Emery and Rimoin's Essential Medical Genetics

Emery and Rimoin's Essential Medical Genetics

Edited by

David L Rimoin†
Steven Spielberg Chair & Director,
Medical Genetics Institute, Cedars-Sinai Medical Center,
Los Angeles, CA, USA

Reed E Pyeritz
William Smilow Professor of Medicine,
Departments of Medicine and Genetics,
Perelman School of Medicine at the University of Pennsylvania,
Philadelphia, PA, USA

Bruce R Korf
Wayne H and Sara Crews Finley Chair in Medical Genetics,
Professor and Chair, Department of Genetics
and Director, Heflin Center for Genomic Sciences,
University of Alabama at Birmingham,
Birmingham, AL, USA

AMSTERDAM • BOSTON • HEIDELBERG • LONDON • NEW YORK
OXFORD • PARIS • SAN DIEGO • SAN FRANCISCO • SINGAPORE
SYDNEY • TOKYO
Academic Press is an imprint of Elsevier

Academic Press is an imprint of Elsevier
The Boulevard, Langford Lane, Kidlington, Oxford, OX5 1GB, UK
225 Wyman Street, Waltham, MA 02451, USA

Notices

Knowledge and best practice in this field are constantly changing. As new research and experience broaden our understanding, changes in research methods, professional practices, or medical treatment may become necessary.

Practitioners and researchers must always rely on their own experience and knowledge in evaluating and using any information, methods, compounds, or experiments described herein. In using such information or methods they should be mindful of their own safety and the safety of others, including parties for whom they have a professional responsibility.

To the fullest extent of the law, neither the Publisher nor the authors, contributors, or editors, assume any liability for any injury and/or damage to persons or property as a matter of products liability, negligence or otherwise, or from any use or operation of any methods, products, instructions, or ideas contained in the material herein.

British Library Cataloguing in Publication Data
A catalogue record for this book is available from the British Library

Library of Congress Cataloging in Publication Data
A catalog record for this book is available from the Library of Congress

ISBN: 978-0-12-407240-4

For information on all Academic Press publications
visit our website at store.elsevier.com

Printed and bound in Italy

13 14 15 16 10 9 8 7 6 5 4 3 2 1

CONTENTS

Preface, xv

Foreword, xvii

Personal memories of David Rimoin, xix

Basic Principles

1 History of Medical Genetics, 3
Victor A McKusick and Peter S Harper

2 Medicine in a Genetic Context, 9
Barton Childs and Reed E Pyeritz

3 Nature and Frequency of Genetic Disease, 12
Bruce R Korf, David L Rimoin, and Reed E Pyeritz

4 Genomics and Proteomics, 15
Raju Kucherlapati

5 Genome and Gene Structure, 16
Daniel H Cohn

6 Epigenetics, 17
Rosanna Weksberg, Darci T Butcher, Daria Grafodatskaya, Sanaa Choufani, and Benjamin Tycko

7 Human Gene Mutation in Inherited Disease: Molecular Mechanisms and Clinical Consequences, 22
Stylianos E Antonarakis and David N Cooper

8 Genes in Families, 25
Jackie Cook

9 Analysis of Genetic Linkage, 28
Rita M Cantor

10 Chromosomal Basis of Inheritance, 30
M Fady Mikhail

11 Mitochondrial Medicine: The Mitochondrial Biology and Genetics of Metabolic and Degenerative Diseases, Cancer, and Aging, 35
Douglas C Wallace, Marie T Lott, and Vincent Procaccio

12 Multifactorial Inheritance and Complex Diseases, 38
Christine W Duarte, Laura K Vaughan, T Mark Beasley, and Hemant K Tiwari

13 Population Genetics, 41
Bronya J B Keats and Stephanie L Sherman

14 Pathogenetics of Disease, 43
Reed E Pyeritz

15 Human Developmental Genetics, 47
Wen-Hann Tan, Edward C Gilmore, and Hagit N Baris

16 Twins and Twinning, 53
Jodie N Painter, Sarah J Medland, Grant W Montgomery, and Judith G Hall

17 **The Molecular Biology of Cancer, 58**
Edward S Tobias

18 **The Biological Basis of Aging: Implications for Medical Genetics, 60**
Junko Oshima, George M Martin, and Fuki M Hisama

19 **Pharmacogenetics and Pharmacogenomics, 63**
Daniel W Nebert and Elliot S Vesell

General Principles

20 **Genetic Evaluation for Common Diseases of Adulthood, 69**
Maren T Scheuner and Shannon Rhodes

21 **Genetic Counseling and Clinical Risk Assessment, 74**
R Lynn Holt and Angela Trepanier

22 **Cytogenetic Analysis, 76**
Nancy B Spinner, Malcolm A Ferguson-Smith, and David H Ledbetter

23 **Diagnostic Molecular Genetics, 78**
Wayne W Grody and Joshua L Deignan

24 **Heterozygote Testing and Carrier Screening, 80**
Matthew J McGinniss and Michael M Kaback

25 **Prenatal Screening for Neural Tube Defects and Aneuploidy, 82**
Amelia L M Sutton and Joseph R Biggio

26 **Techniques for Prenatal Diagnosis, 85**
Lee P Shulman and Sherman Elias

27 **Neonatal Screening, 89**
Richard W Erbe and Harvey L Levy

28 **Enzyme Replacement and Pharmacologic Chaperone Therapies for Lysosomal Storage Disease, 91**
Robert J Desnick, Edward H Schuchman, Kenneth H Astrin, and Seng H Cheng

29 **Gene Therapy: From Theoretical Potential to Clinical Implementation, 93**
Nicholas S R Sauderson, Maria G Castro, and Pedro R Lowenstein

30 **Ethical and Social Issues in Clinical Genetics, 95**
Angus John Clarke

31 **Legal Issues in Genetic Medicine, 100**
Philip R Reilly

Applications to Clinical Problems

32 **Genetics of Female Infertility in Humans, 105**
Bala Bhagavath and Lawrence C Layman

33 **Genetics of Male Infertility, 108**
Csilla Krausz, Chiara Chianese, Ronald S Swerdloff, and Christina Wang

34 **Fetal Loss, 113**
Rhona Schreck and John Williams III

35 **A Clinical Approach to the Dysmorphic Child, 117**
Kenneth L Jones and Marilyn C Jones

36 **Clinical Teratology, 120**
Jan M Friedman and James W Hanson

37 **Neurodevelopmental Disabilities: Global Developmental Delay, Intellectual Disability, and Autism, 124**
John B Moeschler

38 **Abnormal Body Size and Proportion, 131**
John M Graham, Deepika D'Cunha Burkardt, and David L Rimoin

39 **Susceptibility and Response to Infection, 138**
Michael F Murray

40 **Transplantation Genetics, 141**
Steven Ringquist, Ying Lu, Massimo Trucco, and Gaia Bellone

41 **The Genetics of Disorders Affecting the Premature Newborn, 144**
Aaron Prosnitz, Jeffrey R Gruen, and Vineet Bhandari

42 **Disorders of DNA Repair and Metabolism, 146**
Sharon E Plon

Applications to Specific Disorders

Chromosomal Disorders

43 **Autosomal Trisomies, 155**
Cynthia J Curry

44 **Sex Chromosome Abnormalities, 157**
Claus H Gravholt

45 **Deletions and Other Structural Abnormalities of the Autosomes, 161**
Nancy B Spinner, Laura K Conlin, Surabhi Mulchandani, and Beverly S Emanuel

Cardiovascular Disorders

46 **Congenital Heart Defects, 169**
Rocio Moran and Nathaniel H Robin

47 **Inherited Cardiomyopathies, 175**
Polakit Teekakirikul, Carolyn Y Ho, and Christine E Seidman

48 **Heritable and Idiopathic Forms of Pulmonary Arterial Hypertension, 181**
Eric D Austin, John H Newman, James E Loyd, and John A Phillips

49 **Hereditary Hemorrhagic Telangiectasia (Osler – Weber – Rendu Syndrome), 184**
Alan E Guttmacher, Douglas A Marchuk, Scott O Trerotola, and Reed E Pyeritz

50 **Hereditary Disorders of the Lymphatic System and Varicose Veins, 192**
Robert E Ferrell and Reed E Pyeritz

51 **The Genetics of Cardiac Electrophysiology in Humans, 196**
Reed E Pyeritz

52 **Genetics of Blood Pressure Regulation, 201**
Frank S Ong, Kenneth E Bernstein, and Jerome I Rotter

53 **Preeclampsia, 206**
Anthony R Gregg

54 **Common Genetic Determinants of Coagulation and Fibrinolysis, 209**
Angela M Carter, Kristina F Standeven, and Peter J Grant

55 **Genetics of Atherosclerotic Cardiovascular Disease, 213**
Atif N Qasim and Muredach P Reilly

56 **Disorders of the Venous System, 215**
Pascal Brouillard, Nisha Limaye, Laurence M Boon, and Miikka Vikkula

57 **Capillary Malformation/Arteriovenous Malformation, 219**
Nicole Revencu, Laurence M Boon, and Miikka Vikkula

Respiratory Disorders

58 **Cystic Fibrosis, 225**
Garry R Cutting

59 **Genetic Underpinnings of Asthma and Related Traits, 229**
Hakon Hakonarson, Michael E March, and Patrick M A Sleiman

60 **Hereditary Pulmonary Emphysema, 234**
Chad K Oh and Nestor A Molfino

61 **Interstitial and Restrictive Pulmonary Disorders, 237**
William E Lawson and James E Loyd

Renal Disorders

62 **Congenital Anomalies of the Kidney and Urinary Tract, 241**
Grace J Noh, Rosemary Thomas-Mohtat, and Elaine S Kamil

63 **Cystic Diseases of the Kidney, 252**
Angela Sun, Raymond Y Wang, and Dechu P Puliyanda

64 **Nephrotic Disorders, 255**
Hannu Jalanko and Helena Kääriäinen

65 **Renal Tubular Disorders, 257**
Richard E Hillman

66 **Cancer of the Kidney and Urogenital Tract, 259**
Eamonn R Maher

Gastrointestinal Disorders

67 **Gastrointestinal Tract and Hepatobiliary Duct System, 267**
Eberhard Passarge

68 **Bile Pigment Metabolism and Its Disorders, 270**
Namita Roy Chowdhury, Jayanta Roy Chowdhury, and Yesim Avsar

69 **Cancer of the Colon and Gastrointestinal Tract, 272**
C Richard Boland, Barbara Jung, and John M Carethers

Hematologic Disorders

70 **Hemoglobinopathies and Thalassemias, 283**
John Old

71 **Other Hereditary Red Blood Cell Disorders, 290**
Bertil Glader

72 **Hemophilias and Other Disorders of Hemostasis, 292**
Jordan A Shavit and David Ginsburg

73 **Rhesus and Other Fetomaternal Incompatibilities, 298**
Gregory Lau

74 **Leukemias, Lymphomas, and Other Related Disorders, 300**
Yanming Zhang and Janet D Rowley

75 **Immunologic Disorders, 302**
Nancy L Reinsmoen, Kai Cao, and Chih-Hung Lai

76 **Systemic Lupus Erythematosus, 304**
Yun Deng, Bevra H Hahn, and Betty P Tsao

77 **Rheumatoid Disease and Other Inflammatory Arthropathies, 306**
Sarah Keidel, Catherine Swales, and Paul Wordsworth

78 **Amyloidosis and Other Protein Deposition Diseases, 309**
Merrill D Benson

79 **Immunodeficiency Disorders, 312**
Rochelle Hirschhorn, Kurt Hirschhorn, and Luigi D Notarangelo

80 **Inherited Complement Deficiencies, 315**
Kathleen E Sullivan

Endocrinologic Disorders

81 **Disorders of Leukocyte Function, 321**
Harry R Hill, Attila Kumánovics, and Kuender D Yang

82 **Genetic Disorders of the Pituitary Gland, 325**
Amy Potter, John A Phillips, and David L Rimoin

83 **Thyroid Disorders, 330**
Michel Polak and Gabor Szinnai

84 **Parathyroid Disorders, 334**
Geoffrey N Hendy, Murat Bastepe, and David E C Cole

85 **Diabetes Mellitus, 338**
Leslie J Raffel and Mark O Goodarzi

86 **Genetic Disorders of the Adrenal Gland, 350**
Karen Lin-Su, Oksana Lekarev, and Maria I New

87 **Disorders of the Gonads, Genital Tract, and Genitalia, 352**
Joe Leigh Simpson

88 **Cancer of the Breast and Female Reproductive Tract, 361**
Ora Karp Gordon

Metabolic Disorders

89 **Disorders of the Body Mass, 365**
Patricia A Donohoue and Omar Ali

90 **Genetic Lipodystrophies, 367**
Abhimanyu Garg

91 **Amino Acid Metabolism, 372**
Raymond Y Wang, William R Wilcox, and Stephen D Cederbaum

92 **Disorders of Carbohydrate Metabolism, 374**
Priya S Kishnani and Yuan-Tsong Chen

93 **Congenital Disorders of Protein Glycosylation, 376**
Jaak Jaeken

94 **Purine and Pyrimidine Metabolism, 379**
Naoyuki Kamatani, H A Jinnah, Raoul C M Hennekam, and André B P van Kuilenburg

95 **Lipoprotein and Lipid Metabolism, 382**
Robert A Hegele

96 **Organic Acidemias and Disorders of Fatty Acid Oxidation, 384**
Jerry Vockley

97 **Vitamin D Metabolism or Action, 387**
Elizabeth A Streeten and Michael A Levine

98 **Inherited Porphyrias, 391**
R J Desnick and Manisha Balwani, and Karl E Anderson

99 **Inherited Disorders of Human Copper Metabolism, 393**
Stephen G Kaler and Seymour Packman

100 **Iron Metabolism and Related Disorders, 394**
Kaveh Hoda, Christopher L Bowlus, Thomas W Chu, and Jeffrey R Gruen

101 **Mucopolysaccharidoses, 397**
J Ed Wraith

102 **Oligosaccharidoses: Disorders Allied to the Oligosaccharidoses, 399**
Jules G Leroy

103 **Sphingolipid Disorders and the Neuronal Ceroid Lipofuscinoses or Batten Disease (Wolman Disease, Cholesterylester Storage Disease, and Cerebrotendinous Xanthomatosis), 403**
Rose-Mary Boustany, Ibraheem Al-Shareef, and Sariah El-Haddad

104 **Peroxisomal Disorders, 409**
Ronald J A Wanders

Mental and Behavioral Disorders

105 **The Genetics of Personality, 415**
Matt McGue and Lindsay K Matteson

106 **Fragile X Syndrome and X-Linked Intellectual Disability, 417**
Kathryn B Garber, Stephen T Warren, and Jeannie Visootsak

107 **Dyslexia and Related Communication Disorders, 420**
Angela Friend, Bruce F Pennington, Shelley D Smith, and Jeffrey W Gilger

108 **Attention-Deficit/Hyperactivity Disorder, 423**
Stephen V Faraone and Alysa E Doyle

109 **Autism Spectrum Disorders, 425**
Sunil Q Mehta and Daniel H Geschwind

110 **Genetics of Alzheimer Disease, 427**
Adam C Naj, Regina M Carney, Susan E Hahn, Margaret A Pericak-Vance, Michael A Slifer, and Jonathan L Haines

111 **Schizophrenia and Affective Disorders, 429**
Jonathan D Picker

112 **Addictive Disorders, 431**
David Goldman, Paola Landi, and Francesca Ducci

Neurologic Disorders

113 **Neural Tube Defects, 435**
Richard H Finnell, Timothy M George, and Laura E Mitchell

114 **Genetic Disorders of Cerebral Cortical Development, 438**
Ganeshwaran H Mochida, Annapurna Poduri, and Christopher A Walsh

115 **Genetic Aspects of Human Epilepsy, 440**
Asuri N Prasad and Chitra Prasad

116 **Basal Ganglia Disorders, 443**
Andrew B West, Michelle Gray, and David G Standaert

117 **The Hereditary Ataxias, 445**
Puneet Opal and Huda Zoghbi

118 **Hereditary Spastic Paraplegia, 447**
John K Fink

119 **Autonomic and Sensory Disorders, 454**
Felicia B Axelrod

120 **The Phakomatoses, 458**
Susan M Huson and Bruce R Korf

121 **Multiple Sclerosis and Other Demyelinating Disorders, 464**
A Dessa Sadovnick

122 **Genetics of Stroke, 467**
Mateusz G Adamski and Alison E Baird

123 **Primary Tumors of the Nervous System, 469**
Angel A Alvarez and Markus Bredel

Neuromuscular Disorders

124 **Muscular Dystrophies, 473**
Anna Sarkozy, Kate Bushby, and Eugenio Mercuri

125 **Hereditary Motor and Sensory Neuropathies, 476**
Wojciech Wiszniewski, Kinga Szigeti, and James R Lupski

126 **Congenital (Structural) Myopathies, 478**
Heinz Jungbluth and Carina Wallgren-Pettersson

127 **Spinal Muscular Atrophies, 480**
Sabine Rudnik-Schöneborn and Klaus Zerres

128 **Hereditary Muscle Channelopathies, 482**
Frank Lehmann-Horn, Reinhardt Rüdel, and Karin Jurkat-Rott

129 **The Myotonic Dystrophies, 484**
Chris Turner

130 **Hereditary and Autoimmune Myasthenias, 486**
David Beeson

131 **Motor Neuron Disease, 489**
Teepu Siddique, H X Deng, and Senda Ajroud-Driss

Ophthalmologic Disorders

132 **Genetics of Color Vision Defects, 493**
Samir S Deeb and Arno G Motulsky

133 **Optic Atrophy, 496**
Grace C Shih and Brian P Brooks

134 **Glaucoma, 498**
Janey L Wiggs

135 **Defects of the Cornea, 501**
R Krishna Sanka, Elmer Tu, and Joel Sugar

136 **Congenital Cataracts and Genetic Anomalies of the Lens, 503**
Arlene V Drack, Yaron S Rabinowitz, and Edward Cotlier

137 **Hereditary Retinal and Choroidal Dystrophies, 505**
Suma P Shankar

138 **Strabismus, 507**
J Bronwyn Bateman and Sherwin J Isenberg

139 **Retinoblastoma and the RB1 Cancer Syndrome, 508**
A Linn Murphree, Robin D Clark, Linda M Randolph, Uma M Sachdeva, Dan S Gombos, and Joan M O'Brien

140 **Anophthalmia, Microphthalmia, and Uveal Coloboma, 511**
Brian P Brooks

Deafness

141 **Hereditary Hearing Impairment, 517**
Rena Ellen Falk and Arti Pandya

Craniofacial Disorders

142 **Clefting, Dental, and Craniofacial Syndromes, 523**
Jeffrey C Murray and Mary L Marazita

143 **Craniosynostosis, 527**
Ethylin Wang Jabs and Amy Feldman Lewanda

Dermatologic Disorders

144 **Abnormalities of Pigmentation, 541**
Richard A Spritz and Vincent J Hearing

145 **Ichthyosiform Dermatoses, 544**
Howard P Baden and John J DiGiovanna

146 **Epidermolysis Bullosa, 546**
Cristina Has, Leena Bruckner-Tuderman, and Jouni Uitto

147 **Ectodermal Dysplasias, 551**
Dorothy Katherine Grange

148 **Skin Cancer, 557**
Julia A Newton Bishop and Rosalyn Jewell

149 **Psoriasis, 559**
Johann E Gudjonsson and James T Elder

150 **Cutaneous Hamartoneoplastic Disorders, 561**
Katherine L Nathanson

151 **Inherited Disorders of the Hair, 563**
Mazen Kurban and Angela M Christiano

Connective Tissue Disorders

152 **Marfan Syndrome and Related Disorders, 567**
Reed E Pyeritz

153 **Ehlers – Danlos Syndrome, 575**
Peter H Byers

154 **Heritable Diseases Affecting the Elastic Fibers: Cutis Laxa, Pseudoxanthoma Elasticum, and Related Disorders, 579**
Jouni Uitto

Skeletal Disorders

155 **Osteogenesis Imperfecta (and Other Disorders of Bone Matrix), 587**
Craig Munns and David Sillence

156 **Disorders of Bone Density, Volume, and Mineralization, 590**
Maria Descartes and David O Sillence

157 **Chondrodysplasias, 593**
David, Rimoin, Ralph Lachman, and Sheila Unger

158 **Abnormalities of Bone Structure, 595**
William A Horton

159 **Arthrogryposes (Multiple Congenital Contractures), 597**
Judith G Hall

160 **Common Skeletal Deformities, 599**
William A Horton

161 **Hereditary Noninflammatory Arthropathies, 601**
Mariko L Ishimori

Pathways

162 **Pathways—Cohesinopathies, 607**
Matthew A Deardorff and Ian D Krantz

163 **Genes and Mechanisms in Human Ciliopathies, 612**
Dagan Jenkins and Philip L Beales

Index, 615

PREFACE

In the three decades that have elapsed since the first edition of this book was published, the field of medical genetics has experienced explosive growth. The rate of change in the science and application of medical genetics that occurred in the period spanning the first through the fifth editions of this book has markedly accelerated. In the 1980s, genetics was viewed by most practitioners as an important but obscure corner of medicine. Now it is widely recognized that virtually all human disorders have a genetic component, and genetics is viewed as the key basic science in uncovering the mysteries of disease pathogenesis. The tools of molecular genetics have matured remarkably over these past three decades, to the point where the mutations that underlie most single-gene disorders are being reported week by week. This has resulted in insights into disease mechanisms that have solved medical puzzles that have existed for centuries. These advances are also providing new clinical approaches that vastly improve the ability to accurately diagnose these disorders, and in many instances offer new hope for treatment and prevention. Of even greater importance for the medical and public health communities, and the public, is the increasing ability to dissect the genetic contributions to common disorders. The relationships between nuclear and mitochondrial genotypes, epigenetics, environment and chance remain dauntingly complex, but powerful molecular and computational tools are now being applied to these problems. The promise of new approaches to prevention and treatment represents a new paradigm, which includes what some call "personalized medicine," that will transform health care.

The sequencing of the human genome has captured public attention and raised both hopes and concerns. Clinical applications of genomics have expanded dramatically since the last edition of this book, most notably in cytogenomics and whole-exome or whole-genome sequencing. Nevertheless, it remains to be seen how quickly and to what extent the genetic approach will be incorporated into the day-to-day practice of medicine. The complexity of translating scientific developments in genetics to clinical application is being increasingly recognized, particularly with respect to legal, ethical, psychological, economic, and social implications. The need to inform colleagues throughout medicine about advances in genetics and the principles of their clinical application has never been greater.

The continued excitement and progress in our field are reflected in further expansion of this book with several new chapters. The number of genetic disorders understood at the molecular level continues to increase and common disorders not traditionally viewed as "genetic" are included as genetic contributions are coming to light. The shear mass of new information precludes continued publication of the complete chapters in hard copy. The print volume contains short summaries of each chapter along with a few illustrations and recent references of general relevance to the topic. The complete chapters are available on the PPMG6e website. Authors are expected to update the online version at least semiannually.

We continue to be grateful for the comments about previous editions provided by our colleagues. This edition is better for their input, and we accept full responsibility for the deficiencies that remain.

We warmly acknowledge, besides our contributing authors, the assistance of staff at Elsevier Health Sciences and our personal assistants, especially Sue Lief in Los Angeles. We appreciate the continued moral, scholarly, and spiritual support of Professor Alan EH Emery, one of the originators of this book.

Finally, just as this edition was being finalized, the other originator and senior editor of this book, David Rimoin, died after a brief illness. There have been many memorials and tributes to Dr Rimoin, and others will emerge in the future. This sixth edition of *Principles and Practice of Medical Genetics* is one such memorial, to which he made tremendous intellectual and organizational contributions.

Reed E Pyeritz
Bruce R Korf
August 2012

FOREWORD

In the previous five editions of Principles and Practice of Medical Genetics, our late colleague, Victor McKusick, masterfully reviewed the advances that had occurred during the years between the various editions. This sixth edition, which will be published 30 years after the first volumes in 1983, finds the practice of medical genetics in a new environment.

The search for and identification of disease genes have continued to be important in our understanding of genetic disorders. Such discoveries, first reported with the groundbreaking discovery in 1986 of the genes underlying chronic granulomatous disease and Duchenne muscular dystrophy, are now so frequently reported that they will not usually merit publication in a major journal.

Undoubtedly, sequencing of an individual's whole exome or genome is currently forcing dramatic changes in the way genetics is practiced on the clinical and research level. There will be not a single subfield of genetics that will not be profoundly reshaped in the coming years as a consequence of these new technologies. We already observe and welcome fundamental discoveries that revolutionize our understanding of human disease, evolution, and ancestry. Similarly, diagnosis, prognosis and therapeutics have been markedly improved. Even the relationship to our endobiome has to be rethought as a consequence of discoveries in the new field of metagenomics.

In practice, we now increasingly order sophisticated genetic tests to confirm a diagnosis with great accuracy that is already suspected clinically. For example, based on the careful study of the dystrophin gene, we are able not only to confirm a diagnosis but also to predict with considerable accuracy whether a young boy presenting with muscle disease is likely to have the more severe Duchenne muscular dystrophy, or the milder Becker phenotype. This ability to use widely available genetic testing in many clinical situations to dissect out clinical phenotypes with specific gene mutations is invaluable.

The most common genetic disease that we recognize, inherited or acquired, is cancer. Now genetic tests are ordered in families known or thought to be at risk for breast (and other) cancers. Treatment decisions are made for these patients who are found to carry mutant genes that confer increased risk for malignancy, and who might have already presented with cancer. Although we have a vastly greater understanding of tumor biology, as well as predictive laboratory tools, we still lack the ideal cancer treatment, as the tumor in each patient is unique. However, some genetic discoveries related to cancer treatment are among the first to define what many call "personalized medicine".

We are now increasingly using genetic technologies in a more preventive mode across a population, not just in a person in whom we are trying to confirm a diagnosis. The most frequent genetic testing done in the United States and the rest of the developed world in this fashion is that involved with newborn screening. We screen in the public health sector all 4.2 million babies born in the United States for more than 30 disorders, nearly of which are genetically determined. These tests are largely done with metabolite analysis using mass spectroscopy, but there are increasingly direct genetic analyses done as secondary tests. This genetic testing differs in that it is aimed at healthy babies, in an effort to identify and treat serious, often fatal genetic disorders before symptoms appear. The US Center for Disease Control and Prevention has recognized the expansion of newborn screening as one of the most important public health developments of the past 10 years.

The application of next generation sequencing to newborn screening is now in pilot studies and will ultimately be applied at all ages. The actual sequencing, which has been the most daunting issue, is rapidly becoming practical. The costs to sequence the whole exome or genome have dropped markedly, making it feasible to use these resources in an attempt to better understand the outcome of a known condition. For example, by looking across the whole genome of an infant who screens positive for sickle cell disease, it might be possible to identify phenotype-modifying loci and much better predict the course of this severe and variable condition.

Challenges to the broad use of next generation sequencing remain. These include the management of the flood of information that comes from such sequencing and filling gaps in phenotypic annotations of variations in a whole genome. It is likely that we will soon enter the time when information about the person's entire genome will be a part of the medical record and as such a part of medical practice. Such information will require the widespread adoption of the electronic medical record. With the genome sequence in one's medical record one can inquire about the genome over a lifetime, and annotations made to greatly enhance our understanding.

We currently lack the scientific, logistical, ethical and legal framework required for the appropriate and effective use of such information. We will most surely identify modifiers that affect disease states (some of which can possibly be clues to therapies), but we will have a wide array of findings, identifying disorders, which are late in onset and many that lack therapies or any other actionable outcomes. And, while the profession wrestles with issues about what to return of research results and incidental findings from next generation sequencing, several companies have begun offering these genetic tests directly to the consumer.

Health professionals and the general public must have and will benefit from much expanded genetic education. Eventually, everyone will want to understand better their genetic makeup, their risks for disease, how to prevent serious consequences of disease predispositions, and the most appropriate treatments for overt disease. This sixth edition of *Emery and Rimoin's Principles and Practice of Medical Genetics* is one effective mechanism for providing this education.

R. Rodney Howell
Member, Hussman Institute for Human Genomics,
Miller School of Medicine, University of Miami,
Miami, FL, USA

PERSONAL MEMORIES OF DAVID RIMOIN

David and I first met in 1963 when we were both research fellows in Victor McKusick's Unit at Johns Hopkins Hospital. He studied dwarfism and I studied muscular dystrophies. He had an office opposite mine and we shared many experiences and events together until I graduated PhD in 1964 and returned to the UK. David was awarded his PhD a year or two later. We remained on close personal terms thereafter—including a wonderful stay at his home in Beverly Hills when I was Boeckmann Visiting Professor at UCLA. On another occasion we attended a meeting in Moscow (before perestroika and glasnost), which was very challenging for both of us. It was at this meeting when we conceived the notion of editing a new textbook on medical genetics.

Over the years, we worked closely together on *PPMG* and on several occasions he visited Edinburgh where I was then working. In 2006, he was invited to give the annual *Emery Lecture* at Green College, Oxford University. He gave an excellent talk entitled "The skeletal dysplasias: clinical–molecular correlations" which was well-received by everyone.

I shall miss David greatly—his warmth, generosity and friendship. And of course his collaboration with the *Emery & Rimoin's Principles and Practice of Medical Genetics*.

Alan E H Emery
University of Edinburgh;
Honorary Fellow, Green Templeton College,
University of Oxford
June 2012

Basic Principles

History of Medical Genetics, 3
Medicine in a Genetic Context, 9
Nature and Frequency of Genetic Disease, 12
Genomics and Proteomics, 15
Genome and Gene Structure, 16
Epigenetics, 17
Human Gene Mutation in Inherited Disease: Molecular Mechanisms and Clinical Consequences, 22
Genes in Families, 25
Analysis of Genetic Linkage, 28
Chromosomal Basis of Inheritance, 30
Mitochondrial Medicine: The Mitochondrial Biology and Genetics of Metabolic and Degenerative Diseases, Cancer, and Aging, 35
Multifactorial Inheritance and Complex Diseases, 38
Population Genetics, 41
Pathogenetics of Disease, 43
Human Developmental Genetics, 47
Twins and Twinning, 53
The Molecular Biology of Cancer, 58
The Biological Basis of Aging: Implications for Medical Genetics, 60
Pharmacogenetics and Pharmacogenomics, 63

CHAPTER

1

History of Medical Genetics

Victor A McKusick[*]

Formerly University Professor of Medical Genetics, Johns Hopkins University, Physician, Johns Hopkins Hospital, Baltimore, MD, USA

Peter S Harper

Institute of Medical Genetics, School of Medicine, Cardiff University, Heath Park, Cardiff, UK

1.1 A TIMELINE FOR MEDICAL GENETICS (PSH)

Modern medical genetics as a well-defined field of medicine has developed so rapidly since its beginnings half a century ago, that it is often forgotten how far back in time its roots and origins go. It can be reasonably argued that genetics overall was based in considerable measure on problems of human inheritance and inherited disease, and studies of this extend back long before the twentieth century acceptance of Mendelism. Thus medical genetics, when thought of in the widest sense, is perhaps the oldest area of genetics, and certainly not the recent addition that it is sometimes portrayed as.

This "timeline" gives some of what I consider to be the main landmarks along this lengthy course. Not all of these can be considered to be directly part of "medical genetics," even on the broadest definition, but they are all relevant to it in one way or another. I have also included some more general "world events" that have particularly impacted on the development of the field. To economize on space, full references are not given in this table, but they can be found in the web-based version of this chapter. I shall welcome suggestions for other items that might be included in future editions.

The rapid pace of developments in the field gives urgency to preserving the primary records that will be essential if future historians are to be able to make a detailed and unbiased analysis of how medical genetics has developed. These include the personal scientific records and correspondence of key workers across the world, both clinical and scientific; laboratory notebooks; published books; and recorded interviews. Some sources are given in the web-based version of this chapter; many are collected in the Genetics and Medicine Historical Network Website (www.genmedhist.org). Everyone working in the field can contribute to ensuring that our history is fully documented.

For fuller accounts of the history of medical genetics, readers may consult Victor McKusick's masterly chapter in previous print editions of this book (largely conserved in the current web-based edition), or my own book "A Short History of Medical Genetics."

[*]Deceased.

TABLE 1-1 A Timeline For Human and Medical Genetics. Based on the Timeline of the Genetics and Medicine Historical Network (www.genmedhist.org). The Original Version First Appeared in Harper (2008), A Short History of Medical Genetics, OUP

1651	William Harvey's book *De Generatione Animalium* studies the egg and early embryo in different species and states: "Ex ovo omnium" (all things from the egg).
1677	Microscopic observations of human sperm (Leeuwenhoek).
1699	Albinism noted in "Moskito Indians" of Central America (Wafer).
1735	Linnaeus, *Systema Naturae*. First "natural" classification of plants and animals.
1751	Maupertuis proposes equal contributions of both sexes to inheritance and a "Particulate" concept of heredity.
1753	Maupertuis describes polydactyly in Ruhe family; first estimate of likelihood for it being hereditary.
1794	John Dalton. Color blindness described in himself and others. Limited to males. Erasmus Darwin publishes *Zoonomia*. Progressive evolution from primeval organisms recognized.
1803	Hemophilia in males and its inheritance through females described (Otto).
1809	Inherited blindness described in multiple generations (Martin). Lamarck supports evolution, including man, based on inheritance of acquired characteristics.
1814	Joseph Adams, Concepts of "predisposition" and "disposition"; "congenital" and "hereditary."
1852	First clear description of Duchenne muscular dystrophy (Meryon).
1853	Hemophilic son, Leopold, born to Queen Victoria in England.
1858	Charles Darwin and Alfred Russel Wallace. Papers to Linnean Society on Natural Selection.
1859	Charles Darwin publishes *On the Origin of Species*.
1865	Gregor Mendel's experiments on plant hybridization presented to Brunn Natural History Society.
1866	Mendel's report formally published.
1868	Charles Darwin's "provisional hypothesis of pangenesis." Charles Darwin collects details of inherited disorders in *Animals and Plants under Domestication*.
1871	Friedrich Miescher isolates and characterizes "nucleic acid."
1872	George Huntington describes "Huntington's disease."
1882	First illustration of human chromosomes (Flemming).
1885	"Continuity of the germ plasm" (August Weismann).
1887	Boveri shows constancy of chromosomes through successive generations.
1888	Waldeyer coins term "chromosome." Weismann presents evidence against inheritance of acquired characteristics.
1889	Francis Galton's *Law of Ancestral Inheritance*.
1891	Henking identifies and names "X chromosome."
1894	Bateson's book *Material for the Study of Variation*.
1896	EB Wilson's book *The Cell in Development and Inheritance*.
1899	Archibald Garrod's first paper on alkaptonuria.
1900	Mendel's work rediscovered (de Vries, Correns, and Tschermak).
1901	Karl Landsteiner discovers ABO blood group system. Archibald Garrod notes occurrence in sibs and consanguinity in alkaptonuria.
1902	Bateson and Saunders' note on alkaptonuria as an autosomal recessive disorder. Bateson and Garrod correspond. Garrod's definitive paper on alkaptonuria an example of "chemical individuality." Bateson's *Mendel's Principles of Heredity. A Defence* supports Mendelism against attacks of biometricians. Chromosome theory of heredity (Boveri, Sutton).
1903	American Breeders Association formed. Includes section of eugenics from 1909. Cuénot in France shows Mendelian basis and multiple alleles, for albinism in mice. Castle and Farabee show autosomal recessive inheritance in human albinism. Farabee shows autosomal dominant inheritance in brachydactyly.
1905	Stevens and Wilson separately show inequality of sex chromosomes and involvement in sex determination in insects. Bateson coins term "genetics."
1906	First International Genetics Congress held in London.
1908	Garrod's Croonian lectures on "inborn errors of metabolism." Royal Society of Medicine, London, "Debate on Heredity and Disease." Hardy and Weinberg independently show relationship and stability of gene and genotype frequencies (Hardy Weinberg equilibrium).
1909	Bateson's book *Mendel's Principles of Heredity* documents a series of human diseases following Mendelian inheritance. Karl Pearson initiates *The Treasury of Human Inheritance*. Wilhelm Johannsen introduces term "gene."
1910	Thomas Hunt Morgan discovers X-linked "white eye" *Drosophila* mutant. Eugenics Record Office established at Cold Spring Harbor under Charles Davenport.
1911	Wilson's definitive paper on sex determination shows X-linked inheritance for hemophilia and color blindness.
1912	Winiwarter proposes diploid human chromosome number as approximately 47. First satisfactory human chromosome analysis. First International Eugenics congress (London).

TABLE 1-1 A Timeline For Human and Medical Genetics. Based on the Timeline of the Genetics and Medicine Historical Network (www.genmedhist.org). The Original Version First Appeared in Harper (2008), A Short History of Medical Genetics, OUP—*Cont'd*

Year	Event
1913	Alfred Sturtevant constructs first genetic map of *Drosophila* X-chromosome loci. American Genetics Society formed as successor to American Breeders Association.
1914	Boveri proposes chromosomal basis for cancer. (Outbreak of World War I).
1915	J.B.S. Haldane and colleagues show first mammalian genetic linkage in mouse.
1916	Relationship between frequency of a recessive disease and of consanguinity (F. Lenz). Calvin Bridges shows non-disjunction in *Drosophila*.
1918	Anticipation first recognized in myotonic dystrophy (Fleischer). R.A. Fisher shows compatibility of Mendelism and quantitative inheritance.
1919	Hirszfeld and Hirszfeld show ABO blood group differences between populations. Genetical Society founded in UK by William Bateson.
1922	Inherited eye disease volumes of *Treasury of Human Inheritance* (Julia Bell).
1923	Painter recognizes human Y chromosome; proposes human diploid chromosome number of 48.
1927	Hermann Muller shows production of mutations by X-irradiation in *Drosophila*. Compulsory sterilization on eugenic grounds upheld by courts in America (Buck v. Bell).
1928	Stadler shows radiation-induced mutation in maize and barley. Griffiths discovers "transformation" in Pneumococcus.
1929	Blakeslee shows effect of chromosomal trisomy in Datura, the thorn apple.
1930	R.A. Fisher's *Genetical Theory of Natural Selection*. Beginning of major Russian contributions to human cytogenetics. Haldane's book *Enzymes* attempts to keep biochemistry and genetics linked.
1931	Archibald Garrod's second book *Inborn Factors in Disease*. UK Medical Research Council establishes Research Committee on Human Genetics (Chairman J.B.S. Haldane).
1933	Nazi eugenics law enacted in Germany.
1934	Fölling in Norway discovers phenylketonuria (PKU). *Treasury of Human Inheritance* volume on Huntington's disease (Julia Bell). O.L. Mohr's book *Genetics and Disease*. Mitochondrial inheritance proposed for Leber's optic atrophy (Imai and Moriwaki, Japan).
1935	First estimate of mutation rate for a human gene (hemophilia; J.B.S. Haldane). R.A. Fisher (amongst others) suggests use of linked genetic markers in disease prediction.
1937	First human genetic linkage—hemophilia and color blindness (Bell and Haldane). Moscow Medical Genetics Institute closed; director Levit and others arrested and later executed. Destruction of Russian genetics begins. Seventh International Genetics Congress, Moscow canceled. Max Perutz begins crystallographic studies of hemoglobin in Cambridge.
1938	Lionel Penrose publishes "Colchester Survey" of genetic basis of mental handicap.
1939	Seventh International Genetics Congress held in Edinburgh. "Geneticists' Manifesto" issued. (Outbreak of World War II) Cold Spring Harbor Eugenics Record Office closed. Rh blood group system discovered (Landsteiner and Wiener).
1941	Beadle and Tatum produce first nutritional mutants in *Neurospora* and confirm "one gene–one enzyme" principle. Charlotte Auerbach discovers chemical mutagens in Edinburgh (not published until the end of the war).
1943	Nikolai Vavilov, leader of Russian genetics, dies in Soviet prison camp. First American genetic counseling clinic. Mutation first demonstrated in bacteria (Luria).
1944	Schrödinger's book *What Is Life?* provides inspiration for the first molecular biologists. Avery shows bacterial transformation is due to DNA, not protein.
1945	Lionel Penrose appointed as head of Galton Laboratory, London, found modern human genetics as a specific discipline. (Hiroshima and Nagasaki atomic explosions.) Genetic study of effects of radiation initiated on survivors of the atomic explosions (J.V. Neel director).
1946	Penrose's inaugural lecture at University College, London, uses PKU as paradigm for human genetics. John Fraser Roberts begins first UK genetic counseling clinic in London. Sexual processes first shown in bacteria (Lederberg).
1948	Total ban on all genetics (including human genetics) teaching and research in Russia. American Society of Human Genetics founded. H.J. Muller, President.
1949	*American Journal of Human Genetics* begun. Charles Cotterman, first editor. Linus Pauling and colleagues show sickle cell disease to have a molecular basis. J.V. Neel shows it to be recessively inherited. J.B.S. Haldane suggests selective advantage due to malaria. Barr and Bertram (London, Ontario) discover the sex chromatin body.

Continued

TABLE 1-1 A Timeline For Human and Medical Genetics. Based on the Timeline of the Genetics and Medicine Historical Network (www.genmedhist.org). The Original Version First Appeared in Harper (2008), A Short History of Medical Genetics, OUP—*Cont'd*

Year	Event
1950	Curt Stern's Book *Human Genetics*.
	Frank Clark Fraser initiates Medical Genetics at McGill University, Montreal.
1951	Linus Pauling shows triple helical structure of collagen.
	HELA cell line established from cervical cancer tissue of Baltimore patient Henrietta Lacks.
1952	First human inborn error shown to result from enzyme deficiency (glycogen storage disease type 1, Cori and Cori).
	Rosalind Franklin's X-ray crystallography shows helical structure of B form of DNA.
1953	Model for structure of DNA as a double helix (Watson and Crick).
	Bickel et al. initiate dietary treatment for PKU.
	Enzymatic basis of PKU established (Jervis).
	Specific chair in Medical Genetics founded (first holder Maurice Lamy, Paris).
1954	Allison proves selective advantage for sickle cell disease in relation to malaria.
1955	Sheldon Reed's book *Counselling in Medical Genetics*.
	Oliver Smithies develops starch gel electrophoresis for separation of human proteins.
	Fine structure analysis of bacteriophage genome (Benzer).
1956	Tjio and Levan show normal human chromosome number to be 46, not 48.
	First International Congress of Human Genetics (Copenhagen).
	Amniocentesis first validated for fetal sexing in hemophilia (Fuchs and Riis).
1957	Ingram shows specific molecular defect in sickle cell disease.
	Specific Medical Genetics departments opened in Baltimore (Victor McKusick) and Seattle (Arno Motulsky).
1958	First HLA antigen detected (Dausset).
1959	Harry Harris' book *Human Biochemical Genetics*.
	Perutz completes structure of hemoglobin.
	First human chromosome abnormalities identified in:
	Down's syndrome (Lejeune et al.).
	Turner syndrome (Ford et al.).
	Klinefelter syndrome (Jacobs and Strong).
1960	Trisomies 13 and 18 identified (Patau et al. and Edwards et al.).
	First edition of *Metabolic Basis of Inherited Disease*.
	Role of messenger RNA recognized.
	First specific cytogenetic abnormality in human malignancy (Nowell and Hungerford, "Philadelphia chromosome").
	Chromosome analysis on peripheral blood allows rapid development of diagnostic clinical cytogenetics (Moorhead et al.).
	Denver conference on human cytogenetic nomenclature.
	First full UK Medical Genetics Institute opened (under Paul Polani, Guy's Hospital, London).
	First Bar Harbor course in Medical Genetics, under Victor McKusick.
1961	Prevention of rhesus hemolytic disease by isoimmunization (Clarke and colleagues, Liverpool).
	Mary Lyon proposes X-chromosome inactivation in females.
	Cultured fibroblasts used to establish biochemical basis of galactosemia (Krooth and Weinberg), establishing value of somatic cell genetics.
	"Genetic Code" linking DNA and protein established (Nirenberg and Matthaei).
1963	Population screening for PKU in newborns (Guthrie and Susi).
1964	Ultrasound used in early pregnancy monitoring (Donald).
	First journal specifically for medical genetics (*Journal of Medical Genetics*).
	Genetics restored as a science in USSR after Nikita Khrushchev dismissed.
	First HLA workshop (Durham, North Carolina).
1965	High frequency of chromosome abnormalities found in spontaneous abortions (Carr, London, Ontario).
	Human–rodent hybrid cell lines developed (Harris and Watkins).
1966	First chromosomal prenatal diagnosis (Steele and Breg).
	First edition of McKusick's *Mendelian Inheritance in Man*.
	Recognition of dominantly inherited cancer families (Lynch).
1967	Application of hybrid cell lines to human gene mapping (Weiss and Green).
1968	First autosomal human gene assignment to a specific chromosome (Duffy blood group on chromosome 1) by Donahue et al.
1969	First use of "Bayesian" risk estimation in genetic counseling (Murphy and Mutalik).
	First Masters degree course in genetic counseling (Sarah Lawrence College, New York).
1970	Fluorescent chromosome banding allows unique identification of all human chromosomes (Zech, Caspersson and colleagues).
1971	"Two-hit" hypothesis for familial tumors, based on retinoblastoma (Knudson).
	Giemsa chromosome banding suitable for clinical cytogenetic use (Seabright).
	First use of restriction enzymes in molecular genetics (Danna and Nathans).
1972	Population screening for Tay–Sachs disease (Kaback and Zeiger).

TABLE 1-1 A Timeline For Human and Medical Genetics. Based on the Timeline of the Genetics and Medicine Historical Network (www.genmedhist.org). The Original Version First Appeared in Harper (2008), A Short History of Medical Genetics, OUP—*Cont'd*

Year	Event
1973	Prenatal diagnosis of neural tube defects by raised alpha fetoprotein (Brock).
	First Human Gene Mapping Workshop (Yale University).
1975	DNA hybridization (Southern) "Southern blot."
1977	Human beta-globin gene cloned.
1978	Prenatal diagnosis of sickle cell disease through specific RFLP (Kan and Dozy).
	First mutation causing a human inherited disease characterized (beta-thalassemia).
	First birth following in vitro fertilization (Steptoe and Edwards).
1979	Vogel and Motulsky's textbook *Human Genetics, Problems and Approaches*.
1980	Primary prevention of neural tube defects by preconceptional multivitamins (Smithells et al.).
	Detailed proposal for mapping the human genome (Botstein et al.).
1981	Human mitochondrial genome sequenced (Anderson et al.).
1982	Linkage of DNA markers on X chromosome to Duchenne muscular dystrophy (Murray et al.).
1983	First autosomal linkage using DNA markers for Huntington's disease. (Gusella et al.).
1983	First general use of chorionic villus sampling in early prenatal diagnosis.
1984	DNA fingerprinting discovered (Jeffreys).
1985	Application of DNA markers in genetic prediction of Huntington's disease.
	First initiatives toward total sequencing of human genome (US Dept of Energy and Cold Spring Harbor meetings).
1986	Polymerase chain reaction (PCR) for amplifying short DNA sequences (Mullis).
1988	International Human Genome Organisation (HUGO) established.
	US congress funds Human Genome Project.
1989	Cystic fibrosis gene isolated.
	First use of preimplantation genetic diagnosis.
1990	First attempts at gene therapy in immunodeficiencies.
	Fluorescent in situ hybridization introduced to cytogenetic analysis.
1991	Discovery of unstable DNA and trinucleotide repeat expansion (fragile X).
1992	Isolation of *PKU* (phenylalanine hydroxylase) gene (Woo and colleagues).
	First complete map of human genome produced by French *Généthon* initiative (Weissenbach et al.).
1993	Huntington's disease gene and mutation identified.
	BRCA 1 gene for hereditary breast–ovarian cancer identified.
1996	"Bermuda Agreement" giving immediate public access to all Human Genome Project data.
1997	First cloned animal (Dolly the sheep), Roslin Institute, Edinburgh.
1998	Total sequence of model organism *Caenorhabditis elegans*.
	Isolation of embryonic stem cells.
1999	Sequence of first human chromosome (22).
2000	"Draft sequence" of human genome announced jointly by International Human Genome Consortium and by Celera.
	Correction of defect in inherited immune deficiency (SCID) by gene therapy, (but subsequent development of leukemia).
2002	Discovery of microRNAs.
2003	Complete sequence of human genome achieved and published.
2005	Sequencing of chimpanzee genome.
2006	Prenatal detection of free fetal DNA in maternal blood clinically feasible.
2007	First genome-wide association studies giving robust findings for common multifactorial disorders.
2008	First specific individual human genomes sequenced.
2010	Sequencing of Neanderthal genome. Diagnostic use of human exome sequencing.
2011	Modern human genome shown to contain sequence from other ancient hominins (Neanderthal and Denisovan).

FURTHER READING

Adams, J. *A Treatise on the Supposed Hereditary Properties of Diseases*; Callow: London, 1814.

Carlson, E. A. *Mendel's Legacy: The Origin of Classical Genetics*; Cold Spring Harbor Laboratory Press: Cold Spring Harbor, New York, 2004.

Crow, J. F.; Dove, W. F. *Perspectives on Genetics: Anecdotal, Historical and Critical Commentaries, 1987–1998*; The University of Wisconsin Press: Madison, 2000.

Dunn, L. C. *A Short History of Genetics: The Development of Some of the Main Lines of Thought, 1864–1939*; McGraw-Hill: New York, 1965, (Reissued by Iowa State University Press, 1991).

Harper, P. S., Ed. *Landmarks in Medical Genetics: Classic Papers with Commentaries*, Oxford University Press: Oxford, 2004.

Harper, P. S. *First Years of Human Chromosomes: The Beginnings of Human Cytogenetics*; Scion: Oxford, 2006.

Harper, P. S. *A Short History of Medical Genetics*; Oxford University Press, 2008.

Harris, H. *The Cells of the Body: A History of Somatic Cell Genetics*; Cold Spring Harbor Laboratory Press: Cold Spring Harbor, New York, 1995.

Hsu, T. C. *Human and Mammalian Cytogenetics: An Historical Perspective*; Springer-Verlag: New York, 1979.

Judson, H. *The Eighth Day of Creation: Makers of the Revolution in Biology*; Jonathan Cape: London, 1979.

Kevles, D. J. *In the Name of Eugenics: Genetics and the Uses of Human Heredity*; Knopf: New York, 1985.

Lindee, S. *Moments of Truth in Genetic Medicine*; The Johns Hopkins University Press: Baltimore, 2005.

McKusick, V. A. A 60 Year Tale of Spots, Maps and Genes. *Annu. Rev. Genomics Hum. Genet.* **2006**, *7*, 1–27.

McKusick, V. A. Mendelian Inheritance in Man and Its Online Version *OMIM*. *Am. J. Hum. Genet.* **2007**, *80*, 588–604.McKusick, V. A. History of Medical Genetics. In *Emery and Rimoin's Principles and Practice of Medical Genetics*, 5th ed.; Rimoin, D. L.; Connor, J. M.; Pyeritz, P.; Korf, B. R., Eds.; Churchill Livingstone: London, 2007; pp 3–32.

Scriver, C. R.; Childs, B. *Garrod's Inborn Factors in Disease*; Oxford University Press: Oxford, 1989.

Sturtevant, A. H. *A History of Genetics*; Harper and Row: New York, 1965, Reissued 2001 by Cold Spring Harbor Laboratory Press.

CHAPTER 2

Medicine in a Genetic Context[1]

Barton Childs†

Johns Hopkins University School of Medicine, Perelman School of Medicine at the University of Pennsylvania, PA, USA

Reed E Pyeritz

Perelman School of Medicine at the University of Pennsylvania, Philadelphia, PA, USA

Much has been learned in many scientific disciplines through a reductionistic approach. Indeed, our categorical thinking about the field of genetics as applied to medicine, reflected in the organization of this book, is a facile response to the explosion of new information in recent years. However, whatever it is we call medicine has at bottom some consistency and common grounds, one of which is genetics. In the latter half of the previous century, "genetic disease" was partitioned into Mendelian, chromosomal, and multifactorial bins. The distinction between the first two categories has blurred and the latter category has expanded to include more and more conditions. The idea that all diseases are to greater or lesser degree "genetic" is far from a new concept, having arisen several centuries ago. The advent of genomics, which makes possible the study of the genetics of diseases of complex origin in families of patients who have affected relatives, as well as in those who do not, emphasizes that genetic variation underlies the latter no less than the former. The continuity of segregating to nonsegregating familial aggregation is extended to include cases where there is neither segregation nor aggregation. Perhaps we should require a disease to be shown not to be associated with any genetic variation before saying it has no genetic basis.

If genetics is the basic science for medicine, it should be possible to construct a set of principles that characterize disease in a genetic context—that is, a set of generalizations shared by all diseases and framed in genetic terms. And there should be hierarchies of principles, inclusive and of increasing generality and forming a matrix embracing them all. This chapter articulates one such matrix.

In relation to disease, the proximate causes are the products of the variant genes and the experiences of the environment with which they are incongruent. Ultimate or remote causes are the mechanisms of mutation and the causes of fluctuations through time of the elements of the gene pool, including selection, mating systems, founder effects, and drift and the means whereby cultures and social organization evolve. In disease, the variant gene products and the experiences of the environment with which they are incongruent account for characteristic signs and symptoms, but in making available the particular proximate causes assembled by chance in particular patients, it is the remote causes that impart the stamp of individuality to the case.

So the model relates disease to causes, to the gene pool, and ultimately to biological evolution as well as to the evolution of cultures, and to individuality—the latter a consequence of the specificities of both causes. The principles must be seen to be related and interdependent so as to form a network of ideas within which to compose one's thoughts about each specific example of each disease.

In medicine, we tend to think of patients in relation to their disease; that is, as a class of people characterized by the name of the disease. This is what Ernst Mayr called typological thinking. Although patients do differ somewhat one from another, they all share an essence: the disease. In contrast, Mayr proposes population thinking, in recognition that populations consist not of types but of unique individuals. So, in this context, disease has no essence, its variety is imparted by that of the unique individuals who experience it, each in their own private version, and the name of the disease is a convenience, an acknowledgment of the necessity to group patients for logistical purposes.

Medicine is at a peak of success in diagnosis and treatment and moving rapidly to ever new heights of achievement. But all of the changes may not be equally evident. For example, the analysis of pathogenesis, traditionally a top-down process, is beginning to give way to a bottom-up approach in which discovery of variant genes leads to variant protein products and thence to the same

[1]This summary is adapted from the chapter written for the 5th edition of this chapter by Professor Childs (1916–2010).

†Deceased.

molecular analysis of pathogenesis (see Chapter 16). And the genetic heterogeneity and individuality of disease are not easily accommodated in traditional thinking. So we are changing how we look at disease, how we define it and classify it, and the language we use in describing it. For example, genomics and proteomics embody new ways of thinking in the past several decades. And these developments are changing our relationships to biology and to society. Biologists are expressing interest in the fates of the molecules they discover, and the public is becoming aware of what molecular biology and genetics mean to them, as risk factors, for example. So, because this same molecular genetics gives us new insights into the principles that govern—and have always governed—disease, should we not articulate those principles and weave them into our thinking?

If we are to define disease, it must be as loss of adaptation; the open system has had difficulty in maintaining homeostasis. So, our question is, how is this failure of adaptation attained? The straightforward answer is to say that a variation in a homeostatic system was incongruent with its environment, whether within the cell or outside, and the mechanisms for compensation were inadequate, momentarily or permanently, to restore congruence. As a result, other systems were affected, and then still others. But this only tells us that the machine broke down. If we would define disease, we must know what variations can lead to what levels of incongruence. We must know the weaknesses in the evolution of organisms, or if not weaknesses, degrees of flexibility. That is, the origins of human disease lie in both human evolution and in that of the environment with which the human species has evolved to be congruent. And because both biological and cultural evolutions are continuous, though at markedly different rates, congruence must be relative and changing.

In defining disease we must take into account not just the gene that seems most relevant to the phenotype—after which the phenotype may be neglected in the interest of molecular treatment—but also keep alive the relationships of genes (or better, of their products) and phenotypes, the better to grasp the individuality of each, so as to tailor the particularity of the molecular treatment to the biological individuality of a very particular patient. Then, having that principle in mind, the necessity to group patients for treatment can be managed rationally. This rationale underlies the recent surge in interest in "personalizing medicine." So, in the end, how shall we define disease? The definition must include remote causes as well as proximate and the relationship of both to the DNA. And it must be populational in concept, rather than typological, which is to say, it must be nominalist. So, one way of expressing it is as follows: disease is a consequence of incongruence between genetically variable homeostatic systems and the kinds, intensities, and durations of exposures to elements of the environments to which they are called upon to adapt.

Twenty thousand or so genes have been identified, and sooner or later, their products' roles in homeostasis will follow, with obvious benefits for studies of pathogenesis, treatment and prevention. In addition, definitive samples of gene products, useful in defining disease, will be available for characterizing human biological properties hitherto unknown. A few examples of mechanistic—or "how"—questions being asked are as follows:

How variable is the human genome?
Is there an inborn error for every locus?
Are diseases characterized by the qualities of the proteins that are their proximate causes?
Are conserved genes underrepresented or overrepresented in diseases?
What is the role of developmental constraints in fostering or suppressing disease?
Are some proteins more frequently the object of aging processes, or is it random?

Equally intriguing, important and increasingly tractable, are the "why" questions, such as the following:

Why do we have disease?
Why this disease?
Why this person?
Why at this time?

Answering both the "how" and the "why" questions will benefit diagnosis, treatment, and prevention. The logic of prevention is more powerful than that of treatment, but we need both a more comprehensive knowledge of nongenetic proximate causes and, in time, to learn, understand, and adjust to the social dislocations any sudden spate of preventions could bring. But it is in part in the grasp of the possibility and plausibility of both prevention and treatment, and in part in understanding the meaning of individuality in medicine and the virtues of population thinking in relation to it, that we may be able at once to pursue the reductionist path we have so successfully traversed and return to embrace the integration, the humanity, of patients who appeal to us for relief of both the consequences of their molecular incongruities and the injury of the disease to that integrated humanity.

FURTHER READING

1. Childs, B.; Scriver, C. R. Age at Onset and Causes of Disease. *Perspect. Biol. Med.* **1986**, *29*, 437–460.
2. Childs, B. *Genetic Medicine*; Johns Hopkins Press: Baltimore, 1999.
3. Costa, T.; Scriver, C. R.; Childs, B. The Effect of Mendelian Disease on Human Health: A Measurement. *Am. J. Med. Genet.* **1985**, *21*, 231–242.
4. Hayes, A.; Costa, T.; Scriver, C. R.; Childs, B. The Effect of Mendelian Disease on Human Health. II: Response to Treatments. *Am. J. Med. Genet.* **1985**, *21*, 243–255.
5. Green, E. D.; Guyer, M. S. Charting a Course for Henomic Medicine from Basepairs to Bedside. *Nature* **2011**, *470*, 204–213.

6. Jacobs, P. A. The Role of Chromosome Abnormalities in Reproductive Failure. *Reprod. Nutr. Dev.* **1990,** *30* (Supp), 63s–74s.
7. Lander, E. S. Initial Impact of the Sequencing of the Human Genome. *Nature* **2011,** *470,* 187–197.
8. Mayr, E.; Provine, W. B. *The Evolutionary Synthesis*; Harvard University Press: Cambridge, 1980.
9. Scriver, C. R.; Childs, B. *Garrod's Inborn Factors in Disease*; Oxford Press: New York, 1989.
10. Thomas, L. The Future Impact of Science and Technology on Medicine. *Bioscience* **1974,** *24,* 99.
11. Treacy, E.; Childs, B.; Scriver, C. R. Response to Treatment in Hereditary Metabolic Disease: 1993 Survey and 10 Years Comparison. *Am. J. Hum. Genet.* **1995,** *56,* 359–367.
12. Waddington, C. H. *The Strategy of the Genes*; Allen and Unwin: London, 1957.

CHAPTER 3

Nature and Frequency of Genetic Disease

Bruce R Korf

Department of Genetics, Heflin Center for Genomic Sciences, University of Alabama at Birmingham, Birmingham, AL, USA

David L Rimoin†

Medical Genetics Institute, Cedars-Sinai Medical Center, Los Angeles, CA, USA

Reed E Pyeritz

Perelman School of Medicine at the University of Pennsylvania, Philadelphia, PA, USA

Genes are major determinants of human variation. Genome sequencing studies have shown that an individual genome includes 10,000–11,000 differences from the reference genome that changes an amino acid in a protein (non-synonymous change) and an approximately equal number of synonymous variants. Any individual is heterozygous for 50–100 variants that have been associated with genetic disorders (*1*). Genetic variation spans a range from chromosomal changes to single nucleotide changes. The impact of genetically determined characteristics likewise spans a continuum and may be nil at one extreme or lethal at the other.

3.1 CHROMOSOMAL DISORDERS

By definition, a chromosomal disorder is present if there is a visible alteration in the number or structure of the chromosomes. Using routine light microscopy and a moderate level of chromosome banding, the frequency of balanced and unbalanced structural rearrangements in newborns has been estimated at about 9.2 in 1000 (*2*). Some of those with unbalanced rearrangements will have congenital anomalies and/or intellectual disabilities. A proportion of those with balanced changes, will, in adult life, be at increased risk of either miscarriage or having a disabled child. The incidence of aneuploidy in newborns is about 3 in 1000, but the frequency increases dramatically among stillbirths or in spontaneous abortions (*3*). Different types of chromosomal abnormalities predominate in spontaneous abortions as compared with live-born infants. For example, trisomy 16 is the commonest autosomal trisomy in abortions, whereas trisomies for chromosomes 21, 18, and 13 are the only autosomal trisomies occurring at appreciable frequencies in live-born infants. Monosomy for the X chromosome (45,X) occurs in about 1% of all conceptions, but 98% of those affected do not reach term. Triploidy is also frequent in abortions but is exceptional in newborns. The high frequency of chromosomally abnormal conceptions is mirrored by results of chromosome analysis in gametes, which reveal an approximate abnormality rate of 4–5% in sperm (*4*) and 12–15% in oocytes (*5*).

Routine light microscopy cannot resolve small amounts of missing or additional material (less than 4 Mb of DNA). The advent of genomic microarray analysis has revealed a high frequency of submicroscopic deletions and duplications and other copy number variations, including both apparently benign and pathological changes (*6*). Multiple microdeletion and microduplication disorders have been defined in recent years and undoubtedly more await discovery. (see Chapter 29). Such microdeletions, which epistemologically link "chromosomal disorders" with single-gene disorders, account for a proportion of currently unexplained learning disability and multiple malformation syndromes (*7*).

3.2 SINGLE-GENE DISORDERS

By definition, single-gene disorders arise as a result of mutations in one or both alleles of a gene on an autosome

†Deceased.

or sex chromosome or in a mitochondrial gene. There have been many investigations into the overall frequency of single-gene disorders. Many early estimates were misleadingly low due to under-ascertainment, especially of late-onset disorders (e.g. familial hypercholesterolemia, adult polycystic kidney disease, and Huntington disease). Carter (*8*) reviewed the earlier literature and estimated an overall incidence of autosomal dominant traits of 7.0 in 1000 live births, of autosomal recessive traits of 2.5 in 1000 live births, and of X-linked disorders of 0.5 in 1000 live births. This gave a combined frequency of 10 in 1000 live births (1%). At that time, approximately 2500 single-gene disorders had been delineated. The number of recognized Mendelian phenotypes has since almost doubled, and these new entities include several particularly common conditions (e.g. familial breast cancer syndromes, with a combined estimated frequency of 5 in 1000; hereditary nonpolyposis colon cancer syndromes, with a combined frequency of 5 in 1000). In addition, new technologies for DNA analysis have revealed a higher-than-expected frequency of generally asymptomatic people with one or two mutant alleles at a locus (*1*). For example, up to 1% of the population has a mutant allele for von Willebrand factor, but many of these people have few or no symptoms, so again there is the problem of the imprecise and variable boundary between a harmless variant and a clinically important one. Furthermore, DNA analysis has shown that for several important disorders, including myotonic dystrophy and fragile X syndrome, relatives of an affected individual may harbor a premutation that, although not detrimental to the carrier, has the potential for expansion to a full deleterious mutation in an offspring. The prevalence of such premutation carriers may be as high as 1 in 178 females for fragile X syndrome (*9*).

The frequencies of many single-gene disorders show population variation. Geographic variation may be explained by selection or by founder effects or attributed to random genetic drift. Selection has resulted in a carrier frequency of one in three for sickle cell anemia in parts of equatorial Africa, and the Afrikaners of South Africa have a high frequency of variegate porphyria and familial hypercholesterolemia due to a founder effect. The carrier frequency for mutations in *HFE*, one of the genes responsible for hemochromatosis, is 1 in 10 individuals of Celtic ancestry. Undoubtedly, more single-gene disorders are going to be delineated. In theory, at least one per locus will eventually be recognized (about 20,000) minus those with no or a mild phenotype and minus those incompatible with establishment/continuance of a pregnancy. Increasing the total are those loci for which different mutations cause entirely different phenotypes. For example, mutations in LMNA can result in at least 13 distinct disorders (*10*). There is also overlap with the multifactorial category. For example, many patients with acute intermittent porphyria are asymptomatic in the absence of an environmental trigger, and epistatic involvement of other genes is believed to contribute to intrafamilial phenotypic variation for patients with the same mutation. As more gene–environment and gene–gene interactions are identified, the boundary between single-gene disorders and multifactorial disorders will become further blurred.

3.3 MULTIFACTORIAL DISORDERS

Multifactorial disorders result from an interaction of one or more genes with one or more environmental factors. Thus, in effect, the genetic contribution predisposes the individual to the actions of environmental agents. Such an interaction is suspected when conditions show an increased recurrence risk within families, which does not reach the level of risk or pattern seen for single-gene disorders and when identical twin concordance exceeds that for nonidentical twins but is less than 100% (see Chapter 14). For most multifactorial disorders, however, the nature of the environmental agent(s) and the genetic predisposition are currently unclear and are the subject of intensive research efforts. The ability to conduct genome-wide association studies has accelerated progress in this area.

Multifactorial disorders are believed to account for approximately one-half of all congenital malformations and to be relevant to many common chronic disorders of adulthood, including hypertension, rheumatoid arthritis, psychoses, and atherosclerosis (complex common disorders). The former group had an estimated frequency of 46.4 per 1000 in the British Columbia Health Surveillance Registry (*11*). In addition, a multifactorial etiology is suspected for many common psychological disorders of childhood, including dyslexia (5–10% of the population), specific language impairment (5% of children), and attention deficit-hyperactivity disorder (4–10% of children). Hence, the multifactorial disorder category represents the commonest type of genetic disease in both children and adults. Multifactorial disorders also show considerable ethnic and geographic variation. For example, talipes equinovarus is some six times more common among Maoris than among Europeans, and neural tube defects were once 10 times more frequent in Ireland than in North America.

Often ignored, both intellectually and in research, are genotypes that reduce susceptibility to potentially harmful environmental factors. An understanding of alleles that provide protection from disease or increase longevity will yield insight into pathogenesis as well as novel approaches to therapy and prevention.

3.4 SOMATIC CELL GENETIC DISORDERS

Somatic cell mutation is a natural developmental process in the immune system, but is also responsible for a significant burden of genetic disease. This includes somatic or germline mosaicism for single-gene disorders (*12*), as well as mutations that give rise to cancer (see Chapter 21). Cancer cells tend to have accumulated multiple mutations; the first step in the cascade of mutations may be inherited (i.e. involving germ cells and all somatic cells). Carcinogens are important causes of noninherited mutations, and genetic susceptibility is suspected to account for individual variation in risk on exposure. Somatic cell genetic disorders might also be involved in other clinical conditions such as autoimmune disorders and the aging process.

REFERENCES

1. Durbin, R. M.; Abecasis, G. R.; Altshuler, D. L.; Auton, A.; Brooks, L. D.; Durbin, R. M.; Gibbs, R. A.; Hurles, M. E.; McVean, G. A.; Durbin, R. M., et al. A Map of Human Genome Variation from Population-Scale Sequencing. *Nature* **2010,** *467,* 1061–1073.
2. Jacobs, P. A.; Browne, C.; Gregson, N.; Joyce, C.; White, H. Estimates of the Frequency of Chromosome Abnormalities Detectable in Unselected Newborns using Moderate Levels of Banding. *J. Med. Genet.* **1992,** *29,* 103.
3. Hassold, T.; Hunt, P. To Err (Meiotically) Is Human: The Genesis of Human Aneuploidy. *Nat. Rev. Genet.* **2001,** *2,* 280–291.
4. Templado, C.; Vidal, F.; Estop, A. Aneuploidy in Human Spermatozoa. *Cytogenet. Genome Res.* **2011,** *133,* 91–99.
5. Rosenbusch, B. The Incidence of Aneuploidy in Human Oocytes Assessed by Conventional Cytogenetic Analysis. *Hereditas* **2004,** *141,* 97–105.
6. Miller, D. T.; Adam, M. P.; Aradhya, S.; Biesecker, L. G.; Brothman, A. R.; Carter, N. P.; Church, D. M.; Crolla, J. A.; Eichler, E. E.; Epstein, C. J., et al. Consensus Statement: Chromosomal Microarray Is a First-Tier Clinical Diagnostic Test for Individuals with Developmental Disabilities or Congenital Anomalies. *Am. J. Hum. Genet.* **2010,** *86,* 749–764.
7. Girirajan, S.; Eichler, E. E. Phenotypic Variability and Genetic Susceptibility to Genomic Disorders. *Hum. Mol. Genet.* **2010,** *19,* R176–R187.
8. Carter, C. O. Monogenic Disorders. *J. Med. Genet.* **1977,** *14,* 316.
9. Hantash, F. M.; Goos, D. M.; Crossley, B.; Anderson, B.; Zhang, K.; Sun, W.; Strom, C. M. FMR1 Premutation Carrier Frequency in Patients Undergoing Routine Population-Based Carrier Screening: Insights into the Prevalence of Fragile X Syndrome, Fragile X-Associated Tremor/Ataxia Syndrome, and Fragile X-Associated Primary Ovarian Insufficiency in the United States. *Genet. Med.* **2011,** *13,* 39–45.
10. Capell, B. C.; Collins, F. S. Human Laminopathies: Nuclei Gone Genetically Awry. *Nat. Rev. Genet.* **2006,** *7,* 940–952.
11. Baird, P. A.; Anderson, T. W.; Newcombe, H. B.; Lowry, R. B. Genetic Disorders in Children and Young Adults: A Population Study. *Am. J. Hum. Genet.* **1988,** *42,* 677.
12. Youssoufian, H.; Pyeritz, R. E. Mechanisms and Consequences of Somatic Mosaicism in Humans. *Nat. Rev. Genet.* **2002,** *3,* 748–758.

CHAPTER

4

Genomics and Proteomics

Raju Kucherlapati

The last century has witnessed the increasing realization that genetics plays an important role in human health and disease. The principles underlying genetics were discovered by Mendel in 1867, but remained obscure until the beginning of the twentieth century. Archibald Garrod first enumerated the significance of these genetic principles to human disease in 1902.

The initial focus of human genetics was the class of diseases and disorders that are the result of mutations in single genes. Because the inheritance of these disorders follows the rules established by Mendel's discoveries, these disorders are referred to as Mendelian disorders. Because the disorders are the result of mutations in single genes, they are also referred to as single-gene disorders. Much of the twentieth century was devoted to the description of a large number of such disorders. A compilation of these disorders that is available online (Online Mendelian Inheritance in Man [OMIM]) lists more than 10,000 such disorders. Many of these disorders were discovered in newborn children. Although all of these single-gene disorders collectively constitute a significant health burden on the population, the relative rarity of each of the disorders made human genetics an esoteric specialty of medicine. More recent developments resulted in establishing that genetics plays an important role in susceptibility to many disorders and genetic variation may play a modulating role in other disorders.

Tools are now available to identify the location of specific genes within the human genome and to establish what genetic changes might be associated with specific disease. The mapping of genes or diseases to specific locations of the genome is facilitated by the development of genetic and physical maps of the human genome. The construction of genetic maps relies on the use of recombination frequencies between polymorphic markers. During the development of genetic maps several different types of polymorphic markers were used and single nucleotide polymorphisms are currently used extensively in constructing and utilizing genetic maps.

Sequencing of the "first" human genome was accomplished by two methods. One was clone-by-clone sequencing and the second one is "shot-gun" sequencing. To facilitate clone-by-clone sequencing, clone-based physical maps were constructed. During the past 10 years, the cost of sequencing has gone down by several orders of magnitude and it is now possible to sequence an entire human genome for less than $10,000.

There are a number of methods available to identify disease genes. The most current approaches involve sequencing all of the genes (whole-exome sequencing) or whole genome sequencing, identifying all of the significant genetic variants and assessing which of the variants may be responsible for the disease. Use of these methodologies is resulting in rapid advances in identifying human disease genes. Gene sequencing technologies, along with technologies to assess changes in RNA and protein expression, are revolutionizing our understanding of the human genome and the role of individual genes in health and disease.

CHAPTER

5

Genome and Gene Structure

Daniel H Cohn

Department of Molecular, Cell and Developmental Biology, Department of Orthopedic Surgery, UCLA, Los Angeles, CA, USA

The function of the human genome is to transfer information reliably from parent to daughter cells and from one generation to the next. The process of duplicating the genome for both cell division and gametogenesis is *replication*. Replication depends on the double-stranded structure of DNA, where the strands of the parental DNA serve as the template for synthesis of the new strands prior to cell division. Replication occurs with high fidelity, with a mutation rate on the order of 1×10^{-8} per base pair per generation, minimizing the number of new mutations while preserving the evolutionary flexibility that mutations provide.

Transcription represents initiation of the process by which the information encoded in the genome is expressed. Transcription is mediated by RNA polymerases, which read the genome sequence, identify transcription start sites, and synthesize the primary transcript. Gene expression is a highly regulated process, governed by chromatin structure and transcriptional regulatory factors, the largest class of genes present in the human genome. Transcriptional regulators target sequences within genes, including the transcriptional initiation site (promoter) as well as other regulatory sequences such as enhancers and silencers. The regulation of transcription determines the developmental timing, tissue specificity, and level of expression of each gene. Additional regulation can be conferred during the maturation of the primary transcript via capping, splicing, and (for protein coding genes) polyadenylation, yielding an exquisite diversity of mature RNA molecules.

Mature messenger RNAs are acted upon by ribosomes to effect *translation*, synthesizing the wide variety of proteins present in each cell. Translation is also a regulated process, determining the level of synthesis of each protein and, along with the sequence of the encoded protein itself, the subcellular compartment to which the protein is directed. The folding and targeting of each protein is mediated by complex mechanisms that may be specific to each subcellular compartment, determining for example whether proteins are cytoplasmic, secreted, or resident in specific organelles. Posttranslational modifications, including hydroxylation, glycosylation, and phosphorylation, also play a role in the localization and activities of proteins.

The genomes of individuals differ in a multiplicity of ways. During meiosis, *recombination* during male and female gametogenesis reshuffles the parental genomes to generate diversity. The largest number of loci that differs between individuals is single nucleotide polymorphisms. However, a variety of different types of copy number variation account for the greatest number of nucleotide differences between individuals. The aggregate genetic variability between humans dictates the traits we manifest as well as the diseases to which individuals are susceptible.

CHAPTER

6

Epigenetics

Rosanna Weksberg

Genetics and Genome Biology, Research Institute; Clinical and Metabolic Genetics, The Hospital for Sick Children; Institute of Medical Sciences, University of Toronto, Toronto, ON, Canada

Darci T Butcher, Daria Grafodatskaya, and Sanaa Choufani

Genetics and Genome Biology, Research Institute, The Hospital for Sick Children, Toronto, ON, Canada

Benjamin Tycko

Institute for Cancer Genetics, Herbert Irving Comprehensive Cancer Center, Columbia University Medical Center, New York, NY, USA

Despite the tremendous advances in human genetics enabled by the original public and private human genome projects and brought to fruition with high-throughput genotyping and "Nextgen" DNA sequencing, many aspects of human biology still cannot be adequately explained by nucleotide sequences alone. Normal human development requires the specification of a multitude of cell types or organs that depend on transcriptional regulation programmed by epigenetic mechanisms. Epigenetics refers to modifications to DNA and its associated proteins that define the distinct gene expression profiles for individual cell types at specific developmental stages. Disruption of such control mechanisms is associated with a variety of diseases with behavioral, endocrine or neurologic manifestations, and quite strikingly with disorders of tissue growth, including cancer. While the involvement of epigenetic alterations in many of these diseases has been known to specialists for some time, the importance of epigenetics in general clinical medicine has only just begun to emerge. Current research is focused on characterizing *cis*- and *trans*-acting influences of the genetic background on epigenetic marks, delineating cell type- or tissue-specific epigenetic marks in human health and disease, studying the interaction between epigenetic marks and the environment especially with respect to fetal programming and risks for common adult onset disorders, and modulating adverse epigenetic states by drug-based and nutritional therapies.

An epigenetic trait is defined as a "stably heritable phenotype resulting from changes in a chromosome without alterations in the DNA sequence." Epigenetic patterns, essential for controlling gene expression in normal growth and development, are established by a number of mechanisms including DNA methylation at cytosine residues in CpG dinucleotides and covalent modifications of histone proteins, as well as by less well-understood mechanisms controlling long-range chromatin architecture within the cell nucleus. Although DNA methylation and histone modifications are regulated by different sets of enzymes, cross talk between these modifications occurs through interactions of enzymes and other proteins that create and recognize these patterns. The relationship between these two central types of epigenetic modifications is known to be bidirectional, with histone marks being more labile and DNA methylation more stable. Thus DNA methylation can act to "lock in" epigenetic states. However, regulating metastable states of gene expression is so crucial in *trans*-acting influences development and tissue homeostasis that other mechanisms, in addition to histone modifications and DNA methylation, come into play to establish and maintain epigenetic states. Regulatory noncoding RNAs, including small interfering RNA, microRNAs (miRNAs), and long noncoding RNAs, play important roles in gene expression regulation at several levels of transcription, mRNA degradation, splicing, transport, and translation.

Epigenetic mechanisms involve a wide variety of normal developmental processes. Alterations in genes and proteins that are responsible for maintaining epigenomic patterns are the cause of a number of human disorders and diseases. Genomic imprinting, the unequal contribution of maternal and paternal allele to the offspring, caused by DNA methylation contributes to a number of human disorders. Imprinted genes typically function in growth regulation and neurodevelopment, and the corresponding disease phenotypes due to genetic or epigenetic aberrations in these genes indeed entail major abnormalities of intrauterine growth or postnatal cognition and behavior. These disorders include Beckwith–Wiedemann, Russell Silver, Prader–Willi and Angelman syndromes. A number of disorders have been described with mutations or deletions in genes that are important for maintaining normal epigenetic regulation. Loss of function of these genes can disrupt normal establishment, maintenance, or reading of epigenetic marks, thereby resulting in altered chromatin structure and gene expression. Many of these disorders are associated with intellectual disability (ID); other features include facial dysmorphology and various congenital anomalies. These include CHARGE association, and Kleefstra, Sotos, Kabuki, and Rett syndromes.

Epigenetic aberrations can result in a diverse array of non-neoplastic human diseases. Changes in DNA methylation were the first epigenetic alteration identified in cancer, and subsequent work over three decades has shown that both hyper- and hypomethylations are important and pervasive pathogenic mechanisms both in early and late stages of human neoplasia. Not surprisingly, histone modifications and miRNA expression are also altered in cancer. Proteins regulating epigenetic marks, including histone methyltransferases and demethylases, DNA methyltransferases, and chromatin remodeling SWI–SNF complexes are also dysregulated, not just by over- or under-expression but also by recurrent cancer-associated somatic mutations. Two classes of medications, DNA methylation inhibitors and HDAC inhibitors, have been approved by the US Food and Drug Administration as treatment for cancer. The wide array of epigenetic alterations identified in human disease could present valuable targets for approved medications as well as for novel ones that have yet to be developed.

Trans-generational effects of the environment on our epigenome have been documented providing evidence that in humans the mother's diet early in pregnancy can directly affect the programming of the epigenome early in utero that shape our development later in life. In mice and humans, oocytes retrieved following hormonal induction or embryos studied after in vitro culture have shown methylation and/or expression anomalies in several imprinted genes. Increasing attention has recently been focused on reports of increased rates of epigenetic errors in human following infertility/artificial reproductive technology (ART). In particular, two rare disorders exhibited an increased incidence in retrospective studies in children born following infertility/ART.

The role of epigenetic marks in translating the primary genomic sequence has now moved to the forefront of human genetics, with clear implications for our understanding of human development and disease. A number of initiatives have now been implemented to define human epigenetic patterns at high resolution with complete genomic coverage. The technology is now available to investigate multiple tissue-specific epigenomes in humans, and the NIH Roadmap Epigenomics Mapping Consortium (www.roadmapepigenomics.org) was launched to produce a public resource of human epigenomic data to catalyze basic biology and disease-oriented research. Another parallel initiative is the NIH Epigenomics of Health and Disease Roadmap Program, which funds investigator-initiated research. While a good part of what we know so far about epigenetics in disease has come from cancer research, it is telling that most of the initial research grants from this program have been targeted to other complex diseases ranging from Alzheimer disease to adult heart disease and diabetes to autism. These initiatives interface with the International Human Epigenomics Consortium, which was established to accelerate and coordinate epigenomics research worldwide. These data should provide keys to unravel genetic and environmental factors that impinge on epigenomes to affect normal processes such as development and aging and lead to human diseases when these processes go awry.

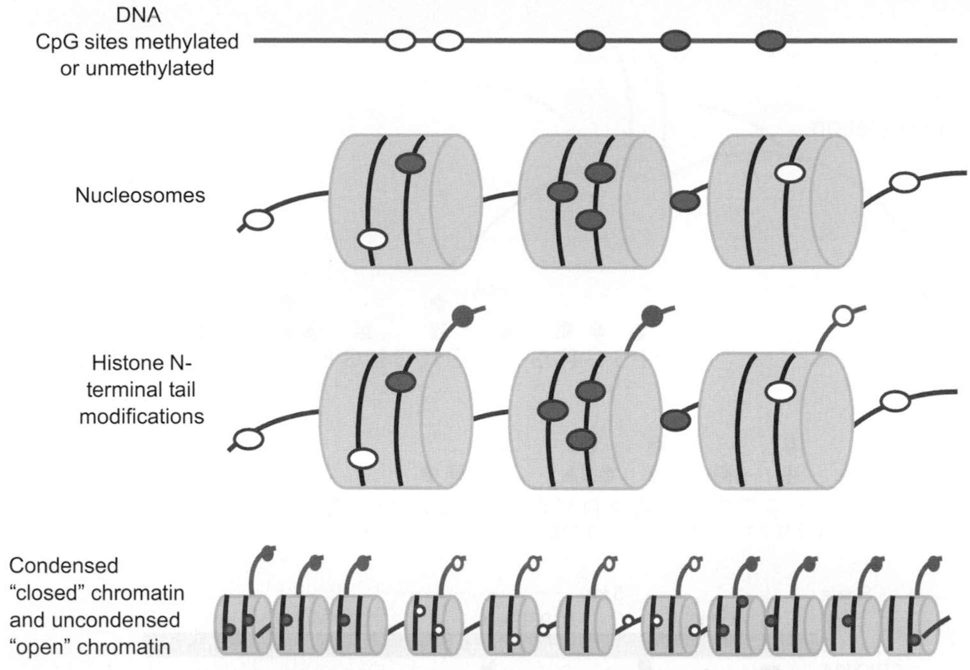

FIGURE 6-1 Epigenetic organization of chromatin: layering of DNA methylation, histone modification to control gene expression. DNA of a gene promoter can be unmethylated (white circles) and in most cases the gene is expressed or the promoter can be methylated (blue circles) and in most cases the gene is not expressed. DNA is not independent of its associated histone proteins. Histone modifications are established and maintained independently or dependently on the DNA methylation state of a region. These protein modifications can activate (open orange circles) or repress (filled orange circles) gene transcription. Although not shown in this figure, but mentioned in the text, additional epigenetic processes, including microRNA and long noncoding RNAs, also contribute to gene regulation. The DNA/histone protein nucleosome core is further compacted to form higher order chromatin structures that also contribute to gene regulation.

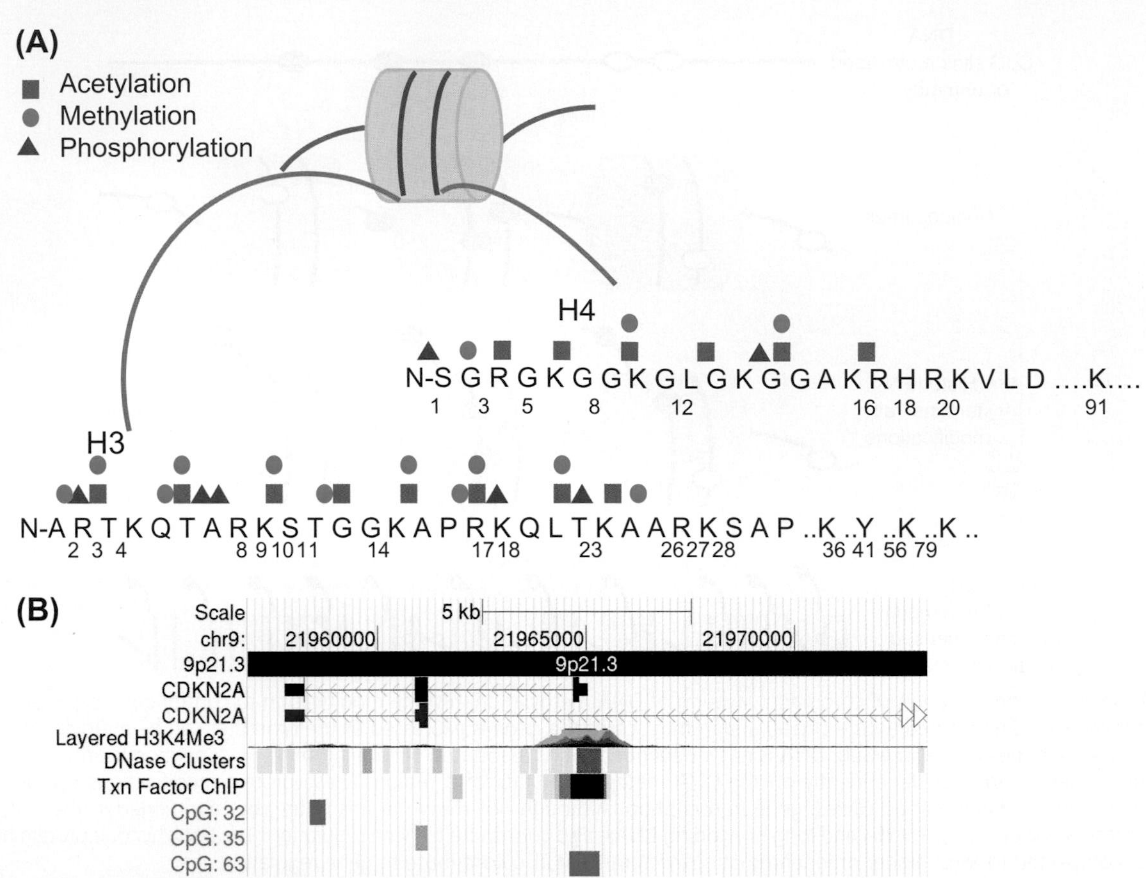

FIGURE 6-2 A. Histone Modifications of Histone H3 and H4 N-Terminal Tails. Posttranslational modifications of N-terminal tails (these can also occur in the C-terminal domain but are not shown here) can occur in combination and are read by the appropriate protein to establish local and global decondensed or open and condensed or closed chromatin states. Ac, acetylation (blue squares); Me, methylation (green circles); P, phosphorylation (red triangles). B. Snapshot from UCSC Genome Browser Representing H3K4 Methylation in the Promoter of the Tumor-suppressor Gene *CDKN2A*. This diagram is an example of epigenetic data available in UCSC genome browser. The description of each genomic feature is shown on the left. Two isoforms of *CDKINC* genes are shown in black and blue, here we focus on a shorter (black) isoform. Enrichment for active histone H3K4me3 is shown by multiple colors in nine cell lines. The peak of H3K4me3 coincides with transcription start site of *CDKN2A*, CpG island (green), as well transcription factor binding sites (TxN factor CHIP) and DNase clusters, which is an indicator of open chromatin. All of these marks—H3K4me3, transcription factor binding as well DNase clusters—indicate that *CDKN2A* is active transcribed in these cell lines. Information about other histone marks and DNA methylation levels is available from at the USCS genome browser under multiple tracks from the Regulation section. This image was downloaded from UCSC genome browser http://genome.ucsc.edu/ (Ref. (*313*).) The ENCODE Regulation data is from Ref. (*314*).

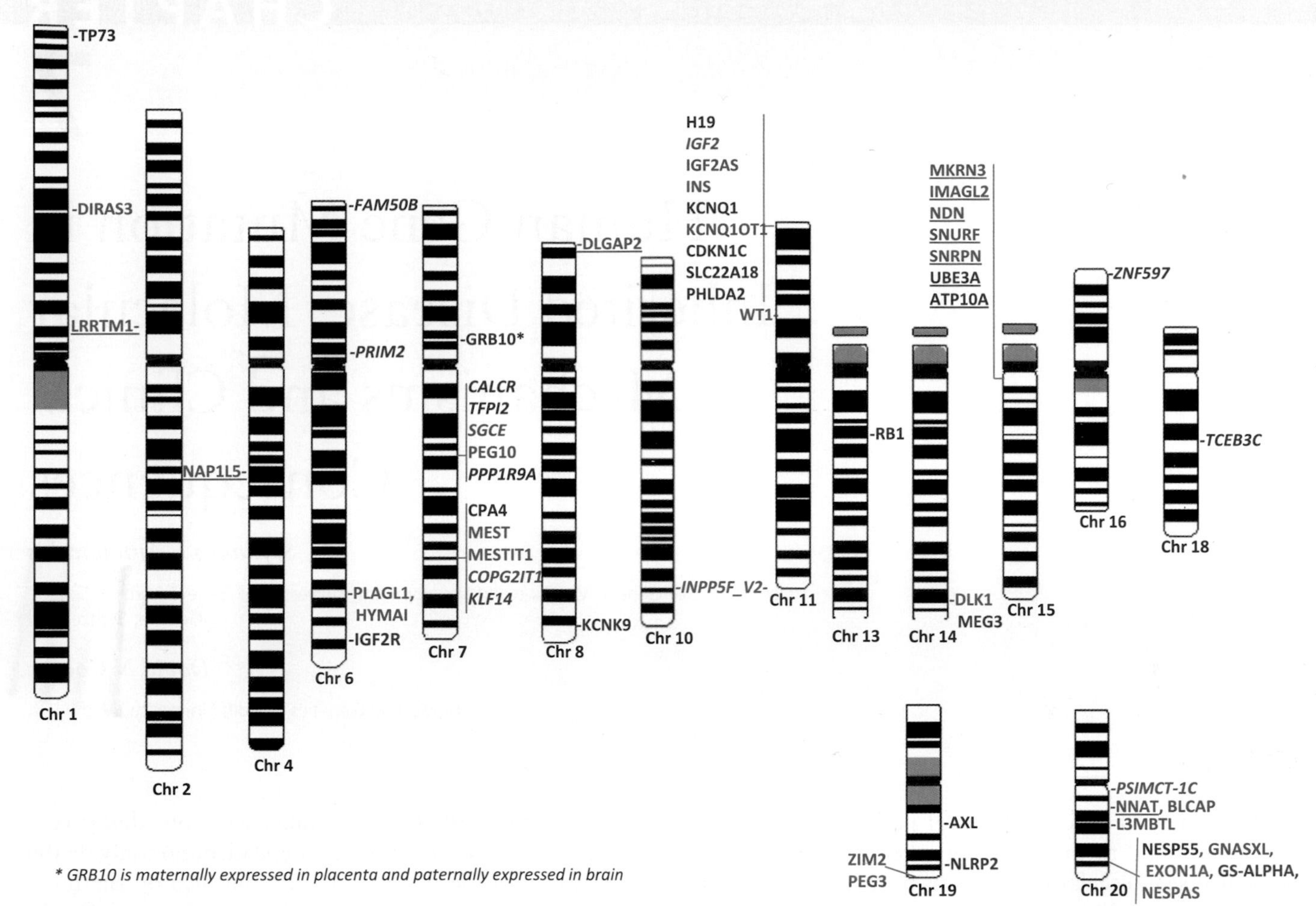

FIGURE 6-3 Ideograms of Human Imprinted Genes. Ideograms were generated using http://www.dna-rainbow.org/ideograms/. Ideogram of each human chromosome known to have an imprinted gene based on the imprinted gene catalog last updated January 2011 (http://igc.otago.ac.nz) and recent literature. The G-bands, areas with proportional more A-T base pairs, are normally colored black in schematic representations. To compare the schematic ideograms with our rendered chromosomes, we colored the A-T bases black and the G-C bases white. Blue areas in the rendered chromosomes identify bases not known yet. Blue genes are paternally expressed, and red genes are maternally expressed. Bold genes are implicated in growth, and underlined genes play roles is neurodevelopment. Genes in italics have no reported function in growth or neurodevelopment.

FURTHER READING

1. Barker, D. J.; Eriksson, J. G.; Forsen, T.; Osmond, C. Fetal Origins of Adult Disease: Strength of Effects and Biological Basis. *Int. J. Epidemiol.* **2002**, *31,* 1235–1239.
2. Berger, S. L.; Kouzarides, T.; Shiekhattar, R.; Shilatifard, A. An Operational Definition of Epigenetics. *Genes Dev.* **2009**, *23,* 781–783.
3. Cedar, H.; Bergman, Y. Linking DNA Methylation and Histone Modification: Patterns and Paradigms. *Nat. Rev. Genet.* **2009**, *10,* 295–304.
4. Ernst, J.; Kellis, M. Discovery and Characterization of Chromatin States for Systematic Annotation of the Human Genome. *Nat. Biotechnol.* **2010**, *28,* 817–825.
5. Espinoza, C. A.; Ren, B. Mapping Higher Order Structure of Chromatin Domains. *Nat. Genet.* **2011**, *43,* 615–616.
6. Ponting, C. P.; Oliver, P. L.; Reik, W. Evolution and Functions of Long Noncoding RNAs. *Cell* **2009**, *136,* 629–641.
7. Portela, A.; Esteller, M. Epigenetic Modifications and Human Disease. *Nat. Biotechnol.* **2010**, *28,* 1057–1068.
8. Sutcliffe, A. G.; Peters, C. J.; Bowdin, S.; Temple, K.; Reardon, W.; Wilson, L.; Clayton-Smith, J.; Brueton, L. A.; Bannister, W.; Maher, E. R. Assisted Reproductive Therapies and Imprinting Disorders—A Preliminary British Survey. *Hum. Reprod.* **2006**, *21,* 1009–1011.
9. Tilghman, S. M. The Sins of the Fathers and Mothers: Genomic Imprinting in Mammalian Development. *Cell* **1999**, *96,* 185–193.
10. van Bokhoven, H.; Kramer, J. M. Disruption of the Epigenetic Code: An Emerging Mechanism in Mental Retardation. *Neurobiol. Dis.* **2010**, *39,* 3–12.
11. Wolffe, A. P.; Matzke, M. A. Epigenetics: Regulation through Repression. *Science* **1999**, *286,* 481–486.

CHAPTER

7

Human Gene Mutation in Inherited Disease: Molecular Mechanisms and Clinical Consequences

Stylianos E Antonarakis

Department of Genetic Medicine and Development, University of Geneva Medical School, Geneva, Switzerland

David N Cooper

Institute of Medical Genetics, Cardiff University, Cardiff, UK

A wide variety of different types of pathogenic mutation have been found to cause human inherited disease, with many diverse molecular mechanisms being responsible for their generation. These types of mutation include single base-pair substitutions in coding, regulatory and splicing-relevant regions of human genes, as well as micro-deletions, micro-insertions, duplications, repeat expansions, combined micro-insertions/deletions (indels), inversions, gross deletions and insertions, and complex rearrangements.

The first description of a heritable human gene mutation at the DNA level was accomplished in 1978: gross deletions of the human α-globin (*HBA*) and β-globin (*HBB*) gene clusters giving rise to α- and β-thalassemia. Over the last 33 years, excess of 120,000 different inherited disease-causing mutations and disease-associated/functional polymorphisms have been identified and characterized in a total of over 4400 human genes. Single base-pair substitutions (67%) and micro-deletions (15.6%) are the most frequently encountered mutations, the remainder comprising an assortment of micro-insertions (6.5%), indels (1.5%), gross deletions (6.6%), gross insertions and duplications (1.4%), inversions, repeat expansions (0.3%), and complex rearrangements (1.0%). The vast majority of known mutations reside within the coding region (86%), the remainder being located in either intronic (11%) or regulatory (3%, promoter, untranslated, or flanking regions) sequences. Mutations may interfere with any stage in the pathway of expression from gene activation to synthesis and secretion of the mature protein product. The question of the proportion of possible mutations within human disease genes that are likely to be of pathological significance is one that is very difficult to address because it is dependent not only on the type and location of the mutation but also on the functionality of the nucleotides involved (itself dependent in part on the amino acid residues that they encode) which is often hard to assess. In addition, some types of mutation are likely to be much more comprehensively ascertained than others, making observational comparisons between mutation types an inherently hazardous undertaking. Different types of human gene mutation may vary in size, from structural variants to single base-pair substitutions, but what they all have in common is that their nature, size, and location are often determined either by specific characteristics of the local DNA sequence environment or by higher order features of the genomic architecture. The human genome is now recognized to contain "pervasive architectural flaws," where certain DNA sequences are inherently mutation prone by virtue of their base composition, sequence repetitivity and/or epigenetic modification. Indeed, the nature, location, and frequency of different types of mutation causing inherited disease are shaped in large part, and often in remarkably predictable ways, by the local DNA sequence environment. The mutability of a given gene or genomic region may also be influenced indirectly by a variety of noncanonical (non-B) secondary structures whose formation is facilitated by the underlying DNA sequence. Since these non-B DNA structures can interfere with subsequent DNA replication and repair and may serve to increase mutation frequencies in generalized fashion (i.e. both in the context of subtle mutations and gross rearrangements), they have the potential to serve as a unifying concept in studies of mutational mechanisms underlying human inherited disease.

A major goal of molecular genetic medicine is to be able to predict the nature of the clinical phenotype through ascertainment of the genotype. However, the extent to which this is feasible in medical genetics is very much disease-, gene-, and mutation-dependent.

Given knowledge of a specific clinical phenotype, to what extent can the underlying causal genotype be inferred? Conversely, given knowledge of a specific genotype, to what extent is it possible to infer the likely clinical phenotypic consequences for the individual concerned, for instance in terms of the penetrance, age of onset, and severity of the disease? This chapter attempts to provide an overview of the nature of the different types of mutation causing human genetic disease and then considers their consequences for the clinical phenotype. We argue that the study of mutations in human genes is of paramount importance not only for understanding the pathophysiology of inherited disorders, but also for optimizing diagnostic testing procedures and guiding the design of new therapeutic strategies.

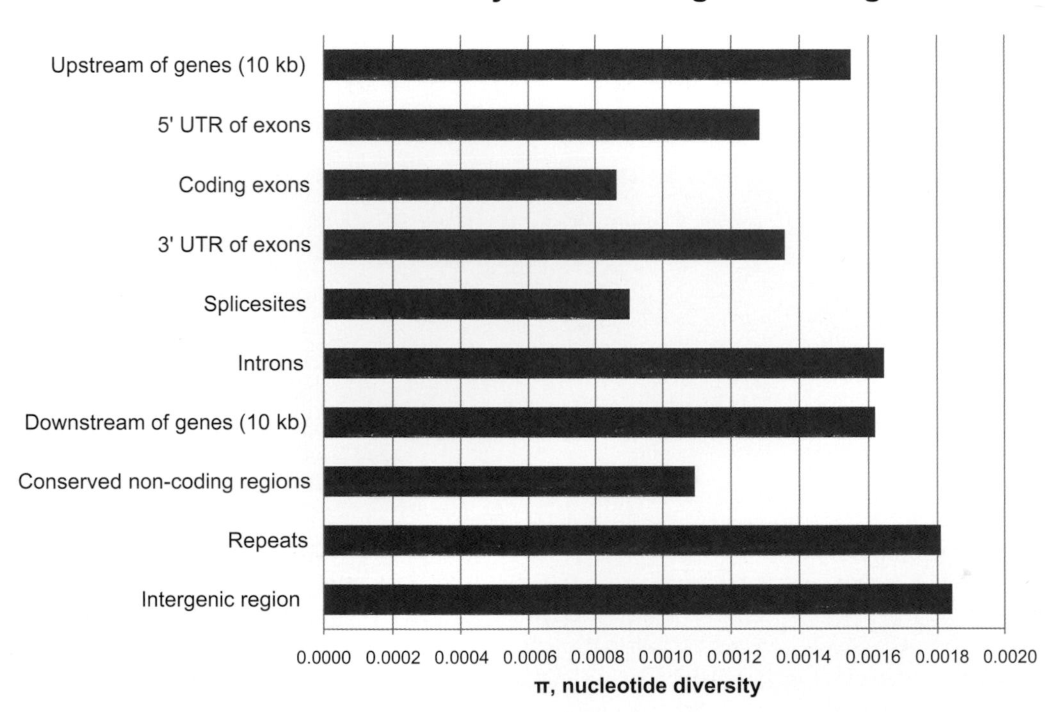

• Genomic variants from CompleteGenomics, 54 unrelated individuals from different ethnic groups (www.completegenomics.com)
• Genomic regions drawn from UCSC genome browser (http://genome.ucsc.edu)

FIGURE 7-1 Nucleotide diversity (equivalent to frequency of polymorphic variants) in different genomic regions. The genomic variants analyzed are from the whole-genome sequences of 54 unrelated human genomes (see text).

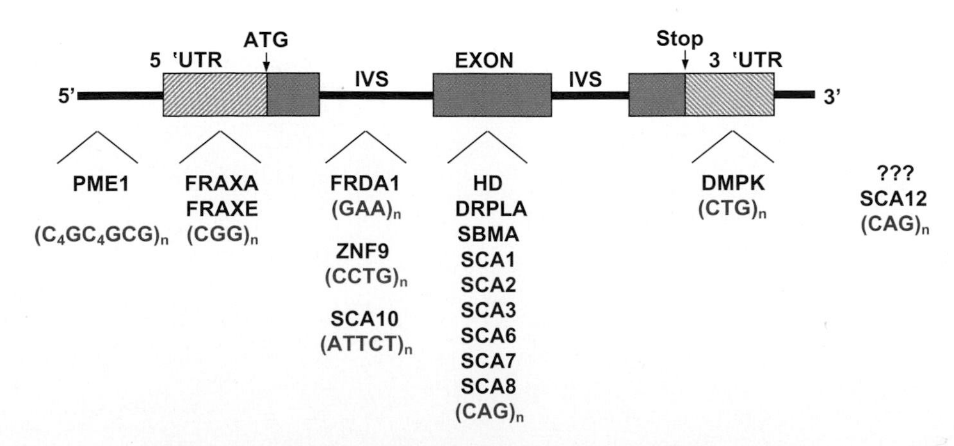

Selected Repeat Expansions in Human Disorders

FIGURE 7-2 Location of the repeat expansion in selected human disorders.

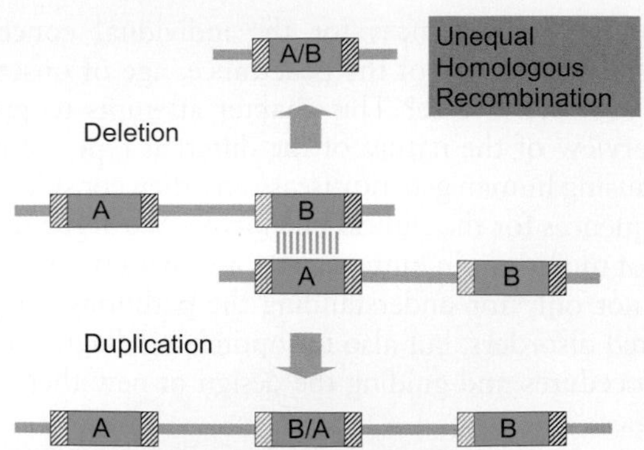

FIGURE 7-3 Homologous unequal recombination between similar regions of sequences A and B. The recombination events cause either deletions or duplications. In case of a deletion, a hybrid sequence is generated with the first part from sequence A and the second from sequence B. The middle sequence in the duplication product is also a hybrid sequence; the first part is from sequence B and the second from sequence A.

CHAPTER

8

Genes in Families

Jackie Cook

Sheffield Regional Genetics Service, Sheffield Children's Hospital, Western Bank, Sheffield, South Yorkshire, UK

8.1 INTRODUCTION

Genetic counseling is the process by which patients and relatives at risk of a disorder that may be hereditary are advised of the consequences of the disorder, the probability of developing and transmitting it, and the ways in which this may be prevented, avoided, or ameliorated. To achieve these aims, an accurate diagnosis and detailed information regarding the family history are essential. The basis for establishing a diagnosis depends on medical history, examination, and investigation. The diagnostic information is combined with information obtained from the family pedigree to determine the mode of inheritance of the disorder and to calculate the risk of recurrence, so that family members can be appropriately counseled.

A family history of a genetic condition may be due to:

1. A mutation within a single nuclear gene
2. A mutation within a mitochondrial gene
3. A contiguous gene deletion or duplication involving anything from a few to a large number of genes
4. A chromosomal rearrangement resulting in unbalanced products at meiosis
5. Multifactorial inheritance

8.2 AUTOSOMAL DOMINANT INHERITANCE

8.2.1 Definition

Autosomal dominant inheritance refers to disorders caused by genes located on the autosomes, thereby affecting both males and females. The disease or mutant alleles are dominant to the wild-type alleles, so the disorder is manifest in the heterozygote (i.e. an individual who possesses both the wild-type and the mutant allele). Autosomal dominant inheritance shows vertical transmission and affects both sexes equally.

8.2.2 Recurrence Risks

Since individuals with autosomal dominant disorders are heterozygous for a mutant and a normal allele, there is a one in two (50%) chance a gamete will carry the normal allele and a one in two (50%) chance a gamete will carry the mutant allele. Assuming that the individual's partner will contribute a normal allele, there is a one in two (50%) chance that the offspring, regardless of sex, will inherit the disorder with each pregnancy.

8.3 AUTOSOMAL RECESSIVE INHERITANCE

8.3.1 Definition

Autosomal recessive inheritance refers to disorders due to genes located on the autosomes, but in which the disease alleles are recessive to the wild-type alleles and are, therefore, not evident in the heterozygous state, only being manifest in the homozygous state. Autosomal recessive disorders affect both males and females and tend to occur in just one generation. The parents of an individual with an autosomal recessive disorder are heterozygous for the disease allele and are usually referred to as being carriers for the disorder.

8.3.2 Recurrence Risks

When two parents carrying the same disease allele reproduce, there is an equal chance that gametes will contain the disease or the wild-type allele. There are four possible combinations of these gametes, resulting in a one in four (25%) chance of having a homozygous affected offspring, a one in two (50%) chance of having a heterozygous unaffected carrier offspring, and a one in four (25%) chance of having a homozygous unaffected offspring.

When an individual with an autosomal recessive disorder has children, they will only produce gametes containing the disease allele. Since it is most likely that their partners will be homozygous for the wild-type allele, the partners will always contribute a normal allele and therefore all the children will be heterozygous carriers and unaffected. If, however, an affected individual has children with a partner who happens to be heterozygous for the disease allele, there will be a 50% chance of transmitting the disorder, depending on whether the partner contributes a disease or a wild-type allele. Such a pedigree is said to exhibit pseudodominance.

8.4 SEX-LINKED INHERITANCE

Strictly speaking, sex-linked inheritance refers to the inheritance patterns shown by genes on the sex chromosomes. If the gene is on the X chromosome, it is said to show X-linked inheritance and, if on the Y chromosome, Y-linked or holandric inheritance.

8.5 X-LINKED RECESSIVE INHERITANCE

8.5.1 Definition

This form of inheritance is conventionally referred to as sex-linked inheritance. It refers to disorders due to recessive genes on the X chromosome. Males have a single X chromosome and are therefore hemizygous for most of the alleles on the X chromosome, so that, if they have a mutant allele, they will manifest the disorder. Females, on the other hand, will usually only manifest the disorder if they are homozygous for the disease allele, and if heterozygous will usually be unaffected. Since it is rare for females to be homozygous for a mutant allele, X-linked recessive disorders usually only affect males with transmission through unaffected female carriers. Male-to-male transmission does not occur.

8.5.2 Recurrence Risks

If a male affected with an X-linked recessive disorder survives to reproduce, he will always transmit his X chromosome with the mutant allele to his daughters, who will be obligate carriers. An affected male will always transmit his Y chromosome to a son, and therefore none of his sons will be affected. A carrier female has one X chromosome with the wild-type allele and one X chromosome with the disease allele; therefore, her sons have a one in two (50%) chance of being affected, while her daughters have a one in two (50%) chance of being carriers.

8.6 X-LINKED DOMINANT INHERITANCE

8.6.1 Definition

X-linked dominant inheritance is an uncommon form of inheritance and is caused by dominant disease alleles on the X chromosome. The disorder will manifest in both hemizygous males and heterozygous females. Random X-inactivation usually means that females are less likely to be severely affected than hemizygous males, unless they are homozygous for the disease allele.

8.6.2 Recurrence Risks

Offspring of either sex have a one in two (50%) chance of inheriting the disorder from affected females. The situation is different for males affected by X-linked dominant disorders, whose daughters will always inherit the gene and whose sons cannot inherit the gene.

8.6.3 X-Linked Dominant Lethal Alleles

In some disorders due to mutant alleles of genes on the X chromosome, affected males are never or very rarely seen (e.g. incontinentia pigmenti and Goltz syndrome). This is thought to be due to a lethal effect of the mutant disease allele in the hemizygous male, resulting in nonviability of the conceptus during early embryonic development. As a consequence, if an affected female were to have children, one would expect a sex ratio of 2:1, female to male, in the offspring and that one half of the females would be affected, while none of the male offspring would be affected. The majority of the mothers of females with these X-linked dominant lethal disorders are generally unaffected, and the disease alleles are therefore thought to arise as new mutations.

8.7 Y-LINKED (HOLANDRIC) INHERITANCE

Y-linked, or holandric, inheritance refers to genes carried on the Y chromosome. They therefore will be present only in males, and the disorder would be passed on to all their sons but never their daughters. Genes involved in spermatogenesis have been mapped to the Y chromosome, but a male with a mutation in a Y-linked gene involved in spermatogenesis would probably be infertile or hypofertile, making it difficult to demonstrate Y-linked inheritance.

8.7.1 Mitochondrial Inheritance

The nuclear chromosomes are not the only source of coding DNA sequences within the cell. Mitochondria possess their own DNA, which, as well as coding for mitochondrial transfer RNA and ribosomal RNA, also carries the genes for 12 structural proteins that are all mitochondrial enzyme subunits. Mutations within these genes have been shown to cause disease (e.g. Leber's hereditary optic neuropathy, LHON). The inheritance pattern of mitochondrial DNA is, however, very different from that of nuclear DNA. Mitochondria are exclusively maternally inherited. Therefore, mitochondrial mutations can only be transmitted through females, although they can affect both sexes equally.

8.7.2 Chromosomal Disorders

Factors that indicate the possibility of a familial chromosomal disorder in a pedigree, such as a balanced reciprocal translocation resulting in unbalanced products at meiosis, include a family history of infertility, multiple spontaneous abortions, and malformed stillbirths or live-born infants with multiple congenital malformations occurring in a pattern that does not conform to that of Mendelian inheritance. The possibility of a subtle rearrangement not detected by routine cytogenetic analysis but leading to recurrent chromosome imbalance in offspring should be borne in mind, and a number of techniques can be used to identify this situation such as fluorescence in situ hybridization (FISH), telomere FISH, comparative genomic hybridization, and microarray analysis.

8.7.3 Multifactorial Inheritance

A large group of relatively common disorders have a considerable genetic predisposition but do not follow clear-cut patterns of inheritance within families. These include many birth defects and chronic diseases of later life. Liability to these disorders appears to be due to the interaction of several genetic and environmental influences. Clusters of cases within a family may simulate a Mendelian pattern of inheritance. It is often difficult to be precise about recurrence risks in multifactorial disorders. The risk is greatest among first-degree relatives, is usually small for second-degree relatives, and often does not exceed the general population risk for third-degree relatives. The risk increases when multiple family members are affected. Risks are also affected by the incidence of the disorder in the general population and the sex of the patient and relatives in disorders that have unequal sex incidence, such as pyloric stenosis and Hirschsprung disease. In these disorders, recurrence is higher in the siblings of the less affected sex, probably reflecting a greater genetic load to cause the disease in that sex. The severity of the disorder also influences risk, for example, the recurrence risk is greater for bilateral cleft lip and palate than for unilateral cleft lip alone.

CHAPTER 9

Analysis of Genetic Linkage

Rita M Cantor

Department of Human Genetics, David Geffen School of Medicine at UCLA, Los Angeles, CA, USA

Linkage analysis is a well-established genetic analysis method used to the map genes for heritable traits to their chromosome locations. It is part of a larger process that has been referred to as "reverse genetics," because the approach works in the reverse order from our model of how genes operate, biologically. That is, while genes act in a forward fashion to produce a trait, reverse genetics starts with the trait, and linkage analysis, along with other analytic methods, is used in a process to identify the genes that cause the trait. Reverse genetics became feasible in the 1990s, when a very extensive panel of multi-allelic markers that spanned the human genome was established. During the last 15 years, the markers have evolved from multi-allelic to 2.4 million bi-allelic single nucleotide polymorphisms (SNPs), where the inter-marker spacing is very dense and the whole genome is analyzed with an appropriate set of markers to conduct a full genome scan. To accomplish this, the markers are genotyped and statistically tested in a study sample of pedigrees that have been ascertained for individuals showing the trait of interest. Those chromosome regions showing the strongest statistical evidence of linkage and exceeding a predetermined threshold localize the trait genes. Using the advances made by the Human Genome Project, reverse genetics has been very effective in identifying the genes causing rare Mendelian disorders that result from the effects of a single gene within pedigrees.

In this chapter, the necessary concepts and methods to understand and interpret the process of linkage analysis are explained. These include the biological concept of chromosome recombination in meiosis, which allows one to assess the distance between trait genes and markers along a chromosome, and the statistical methods of likelihood, maximum likelihood estimation, and LOD scores, which allow one to test and assess the strength of the evidence supporting the linkage.

Linkage analysis for humans was originally developed for binary traits in large pedigrees. The statistics supporting this approach are presented in detail. The parametric linkage method has been extended to include traits with genotypes that are not fully penetrant and for which there is locus heterogeneity. These extensions are discussed. Multipoint linkage analysis has been very effective in localizing genes within linked regions, and this method is described. More recently, model-free linkage methods have been developed for traits with complex and unknown genetic models. The central principles of these methods are explained. Traits such as height and blood pressure are quantitative, continuous, and possibly normally distributed. They may contain more linkage information than those that are dichotomized. Methods for their analysis are discussed. The design of a good linkage study involves the consideration and coordination of all of these factors.

Well-designed computer software is essential for linkage analyses. Options that conduct parametric linkage analyses in large pedigrees, as well as those that incorporate all of the extensions and alternative methods discussed are presented in tables and a list of websites. Extra features of the software are also presented. Further Reading for the range of linkage approaches is given.

It should also be noted that currently many gene identification studies are bypassing the traditional early steps of reverse genetics, such as linkage analysis to begin with a genome-wide association study. Using this approach, pedigrees have been reduced to single individuals, genotyping is done with very dense SNP marker panels, and genes are identified by testing individual SNPs using the statistical method of association analysis. The relation of this linkage disequilibrium approach is compared to that of linkage analysis. The future of linkage analysis is discussed in the context of exome sequencing for rare variants.

FURTHER READING

1. Almasy, L.; Blangero, J. Multipoint Quantitative-Trait Linkage Analysis in General Pedigrees. *Am. J. Hum. Genet.* **1998**, *62*, 1198–1211.
2. Haseman, J. K.; Elston, R. C. The Investigation of Linkage between a Quantitative Trait and a Marker Locus. *Behav. Genet.* **1972**, *2*, 3–19.
3. Hirschhorn, J. N.; Daly, M. J. Genome-Wide Association Studies for Common Diseases and Complex Traits. *Nat. Rev. Genet.* **2005**, *6*, 95–108.
4. Kruglyak, L.; Lander, E. S. Complete Multipoint Sib-Pair Analysis of Qualitative and Quantitative Traits. *Am. J. Hum. Genet.* **1995**, *57*, 439–454.

5. Morton, N. E. Sequential Tests for the Detection of Linkage. *Am. J. Hum. Genet.* **1955,** *7,* 277–318.
6. Nyholt, D. R. Invited Editorial: All LODs Are Not Created Equal. *Am. J. Hum. Genet.* **2000,** *67,* 282–288.
7. Ott, J. *Analysis of Human Genetic Linkage*; Johns Hopkins University Press: Baltimore, 1999.
8. Risch, N. Linkage Strategies for Genetically Complex Traits. I. Multilocus Models. *Am. J. Hum. Genet.* **1990,** *46,* 222–228.
9. Risch, N. Linkage Strategies for Genetically Complex Traits. II. The Power of Affected Relative Pairs. *Am. J. Hum. Genet.* **1990,** *46,* 229–241.
10. Sobel, E.; Lange, K. Descent Graphs in Pedigree Analysis: Applications to Haplotyping, Location Scores, and Marker-Sharing Statistics. *Am. J. Hum. Genet.* **1996,** *58,* 1323–1337.

CHAPTER
10
Chromosomal Basis of Inheritance

M Fady Mikhail

Cytogenetics Laboratory; Department of Genetics, University of Alabama at Birmingham, Birmingham, AL, USA

The human genome is packaged into a set of chromosomes, which are thus the vehicles of inheritance as they contain virtually the entire cellular DNA with the exception of the small fraction present in the mitochondria. Chromosomes are derived in equal numbers from the mother and father. Each ovum and sperm contains a set of 23 different chromosomes, which is the haploid number of chromosomes in humans. The diploid fertilized egg and virtually every cell of the body arising from it has two haploid sets of chromosomes, resulting in the diploid human chromosome number of 46. The human karyotype consists of 22 pairs of autosomes and a pair of sex chromosomes. The haploid human genome consists of $\sim 3 \times 10^9$ base pairs (bp) of DNA. The total length of the diploid human genome is about 2 m. As the cell nucleus is no more than 10 µm in diameter, it is necessary to fold and compact this DNA, which is accomplished by packaging it in a hierarchy of levels into chromosomes of manageable size. The first level of packaging, and thus the fundamental unit of chromosome organization, is a regularly repeating protein-DNA complex called the nucleosome. The nucleosome consists of a cylindrical core of about 11 nm in diameter made up of two molecules each of the four core histones (H2A, H2B, H3, and H4) with 147 bp of DNA wrapped around it. A "linker" DNA connects adjacent nucleosomes. Each nucleosome is also associated with a molecule of histone H1, which changes the path of the DNA as it exits from the nucleosome, and plays a role in further condensation of chromosomal DNA. The next higher level of packaging is the chromatin fiber, which is a superhelix 30 nm in diameter composed of nucleosomes and histone H1. The 30-nm fiber is the basic component of interphase chromatin and metaphase chromosomes. The "solenoid" and the "zigzag" models have been proposed for the formation of the 30-nm chromatin fiber. At the final level of packaging, the 30-nm chromatin fiber is arranged into loops that radiate from a core or scaffold of the metaphase chromosome. The details of the higher order structure of chromosomes are not well understood at the molecular level. Chromatin is generally classified into euchromatin and heterochromatin. Euchromatin consists of active genes; however, not all genes in euchromatic regions are active at any given time. Euchromatin is dispersed in the interphase nucleus and replicates its DNA early in the S phase of the cell cycle. Heterochromatin consists predominantly of inactive genetic material, replicates its DNA late in the S phase, and is condensed in the interphase nucleus.

Cell division and proliferation are central to growth and development of multicellular organisms. The major events in the cell cycle are replication and segregation of chromosomes. Cell division also ensures proper segregation and partitioning of the genetic material into daughter cells, thus providing the basis for Mendelian laws of inheritance. Knowledge of the essential features of mitosis and meiosis is crucial for understanding the Mendelian laws of inheritance, construction of genetic maps, and the origin of chromosome aberrations. In somatic cells and in cells of the germline prior to the time they undergo their first specialized meiotic divisions, nuclear division takes place by mitosis. During mitosis, each chromosome divides into two daughter chromosomes (sister chromatids), one of which segregates into each daughter cell. Therefore, the number of chromosomes per nucleus remains unchanged producing daughter cells with identical chromosome constitutions. In cells with a generation time of 18–24 h, mitosis takes about 1–2 h and is divided into five major stages: prophase, prometaphase, metaphase, anaphase, and telophase. Cytokinesis, the division of the cytoplasm, follows telophase and leads to the formation of two genetically identical daughter cells. Meiosis on the other hand is a specialized cell division in germ cells, which generates gametes with the haploid set of 23 chromosomes. The final gametic set includes single representatives of each of the 23 chromosome pairs selected at random. The details of meiosis and gamete formation are somewhat different in males and females, but the basic features are the same in both and are of fundamental importance. Meiosis accounts for the major principles of Mendelian genetics: segregation, independent assortment, and recombination of linked genes. Recombination or crossing over is the exchange of genetic material between homologous non-sister chromatids, a process that adds to genetic diversity by generating new combinations of genes. Meiosis consists of two cell divisions (meiosis I and II). Meiotic prophase I is rather prolonged and can be subdivided into five stages on the basis of condensation of chromosomes and the extent of homologous pairing: leptotene, zygotene, pachytene, diplotene, and diakinesis. The meiosis II division resembles an ordinary mitotic division, except for the presence of a single set of 23 duplicated chromosomes, each with two chromatids held together at their centromeres. At the end of the two meiotic divisions, each primary spermatocyte or oocyte has given rise to four haploid products.

Technical innovations in the past 60 years have revolutionized the study of human chromosomes. Chromosomes are normally visible only during cell division as they become condensed in preparation for orderly division. Based on the position of their centromeres, human chromosomes are classified as metacentric, in which the centromere is at or near the middle of the chromosome; submetacentric, in which the centromere is located significantly off center; or acrocentric, in which the centromere is very close to one end. For all categories, the short arm of the chromosome is referred to as "p" for petite, and the long arm as "q." Human chromosomes are assigned to groups A through G according to their general size and position of the centromere. Techniques such as Giemsa (G)-banding and reverse (R)-banding produce the full range of bands along each chromosome, allowing identification of individual human chromosomes. Other banding techniques produce much more restricted staining of specific subsets of chromosome bands and include centromere (C)-banding and NOR-banding.

The gap between light microscope resolution of chromosome structure and the gene was bridged by the introduction of several molecular cytogenetic techniques. Fluorescence in situ hybridization (FISH) involves hybridizing a fluorescently labeled single-stranded DNA probe to denatured chromosomal DNA on a microscope slide preparation of metaphase chromosomes and/or interphase nuclei prepared from the patient's sample. After overnight hybridization, the slide is washed and counterstained with a nucleic acid dye, allowing the region where hybridization has occurred to be visualized using a fluorescence microscope. FISH is now widely used for clinical diagnostic purposes. One type of FISH that has the potential to reveal chromosomal imbalances across the genome is comparative genomic hybridization (CGH). In CGH, DNA specimens from patient and normal control are differentially labeled with two different fluorescent dyes and hybridized to normal metaphase chromosome spreads. Difference between the fluorescent intensities of the two dyes along the length of any given chromosome will reveal gains and losses of genomic segments The latest addition to molecular cytogenetic techniques is array CGH, where CGH is applied to an array of DNA targets (probes) each representing a part of the human genome and fixed to a solid support. Like CGH, array CGH directly compares DNA content between two differentially labeled DNA specimens (a test and a normal control), which are labeled and co-hybridized onto the array. In the past few years, high-resolution whole-genome coverage array CGH platforms have been increasingly used in clinical molecular cytogenetic labs. These provide a relatively quick method to scan the entire genome for gains and/or losses with significantly high resolution and greater clinical abnormality yield than was previously possible. This led to the identification of novel genomic disorders in patients with autism spectrum disorders, developmental delay, mental retardation, and/or multiple congenital anomalies.

Genomic imprinting refers to a process by which maternal and paternal alleles of specific genes or chromosomal regions are differentially marked during gametogenesis such that they are expressed differently in the embryo. One allele of the imprinted gene is usually active, while the other is inactive. Thus the paternal and maternal copies are not functionally equal for all genes, and therefore both a maternal copy and a paternal copy are required for normal development. Imprinting is known to affect only a small number of genes and chromosomal regions in the human genome; however, imprinted genes contribute significantly to genetic disease.

Human cytogenetics has advanced during the past four decades because of continuing technical advances and the high incidence of chromosome abnormalities in the human population. Constitutional chromosome aberrations have a significant impact as causes of pregnancy wastage, congenital malformations, abnormalities of sex differentiation, mental retardation, and behavior problems. Whereas acquired chromosomal changes play a significant role in carcinogenesis and in tumor progression. Most chromosomal abnormalities exert their phenotypic effects by increasing or decreasing the quantity of genetic material. Chromosomal abnormalities can be divided into numerical and structural abnormalities. Structural changes such as translocations and inversions pose a much more serious familial recurrence risk for chromosome abnormalities. This is due to aberrant segregation of chromosomes during meiosis in clinically normal carriers of these balanced rearrangements.

Much remains to be learned about the molecular aspects of chromosome structure, function, and behavior. It is anticipated that the human genome sequence and its functional characterization will provide the tools with which to approach these problems and define a new frontier for the role of chromosomes in human disease.

TABLE 10-1 Characteristics of Chromosome Bands

Characteristic	Q- or G-Bands	R-Bands	C-Bands
Location	Chromosome arms	Chromosome arms	Centromeres, distal Yq
Type of DNA sequence	Repetitive, some unique	Unique, some repetitive	Highly repetitive satellite
Base composition	AT-rich	GC-rich	AT-rich, some GC-rich
5-Methylcytosine content	Low	Moderate	High
Type of chromatin	Heterochromatin	Euchromatin	Heterochromatin
Replication	Mid to late S phase	Early S phase	Late S phase
Transcription	Low	High	Absent
Gene density	Low	High	Absent
CpG-rich islands	Few	Many	Absent
Repeats	LINE-rich	SINE-rich	–
Acetylated histones	Low	High	Absent

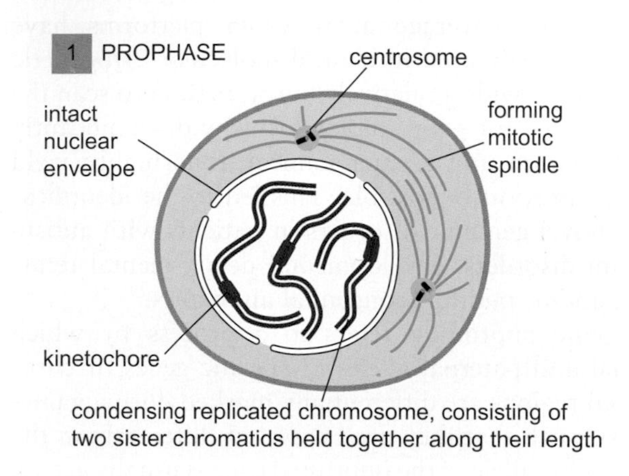

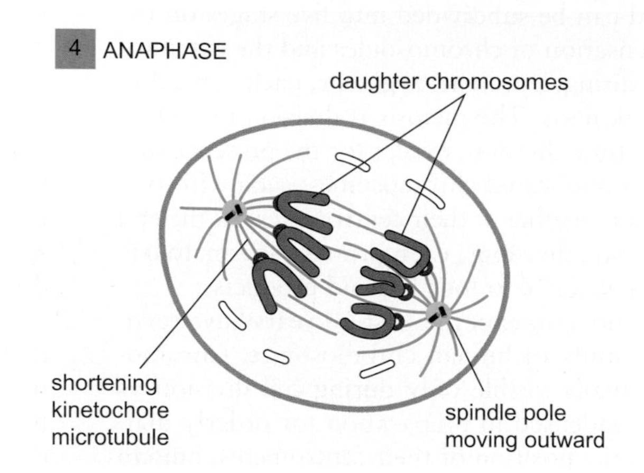

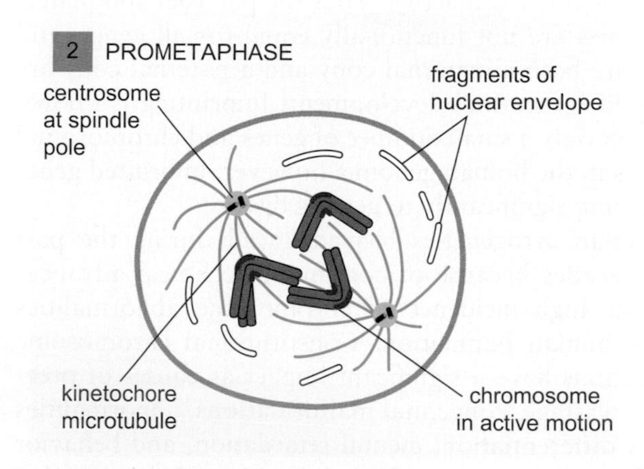

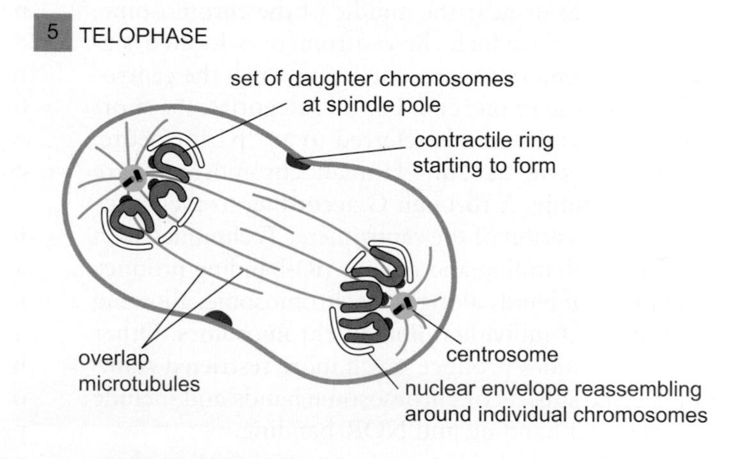

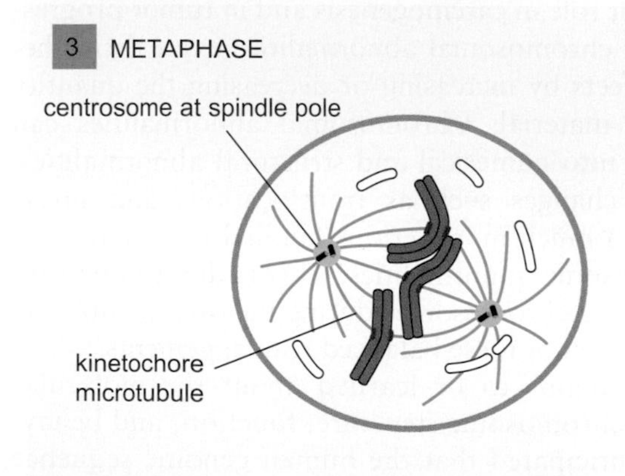

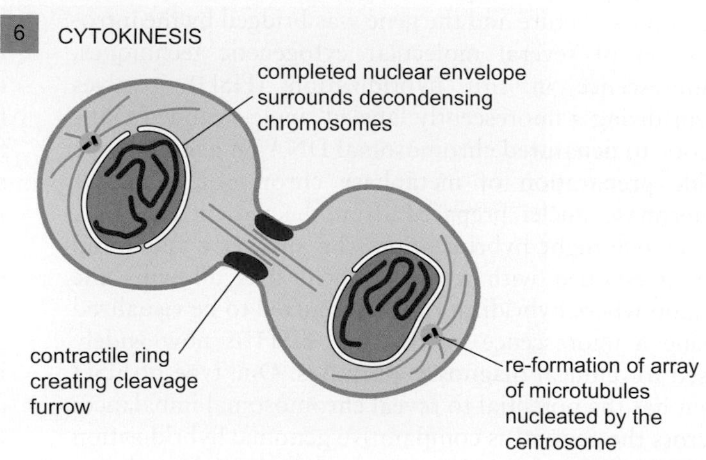

FIGURE 10-1 Diagrammatic representation of the stages of mitosis. *(Alberts, B., Johnson, A., Lewis, J., et al. Part IV: Internal Organization of the Cell, Chapter 18: The Mechanics of Cell Division, An Overview of M Phase, Panel 18-1: The Principal Stages of M Phase. In* Molecular Biology of the Cell, *4th ed.; Garland Science: New York, 2002.)*

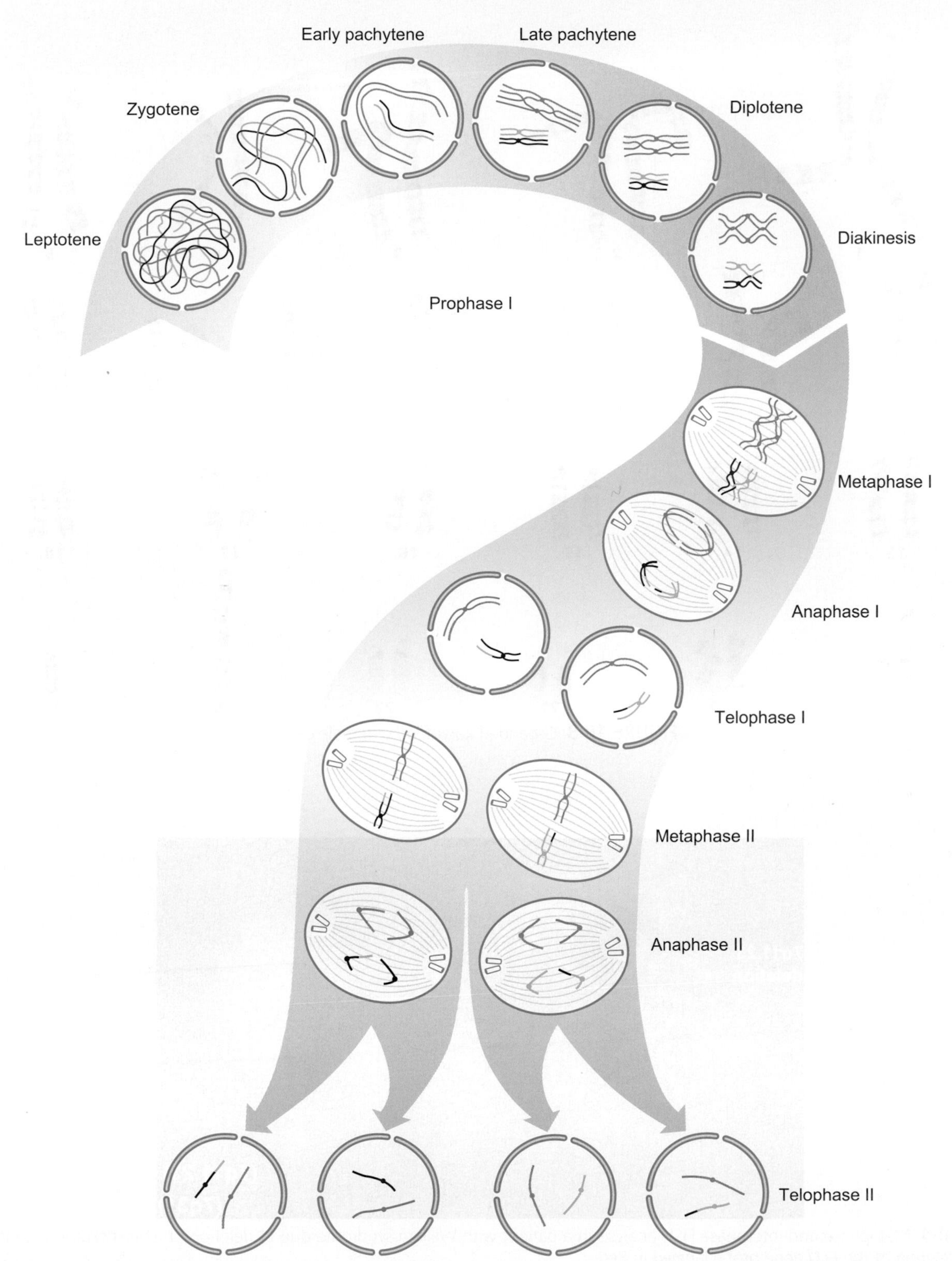

FIGURE 10-2 Diagrammatic representation of the stages of meiosis. *(Turnpenny, P. D., Ellard, S., Chapter 3: Chromosomes and Cell Division, Figure 3.19: Stages of Meiosis, In* Emery's Elements of Medical Genetics, *12th ed.; Churchill Livingstone: Elsevier, 2005.)*

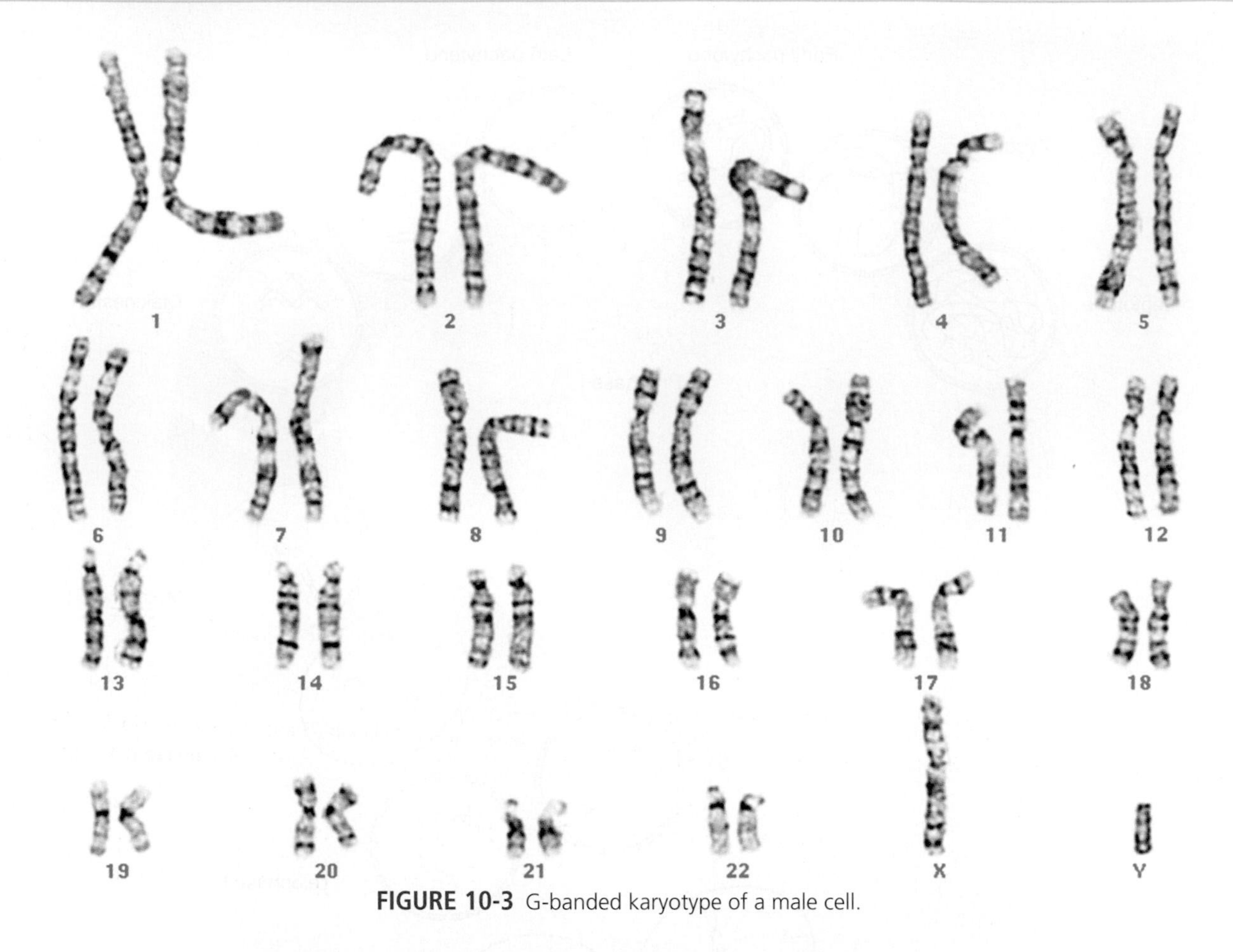

FIGURE 10-3 G-banded karyotype of a male cell.

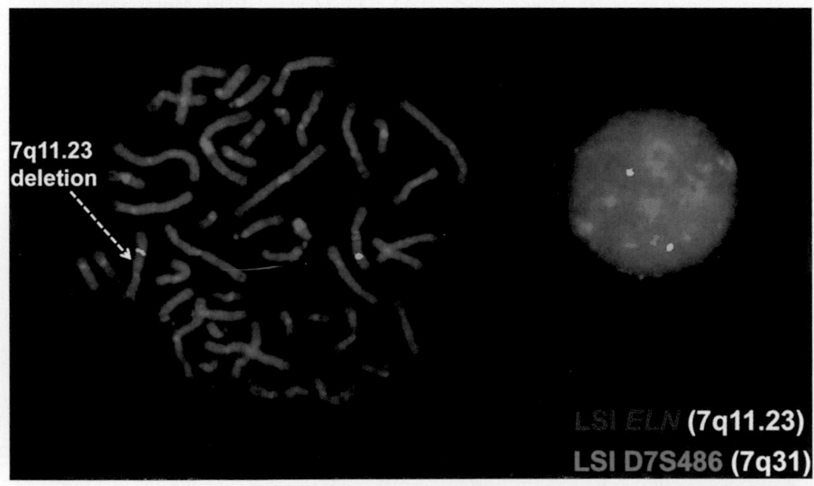

FIGURE 10-4 Metaphase and interphase FISH analysis in a patient with William syndrome due to deletion on chromosome 7 band q11.23. Note the deletion of the *ELN* gene probe labeled in Red.

CHAPTER

11

Mitochondrial Medicine: The Mitochondrial Biology and Genetics of Metabolic and Degenerative Diseases, Cancer, and Aging

Douglas C Wallace and Marie T Lott

Center for Mitochondrial and Epigenomic Medicine (CMEM), Children's Hospital of Philadelphia, Research Institute, Philadelphia, PA, USA

Vincent Procaccio

Biochemistry and Genetics Department, National Center for Neurodegenerative and Mitochondrial Diseases, CHU Angers, Angers, France

In the 14 years, since the first disease-causing mitochondrial DNA (mtDNA) mutations were reported, hundreds of pathogenic mutations have been identified in both the mtDNA and the nuclear DNA (nDNA)-encoded mitochondrial genes. Concurrently, major advances have been made in elucidating the molecular structure of the mitochondrial oxidative phosphorylation (OXPHOS) complexes revealing the central function of the mtDNA-encoded proteins in OXPHOS, providing insight into the assembly of the complexes, and revealing the cellular role of mitochondrial bioenergetics and metabolism in cellular physiology. The coalescence of these genetic and biochemical advances has resulted in a new synthesis that posits that much of the etiology of the common, "complex," metabolic and degenerative diseases, aging, and cancer is the product of mitochondrial bioenergetic dysfunction.

Recently, deleterious, mtDNA mutations have been observed in all of the genes of the maternally inherited mtDNA including the *rRNA* and *tRNA* genes for mitochondrial proteins synthesis and the 13 polypeptide genes, all of which are central subunits of OXPHOS, seven of the 45 polypeptides of complex I (NADH dehydrogenase), one of the 11 of complex III (bc_1 complex), three of the 13 of complex IV (cytochrome c oxidase) and two of the approximate 17 of complex V (ATP synthase). Mutations in *mtDNA*, *tRNA* and *rRNA* genes generally cause multisystem disease presenting with symptoms that can include diabetes mellitus; myoclonic epilepsy and ragged red fiber disease; and mitochondrial encephalomyopathy, lactic acidosis and stroke-like episodes (MELAS), etc. Mutations in mtDNA polypeptide genes can cause either stereotypic clinical presentations such as Leber hereditary optic neuropathy (LHON) or Leigh syndrome or more complex multisystem presentations such as MELAS. Mutations also accumulate in cells and tissues with aging and degenerative diseases. A significant portion of these mutations affect the mtDNA control region which regulates mtDNA transcription and replication. The accumulation of these somatic mutations has been proposed to be the aging clock.

Ancient, polymorphic variants have also been observed in the mtDNA, and these have been found to correlate highly with the geographic origin of indigenous populations. Many of these mutations alter highly evolutionarily constrained mtDNA nucleotide positions and found regional mtDNA lineages called haplogroups. Recent biochemical studies indicate that these variants alter mitochondrial function. Therefore, it has been hypothesized that the population variants were adaptive by altering the efficiency of mitochondrial energy metabolism and thus permitted populations to become established in different selective environments. Analysis of the mtDNA haplogroup distributions of patients harboring various pathogenic mtDNA mutations has revealed that the background mtDNA haplogroup can significantly affect the penetrance of the milder mtDNA LHON mutations. Hence, functional haplogroup variants are both adaptive and clinically relevant.

Mutations in approximately 1000–2000 nDNA-encoded mitochondrial proteins have been identified in OXPHOS structural genes, OXPHOS assembly genes, metabolic genes, and biogenesis genes. The identification of *nDNA* gene mutations in patients with complex I defects has revealed many assembly factors permitting the elucidation of the complex steps required to generate this enormous enzyme. Mutations in metabolic genes have revealed new biochemical pathways and mutations in nDNA encoded in mitochondrial biogenesis genes have been found to cause a plethora of clinical presentations. For example, mutations in the mitochondrial DNA polymerase (mtpolymerase γ) have a carrier frequency of about 1/50, destabilize the mtDNA, and can have different severity and result in very different manifests ranging from lethal Aplers syndrome to mild ophthalmoplegia and psychiatric disorders. Finally, subclinical nDNA-encoded mitochondrial gene variants can result in severe OXPHOS disease if combined with incompatible subclinical mtDNA variants.

This growing understanding of familial mitochondrial diseases and mtDNA population genetics has offered a bold new perspective on the common, "complex" diseases. An mtDNA lineage harboring a $tRNA^{Gln}$ variant was reported in 1993 to be associated with 3–7% of late-onset Alzheimer disease (AD) and Parkinson disease (PD). Subsequent studies have shown that somatic mtDNA mutations are systemically increased in AD patients and these affect brain mtDNA transcription and copy number. Hence, partial mitochondrial defects associated with regional mtDNA variants may be exacerbated by the secondary accumulation of somatic mtDNA mutations resulting in neurological dysfunction. This concept for AD has been strongly validated by the cloning of *nDNA* genes that predispose to PD, many of which have been found to be involved in maintaining mitochondrial functional integrity. Therefore, mitochondrial medicine can provide a conceptual framework for investigating a wide range of metabolic and degenerative diseases, autism being one example.

As information on mitochondrial biology and genetics has advanced, new information has permitted reassessment of Warburg's hypothesis that mitochondrial dysfunction is important in cancer. These investigations have been fueled by the discovery that mutations in mitochondrial OXPHOS and tricarboxylic acid cycle enzymes can cause cancer. For example, mutations in OXPHOS complex II (succinate dehydrogenase) genes generate paragangliomas. However, the broad role of mitochondrial dysfunction in cancer is being appreciated as large numbers of mtDNA mutations are being discovered in cancer cells, mitochondrial functions are shown to be critical for cancer cell metabolism, and mitochondrial stress-activated retrograde signaling is important in tumor progression and invasiveness.

Our growing understanding of the importance of mitochondrial function in human health and disease has made the development of therapeutics for treating mitochondrial dysfunction critical. Therapeutic approaches include both pharmacological and nutraceutical metabolic treatments and somatic and germ line genetic therapies. Metabolic therapeutics have focused on increasing mitochondrial energy output and mitochondrial biogenesis, reducing the levels and adverse effects of mitochondrial oxidative stress, and modulating apoptosis through manipulation of the mitochondrial permeability transition pore. Somatic gene therapeutics have focused on transduction of normal copies of mutant nDNA-encoded mitochondrial genes, mtDNA heteroplasmy shifting, direct manipulation of mtDNA and mitochondrial biogenesis, and introduction of corrected copies of *mtDNA* genes into the nucleus with the normal protein transferred back into the mitochondria (allotropic gene expression). However, the boldest therapeutic approach being explored for maternally inherited mtDNA diseases is the transfer of the spindle from the mutant mtDNA-containing oocyte of the affected mother to the oocyte of a normal donor followed by in vitro fertilization thus eliminating the mother's mutant mtDNAs.

These advances have generated a revolutionary new synthesis for the etiology of the metabolic and degenerative diseases, cancer, and aging. Contrary to the traditional focus on organ-specific anatomical medicine, the new synthesis takes a systemic bioenergetic perspective leading to mitochondrial medicine. The bioenergetic perspective envisions that all of the common, "complex" diseases have a common, central bioenergetic etiology. Mitochondrial bioenergetics can be perturbed by alterations in nDNA-encoded mitochondrial genes or changes in *nDNA* gene expression via the epigenome, mutations in the mtDNA are either recent deleterious or ancient adaptive, and alterations in the availability and demands on the use of dietary calories. Severe mitochondrial mutations will generate symptoms directly. However, mild mitochondrial defects will increase the damage rate of the mtDNA and thus speed up the aging clock. The accumulating somatic cell mtDNA damage will progressively erode inherited partial mitochondrial defects until the mitochondrial energy output falls below the minimum necessary for the most mitochondrial energy dependent tissues to function, that is the central nervous system, heart, muscle, renal, and endocrine system. This explains the delayed onset and progressive course of the common, "complex" diseases and their variable penetrance and expressivity. Mitochondrial dysfunction will also disrupt the flow of energy through the body resulting in diabetes, obesity, and other manifestations of the metabolic syndrome. The mitochondria are the most prevalent bacteria in our bodies and the mtDNA and mitochondrial N-formyl-polypeptides will prime the immune system like other bacteria and thus may initiate autoimmune disease. Finally, cancer cells are limited because of rapid growth for both biogenesis substrates and energy, both of which require the modulation of mitochondrial function. Hence, the mitochondrial bioenergetic perspective places all of the complex diseases within a common nonanatomical and non-Mendelian pathophysiological and genetic system.

CHAPTER

12

Multifactorial Inheritance and Complex Diseases

Christine W Duarte, Laura K Vaughan, T Mark Beasley, and Hemant K Tiwari

Department of Biostatistics, Section on Statistical Genetics, University of Alabama at Birmingham, Birmingham, AL, USA

Multifactorial diseases are caused by multiple genetic factors with possible contributions or interactions with environmental factors. The study of mutlifactorial disease involves several steps, which include (1) determination of the heritability of a trait or disease; (2) segregation analysis to determine the type of genetic model; (3) linkage analysis to determine the rough genomic locations of the genetic loci; (4) association analysis or fine mapping to determine the actual causal variants; and (5) replication, functional, or other studies to confirm the initial findings. With the conclusion of the HapMap project, which surveyed the set of common single nucleotide polymorphisms (SNPs) that exist in a diverse set of human populations, and with advances in affordable high throughput genotyping technology, population-based association studies have become the most popular method for discovery of genes associated with complex traits and disease.

Indeed, the HapMap project contributed to a revolution in how common complex diseases are studied. The data gathered by the HapMap project revealed genome-wide patterns of association between SNPs (LD) and the differences in those LD patterns and allele frequencies across populations. Advances in affordable high throughput genotyping technology combined with advances in the control of population stratification in association studies in population data (*1–9*) enabled researchers to perform large-scale genome-wide association studies (GWAS). As of August 19, 2011, there were 990 GWAS publications on a variety of different diseases (see http://www.genome.gov/26525384, for details).

The GWAS analysis protocol is well established. GWAS involves several phases: determination of study design (population data, case-control, or familial data), deciding on the genotypic platform, quality control (QC) of the genotypic data both at the genotyping laboratory and statistical methods for QC after the data are delivered to the analyst, using an appropriate association test in the discovery set, replication in a validation set, and finally functional analysis. Balding (*10*) provided a tutorial on statistical methods for population association studies. Other reviews have been published on topics such as population stratification, selection of controls and cases, and overall methodological issues with GWAS (*11–20*).

More recently, direct sequencing has become a popular technique for discovery of associated genetic variants for disease. Such technologies overcome some of the shortcomings of GWAS methods such as ascertainment bias in the set of currently available SNPs, the ability to assay rare or private variants, and greater ability to search for variants other than SNPs such as copy number variants or CNVs, insertions or deletions or indels, inversions, etc. Whole exome sequencing, in which only the sequence of exons is assayed, has been used to discover causal mutations in a number of Mendelian disorders, such as Miller syndrome (*21*) and hereditary spastic paraparesis (*22*). Prior bioinformatic processing may be used for a variety of purposes such as (1) to filter for variants that are identical by descent if family data are present (*23*); (2) to filter out common variants using dbSNP (*24*), if only rare variants are desired; and (3) to predict functional variants (nonsense, missense, splice site variants, indels, frameshift mutations, etc.) using tools such as SIFT and PolyPhen (*24,25*). New statistical methods for summarizing the effect of multiple rare variants at a single gene can be applied, including the cohort allelic sums test method (*26*), the combined multivariate and collapsing method (*27*), methods that weight the counts of each variant using the estimated standard deviation of the total number of mutations (*28,29*), and a method which models these weights in a flexible Bayesian framework (*30*).

Interpretation of genetic studies in the context of prior biological knowledge or other cross-platform functional characterization is becoming a popular way to increase the power and interpretability of genetic studies. One class of methods termed "Pathway Analysis" typically examine test statistics to determine if the members of group of genes are enriched for association with a trait (*31,32*), or to test if the group itself is associated with the trait (*33,34*). For additional information on these methods, see Refs. (*9,35–41*).

The study of multifactorial diseases is a challenging field that has undergone great change in the past two decades and is likely to undergo yet more change in the years to come. Fundamental biological questions such as the common variant hypothesis as well as practical public health problems are as yet unresolved by current studies.

REFERENCES

1. Dadd, T.; Weale, M. E.; Lewis, C. M. A Critical Evaluation of Genomic Control Methods for Genetic Association Studies. *Genet. Epidemiol.* **2009 May,** *33* (4), 290–298, Review.
2. Devlin, B.; Roeder, K. Genomic Control for Association Studies. *Biometrics* **1999 Dec,** *55* (4), 997–1004.
3. Devlin, B.; Bacanu, S. A.; Roeder, K. Genomic Control to the Extreme. *Nat. Genet.* **2004 Nov,** *36* (11), 1129–1130.
4. Patterson, N.; Price, A. L.; Reich, D. Population Structure and Eigen Analysis. *PLoS Genet.* **2006 Dec,** *2* (12), e190.
5. Price, A. L.; Patterson, N. J.; Plenge, R. M.; Weinblatt, M. E.; Shadick, N. A.; Reich, D. Principal Components Analysis Corrects for Stratification in Genome-Wide Association Studies. *Nat. Genet.* **2006 Aug,** *38* (8), 904–909.
6. Pritchard, J. K.; Rosenberg, N. A. Use of Unlinked Genetic Markers to Detect Population Stratification in Association Studies. *Am. J. Hum. Genet.* **1999 Jul,** *65* (1), 220–228.
7. Pritchard, J. K.; Stephens, M.; Rosenberg, N. A.; Donnelly, P. Association Mapping in Structured Populations. *Am. J. Hum. Genet.* **2000 Jul,** *67* (1), 170–181.
8. Redden, D.; Divers, J.; Vaughan, L.; Tiwari, H.; Beasley, T.; Fernandez, J.; Kimberly, R.; Feng, R.; Padilla, M.; Lui, N., et al. Regional Admixture Mapping and Structured Association Testing: Conceptual Unification and an Extensible General Linear Model. *PLoS Genet.* **2006 Aug 25,** *2* (8), e137.
9. Thomas, D. C.; Conti, D. V.; Baurley, J.; Nijhout, F.; Reed, M.; Ulrich, C. M. Use of Pathway Information in Molecular Epidemiology. *Hum. Genomics* **2009 Oct,** *4* (1), 21–42, Review.
10. Balding, D. J. A Tutorial on Statistical Methods for Population Association Studies. *Nat. Rev. Genet.* **2006 Oct,** *7* (10), 781–791, Review.
11. Donnely, P. Progress and Challenges in Genome-Wide Association Studies in Humans. *Nature* **2008 Dec 11,** *456* (7223), 728–731.
12. Hirschhorn, J. N.; Daly, M. J. Genome-Wide Association Studies for Common Diseases and Complex Traits. *Nat. Rev. Genet.* **2005 Feb,** *6* (2), 95–108, Review.
13. Laird, N. M.; Lange, C. Family-Based Designs in the Age of Large-Scale Gene-Association Studies. *Nat. Rev. Genet.* **2006 May,** *7* (5), 385–394, Review.
14. Manolio, T. A. Genomewide Association Studies and Assessment of the Risk of Disease. *N. Engl. J. Med.* **2010 Jul 8,** *363* (2), 166–176, Review.
15. Manolio, T. A.; Collins, F. S.; Cox, N. J.; Goldstein, D. B.; Hindorff, L. A.; Hunter, D. J.; McCarthy, M. I.; Ramos, E. M.; Cardon, L. R.; Chakravarti, A., et al. Finding the Missing Heritability of Complex Diseases. *Nature* **2009 Oct 8,** *461* (7265), 747–753, Review.
16. McCarthy, M. I.; Abecasis, G. R.; Cardon, L. R.; Goldstein, D. B.; Little, J.; Ioannidis, J. P.; Hirschhorn, J. N. Genome-Wide Association Studies for Complex Traits: Consensus, Uncertainty and Challenges. *Nat. Rev. Genet.* **2008 May,** *9* (5), 356–369, Review.
17. Psychiatric GWAS Consortium Coordinating Committee; Cichon, S.; Craddock, N.; Daly, M.; Faraone, S. V.; Gejman, P. V.; Kelsoe, J.; Lehner, T.; Levinson, D. F.; Moran, A.; Sklar, P., et al. Genomewide Association Studies: History, Rationale, and Prospects for Psychiatric Disorders. *Am. J. Psychiatry* **2009 May,** *166* (5), 540–556, Epub 2009 Apr 1.
18. Stranger, B. E.; Stahl, E. A.; Raj, T. Progress and Promise of Genome-Wide Association Studies for Human Complex Trait Genetics. *Genet.* **2011 Feb,** *187* (2), 367–383, Epub 2010 Nov 29. Review.
19. Tiwari, H. K.; Barnholtz-Sloan, J.; Wineinger, N.; Padilla, M. A.; Vaughan, L. K.; Allison, D. B. Review and Evaluation of Methods Correcting for Population Stratification with a Focus on Underlying Statistical Principles. *Hum. Hered.* **2008,** *66* (2), 67–86.
20. Witte, J. S. Genome-Wide Association Studies and Beyond. *Annu. Rev. Public Health* **2010 Apr 21,** *31,* 9–20.
21. Ng, S. B.; Buckingham, K. J.; Lee, C.; Bigham, A. W.; Tabor, H. K.; Dent, K. M.; Huff, C. D.; Shannon, P. T.; Jabs, E. W.; Nickerson, D. A., et al. Exome Sequencing Identifies the Cause of a Mendelian Disorder. *Nat. Genet.* **2010 Jan,** *42* (1), 30–35.
22. Erlich, Y.; Edvardson, S.; Hodges, E.; Zenvirt, S.; Thekkat, P.; Shaag, A.; Dor, T.; Hannon, G. J.; Elpeleg, O. Exome Sequencing and Disease-Network Analysis of a Single Family Implicate a Mutation in KIF1A in Hereditary Spastic Paraparesis. *Genome Res.* **2011 May,** *21* (5), 658–664.
23. Rödelsperger, C.; Krawitz, P.; Bauer, S.; Hecht, J.; Bigham, A. W.; Bamshad, M.; de Condor, B. J.; Schweiger, M. R.; Robinson, P. N. Identity-by-Descent Filtering of Exome Sequence Data for Disease-Gene Identification in Autosomal Recessive Disorders. *Bioinformatics* **2011 Mar 15,** *27* (6), 829–836.
24. Ng, S. B.; Turner, E. H.; Robertson, P. D.; Flygare, S. D.; Bigham, A. W.; Lee, C.; Shaffer, T.; Wong, M.; Bhattacharjee, A.; Eichler, E. E., et al. Targeted Capture and Massively Parallel Sequencing of 12 Human Exomes. *Nature* **2009 Sep 10,** *461* (7261), 272–276.
25. Sunyaev, S.; Ramensky, V.; Koch, I.; Lathe, W., 3rd; Kondrashov, A. S.; Bork, P. Prediction of Deleterious Human Alleles. *Hum. Mol. Genet.* **2001 Mar 15,** *10* (6), 591–597.
26. Morgenthaler, S.; Thilly, W. G. A Strategy to Discover Genes That Carry Multi-Allelic or Mono-Allelic Risk for Common Diseases: A Cohort Allelic Sums Test (CAST). *Mutat. Res.* **2007 Feb 3,** *615* (1–2), 28–56.
27. Li, B.; Leal, S. M. Methods for Detecting Associations with Rare Variants for Common Diseases: Application to Analysis of Sequence Data. *Am. J. Hum. Genet.* **2008 Sep,** *83* (3), 311–321.
28. Madsen, B. E.; Browning, S. R. A Groupwise Association Test for Rare Mutations Using a Weighted Sum Statistic. *PLoS Genet.* **2009 Feb,** *5* (2), e1000384.
29. Price, A. L.; Kryukov, G. V.; de Bakker, P. I.; Purcell, S. M.; Staples, J.; Wei, L. J.; Sunyaev, S. R. Pooled Association Tests for Rare Variants in Exon-Resequencing Studies. *Am. J. Hum. Genet.* **2010 Jun 11,** *86* (6), 832–838.
30. Yi, N.; Zhi, D. Bayesian Analysis of Rare Variants in Genetic Association Studies. *Genet. Epidemiol.* **2011 Jan,** *35* (1), 57–69.
31. Subramanian, A.; Kuehn, H.; Gould, J.; Tamayo, P.; Mesirov, J. P.GSEA-P: A Desktop Application for Gene Set Enrichment Analysis. *Bioinformatics* **2007 Dec 1,** *23* (23), 3251–3253.

32. Subramanian, A.; Tamayo, P.; Mootha, V. K.; Mukherjee, S.; Ebert, B. L.; Gillette, M. A.; Paulovich, A.; Pomeroy, S. L.; Golub, T. R.; Lander, E. S., et al. Gene Set Enrichment Analysis: A Knowledge-Based Approach for Interpreting Genome-Wide Expression Profiles. *Proc. Natl. Acad. Sci. U.S.A* 2005 **Oct 25,** *102* (43), 15545–15550.
33. Chen, L. S.; Hutter, C. M.; Potter, J. D.; Liu, Y.; Prentice, R. L.; Peters, U.; Hsu, L. Insights into Colon Cancer Etiology Via a Regularized Approach to Gene Set Analysis of GWAS Data. *Am. J. Hum. Genet.* 2010 **Jun 11,** *86* (6), 860–871.
34. Chen, X.; Wang, L.; Hu, B.; Guo, M.; Barnard, J.; Zhu, X. Pathway-Based Analysis for Genome-Wide Association Studies Using Supervised Principal Components. *Genet. Epidemiol.* 2010 **Nov,** *34* (7), 716–724.
35. Cantor, R. M.; Lange, K.; Sinsheimer, J. S. Prioritizing GWAS Results: A Review of Statistical Methods and Recommendations for Their Application. *Am. J. Hum. Genet.* 2010 **Jan,** *86* (1), 6–22, Review.
36. Holmans, P. Statistical Methods for Pathway Analysis of Genome-Wide Data for Association with Complex Genetic Traits. *Adv. Genet.* 2010, *72,* 141–179.
37. Hong, M. G.; Pawitan, Y.; Magnusson, P. K.; Prince, J. A. Strategies and Issues in the Detection of Pathway Enrichment in Genome-Wide Association Studies. *Hum. Genet.* 2009 **Aug,** *126* (2), 289–301.
38. Wang, L.; Jia, P.; Wolfinger, R. D.; Chen, X.; Zhao, Z. Gene Set Analysis of Genome-Wide Association Studies: Methodological Issues and Perspectives. *Genomics* 2011 **Jul,** *98* (1), 1–8.
39. Wang, L.; Jia, P.; Wolfinger, R. D.; Chen, X.; Grayson, B. L.; Aune, T. M.; Zhao, Z. An Efficient Hierarchical Generalized Linear Mixed Model for Pathway Analysis of Genome-Wide Association Studies. *Bioinformatics* 2011 **Mar 1,** *27* (5), 686–692.
40. Voight, B. F.; Pritchard, J. K. Confounding from Cryptic Relatedness in Case-Control Association Studies. *PLoS Genet.* 2005 **Sep,** *1* (3), e32.
41. Wang, W. Y.; Barratt, B. J.; Clayton, D. G.; Todd, J. A. Genome-Wide Association Studies: Theoretical and Practical Concerns. *Nat. Rev. Genet.* 2005 **Feb,** *6* (2), 109–118, Review.

CHAPTER

13

Population Genetics

Bronya J B Keats

Research School of Biology, Australian National University, Canberra, ACT, Australia

Stephanie L Sherman

Department of Human Genetics, Emory University School of Medicine, Atlanta, GA, USA

The principles of population genetics attempt to explain the genetic diversity in present populations and the changes in allele and genotype frequencies over time. Population genetic studies facilitate the identification of alleles associated with disease risk and provide insight into the effect of medical intervention on the population frequency of a disease. Allele and genotype frequencies depend on factors such as mating patterns, population size and distribution, mutation, migration, and selection. By making specific assumptions about these factors, the Hardy–Weinberg law, a fundamental principle of population genetics, provides a model for calculating genotype frequencies from allele frequencies for a random mating population in equilibrium.

Hardy–Weinberg equilibrium at a single autosomal locus is established in one generation of random mating; if the allele frequencies are p and q, then the genotype frequencies are p^2, 2pq, and q^2. Thus, the carrier frequency for a rare, autosomal recessive disease is approximately twice the square root of the disease frequency in the population. Unlike an autosomal locus, equilibrium at an X-linked locus is not reached in one generation; it takes several generations and the equilibrium allele frequency is two-thirds of the frequency in females plus one-third of the frequency in males. A population that is in equilibrium with respect to two loci considered jointly must be in equilibrium with respect to each locus separately, but the converse is not true. The approach to equilibrium may be much longer for two loci considered jointly; if recombination between them is rare, then thousands of generations are required to reach equilibrium. The linkage disequilibrium coefficient is a measure of association between the alleles at the two loci.

If mating is not at random, the allele frequencies are not affected, but the genotype frequencies are not in Hardy–Weinberg proportions. Examples of non-random mating in a population are inbreeding, assortative mating, and stratification. Their effect is to increase the frequency of homozygotes and decrease the frequency of heterozygotes in the population.

Evolutionary forces such as random genetic drift, mutation, selection, and migration change the allele frequencies in a population. Important examples of each of these forces have been documented in human populations, and their effects are becoming better understood as knowledge of the genetic structure of populations at the DNA level increases. Mutations are the source of variation in a population and lead to changes in allele frequencies and increased heterozygosity. However, mutation pressure alone is a very weak evolutionary force. Migration (gene flow), like mutation, increases heterozygosity in a population. It can lead to substantial changes in allele frequencies over short periods of time.

Random genetic drift is the change in allele frequencies that occurs from one generation to the next in small populations by chance. The eventual result of random drift is fixation or loss of each allele in the initial population. Like inbreeding, random genetic drift can lead to an excess of homozygotes at the expense of heterozygotes. Founder effect is a special case of random genetic drift in which population size is severely reduced by such events as famine, disease epidemics, or migration of a small subset of individuals to a new homeland.

The effect of selection on allele frequencies depends on the relative fitnesses of the genotypes. In general, selection will lead to the eventual loss of an allele, but a large number of generations of selection may be required. However, if selection favors the heterozygote over both homozygotes, equilibrium allele frequencies will be reached.

Ethnic variation in allele frequencies is found throughout the genome, and by examining this genetic diversity, evolutionary patterns can be inferred, and variants contributing to the cause of common diseases can be identified. As a result of major international initiatives, most recently the 1000 Genomes Project, nearly 20 million single nucleotide polymorphisms and copy number variants have been generated and cataloged. These extensive datasets provide the population geneticist with a huge set of densely mapped polymorphisms for reconciling genome variation with population histories of bottlenecks, admixture, and migration, and for revealing evidence of natural selection.

The availability of massive databases of genetic variation and automated technology for genotyping, sequencing, and bioinformatic analysis is significantly enhancing collaborative efforts between population geneticists and molecular geneticists, and advancing the understanding of many diseases. As we would anticipate, thousands of interesting new questions are arising and the tools to answer many of them are now at our fingertips.

FURTHER READING

1. Hindorff, L. A.; Sethupathy, P.; Junkins, H. A., et al. Potential Etiologic and Functional Implications of Genome-Wide Association Loci for Human Diseases and Traits. *Proc. Natl. Acad. Sci. U.S.A.* **May 27, 2009**.
2. Maniolo, T.; Brooks, L. D.; Collins, F. S. A HapMap Harvest of Insights into the Genetics of Common Disease. *J. Clin. Invest.* **2008**, *118*, 1590–1605.
3. Psychiatric GWAS Consortium Coordinating Committee. Genomewide Association Studies: History, Rationale, and Prospects for Psychiatric Disorders. *Am. J. Psychiatry* **2009**, *166*, 540–556.

CHAPTER

14

Pathogenetics of Disease

Reed E Pyeritz

Perelman School of Medicine at the University of Pennsylvania, Philadelphia, PA, USA

The great ease with which molecular information can be collected on the genomes of higher organisms will tempt many. We can inevitably expect vast compendia of sequences but, without functional reference, these compendia will be uninterpretable, like an undeciphered ancient language. Many people and many computers will play games with these sequences, but we will have to find out by experiment what the sequences do and how the products they make participate in the physiology and development of the organism. Thus, although the analysis of the genotype has been taken care of, we still need better ways of analyzing phenotypes. Many of us are ultimately interested in the causal analysis of development and the reduction of the complex phenotypes of higher organisms to the level of gene products. This is still the major problem of biology. We must understand what cells can do because all of what we are is generated by cells growing, moving and differentiating.

Sydney Brenner, 1973

The effort spent on the identification of genes is likely to prove only a small fraction of that required to work out their normal function in the tissues in which they are expressed. Yet that is where clues to the treatment and prophylaxis of disease are most likely to arise.

John Maddox, 1994

14.1 INTRODUCTION

The foregoing quotations emphasize that the general theme of this chapter—all that occurs between the gene and the bedside—is not a new one. The true promise of the Human Genome Project only began to be realized when our genome was sequenced, and the really hard work remains. Etiology is the study of the causes of a phenomenon and, in the medical context, of disease. Its method is to discover the association between factors that are thought to be causes and certain features that we wish to explain, such as a syndrome or a common disease. The goal and method of the discipline are strictly empirical, with at best minor interest in discerning the actual mechanisms involved. For instance, there is a heavily documented etiologic relationship between the total cholesterol level in plasma and the incidence of atherosclerosis. However, whether the former directly causes the latter or the relationship is more oblique is as yet unresolved. Pathogenesis is the study of the mechanisms by which the etiologic factors are converted into disease states. Genetic etiology is a more specialized topic that deals with the properties of the genetic causal factors of disease and how they behave. Mendel's laws, which were formulated before anything was known of the genes and their mechanisms, are arguably the high point of genetic etiology. A positive family history for early coronary artery disease is widely recognized as an important risk factor in the cause of myocardial infarction, with no appeal to explicit mechanisms, even genetic. Pathogenetics, a condensation of "genetic pathogenesis," is the study of how anomalies in the genome are converted into phenotypic disorders.

In this chapter, the emphasis is on disease. However, we stress the relationship of genetic processes to ordinary developmental mechanisms and maintenance of the healthy body, known as orthogenetics. Precise definition of a disorder, although an important question in its own right, is a luxury that we cannot afford here. For this reason, orthogenetics and pathogenetics are best seen as parts of a single continuous field of inquiry. Understanding the phenotype comes from the interactions among the insights from many disciplines.

Part of the notorious "nature-nurture" vision of medicine (i.e. the image that disease is in origin either genetic or environmental or between them) was that it was suggested for making sure that nothing is overlooked. We have serious doubts that it was ever an illuminating axis. One merely has to think of whether scurvy is a vitamin deficiency or (so far as we know) an inborn error of metabolism. In a fair-skinned population at the equator, sunburn is almost purely environmental. In a mixed population of pure fair-skinned and pure black-skinned individuals, it will be almost purely genetic. There is also the hidden and unsatisfactory notion of anomalous division into classes that are exclusive and exhaustive. However, to define a trait as "multifactorial" requires scarcely any proof of etiology at all. As many chapters in this book illustrate, no condition is purely "genetic" (or "environmental"), even those displaying Mendelian inheritance.

14.2 PATHOGENETICS OF MENDELIAN DISORDERS

Two pervasive characteristics of Mendelian disorders are "variability" and "pleiotropy." In the strict sense, the former describes variation in phenotype among people (often relatives) who have exactly the same mutant genotype. The latter reflects multiple (especially seemingly unrelated) phenotypic features due to the same genotype. Neither concepts can be understood based on etiology alone, and crucially depend on the interplay of the various factors that influence development of the disorder in addition to the action (or inaction) of the mutant gene.

When the defect in a genetic disorder is described at the level of the mutation, the field of inquiry is etiology. Anything more remote from the mutation represents phenotype and at least the first layer of complexity in the pathogenetics (Table 14-1). Thus, the resolution or sensitivity of the methods being brought to bear on the investigation of how the disorder arises determines how closely the mutation can be approached.

It is more feasible and instructive to focus on the stable product of most genes, the protein, and describe the types of molecular pathology that arises from mutation. In the most fundamental terms, a mutation can affect the quantity of a protein, the quality of a protein, and occasionally both aspects. The quantity of a protein synthesized by a gene is regulated at the level of transcription, by promoters, enhancers, noncoding RNAs, and other locus control elements, and at the level of translation (see Chapter 5). Mutations in any of a number of sites *cis* and *trans* to the gene of interest can affect the amount of protein produced. Usually, but not always, production from mutant alleles is decreased.

A change in the primary structure, the amino acid sequence, of a protein may alter its function (i.e. the quality of the protein). The study of diverse variants of the same protein has greatly advanced the understanding of molecular pathogenetics and given a certain sophistication to the "new" field of endeavor, investigating authentic relationships between genotype and phenotype. How the quality of a protein can be affected depends, in the first instance, on its normal function.

For example, a change in primary sequence might affect the stability of the protein and lead to enhanced (or retarded) degradation. In some situations, the amount of the mutant protein is crucial to the severity of the phenotype, especially in the dominant-negative scenario.

Proteins can be classified into three classes based on function: (1) those whose essential functions involve interactions with small molecules, such as enzymes, receptors, and transporters; (2) those that perform regulatory roles, such as transcription factors and hormones; and (3) those that function in complex systems, often in a structural role and often in association with other proteins.

Within each of the three classes of proteins, a mutation can have one of four consequences: quantitative increase or decrease in function, or qualitative gain or loss of function. Each of these consequences can have a number of molecular explanations.

A quantitative increase in function can be due to a regulatory mutation. An example is loss of sensitivity to inhibition, such as by a repressor molecule. A mutation could also affect the active site of an enzyme, such that its V_{max} was increased, or the binding site of a hormone, such that the K_M was lowered. Many disorders associated with expansion of unstable trinucleotide repeats, such as Huntington disease, share gain-of-function pathogenesis.

A quantitative decrease in function could operate by the converse of any of the afore-mentioned mechanisms. The extreme of the spectrum of decreased function is loss of function, perhaps the easiest to conceptualize, and certainly the most prevalent consequence. For example, most inborn errors of metabolic pathways result from an enzymatic failure. The enzymopathy can be due to a mutation in or around the locus encoding that enzyme, resulting in a qualitative or quantitative defect as described earlier; to abnormal posttranslational processing of the nascent enzyme; to abnormal subcellular localization or extracellular trafficking; to altered affinities for substrates or cofactors; or to altered responsiveness to allosteric regulators of activity. Other examples of loss-of-function phenotypes include familial hypercholesterolemia due to many of the defects in the low-density lipoprotein receptor and cancers due to defects in tumor suppresser genes, such as the retinoblastoma or neurofibromatosis type 1 genes. Strains of mice-bearing gene "knockouts" represent specified loss-of-function mutations; these are especially popular tools for studying development and neoplasia.

Quantitative and qualitative loss of function clearly overlaps. A mutation that reduces the ability of an enzyme to bind substrate also might lead to enhanced degradation and a reduced steady-state amount of the protein molecule.

Mutations that cause a gain in function, that is, a function not intrinsic to the wild-type protein, are less common.

The diverse familial amyloidoses are examples, in which a change in amino acid sequence of one or another protein (e.g. transthyretin) results in enhanced stability of the protein and abnormal tissue deposition (see Chapter 83).

The least commonly recognized molecular phenotype, also qualitative, is a change in function. One example is the product of the fusion of the *BCR* and the *ABL* genes in chronic myelogenous leukemia (see Chapter 79). Another example is the p53 protein, which when mutated in some ways, assumes regulatory capabilities foreign to the normal product.

As useful as these protein phenotypes are for classification (and education), there are limitations in making the intellectual leap to the next level of pathogenetic complexity. For example, gene knockout mutations are relatively easy to generate in mice, and increasingly in other species. Many investigators see this technique as a facile way to isolate the physiologic role of a particular gene product, to generate an animal model of a given disease, or to serve as the background strain into which a defined mutation is introduced. There is no question that the approach has been brilliantly successful in a number of instances. However, the pitfalls have been underemphasized.

The actual effect of the mutation may be loss of function at the protein level but gain of function at the cellular level. For example, Rett syndrome (OMIM*312750) is pleiotropic, severe neurologic disorder that primarily affects girls. The cause is mutation in *MECP2*, which encodes a protein that represses transcription of other genes. Mutations that inactivate the MECP2 protein result in enhanced or inappropriate production of proteins in various tissues, most obviously the brain.

Progress in understanding variability, of great interest in many clinical situations, has been achieved for a number of disorders, such as cystic fibrosis and familial breast and ovarian cancer due to mutation in *BRCA1* and *BRCA2*. A variety of techniques can be used. The more that is known about pathogenesis of the disorder of interest, the easier it is to select candidate genes, variations that might impact the fundamental mutation, or to validate loci identified by genome-wide association studies.

14.3 PATHOGENETICS OF COMMON DISORDERS

The major progress in the past 5 years in understanding both etiology and pathogenetics has occurred through GWAS. Exciting advances have been made despite the undoubted heterogeneity in causation that underlies the thousands of patients required for discovery and replication studies. Over 1100 loci involving over 165 disorders have been identified. These loci were selected because of relatively common genotypic variation in the general population. An optimistic analysis suggests that, for many disorders, close to 50% of the heritability has been identified. Nonetheless, an important "missing heritability" exists, which may represent rare variants. A great many of the associated loci are likely involved in the pathogenetic nexus of the disease. As such, they represent interesting targets for, among other things, drug discovery. Some associated loci are not near protein-coding sequences. Whether they are linked to the sites encoding the variety of regulatory RNAs, or represent remote control elements, is an area of intense investigation.

Another approach to the pathogenetics of common diseases is to identify endophenotypes, which represent reliable, clinically identifiable indications of the latent disease process. Various imaging modalities are common tools. So too are biomarkers, which appear in blood, CSF, or tissue before clear evidence of the disease emerges. One benefit of preclinical markers of disease is the facilitation of family studies, because many disorders only become clinically evident in late adulthood.

14.4 CONCLUSIONS

The prognosis of a disease is largely a matter of pathogenesis. For instance, its age of onset, the rapidity of its course, and the vulnerable points at which disease and complications may occur all depend on details that in principle, as much as in fact, may be difficult to infer even from the most detailed knowledge of the basic defect. Some knowledge of the prognosis may come from "black-box" empirical inquiries—the natural history of myotonic dystrophy, for instance—but this course calls for extensive data, and there may be disturbing discrepancies between one study and another that are not readily reconciled. If the pathogenesis is understood, even partially, more incisive methods may be available, including direct measurements of the progress of components of the disease. For instance, the pathogenesis of familial polyposis coli is not clearly established, but currently the course of this disease and its response to treatment are easier to study than Alzheimer dementia. Refined studies at the molecular level make for very precise statements about etiology. It is tempting, but rather treacherous, to view pathogenesis in the same way. But where the concern lies in either the assessment of morbidity or the study of the population and eugenic behavior of the mutant, to attach too much weight to refined biochemistry may push the precision of the statement at the expense of its significance. The overt clinical pattern and the target of selection are very coarse matters; the many modifying factors, which to the basic scientist are largely a nuisance, may have important attenuating effects on the course disease.

Many advances in therapeutics have resulted from largely empirical reasoning as to choosing an approach, but from an understanding of natural history in judging whether the therapy was successful. A more rational approach to targeting therapy is based on an understanding of pathogenesis. Some fondly held the hope of circumventing "indirect" therapies for genetic disorders by simply replacing the defective gene. But considerable experience has amply shown the general fallacy in this approach. Until the molecular pathogenesis of a disorder is elucidated, the effects of simply adding back, or even replacing, a gene that should have been functioning perhaps from conception will be as empirical as anything physicians had available in the eighteenth century.

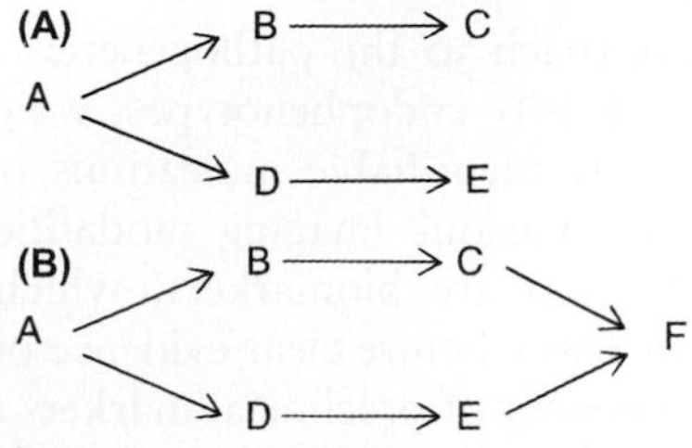

FIGURE 14-1 Metabolic pathways with branches. A. Open branched pathway. B. Closed branched pathway.

TABLE 14-1	Exploration of the Phenotype by Increasing Level of Complexity
mRNA	
Translated protein	
Posttranslationally modified protein	
Localization of gene product	
Macromolecular aggregate	
Cellular metabolism	
Tissue/organ function or structure	
Clinical manifestations	

FURTHER READING

1. Childs, B. *Genetic Medicine: A Logic of Disease*; Johns Hopkins University Press: Baltimore, 1989.
2. Cutting, G. R. Modifier Genes in Mendelian Disorders: The Example of Cystic Fibrosis. *Ann. N.Y. Acad. Sci.* **2010,** *1214,* 57–69.
3. Dietz, H. C. New Therapeutic Approaches to Mendelian Disorders. *N. Engl. J. Med.* **2010,** *363,* 852–863.
4. Green, E. D.; Guyer, M. S. Charting a Course for Genomic Medicine from Base Pairs to Bedside. *Nature* **2011,** *470,* 204–213.
5. Lander, E. S. Initial Impact of the Sequencing of the Human Genome. *Nature* **2011,** *470,* 187–197.
6. Milne, R. L.; Antoniou, A. C. Genetic Modifiers of Cancer Risk for *BRCA1* and *BRCA2* Mutation Carriers. *Ann. Oncol.* **2011,** *22* (Suppl. 1), 11–17.
7. Murphy, E. A.; Pyeritz, R. E. Homeostasis. VII. A Conspectus. *Am. J. Med. Genet.* **1986,** *24,* 735–751.
8. Pyeritz, R. E. Pleiotrophy Revisited. *Am. J. Med. Genet.* **1989,** *34,* 124–134.
9. Visel, A.; Zhu, Y.; May, D., et al. Targeted Deletion of the 9p21 Non-Coding Coronary Artery Disease Risk Interval in Mice. *Nature* **2010,** *464,* 409–413.
10. Wang, T. J.; Larson, M. G.; Vasan, R. S., et al. Metabolite Profiles and the Risk of Developing Diabetes. *Nat. Med.* **2011,** *17,* 448–454.
11. Wu, Y.; Le, W.; Jankovic, J. Preclinical Biomarkers of Parkinson Disease. *Arch. Neurol.* **2011,** *68,* 22–30.

CHAPTER 15

Human Developmental Genetics

Wen-Hann Tan

Division of Genetics, Children's Hospital Boston, Boston, MA, USA

Edward C Gilmore

Department of Pediatrics, Case Western Reserve University, Cleveland, OH, USA

Hagit N Baris

Institute of Medical Genetics, Rabin Medical Center—Beilinson Campus, Petach-Tikva, Israel

15.1 TIMING OF NORMAL HUMAN DEVELOPMENT

The development of the human embryo (i.e. embryogenesis) occurs in the first 8–9 weeks from fertilization to the beginning of the fetal period, and can be divided into 23 "Carnegie stages." These are based on the classic collection of human embryos accumulated at the Carnegie Institution decades ago. These stages are based primarily on the morphological characteristics rather than the age or size of the embryo, although each stage corresponds to an approximate post-fertilization/postconception age (NOT gestation age, which begins from the last menstrual period, approximately 2 weeks prior to fertilization in women with regular menstrual cycles) and size in terms of crown-rump length. The Carnegie stages, with the key events at each stage and the approximate postconception age in days, are shown in Table 15-1.

The first 4 weeks of development, from fertilization until the end of gastrulation (Carnegie stage 13) is also known as "blastogenesis." Organogenesis begins at the end of blastogenesis and is characterized by morphogenesis ("formation of organs and other body parts") and histogenesis ("differentiation of cells and tissues"). The fetal period, which begins after Carnegie stage 23 and continues through delivery, is the period of continuing growth and differentiation of the organs formed during embryogenesis.

15.2 THE CONCEPT OF DEVELOPMENTAL FIELDS AND FIELD DEFECTS

Morphogenesis is believed to occur within developmental fields, defined as a "region or a part of the embryo which responds as a coordinated unit to embryonic induction and results in complex or multiple anatomic structures." The primary field, which comprises the entire early embryo, is established through blastogenesis. The primary field is then subdivided through the process of pattern formation into progenitor fields, which give rise to the primordia of all morphological structures. Subsequent differentiation results in secondary or epimorphic fields, which form the "irreversibly determined final structure(s)." Histogenesis occurs much later in development, usually after pattern formation. Most organs that are malformed are histologically normal and not at risk of developing malignancies; on the other hand, abnormal histogenesis in a dysmorphogenetic organ might result in malignancy.

Abnormal differentiation during organogenesis usually leads to a "monotopic" field defect and hence an anomaly in a single structure or organ. In contrast, abnormal blastogenesis usually results in a "polytopic" field defect in which multiple structures/organs in different parts of the body that share the same primary developmental field are malformed. Multiple congenital anomalies may also be part of a sequence, a syndrome, or an association. A sequence is defined as a pattern of multiple anomalies derived from a single known or presumed prior anomaly or mechanical factor. A syndrome is defined as a pattern of multiple anomalies thought to be pathogenetically related and not known to represent a single sequence or a polytopic field defect. An association is defined as a nonrandom occurrence in two or more individuals of multiple anomalies not known to be a polytopic field defect, sequence, or syndrome and refers solely to statistically, not pathogenetically, or causally related anomalies (see Chapter 43). The term association should be limited to the idiopathic statistical occurrence of multiple congenital anomalies apparently of non-blastogenetic origin. Hence, the VATER association (OMIM# 192350), which is postulated to be due to a blastogenetic defect, should be considered as a polytopic field defect instead of an association.

15.3 A LIMITED REPERTOIRE OF DEVELOPMENTAL GENES AND PATHWAYS

Embryogenesis and the development of various organ systems are temporally and spatially regulated by a common set of genes, which are often transcription factors, as well as developmental pathways, many of which regulate transcription factors. The same gene may be expressed in different tissues at different developmental stages or regulate different developmental pathways and hence affect the development of different organ systems. Some developmental pathways interact with one another. There are also parallel pathways leading to "redundancy" due to the presence of paralogous genes (i.e. genes with very similar nucleotide sequences and are presumably derived from the same ancestral gene by gene/chromosome duplication).

Studies in animal models have shown that in early embryogenesis, Oct4 (also known as Pou5f1 or Oct3), Nanog, and Sox2 confer pluripotency and are required for the formation of the inner cell mass, which will develop into embryo; Sall4 activates the expression of these three genes and suppresses the expression of *Cdx2*, which promotes the formation of the trophectoderm; the trophectoderm develops into the embryonic component of the placenta. Gastrulation, the process through which all three embryonic germ layers (endoderm, mesoderm, and ectoderm) are formed, requires the expression of NODAL (a member of the TGF-β protein superfamily), specific fibroblast growth factors such as FGF8, the SNAI group of zinc finger transcription factors, the T-box transcription factor EOMES, and the basic helix-loop-helix transcription factors, MESP1 and MESP2. Following gastrulation and the formation of the notochord (a solid midline dorsal structure involved in the formation of the central nervous system) from mesodermal cells, the developing embryo undergoes somitogenesis, the process through which the paraxial mesoderm that flanks the notochord undergoes segmentation in the cranio-caudal axis to form somites, which eventually give rise to the axial skeleton, skeletal muscles and associated tissues. Somitogenesis is regulated by oscillating expression of specific genes in the Notch signaling pathway including *HES7*, *LNFG*, and *DLL3* (all of which are part of the "segmentation clock") and by expression of *FGF8*, *MESP1*, and *MESP2* (Figure 15-1). Thus, some of the genes that regulate gastrulation also regulate somitogenesis. Moreover, somites are formed only when these segmentation clock genes are in a specific phase and *FGF8* expression is below a specific threshold. This highlights the importance of spatial and temporal regulation of developmental gene expression.

It is notable that loss-of-functional mutations in any of the three Notch signaling pathway genes that are part of the segmentation clock lead to one of the types of spondylocostal dysostosis, while loss-of-functional mutations in *MESP2* lead to spondylothoracic dysostosis, both of which are skeletal dysplasias characterized by contiguous segmentation defects of the vertebrae.

The establishment of left–right asymmetry in the embryo is regulated by the differential expression between the left and right lateral plate mesoderm of specific genes including *Nodal* (also involved in gastrulation), *Lefty2*, and *Pitx2*, at least in the mouse model; it is also dependent on cilia-mediated flow of morphogen-containing "nodal vesicular parcels" at the primitive node and hence requires the expression of genes involved in the assembly and function of the cilia. Reference (*12*) summarizes the human genetic syndromes that result from mutations in the genes implicated in left–right asymmetry.

15.4 REGULATION OF GENE EXPRESSION IN DEVELOPMENT

The tight spatial and temporal regulation of developmental genes allows the same gene to be involved in different developmental processes from embryogenesis through organogenesis. Moreover, since the genes in almost all cells in the body are identical (i.e. "genomic equivalence"), the genes must be differentially expressed in the different cells to enable the cells to differentiate into different tissues and organs. Differential gene expression may be controlled at the level of transcription, posttranscription (i.e. before translation of mRNA), translation, and posttranslation. Transcriptional regulation may be accomplished through epigenetic mechanisms (e.g. modification of histones by acetylation or methylation, or methylation of DNA residues) or genetic mechanisms (e.g. promoter, enhancer, silencer, and insulator DNA sequences that often lie outside of the coding sequences of the genes that they regulate). The transcribed pre-mRNA may be modified by RNA editing followed by alternative splicing before it is translated. Translation of the processed mRNA is regulated by a set of eukaryotic initiation factors (eIFs). In addition, microRNAs (miRNA) can inhibit the initiation of translation, the elongation of the primary polypeptide, or the degradation of mRNA by deadenylation. As the primary polypeptide is folded and assembled into its final quaternary structure, various posttranslational modifications may occur, such as phosphorylation, ubiquitination, sumoylation, acetylation (of lysine residues), methylation (of lysine and arginine residues), hydroxylation, and glycosylation. Defects in these processes can result in a wide variety of genetic syndromes including Rubinstein–Taybi syndrome (mutation in histone acetyltransferases), immunodeficiency-centromeric instability-facial anomalies (ICF) syndrome (mutation in a DNA methyltransferase), dyschromatosis symmetrica hereditaria (mutation in a RNA-editing gene), myotonic dystrophy (aberrant alternative splicing), and congenital disorders of glycosylation due to abnormal posttranslational glycosylation.

15.5 ORGANOGENESIS

Development of the human brain begins with the formation of the three primary brain vesicles by day 28 in the most cranial part of the neural tube, which is derived from the ectoderm. The secondary vesicles are formed in the 5th week of development. These vesicles and the brain structures that derive from them are shown in Figure 15-2.

The basic processes involved in brain development are, in order, tissue/cellular specification, proliferation followed by final differentiation, neuronal migration, axonal outgrowth/target finding and finally synaptogenesis/network modification. The timing of these processes is shown in Table 15-2.

Multiple genes and gene pathways regulate the differentiation of specific brain regions. Genes that induce the development of the forebrain include members of the fibroblast growth factor (FGF) family, the *Wnt* gene family, the *BMP* (bone morphogenetic protein) genes, and *SHH* (sonic hedgehog). Genes that regulate the development and segmentation of the hindbrain include the *HOX* gene family.

Formation of the embryonic heart begins with the fusion of a pair of heart tubes derived from the splanchnic mesoderm, and it starts to beat at around 22–23 days even before it has fully formed, making it the first functional organ in the embryo. It then undergoes looping which results in left–right asymmetry, followed by septation leading to the formation of the four chamber heart. The genetic and epigenetic regulation of cardiogenesis have recently been reviewed by López-Sánchez et al. and van Weerd et al.

The epithelial lining of the gastrointestinal tract comprises cells derived from the endoderm, but the surrounding smooth muscle and connective tissue are derived from the mesoderm, and the enteric nervous system is derived from the ectoderm. The genes involved in the development of the intestines have been reviewed by Noah et al.

The kidneys are derived from the intermediate mesoderm. The genes involved in the development of the kidneys have been reviewed by Potter et al., and Song and Yosypiv recently reviewed the genetic bases of congenital renal malformations.

The bones and connective tissue of the limbs are derived from the lateral plate mesoderm. Members of the *FGF* gene family induce the development of the limb bud, at the tip of which is the apical ectodermal ridge where specific *FGF* genes are expressed and promote the longitudinal outgrowth of the developing limb. The expression of *SHH* in the zone of polarizing activity in the caudal part of the limb bud determines the anterior–posterior axis (i.e. digit 1 vs. digit 5) of the developing limbs. The dorsal-ventral axis of the limbs is specified various genes including *LMX1B* and *WNT7A*. In addition, some members of the *HOX* and *T-box* (*TBX*) gene families also regulate limb development.

15.6 CONCLUSION

Human development requires the precise regulation of a large but finite number of genes in the different developing tissues and organs, both in terms of where the genes are expressed and when they are expressed. Malformations in one or more organ systems occur when these genes are mutated and thereby disrupting the evolutionary-conserved sequence of development.

TABLE 15-1 Carnegie Stage Table

Stage	Days (Approx)	Size (mm)	Images (Not to Scale)	Events
1	1 (week 1)	0.1–0.15		fertilized oocyte, zygote, pronuclei
2	2–3	0.1–0.2		morula cell division with reduction in cytoplasmic volume, blastocyst formation of inner and outer cell mass
3	4–5	0.1–0.2		loss of zona pellucida, free blastocyst
4	5–6	0.1–0.2		attaching blastocyst

Continued

TABLE 15-1 Carnegie Stage Table—*Cont'd*				
Stage	**Days (Approx)**	**Size (mm)**	**Images (Not to Scale)**	**Events**
5	7–12 (week 2)	0.1–0.2		implantation
6	13–15	0.2		extraembryonic mesoderm, primitive streak, gastrulation
7	15–17 (week 3)	0.4		gastrulation, notochordal process
8	17–19	1.0–1.5		primitive pit, notochordal canal
9	19–21	1.5–2.5		Somitogenesis **Somite Number 1–3** neural folds, cardiac primordium, head fold
10	22–23 (week 4)	2–3.5		**Somite Number 4–12** neural fold fuses
11	23–26	2.5–4.5		**Somite Number 13–20** rostral neuropore closes
12	26–30	3–5		**Somite Number 21–29** caudal neuropore closes
13	28–32 (week 5)	4–6		**Somite Number 30** leg buds, lens placode, pharyngeal arches
14	31–35	5–7		lens pit, optic cup
15	35–38	7–9		lens vesicle, nasal pit, hand plate
16	37–42 (week 6)	8–11		nasal pits moved ventrally, auricular hillocks, foot plate
17	42–44	11–14		finger rays
18	44–48 (week 7)	13–17		ossification commences
19	48–51	16–18		straightening of trunk

TABLE 15-1 Carnegie Stage Table—*Cont'd*

Stage	Days (Approx)	Size (mm)	Images (Not to Scale)	Events
20	51–53 (**week 8**)	18–22		upper limbs longer and bent at elbow
21	53–54	22–24		hands and feet turned inward
22	54–56	23–28		eyelids, external ears
23	56–60	27–31		rounded head, body and limbs

Source: http://php.med.unsw.edu.au/embryology/.

TABLE 15-2 Timing of Human Brain Development

Event	Gestation Time
Neural plate forms	16th day
Longitudinal groove forms	20th day
Neural tube closure start at 2nd–7th somite	22nd day
Neuromeres start to form in hindbrain	22nd day
Forebrain induction	3rd week
Cerebellum formation starts	4th week
Purkinje cells are born	5–6 weeks
Cerebral cortex start to expand	6 weeks
Ganglionic eminences form	6–7 weeks
Cerebral cortex preplate forms	6–7 weeks
Cortical plate (CP) forms	7–10 weeks
Cerebral cortical neurogenesis	7–18 weeks
Neuronal migration into cortical plate (including GABAergic neurons)	7–28 weeks
Corpus colossal fiber start to cross midline	12 weeks
Myelination-	
Most of cerebral cortex	1st year of life
Cerebral commissures	1st half of 1st decade
Intra-cortical association	Through 2nd decade
Synapse dynamics (visual cortex)	
Formation	3rd Trimester to 5 years
Pruning	5 years to 2nd decade

Adapted from Sidman, R. L.; Rakic, P. Development of the human central nervous system. In *Histology and histopathology of the nervous system*, Haymaker, W.; Adams, R. D., Eds.; Charles C Thomas: Springfield, IL, 1982; Vol. 1, pp 3–145; Huttenlocher, P. R., Morphometric study of human cerebral cortex development. *Neuropsychologia* 1990, *28* (6), 517–527; Rabinowicz, T.; de Courten-Myers, G. M.; Petetot, J. M.; Xi, G.; de los Reyes, E., Human cortex development: estimates of neuronal numbers indicate major loss late during gestation. *J. Neuropathol. Exp. Neurol.* 1996, *55* (3), 320–328; Yakovlev, P. I.; Lecours, A. R., The Myelogenic cycles of regional maturation of the brain. In *Regional Development of the Brain in Early Life*, Minkowski, A., Ed. Blackwell Scientific Publications: Oxford, 1967; pp 3–70; ten Donkelaar, H. J.; Lammens, M.; Wesseling, P.; Thijssen, H. O.; Renier, W. O., Development and developmental disorders of the human cerebellum. *J. Neurol.* 2003, *250* (9), 1025–1036.

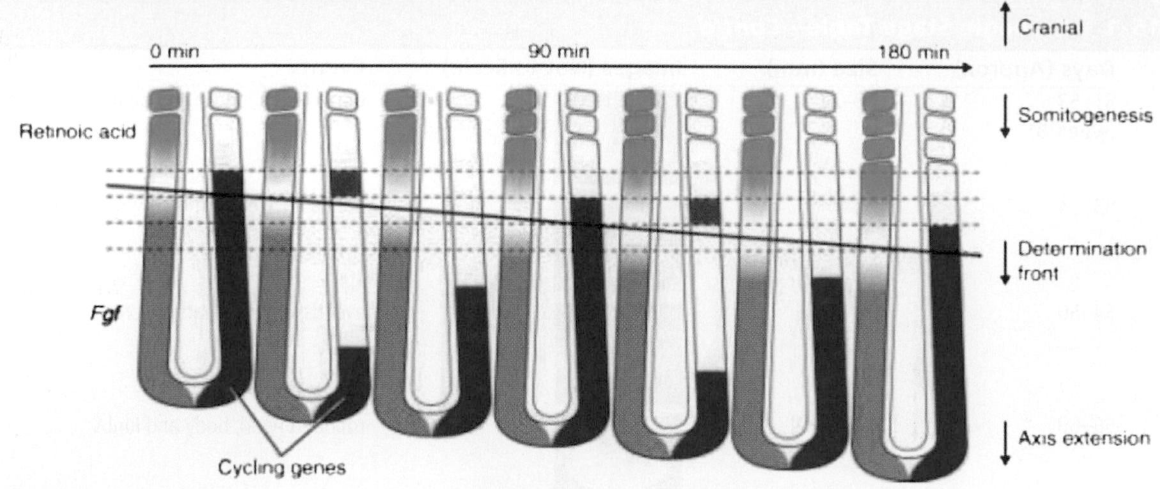

FIGURE 15-1 Clock-and-wavefront model of somitogenesis. Diagrams of the caudal end of chick embryos during two rounds of somitogenesis. Retinoic acid (blue) and Fgf8 (gray) gradients move caudally as the embryo elongates (axis extension) during somitogenesis. In chick, a somite pair forms every 90 min, which constitutes the length of the clock cycle. Expression of cycling genes (red) extends from caudal to cranial, and when expression of these genes spreads cranially to cross the threshold level of *Fgf8* signaling (called the determination wavefront; diagonal line), somites are established (indicated by expression of *Mesp* genes; purple). *(Schoenwolf, G. C.; Bleyl, S. B.; Brauer, P. R.; Francis-West, P. H.; Larsen's Human Embryology, 4th ed.; Churchill Livingstone: Philadelphia, 2008.)*

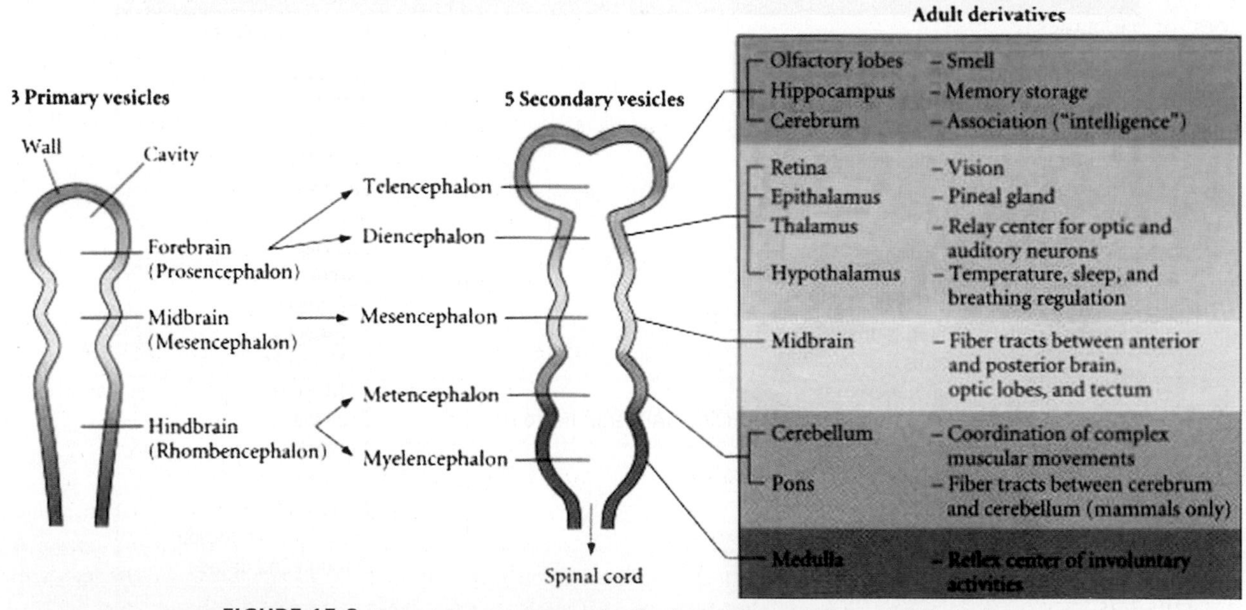

FIGURE 15-2 The primary and secondary brain vesicles and their adult derivatives.

GENERAL REFERENCES

1. Epstein, C. J.; Erickson, R. P.; Wynshaw-Boris, A. *Inborn Errors of Development: The Molecular Basis of Clinical Disorders of Morphogenesis*, 2nd ed.; Oxford University Press: New York, 2008.
2. Gilbert, S. F. *Developmental Biology*, 9th ed.; Sinauer Associates, Inc.: Sunderland, MA, 2010.
3. Hogg, M.; Paro, S.; Keegan, L. P.; O'Connell, M. A. RNA Editing by Mammalian ADARs. *Adv. Genet.* **2011,** *73,* 87–120.
4. Hill, M. A. UNSW Embryology. http://php.med.unsw.edu.au/embryology/ (accessed March 18, 2012).
5. Lopez-Sanchez, C.; Garcia-Martinez, V. Molecular Determinants of Cardiac Specification. *Cardiovasc. Res.* **2011,** *91* (2), 185–195.
6. Martinez-Frias, M. L.; Frias, J. L.; Opitz, J. M. Errors of Morphogenesis and Developmental Field Theory. *Am. J. Med. Genet.* **1998,** *76* (4), 291–296.
7. Noah, T. K.; Donahue, B.; Shroyer, N. F. Intestinal Development and Differentiation. *Exp. Cell Res.* **2011,** *317* (19), 2702–2710.
8. Opitz, J. M.; Zanni, G.; Reynolds, J. F., Jr.; Gilbert-Barness, E. Defects of Blastogenesis. *Am. J. Med. Genet.* **2002,** *115* (4), 269–286.
9. Potter, S. S.; Brunskill, E. W.; Patterson, L. T. Defining the Genetic Blueprint of Kidney Development. *Pediatr. Nephrol.* **2011,** *26* (9), 1469–1478.
10. Sayed, D.; Abdellatif, M. MicroRNAs in Development and Disease. *Physiol. Rev.* **2011,** *91* (3), 827–887.
11. Schoenwolf, G. C.; Bleyl, S. B.; Brauer, P. R.; Francis-West, P. H. *Larsen's Human Embryology*, 4th ed.; Churchill Livingstone: Philadelphia, 2008.
12. Sutherland, M. J.; Ware, S. M. Disorders of Left–Right Asymmetry: Heterotaxy and Situs Inversus. *Am. J. Med. Genet. C Semin. Med. Genet.* **2009,** *151C* (4), 307–317.

CHAPTER

16

Twins and Twinning

Jodie N Painter, Sarah J Medland and Grant W Montgomery

Queensland Institute of Medical Research, Herston, Queensland, Australia

Judith G Hall

Department of Medical Genetics, British Columbia's Children's Hospital, Vancouver, Canada

Humans have always been fascinated by twins although throughout history different cultures have had different attitudes regarding them, considering them as signs of either good or bad times to come. Nowadays, as scientific and medical knowledge have increased, the opportunity to study similarities and differences within twin pairs has lead to numerous scientific advances (*1*), particularly in the fields of epidemiology and genetics.

Twins can be either dizygotic ("fraternal") or monozygotic ("identical"). Dizygotic twins (DZ) result when two different ova are released (multiple ovulation) and are fertilized by two different sperm, growing and developing at the same time in the same uterus. Monozygotic twins (MZ) result from one ovum fertilized by one sperm that then divides to form two embryos. The most reliable method for determining zygosity is the analysis of DNA.

The incidence of MZ twins seems constant throughout the world at 3–4 in every 1000 births while the incidence of DZ twins varies from population to population, from 2 to 7 per 1000 births in Japan through to 45–50 per 1000 births in Nigeria (*1,2*). The incidence of both MZ and DZ twins has increased over the last few decades, to almost 1 in 36 live births. This mostly represents an increase in DZ twins, due primarily to increased numbers of older mothers and to the use of assisted reproductive technologies (ARTs) (*2*). Interestingly, MZ twinning is also increased with the use of ARTs (*3*).

There are a number of different types of placentation according to the number of chorionic membranes and amniotic sacs that are present. Placentation in MZ twins is thought to correspond to the stage in embryonic life during which the twinning event occurred (Table 16-2; Figure 16-1). MZ twins may have separate or contiguous placentas and may be monochorionic monoamniotic, monochorionic diamniotic, or dichorionic diamniotic (see Table 16-2). DZ twins typically have two placentas, with two chorions and two amnions (diamniotic dichorionic), although they may fuse and look like one.

Both MZ and DZ twins have an increased risk for structural defects compared to singletons. Deformations are presumably related to external pressure caused by two growing fetuses sharing the space usually meant for one. Structural defects in MZ twins are 3 times more frequent than among DZ twins and 2–3 times more frequent than in singletons. MZ twins are also known to have a higher incidence of all types of congenital anomalies (see Table 16-1), some of which are unique to the MZ twinning process itself and thought to be due to defects in the completion of the development of the embryo (*5*, see Table 16-3).

The etiology of MZ twinning in humans is unknown, but several mechanisms have been proposed, including delayed fertilization or implantation, abnormalities or rupture of the zona pellucida, congenital anomalies, chromosomal abnormalities and epigenetic events such as genomic imprinting and X-inactivation (*1,4*). Women with a history of spontaneous DZ twinning have increased incidence of multiple follicle growth, and higher levels of follicle-stimulating hormone (FSH) and luteinizing hormone than those who deliver singletons (*2*). Increased DZ twinning in older women is thought to result from depletion of the ovarian follicle pool and the subsequent rise in FSH concentrations (*5*).

There are a number of reports of MZ twinning occurring in families (*6*), suggested to be related to an inherited defect in the zona pellucida leading to early hatching, or to environmental or epigenetic factors. However, investigation of familial MZ twinning is hampered by the possibility of ascertainment bias as such studies typically only include families containing multiple MZ twin pairs. Familial DZ twinning is firmly established, most likely due to an inherited predisposition to multiple ovulations, with the risk of having twins being up to 2.5 times higher for a woman with a sister with DZ twins than it is for the general population (*2*).

A number of studies have attempted to determine genes underlying the predisposition to DZ twinning, including family-based linkage studies and the screening of biologically plausible candidate genes. At present only mutations in the growth differentiation factor 9 (*GDF9*) gene appear to increase the chance of DZ twinning, although such mutations are rare (*7*). Such studies have confirmed that DZ twinning is a complex trait likely to be influenced by multiple genes.

Classic twin studies (*8*) rely on the assumption that MZ twins are genetically identical whereas DZ twins are not, while both are exposed to the same environmental influences. Hence any discordance between MZ twins is expected to be due to environmental influences, while differences between DZ twins are attributed to a combination of both genetic and environmental factors. Studies of MZ versus DZ twins have helped to establish the extent to which genetic versus environmental factors contribute to traits and disorders with multifactorial inheritance (termed "complex" traits or diseases) (*1*).

Studying MZ twins has revealed that a number of neurological, immunological, and structural aspects of embryogenesis appear to be independent of genetic factors. There are also increasing reports of discordance for phenotypic and genetic factors between MZ twins, most clearly seen when twins are discordant for disease (see Table 16-5). Recent advances in genetic technology have revealed further MZ twin discordance at the DNA, RNA and epigenetic levels (*9,10*).

In the field of twins and twinning, there is still a great deal to be learned. The development of new cytogenetic techniques, the use of prenatal diagnosis such as chorionic villi sampling, amniocentesis, and ultrasound examination in humans, as well as embryopathology, histology and molecular genetic advances all give clues to the increased understanding of MZ and DZ twins and the twinning process itself.

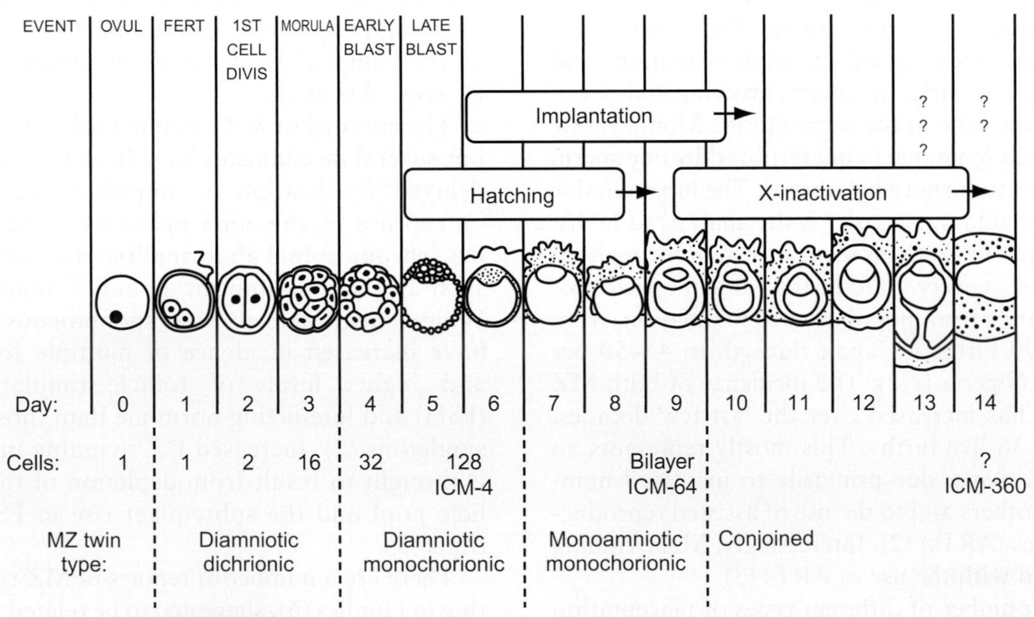

FIGURE 16-1 Process of monozygotic twinning during postfertilization.

TABLE 16-1 Structural Defects in Monozygotic Twins

Associated with the Twinning Process Due to Incomplete Splitting of the Embryo	Due to Shared Vascular Connections or to *in Utero* Death of Second Twin	Due to Fetal Constraint or Crowding *in Utero*
Conjoined twins	Acardia (TRAP sequence)	Craniosynostosis
Fetus *in fetu*	Asplenia	Positional defects of the foot
Fetus papyraceous	Microcephaly, hydrocephaly	Bowing of the limbs
	Intestinal atresia	Some contractures
	Aplasia cutis	
	Terminal limb defects	
	Gastroschisis	
	Disseminated intravascular coagulation	

TABLE 16-2 Types of Monozygotic Twins According to Placentation

Types of Placentation in MZ Twins	Time of Division After Fertilization (Days)	MZ Twin Pregnancies Surviving at Birth (%)
Dichorionic diamniotic	<3	25
Monochorionic diamniotic	4–6	70–75
Monoamniotic monochorionic	7–13	1–2

TABLE 16-3 Anomalies Exclusive to Monozygotic Twins

Anomaly	Description
Fetus *in fetu*	Small parasitic dead twin attached to a normal twin. Often confused with a tumor. Generally located at the origin of the superior mesenteric vessels. Other sites have been reported.
Fetus papyraceous	Mummified dead fetus usually attached to the placenta and present with a normal or more viable twin.
Acardia	Twin with an absent or rudimentary or nonfunctioning heart and whose circulation has been sustained by a normal twin. Associated with a higher rate of chromosomal anomalies (1 in 35,000 births with excess of females).
Conjoined twins	Incomplete twins resulting from an abnormality of the twinning process. They are derived from a single zygote and are always of the same sex. Incidence varies from one to 20,000 to one to 100,000. Females make up 80% of conjoined twins.

TABLE 16-4 Sex Ratio in Twins

	DZ Twins and Singletons	All MZ Twins	Conjoined Twins	Sacral Teratomas
Sex ratio (M/(M + F)	0.514	0.496	0.23	0.25

Data from James (*167*).

TABLE 16-5 Examples of Reported Discordance in Monozygotic Twins
Acardiac
Adrenal hyperplasia
Aging
Aglossia-adactylia
Aicardi syndrome
Alagille syndrome
Alzheimer disease
Amniotic bands
Amyoplasia
Amyotrophic lateral sclerosis
Anorchia
Asplenia
Asthma
Attention deficit disorder
Autism
Basal cell carcinoma
Beckwith–Wiedemann syndrome
Behçet's syndrome
Biliary atresia
Body mass index/oOverweight
Body stalk
Breast cancer
Cavum septum pellucidum
Cerebral hemisphere
Cerebral palsy
Choanal atresia
Chromosomes 1 and 4
Chronic fatigue syndrome
Chronic periodontitis
Cleft lip ± cleft palate
Cloverleaf skull
Coeliac disease
Congenital heart disease
Copy number variation
Corpus callosum
Costello syndrome
Cutaneous mastocytosis
Cutis laxa
Cystitis
Deletion 22q
Developmental coordination disorder
Diabetes
Duane retraction syndrome
Duchenne muscular dystrophy
Endocardial fibroelastosis
Epilepsy
Factor IX deficiency
Fibular aplasia
Fragile X syndrome
Frontonasal dysplasia
G syndrome
Genital anomalies
Gerstmann–Sträussler–Scheinker disease
Goldenhar syndrome
Gonadal dysgenesis
Hair whorls
Handedness
Hirschsprung disease
Hunter disease
Huntington disease
Hydranencephaly
Hypertrophic cardiomyopathy
Hypothyroidism
Infantile spasms
Inflammatory bowel disease (Crohn's Disease, ulcerative colitis)

TABLE 16-5 Examples of Reported Discordance in Monozygotic Twins—cont'd
Joint mobility
Kabuki syndrome
Kallmann syndrome
Kleeblattschädel anomaly
Leukemia
Lipodystrophy
McCune–Albright syndrome
Macular degeneration
Megacystis–microcolon–intestinal hypoperistalsis syndrome
Mental retardation
Migraines
Monosomy 11p
Multiple sclerosis
Myasthenia gravis
Neural tube defects
Neuroblastoma
Oculo–oto–radial syndrome
Oral–facial–digital syndrome, type 1
Parkinson disease
Polydactyly
Primary biliary cirrhosis
Proteus syndrome
Pyloric stenosis
Renal agenesis
Retinitis pigmentosa
Rhabdoid tumor
Rheumatoid arthritis
Ring chromosome 19 mosaicism
Rubinstein–Taybi syndrome
Russell–Silver syndrome
Sex
Schimmelpenning–Feuerstein–Mims syndrome
Schizophrenia/bipolar disorder
Scheuermann disease
Scleroderma
Seizures
Sirenomelia
Skeletal dysplasia
Sotos syndrome
Spinocerebellar ataxia
Strabismus
Sudden infant death syndrome
Suicide
Symbrachydactyly
Synaesthesia
Systemic sclerosis
Teratoma
Testicular cancer
Thyroid dysgenesis
Toxoplasmosis infection
Trisomy 1
Trisomy 13 (Patua syndrome)
Trisomy 18 (Edward syndrome)
Trisomy 21 (Down syndrome)
Tuberous sclerosis
Turner syndrome
Urinary tract anomalies
Vaginal dysgenesis
van der Woude syndrome
VATER association
Vitiligo
von Hippel–Lindau syndrome

REFERENCES

1. Hall, J. G. Twinning. *Lancet* **2003,** *362,* 735–743.
2. Hoekstra, C.; Zhao, Z. Z.; Lambalk, C. B.; Willemsen, G.; Martin, N. G.; Boomsma, D. I.; Montgomery, G. W. Dizygotic Twinning. *Hum. Reprod. Update* **2008,** *14,* 37–47.
3. Aston, K. I.; Peterson, C. M.; Carrell, D. T. Monozygotic Twinning Associated with Assisted Reproductive Technologies: A Review. *Reproduction* **2008,** *136,* 377–386.
4. Machin, G. Non-identical Monozygotic Twins, Intermediate Twin Types, Zygosity Testing, and the Non-random Nature of Monozygotic Twinning: A Review. *Am. J. Med. Genet. C Semin. Med. Genet.* **2009,** *151C,* 110–127.
5. Beemsterboer, S. N.; Homburg, R.; Gorter, N. A.; Schats, R.; Hompes, P. G.; Lambalk, C. B. The Paradox of Declining Fertility but Increasing Twinning Rates with Advancing Maternal Age. *Hum. Reprod.* **2006,** *21,* 1531–1532.
6. Cyranoski, D. Two by Two. *Nature* **2009,** *458,* 826–829.
7. Palmer, J. S.; Zhao, Z. Z.; Hoekstra, C.; Hayward, N. K.; Webb, P. M.; Whiteman, D. C.; Martin, N. G.; Boomsma, D. I.; Duffy, D. L.; Montgomery, G. W. Novel Variants in Growth Differentiation Factor 9 in Mothers of Twins. *J. Clin. Endo. Metab.* **2006,** *91,* 4713–4716.
8. Boomsma, D.; Busjahn, A.; Peltonen, L. Classical Twin Studies and Beyond. *Nat. Rev. Genet.* **2002,** *3,* 872–882.
9. Shur, N. The Genetics of Twinning: From Splitting Eggs to Breaking Paradigms. *Am. J. Med. Genet. C* **2009,** *151,* 105–109.
10. Zwijnenburg, P. J.; Meijers-Heijboer, H.; Boomsma, D. I. Identical but Not the Same: The Value of Discordant Monozygotic Twins in Genetic Research. *Am. J. Med. Genet. B Neuropsychiatr. Genet.* **2010,** *153B,* 1134–1149.

CHAPTER

17

The Molecular Biology of Cancer

Edward S Tobias

Wolfson Medical School Building, Glasgow, UK

As reviewed in this chapter, we now know that germline and somatic mutations in two major classes of genes, that is, oncogenes and tumor suppressor genes (which include several DNA damage recognition and repair genes) play a primary role in the pathogenesis of cancer. However, as noted at the outset of this chapter, given the enormous progress that has been made in furthering our understanding of the genetic basis of cancer, it is not possible to review in detail all of the various mutations in oncogenes and tumor suppressor genes that have been identified in human cancers (continually updated lists of human mutations are accessible through the Human Gene Mutation Database at http://www.hgmd.cf.ac.uk/ and the catalog of identified cancer-related genes is provided at the Cancer Genome Project website http://www.sanger.ac.uk/genetics/CGP/Census/). Nor is it possible to describe all of the studies that have been carried out to address the cellular functions of various oncogene and tumor suppressor gene protein products. Rather, the primary goals of this chapter are (1) to provide an overview of the experimental strategies used to characterize specific genetic alterations in cancer; (2) to review some of the most prevalent and/or most informative genetic alterations in human cancer that have been elucidated; (3) to highlight some of the relationships between specific genetic alterations and clinical aspects of cancer; and (4) to review the current thinking regarding the normal cellular functions of the genes and proteins involved.

Undoubtedly, many additional oncogenes, tumor suppressor genes and, especially, risk-modifying genes remain to be identified. In addition, the relationships between dietary and environmental factors and genetic alterations in cancer will need to continued for investigation. Further studies will no doubt be facilitated by the elucidation of the sequence of the human genome, the continued identification of polymorphic DNA markers (especially single nucleotide polymorphisms) and several technological advances. One of the latter is massively parallel DNA sequencing (next generation sequencing) that has greatly assisted the ongoing systematic DNA sequencing of thousands of individual cancers to identify all the cancer-associated mutations.

A previous development was the use of complementary DNA microarrays for transcriptional profiling, whereby the expression levels of thousands of gene sequences could be compared between different tissues and patients. The application of this technology permitted the molecular subclassification of various tumor types, such as diffuse large B-cell lymphoma and malignant melanoma. Furthermore, despite the current complexity of the analyses, such methodology has already been demonstrated to be capable of assisting in the classification of tumors for diagnostic and prognostic purposes. A more recent advance, however, has been the development of methods for the detection of levels of microRNAs, a class of small nonprotein-coding RNAs that play important roles in the regulation of gene expression. As was found for messenger RNA, abnormalities in the relative levels of individual microRNAs are increasingly being found to be associated with particular cancer types. Moreover, the stability of these microRNAs in body fluids, such as serum, raises the future possibility of using microRNA detection as a means of detecting or diagnosing tumors or monitoring the effectiveness of cancer therapy.

In addition, important developments have also been made in relation to technologies such as array comparative genome hybridization and multiplex ligation-dependent probe amplification, which detect DNA amplifications and deletions, in addition to proteomics and high-throughput screening.

A cause for optimism is that the continued increase in our understanding of the molecular biology of cancer and improved methods of analysis and drug development have facilitated a concomitant increase in the variety and effectiveness of molecularly targeted therapeutic agents. Such agents were originally exemplified by imatinib mesylate

(Gleevec), trastuzumab (Herceptin), and gefitinib (Iressa). They have now been followed by the PARP inhibitors, the results of which are particularly encouraging. In addition, small molecule inhibitors of several mutated proteins such as BRAF, EGFR, ERBBS, KIT, PDGFRA, PML-RARA, MET, and ALK have already been used in clinical trials and it is likely that many more such potential targeted therapies will be developed in the near future.

CHAPTER

18

The Biological Basis of Aging: Implications for Medical Genetics

Junko Oshima and George M Martin

Department of Pathology, University of Washington, Seattle, WA, USA

Fuki M Hisama

Division of Medical Genetics, Department of Medicine, University of Washington, Seattle, WA, USA

18.1 WHAT IS AGING?

Simply put, aging at the population level is the exponential increase in the force of mortality over unit time, a concept central to the life insurance industry. The Gompertz–Makeham equation expresses both age-independent and age-dependent mortality, the latter exhibiting exponential kinetics over the adult (Figures 18-1 and 18-2). Although longitudinal studies have shown striking variation in physiologic parameters such as renal function with aging, there is little understanding of allelic variants that contribute to superior function during aging. An exception may be the Apolipoprotein E2 (ApoE2) allele, which affords some protection against Alzheimer disease.

18.2 WHY DO WE AGE?

There are two major schools of thought as to why we age. The first implicates the accumulation of constitutional mutations that did not reach phenotypic expression until an age beyond the force of natural selection. The second has been termed antagonistic pleiotropy, meaning that alleles may be selected because they enhance fitness in youth, but paradoxically contribute to deleterious phenotypes later in life. Evolutionary formulations of the nature of aging also implicate additional factors, including stochastic processes, multiple contributing genes (on the order of up to 7% of loci in the human genome), multiple mechanisms, and so forth. Yet, despite wide variation in species as far as life span, and reproductive strategies, there is strong evidence for plasticity. A single environmental variable, dietary restriction, has been shown to enhance the life spans and healthspans of multiple species. "Leaky" mutations in neuroendocrine signal transduction pathways involving insulin-like growth factor receptors and the nuclear translocation of a transcription factor can also result in *C. elegans*, *D. melanogaster* substantial extensions of mean and maximum life spans of such diverse organisms as and *Mus musculus domesticus*.

18.3 HOW DO WE AGE?

There is no consensus as to how we age. One school of thought focuses on alterations in proteins, including the posttranslational modifications in proteins of aging tissue. An example is glycation, the nonenzymatic reaction of glucose with proteins and nucleic acids. Amyloids are a group of proteins that accumulate in different tissues as extracellular protein aggregates, and are implicated in Alzheimer disease. Alterations in membrane lipids including peroxidation (an integral component of the free radical theory of aging) could contribute to age-related cellular dysfunction. Alterations in DNA have also been extensively studied, including the accumulation of somatic mutations in nuclear DNA, epigenetic modifications, mitochondrial DNA, and regulation of telomere length, the specialized structures "capping" the ends of linear chromosomes.

18.4 PROGEROID SYNDROMES OF HUMANS

Although mutations in no single gene have been found to reproduce all aspects of the human aging phenotype (*90*), there are a number of "segmental progeroid mutations" including adult-onset progeria (Werner syndrome), an autosomal recessive disorder caused by mutations in a RecQ helicase, and Hutchinson–Gilford syndrome, a childhood disorder caused by Lamin A/C mutations.

18.5 HUMAN ALLELIC VARIANTS HOMOLOGOUS TO PRO-LONGEVITY GENES IN MODEL ORGANISMS

Although the genetic basis of extreme longevity is unknown, the number of centenarians in the population is growing. The investigation of allelic variants that increase the likelihood of exceptional late-life phenotypes represents a departure from the traditional model of medical geneticists' focus on diseases.

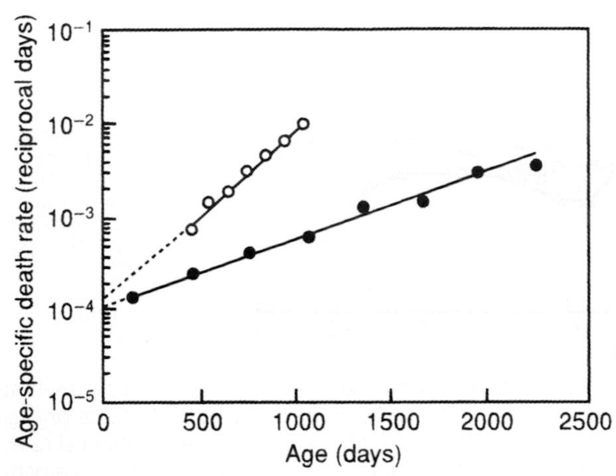

FIGURE 18-1 Gompertz function plot of the age-specific mortality rates for combined sexes of two different murine species of contrasting maximum life span potentials but of comparable size. Both species were wild type and randomly bred from small cohorts captured near the Argonne National Laboratories (Argonne, IL) by the late George A Sacher. They were housed under essentially identical conditions (caging, bedding, humidity, temperature, diet) in adjacent animal rooms with no special efforts to establish specific pathogen-free conditions (G.A. Sacher, personal communication to G.M. Martin). The longer lived species (●), *Peromyscus leucopus*, was found to have a maximum life span of about 8 years, approximately twice that of *Mus musculus* (m). *(From Sacher G. A. Evolution of Longevity and Survival Characteristics in Mammals. In* The Genetics of Aging; *Schneider, E. L., Ed.; Plenum Press: New York, 1978.)*

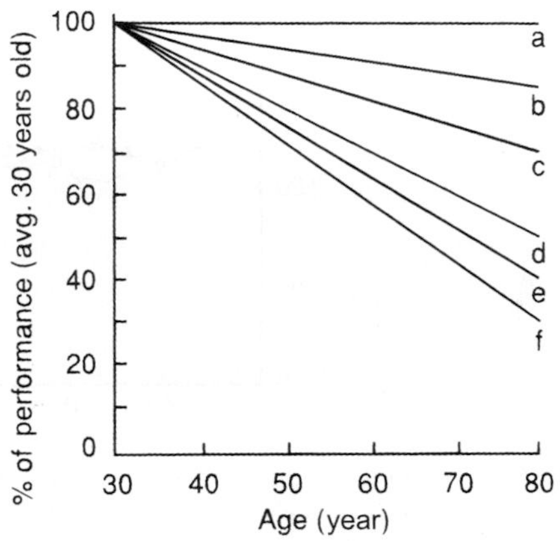

FIGURE 18-2 Linear declines of functional assays for several different human physiological parameters as studied cross-sectionally by the late Nathan W Shock and colleagues. The values are expressed as percentages of the average performances of healthy 20- to 35-year-old male subjects. (a) Fasting blood glucose. (b) Nerve conduction velocity. (c) Cardiac index (resting). (d) Vital capacity and renal blood flow. (e) Maximum breathing capacity. (f) Maximum work rate and maximum oxygen uptake. *(Adapted from Schock N. W. Systems Integration, In* Handbook of the Biology of Aging; *Finch C. E., Hayflick, L., (Eds.); Van Nostrand-Reinhold: New York, 1977.)*

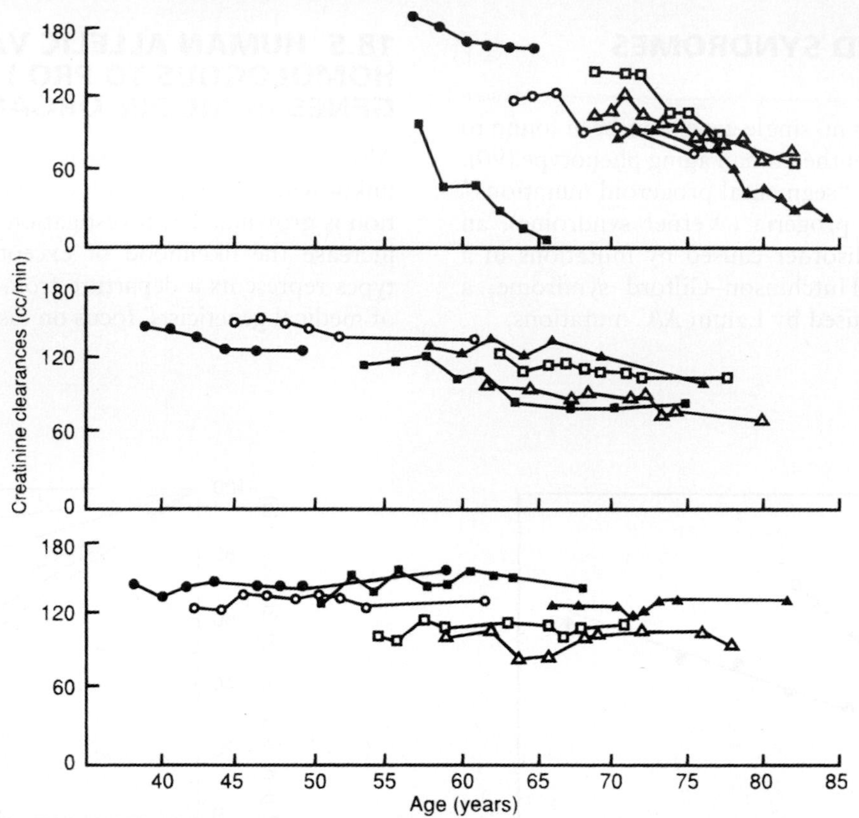

FIGURE 18-3 Longitudinal studies of creatinine clearance (an approximate measure of the glomerular filtration rate) for a representative sample of a subset of 446 clinically normal male volunteers in the Baltimore Longitudinal Study of Aging of the National Institute on Aging followed between 1958 and 1981. The results could be classified in one of three major patterns. The top panel illustrates substantial rates of decline in this measure of renal function for six representative subjects who were followed for 8–14 years. The middle panel illustrates a pattern of slight, but significant decline for six representative subjects followed for 11–22 years. For the six representative subjects in the bottom panel, who were followed for periods of 15–21 years, there were no apparent declines in this measure of renal function. *(From Lindeman R. D., et al. Longitudinal Studies on the Rate of Decline in Renal Function with Age.* J. Am. Geriatr. Soc. *1985,* 33, *278–285.)*

FURTHER READING

1. Austad, S. N. *Why We Age*; John Wiley and Sons: New York, 1997.
2. Corder, E. H.; Saunders, A. M.; Strittmatter, W. J.; Schmechel, D. E.; Gaskell, P. C.; Small, G. W.; Roses, A. D.; Haines, J. L.; Pericak-Vance, M. A. Gene Dose of Apolipoprotein E Type 4 Allele and the Risk of Alzheimer Disease in Late Onset Families. *Science* **1993,** *261,* 921–923.
3. Eriksson, M.; Brown, W. T.; Gordon, L. B.; Glynn, M. W.; Singer, J.; Scott, L.; Erdos, M. R.; Robbins, C. M.; Moses, T. Y.; Berglund, P., et al. Recurrent de novo Point Mutations in Lamin A Cause Hutchinson-Gilford Progeria Syndrome. *Nature* **2003,** *423,* 293–298.
4. Fontana, L.; Partridge, L.; Longo, V. D. Extending Healthy Life Span—From Yeast to Humans. *Science* **2010,** *328,* 321–326.
5. Kleindorp, R.; Flachsbart, F.; Puca, A. A.; Malovini, A.; Schreiber, S.; Nebel, A. Candidate Gene Study of FOXO1, FOXO4 and FOXO6 Reveals No Association with Human Longevity in Germans. *Aging Cell* **2011,** *10,* 622–628.
6. Shay, J. W.; Wright, W. E. Senescence and Immortalization: Role of Telomeres and Telomerase. *Carcinogenesis* **2005,** *26,* 867–874.
7. Yu, C. E.; Oshima, J.; Fu, Y. H.; Wijsman, E. M.; Hisama, F.; Alisch, R.; Matthews, S.; Nakura, J.; Miki, T.; Ouais, S., et al. Positional Cloning of the Werner's Syndrome Gene. *Science* **1996,** *272,* 258–262.

CHAPTER

19

Pharmacogenetics and Pharmacogenomics

Daniel W Nebert

Department of Environmental Medicine and Center for Environmental Genetics (CEG); Department of Pediatrics & Molecular Developmental Biology, Division of Human Genetics, University of Cincinnati Medical Center, Cincinnati, OH, USA

Elliot S Vesell

Department of Pharmacology, Penn State College of Medicine, Hershey, PA, USA

Pharmacogenetics is defined as "the study of heritable variability in drug response" or simply "gene–drug interactions." Responses to drugs often differ among individuals. Each person has a unique "pharmacogenetic profile"—just as each of us has our own distinct pattern of microsatellite differences, single nucleotide polymorphisms (SNPs), or thumbprint. In large part, our genetic makeup determines our drug response. However, drug response is very complex, being influenced by (a) numerous events that are manifest by our genomes (some known, others not yet understood), (b) many environmental effects (e.g. diet, cigarette smoking, drug–drug interactions, exposure to occupational chemicals, and other environmental pollutants), and (c) endogenous factors (e.g. age, gender, exercise, various disease states, status of renal function, function of other organs, etc.).

Pharmacogenomics is defined as "the study of how drugs interact with the *total expression output of the genome*, to influence biological pathways and processes." This field, a direct byproduct of The Human Genome Project, should help in identifying new drug targets and, thus, in designing new drugs. The terms "pharmacogenetics" and "pharmacogenomics" are often used interchangeably, but they should not be. Recently, pharmacogenetics/pharmacogenomics has become synonymous with "individualized drug therapy," a major subset within the broader field of "personalized medicine."

Topics covered in the first half of the chapter include pharmacokinetics (PK) and pharmacodynamics (PD), therapeutic index (or "window"), human monogenic diseases, definition of allelic variants and the Hardy–Weinberg equilibrium, high-penetrance predominantly monogenic (hPpM) traits, drug efficacy vs therapeutic failure versus adverse drug reactions (ADRs), human complex diseases, drug metabolism enzymes (DMEs) and DME-related transporters, genetic contribution to plasma clearance of a drug, extrahepatic pharmacogenetic differences, DME endogenous functions, ethnic differences in gene–drug interactions, and four examples of hPpM pharmacogenetic polymorphisms [*N*-acetylation (*NAT2* gene), debrisoquine/sparteine oxidation (*CYP2D6*), thiopurine methyltransferase (*TPMT*), and glutathione *S*-transferase null alleles (*GSTM1**0*, *GSTT1**0*)].

Topics covered in the second half of the chapter include analysis of "SNP fever," results from the Encyclopedia of DNA Elements (ENCODE) Project, update on genome-wide association (GWA) studies, copy number variants, examples of pharmacogenomic GWA studies, the dilemma of "missing heritability" in GWA studies of multiplex phenotypes, our increasing appreciation of the epigenome, DNA methylation, RNA interference and microRNAs, chromatin remodeling, histone modifications, decanalization (cryptic genetic variation), and possible assays to complement genomics (transcriptomics, proteomics, and metabolomics).

The human genome is incredibly complex, far more than most of us could ever have imagined. Consequently, phenotyping assays remain superior to DNA-genotyping assays as a means of assessing most pharmacogenetic and pharmacogenomic disorders. Statistical interaction is a population-level concept, whereas for the individual, we need to understand biological interaction or function; therefore, DNA variants showing statistical significance in a large cohort often have an effect size too small to predict in, or provide benefit to, the individual patient.

In this chapter we have categorized genes into three types: those associated with monogenic traits, hPpM disorders, and complex diseases. The *hPpM* genes are principally responsible for changes in urine or plasma drug or metabolite levels, as well as therapeutic failure. Drug efficacy and ADRs usually represent multiplex phenotypes that are conceptually not different from complex diseases. One gene generally contributes 90–98% to a monogenic disease, whereas one hPpM gene might contribute 10–25% to an hPpM trait; in contrast, one gene might contribute 0.001–1.0% to multiplex phenotypes such as complex diseases, drug efficacy, or risk of ADRs.

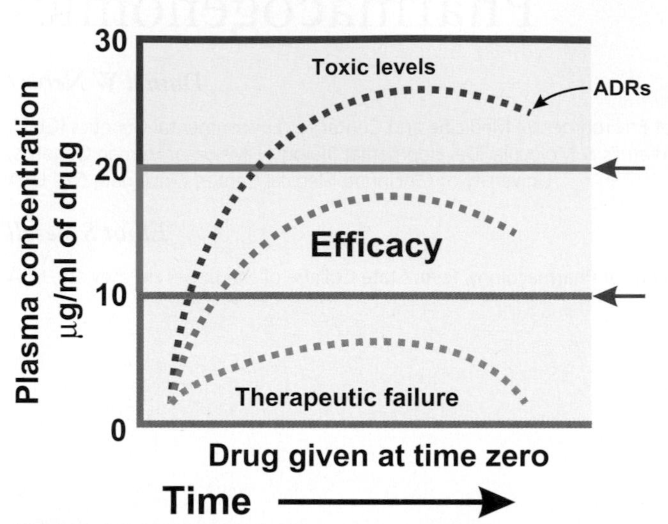

FIGURE 19-1 Theoretical plasma concentration curves for any drug, as a function of time after oral administration of the dose. In this hypothetical case, the horizontal line (at 10 µg/mL) is the minimum *effective concentration*; the horizontal line (at 20 µg/mL) is the minimum *toxic concentration*.

GWA studies have been extremely useful in identifying genes participating in multiplex phenotypes; these might lead to the development of novel drug targets. After more than 1300 GWA publications, we now realize that the contribution of the total number of genes associated with any multiplex phenotype is between 5% and perhaps 60%. The remaining 40–95% is referred to as "missing heritability."

It is likely that epigenetics is a main contributor to this missing heritability. Epigenetics is the study of chromosomal events that alter phenotype, but in which there are no changes in DNA sequence (i.e. no mutations); intriguingly, in some instances these epigenetic effects can be passed on to subsequent generations. Epigenetics includes the processes of DNA methylation, RNA-interference processes, histone modifications, chromatin remodeling, and decanalization (also called cryptic genetic variation).

Urinary metabolite profiles in the individual patient are expected to reflect, at that particular moment in time, the combination of all genetic plus all epigenetic influences. Metabolomics might be regarded as being similar to a "liver profile" test in clinical pathology, except

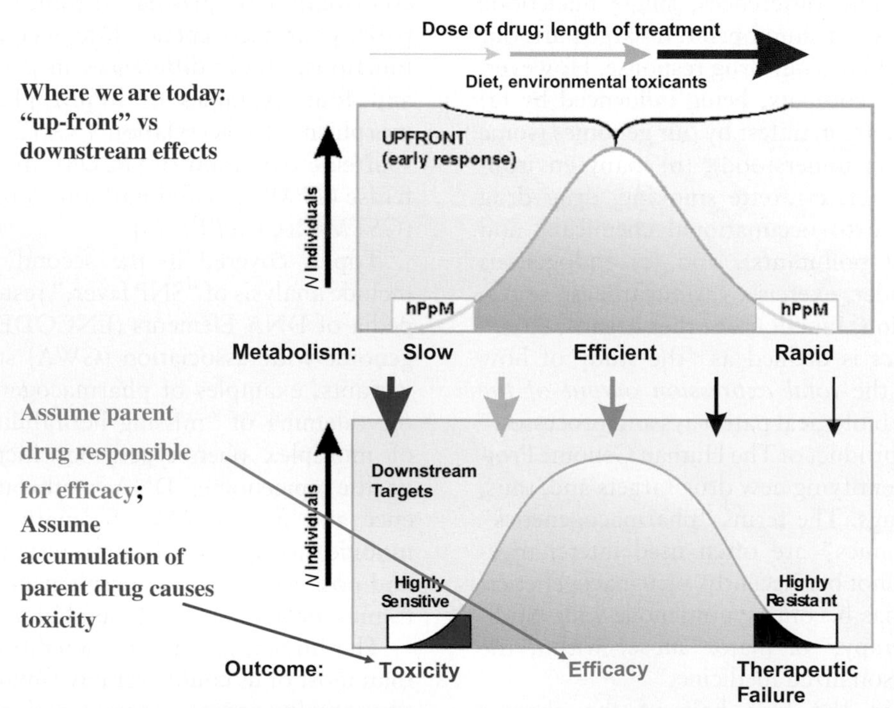

FIGURE 19-2 Two-tier system to explain drug toxicity vs efficacy vs therapeutic failure. "Up-front" early responses include the *hPpM* genes and gene products, which are PK processes. Downstream targets include dozens or hundreds of others genes and gene products, which are PD processes. *(Modified from Nebert, D. W.; Zhang, G.; Vesell, E. S. From Human Genetics and Genomics to Pharmacogenetics and Pharmacogenomics: Past Lessons, Future Directions.* Drug Metab. Rev. ***2008,*** 40 (2), *187–224.)*

that metabolomics includes measurement of metabolites present at much lower concentrations and, accordingly, provides several orders of magnitude greater sensitivity (femtomolar to attomolar range).

Transcriptomics, proteomics, and metabolomics might complement genomic (DNA sequence) assays in achieving improved success in personalized medicine, a large subset of which is individualized drug therapy. Given all the complexity of the human genome, it is now easy to realize how difficult it would ever be to predict a common disease (personalized medicine) or to predict in advance "which drug and dose to give" (individualized drug therapy) to the individual patient.

TABLE 19-1 Predominantly Monogenic Pharmacogenetic Disorders that Have Been Characterized[a]

Disorder	Major Gene Known to be Responsible
Phenylthiourea—nontaster	*TAS2R1*
Hypocatalasemia	*CAT*
Atypical serum cholinesterase	*BCHE*
Glucose-6-phosphate dehydrogenase deficiency	*G6PD*
Isoniazid slow *N*-acetylation	*NAT2*
Fish-odor syndrome trimethylaminuria	*FMO3*
Debrisoquine/sparteine oxidation poor metabolizer	*CYP2D6*
Serum paraoxonase low activity	*PON1*
Thiopurine methyltransferase deficiency	*TPMT*
Sensitivity to alcohol	*ALDH2*
S-mephenytoin oxidation deficiency	*CYP2C19*
Sulfotransferase deficiency	*SULT1A1, SULT1A2*
Coumarin, nicotine oxidase deficiency	*CYP2A6*
P-glycoprotein transporter defect	*ABCB1*
Malignant hyperthermia	*RYR1*
Quinone oxidoreductase defect	*NQO1*
Peptide transporter defect	*TAP2*
Phenytoin, warfarin oxidation defect	*CYP2C9*
Debrisoquine ultra-metabolizers	*CYP2D6*1XN*
Epoxide hydrolase deficiency	*EPHX1*
Glutathione *S*-transferase null alleles	*GSTM1*0, GSTT1*0*
Long-QT syndrome	*KCNH2*
Dihydropyrimidine dehydrogenase deficiency	*DPYD*
Chlorzoxazone hydroxylation defect	*CYP2E1*
Peptide transporter defect	*TAP1*
Sulfonylurea receptor defect	*ABCC8*
Calcium channel defect	*CACNA1A*
Androstane glucuronosyl conjugation	*UGT2B4*
Congenital long-QT syndrome	*SCN5A*
S-oxazepam glucuronosyl conjugation	*UGT2B7*
Paclitaxel hydroxylase deficiency	*CYP2C8*
Chlorpyrifos oxidation deficiency	*CYP3A4*
Acrodermatitis enteropathica	*SLC39A4*
Nifedipine oxidation deficiency	*CYP3A5*
Cyclophosphamide metabolism deficiency	*CYP2B6*
Hyperinsulinemic hypoglycemia	*SLC16A1*
Hereditary folate malabsorption	*SLC46A1*

[a]This list is not meant to be all-inclusive. In each case, compared with the consensus allele, one or more variant alleles lead to a defective gene product, resulting in decreased metabolism, transporter or receptor activity, or channel function. The clinical consequence in most homozygous affected subjects is toxicity, due to drug accumulation with enhanced drug activity. Occasionally, decreased drug activity (*therapeutic failure*) ensues if the variant reflects ultra-rapid drug metabolism or if, for activity, the drug requires metabolic conversion to an active form and this conversion is decreased in the variant. We realize that some of the traits listed here might concern primarily environmental toxicants (e.g. *TAS2R1*, *CAT*, *FMO3*, and *PON1*) rather than prescribed drugs.

TABLE 19-2 List of Recent Associations Between Pharmacogenomics Traits and One or Several Genes

Gene (or Allele)	Drug
TPMT	Mercaptopurine-, thioguanine-induced toxicity
ESR1	Tamoxifen resistance
BCR-ABL	Imatinib, Dasatinib, Nilotinib resistance
UGT1A1	Irinotecan, Nilotinib resistance
ERBB2	Lapatinib, Trastuzumab resistance
*HLA-B*5701, HLA-DR7, HLA-DQ3*	Abacavir-induced hypersensitivity
EGFR	Cetuximab, Erlotinib, Gefitinib, Panitumumab resistance
CYP2C9, VKORC1, CYP4F2	Warfarin relative resistance
*HLA-B*1502*	Carbamazepine-induced SJS-TEN[a] in Asians
KRAS	Cetuximab, Panitumumab resistance
KIT	Imatinib resistance
CYP2D6	Codeine toxicity in ultrarapid metabolizers
*HLA-B*5701*	Floxacillin-induced hepatic injury
IL28B	Response to HCV infection by peg-Interferon-α-2b or α-2a combined with Ribavirin
CYP2C19	Clopidogrel response
*HLA-DRB1*1501, HLA-DQB1*0602, HLA-DRB5*0101, HLA-DQA1*0102*	Lumiracoxib-induced hepatic injury
*HLA-A*3101*	Carbamazepine-induced hypersensitivity; SJS-TEN in Europeans
CYP1A2, AHR	Caffeine addiction

[a]SJS-TEN, Stevens–Johnson syndrome-toxic epidermal necrolysis.

FURTHER READING

1. Nebert, D. W.; Zhang, G.; Vesell, E. S. From Human Genetics and Genomics to Pharmacogenetics and Pharmacogenomics: Past Lessons, Future Directions. *Drug Metab. Rev.* **2008,** *40,* 187–224.
2. Wang, L.; McLeod, H. L.; Weinshilboum, R. M. Genomics and Drug Response. *N. Engl. J. Med.* **2011,** *364,* 1144–1153.

General Principles

Genetic Evaluation for Common Diseases of Adulthood, 69
Genetic Counseling and Clinical Risk Assessment, 74
Cytogenetic Analysis, 76
Diagnostic Molecular Genetics, 78
Heterozygote Testing and Carrier Screening, 80
Prenatal Screening for Neural Tube Defects and Aneuploidy, 82
Techniques for Prenatal Diagnosis, 85
Neonatal Screening, 89
Enzyme Replacement and Pharmacologic Chaperone Therapies for Lysosomal Storage Disease, 91
Gene Therapy: From Theoretical Potential to Clinical Implementation, 93
Ethical and Social Issues in Clinical Genetics, 95
Legal Issues in Genetic Medicine, 100

CHAPTER 20

Genetic Evaluation for Common Diseases of Adulthood

Maren T Scheuner and Shannon Rhodes

Common diseases are typically chronic conditions that develop over decades, usually occurring in adulthood due to genetic and environmental risk factors, including exposures, infectious agents, and cultural and behavioral factors, such as diet and exercise (i.e. multifactorial etiology). Rarely, a common chronic disease of adulthood can occur as the manifestation of a Mendelian disorder (*1*).

Genetic tests are increasingly available for both Mendelian and multifactorial forms of common diseases. Genetic test results have the potential to impact health outcomes by improving our ability to diagnosis, treat, and prevent common disease. Results of genetic tests are also used to inform reproductive and personal decisions made by patients, such as decisions about career, finances, and marriage. However, genetic testing for common diseases in clinical practice is generally limited to testing for Mendelian disorders. Very little is understood regarding the interactions between genetic and non-genetic risk factors underlying multifactorial diseases, and attempts to build genetic risk models have failed to show that additional genetic information can substantially improve the prediction of risk for common diseases (*2,3*), and evidence for improved health outcomes is lacking (*4*). Thus, to date, multiplex testing for common diseases has limited proven utility (*5*) and family history remains the best strategy for assessing genetic risk for these disorders (*6*). Family history is also key to identifying individuals at risk for Mendelian forms of common diseases (*1*).

Outcomes relating to genetic evaluation for common diseases can be characterized as psychological, cognitive, behavioral, or clinical (*7*). Generally psychological and cognitive outcomes related to genetic counseling and testing for common diseases are positive (*8–12*). Behavioral responses to genetic risk information among individuals at risk for cancer found mixed results, with some studies showing no change in behavioral outcomes, while others reported increased participation in cancer screening (*13*). A few studies have documented lifestyle changes after receiving genetic test results (*14,15*). Clinical outcomes, such as reduced incidence of disease after a recommended intervention, are less well studied.

Clinical genetic evaluation for common disease should be considered for individuals with a strong familial risk or when a Mendelian disorder is suspected. Genetic evaluation for common diseases is comprised of several components including (1) genetic risk assessment and diagnosis through clinical assessment, family history assessment, and genetic testing; (2) recommendations for management and prevention options appropriate given a genetic risk or diagnosis; and (3) genetic counseling and education. Primary care clinicians are best suited to identify patients with increased familial risk, with referral of high-risk cases or those suspected of having a Mendelian disorder. However, barriers include clinicians' lack of time and perceived lack of knowledge and skills to collect and interpret family history, and lack of sufficient evidence on how to perform familial risk assessment and the effects of familial risk on clinical outcomes (*16,17*). Electronic family history tools have the potential to mitigate these barriers.

Knowledge of genetic susceptibility to common diseases may identify important biologic differences that could lead to better disease management and prevention through the use of targeted therapies and enhanced screening and prevention strategies. Prevention strategies for common chronic diseases include targeted lifestyle changes, screening at earlier ages, more frequently and with more intensive methods than used for average risk individuals, use of chemoprevention, and, for those at highest risk, prophylactic procedures and surgeries. However, most guidelines are based on clinical observation and expert opinion, and outcomes research is needed that assesses their clinical utility. In the absence of clinical guidelines for common chronic diseases associated with

Mendelian disorders, clinicians can suggest management and prevention strategies that have been proven effective for the general population.

Genetic counseling is critical for delineating a patient's motivation for genetic evaluation and likely responses to learning of a genetic risk or diagnosis. Through genetic counseling, patients will be educated about the role of genetic and non-genetic risk factors for disease; basic concepts of genetics, such as inheritance patterns, penetrance, and variable expressivity; and the options for treatment, prevention, or risk factor modification tailored to the genetic risk or diagnosis. Genetic counseling also ensures the opportunity to provide informed consent, including discussion of the potential benefits, risks and limitations of testing, implications for family members, and the alternative of not testing.

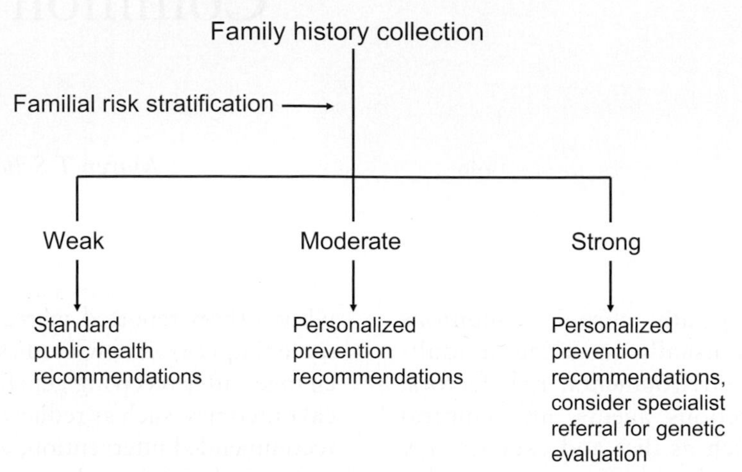

FIGURE 20-1 Family history collection followed by risk stratification that recognizes characteristics that increase disease risk (e.g. early age at diagnosis, two or more close relatives affected with a disease or a related condition, a single family member with two or more related diagnoses, multifocal or bilateral disease, and occurrence of disease in the less often affected sex) can guide risk-specific recommendations for disease management and prevention. Standard public health messages would be appropriate for individuals with a weak familial risk and the absence of significant personal risk factors. Personalized prevention recommendations should be provided to individuals with moderate or strong familial risk, which should include emphasis of public health messages, and consideration of earlier and more frequent screening and use of chemoprevention when available. Referral for genetic evaluation by a geneticist or other specialist should be considered for individuals with strong familial risk or when a Mendelian disorder is suspected.

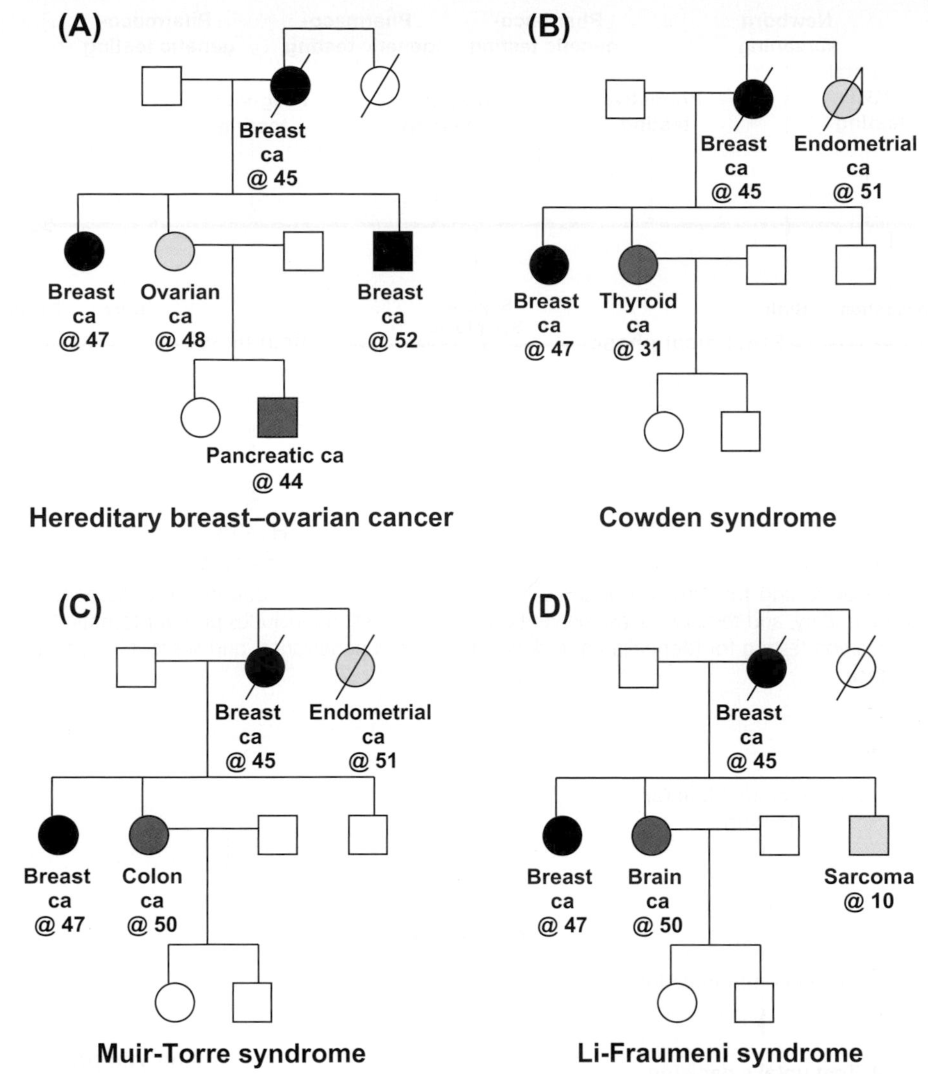

FIGURE 20-2 Each pedigree depicts a strong familial risk for breast cancer. However, by recognition of the patterns of cancer in the family, a more accurate diagnosis can be made. Pedigree A features early-onset breast and ovarian cancer and is most consistent with hereditary breast–ovarian cancer syndrome, which is almost always due to *BRCA1* or *BRCA2* gene mutations. In this case, a *BRCA2* gene mutation is likely given the family history of male breast cancer and pancreatic cancer. Pedigree B features early-onset breast, thyroid, and endometrial cancer and is most consistent with Cowden syndrome due to *PTEN* gene mutations. Pedigree C features early-onset breast, colon, and endometrial cancer and suggests the possibility of Muir–Torre syndrome, a variant of hereditary nonpolyposis colon cancer due to mutations in mismatch repair genes. The early-onset breast cancer, brain tumor, and childhood sarcoma in pedigree D is consistent with Li–Fraumeni syndrome due to *TP53* gene mutations. Multiple primary cancers are common among individuals with Li–Fraumeni syndrome.

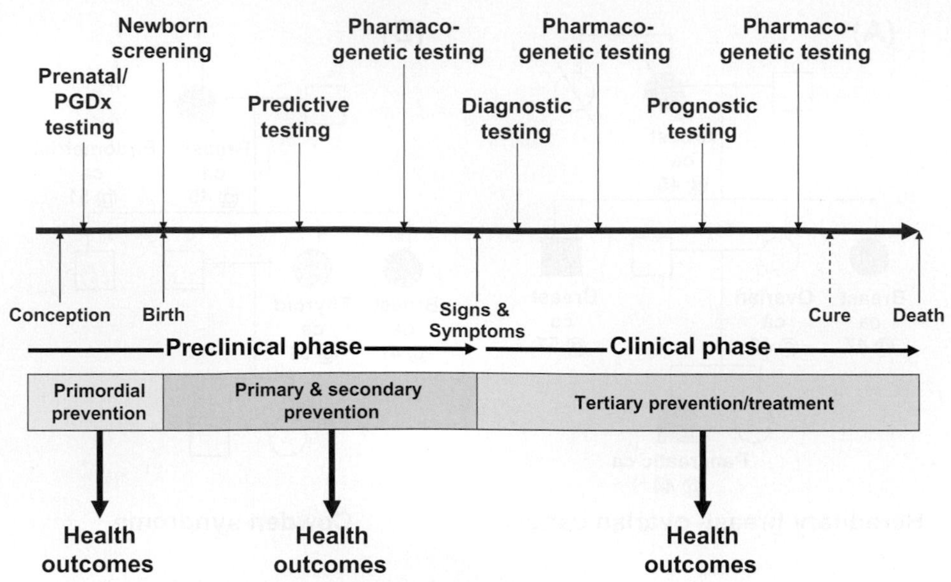

FIGURE 20-3 Currently, genetic testing for common diseases of adulthood has a role throughout the lifespan, including testing that can impact primordial, primary, secondary, and tertiary disease prevention. Predictive testing includes presymptomatic testing for highly penetrant Mendelian disorders, predisposition testing for Mendelian disorders with reduced penetrance, and susceptibility testing for common diseases due to multifactorial inheritance.

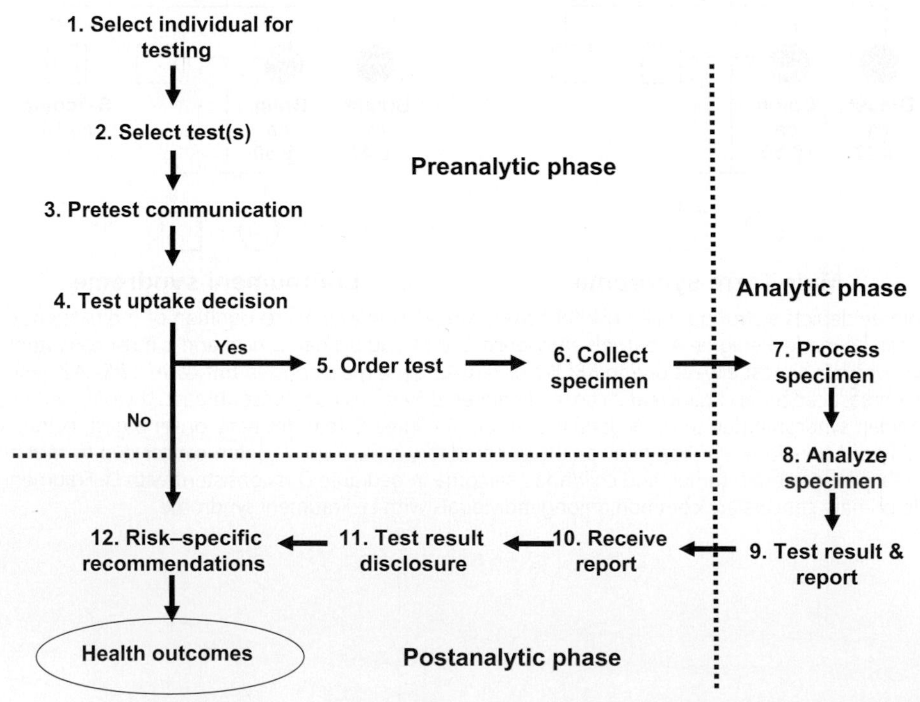

FIGURE 20-4 A framework depicting the delivery of genetic tests according to the pre-analytic, analytic, and post-analytic phases of the genetic testing process is shown. The specific actions of each step will vary, depending in large part on the indication for genetic testing. Yet in general, all of the steps should occur for any given clinical scenario.

REFERENCES

1. Scheuner, M. T.; Yoon, P. W.; Khoury, M. J. Contribution of Mendelian Disorders to Common Chronic Disease: Opportunities for Recognition, Intervention, and Prevention. *Am. J. Med. Genet. C Semin. Med. Genet.* **2004,** *125C* (1), 50–65.
2. Paynter, N. P., et al. Association between a Literature-Based Genetic Risk Score and Cardiovascular Events in Women. *JAMA* **2010,** *303* (7), 631–637.
3. Meigs, J. B., et al. Genotype Score in Addition to Common Risk Factors for Prediction of Type 2 Diabetes. *N. Engl. J. Med.* **2008,** *359* (21), 2208–2219.
4. Palomaki, G. E., et al. Use of Genomic Profiling to Assess Risk for Cardiovascular Disease and Identify Individualized Prevention Strategies—A Targeted Evidence-Based Review. *Genet. Med.* **2010,** *12* (12), 772–784.
5. Khoury, M. J., et al. An Epidemiologic Assessment of Genomic Profiling for Measuring Susceptibility to Common Diseases and Targeting Interventions. *Genet. Med.* **2004,** *6* (1), 38–47.

6. Valdez, R., et al. Family History in Public Health Practice: A Genomic Tool for Disease Prevention and Health Promotion. *Annu. Rev. Public Health* **2010,** *31,* 69–87.
7. Scheuner, M. T.; Sieverding, P.; Shekelle, P. G. Delivery of Genomic Medicine for Common Chronic Adult Diseases: A Systematic Review. *JAMA* **2008,** *299* (11), 1320–1334.
8. Butow, P. N., et al. Psychological Outcomes and Risk Perception after Genetic Testing and Counselling in Breast Cancer: A Systematic Review. *Med. J. Aust.* **2003,** *178* (2), 77–81.
9. Schlich-Bakker, K. J.; ten Kroode, H. F.; Ausems, M. G. A Literature Review of the Psychological Impact of Genetic Testing on Breast Cancer Patients. *Patient Educ. Couns.* **2006,** *62* (1), 13–20.
10. Collins, V., et al. Cancer Worries, Risk Perceptions and Associations with Interest in DNA Testing and Clinic Satisfaction in a Familial Colorectal Cancer Clinic. *Clin. Genet.* **2000,** *58* (6), 460–468.
11. Pieterse, A. H., et al. Cancer Genetic Counseling: Communication and Counselees' Post-Visit Satisfaction, Cognitions, Anxiety, and Needs Fulfillment. *J. Genet. Couns.* **2007,** *16* (1), 85–96.
12. Green, R. C., et al. Disclosure of APOE Genotype for Risk of Alzheimer's Disease. *N. Engl. J. Med.* **2009,** *361* (3), 245–254.
13. Marteau, T. M.; Lerman, C. Genetic Risk and Behavioural Change. *BMJ* **2001,** *322* (7293), 1056–1059.
14. McBride, C. M., et al. The Behavioral Response to Personalized Genetic Information: Will Genetic Risk Profiles Motivate Individuals and Families to Choose More Healthful Behaviors? *Annu. Rev. Public Health* **2010,** *31,* 89–103.
15. McBride, C. M., et al. Incorporating Genetic Susceptibility Feedback into a Smoking Cessation Program for African–American Smokers with Low Income. *Cancer Epidemiol. Biomarkers Prev.* **2002,** *11* (6), 521–528.
16. Rich, E. C., et al. Reconsidering the Family History in Primary Care. *J. Gen. Intern. Med.* **2004,** *19* (3), 273–280.
17. Wilson, B. J., et al. Systematic Review: Family History in Risk Assessment for Common Diseases. *Ann. Intern. Med.* **2009,** *151* (12), 878–885.

CHAPTER

21

Genetic Counseling and Clinical Risk Assessment

R Lynn Holt

Clinical and Diagnostic Sciences; Genetic Counseling Program, University of Alabama at Birmingham, School of Health Professions, Birmingham, AL, USA

Angela Trepanier

Center for Molecular Medicine and Genetics; Genetic Counseling Graduate Program, Wayne State University School of Medicine, Detroit, MI, USA

Genetic counseling aims to translate scientific knowledge and medical information to individuals in a meaningful and thoughtful manner. It is a multistep process that involves medical evaluation, education, and communication. The term was first coined by Dr. Sheldon Reed in 1947 and the definition has evolved since that time. Despite significant changes in the scope of medical genetics services over the last 60 years, promoting understanding and adaptation remain fundamental goals of genetic counseling.

The current definition is stated as "Genetic counseling is the process of helping people understand and adapt to the medical, psychological, and familial implications of genetic contributions to disease." This process integrates the following:

- Interpretation of family and medical histories to assess the chance of disease occurrence or reoccurrence.
- Education about inheritance, testing, management, prevention, resources, and research.
- Counseling to promote informed choices and adaptation to the risk or condition.

The goals and aims of genetic counseling have evolved as the profession and applications of genetics in medicine have grown. Historically, the field was based on the principle of nondirectiveness with the goal of providing information to clients in a value-neutral manner that would permit clients to make decisions based on their own values and circumstances. However, critics of nondirectiveness argue that it may actually deter genetic counselors from actively engaging clients in an exploration of their situations and that it may not be applicable to all genetic counseling situations (e.g. cardiovascular or cancer genetics). Furthermore, although some consider nondirectiveness a practice model, it is not. Using it in this way may actually be impeding the development of real practice models. Models serve an important function in that they lay out the tenets of a profession, the goals, and the techniques used to achieve them and they provide a basis for evaluating the effectiveness counseling strategies.

Actual models of practice including the teaching model, counseling model, and the recently developed reciprocal engagement model have been explored. The development, vetting, and implementation of such models is a critical step in determining which strategies are most effective in achieving the goals of genetic counseling and in providing services in an evidence-based manner. The genetic counseling process is multistep and is accomplished through five basic tasks: information gathering, risk assessment, information giving, psychosocial assessment and counseling, and management/follow-up. Information gathering includes collecting the relevant medical and family history facts to establish a diagnosis or for risk assessment purposes. Risk assessment includes the application of Mendelian inheritance patterns as well as mathematical principles to determine an accurate and individualized risk for a disease or recurrence of disease. Family history collection and interpretation are an essential part of this process and risk assessment. Information giving is based on the needs of the client and may include (but is not limited to) client education about a diagnosis or differential diagnosis, informed consent, medical management, and resources.

Psychosocial support and counseling are indispensable components of the genetic counseling process due to the complex emotionally reactions that accompany genetic disease and disability. This part of the process is also important to promote client adaptation. The fifth and final step is management/follow-up. This provides the client with a mutually agreed upon plan for management of a genetic condition or risk and return visits to the genetics clinic for future evaluations and additional counseling and education as the needs of the client and family change over time. Other essential components of follow-up include communication with other members of a client's medical team, coordination of any additional referrals for specialists or testing, in some cases disclosure of test results, and when relevant facilitation of communication with at-risk relatives.

Currently, the primary practice areas of genetic counseling include reproductive genetics, pediatric genetics, cancer genetics, and neurogenetics. However, genetics services are rapidly expanding to include risk assessment and counseling for additional complex adult onset disorders such as heart disease, stroke, and diabetes, hypertension, and Alzheimer disease. The primary goal of many of these evaluations is to use family history information, with or without genomic testing, to establish a person's risk level—high, moderate, or low (population risk)—for the purpose of developing and effectively communicating a personalized medical management plan (personalized medicine) that results in improved health outcomes for the individual and family. Genetic counseling services are provided by a variety of health care professionals in a number of different settings. In some settings, like pediatric genetics, metabolic disease, and specialty disease clinics, services are often provided by the genetics team. This team is comprised of professionals trained in various aspects of genetics including clinical (MD) geneticists, genetic counselors, genetic nurses, and medical (PhD) geneticists with laboratory support from cytogeneticists, molecular geneticists, and biochemical geneticists. Other key professionals such as nurses, dieticians, social workers, and/or psychologists can also be members of the team. Each team member has his/her own set of skills and corresponding role on the team although there can be some overlapping functions. Various team members may contribute to the interpretation/risk assessment, education, and counseling functions of genetic counseling. This role is not limited to genetic counselors.

In reproductive genetics, genetic counseling services are often provided by genetic counselors or genetic nurses working with geneticists or with physicians boarded in obstetrics or maternal fetal medicine specialists. Similarly, genetic counselors or genetic nurses may work with geneticists or with oncologists, surgeons, or gastroenterologists (cancer genetics), cardiologists (cardiovascular genetics), or neurologists (neurogenetics).

Future service lines for genetic counseling are almost limitless as the field of medical genetic evolves. Current drivers in the field include direct to consumer genetic testing, advances in genetic technology including exome sequencing and full genome sequencing, and the discovery of the genetic contribution to multifactorial conditions such as heart disease and diabetes. In many ways, the growth in technology has outpaced current medical models to implement the information that is generated. Among the many questions that come with this new technology is the question of how will people use this information. Genetic counseling will play an important role in addressing this question.

REFERENCE

1. National Society of Genetic Counselors' Definition Task Force. A New Definition of Genetic Counseling: National Society of Genetic Counselors' Task Force Report. *J. Genet. Couns.* **2006,** *15* (2), 77–83.

FURTHER READING

1. Bennett, R. L.; Steinhaus-French, K. A.; Resta, R. G.; Doyle, D. L. Recommendations for Standardized Pedigree Nomenclature: Update and Assessment of the Recommendations of the National Society of Genetic Counselors. *J. Genet. Couns.* **2008,** *17* (5), 424–433.
2. Bennett, R. L. *The Practical Guide to the Genetic Family History* 2nd ed.; Wiley Blackwell: Hoboken, NJ, 2010.
3. Hadley, D. W.; Letocha Ersig, A. D.; Holohan Quattrocchi, M. K. Guidelines and Policies on Genetic Testing in Children and Families. In *Handbook of Genomics and the Family, Issues in Clinical Child Psychology*; Tercyak, K. P., Ed.; Springer: New York, 2010.
4. Hampel, H.; Grubs, R. E.; Walton, C. S.; Nguyen, E.; Breidenbach, D. H.; Nettles, S. The American Board of Genetic Counseling 2008 Practice Analysis Advisory Committee, Genetic Counseling Practice Analysis. *J. Genet. Couns.* **2009,** *18,* 205–216.
5. Hodge, S. E. A Simple, Unified Approach to Bayesian Risk Calculations. *J. Genet. Couns.* **1998,** *7* (3), 235–261.
6. LeRoy, B. S.; McCarthy Veach, P.; Bartels, D. M. *Genetic Counseling Practice Advanced Concepts and Skills*; Wiley-Blackwell: Hoboken NJ, 2010.
7. McCarthy Veach, P.; LeRoy, B.; Bartels, D. *Facilitating the Genetic Counseling Process, A Practice Manual*; Springer-Verlag: New York, 2003.
8. National Society of Genetic Counselors' Definition Task Force. A New Definition of Genetic Counseling: National Society of Genetic Counselors' Task Force Report. *J. Genet. Couns.* **2006,** *15* (2), 77–83.
9. Resta, R. G., Ed. *Psyche and Helix, Psychological Aspects of Genetic Counseling, Essays by Seymour Kessler, Ph.D.*; Wiley-Liss: New York, 2002.
10. Schneider, K. A. *Counseling About Cancer, Strategies for Genetic Counseling* 2nd ed.; Wiley-Liss: New York, 2002.
11. Smith, D. H.; Quaid, K. A.; Dworkin, R. B.; Gramelspacher, G. P.; Granbois, J. A.; Vance, G. H. *Early Warning*; Indiana University Press: Bloomington, IN, 1998.
12. Uhlmann, W. R.; Schuette, J. L.; Yashar, B. M. *A Guide to Genetic Counseling* 2nd ed.; Wiley-Liss: New York, 2009.
13. Young, I. D. *Introduction to Risk Calculation in Genetic Counseling* 2nd ed.; Oxford University Press: New York, 1999.

CHAPTER

22

Cytogenetic Analysis

Nancy B Spinner

Abramson Research Center, The Children's Hospital of Philadelphia; Perelman School of Medicine at the University of Pennsylvania, Philadelphia, PA, USA

Malcolm A Ferguson-Smith

Department of Veterinary Medicine, Cambridge Resource Centre for Comparative Genomics, Cambridge, UK

David H Ledbetter

Geisinger Health System, Danville, PA, USA

Cytogenetics traditionally refers to the study of chromosomes by microscopy following the application of banding techniques, permitting identification of abnormalities of chromosome number, loss or gain of chromosomal material or positional changes. Advances over the past 30 years have resulted in an increased reliance on molecular techniques, such that the current field is a hybrid of microscopic and molecular-based technologies. Because chromosomes carry the genetic material, an understanding of their behavior, structure, and function is essential for an understanding of genetics.

Visible or submicroscopic changes in the number and structure of chromosomes have been associated with a broad spectrum of disease, so the analysis of chromosomes has been an important tool in medical genetics for a number of decades. Cytogenetic analysis for medical diagnostics began shortly after the realization, in 1956, that humans have 46 chromosomes. This landmark was achieved, thanks to the introduction of a hypotonic solution into the protocol for preparation of chromosomes, which allowed for adequate dispersal of the chromosomes to allow counting and further analysis. This discovery was followed in 1959 by the first demonstration of a human syndrome being caused by an abnormality in chromosome number, with the recognition that Down syndrome is caused by trisomy of chromosome 21. Simultaneously, the role of chromosome abnormalities was discovered in cancer cells, with the recognition of the Philadelphia chromosome (a derivative chromosome 22 from a translocation between chromosomes 9 and 22), in patients with chronic myeloid leukemia. Over the next two decades, there has been a steady stream of reports identifying patients with constitutional and somatic chromosomal abnormalities including aneuploidy, deletions and duplications, translocations and other structural rearrangements.

In the late 1960s, techniques for the banding of human chromosomes were introduced and further refined during the 1970s, which lead to the establishment of a standard chromosome banding pattern that became the criterion for chromosome identification. Following the early discoveries of chromosome abnormalities in patients with specific constitutional and somatic abnormalities, techniques for chromosome identification continued to improve, with karyotypes going from less than 400 bands up to high resolution karyotypes with more than 1000 bands.

The introduction of in situ hybridization, first using radioactivity in the early 1970s and later using fluorescence, expanded the repertoire of the cytogeneticist and allowed for the integration of the incredible advances that were being made at the molecular level. Fluorescence in situ hybridization (FISH), brought into common use in the mid 1980s, had an immediate strong following, as it is a relatively straightforward technique that served as a bridge between DNA analysis and cytogenetics and had an enormous impact on our ability to diagnose chromosome aberrations not identifiable by banding such as small markers, and small deletions or duplications.

The next major advance in cytogenetics was the introduction of comparative genomic hybridization, first using metaphase chromosomes, and then arrays. This extraordinary technology allows interrogation of the entire genome at much higher resolution than cytogenetic analysis by G-banding, and it has lead to the identification of an entire new generation of deletion and duplication syndromes. Array-based genomic tests have also lead to the recognition that normal individuals have many small deletions and duplications, necessitating thorough analysis of these copy number variants to distinguish the pathogenic from benign. Array-based technologies have evolved over the past several years to incorporate genotype information, along with copy number information, expanding the possible diagnoses to include uniparental disomy and regions of homozygosity. The utilization of array-based techniques, coupled with the available data from the Human Genome Project that lead to the creation of databases of normal and pathogenic genomic alterations, have resulted in an extremely powerful diagnostic tool. However, array-based techniques only identify imbalances resulting in deletions or duplications, so balanced rearrangements such as translocation, inversions, or rings may not be identified, and require a standard chromosome analysis.

In this chapter, we describe the most useful techniques of those now available for chromosome analysis and provide a historical context for the present state of the field. Array-based techniques have become the first line of investigation for analysis of chromosome abnormalities, with cytogenetic and FISH tools utilized to help interpret the results of imbalances diagnosed by array.

CHAPTER

23

Diagnostic Molecular Genetics

Wayne W Grody and Joshua L Deignan

Departments of Pathology and Laboratory Medicine, Pediatrics, and Human Genetics, UCLA School of Medicine, UCLA Medical Center, Los Angeles, CA, USA

There can be no question that we are in the midst of a revolution in clinical medicine, as profound in its own way as the early revelations of human anatomy and physiology and the germ theory of disease. The twenty first century has already been christened the age of molecular medicine—the molecule in question being, of course, DNA. Molecular biology, which came to dominate basic life sciences research over the second half of the twentieth century, has now become firmly ensconced in clinical medicine as well. The development of robust technologies such as the polymerase chain reaction (PCR) and massively parallel DNA sequencing, and the extraordinary productivity of the Human Genome Project in elucidating medically relevant targets for these techniques, have firmly established patient DNA as a powerful "analyte" in the clinical laboratory and as a valid substrate for therapeutic manipulation in the clinic.

While all areas of medicine have been impacted by these developments, it is medical genetics, as the specialty most directly connected to the human genome, that is the first discipline to be entirely transformed by them. And if the promise of gene therapy continues to remain frustratingly elusive through successive editions of this book, such is not the case for gene-based diagnostics. In no other sector of medical practice does at least the question, if not always the actual execution, of molecular testing arise with virtually every patient seen. Not only is medical genetics by definition concerned with inherited alterations in the patient's genome that are detectable in this way, but much of its attention is directed at ascertaining recessive carrier states, fetal conditions that will not be expressed until after birth, and predisposition to adult-onset disorders before they are symptomatic; such diagnoses often exhibit no other evidence of their presence than at the DNA level. For these reasons, medical geneticists and genetic counselors, as a group, tend to be significantly more facile and familiar with the applications and interpretation of molecular genetic tests than other clinicians. Nevertheless, keeping up with a field as rapidly evolving as this one, in which new techniques and disease genes (not to mention their associated ethical and societal dilemmas) are reported every week, can be a challenge for even the most dedicated specialist. That is why close two-way communication between the clinic and the laboratory is so essential, more so in medical genetics practice than in any other.

Molecular genetics has a unique range of indications, most of which are quite different from the uses of traditional clinical laboratory testing and even molecular biologic testing in other disease classes (e.g. infectious disease, cancer). Most notably, as mentioned earlier, these procedures are often performed on healthy people who have no other signs or symptoms of the disease being tested. Even if they are symptomatic, the assay is not directed at a discrete anatomic site, as would be the case with a DNA test for an infection or a tumor. Since genetic disorders reflect heritable mutations in the germline, the test in most cases can be performed on any accessible body fluid or tissue. The most common specimen is whole blood, though saliva, buccal swabs, dried blood spots, urine, and other specimens can be used, while amniocytes or chorionic villus samples (CVS) and, most recently, fetal DNA in maternal blood are collected for prenatal diagnosis. Additionally, the technical approaches as well as the psychosocial and ethical implications of molecular genetic tests may vary substantially depending on the reason for testing.

Just as many of the applications are unique, so too the types of patient samples collected for molecular genetic testing may be different from those obtained for other types of clinical laboratory testing. Since germline mutations are present in every cell of the body, site-specific biopsy is not required and simple phlebotomy will suffice for most purposes (this is in contrast to the approach for laboratory diagnosis of infectious and neoplastic diseases, where a specific lesion must be sampled). Alternatively, the great sensitivity of PCR allows for genetic testing to be performed on minute samples of any readily accessible tissue or body fluid, including saliva and urine. Newborn DNA screening can easily be performed on the same blood spots collected for biochemically based screening. For prenatal diagnosis, either amniocytes or CVS can be used, since both contain the full complement of fetal DNA. In addition, PCR-based single-cell genetic analysis now allows for preimplantation diagnosis on individual embryonic blastomeres, as discussed earlier, and most recently prenatal diagnosis targeting fetal DNA circulating in the maternal blood has become practicable, thus potentially obviating the need for any sort of invasive fetal sampling.

Once in hand, the specimen can be subjected to any of the molecular genetic techniques in the current armamentarium. The choice of technique will depend on the nature of the disease gene being studied (especially its size and mutational heterogeneity), the purpose of the test, and to some extent the condition of the specimen. Specimens that are extremely small, fixed, or degraded will require some sort of PCR-based analysis, since that technique is the most tolerant of suboptimal samples. Once amplified to an abundant amount, the DNA is then amenable to further analysis by other techniques such as allele-specific oligonucleotide hybridization or DNA sequencing. Southern blotting, on the other hand, which is still used for certain tests involving large genomic DNA targets (such as the fragile X full mutation, though PCR-based methods have recently been developed), requires intact DNA of high molecular weight in pristine enough condition to be digested efficiently with restriction endonucleases and to produce reproducible patterns on gel electrophoresis. This usually means at least a few milliliters of whole blood, or one or two dishes of cultured amniocytes, transported to the lab in a timely fashion. In either case, the first step is typically the isolation of DNA from the specimen, for which a variety of commonly used methods and commercial reagents are available.

We should always remain aware of the limitations of molecular genetic testing, no matter how sophisticated the technology. Single genes and mutations rarely act entirely alone, and complex disorders clearly involve hundreds or perhaps thousands of genes all interacting at the DNA, RNA, and protein levels and with the outside environment. Thus, no DNA test, no matter how comprehensive, can ever tell the whole story. As the initial phase of the Human Genome Project is completed and we move on to the "post-genomic" or "proteomics" era, we will need to begin to think of genetic disease pathogenesis on a more "three-dimensional" level, taking into account the many interacting gene products in the nucleus and cytoplasm, as opposed to merely the "one-dimensional" string of nucleotides on a stretch of DNA. Whether tests of such phenomena, once they arrive, will be part of the domain of the molecular genetics laboratory or will move into a new "proteomics" laboratory section is anyone's guess. But as we await such advances, we can remain confident that the molecular genetic tests of today, and those added to the menu with each passing month, already offer a tremendous and irreplaceable service to many patients and their families, providing information and choices for their own lives and those of their offspring that had never been available before, in any form, in the entire history of medicine.

FURTHER READING

1. American Society of Clinical Oncology. American Society of Clinical Oncology Policy Statement Update: Genetic Testing for Cancer Susceptibility. *J. Clin. Oncol.* **2003,** *21,* 2397–2406.
2. Billings, P. R. Genetic Nondiscrimination. *Nat. Genet.* **2005,** *37,* 559–560.
3. Das, S.; Bale, S. J.; Ledbetter, D. H. Molecular Genetic Testing for Ultra Rare Diseases: Models for Translation from the Research Laboratory to the CLIA-Certified Diagnostic Laboratory. *Genet. Med.* **2008,** *10,* 332–336.
4. Grody, W. W. Cystic Fibrosis: Molecular Diagnosis, Population Screening, and Public Policy. *Arch. Pathol. Lab. Med.* **1999,** *123,* 1041–1046.
5. Grody, W. W.; Cutting, G.; Klinger, K., et al. Laboratory Standards and Guidelines for Population-Based Cystic Fibrosis Carrier Screening. *Genet. Med.* **2001,** *3,* 149–154.
6. Lo, Y. M.; Chan, K. C.; Sun, H.; Chen, E. Z.; Jiang, P.; Lun, F. M.; Zheng, Y. W.; Leung, T. Y.; Lau, T. K.; Cantor, C. R., et al. Maternal Plasma DNA Sequencing Reveals the Genome-Wide Genetic and Mutational Profile of the Fetus. *Sci. Transl. Med.* **2010,** *2,* 61ra91.
7. Monaghan, K. G.; Feldman, G. L.; Palomaki, G. E.; Spector, E. B.; Ashkenazi Jewish Reproductive Screening Working Group. Molecular Subcommittee of the ACMG Laboratory Quality Assurance Committee Technical Standards and Guidelines for Reproductive Screening in the Ashkenazi Jewish Population. *Genet. Med.* **2008,** *10,* 57–72.
8. Palomaki, G. E.; Kloza, E. M.; Lambert-Messerlian, G. M.; Haddow, J. E.; Neveux, L. M.; Ehrich, M.; van den Boom, D.; Bombard, A. T.; Deciu, C.; Grody, W. W., et al. DNA Sequencing of Maternal Plasma to Detect Down Syndrome: An International Clinical Validation Study. *Genet. Med.* **2011,** *13,* 913–920.
9. Richards, C. S.; Bale, S.; Bellissimo, D. B.; Das, S.; Grody, W. W.; Hegde, M. R.; Lyon, E.; Ward, B. E. Molecular Subcommittee of the ACMG Laboratory Quality Assurance Committee ACMG Recommendations for Standards for Interpretation and Reporting of Sequence Variations: Revisions 2007. *Genet. Med.* **2008,** *10,* 294–300.
10. Strom, C. M.; Crossley, B.; Buller-Buerkle, A.; Jarvis, M.; Quan, F.; Peng, M.; Muralidharan, K.; Pratt, V.; Redman, J. B.; Sun, W. Cystic Fibrosis Testing 8 Years on: Lessons Learned from Carrier Screening and Sequencing Analysis. *Genet. Med.* **2011,** *13,* 166–172.
11. ten Bosch, J. R.; Grody, W. W. Keeping up with the Next Generation: Massively Parallel Sequencing in Clinical Diagnostics. *J. Mol. Diagn.* **2008,** *10,* 484–492.
12. Watson, M. S.; Cutting, G. R.; Desnick, R. J., et al. Cystic Fibrosis Couple Carrier Screening: 2004 Revision of the American College of Medical Genetics Mutation Panel. *Genet. Med.* **2004,** *6,* 387–391.

CHAPTER

24

Heterozygote Testing and Carrier Screening

Matthew J McGinniss
Molecular Diagnostics, Caris Life Sciences, Phoenix, AZ, USA

Michael M Kaback
Departments of Pediatrics and Reproductive Medicine, University of California, San Diego, CA, USA

A list of relatively "common" autosomal recessive disorders (Table 24-1) seen in the defined subpopulations in the United States is presented with data indicating the respective carrier frequency and newborn disease incidence in those ethnic groups. The fact that selected genetic diseases occur predominantly in certain ethnic, religious, or racial groups is not surprising when one considers the relatively high degree of inbreeding seen in defined subpopulations. In addition to inbreeding, in some situations, selective environmental factors may have existed at some point in history that provided a biological (reproductive) advantage to carriers of the recessive gene (e.g. relative resistance to malaria in individuals who are heterozygous for sickle cell hemoglobin, β-thalassemia, or G6PD deficiency). Because of this selective effect, the gene becomes "enriched" from one generation to the next in that population.

It should be emphasized that while many of these diseases are relatively rare, the frequency of carriers can be quite high. For example, the disease incidence of cystic fibrosis (CF) among Caucasians of Northern European ancestry is approximately 1 in 2500 newborns, yet nearly 1 in 25 such individuals are carriers of a CF mutation. Similarly, the disease incidence for Tay–Sachs disease (TSD) among Ashkenazi Jewish newborns is about 1 in 3600, while the carrier rate in this population is approximately 1 in 30. Because of these population distributions and the availability of relatively simple, accurate, and inexpensive carrier detection methods, it is possible to screen individuals in these groups and identify persons and, more critically, couples at risk for homozygous disease in their offspring before affected children are born. With comprehensive genetic counseling, and the important options that prenatal diagnosis can provide, many "at-risk" families, identified through screening such subpopulations, might choose to have only children unaffected with the disorder for which they are found at risk.

For those relatively "common" autosomal disorders that occur within defined ethnic groups, there is usually no known prior history of the disease in either side of the family. Such a positive history may be present in only 20% of instances where a child with such a disorder is diagnosed. For this reason, clinicians should consider carrier screening for all individuals in these subpopulation groups where heterozygote detection is readily available. Certainly, such testing and its implications should be thoroughly discussed with the patients—or they can be referred to appropriate regional agencies for such services. Not only is this considered optimal preventive medicine, but has become the standard of care in many cases.

The implementation of carrier screening programs must rely upon education and standardization. Both the affected population and the health care providers need education about genetic screening programs. This is amply illustrated by the cumulative experiences of the TSD, β-thalassemia, and CF carrier screening programs. Even if the individual tested receives a negative screening result, it will still be appropriate for the physician to order a diagnostic test if warranted by the clinical circumstances. Subpopulations affected by targeted carrier screening programs also need to be reliably informed about the mutation detection rates and overall clinical utility to ensure that genetic counseling is effective. Standardization of analytical testing platforms will play an increasingly important role as carrier screening programs expand and new ones develop. In addition, carrier screening programs can borrow from the successes of newborn screening programs that exist in the United States and those around the world.

TABLE 24-1 Frequency and Incidence for Selected Autosomal Recessive Disorders in Defined Ethnic Groups in the United States

Disease	Ethnic Group	Gene Frequency	Carrier Frequency	"At-Risk" Couple Frequency[a]	Disease Incidence in Newborns
Sickle cell anemia	African-Americans	0.040	0.080	1:150	1:600
Tay–Sachs disease	Ashkenazi Jews	0.016	0.032	1:900	1:3600
β-Thalassemia	Greeks, Italians	0.016	0.032	1:900	1:3600
α-Thalassemia	Southeast Asians and Chinese	0.020	0.040	1:625	1:2500
Cystic fibrosis	Northern Europeans	0.020	0.040	1:625	1:2500
Spinal muscular atrophy	Asian	0.010	0.020	1:2500	1:10,000
Phenylketonuria	Northern Europeans	0.008	0.016	1:4000	1:16,000

[a]Likelihood that both members of a couple are heterozygous for the same recessive allele (assuming nonconsanguinity and that both are of the same ethnic group).

FURTHER READING

1. Cao, A.; Cristina Rosatelli, M.; Galanello, R. Control of β-Thalassaemia by Carrier Screening, Genetic Counselling and Prenatal Diagnosis: The Sardinian Experience. In *Ciba Foundation Symposium 197-Variation in the Human Genome*; Chadwick, D.; Cardew, G., Eds.; John Wiley & Sons: Chichester, UK, 2007.
2. Desnick, R. J.; Kaback, M. M. *Tay–Sachs Disease (Advances in Genetics, Volume 44)*; Academic Press: San Diego, CA, 2001.
3. Harper, P. *Practical Genetic Counseling*; 7th edn.; Oxford University Press: Oxford UK, 2010.
4. McKinlay Gardner, R. J.; Sutherland, G. R. *Chromosome Abnormalities and Genetic Counseling*; Oxford Monographs on Medical Genetics No. 46; 3rd ed.; Oxford University Press: Oxford UK, 2004.
5. Young, I. D. *Introduction to Risk Calculation in Genetic Counseling*; 3rd ed.; Oxford University Press: Oxford UK, 2007.

HELPFUL WEBSITES

The GeneTests website provides current, authoritative information on genetic testing and its use in diagnosis, management, and genetic counseling. GeneTests promotes the appropriate use of genetic services in patient care and personal decision making. http://www.ncbi.nlm.nih.gov/sites/GeneTests/servlet/access.

Online Mendelian Inheritance in Man (OMIM) is a comprehensive, authoritative, and timely compendium of human genes and genetic phenotypes. The full text, referenced overviews in OMIM contain information on all known Mendelian disorders and over 12,000 genes. It is updated daily, and these entries contain links to other genetics resources. http://www.ncbi.nlm.nih.gov/omim.

The American College of Medical Genetics (ACMG) defines and promotes excellence in medical genetics practice and the integration of translational research into practice. In addition, they promote and provide medical genetics education, clinical practice guidelines, and laboratory standards and guidelines. http://www.acmg.net.

National Tay–Sachs and Allied Diseases Association (NTSAD) was founded by concerned parents with children affected by TAD or a related genetic disorder, including all the lysosomal storage diseases and leukodystrophies. NTSAD was an early pioneer in the development of community education about TAD, carrier screening programs, and laboratory quality control programs—thereby ensuring that those being tested received accurate and reliable information and test results. http://www.ntsad.org/.

The Cystic Fibrosis Foundation is the world's leader in the search for a cure for CF. The Foundation funds more CF research than any other organization, and nearly every CF drug available today was made possible because of the Foundation's support. http://www.cff.org/.

CHAPTER

25

Prenatal Screening for Neural Tube Defects and Aneuploidy

Amelia L M Sutton and Joseph R Biggio

Department of Obstetrics and Gynecology, Division of Maternal Fetal Medicine; Department of Genetics, University of Alabama, Birmingham, AL, USA

Neural tube defects (NTDs) and aneuploidies are the major causes of perinatal death and childhood morbidity. Routine screening for these anomalies has become a standard part of prenatal care. Improvements in screening tests for aneuploidy have led to a dramatic shift in the practice patterns and a decline in the number of diagnostic tests. A combination of folate supplementation for primary and secondary prevention and screening with maternal serum alpha-fetoprotein (MSAFP) followed by comprehensive ultrasound evaluation for screen-positive women has reduced the birth incidence of NTDs. Recent advances in combined first- and second-trimester screening strategies, utilizing both biochemical and ultrasound evaluation, have increased the detection rates of chromosomal abnormalities to such levels that many women elect to forego prenatal diagnostic procedures resulting in reduction of rates of these procedures by more than 50%.

25.1 SCREENING FOR AND DIAGNOSIS OF NEURAL TUBE DEFECTS

Worldwide, the incidence of NTDs is approximately 1–10 per 1000 births, with some geographic variation. Adequate maternal folate levels are essential for normal fetal neural tube development. Widespread dietary supplementation and fortification of food staples with folate in the United States produced a 23% reduction in spina bifida in less than a decade. Approximately 90–95% of NTDs occur in women with no family history of the condition.

MSAFP levels are elevated in most of the pregnancies affected by NTDs. Traditionally, women with elevated MSAFP were offered genetic amniocentesis with measurement of AFP and acetylcholinesterase in the amniotic fluid. Advances in ultrasound imaging have dramatically changed the evaluation of screen-positive gestations. AFP is a major protein in the fetal circulation with concentrations 10^6-fold higher in the fetal circulation compared to the maternal circulation and accumulates in the amniotic fluid with advancing gestation. AFP reaches the maternal serum by a combination of transplacental and transamniotic diffusion. The serum concentration of AFP is gestational age dependent and peaks in the third trimester; therefore, an accurate estimation of gestational age is essential for interpretation of the concentration. Concentrations are usually expressed as multiples of the median (MoM) of unaffected pregnancies at the specific gestational age and comparisons are then made using MoMs.

Defects in fetal skin, including open NTDs and ventral wall abnormalities, cause increased amniotic fluid AFP concentrations by a direct transudative process. Fetal renal disease, intrauterine demise, and contamination of the amniotic fluid with fetal blood also result in elevated AFP concentrations. Although there is considerable overlap between affected and unaffected pregnancies, a cutoff of 2.5 MoM results in a detection rate of 95% for anencephaly and 80–85% for spina bifida with a 3% false-positive rate. Detection is enhanced when pregnancy dating criteria are based on a biparietal diameter measurement. Serum screening for NTDs using MSAFP is not typically effective until after 14 weeks of gestation and is typically performed at 15–20 weeks in the window typically utilized in screening programs for aneuploidy. Serum marker levels can be affected by multiple factors including maternal weight, ethnicity, and the presence of insulin-dependent diabetes. Multiple gestations pose a diagnostic dilemma in serum screening for NTDs. In general, serum biomarkers, including AFP, in twin pregnancies are approximately twice those of singletons. Detection rates of NTDs are significantly lower in twins than in singletons. Using a cutoff of 4.0 MoM provides detection rates around 90–95% for anencephaly and 65–80% for open spina bifida with a 7–10% false-positive rate.

Ultrasound screening for major fetal anomalies in the second trimester has become a routine part of prenatal care in many countries. Anencephaly results from failed closure of the cephalic portion of the neural tube and disrupted cranial development resulting in exencephaly, the precursor of anencephaly. Ultrasound findings include decreased crown rump length (CRL), absent calvaria, extruding lobulated cerebral tissue (exencephaly) or absent neural tissue, and abnormal head shape with the eyes delineating the upper portion of the fetal face. Open spina bifida is detected by both cranial and spinal abnormalities on ultrasound. The two major cranial anomalies, the so-called "lemon" and "banana" signs, are recognized in more than 90% of fetuses with open NTDs. The lemon sign describes the flattening or concavity of the frontal bones of the skull visible in the transverse plane. The banana sign signifies the caudal displacement of the cerebellum with alignment of the cerebellar hemispheres as part of type II Arnold–Chiari malformation. Sonographic identification of the area of spinal dysraphism is often more difficult than detecting cranial abnormalities in spina bifida.

Traditionally, women with elevated MSAFP levels were further evaluated by amniocentesis so that amniotic fluid levels of AFP and acetylcholinesterase were measured to confirm the diagnosis of a NTD. Elevated amniotic fluid acetylcholinesterase, defined as an amniotic fluid AFP concentration greater than 2.0 MoM for the gestational age, detects essentially 100% of anencephaly and spina bifida cases with a false-positive rate of less than 1%. Ultrasound technology has advanced significantly since the introduction of MSAFP screening and carries detection rates for NTDs approaching that of amniocentesis.

25.2 PRENATAL SCREENING FOR ANEUPLOIDY

The association between advanced maternal age and an increased risk of Down syndrome was first described in the 1930s. In the early years of prenatal screening, maternal age primarily was used to select pregnancies at high risk for aneuploidy. Diagnostic testing, by amniocentesis or chorionic villus sampling (CVS), was offered to all women over a specific age cutoff, typically age 35; however, evidence has emerged that maternal age is not adequate to screen for aneuploidy. The American College of Obstetricians and Gynecologists (ACOG) now recommends offering prenatal screening for aneuploidy to women of all ages. ACOG further recommends that all women should have the option to undergo invasive testing with either CVS or amniocentesis, irrespective of screening results, for chromosomal abnormalities.

25.3 MATERNAL SERUM SCREENING

Precise estimation of gestational age is a prerequisite for valid screening with serum markers and these analyte concentrations are expressed as MoMs. A likelihood ratio for serum markers may be calculated by dividing the height of the affected pregnancy frequency distribution curve at a given MoM value by the height of the distribution curve of unaffected pregnancies at the same MoM. This likelihood ratio can then be multiplied by the age-specific risk to estimate the adjusted risk. In women carrying a fetus with trisomy 21, the concentrations of MSAFP tend to be 30% and unconjugated estriol (uE_3) 20% lower than the median, while human chorionic gonadotropin (hCG) and inhibin-A (InhA) concentrations are twofold higher. The combination of uE_3, MSAFP, hCG (the so-called "triple screen"), and maternal age results in a detection rate of greater than 75% with a false-positive rate of 3–5% and addition of InhA increases the detection rate by 15–20%.

The three analytes—AFP, hCG, and uE3—are decreased in pregnancies affected by trisomy 18, achieving a detection rate of 58–80% at a false-positive rate of just 0.3–0.6%. Similar to the second trimester, hCG levels are elevated in the first trimester in pregnancies with Down syndrome, while concentrations of pregnancy-associated plasma protein-A (PAPP-A) are 50–60% reduced. In combination with maternal age, hCG and PAPP-A have a detection rate of approximately 67%. In affected twin pregnancies, however, analyte levels are not consistent. The overall detection rate for aneuploidy in twin gestations is only 50–55% with second-trimester serum screening. A number of factors, including ethnicity and diabetes, affect serum marker concentrations. hCG levels are higher in black women as compared to Caucasian women, but uE_3 levels are similar among the two groups. Furthermore, AFP levels are also higher in black women; thus, race-specific medians have been developed for Caucasian and African–American women. AFP levels are significantly reduced in women with preexisting diabetes as compared to healthy controls. Additionally, uE_3 and PAPP-A levels are slightly lower, and InhA levels are higher in diabetic women. Most programs make adjustments in the standard risk calculations in diabetic women.

25.4 ULTRASOUND SCREENING FOR ANEUPLOIDY

Mid trimester thickening of the nuchal fold was first described as sonographic marker of Down syndrome. Using a cutoff of 6 mm or more for the nuchal fold has been demonstrated to allow detection of approximately 40% of fetuses with trisomy 21. The ratio of the femur or humerus length to the biparietal diameter is used to detect the relatively shortened femurs. The combination of a thickened nuchal fold and relatively short femur or humerus has a sensitivity of 75% and specificity of 98% for Down syndrome. A number of other ultrasound markers have been described, including choroid plexus cysts, renal pelvic dilation, echogenic intracardiac foci, sandal toe deformity, and a number of others. Multiple scoring systems have been developed to quantify the risk of aneuploidy with a combination of markers. By integrating maternal age into the scoring system, further refinement of the method can be used to modify the *a priori* risk of aneuploidy. More than one sonographic marker is associated with over a fivefold increase in the risk of Down syndrome, whereas a normal ultrasound decreases the chance of Down syndrome by approximately one-half.

The nuchal translucency (NT) is a term used to describe a collection of fluid behind the fetal neck recognized during the first trimester. Fetuses with trisomy 21, as well as other chromosome abnormalities, have measurements that are two- to threefold larger than that of unaffected fetuses. An increased NT has also been identified in a multitude of single gene disorders and structural malformations. The NT is measured between 11 and 13 6/7 weeks at a corresponding CRL of 45–84 mm. The NT measurement may be expressed as the delta NT (the difference in millimeters between the normal median for the CRL and the measured NT) or in MoM by dividing the measured NT by the normal median at the gestational age allowing for an age-adjusted risk assessment, similar to biochemical screening. Using an increased NT as the sole criteria for fetal aneuploidy screening detects approximately 70–75% of fetuses with trisomy 21, and when combined with the maternal age-associated *a priori* risk, a detection rate in excess of 80% is achievable. Measurements are independent of maternal serum levels of hCG and PAPP-A; therefore, these markers can be combined into a single screening test. The combination of free β-hCG or hCG, PAPP-A, and NT with maternal age is able to detect approximately 80–85% of fetuses with trisomy 21 at a false-positive rate of 5%. Other ultrasound markers can also be incorporated into the algorithm in the first trimester including tricuspid regurgitation and absence of the nasal bone.

By combining the ultrasound with biochemical analysis, high detection rates with low false-positive rates can be achieved. The First- and Second-Trimester Evaluation of Risk (FASTER) trial compared combinations of screening strategies in nearly 40,000 women, including sequential (first- and second-trimester results reported independently) *vs.* integrated (first- and second-trimester results reported as single risk assessment) screening. These combined screening strategies that incorporate both first- and second-trimester components achieve the highest detection rates, up to 96% if the NT is included, at a false-positive rate of 5% or less.

FURTHER READING

Cheschier, N. ACOG Practice Bulletin. Neural Tube Defects. Number 44, July 2003. (Replaces Committee Opinion Number 252, March 2001); *Int. J. Gynaecol. Obstet.* **2003,** *83* (1), 123–133.

Adzick, N. S.; Thom, E. A.; Spong, C. Y.; Brock, J. W., 3rd; Burrows, P. K.; Johnson, M. P., et al. A Randomized Trial of Prenatal Versus Postnatal Repair of Myelomeningocele. *N. Engl. J. Med.* **2011,** *364* (11), 993–1004.

Wald, N. J.; Kennard, A.; Hackshaw, A.; McGuire, A. Antenatal Screening for Down's Syndrome. *Health Technol. Assess.* **1998,** *2* (1), 1–112, i–iv.

Ghi, T.; Pilu, G.; Falco, P.; Segata, M.; Carletti, A.; Cocchi, G., et al. Prenatal Diagnosis of Open and Closed Spina Bifida. *Ultrasound Obstet. Gynecol.* **2006,** *28* (7), 899–903.

Spencer, K. Aneuploidy Screening in the First Trimester. *Am. J. Med. Genet. C Semin. Med. Genet.* **2007,** *145C* (1), 18–32.

Nicolaides, K. H. Screening for Fetal Aneuploidies at 11 to 13 Weeks. *Prenat. Diagn.* **2011,** *31* (1), 7–15.

Wald, N. J.; Kennard, A.; Hackshaw, A. K. First Trimester Serum Screening for Down's Syndrome. *Prenat. Diagn.* **1995,** *15* (13), 1227–1240.

Nicolaides, K. H. Nuchal Translucency and Other First-Trimester Sonographic Markers of Chromosomal Abnormalities. *Am. J. Obstet. Gynecol.* **2004,** *191* (1), 45–67.

Malone, F. D.; Canick, J. A.; Ball, R. H.; Nyberg, D. A.; Comstock, C. H.; Bukowski, R., et al. First-Trimester or Second-Trimester Screening, or Both, for Down's Syndrome. *N. Engl. J. Med.* **2005,** *353* (19), 2001–2011.

Cicero, S.; Avgidou, K.; Rembouskos, G.; Kagan, K. O.; Nicolaides, K. H. Nasal Bone in First-Trimester Screening for Trisomy 21. *Am. J. Obstet. Gynecol.* **2006,** *195* (1), 109–114.

Lo, Y. M.; Tein, M. S.; Lau, T. K.; Haines, C. J.; Leung, T. N.; Poon, P. M., et al. Quantitative Analysis of Fetal DNA in Maternal Plasma and Serum: Implications for Noninvasive Prenatal Diagnosis. *Am. J. Hum. Genet.* **1998,** *62* (4), 768–775.

CHAPTER 26

Techniques for Prenatal Diagnosis

Lee P Shulman

Department of Obstetrics and Gynecology; Division of Clinical Genetics, Feinberg School of Medicine, Northwestern University, Chicago, IL, USA

Sherman Elias

Department of Obstetrics and Gynecology, Feinberg School of Medicine, Northwestern University; Prentice Women's Hospital, Chicago, IL, USA

26.1 INTRODUCTION

The cornerstone of prenatal diagnosis has been the direct evaluation of fetal tissue. Advances resulting from the Human Genome Project have increased the number of fetal abnormalities amenable to invasive prenatal diagnosis. In addition, in the past few decades, there has been an expansion of our capabilities to assess the risk for fetal abnormalities by noninvasive screening modalities including high-resolution ultrasonography and maternal serum analyte screening. Effective screening protocols reduce the number of invasive procedures in women carrying unaffected fetuses. Recent data showing improved safety of chorionic villus sampling (CVS) and amniocentesis now raise the possibility of offering invasive prenatal diagnosis to all women in the first and second trimesters, an approach to prenatal care that may be considerably expanded with the application of microarrays to evaluating the fetal genome.

26.2 AMNIOCENTESIS

Amniocentesis, the aspiration of amniotic fluid, has traditionally been performed at and after 15–17 weeks gestation (menstrual weeks). Amniocentesis is routinely performed in an outpatient facility. An ultrasound examination should be done immediately before the procedure to evaluate fetal number and viability, perform fetal biometric measurements to confirm gestational age, establish placental location, and estimate amniotic fluid volume. Any procedure that involves passing a device into an organ, especially the pregnant uterus, carries a risk. Amniocentesis is no exception. Amniocentesis carries potential danger to both mother and fetus. Maternal risks are quite low, with symptomatic amnionitis occurring only rarely. Minor maternal complications such as transient vaginal spotting and minimal amniotic fluid leakage occur in 1% or less of cases, but these are almost always self-limited in nature. Other very rare complications include intra-abdominal organic injury or hemorrhage. We believe it wise to continue to counsel that the risk of pregnancy loss secondary to amniocentesis is 0.5% over baseline, or perhaps slightly less at centers with experienced operators.

26.3 CHORIONIC VILLUS SAMPLING

Because amniocentesis is most commonly performed in the mid-second trimester (15–16 weeks), fetal diagnosis cannot usually be established prior to 17–18 weeks gestation. A technique that could be performed during the first trimester would be highly desirable to reduce the psychological stress of awaiting results until midpregnancy and to allow a safer method of pregnancy termination, should an abnormality be detected. CVS is such a technique. Chorionic villi can be obtained by transcervical, transabdominal, or transvaginal aspiration, depending on the location of the placenta and maternal intra-abdominal anatomy. Recently, Caughey and Colleagues (*85*) reported a retrospective cohort study of all amniocentesis and CVS procedures resulting in a normal karyotype from 1983 to 2003 at the University of California, San Francisco. In a comparison of 9886 CVS and 30,893 amniocentesis procedures performed during the study period, the overall loss rates were 3.12% for CVS and 0.83% for amniocentesis ($P < 0.001$). However, in the most recent time period, 1998–2003, there was no difference between the two procedures (adjusted odds ratio 1.03, 95% CI 0.23–4.52). We conclude that clinical judgment and patient individualization in choosing the optimal approach for CVS increases safety. For example, some technically difficult transcervical CVS procedures (e.g. fundal placentas) should be avoided in favor of a technically more facile transabdominal approach. The converse is also true.

26.4 FETAL BLOOD SAMPLING

Many of the indications for detecting fetal abnormalities that previously required fetal blood sampling are now performed by amniocentesis or CVS using DNA analysis; nonetheless, the continuing value of fetal blood sampling lies not only with those few remaining diagnostic indications but also in providing the potential for drug therapy. For example, fetal arrhythmias have been treated with direct administration of antiarrhythmic medications, and fetal paralysis may be induced to facilitate invasive procedures such as fetal transfusions or for magnetic resonance imaging.

26.5 FETAL TISSUE SAMPLING

Advances in gene and protein identification as a result of the Human Genome Project and other public and private genetic initiatives has considerably reduced the need for fetal tissue biopsy for the prenatal detection of genetic disorders. Conditions such as fetal genodermatoses and Duchenne muscular dystrophy once required fetal skin and muscle biopsy, respectively, for direct pathological evaluation. However, most of these conditions have been associated with specific gene mutations and are now amenable to diagnosis by CVS, amniocentesis preimplantation genetic diagnosis.

26.6 SUMMARY

While many of the fetal conditions that required fetal tissue sampling have now been characterized by specific gene mutations that allow for less invasive techniques for accurate prenatal diagnosis. Nonetheless, the ability to access the fetal compartment and obtain fetal tissue in a safe and tolerable fashion will likely be needed until a reliable approach to obtaining fetal cells from maternal blood is developed. Until then, CVS and amniocentesis continue to provide relatively safe and highly effective approaches for prenatal diagnosis that will allow for any and all diagnostic procedures amenable for the detection of fetal chromosome, genetic, and genomic abnormalities.

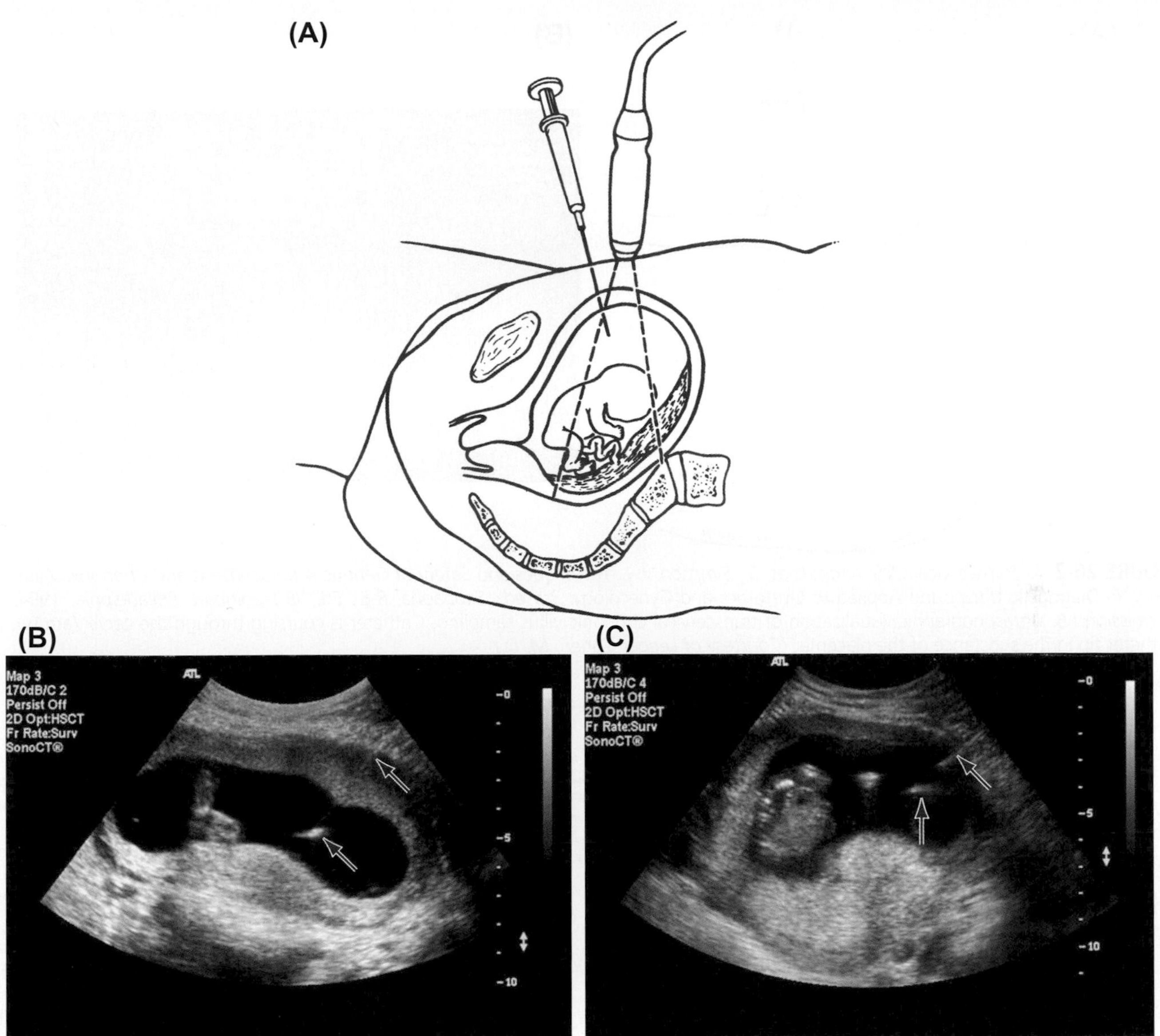

FIGURE 26-1 A. Amniocentesis performed concurrently with ultrasound. *(From Simpson, J. L.; Elias, S. Prenatal Diagnosis of Genetic Disorders. In* Maternal-Fetal Medicine: Principles and Practice; *Creasy, R. K.; Resnik, R. Eds.; WB Saunders: Philadelphia, 1994, with permission.)* B. Ultrasonographic visualization of a transplacental amniocentesis. Thin arrow points to the needle shaft within the placenta and the wall of the uterus, and the thick arrow points to the needle tip within the amniotic cavity. (*Courtesy of Leeber Cohen, M. D.*) C. Ultrasonographic visualization of non-transplacental amniocentesis. Thin arrow points to the needle shaft within the wall of the uterus, and the thick arrow points to the needle tip within the amniotic cavity. *(Courtesy of Leeber Cohen, M. D.)*

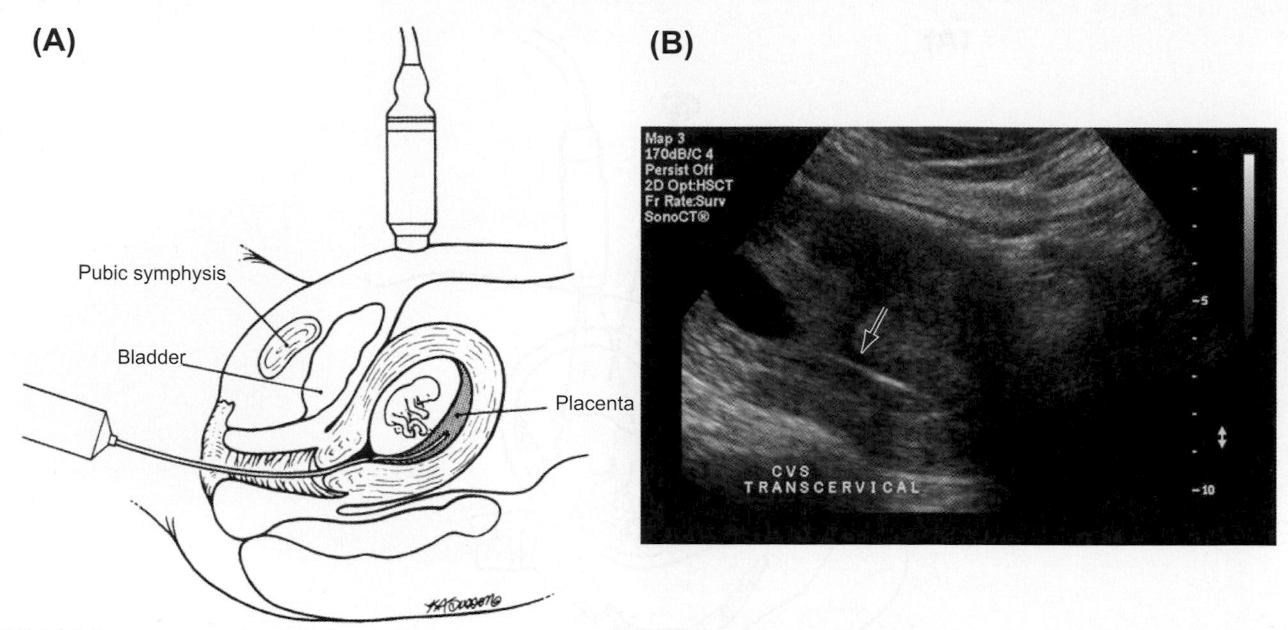

FIGURE 26-2 A. Transcervical CVS. *(From Elias, S.; Simpson, J. L. Techniques and Safety of Genetic Amniocentesis and Chorionic Villus Sampling. In* Diagnostic Ultrasound Applied to Obstetrics and Gynecology, *3rd ed.; Sabbagha, R.E., Ed.; JB Lippincott: Philadelphia, 1994, with permission.)* B. Ultrasonographic visualization of transcervical chorionic villus sampling. Catheter is coursing through the cervix (arrow), with catheter tip in the substance of the placenta. *(Courtesy of Leeber Cohen, M. D.)*

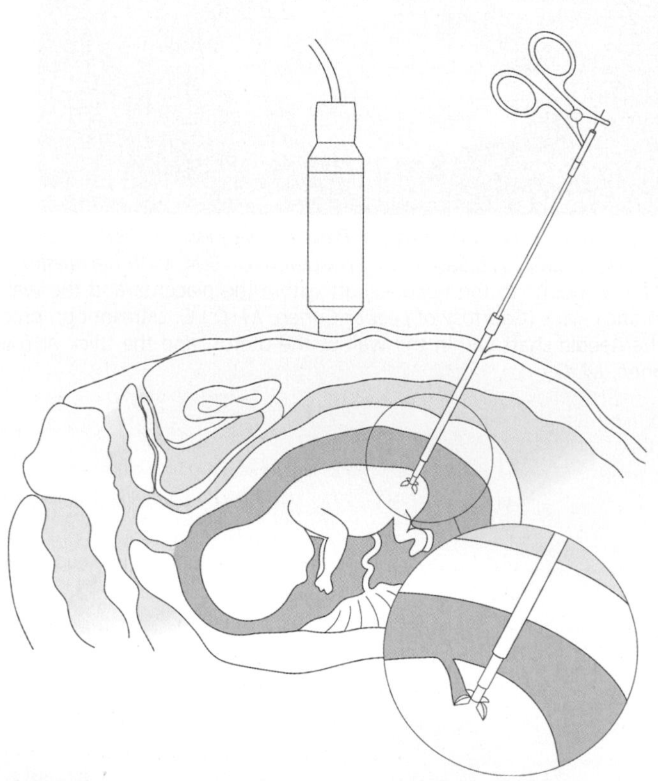

FIGURE 26-3 Fetal skin sampling. *(From Elias, S; Emerson, D. S.; Simpson, J. L., et al. Ultrasound-Directed Fetal Skin Sampling for Prenatal Diagnosis of Genodermatoses.* Obstet. Gynecol. ***1994***, 83, *337, with permission.)*

FURTHER READING

1. Eddleman, D.; Malone, F.; Sullivan, L., et al. Pregnancy Loss Rates After Midtrimester Amniocentesis by the First and Second Trimester Evaluation of Risk (FASTER) Trial Research Consortium. *Obstet. Gynecol.* **2006**, *109* (2, Part 1), 1067.
2. Rhoads, G.; Jackson, L.; Schlesselman, S., et al. The Safety and Efficacy of Chorionic Villus Sampling for Early Prenatal Diagnosis of Cytogenetic Abnormalities. *N. Engl. J. Med.* **1989**, *320*, 609.
3. Jackson, L.; Zachary, J.; Desnik, R., et al. A Randomized Comparison of Transcervical and Transabdominal Chorionic Villus Sampling. *N. Engl. J. Med.* **1992**, *327*, 594.
4. Shulman, L.; Simpson, J.; Elias, S., et al. Transvaginal Chorionic Villus Sampling Using Transabdominal Ultrasound Guidance: A New Technique for First-Trimester Prenatal Diagnosis. *Fetal. Diagn. Ther.* **1993**, *8*, 144.
5. WHO/PAHO Consultation on CVS Evaluation of Chorionic Villus Sampling Safety. *Prenat. Diagn.* **1999**, *19*, 97.
6. Caughey, A.; Hopkins, L.; Norton, M. Chorionic Villus Sampling Compared with Amniocentesis and the Difference in the Rate of Pregnancy Loss. *Obstet. Gynecol.* **2006**, *108*, 612.
7. Daffos, F. Fetal Blood Sampling. *Annu. Rev. Med.* **1989**, *40*, 319.
8. Elias, S.; Annas, G. Generic Consent for Genetic Screening. *N. Engl. J. Med.* **1994**, *330*, 1611.
9. Elias, S.; Emerson, D.; Simpson, J., et al. Ultrasound-Directed Fetal Skin Sampling for Prenatal Diagnosis of Genodermatoses. *Obstet. Gynecol.* **1994**, *83*, 337.
10. Evans, M.; Hoffman, E.; Cadrin, C., et al. Fetal Muscle Biopsy: Collaborative Experience with Varied Indications. *Obstet. Gynecol.* **1994**, *84*, 913.
11. Nicolaides, K.; Azar, G.; Byrne, D., et al. Fetal Nuchal Translucency: Ultrasound Screening for Chromosomal Defects in First Trimester of Pregnancy. *BMJ* **1992**, *304*, 867.
12. Malone, F.; Canick, J.; Ball, R., et al. First-Trimester or Second-Trimester Screening, or Both, for Down's Syndrome. *N. Engl. J. Med.* **2005**, *353*, 2001.
13. Ehrich, M.; Deciu, C.; Zwiefelhofer, T., et al. Noninvasive Detection of Fetal Trisomy 21 by Plasma DNA Sequencing in a Clinical Setting. *Am. J. Obstet. Gynecol.* in press.

CHAPTER

27

Neonatal Screening

Richard W Erbe

Departments of Pediatrics and Medicine, State University of New York; Division of Genetics, The Women and Children's Hospital of Buffalo, Buffalo, NY, USA

Harvey L Levy

Pediatrics, Harvard Medical School; Medicine/Genetics, Boston Children's Hospital, Boston, MA, USA

Neonatal screening dates from the 1960s when the methods developed by Dr. Robert Guthrie were applied to phenylketonuria (PKU) on a population-wide basis. Guthrie's bacterial assays of a single analyte for each newborn screened were gradually expanded to encompass disorders such as maple syrup urine disease, homocystinuria, and tyrosinemia. Further, exploiting the blood spots with additional analytical methods, including enzyme activity assays, immunoassays, and electrophoresis, expanded the target disorders to include hypothyroidism, congenital adrenal hyperplasia, the sickling hemoglobinopathies, cystic fibrosis (CF), and others. All of these new measures were based on Guthrie's simple and effective filter paper blood spot method for specimen collection, transport, and analysis. State programs continued to test newborn specimens in a designated state laboratory, with communication of the test results to specified physicians. This state-by-state system of organization led to great variability in the nature and number of screened disorders. The more recent introduction of tandem mass spectrometry (MS/MS) substantially expanded the capability of screening for fatty acid oxidation defects, organic acidemias, and additional amino acid disorders, among others. Moreover, MS/MS methods can screen for many disorders from each Guthrie specimen. A committee of the American College of Medical Genetics (ACMG) recommended that all newborn screening programs include at least 29 "core" disorders and all programs in the United States now adhere to this recommendation. An additional 25 "secondary target" disorders would be detected in the differential diagnosis of the core disorders, resulting in a substantially larger number of screened disorders and much greater uniformity of coverage among different state programs.

Regarding MS/MS testing, tables are provided that list all disorders identifiable by newborn screening, their major symptoms, the markers analyzed by MS/MS, and common causes of false-positive results. This screening has focused primarily on detecting an increase in an analyte as measured by MS/MS, although several recent applications have used the reduction of an analyte occurring within a series of closely related reactions such as in the urea cycle to identify a disorder. Further examples are provided to illustrate other types of analyses that continue to be used for primary screens, including enzyme activity assays in galactosemia and biotinidase deficiency; immunoassays for congenital hypothyroidism, congenital adrenal hyperplasia, and CF; isoelectric focusing or hemoglobin electrophoresis for the sickling hemoglobinopathies; and DNA-based gene tests as second-level screens for medium chain acyl-CoA dehydrogenase deficiency (MCADD), galactosemia, biotinidase deficiency, and CF.

Expanded newborn screening has promoted the development of regional, national, and international collaborations aimed at refining screening cutoffs in order to improve diagnostic accuracy and reduce false positives. The cutoff or threshold level selected possible metabolic disorders is usually a level that presumably will be exceeded by all affected infants to maximize test sensitivity while achieving appropriate positive predictive values (i.e. the proportion of persons with positive test results who actually have the disorder). In some instances, ratios of some of the multiple analytes simultaneously measured in each blood spot can be used to improve accuracy, such as the phenylalanine to tyrosine ratio in cases of hyperphenylalaninemia or the C8-to-C6 ratio in MCADD. Collection of detailed clinical data aims to refine distinctions between variability related to the analytical methods and biological and pathological variations in the newborn. Indeed, the majority of positive screening results are not due to a metabolic or other disorder. Some result from mild, transient elevations in blood amino acids or acylcarnitines in otherwise normal newborns or occur in premature infants receiving hyperalimentation or with liver dysfunction, or after transfusion. Other problems arise from improper specimen handling, from failure to collect the Guthrie specimen from transferred newborns or from increasing instances of "opting out." Confirmatory testing is essential after an abnormal screening result in order to reach a diagnosis. Screening abnormalities may have more than one possible genetic or nongenetic cause. Confirmation of PKU, galactosemia, and other inborn errors of metabolism requires not only additional diagnostic tests but also rigorous clinical diagnostic evaluation before an appropriate plan of management can be formulated. DNA testing for confirmation of diagnosis and identification of mutant alleles is increasingly available. Regardless of the subsequent test results, a patient with the clinical abnormalities that are characteristics of a screened disorder should be managed as potentially affected until the status is resolved.

Screening, diagnosis, and management of inborn errors of metabolism are complex. Sources of guidelines and information are increasingly available online. For example, the ACMG has developed "ACT Sheets," succinct descriptions of the possible disorder, and an algorithm for follow-up testing and interpretation of test results. More detailed but focused explanations of the metabolic disorders, along with recommended testing and genetic implications for relatives are available online at GeneReviews.

The prominence of newborn screening has led to a number of highly contested issues. One involves the retention of the newborn Guthrie specimen. A number of programs retain these specimens for decades or indefinitely. The ability to sequence the DNA for a specific gene or for parts or all of the genome has led to concern over potential misuse of the stored specimen and calls for protective measures. Another issue involves the addition of disorders to the screening menu. How much knowledge of natural history should we know and how effective should therapy be before a disorder is added? The ACMG, through the federally funded Newborn Screening Translational Research Network, is embarking on a multi-year study to address these and other questions.

CHAPTER

28

Enzyme Replacement and Pharmacologic Chaperone Therapies for Lysomal Storage Disease

Robert J Desnick and Edward H Schuchman

Department of Genetics and Genomic Sciences, Mount Sinai School of Medicine, New York, NY, USA

Kenneth H Astrin

Department of Preclinical Sciences, New York College of Podiatric Medicine, New York, NY, USA

Seng H Cheng

Rare Diseases Science, Genzyme, a Sanofi Company, Framingham, MA, USA

The lysosomal storage diseases are a group of disorders that result as a consequence of a deficiency or loss of one or more of the approximately 40 lysosomal enzymes. These abnormalities invariably cause the progressive aberrant accumulation of uncatabolized substrates in the lysosomes with consequent distention of the organelle and resultant cellular damage. Patients can present with a wide range of clinical manifestations depending on the nature of the offending substrate that accumulates in the lysosomal compartment. These include hepatosplenomegaly, cardiomegaly, skeletal deformities, renal failure, neuromuscular disease, cardiorespiratory failure, ocular opacities, neurodegenerative disease, and psychomotor retardation. Severity and onset of disease are correlated with the amount of residual cellular enzymatic activity with patients harboring mutations that result in lower enzyme function typically presenting with more rapidly progressive disease and a shortened life span.

Following the first demonstration of a successful reduction to practice the concept of enzyme replacement therapy for Gaucher disease, this therapeutic paradigm has been applied to the management of several other lysosomal storage diseases. Presently, clinically approved enzyme preparations are available for Gaucher, Fabry, MPS I, II and VI, and Pompe diseases. With the advent of new and more efficient manufacturing processes, alternate enzyme formulations or biosimilars are also in development for these indications. Additionally, enzyme replacement-based therapies are in clinical development for Niemann-Pick B and MPS IV diseases. While these therapies are clearly effective and arguably transformative to the patients, the treatments are intrusive, requiring weekly or biweekly infusions of the enzyme preparations for the lifetime of the patients. Moreover, they do not adequately address all the disease manifestations, particularly those with central nervous system (CNS) involvement. It is for these reasons that improved therapies are continually being sought and developed for this group of disorders.

An exciting development of late is the concept of substrate reduction therapy, at least for a subset of the lysosomal storage diseases. In contrast to enzyme replacement therapy which seeks to augment the enzyme activity that is deficient, substrate reduction therapy aims to lower the amount of substrate that is synthesized by the affected cells. This is achieved by inhibiting the enzymes that are engaged in the biosynthesis of the offending substrates. In cases where patients retain residual enzymatic activity, the deployment of substrate reduction therapy can serve to abate the burden of lysosomal accumulation of the unmetabolized substrates. Presently, one such effector molecule has been approved for use in Gaucher type 1 disease and another is in late-stage clinical testing. As the small molecule drug approved for use in Gaucher disease inhibits a key enzyme in the synthesis of a number of glycosphingolipids, it has the potential of being effective for treating other glycosphingolipidoses as well. This therapeutic approach offers the added benefits of being orally available and the ability to treat the CNS. Consequently, it has also been evaluated for use in treating neuronopathic lysosomal storage diseases. Presently, substrate reduction therapy is also approved for patients with Niemann-Pick type C disease.

Yet another emerging therapeutic approach is pharmacological chaperone therapy, which seeks to rescue misfolded mutant lysosomal proteins through the use of low-molecular weight drugs that bind specifically to the enzymes. By stabilizing the folding intermediates of the mutant enzymes, it is the hope that this will redirect them away from the endoplasmic reticulum-associated degradation pathway that targets them for degradation and into the lysosomal compartment. By chaperoning the delivery of mutant enzymes that retain residual catalytic activity to this organelle, it is anticipated this will reduce the burden of lysosomal storage. This therapeutic strategy will require the development of drug entities that are tailored to each enzyme and then to a subset of the mutant enzymes (ideally those that retain residual activity). As these are small molecule drugs, this approach also offers the potential to address the neuronopathic disease. Moreover, as the biodistribution of these molecular entities are likely to be different to enzymes, it is possible that they be used as an adjuvant therapy to existing treatments. Presently, pharmacologic chaperone therapy is in clinical testing for Fabry disease.

In addition to these approaches, several other emerging technology platforms are under active investigation. These include, for example, gene and cell-based therapies that are in late-stage preclinical development for a number of lysosomal storage diseases. It is evident that over the next decade, patients with this group of disorders will have increasing therapeutic options for managing their disease. It is also anticipated that with some of the lysosomal storage diseases, different treatment modalities will be used in combinations to effect optimal clinical outcomes.

CHAPTER

29

Gene Therapy: From Theoretical Potential to Clinical Implementation

Nicholas S R Sauderson, Maria G Castro, and Pedro R Lowenstein

Department of Neurosurgery and Department of Cell and Developmental Biology, The University of Michigan Medical School, Ann Arbor, MI, USA

This chapter explains the theory and practice of the art of introducing new genetic elements to somatic cells to treat human disease. In the early 1970s, the field grew around the idea of using gene replacement for the treatment of genetic diseases. A wide range of gene transfer vectors have been developed over more than 40 years for this purpose, and a wide range of diseases have been targeted, not only genetic diseases, but also non-inherited diseases, including cancer. This chapter focuses on viral vectors as tools and on diseases of the CNS, particularly brain cancer, as targets.

The most commonly used viruses for vector development are retroviruses, including lentiviruses, herpesviruses, adenoviruses, and adeno-associated viruses. These have all been engineered by the removal of genetic sequences necessary for the virus' own replicative cycle so that the resulting viral vector is incapable of replicating in a normal human cell. Modified cell lines that express the missing proteins are used to grow and expand the engineered viral vectors, which can be purified to high titers, largely free from wild-type viruses and other undesirable contaminants. Potentially therapeutic transgenes can then be inserted, and after infection of the target cells, the vectors express the therapeutic products under the spatiotemporal control of specific promoters. In addition to cell-type specificity, production of transgenes can also be tightly regulated using expression control mechanisms such as the tetracycline-inducible promoter.

The simplest scenario for gene therapy is perhaps that of a genetic disorder in which a single gene is missing or defective in the patient's genome, and its complementation by a transgene expressed from a viral vector is sufficient to cure the disease. This concept has now advanced to a stage where clinical success has been achieved in the treatment of genetic immunodeficiencies, adrenoleukodystrophies, and retinal degeneration, all caused by mutations in a single gene.

Over the years, the concept of gene therapy was expanded to the proposed use of genes as pharmacological compounds. As such, genes were proposed to be used to deliver a product not because it was missing but because it would have a desired pharmacological effect. Though conceptually less straightforward, gene therapy is now being used to treat various neurodegenerative disorders for which the cause is not a simple genetic lesion; thus a cure is currently not a defined goal for gene therapy of such diseases, as the goal is symptomatic relief. For example, some of the symptoms of Parkinson disease are caused by low levels of dopamine in the striatum and therefore can be partially addressed by expression from viral vectors of the enzymes responsible for synthesizing this neurotransmitter. Alzheimer disease is associated with loss of basal forebrain cholinergic neurons, and viral vectors have been used to express various neurotrophic factors in an attempt to slow this process.

Yet more challenging are gene therapy approaches to treating cancer. Cancer is thought to be essentially a genetic disorder, and so in principle ought to be amenable to gene therapy. An obvious strategy for tackling cancers associated with the loss of tumor suppressor genes such as p53, PTEN, Rb, etc. would be to use viral vectors to replace the missing gene. This approach has shown little promise in practice because of the difficulty of transducing every single cell manifesting the deficit. More successful at the preclinical level has been the use of vectors encoding enzymes capable of activating pro-drugs to cytotoxic metabolites. For example, the thymidine kinase enzyme from Herpes simplex virus can be expressed in tumor cells, and is then capable of phosphorylating the pro-drug ganciclovir. The phosphorylated form is incorporated into the DNA of replicating cells, interferes with DNA elongation and causes the death of the cell. There appears to be some degree of "bystander effect," meaning that compared with the replacement of defective tumor suppressor genes, a smaller percentage of transformed cells would need to be transduced in order to effect a cure, but once again, results in clinical trials so far have been disappointing.

In view of the likely failure of any strategy that is dependent on transducing all or most transformed cells, considerable efforts have also been invested in developing gene therapy strategies for enhancing the immune response against cancer cells. Cancer patients often possess antitumor T cells and antibodies, so it seems reasonable to imagine that with the right kind of stimulation, this partial immune response might be enhanced to the point that it becomes able to clear the cancer, just as viral infections are routinely cleared by the immune system. Approaches aimed at stimulating the activity of various immune cells include those based on inducing the expression of immunomodulatory cytokines such as interleukins, tumor necrosis factor-α, interferons, granulocyte/macrophage colony stimulating factor, and Fms-like tyrosine kinase ligand. Alternatively, various attempts have been made to interfere with the activities of the numerous components of the immune system that normally function to suppress the immune response, such as regulatory T cells and CTLA-4.

A fourth approach is based on interfering with angiogenesis. It is widely believed that tumors cannot grow beyond a certain size without the formation of new blood vessels to supply the new cells with nutrients and oxygen, and therefore the expression of proteins that interfere with the signaling pathways normally involved in angiogenesis ought to have an antitumor effect based on starving the tumor. Attention has focused on the growth factors involved in angiogenesis, such as vascular epithelial growth factor, epidermal growth factor and basic fibroblast growth factor.

Finally, though not strictly gene therapy, several attempts have been made to capitalize on the propensity of viruses to kill their host cells, by developing viruses that replicate preferentially in tumor cells. One of the earliest was ONYX-015, an adenovirus engineered to lack the E1B-55kDa gene. This deficit prevents the virus from replicating in healthy cells, but the virus is able to replicate and kill cells that lack the tumor suppressor gene p53. Since this gene is commonly deleted or defective in human cancers, this confers tumor-specificity on the virus' killing ability. In addition, such replication competent vectors are being engineered to carry further genes that will aid in the destruction of cancer cells.

Currently, gene therapy is a clinical therapeutic tool used to treat a small number of monogenic disorders, but has still to achieve clinical efficiency in the treatment of non-inherited complex diseases such as neurodegenerative disorders and cancer. A small number of adverse effects have been observed, including the deaths of two patients, but overall the safety record of gene therapy is very good. Reasons for the tragic failures have been closely examined, and improved methodologies developed to prevent their recurrence. Perhaps the best speculation for the future of the field is that gene therapies will start to show dramatic successes in targeting the more complicated diseases, once our understanding of the diseases themselves improves sufficiently to give gene therapists better-defined targets.

CHAPTER 30

Ethical and Social Issues in Clinical Genetics

Angus John Clarke

Institute of Medical Genetics, School of Medicine, Cardiff, Wales, UK

Clinical genetic practice raises a wide range of ethical and social issues not only for professionals but also for our patients and clients and for society at large. These include the proper scope of prenatal diagnosis and the selective termination of pregnancies, the communication of sensitive information within families, the making of decisions on behalf of children or incompetent adults, and the potentially competing obligations of professionals to their patients or clients, their patients' families, and to the society more broadly. And what obligations do we professionals have to ensure the just treatment by society of individuals with physical or cognitive impairments?

The pace of technical developments makes these issues more difficult to assess because of the rapidly increasing quantities of genetic data that can now be generated about individuals, especially when the interpretation of so much of this is uncertain and is likely to remain somewhat provisional for years or even decades. Established applications now include (i) decisions about disease prevention or the surveillance for complications of disease in those at increased risk of a familial disorder, (ii) reproductive decisions made either before conception (from identifying the healthy carriers of recessive disorders) or after a conception (considering whether to establish or continue with a pregnancy when the child would be likely to be affected by a genetic condition), and (iii) guidance on the choice of therapies, especially in the treatment of malignancy. In addition to these applications, however, there are many other, often poorly validated, applications being promoted outside regular health care, such as ancestry and relationship (especially paternity) testing, predictions relevant to non-health matters (such as personality, intelligence, and sporting prowess), or decisions about lifestyle and dietary choice. Other governmental uses of genetic testing operate in the context of identity testing for forensics and immigration.

The technical developments in genome-based testing are changing the shape of the ethical and social issues confronting us in clinical genetics because the contexts within which clinicians and clients/patients/families are operating are, for the moment, unfamiliar. We are yet to develop common professional habits of practice that allow all parties to feel "safe."

30.1 GOALS AND OUTCOMES OF GENETIC SERVICES

One important area concerns the goals of genetic services, as formulated in measures of outcome. One must ensure that these do not promote either a cost-driven or a consumer-led eugenic program, perhaps unwittingly, as may happen if success is measured in terms of birth rates or terminations of pregnancy. This concern merges with the need for a sensitive and self-consciously critical approach to how services can be shaped or driven by the targets against which they are held accountable. An inappropriate method of evaluating clinical services could, at the microlevel, drive the clinical interaction between the professional (doctor or counselor) and the patient (or client), leading the professional to be inappropriately directive in shaping the client's decisions.

30.2 DIAGNOSTIC GENETIC TESTS

Diagnostic genetic tests—performed to find the cause of a presenting medical complaint—can confront families with difficult ethical choices to be made, usually when there are implications for others in the family. The parents of a child diagnosed with Duchenne muscular dystrophy, for example, may be faced with decisions about prenatal diagnosis in a pregnancy and about their obligation to give this information to the mother's sisters and cousins, who may be carriers. If this information is not passed on appropriately, the professionals may consider whether they have an obligation to transmit this information even against the wishes of the boy's parents.

30.3 PREDICTIVE GENETIC TESTS

Predictive genetic testing may be sought by those at high risk of inheriting a serious disorder, perhaps a familial cancer or an inherited neurodegeneration. If there are no useful medical interventions for those shown to have the altered gene, as currently for Huntington's disease, then many choose not to be tested and the genetic counseling for those who seek testing involves a process designed to help them think through the implications in advance. The major issues to be considered include the psycho-emotional impact on the client, questions of employment and life and health insurance, the family and social dimension such as the impact on a partner and decisions about reproduction, and the wider family. Where there are major health benefits, as in some hereditary cancers, the counseling helps to prepare the clients but the process is more straightforward. For other conditions, there may be potential clinical advantages of testing but these may be less clear and the decision may be more finely balanced, as for some women at risk of familial breast cancer.

In this realm of predictive testing—and potentially in the less accurate realm of susceptibility testing—those tested may wish to keep the results of testing, and the fact of having been tested, from powerful Third Parties, especially in relation to life and health insurance and current or prospective employers. There is some debate as to whether genetic privacy is, or can remain, a credible goal; in that case the efforts to maintain confidentiality may give way to strengthened efforts at preventing the unfair use of genetic information. The definition of genetic information in such Genetic Non-Discrimination legislation then becomes crucial; does it include a state of ill health caused by a genetic disorder? Does it include family history as well as genetic test results? Professionals will usually want to protect their patients' privacy from Third Parties unless safety is at stake (e.g. if the patient is unfit to drive).

Within the family, the issue of the confidentiality of genetic information of any sort, including test results, remains important in predictive and diagnostic settings. In practice, professionals almost always respect this but they may work with families to help them pass difficult information to relatives. It is very uncommon for professionals to force the disclosure of genetic information to family members against the wishes of their client.

30.4 POPULATION SCREENING

Population screening for genetic disorders is a very different context within which genetic tests may be made available. Instead of a personal or family history of a genetic condition having caused concern that prompted a clinical referral, those at population (low) risk of a problem may have anxieties provoked by the screening program that may not be allayed by a low risk result, as screening tests are usually not 100% sensitive. The way in which the screening test is offered may have a major influence on uptake, so that the quality of information provided and of consent obtained may both be open to question. Personal commitment to the screening program by staff may appropriately translate into efforts to maximize uptake (as with newborn screening for treatable disorders), but this would be thoroughly inappropriate in the setting of reproductive decisions, such as antenatal screening for Down syndrome. Indeed, the simple fact of being offered a population screening test can itself be understood as a recommendation by the staff that patients accept the offer. This can be termed "structural nondirectiveness." In such a setting, the health practitioners must work hard to ensure that their patients understand that the offer of a test is not a recommendation but a decision for them to make. It may be possible to assist this process, in turn, by altering certain structural features of the program.

30.5 NEWBORN SCREENING

Newborn screening programs have usually been introduced for the benefit of those children whose health and well-being will be greatly improved by early treatment—before they would be likely to have their disorder diagnosed in other ways. This is certainly true for phenylketonuria and hypothyroidism. As screening expands to include more disorders, the future health of the child may no longer be the primary consideration but perhaps broader family or societal interests (early diagnosis leading to lower cost of health care or enabling informed reproductive decisions) or the program may be driven by the technology, perhaps in an unhelpful manner. Equally, the test assay employed may affect the program's goals and outcomes. Thus, if the goal of screening for cystic fibrosis ceases to be the early diagnosis and treatment of infants affected by the clinical disorder cystic fibrosis but instead becomes the identification of all infants with two variants in the *CFTR* gene, many clinically unnecessary interventions may be triggered and the family may suffer practical and emotional disruption, as when boys are identified as likely to suffer from male infertility in adulthood.

30.6 ANTENATAL SCREENING

Antenatal screening raises many of the long-standing concerns familiar from the consideration of eugenic programs. The concern for the quality of the "race," the "nation," or the "population" may be seen as the rationale for screening, and the motivation for this may be idealistic or emotional (with an underlying patriotism or racism), economic (aiming to minimize expenditure on nonproductive citizens), or driven by consumer pressure, itself a product of other factors and forces. The offer of screening may be routinized so that little consideration of the purpose and possible consequences of participation may have occurred. If an initial screening test indicates an increased risk of a chromosomal or fetal structural problem, for example, it may then be difficult for the woman/couple to decline further investigations suggested by the practitioners to clarify the situation and resolve the uncertainties that have been generated. Some of the further investigations likely to be triggered include chorionic villus biopsy or amniocentesis, so that a number of miscarriages will result.

It would consume more time than is available in most antenatal clinics to consider in any depth the social context within which screening is offered. Not only would the woman's knowledge and experience of Down syndrome, spina bifida, and other conditions be discussed—and further information provided—but also the level of care and support required by an affected child and the level of support from health and social services that is provided by the state or insurance schemes. It would also be possible to address additional topics such as the experience of family life with an affected child and the social stigmatization and isolation that can be experienced. These factors may all be relevant to the decisions made about screening and some women/couples will also want counseling to support their coming to a decision in the light of their ethical values and/or religious beliefs.

When faced by a firm diagnosis of a fetal chromosome anomaly or malformation, the parent(s) will be confronted by the same questions and issues as in the somewhat hypothetical, "perhaps" mode of a screening risk estimate but now fully "for real." Progress through the screening system can be experienced as a series of small decisions, sometimes made by default and each deferred to the last possible moment. These deferrals can be further extended if even the final results of the diagnostic testing express substantial uncertainty. A decision may then have to be made as to whether to continue or terminate the pregnancy in the face of incomplete information.

Beyond the few individuals caught up in any one decision, there are wider social issues too. Thus, is society causing justified sadness, anger, or resentment among those affected by any of the conditions likely to be detected by antenatal screening? While the rhetoric of screening does seek to keep attitudes to antenatal screening (and making diagnoses) as distinct, this separation looks unconvincing from the perspective of a midwife or a clinical geneticist. It is not unreasonable for someone with Down syndrome to resent antenatal screening that targets his/her condition in particular, especially if one of the major factors underlying this seems to be the drive to avoid the birth of children with conditions that require costly care, whether the "burden" would fall on state or family.

30.7 FREE FETAL DNA AND NONINVASIVE PRENATAL DIAGNOSIS

Developments in prenatal diagnostic methods, using genomic sequencing technologies to identify fetal DNA sequences in maternal blood, are expected soon to be applied to routine antenatal screening. Once this occurs, it will be possible routinely to identify fetal chromosome anomalies at 7–8 weeks of gestation. Down syndrome could be (more or less) eliminated. The social consequences of this, with most affected children then being born as a positive choice rather than as a failure of screening or a reluctance to enter screening that might lead to a late termination of pregnancy, are difficult to foresee but are likely to become manifest only slowly, over a generation or more.

30.8 GENOMIC CARRIER SCREENING FOR CARRIERS OF AUTOSOMAL RECESSIVE DISORDERS

Another application of high-throughput sequencing technologies has emerged as screening for carrier status of many autosomal recessive (AR) disorders. Within families already known to carry a particular AR disorder because of the birth of an affected child, couples can usually already avoid the birth of further affected children through carrier testing and, for couples where both are carriers, the application of prenatal diagnosis. The new technologies open this possibility—avoiding the birth of affected children—to everyone, not only to those who have already had an affected child in the family. This development may well be welcomed by many, especially where recessive disease is especially prevalent, as in communities that practice customary consanguineous marriage. The interaction between these new technological possibilities and social custom—not only consanguinity nut also arranged marriage—is likely to be interesting and complex and will take some decades to work through.

30.9 CROSS-CULTURAL GENETIC COUNSELING

Linguistic difference between professionals and clients or patients will often act as a barrier to accessing services and may make it more difficult for a client to be given a high-quality service. This becomes a particular problem when there are preexisting differences of power, wealth, and status between communities and the less powerful community is further disadvantaged by poor health care. The use of informal or ad hoc interpreters in clinic—from the client's own family or community—may also undermine privacy and lead to the prospect of social coercion.

In addition to language, differences in religion may complicate such acts of inter-community communication. Stereotyping of the clients' community by professionals may be a major problem reinforcing barriers to uptake (e.g. assumptions such as "Oh, they would not want to do that"). Patterns of decision making within a family or community may differ from those in the professionals' community, who may then experience a tension between respect for the client's community and a wish to promote the autonomy of the individual client. These tensions may be especially difficult if the status of women in the client's community is perceived by the professionals as being inferior and in need of strengthening. The boundary between professional support for the individual client and social action with a goal of political change could easily be transgressed.

30.10 LABELING AND NAMING OF SYNDROMES

Attaching a diagnostic label to a patient can be a delicate process, especially in dysmorphology when it may involve physical examination, photography, and discussion about the affected person's physical features that could easily cause offense in the patient or their relatives. It is important for professionals to approach the question of physical features and appearances with tact and delicacy.

Naming a genetic syndrome can be another delicate task. Examples of unfortunate labels that have dropped out of professional favor because of the offense they caused in some families are CATCH-22 syndrome and LEOPARD syndrome, because of how the labels added insult to the injury of having the condition; the condition once named after the physicians who described it, Hallervordern and Spatz, is now given a biochemical name as the two physicians conducted their studies on the cadavers of children they had identified and selected to be killed for these very studies.

30.11 DUTY TO RECONTACT

There has long been discussion about how actively genetics services should strive to ensure that patients seen in the past should be recontacted when new information comes to light about their diagnosis. There is a balance between making unsolicited and perhaps unwelcome approaches to former patients on the one hand and denying them the benefits of new knowledge on the other. One powerful reason for caution has been the concern that a regular policy of active recontact could lead to an unsustainable—because too onerous and costly—standard of care becoming legally enforceable. With genomic advances, however, these pressures may become stronger as new situations arise: (i) more may now be known about a diagnosis made in the past; (ii) a genetic test carried out in the past may now be interpreted more accurately, as when variants of uncertain significance do become interpretable, when those tested in the past may expect to be given updated interpretations; or (iii) a new test becomes applicable that might establish a precise diagnosis for a patient. It is unclear how genetic services will cope with these pressures if the likely burden of recontacting former clients increases dramatically.

30.12 GENOMIC SCREENING FOR COMMON COMPLEX DISEASES

Genome-wide association studies (GWAS) have identified genetic variants (usually *single-nucleotide polymorphisms* (SNPs)) at many sites that show slight association with disease (e.g. with an odds ratio of 1.2:1) but that may be highly significant statistically. Such findings typically account for only a modest fraction of the genetic contribution to the disease, so that even the application of many of these SNPs to calculating disease risk will be of little clinical utility. Direct-to-consumer corporations do, however, market such tests, often describing them as "lifestyle" tests rather than medical tests, so as to avoid regulatory interference. One major concern about such commercial enterprises is that they may not explain adequately to potential participants that the appropriate genetic investigation in the face of a clear family history of disease may not be a GWAS panel but rather a set of Mendelian gene mutation searches. Another possible problem of "clinically oriented" GWAS are the paradoxical behavioral responses that may not always encourage risk-reducing behaviors.

30.13 RESEARCH

30.13.1 Consent and Feedback

Consent has to be obtained for biomedical research to be compatible with the professional ethics of clinical researchers and with the tolerance and acceptance of the society at large. The semi-redundant concept of informed consent (consent cannot be "uninformed") is the accepted standard for good practice, with necessary exceptions for worthwhile research that cannot be performed on this basis, in which situations consent can be assumed (e.g. with research on the unconscious) or provided by others (in the case of children or the incompetent). Difficult cases abound in this area and some principles require further development, as with group consent or community consent for some types of research conducted within illiterate communities in developing countries.

In the traditional, hypothesis-driven type of research that addresses very specific questions, families who participated in genetics research could often be given any specific results that emerged such as the name of the gene or mutation identified as relevant to their condition. In genetic association studies, in contrast, the passing to an individual of their results is much more problematic. This is because their own results may have no meaning except when pooled with the results of many others, because the tests were not performed within an accredited laboratory to diagnostic standards or because the mass of results would be overwhelming and simply uninterpretable. Many clinical researchers therefore prefer to give a summary of the project's results to participants but not their own personal results, although some commentators disagree.

30.13.2 Confidentiality

Information about research participants, it is usually assumed, will be treated with great respect and care will be taken to ensure that personal information about a patient is not released inappropriately. In the context of biobank research, however, there are particular difficulties in ensuring that personal information about participants remains confidential. One challenge is that research projects now generate massive quantities of sequence information and this may be accessible through the Internet. Such information is intrinsically identifying and it has emerged that sequence information about specific individuals can be extracted from the databanks of raw project data that some funding bodies insist upon. One response to this is to tighten the security net; another is not to guarantee genetic privacy to research participants but instead to focus more on legal methods of preventing the abuse of personal information once it has been obtained.

30.13.3 Geneticization

Research on genetics could be seen as emphasizing the importance of genetic factors in disease causation, at the expense of other—perhaps more remediable—environmental factors. While genetic factors are the principal factors underlying some conditions, many common, complex disorders are the result of interacting genetic and environmental factors. It is therefore perhaps somewhat paradoxical—or at least unfortunate—that the genetic factors may be more amenable than environmental factors to investigation and research, whereas intervening to modify these same genetic factors may be rather more difficult than modifying at least some environmental factors.

30.13.4 Effective Eugenics?

The arrival of genomic technologies that permit low-cost genome-wide analyses shows great potential for the delivery of treatments for the rare inherited disorders as well as the common disorders of our time. In addition, however, they also permit interventions at the population level, such as preconception or antenatal carrier screening and noninvasive prenatal diagnostic testing, based on the fetal DNA present in maternal blood from early in pregnancy. Such developments have the power to make substantial changes in the composition of human populations; the improvements to human populations that were sought by the eugenicists of the past now seem to be attainable. What should we do with this new knowledge and these new technologies?

FURTHER READING

1. Bell, C. J.; Dinwiddie, D. L.; Miller, N. A.; Hateley, S. L., et al. Carrier Testing for Severe Childhood Recessive Diseases by Next-Generation Sequencing. *Sci. Transl. Med.* **2011,** *3,* 65ra4. DOI: http://dx.doi.org/10.1126/scitranslmed.3001756.
2. British Society for Human Genetics. Genetic Testing of Children: Report of a Working Party of the British Society for Human Genetics. Birmingham, England, 2010.
3. Corrigan, O. Empty Ethics: The Problem of Informed Consent. *Sociol. Health Illn.* **2003,** *25,* 768–792.
4. Forrest, K.; Simpson, S. A.; Wilson, B. J.; van Teijlingen, E. R., et al. To Tell or Not to Tell: Barriers and Facilitators in Family Communication About Genetic Risk. *Clin. Genet.* **2003,** *64,* 317–326.
5. Forrest Keenan, K.; van Teijlingen, E.; McKee, L.; Miedzybrodzka, Z.; Simpson, S. A. How Young People Find Out About Their Family History of Huntington's Disease. *Soc. Sci. Med.* **2009,** *68,* 1892–1900.
6. Geelen, E.; Van Hoyweghen, I.; Doevendans, P. A.; Marcelis, C. L.; Horstman, K. Constructing "Best Interests": Genetic Testing of Children in Families with Hypertrophic Cardiomyopathy. *Am. J. Med. Genet. A* **2011 Aug,** *155* (8), 1930–1938.
7. Hallowell, N.; Foster, C.; Eeles, R.; Ardern-Jones, A.; Murday, V.; Watson, M. Balancing Autonomy and Responsibility: The Ethics of Generating and Disclosing Genetic Information. *J. Med. Ethics* **2003,** *29,* 74–83.
8. Khoury, M. J.; Coates, R. J.; Evans, J. P. Evidence-Based Classification of Recommendations on Use of Genomic Tests in Clinical Practice: Dealing with Insufficient Evidence. *Gen. Med.* **2010,** *12* (11), 680–683.
9. Lucassen, A.; Parker, M. Confidentiality and Serious Harm in Genetics—Preserving the Confidentiality of One Patient and Preventing Harm to Relatives. *Eur. J. Hum. Genet.* **2004,** *12,* 93–97.
10. McAllister, M.; Wood, A. M.; Dunn, G.; Shiloh, S.; Todd, C. The Genetic Counseling Outcome Scale: A New Patient-Reported Outcome Measure for Clinical Genetics Services. *Clin. Genet.* **2011,** *79,* 413–424.
11. Pyeritz, R. E. The Coming Explosion in Genetic Testing—Is There a Duty to Recontact? *N. Engl. J. Med.* **2011,** *365,* 1367–1369.
12. Raymond, F. L.; Whibley, A.; Stratton, M. R.; Gecz, J. Lessons Learnt from Large-Scale Exon Re-Sequencing of the X Chromosome. *Hum. Mol. Genet.* **2009,** *18* (R1), R60–R64.
13. Rehmann-Sutter, C.; Müller, H. J., Eds. *Disclosure Dilemmas. Ethics of Genetic Prognosis After the 'Right to Know/Not to Know' Debate*; Ashgate Publishing Ltd: Farnham, England and Burlington, USA, 2009.
14. Sarangi, S.; Bennert, K.; Howell, L.; Clarke, A.; Harper, P.; Gray, J. (Mis)alignments in Counselling for Huntington's Disease Predictive Testing: Clients' Responses to Reflective Frames. *J. Genet. Couns.* **2005,** *14,* 29–42.
15. Scully, J. L.; Porz, R.; Rehman-Sutter, C. 'You Don't Make Genetic Test Decisions from One Day to the Next'—Using Time to Preserve Moral Space. *Bioethics* **2007,** *21,* 208–217.
16. Shaw, A.; Hurst, J. A. "What Is This Genetics, Anyway?" Understandings of Genetics, Illness Causality and Inheritance Among British Pakistani Users of Genetic Services. *J. Genet. Couns.* **2008 Aug,** *17* (4), 373–383.
17. Shirts, B. H.; Parker, L. S. Changing Interpretations, Stable Genes: Responsibilities of Patients, Professionals, and Policy Makers in the Clinical Interpretation of Complex Genetic Information. *Genet. Med.* **2008,** *10* (11), 778–783.
18. Wright, C., 2009. Cell-Free Fetal Nucleic Acids for Non-Invasive Prenatal Diagnosis: Report of the UK expert working group, PHG Foundation. Available online at: http://www.phgfoundation.org/reports/4985/.

CHAPTER

31

Legal Issues in Genetic Medicine

Philip R Reilly

Venture Partner, Third Rock Ventures, Boston, MA, USA

The key legal issues in genetic medicine generally address the problem of the proper use of genetic information. Most of the growing number of clinical interactions that involve the generation and use of genetic information can be analyzed in the context of the long-established legal framework for medical malpractice. To prove a wrong, the plaintiff must establish the existence of a duty, the breach thereof, and a consequent harm recognized by the courts. Because genetic information is inherently familial, genetic testing and counseling pose new challenges for the law as relates to the expectation of confidentiality, problems explored under the notion of the duty to warn. Most genetic counselors are not yet independent practitioners, but with the advent of state licensure systems, they may soon be able to act as independent professionals.

The results of prenatal screening and testing often determine a decision about pregnancy termination. Thus, the ever-contentious constitutional debate over abortion is important to clinical genetics. Despite the unending assault on *Roe v. Wade* since 1973 and an ascendant judicial philosophy that the boundaries of the right to obtain an abortion should be determined by the states, the core principle that the right to privacy encompasses a rationally regulated right to terminate a pregnancy is likely to survive.

Information derived from genetic testing is now potentially of relevance in decisions about adoption, and lawsuits over failure to disclose health issues in babies placed for adoption have occurred. Genetic testing soon may become more widely used in screening sperm and egg donors, two activities that remain largely unregulated in the United States. DNA forensic testing to ascertain relationship has been used in resolving some immigration problems and could become more widely used in the future.

Although the major reason to turn to surrogate mothers is usually infertility, some prospective parents do so due to a genetic burden carried by one or both of them. The law of surrogacy is unsettled in the United States; some states attempt to regulate the practice; others do not. Key questions—which generally focus on custody battles—have been decided in the courts.

Despite some appellate court decisions, the legal status of frozen embryos remains murky in the United States. Fortunately, bitter disputes among divorcing couples over custody and use of embryos are becoming of only historical interest. For some years now, most fertility clinics have been careful to establish the plan of disposition by written agreement before the embryos are created.

Newborn screening (growing in scope) is conducted pursuant to mandatory laws (some of which permit religious exemptions). In the United States, lawsuits have successfully challenged the right of the states to retain samples and use them for research purposes. The constitutionality of mandatory newborn screening has not been directly challenged in the United States, but the Supreme Court of Ireland has recognized the right of a family to refuse screening.

Prenatal screening and carrier testing are not (with rare exceptions) conducted under government programs. Testing or counseling errors in those contexts would be resolved under traditional precepts of tort law.

Genetic discrimination (the denial of some economic or social right) because of the results of a genetic test has been a topic of great concern since the 1980s. The enactment of The Genetic Information and Non-Discrimination Act of 2008 in the United States has alleviated much of that concern. In most countries other than the United States, fear of genetic discrimination has not been a major social concern largely because health care is guaranteed under a single payer system.

In general, although it may have jurisdiction to do so, the FDA has not chosen to closely regulate genetic testing. Labs need only be licensed under the appropriate federal law; if licensed, they may offer tests that they develop without seeking further regulatory approval. Offers to provide genetic test results directly to the consumer rather than through a physician began about a decade ago. A few states such as New York strictly regulate such activities; most states do not. Direct to consumer genetic testing is likely to grow; if so it is possible that there will be more regulatory oversight.

Over the last 20 years, genetic testing has become a regular part of research involving human subjects. This has led institutional review boards (and on occasion the courts) to face new issues in informed consent such as ownership of tissue, dissemination of newly discovered information, and determining rights (if any) to intellectual property. The law seems settled that ownership of tissue can be transferred by donation and that the gift is irrevocable. Less clear is the nature of redress should tissue be used beyond the agreement upon which transfer was based.

There is an ongoing international debate about patenting DNA sequence information. The general trend is that the patent offices and the courts are narrowing the dimensions of patentability, but in the United States patents based on the correlation of sequence information with other data are still being issued. Two major lawsuits have challenged patentability in this area.

The burgeoning biotech industry is heavily focused on genetic information. A significant percentage of newly approved drugs are for rare genetic (orphan) diseases. The Orphan Drug Act has greatly stimulated research in this sector, but it has also raised questions as to whether it is anticompetitive and actually may be chilling research.

FURTHER READING

1. American College of Medical Genetics and Genomics Special Issue: Managing Incidental Findings and Research Results in Genomic Research Involving Biobanks and Archived Data Sets. *Genet. Med.* **2012,** *14* (4), 355–496.
2. Andrews, L. B.; Mehlman, M. J.; Rothstein, M. A. *Genetics, Ethics, Law, and Policy*, 3rd ed.; West Group, 2010.
3. Clayton, E. W.; Steinberg, K. K.; Khoury, M. J., et al. Informed Consent for Genetic Research on Stored Tissue Samples: Consensus Statement. *JAMA* **1995,** *274,* 1786–1792.
4. Cook-Deegan, R.; Heaney, C. Patents in Genomics and Human Genetics. *Annu. Rev. Genomics Hum. Genet.* **2010,** *11* (17), 1–43.
5. Greeley, H. T. The Uneasy Ethical and Legal Underpinnings of Large-Scale Genomic Biobanks. *Annu. Rev. Genomics Hum. Genet.* **2007,** *8,* 343–362.
6. Knoppers, B. M.; Bordet, S.; Isasi, R. M. Pre-Implantation Genetic Diagnosis: An Overview of Socio-Ethical and Legal Considerations. *Annu. Rev. Genomics Hum. Genet.* **2006,** *7,* 201–222.
7. Rothstein, M. A. Putting the GINA in Context. *Genet. Med.* **2008,** *10,* 65–66.
8. Skeene, L. Ownership of Human Tissue and the Law. *Nat. Rev. Genet.* **2002,** *3,* 145–148.
9. World Health Organization. *Genetics, Genomics, and the Patenting of DNA: A Review of Potential Implications for Health in Developing Countries*; WHO, 2005.

Applications to Clinical Problems

Genetics of Female Infertility in Humans, 105
Genetics of Male Infertility, 108
Fetal Loss, 113
A Clinical Approach to the Dysmorphic Child, 117
Clinical Teratology, 120
Neurodevelopmental Disabilities: Global Developmental Delay, Intellectual Disability, and Autism, 124
Abnormal Body Size and Proportion, 131
Susceptibility and Response to Infection, 138
Transplantation Genetics, 141
The Genetics of Disorders Affecting the Premature Newborn, 144
Disorders of DNA Repair and Metabolism, 146

CHAPTER

32

Genetics of Female Infertility in Humans

Bala Bhagavath

Division of Reproductive Endocrinology and Infertility, University of Rochester, Rochester, NY, USA

Lawrence C Layman

Section of Reproductive Endocrinology, Infertility, & Genetics, Department of Obstetrics and Gynecology, Institute of Molecular Medicine and Genetics, Neuroscience Program, Medical College of Georgia, Georgia Health Sciences University, Augusta, GA, USA

Normal pubertal development and fertility depend on the integrated action of the hypothalamic–pituitary–gonadal–outflow tract. Infertility affects 10–15% of all the couples and may be due to a male factor, a female factor, or a combination of both. Encountered patients with reproductive dysfunction are much more commonly eugonadal than hypogonadal. Hypogonadism is further classified as hypogonadotropic (low gonadotropins), which indicates a hypothalamic-pituitary defect, or hypergonadotropic (high gonadotropins) in the presence of a gonadal etiology. The genes involved in these conditions have been somewhat arbitrarily categorized as causing hypothalamic, pituitary, gonadal, and outflow tract abnormalities. Of note, most mutations have been identified in the more severe, but less common hypogonadal individuals rather than the much more common eugonadal patients.

32.1 HYPOGONADOTROPIC HYPOGONADISM

32.1.1 Hypothalamic Causes of Hypogonadism

Mutations in at least 16 genes have been implicated in impaired hypothalamic function, 15 of which cause idiopathic hypogonadotropic hypogonadism (IHH)/Kallmann syndrome (KS) in females. The inheritance of IHH/KS is not completely understood, but current evidence suggests autosomal dominant (*FGFR1, FGF8, CHD7, WDR11*), autosomal recessive (*KISS1R, LEP, LEPR, PCSK1, PROKR2, PROK2, NELF, TAC3, TACR3, NELF, GNRH1*), or X-linked recessive (*NR0B1*) forms. Mutations in the first identified gene (*KAL1*) have been found only in males. Digenic mutations have been reported in 3–10% of patients. Many of the IHH/KS genes include ligand/receptor pairs—*FGF8/FGFR1, LEP/LEPR, PROK2/PROKR2, TAC3/TACR3*, and *GNRH1/GNRHR*. Mutations in some genes have been associated with additional nonreproductive phenotypic features such as synkinesia, dental agenesis, and midfacial defects, and some features of CHARGE syndrome.

32.1.2 Pituitary Causes of Hypogonadism

Mutations in at least nine genes result in impaired pituitary action. Three (*GNRHR, LHB*, and *FSHB*) affect only gonadotropin function, and they are inherited in an autosomal recessive fashion, while others affect other pituitary functions. Combined pituitary hormone deficiency (CPHD) is defined as the deficiency of growth hormone plus at least one additional pituitary hormone—adrenocorticotropic hormone, thyroid stimulating hormone, prolactin, follicle stimulating hormone, or luteinizing hormone. At least six different genes (*PROP1, HESX1, LHX3, LHX4, SOX2* and *SOX3)* are involved in the pathophysiology of CPHD.

32.2 HYPERGONADOTROPIC HYPOGONADISM

32.2.1 Chromosome Abnormalities

About 90% of females with a 45,X karyotype with or without mosaicism lack pubertal development and have hypergonadotropic hypogonadism. The short stature of Turner syndrome is likely explained at least in part by haploinsufficiency of *SHOX*. Approximately 10–15%

of phenotypic females with gonadal failure and a 46,XY karyotype (Swyer syndrome) have SRY mutations, but the etiology for most patients remains unknown.

32.2.2 Single Gene Disorders Associated with Ovarian Failure

Surprisingly, mutations in only several genes on the X chromosome, including *FMR1* (most commonly) and *BMP15* have been reported. Autosomal genes have been found more commonly. *FSHR* gene mutations have been identified most commonly in Finland and usually result in primary amenorrhea with or without breast development. Other rarer single gene defects include *NOBOX* and *FIGLA*. Mutations in some genes cause other phenotypic effects: *AIRE* (autoimmune polyglandular syndrome type 1 with hypoparathyroidism, adrenal failure, and gonadal failure); *FOXL2* (blepharophimosis-ptosis-epicanthus syndrome); *NR5A1* (gonadal failure with or without adrenal failure); *GALT* (galactosemia); and *EIF2B* (leukoencephalopathy with vanishing white matter). Several enzymes (*CYP17A1* and *CYP19A1*) in the steroid pathway have also been implicated in hypergonadotropic hypogonadism.

32.3 EUGONADAL CAUSES OF INFERTILITY

Polycystic ovary syndrome is the most common cause in this category but the molecular basis for this condition is poorly understood. Two disorders, complete androgen insensitivity syndrome (CAIS) and Müllerian aplasia (MA), represent the most common congenital obstructive forms of eugonadal infertility. CAIS is caused by *AR* mutations in most patients. In contrast, *WNT4* mutations only rarely cause MA suggesting the presence of yet to be identified genes. It is interesting that some patients with maturity onset diabetes of the young (MODY5) due to *HNF1B* intragenic deletions have MA.

32.4 CONCLUSION

Increasingly, additional genes are being identified to cause the most severe forms of infertility in females. These findings have improved our understanding of normal physiologic processes regulating reproduction; however, genes important in the most common types of female reproductive dysfunction remain to be discovered.

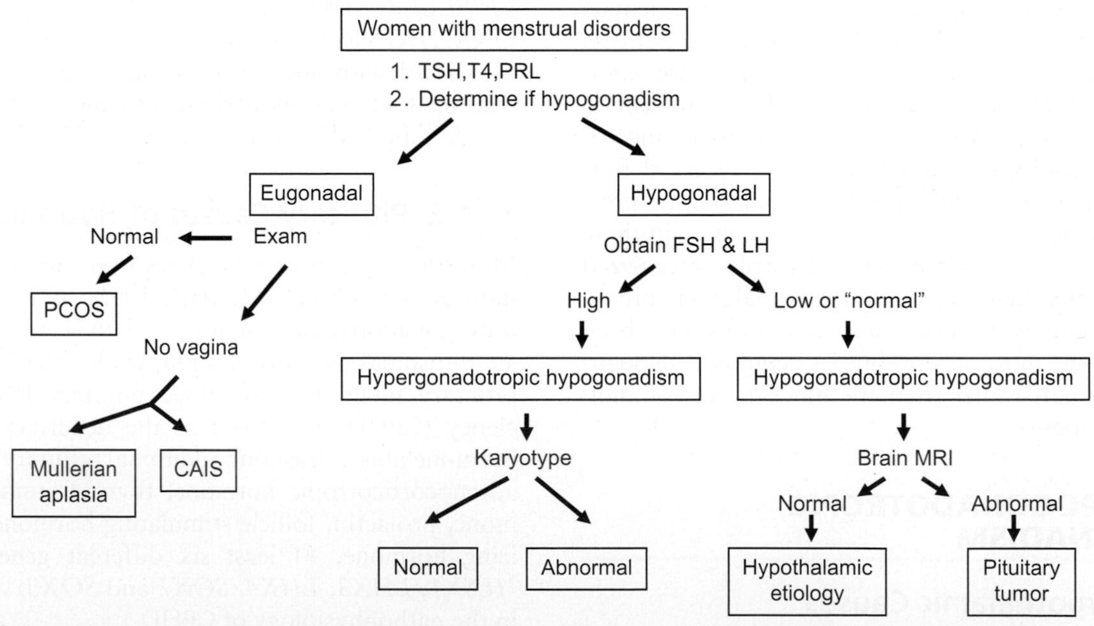

FIGURE 32-1 An overview of the diagnostic steps in females with reproductive dysfunction. The two most common causes of outflow obstruction causing primary amenorrhea are complete androgen insensitivity syndrome (CAIS) and Müllerian aplasia (Mayer–Rokitansky–Kuster–Hauser syndrome). PCOS = polycystic ovary syndrome; T4 = thyroxin; TSH = thyroid stimulating hormone; PRL = prolactin. CDP = constitutional delay of puberty; IHH = idiopathic hypogonadotropic hypogonadism.

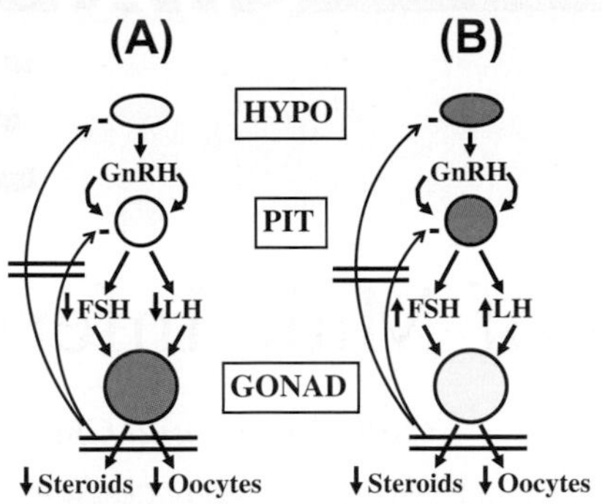

FIGURE 32-2 The hypothalamus–pituitary–gonadal axis is shown. (A) Hypogonadotropic hypogonadism is depicted. Low estradiol fails to trigger a rise in GnRH and FSH indicating that the defect is at the hypothalamus or pituitary (the hypothalamus and pituitary are shaded lighter to reflect hypofunction). (B) Hypergonadotropic hypogonadism is shown. Low estradiol elicits an appropriate rise in hypothalamic GnRH and pituitary gonadotropins indicating the defect is the lack of a gonadal response (the ovary is shaded lighter to represent hypofunction).

TABLE 32-1 Diagnostic Categories of Female Infertility

I. Hypogonadism
- A. Hypogonadotropic hypogonadism
- B. Hypergonadotropic hypogonadism
  - 1. Abnormal chromosomes
    - Females: 45,X (with or without mosaicism); 46,XY
    - Males: 47,XXY; 46,XY
  - 2. Normal chromosomes

II. Eugonadism
- 1. Ovulatory disorders—PCOS
- 2. Endometriosis
- 3. Disorders of the Genital Tract

TABLE 32-2 Prevalence of Diagnostic Categories of both Primary Amenorrhea and Secondary Amenorrhea in Females Are Shown

	Primary Amenorrhea (%)	Secondary Amenorrhea (%)
I. Hypogonadism		
A. Hypergonadotropic	43	11
1. Abnormal chromosomes	27	0.5
2. Normal chromosomes	16	10
B. Hypogonadotropic	31	42
1. Reversible	19	39
2. Irreversible	12	3
II. Eugonadism	26	46
1. Ovulation disorder	8	39
2. Genital tract obstruction	18	7

FURTHER READING

1. Aittomaki, K.; Lucena, J. L.D.; Pakarinen, P., et al. Mutation in the Follicle-Stimulating Hormone Receptor Gene Causes Hereditary Hypergonadotropic Hypogonadism. *Cell* **1995**, *82*, 959–968.
2. Biason-Lauber, A.; Konrad, D.; Navratil, F.; Schoenle, E. J. A WNT4 Mutation Associated with Müllerian-duct Regression and Virilization in a 46, XX Woman. *N. Engl. J. Med.* **2004**, *351*, 792–798.
3. Gottlieb, B.; Pinsky, L.; Beitel, L. K.; Trifiro, M. Androgen Insensitivity. *Am. J. Med. Genet.* **1999**, *89*, 210–217.
4. Kim, H. G.; Bhagavath, B.; Layman, L. C. Clinical Manifestations of Impaired GnRH Neuron Development and Function. *Neurosignals* **2008**, *16*, 165–182.
5. Kim, H. G.; Kurth, I.; Lan, F., et al. Mutations in CHD7, Encoding a Chromatin-Remodeling Protein, Cause Idiopathic Hypogonadotropic Hypogonadism and Kallmann Syndrome. *Am. J. Hum. Genet.* **2008**, *83*, 511–519.
6. Layman, L. C. Human Gene Mutations Causing Infertility. *J. Med. Genet.* **2002**, *39*, 153–161.
7. Layman, L. C. Genetic Diagnosis of Hypogonadotropic Hypogonadism and Kallmann Syndrome. In *Genetic Diagnosis of Endocrine Disorders*; Weiss, R. E.; Refetoff, S., Eds.; Academic Press: Amsterdam, 2010; pp 217–225.
8. Lindner, T. H.; Njolstad, P. R.; Horikawa, Y.; Bostad, L.; Bell, G. I.; Sovik, O. A Novel Syndrome of Diabetes Mellitus, Renal Dysfunction and Genital Malformation Associated with a Partial Deletion of the Pseudo-POU Domain of Hepatocyte Nuclear Factor-1Beta. *Hum. Mol. Genet.* **1999**, *8*, 2001–2008.
9. Lourenco, D.; Brauner, R.; Lin, L., et al. Mutations in NR5A1 Associated with Ovarian Insufficiency. *N. Engl. J. Med.* **2009**, *360*, 1200–1210.
10. McCabe, M. J.; Alatzoglou, K. S.; Dattani, M. T. Septo-Optic Dysplasia and Other Midline Defects: The Role of Transcription Factors: HESX1 and Beyond. *Best Pract. Res. Clin. Endocrinol. Metab.* **2011**, *25*, 115–124.
11. Nelson, L. M. Clinical Practice. Primary Ovarian Insufficiency. *N. Engl. J. Med.* **2009**, *360*, 606–614.
12. Sykiotis, G. P.; Plummer, L.; Hughes, V. A., et al. Oligogenic Basis of Isolated Gonadotropin-Releasing Hormone Deficiency. *Proc. Natl. Acad. Sci. U.S.A.* **2010**, *107*, 15140–15144.

CHAPTER 33

Genetics of Male Infertility

Csilla Krausz and Chiara Chianese

Unit of Sexual Medicine and Andrology, Department of Clinical Physiopathology, University of Florence, Florence, Italy

Ronald S Swerdloff and Christina Wang

Division of Endocrinology, Department of Medicine, Harbor-UCLA Medical Center and Los Angeles Biomedical Research Institute, Torrance, CA, USA

Approximately 7% of all men are confronted with fertility problems, the etiology of which can be ascribed to different factors including genetic anomalies. With the advancement and wide spread use of molecular genetic tools, a number of genetic factors can now be easily identified and some of them are currently part of the diagnostic work up of selected groups of subfertile patients. These factors can be transmitted as autosomal recessive, autosomal dominant, X-linked, and Y-linked traits. With the introduction and utilization of assisted reproductive techniques throughout the world for the treatment of spermatogenesis defects in the male partner, many previously infertile or subfertile men can now father biological children. However, when a genetic factor is the cause of male infertility, the potential risk of transmitting genetic defects to the offspring must be taken into consideration during the counseling of the subfertile couple.

33.1 CHROMOSOME ANOMALIES

Studies published to date lead to the general statement that the more severe the testicular phenotype, the higher is the frequency of chromosomal abnormalities. Patients with less than 10 million spermatozoa/ml ejaculate have a ten times higher incidence (4%) of being carriers of autosomal abnormalities compared to the general population. Robertsonian translocations, reciprocal translocations, paracentric inversions, and marker chromosomes are structural abnormalities most frequently found in oligozoospermic men. The incidence of abnormal karyotype in azoospermic men is 15% mainly related to the presence of an extra X chromosome leading to Klinefelter syndrome. In the group of numerical sex chromosome aberrations (sex chromosome aneuploidy), Klinefelter syndrome (47, XXY or mosaic) is the most frequent and large majority of these patients are azoospermic. However, testicular biopsy followed by ICSI permits a moderate testicular sperm recovery rate that may allow Klinefelter patients to generate their own genetic children. Impaired sperm production due to both numerical and structural chromosomal anomalies has been suggested as a structural effect of alterations occurring in chromosome synapsis during meiosis. Microdeletions of the Y chromosome AZF regions represent an important cause of male infertility and have significant implications in prognosis and outcome of the progeny. Complete AZF deletions have a clear cause–effect relationship with defective spermatogenesis leading to infertility and partial deletions (gr/gr deletions) constitute an important risk factor for spermatogenic impairment and subfertility. AZF deletion screening has also a prognostic value for testicular sperm retrieval in azoospermic men depending on the subtype of AZF deletions. Both karyotype analysis and Y chromosome microdeletions screening have become part of the diagnostic work-up of oligo/azoospermic men (karyotype should be performed starting from <10 million spermatozoa/ml ejaculate, whereas Y microdeletion screening in men with <5 million/ml ejaculate). These genetic tests are relevant also for genetic counseling and are important to predict health consequences for the future offspring.

33.2 GENE DEFECTS INVOLVED IN ENDOCRINE FORMS OF INFERTILITY

Mutations of the *AR* gene result in a variable phenotype from a phenotypic female in the complete androgen insensitivity syndrome to an underandrogenized male with ambiguous genitalia in the partial androgen insensitivity syndrome. However, they can also be associated with a normal male phenotype with defective spermatogenesis. The expansion of the polyglutamine repeats (CAG)n in the *AR* gene has been the object of several studies, but a clear role in male infertility has not been demonstrated.

Congenital hypogonadotropic hypogonadism (cHH) generally manifests as pubertal delay and low levels of gonadotropins with consequent azoospermia (Table 33-1). cHH in association with decreased sense of smell (anosmia or hypo-osmia) is known as Kallmann syndrome. Normosmic cHH and Kallmann syndrome both present as sporadic and inherited cases, with autosomal recessive, autosomal dominant, and X-linked modes of inheritance. Genes associated with cHH can be classified according to whether the proteins they encode are involved in the development and migration of GnRH neurons from the nasal placode to the hypothalamus, regulation of GnRH secretion, or GnRH action (see Table 33-2) for details). Only one single mutation has been described in the *LH*-beta gene, while for the *LH* receptor gene both inactivating and activating mutations have been described, respectively associated with impaired testosterone response and familial precocious male puberty. Mutations in the *FSH* and *FSHR* genes have been rarely reported in males and their phenotypic consequences are variable from azoospermia to oligozoospermia.

33.3 MONOGENIC DEFECTS IN POST-TESTICULAR AND PRIMARY TESTICULAR FORMS OF MALE INFERTILITY

Congenital bilateral absence of the vas deferens is a form of post-testicular azoospermia. The congential unilateral absence of vas deferens (CUAVD) is instead associated with oligozoospermia. Both conditions are transmitted in an autosomal recessive fashion and are caused by mutations in *CFTR*. Rarely, CUAVD is associated with renal agenesis and independent from the *CFTR* mutation. In primary testicular alterations, a large number of genetic variants, mainly in autosomes, have been proposed as risk factors for male infertility. In many cases, only sporadic data are available and when more studies are published on the same polymorphism results are often contradictory. Recurrent causative mutations with cause–effect relationship have not been reported in primary testicular failure, therefore in clinical practice only the *CFTR* mutation is assessed in CUAVD patients.

33.4 SYNDROMIC MONOGENIC DEFECTS

Bardet–Biedl syndrome (MIM *209900) is inherited as an autosomal recessive disease and is related to male infertility since hypogenitalism is one of the typical symptoms. Fourteen genes have so far been implicated at different chromosomes. *Prader–Willi syndrome* (MIM *176270) is a neurobehavioral disease that affects both males and females. Hypogonadism is a symptom and presents in the form of cryptorchidism and scrotal hypoplasia. *Primary Ciliary Dyskinesia* (MIM *242650) is also known as immotile cilia syndrome and male patients are also infertile because of the immotility or poor mobility of spermatozoa as a consequence of structural and functional abnormalities in the central part of the sperm flagella. *Noonan syndrome* (MIM *163950) is an autosomal dominant disorder and about half of male patients suffer from azoospermia or oligozoospermia resulting from bilateral cryptorchidism. *Myotonic dystrophy* (MIM *160900) is an autosomal dominant disease caused by a trinucleotide (CTG) repeat expansion in the *DMPK* gene and the majority of male patients also present progressive testicular tubular atrophy.

33.5 CONCLUSION

Thousands of genes are involved in spermatogenesis but only a fraction has been analyzed. The first and only currently available genome wide association study in relationship with male infertility has identified a number of potential SNPs as risk factors; however, their role was not confirmed in a large follow-up study. This unsuccessful attempt with such an innovative approach implies that major advances can be expected only with the use of whole genome sequencing in large well-selected study populations. In the era of ICSI, the definition of genetic causes is crucial, because at least 50% of infertile men have an unknown etiology and are likely to be carriers of unknown genetic anomalies. In addition to the genetics of male subfertility, an innovative future direction is represented by studies on epigenetic modifications, the alteration of which presumably plays a role in male infertility.

TABLE 33-1 Classification of Genes and Phenotypes Associated with Congenital Hypogonadotropic Hypogonadism

Gene	Locus	OMIM	Product	Inheritance	Associated Phenotype
Development and Migration of GnRH Neurons					
KAL1	Xp22.3	308700	Anosmin-1	X-linked	Hyposmia/anosmia, unilateral renal agenesis, midline facial defects (cleft lip/palate), synkinesia, oculomotor abnormalities, short metacarpals, sensorineural hearing loss, cerebellar ataxia, gut malrotations
FGFR1 (*KAL-2*)	8q11.2-p11.1	136350	Fibroblast growth factor receptor- 1	AD, Digenic	From anosmia to normosmia, variable severity of hypogonadism, cleft lip/palate, synkinesia, dental agenesis
FGF8	10q24	600483	Fibroblast growth factor 8	AD,AR, Digenic	Hyposmia/anosmia Cleft lip/palate, synkinesia
PROK2	3p21.1	607002	Prokineticin-2	AD/AR	Sleep disorders, obesity
PROKR2	20p13	244200	Prokineticin-2 receptor	AD,AR, Digenic	From anosmia to normosmia
CHD7	8q12.1	608892	Chromodomain-helicase-DNA-binding protein 7	AD	Part of CHARGE syndrome (coloboma, heart disease, choanal atresia, retarded growth and development and/or central nervous system anomalies, genital anomalies and/or hypogonadism, ear anomalies and/or deafness), from anosmia to normosmia
NELF	9q34.3	608137	Nasal embryonic LHRH factor (for neurons and axonal outgrowth)	Digenic	Hyposmia/anosmia
WDR11	10q26	606417	Member of the WD repeat-containing protein family	AD?	None reported
GnRH Secretion					
KISS1[a]	1q32	603286	Kisspeptin		None reported
KISS1R	19p13.3	604161	KISS1 receptor	AR	Normosmia
LEP	7q31.2	164160	Leptin	AR	Severe Obesity
LEPR	1p31	601007	Leptin receptor	AR	
TAC3	12q13–q21	162330	Neurokinin B	AD.AR	Microphallus, cryptorchidism
TACR3	4q25	162332	Receptor for TAC3		
PCSK1	15q15–q21	162150	Protein convertase subtilisin/kexin-type 1	AR	Obesity, variably impaired processing of several pro-hormones including those of LH, FSH, proinsulin and pro-opiomelanocortin, hormone levels either low or immuno-logically measurable with low bioactivity
NROB1	Xp21.3–p21.2	300473	Dosage-sensitive sex reversal-adrenal hypoplasia congenita (DAX1)		Adrenal failure
GNRH1	8p21–p11.2	152760	Gonadotropin-releasing hormone	AR	None reported
GnRH Action					
GNRHR	4q21.2	138850	GnRH receptor	AR, Digenic	Microphallus, cryptorchidism

[a]*KISS1* gene mutations have been described only in association with central precocious puberty (CPP).

TABLE 33-2 Genes associated with cHH

	Gene	Locus	OMIM	Product	Inheritance	Associated phenotype
Development and Migration of GnRH neurons	KAL1	Xp22.3	308700	Anosmin-1	X-linked	Hyposmia/anosmia, unilateral renal agenesis, midline facial defects (cleft lip/palate), synkinesia, oculomotor abnormalities, short metacarpals, sensorineural hearing loss, cerebellar ataxia, gut malrotations
	FGFR1 (KAL-2)	8q11.2-p11.1	136350	Fibroblast growth factor receptor- 1	AD, Digenic	From anosmia to normosmia, variable severity of hypogonadism, cleft lip/palate, synkinesia, dental agenesis
	FGF8	10q24	600483	Fibroblast growth factor 8	AD,AR, Digenic	Hyposmia/anosmia Cleft lip/palate, synkinesia
	PROK2	3p21.1	607002	Prokineticin 2	AD/AR	Sleep disorders, obesity
	PROKR2	20p13	244200	Prokineticin-2 receptor	AD,AR, Digenic	From anosmia to normosmia
	CHD7	8q12.1	608892	Chromodomain helicase DNA-binding protein-7	AD	Part of CHARGE syndrome (coloboma, heart disease, choanal atresia, retarded growth and development and/or central nervous system anomalies, genital anomalies and/or hypogonadism, ear anomalies and/or deafness), from anosmia to normosmia
	NELF	9q34.3	608137	Nasal embryonic LHRH factor (for neurons and axonal outgrowth)	Digenic	Hyposmia/anosmia
	WDR11	10q26	606417	member of the WD repeat-containing protein family	AD ?	None reported
GnRH secretion	KISS1*	1q32	603286	Kisspeptin		None reported
	KISS1R	19p13.3	604161	KISS1 receptor	AR	Normosmia
	LEP	7q31.2	164160	Leptin	AR	Severe Obesity
	LEPR	1p31	601007	Leptin receptor	AR	
	TAC3	12q13-q21	162330	Neurokinin B	AD.AR	Microphallus, cryptorchidism
	TACR3	4q25	162332	Receptor for TAC3		
	PCSK1	15q15-q21	162150	Protein convertase subtilisin/kexin-type 1	AR	Obesity, variably impaired processing of several pro-hormones including those of LH, FSH, pro-insulin and pro-opiomelanocortin, hormone levels either low or immunologically measurable with low bioactivity
	NROB1	Xp21.3-p21.2	300473	Dosage-sensitive sex reversal-adrenal hypoplasia congenita (DAX1)		Adrenal failure
	GNRH1	8p21-p11.2	152760	Gonadotropin releasing hormone	AR	None reported
GnRH action	GNRHR	4q21.2	138850	GnRH receptor	AR, Digenic	Microphallus, cryptorchidism

*KISS1 gene mutations have been described only in association with central precocious puberty (CPP).

FURTHER READING

1. Aston, K. I.; Carrell, D. T. Genome-Wide Study of Single-Nucleotide Polymorphisms Associated with Azoospermia and Severe Oligozoospermia. *J. Androl.* **2009**, *30*, 711–725.
2. Aston, K. I.; Krausz, C.; Laface, I., et al. Evaluation of 172 Candidate Polymorphisms for Association with Oligozoospermia or Azoospermia in a Large Cohort of Men of European Descent. *Hum. Reprod.* **2010**, *25*, 1383–1397.
3. Bianco, S. D.; Kaiser, U. B. The Genetic and Molecular Basis of Idiopathic Hypogonadotropic Hypogonadism. *Nat. Rev. Endocrinol.* **2009**, *5*, 569–576.
4. Carrell, D. T.; Hammoud, S. S. The Human Sperm Epigenome and Its Potential Role in Embryonic Development. *Mol. Hum. Reprod.* **2010**, *16*, 37–47.
5. Davis-Dao, C. A.; Tuazon, E. D.; Sokol, R. Z., et al. Male Infertility and Variation in CAG Repeat Length in the Androgen Receptor Gene: A Meta-Analysis. *J. Clin. Endocrinol. Metab.* **2007**, 924319–924326.
6. Forti, G.; Corona, G.; Vignozzi, L., et al. Klinefelter's Syndrome: A Clinical and Therapeutical Update. *Sex Dev.* **2010**, *4*, 249–258.
7. Huhtaniemi, I.; Alevizaki, M. Mutations Along the Hypothalamic-Pituitary-Gonadal Axis Affecting Male Reproduction. *Reprod. Biomed. Online* **2007**, 15622–15632.
8. Krausz, C.; Chianese, C.; Giachini, C., et al. The Y Chromosome-Linked CNVs and Male Fertility. *J. Endocrinol. Invest.* **2011**.
9. Krausz, C. Male Infertility: Pathogenesis and Clinical Diagnosis. *Best Pract. Res. Clin. Endocrinol. Metab.* **2011**, *25*, 271–285.
10. Larriba, S.; Bonache, S.; Sarquella, J., et al. Molecular Evaluation of CFTR Sequence Variants in Male Infertility of Testicular Origin. *Int. J. Androl.* **2005**, *28*, 284–290.
11. Nuti, F.; Krausz, C. Gene Polymorphisms/Mutations Relevant to Abnormal Spermatogenesis. *Reprod. Biomed. Online* **2008**, *16*, 504–513.
12. Simoni, M.; Bakker, E.; Krausz, C. EAA/EMQN Best Practice Guidelines for Molecular Diagnosis of Y-Chromosomal Microdeletions. State of the Art 2004. *Int. J. Androl.* **2004**, *27*, 240–249.
13. Vincent, M. C.; Daudin, M.; De, M. P., et al. Cytogenetic Investigations of Infertile Men with Low Sperm Counts: A 25-Year Experience. *J. Androl.* **2002**, *23*, 18–22, Review.

CHAPTER

34

Fetal Loss

Rhona Schreck

Division of Molecular Pathology, Medical Genetics & Pathology and Laboratory Medicine, Cedars-Sinai Medical Center, Los Angeles, CA, USA

John Williams III

Division of Maternal-Fetal Medicine, Department of Obstetrics and Gynecology, Cedars-Sinai Medical Center, Los Angeles, CA, USA

About 15% of recognized pregnancies result in miscarriage, usually during the first 13 weeks of pregnancy, with the incidence of early pregnancy loss being strongly correlated with maternal age and a history of prior pregnancy loss. An additional 22% of conceptions are lost between implantation and clinical recognition of pregnancy (preclinical losses), bringing the total rate of early pregnancy loss to nearly 40% of conceptions. The incidence of these early losses is directly correlated with the age of the mother (as is the risk of aneuploidy), as the incidence of pregnancy loss rate increases threefold over the mean in women who are between 34 and 39 years of age, and sixfold in women over 40 years of age.

Early studies using chromosome analysis demonstrated an unbalanced karyotype in 60% of miscarriage samples. This should be considered a minimum estimate of the incidence of abnormal karyotypes in early pregnancy loss, as many of the "normal" results represent the growth of maternal tissue when the fetal cells were no longer viable and older cytogenetic techniques would have missed more subtle chromosome imbalances. More recent studies, by analysis of chorionic villi from miscarriage samples, have shown that as many as 80% of lost pregnancies harbor a chromosome abnormality. The most common abnormalities seen are trisomy (22%), triploidy (8%), and sex chromosome monosomy (18%).

The fact that many human conceptions demonstrate chromosome abnormalities is supported by a variety of methodologies used to study embryos generated by in vitro fertilization. The highest abnormality rates have been reported by fluorescence in situ hybridization (up to 83%), but these studies often include observations of monosomy and nullosomy, which may be technical artifacts. Array CGH and SNP-micro array studies have also reported high levels of karyotypic abnormalities in early embryos. However, these studies often show embryonic mosaicism, a post-fertilization event, which may not occur in naturally conceived embryos and which have been show to often spontaneously resolve.

While fetal chromosome abnormalities are the most common reason for pregnancy loss, many other factors have been proposed, particularly in couples experiencing recurrent pregnancy loss (three or more losses with the same partner). Many of the proposed causes of pregnancy loss have not been rigorously substantiated. Most of the nonchromosomal, confirmed factors contributing to pregnancy loss are relatively rare, and affect only a small minority of couples.

In about 5% of couples experiencing repeated pregnancy loss, one partner is found to have a balanced chromosome rearrangement that leads to the production of genetically unbalanced gametes and pregnancies. Hence, parental chromosome analysis is recommended in couples with repeated abortions. The yield of this testing will be higher in couples whose fetuses have been observed to have structural chromosome abnormalities, which occur in less than 2% of miscarriages. In couples where each is a heterozygote for a mutation in the same gene and the homozygous state is lethal (e.g. some X-linked disorders, α-thalassemia, and spinal muscular atrophy) can present with recurrent pregnancy loss. Various genetic mutations associated with a risk for venous or arterial thrombosis have been seen at higher frequencies in women with repeated pregnancy loss. Women with uncontrolled diabetes are also at increased risk for pregnancy loss, as are those with inadequately treated hypothyroidism. Uterine abnormalities, particularly the presence of an intrauterine septum, have been associated with a higher pregnancy loss rate, which can be reduced by surgical correction of the defect. Rarely, intrauterine infection has been show to be the cause of pregnancy loss, especially losses occurring later in pregnancy and involving infection of the fetus.

There are very rare examples of teratogens (specifically ionizing radiation and folic acid antagonists used as abortifacients) inducing pregnancy loss. Interestingly, just as advanced maternal age is a major risk for pregnancy loss, with the incidence of miscarriage increasing from 10% in women 20–24 years of age to 90% in women over 45 years of age, a history of prior pregnancy loss is associated with a lower success rate in subsequent pregnancies, most likely due to the presence of one of the factors already described.

In summary, pregnancy loss is most often caused by the presence of a chromosome abnormality in the fetus, which emphasizes the value of karyotype analysis of miscarriage samples. In individuals with the loss of chromosomally normal fetuses (where maternal cell contamination of the sample has been excluded), recurrent loss or young maternal age, other medical investigations, including genetic testing, endocrine testing, and anatomical evaluation of the mother may provide information relating to the etiology of pregnancy loss.

TABLE 34-1 Evaluation and Management of Recurrent Early Pregnancy Loss

Etiology	Prevalence	Diagnostic Studies	Treatment
Parental chromosome rearrangement	5–8%	Karyotype of both partners	Genetic counseling PGD? Donor gametes?
Uterine anatomy	15–20%	Hysterography Hysterosalpingography Saline infusion sonohysterography	Hysteroscopic metroplasty Hysteroscopic myomectomy Lysis of adhesions
Immunologic	15–20%	Lupus anticoagulant Anticardiolipin IgG/IgM B_2-glycoprotein-1 IgG/IgM Phosphatidylserine IgG/IgM	Low-dose aspirin Heparin/Enoxaparin
Thrombophilia	8–12%	Factor V Leiden mutation Prothrombin gene mutation Fasting homocysteine level Antithrombin III activity Protein C activity Protein S activity	Heparin/Enoxaparin Folic acid
Endocrinologic	8–12%	Mid-luteal phase endometrial biopsy Mid-luteal phase progesterone Thyroid stimulating hormone Prolactin level Fasting glucose and insulin	Progesterone Levothyroxine Cabergoline/bromocriptine Metformin/insulin
Microbiologic	8–10%	Endometrial biopsy Vaginal/cervical cultures	Appropriate antibiotics
Environmental	5%	Review exposure to alcohol, tobacco and caffeine Review exposure to environmental chemicals and toxins	Eliminate exposure

TABLE 34-2 Factors that Contribute to Vascular Problems in the Placenta

Autoantibodies	Anticardiolipin Lupus anticoagulant Antiphospholipid syndrome Systemic lupus erythematosus
Clotting factors	Factor V Leiden Factor XII Protein C Prothrombin
Metabolic problems	Hyperhomocyst(e)inemia Methylenetetrahydrofolate reductase deficiency
Other	45,X Karyotype

TABLE 34-3 Factors Definitively Associated with Early Pregnancy Loss

Unbalanced karyotype in fetus	>60%
Parental chromosome rearrangement	5%
Mendelian disorder in fetus	Rare—family dependent
Uterine abnormalities	~20–75%
Diabetes mellitus (poorly controlled)	Rel. Risk: 1.9–5.8

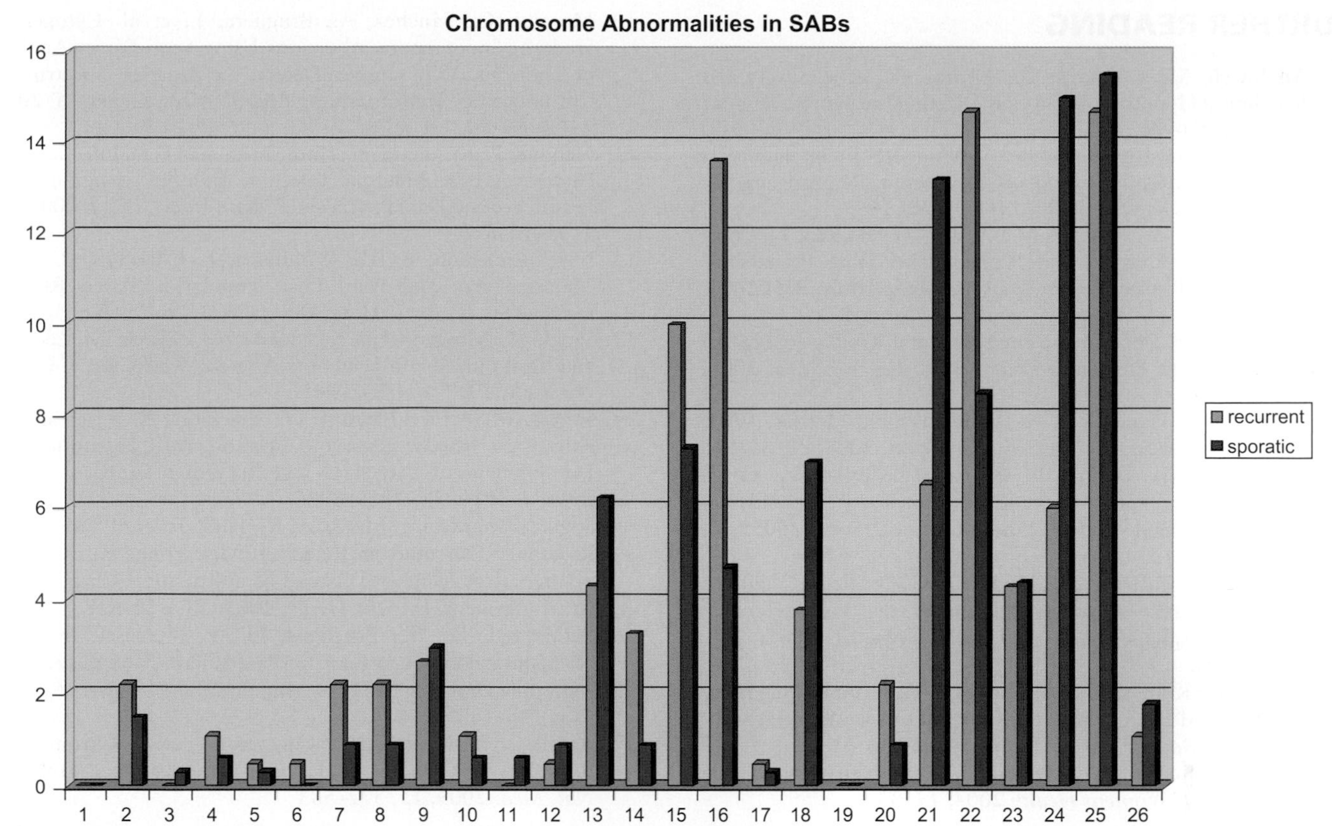

FIGURE 34-1 Relative frequency of chromosome abnormalities observed in cytogenetically abnormal POCs from women with a reported history of recurrent pregnancy loss compared to those with reported sporadic pregnancy loss. Mean maternal age in the recurrent group was 37.3 years and 36.2 years in the sporadic group. Number of chromosome involved presented across the X axis with 23 = double trisomy, 24 = monosomy X, 25 = triploidy, and 26 = tetraploidy. The abnormalities which are considered viable (trisomy 13, 18, 21 and monosomy X) are all more frequent in the group with sporadic losses, with trisomies 15, 16 and 22 being more prevalent among those with recurrent loss.

FURTHER READING

1. Abalovich, M.; Guitierrez, S.; Alcaraz, G., et al. Overt and Subclinical Hypothyroidism Complicating Pregnancy. *Thyroid* **2002**, *12*, 63–68.
2. Escudero, T.; Estop, A.; Fischer, J.; Munne, S. Preimplantation Genetic Diagnosis for Complex Chromosome Rearrangements. *Am. J. Med. Genet. A* **2008**, *146A*, 1662–1669.
3. Evers, I. M.; de Valk, H. W.; Visser, G. H. Risk of Complications of Pregnancy in Women with Type 1 Diabetes: Nationwide Prospective Study in the Netherlands. *BMJ* **2004**, *328*, 915.
4. Gardella, J. R.; Hill, J. A. Environmental Toxins Associated with Recurrent Pregnancy Loss. *Semin. Reprod. Med.* **2000**, *18*, 407–424.
5. Hunt, P. A.; Hassold, T. J. Human Female Meiosis: What Makes a Good Egg Go Bad? *Trends Genet.* **2008**, *24*, 86–93.
6. Jensen, D. M.; Damm, P.; Moelsted-Pedersen, L., et al. Outcomes in Type 1 Diabetic Pregnancies: A Nationwide, Population-Based Study. *Obstet. Gynecol. Surv.* **2005**, *60*, 279–280.
7. Key, T. C.; Giuffrida, R.; Moore, T. R. Predictive Value of Early Pregnancy Glycohemoglobin in the Insulin-Treated Diabetic Pregnancy. *Am. J. Obstet. Gynecol.* **1987**, *156*, 1096–1100.
8. Kline, J. K.; Kinney, A. M.; Levin, B., et al. Trisomic Pregnancy and Elevated FSH: Implications for the Oocyte Pool Hypothesis. *Hum. Reprod.* **2011**, *26*, 1537–1550.
9. Lin, P. C. Reproductive Outcomes in Women with Uterine Anomalies. *J. Womens Health* **2004**, *13*, 33.
10. Morales, C.; Sánchez, A.; Bruguera, J., et al. Cytogenetic Study of Spontaneous Abortions Using Semi-Direct Analysis of Chorionic Villi Samples Detects the Broadest Spectrum of Chromosome Abnormalities. *Am. J. Med. Genet. A* **2008**, *146A*, 66–70.
11. Nunoue, T.; Kusuhara, K.; Hara, T. Human Fetal Infection with Parvovirus B19: Maternal Infection Time in Gestation, Viral Persistence and Fetal Prognosis. *Pediatr. Infect. Dis. J.* **2002**, *21*, 1133–1136.
12. Nybo Anderson, A. M.; Wohlfahrt, J.; Christens, P., et al. Maternal Age and Fetal Loss: Population Based Register Linkage Study. *Br. J. Med.* **2000**, *320*, 1708–1712.
13. Sanghi, A.; Morgan-Capner, P.; Hesketh, L.; Elstein, M. Zoonotic and Viral Infection in Fetal Loss After 12 Weeks. *Br. J. Obstet. Gynaecol.* **1997**, *104*, 942–945.
14. Santos, M. A.; Teklenburg, G.; Macklon, N. S., et al. The Fate of the Mosaic Embryo: Chromosomal Constitution and Development of Day 4, 5 and 8 Human Embryos. *Hum. Reprod.* **2010**, *25*, 1916–1926.
15. Sugiura-Ogasawara, M.; Aoki, K.; Fujii, T., et al. Subsequent Pregnancy Outcomes in Recurrent Miscarriage Patients with a Paternal or Maternal Carrier of a Structural Chromosome Rearrangement. *J. Hum. Genet.* **2008**, *53*, 622–628.
16. Sutherland, H. W.; Pritchard, C. W. Increased Incidence of Spontaneous Abortion in Pregnancies Complicated by Maternal Diabetes Mellitus. *Am. J. Obstet. Gynecol.* **1986**, *155*, 145–148.
17. Vanneste, E.; Voet, T.; Le Caignec, C., et al. Chromosome Instability Is Common in Human Cleavage-Stage Embryos. *Nat. Med.* **2009**, *15*, 577–583.

CHAPTER

35

A Clinical Approach to the Dysmorphic Child

Kenneth L Jones and Marilyn C Jones

Department of Pediatrics, University of California, San Diego, CA, USA

The purpose of this chapter is to present a clinical approach to the child with structural defects. The approach is predicated on the concept that the nature of the structural defects presents clues to the time of onset, mechanism of injury, and probable etiology of the problem, all of which determine the direction of the evaluation. It presumes that the dysmorphic child represents an experiment in human development, which, if interpreted correctly, can provide answers regarding the etiology of various structural defects, as well as permit insights into mechanisms of normal and abnormal morphogenesis. The precise cause of many malformations and malformation syndromes is not known. However, careful clinical evaluation in combination with an expanded range of cytogenetic and molecular testing has allowed the elucidation of the mechanism underlying a growing list of clinical disorders. The separation between genetic and environmental factors as well as cytogenetic and single-gene abnormalities is somewhat arbitrary. However, the approach is intended to be practical and to facilitate detection and prevention of human malformations.

A method of approach to children with structural defects is set forth diagrammatically Figure 35-1. Although the lists of exceptions are growing, a history and physical examination usually makes it possible to determine if the structural abnormality is of prenatal or postnatal onset. In this chapter, "prenatal onset" designates structural abnormalities that are present at birth, and "postnatal onset" designates structures that have previously developed and differentiated normally. Whereas the genetic alteration responsible for many of the disorders included under postnatal onset structural defects is present at the time of conception, the structural manifestations of that genetic alteration do not become obvious until postnatal life.

Once a given problem has been determined to be of prenatal onset, a distinction should be made between those that are single primary defects in development and those that are multiple malformation syndromes. Although the concepts are not totally analogous, separation of prenatal problems into these two categories permits some practical generalizations that can be extremely helpful in counseling about recurrence risk.

Conceptually, "single primary defect in development" is an anatomic or morphogenetic designation. In most cases, the defect involves only a single structure, and the child is otherwise completely normal. For single defects, the specific etiology is unknown, making definitive recurrence risk counseling difficult. However, most single primary defects are explained on the basis of multifactorial inheritance, which is thought to carry a recurrence risk for first-degree relatives of between 2% and 5%. Single primary defects can be subcategorized according to the nature of the error in morphogenesis that has produced the observed structural defect. Thus, single primary defects involve either malformation, deformation, disruption, or dysplasia of the developing structure. Elucidation of which mechanism has produced the birth defect is helpful in predicting prognosis as many deformations improve spontaneously or with postural intervention. Moreover, some dysplasias carry a risk for malignant transformation.

In contrast to the anatomic concept of the single primary defect in development, the designation "multiple malformation syndrome" indicates that the observed structural defects all have the same cause. The defects themselves usually include a number of anatomically unrelated errors in morphogenesis. Multiple malformation syndromes are caused by gross chromosomal abnormalities, chromosomal microdeletions and duplications, teratogens, and single-gene defects usually inherited in Mendelian patterns. Recurrence risk depends on an accurate diagnosis and ranges from zero in cases that represent fresh gene mutations or are caused by one-time teratogen exposures to 100% for the unusual case of a child with the Down syndrome in which one parent is a balanced 21/21 translocation carrier.

The category of multiple malformation syndromes includes patients in whom a primary developmental anomaly of two or more systems has occurred, all of which are thought to be due to a common etiology. Other than Down syndrome, which has an incidence of 1:660, and XXY syndrome (1:500 males), few of these disorders occur more frequently than 1 in 3000 live births.

Multiple malformation syndromes can be categorized on the basis of etiology. Multiple malformation syndromes for which a cause has not yet been identified will likely resolve into one of these categories when the etiology becomes known.

A clinical approach to children with structural defects has been set forth. It is based on the concept that the diagnostic evaluation should be directed by the nature of the structural defects. The ultimate goal of this approach is a specific overall diagnosis. The purpose of a diagnosis is to address the two questions facing all parents when a child with a birth defect is born: what does this condition mean to my child and what are my chances of having another child with a similar problem? An accurate overall diagnosis is critical to understanding the natural history of the disorder such that associated complications may be anticipated through screening, appropriate interventions offered, and precise reproductive counseling be provided. The birth of a child with congenital anomalies usually places severe stress on a family eliciting all of the stages of a grief response. Understanding the cause and implications of the condition is often helpful in mitigating this stress. When an overall diagnosis is lacking, a better understanding of the nature and onset of the problem is commonly possible. That in itself can often be helpful to parents and to all others dealing with children who have structural defects. Lastly, reaching an overall diagnosis may take time as some characteristic developmental and behavioral findings may not manifest until an individual is older. Re-evaluation is a critical component of this approach.

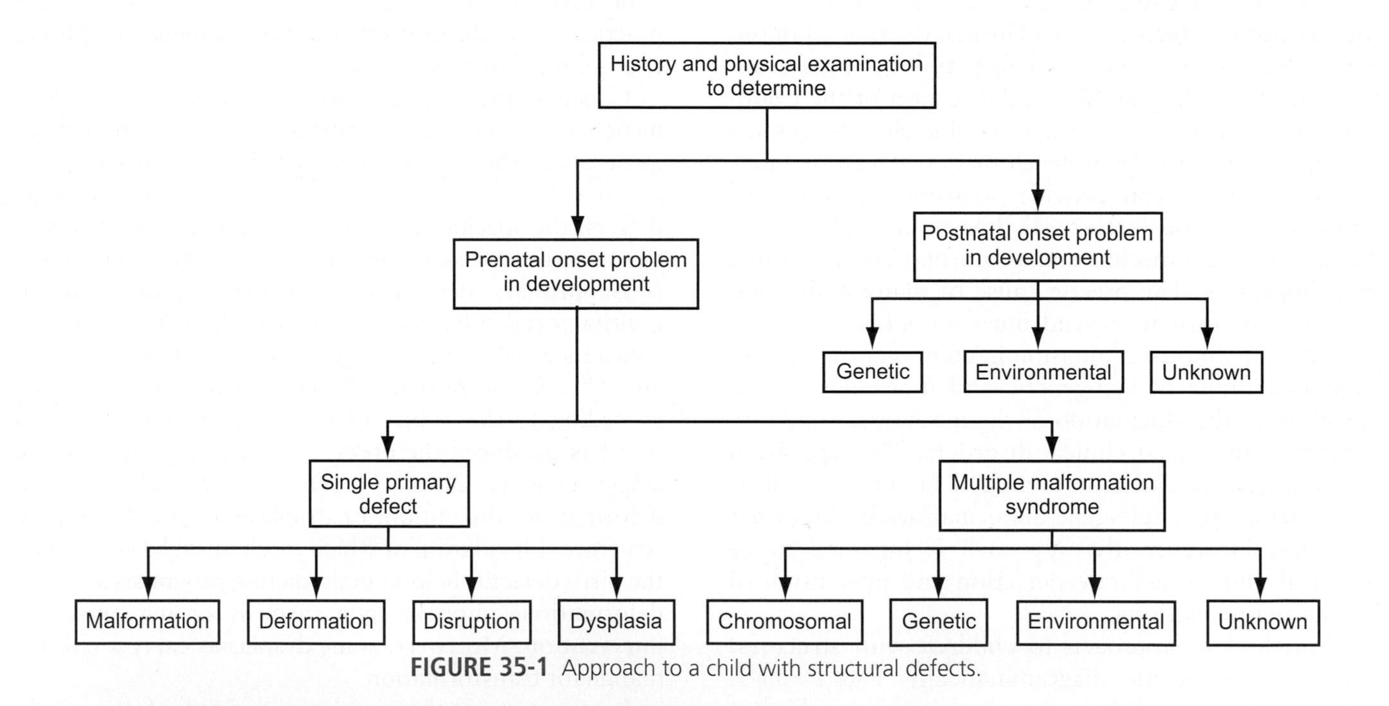

FIGURE 35-1 Approach to a child with structural defects.

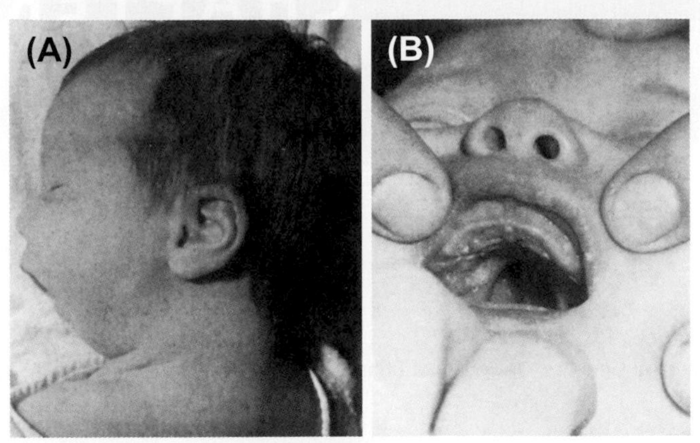

FIGURE 35-2 Infant with the Robin malformation sequence. A. Micrognathia. B. U-shaped palatal cleft.

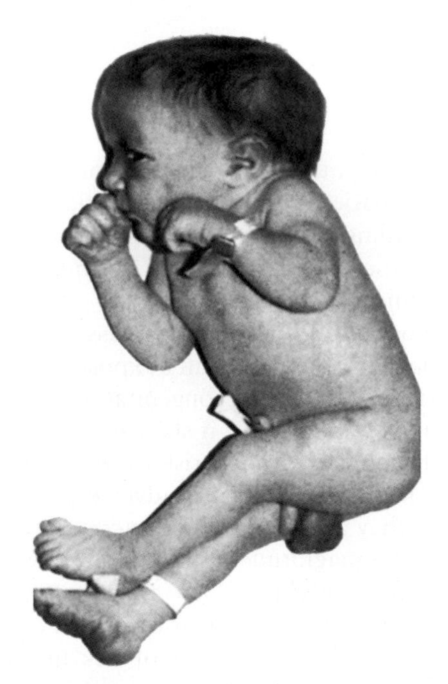

FIGURE 35-3 Newborn infant with breech deformation sequence. Note the deformed cranial shape and positional deformities at the hips and feet.

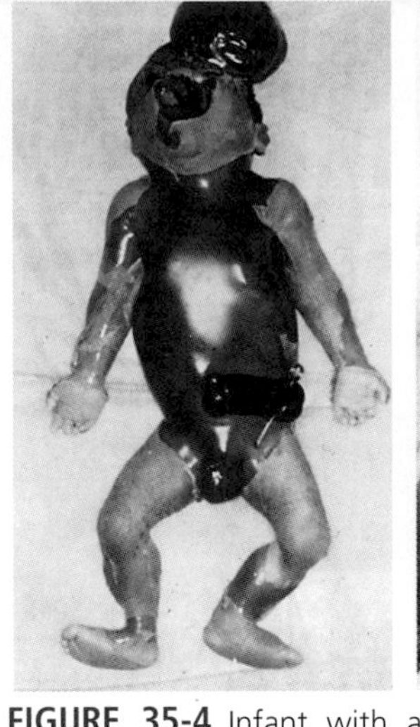

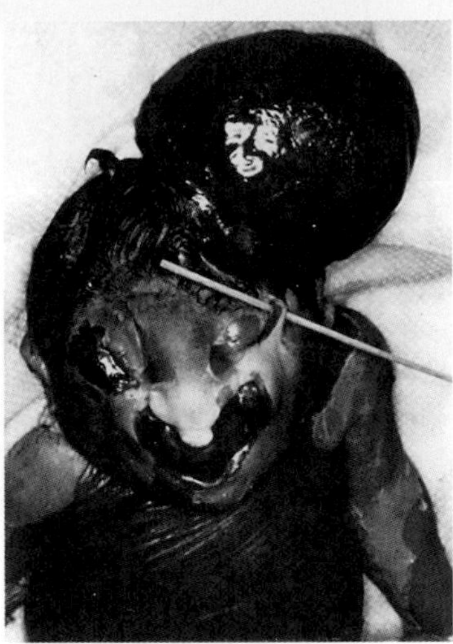

FIGURE 35-4 Infant with amniotic band disruption sequence. Note the asymmetrical encephalocele, severe disruption of facial development, and digital anomalies.

TABLE 35-1 Common Single Primary Defects in Development
Malformation: Cleft lip ± cleft palate, cleft palate, cardiac septal defects, defect in neural tube closure
Deformation: Congenital hip dislocation, talipes equinovarus
Dysplasia: pyloric stenosis

FURTHER READING

1. Graham, J. M. *Smith's Recognizable Patterns of Human Deformation*, 3rd ed.; Elsevier-Saunders: Philadelphia, 2007.
2. Hennekam, R. C. M.; Krantz, I. D.; Allanson, J. E. *Gorlin's Syndromes of the Head and Neck*, 5th ed.; Oxford University Press: New York, 2010.
3. Jones, K. L. *Smith's Recognizable Patterns of Human Malformation*, 6th ed.; Elsevier-Saunders: Philadelphia, 2006.
4. Reprotox. 2011. Available at http://www.reprotox.org/Default.aspx (requires subscription).
5. Shepard, T. H.; Lemire, R. J. *A Catalog of Teratogeneic Agents*, 12th ed.; Johns Hopkins University Press: Baltimore, 2007.
6. Spranger, J.; Benirschke, K.; Hall, J. G., et al. Errors of Morphogenesis: Concepts and Terms. *J. Pediatr.* **1982,** *100,* 160–165.
7. Winter, RM.; Baraister, M. 2011. London Dysmorphology Database. http://lmdatabases.com/ (requires subscription).

CHAPTER

36

Clinical Teratology

Jan M Friedman

Department of Medical Genetics, University of British Columbia, Vancouver, BC, Canada

James W Hanson

Center for Developmental Biology and Perinatal Medicine, Eunice Kennedy Shriver, National Institute of Child Health and Human Development, National Institutes of Health, Rockville, MD, USA

Exposures that can cause permanent structural or functional abnormalities in an exposed embryo or fetus are called "teratogenic." Teratogenicity is a property of an *exposure*, and agents that are teratogenic exhibit this potential only under certain conditions. One factor that influences teratogenicity is the chemical and/or physical nature of the agent itself. Some agents are inherently more risky than others. Another critical factor is the developmental stage of the embryo at the time of exposure. The most sensitive period to alter embryonic development is from about 2 weeks after conception to the 8th week after conception for most, but certainly not all, teratogenic exposures. Dose is also a critical feature of any teratogenic exposure—teratogenic effects occur only when the dose exceeds a certain threshold. Exposures to agents that are generally considered to be safe may have adverse effects on the embryo or fetus if given in doses high enough to produce maternal toxicity. The route of exposure is also of importance—there is unlikely to be a risk associated with any agent when the exposure occurs by a route that does not permit systemic absorption. A few agents, such as ionizing radiation, have direct access to the embryo, whereas others do not reach the embryo until after extensive metabolism by the mother. The teratogenicity of agents that are metabolized by the mother may depend on whether teratogenic metabolites reach the embryo or fetus in sufficient quantities to produce adverse effects. This, in turn, depends on a number of factors including route of entry, physical properties of the agent, maternal dose, amount of systemic absorption, and maternal metabolic capacity. The teratogenicity of an exposure may also be influenced by both the maternal and the fetal genotypes.

Providing teratogen risk counseling involves identifying and estimating the magnitude of risk, and this information must be communicated to the patient in a way that allows her to make informed decisions about the management of her pregnancy. The approach varies from patient to patient and depends on many factors, including the patient's cultural and social background, her understanding of the counselor's language, her level of general and scientific knowledge, and her commitment to the pregnancy.

The purpose of teratogenic risk assessment is to determine whether a pregnant woman's exposure increases her risk of having a child with congenital anomalies above the risk that she would have if she were unexposed. It is important for any teratogenic risk to be presented in the context of these other risks for adverse pregnancy outcomes when they are present. The uncertainty that often exists about the magnitude or nature of the risk complicates teratogen counseling—there are very few exposures for which the available information is sufficient to estimate the magnitude and severity of risk in a particular pregnancy with any confidence.

Clinical assessment of human teratogenic risk requires careful interpretation of data obtained from several kinds of studies. These include experiments done in laboratory animals, case reports, case series, pregnancy registries, randomized controlled trials, cohort studies, case–control studies, record linkage studies, and ecological studies. Meta-analysis provides a systematic approach to identifying, evaluating, synthesizing, and combining the results of epidemiological studies. An alternative way of interpreting multiple clinical teratology studies available on a particular exposure is through expert consensus. It is critically important that observed associations make biological sense.

Teratogenic exposures may be grouped into four major categories on the basis of the kinds of agents involved:

- *Infectious agents* such as rubella, cytomegalovirus, varicella-zoster virus, parvovirus B19, *Treponema pallidum* (syphilis), or *Toxplasma*

- *Physical agents* including ionizing radiation, hyperthermia, mechanical constraint, and early invasive prenatal diagnosis
- *Maternal metabolic factors*, which include inadequate folic acid intake, maternal obesity, maternal auto-antibodies of various kinds, maternal phenylketonuria, and, most importantly, maternal diabetes mellitus
- *Drug and chemical agents*

The adverse fetal effects of organic mercury compounds or PCBs demonstrate that some environmental pollutants can be potentially teratogenic, but there is no compelling evidence that human exposures to other environmental contaminants are teratogenic at doses that are usually encountered.

Among the host of "recreational" drugs that individuals may use or abuse, alcohol and tobacco stand out as the most important public health problems for the developing embryo or fetus. Prenatal exposure to ethanol can result in a wide spectrum of effects on the embryo and fetus, ranging from severely affected individuals with the fetal alcohol syndrome, through less severe but much more frequent alcohol-related birth defects and neurocognitive deficits. The frequency and severity of adverse outcomes appear to be dose related.

Maternal smoking during pregnancy interferes with fetal growth in a dose-related fashion. Women who smoke heavily during pregnancy also have higher than expected rates of spontaneous abortion, late fetal death, neonatal death, and prematurity. Adverse effects have also been observed in the children of women who abused cocaine or toluene (by inhalation) during pregnancy. There is no convincing evidence linking maternal caffeine use during pregnancy and congenital anomalies in humans.

Prescription drugs treatments are prescribed by physicians and which physicians, therefore, bear a particular responsibility. Treatment with some prescription medications in conventional therapeutic doses is potentially teratogenic in humans. Other drugs can be used safely during pregnancy, and in some instances such treatment is beneficial to both the mother and the fetus. Unfortunately, however, available data are insufficient to determine whether maternal treatment with most prescription drugs during pregnancy poses a substantial teratogenic risk.

The most dramatic epidemic of drug-induced birth defects ever recognized was caused by widespread use of thalidomide by pregnant women. Thalidomide embryopathy is a very unusual pattern of congenital anomalies that includes phocomelia and other limb reduction malformations, anomalies of the external ear, eyes, and heart. Other maternal treatments early in pregnancy that may produce characteristic patterns of congenital anomalies (i.e. syndromes) include isotretinoin, aminopterin or methotrexate, warfarin anticoagulants, methimazole or carbimazole, and anticonvulsants (especially, valproic acid, carbamazepine, phenytoin, trimethadione, or phenobarbital). Major malformations occur more often in infants whose mothers are treated with valproic acid during the first trimester of pregnancy than in infants of women who are treated with most other anticonvulsants. The risk is higher in the children of women who require treatment early in pregnancy with multiple anticonvulsants than in those who are adequately treated with a single medicine.

Maternal treatment with androgens during pregnancy can cause masculinization of the external genitalia of female fetuses. The daughters of women who were treated with diethylstilbestrol during pregnancy have a greatly increased risk of gross structural anomalies of the uterus and vagina and of developing clear cell adenocarcinoma of the vagina or cervix.

Maternal treatment with some other prescription medications later in pregnancy poses a serious risk to the fetus. For example, maternal treatment with tetracyclines during the second and third trimesters may produce staining of the infant's primary teeth, and maternal treatment with angiotensin converting enzyme inhibitors or angiotensin II receptor inhibitors later in pregnancy may cause oligohydramnios and fetal joint contractures or death, as well as neonatal renal failure or hypotension in surviving infants. Maternal treatment with indomethocin later in pregnancy has been associated with decreased fetal urinary output and oligohydramnios, as well as with premature closure of the ductus arteriosus in the fetus and consequent persistent pulmonary hypertension in the infant.

Maternal treatment with radioactive iodine after the first trimester can cause hypothyroidism and consequent intellectual disability in the children. Fetal goiter can be produced by maternal treatment during the second or third trimester of pregnancy with iodides or with amiodarone, which contains large amounts of iodine.

No paternal exposure to any chemical or physical agent has been convincingly shown to increase the risk of birth defects in subsequently conceived children. The available evidence also suggests that the risk of congenital anomalies is not measurably increased among the children of women who have had preconceptional exposures to any chemical or physical agent in comparison to unexposed women, although the available data are insufficient to characterize the reproductive toxicity of most exposures with any confidence.

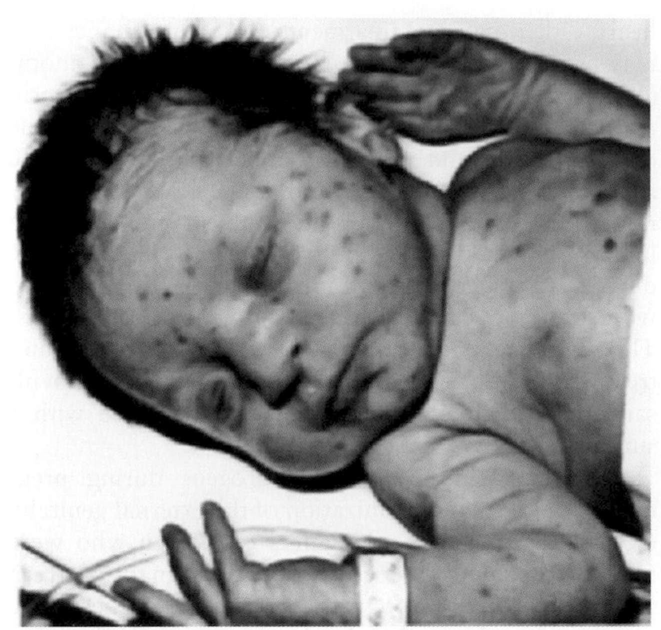

FIGURE 36-1 A child with rubella embryopathy. Note the "blueberry muffin" petechial rash. (*Photograph courtesy of Susan Reef, MD, National Congenital Rubella Syndrome Registry, Centers for Disease Control and Prevention.*)

TABLE 36-1 Mechanisms of Teratogenesis
Alterations of the cytoskeleton
Alterations of the integrity of intracellular organelles
Altered energy sources[a]
Altered membrane characteristics[a]
Altered nucleic acid integrity or function[a]
Chromosomal nondisjunction and breaks[a]
Disturbances of intracellular or intercellular signaling
Dysfunction of molecular chaperones
Effects of mechanical forces on embryogenesis
Effects of small regulatory RNAs
Effects on the distribution of molecules into subcellular compartments
Enzyme inhibitions[a]
Epigenetic control of gene expression
Genomic imbalance resulting from copy number changes
Lack of precursors and substrates needed for biosynthesis[a]
Mitotic interference[a]
Mutation[a]
Osmolar imbalance[a]
Perturbations of the extracellular matrix

[a]Included among the mechanisms presented by Wilson (1977).

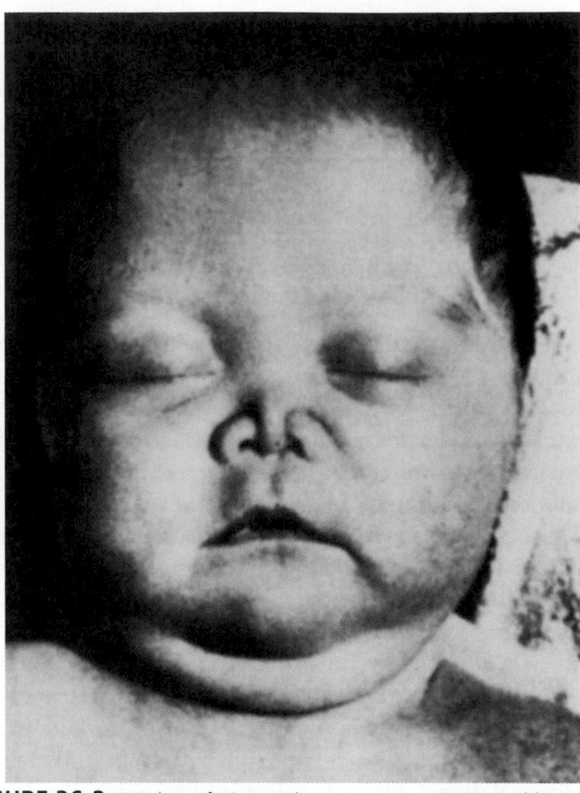

FIGURE 36-2 Fetal warfarin syndrome, note severe nasal hypoplasia. (*From Pauli et al. with permission.*)

TABLE 36-2 Characterization of Teratogenic Exposures
Agent
Nature of the chemical, physical, or infectious agent
Inherent developmental toxicity
Capacity to produce other kinds of toxicity in the mother
Dosage to embryo or fetus
Single, repeated, or chronic exposure
Duration of exposure
Maternal dose
Maternal route of exposure
Maternal absorption
Maternal metabolism and clearance
Placental transfer
Fetal metabolism and clearance
Time of exposure in pregnancy expressed in gestational weeks (or days)
Between conception and onset of embryogenesis
Embryogenesis
Fetal period
Other factors
Genetic susceptibility of mother
Genetic susceptibility of the fetus
Other concurrent exposures
Maternal illness or other condition associated with exposure
Availability of tests to quantify the magnitude of maternal exposure

TABLE 36-3 Characterization of Teratogenic Effects for Counseling

General effects
Alterations of morphogenesis
Alterations of CNS function
Other functional impairments
Death of the conceptus, embryo, or fetus
Prenatal-onset growth deficiency
Carcinogenesis
Specific effects
Recognizable syndrome
Other distinctive features
Magnitude of risk
Absolute
Relative
Prenatal diagnosis
Detailed ultrasound examination
Amniocentesis or other invasive method
Availability
Reliability
Utility

FURTHER READING

1. Briggs, G. G.; Freeman, R. K.; Yaffe, S. J. *Drugs in Pregnancy and Lactation*, 9th ed.; Lippincott Williams & Wilkins: Philadelphia, 2011.
2. Jones, K. L. *Smith's Recognizable Patterns of Human Malformations*, 6th ed.; Elsevier Saunders: Philadelphia, 2006.
3. Koren, G. *Medication Safety in Pregnancy and Breastfeeding*; McGraw-Hill: New York, 2007.
4. Remington, J. S.; Klein, J. O.; Wilson, C. B.; Nizet, V.; Maldonado, Y. A. *Infectious Diseases of the Fetus and Newborn Infant*, 7th ed.; Elsevier Saunders: Philadelphia, 2011.
5. Schaefer, C.; Peters, P.; Miller, R. K. *Drugs during Pregnancy and Lactation: Treatment Options and Risk Assessment*, 2nd ed.; Elsevier: Amsterdam, 2007.
6. Schardein, J. L.; Macina, O. T. *Human Developmental Toxicants: Aspects of Toxicology and Chemistry*; Taylor & Francis: Boca Raton, Florida, 2007.
7. Shepard, T. H. *Catalog of Teratogenic Agents*, 13th ed.; Johns Hopkins University Press: Baltimore, 2010.

CHAPTER 37

Neurodevelopmental Disabilities: Global Developmental Delay, Intellectual Disability, and Autism

John B Moeschler

Section of Medical Genetics, Department of Pediatrics, Dartmouth Medical School and Dartmouth–Hitchcock Medical Center, Lebanon, NH, USA

This chapter addresses the genetic heterogeneity in intellectual disabilities and autism spectrum disorders (ASD). The number of genes identified that cause intellectual disability suggests that an intellectual disability phenotype can emerge as the final common pathway of many different types of abnormal cellular processing and that no one overriding mechanism is likely to be the cause. There are over 2500 genes or phenotypes listed in OMIM with "mental retardation" as the search term, and it is estimated that one-third of entire genome is expressed in the human brain. Mutations in more than 450 genes have been identified to cause intellectual disabilities and related cognitive disorders. There are more than 200 X-linked genes, many cloned or mapped, that when mutated cause intellectual disability. There are many inferred mechanisms by which pathogenic mutations affect CNS function, i.e. cause intellectual disability. Intellectual disability can result from a wide range of protein abnormalities. The following categories of mechanism or dysfunction: enzymatic, mediators of signal transduction, binding proteins, transporter proteins, cell adhesion molecules, structural molecules, motor proteins, tRNAs, apoptosis regulators, chaperones, and enzyme regulators.

The clinical application of the chromosomal microarray has replaced the standard karyotype as the first-line clinical test for ID. The diagnostic rates among study populations with intellectual disability range from 5 to 20%, about twice the rate of conventional karyotype. Most of these cases are micro-deletion/duplications with many recurrent patients having similar recognizable phenotypes. However, many of the microdeletions/duplications are rare or unique. Consequently, there are several efforts to catalog patient phenotypes associated with copy number abnormalities.

Autosomal recessive forms of intellectual disability are not well studied; much more is understood about X-linked genes causing disability. Only three different genes for autosomal recessive intellectual disability have been identified thus far including forms that are associated with microcephaly.

Approximately 200 X-linked genes cause intellectual disability; about half have been identified and another 20% are regionally mapped. All genes responsible for XLID families reported before the re-discovery of fragile X syndrome have now been identified, and, most, if not all, the high-prevalence genes, that is, those of a prevalence of ~1% or greater. Nevertheless, there remain many genes unmapped and mapped but of unclear function, and together they comprise a substantial proportion of X-linked intellectual disability. There are at least two curated online resources for X-linked genes causing intellectual disability.

At least 103 specific genes and 44 genomic loci have been reported in patients with ASD or autistic behavior. These genes and loci have all been causally implicated in intellectual disability, indicating that these two neurodevelopmental disorders share common genetic bases. The genetic architecture of ASDs is highly heterogeneous. About 20% of individuals with an ASD have an identified genetic etiology. ASDs can be due to mutations of single genes involved in autosomal dominant, autosomal recessive, and X-linked disorders. The most common single gene mutation in ASDs is fragile X syndrome (*FMR1*), present in about 2% of cases. Other monogenic disorders described in ASDs include tuberous sclerosis (*TSC1*, *TSC2*), neurofibromatosis

(*NF1*), Angelman syndrome (*UBE3A*), Rett syndrome (*MECP2*), and *PTEN* mutations in patients with macrocephaly and autism. Mutations have been identified in synaptic genes, including *NLGN3*, *NLGN4X*, *SHANK3*, and *SHANK2*. Chromosome microarray studies have revealed submicroscopic deletions and duplications, called copy number variation, affecting many loci and including de novo events in 5–10% of ASD cases.

The inherited metabolic conditions that cause intellectual disability typically are associated with other neurological or systemic signs and symptoms. In several series of patients presenting with intellectual disabilities, the incidence of diagnosed metabolic disorders ranges from 1% or less to 5%. While this incidence is lower than genomic or single gene conditions, the potential for response to treatment is relatively better. In one systematic review of the literature, 81 inborn errors of metabolism that were responsive to treatment and associated with ID were identified; of these, 50 were diagnosed with routine metabolic screening tests.

CNS malformations are a common finding in those with ID. Abnormal findings on MRI are seen in approximately 30% of children with ID. The mean rate of abnormalities found on brain imaging is estimated to be 30%. Brain imaging is often useful in guiding the medical geneticist in the etiological investigations based on the specific CNS abnormalities noted on imaging.

TABLE 37-1 The Medical Genetics Evaluation for Global Developmental Delays or ID

Medical genetics evaluation process:
1. Clinical history
2. Family history
3. Physical examination (especially for minor anomalies)
4. Neurological examination
5. Specific confirmatory genetic tests for suspected syndromes
6. Microarray CGH
8. Fragile X-molecular genetic testing
9. Metabolic screening in all[a]
10. Targeted MRI brain imaging

[a]see Table 37-5.

Source: Adapted from Moeschler, J. B.; Shevell, M. *Pediatrics* 2006, *117* (6), 2304–23016; Moeschler, J. B. *Curr. Opin. Neurol.* **2008**, *21* (2), 117–122; Michelson, D. J.; Shevell, M. I.; Sherr, E. H., et al. *Neurology* **2011**, *77* (17), 1629–1635; Van Karnebeek C. S.; Stockler, I. S. Evidence-Based Approach to Identify Treatable Metabolic Diseases Causing Intellectual Disability. Paper presented at Annual Conference of the American College of Medical Genetics, 2011.

TABLE 37-2 Recurrent Interstitial CMA Deletions and Duplications in ID

Name	Size (Mb)[a]	LCR	MIM	Clinical Features
1q21.1 microdeletion	1.1	+	612474	Mild-to-moderate MR, MC, cardiac abnormalities, cataracts, clear incomplete penetrance
1q21.1 microduplication	1.1	+	612475	Autism or autistic behaviors, mild-to-moderate MR, microcephaly, mild FD
1q41q42 microdeletion	1.2	–	–	MR, seizures, various dysmorphisms, cleft palate, diaphragmatic hernia
2p15q16.1 microdeletion	3.9	–	–	MR, MC, receding forehead, ptosis, telecanthus, short palpebral fissures, downslanting palpebral fissures, broad/high nasal bridge, long/straight eyelashes, smooth and long philtrum, smooth upper vermillion border, everted lower lip, high narrow palate, hydronephrosis, optic nerve hypoplasia
3q29 microdeletion	1.6	+	609425	MR, mild FD, including high nasal bridge and short philtrum
3q29 microduplication	1.6	+	611936	Mild/moderate MR, MC, obesity
7q11.23 microduplication	1.5	+	609757	MR, speech and language delay, autism spectrum disorders, mild FD
9q22.3 microdeletion	6.5	–	–	MR, hyperactivity, overgrowth, trigonocephaly, macrocephaly, FD
12q14 microdeletion	3.4	–	–	Mild MR, failure to thrive, proportionate short stature and osteopoikilosis
14q11.2 microdeletion	0.4	–	–	MR, widely spaced eyes, short nose with flat nasal bridge, long philtrum, Cupid's bow of the upper lip, full lower lip, auricular anomalies
15q13.3 microdeletion	1.5	+	612001	MR, epilepsy, hypotonia, short stature, microcephaly and cardiac defects
15q24 microdeletion	1.7	+	–	MR, growth retardation, MC, digital abnormalities, genital abnormalities, hypospadias, loose connective tissue, high frontal hairline, broad medial eyebrows, downslanted palpebral fissures, long philtrum

TABLE 37-2 Recurrent Interstitial CMA Deletions and Duplications in ID—*Cont'd*

Name	Size (Mb)[a]	LCR	MIM	Clinical Features
16p11.2 microdeletion/duplication	0.6	+	611913	Association with MR, autism, schizophrenia
16p11.2p12.2 microdeletion	7.1	+	–	MR, flat facies, downslanting palpebral fissures, low-set and malformed ears, eye anomalies, orofacial clefting, heart defects, frequent ear infections, short stature, minor hand and foot anomalies, feeding difficulties, hypotonia
16p13.1 microduplication	1.6	+	–	Association with autism, significance uncertain
16p13.1 microdeletion	1.6	+	–	MR, MC, epilepsy, short stature, phenotypic variability
17p11.2 microduplication[b]	3.7	+	610883	MR, infantile hypotonia, failure to thrive, autistic features, sleep apnea, and structural cardiovascular anomalies
17q21.31 microdeletion	0.5	+	610443	MR, hypotonia, long hypotonic face with ptosis, large and low set ears, tubular or pear shaped nose with bulbous nasal tip, long columella with hypoplastic alae nasi, broad chin
19q13.11 microdeletion	0.7	–	–	MR, pre- and postnatal growth retardation, primary microcephaly, hypospadias, ectodermal dysplasia including scalp aplasia, dysplastic nails and dry skin
22q11.2 microduplication	3.7	+	608363	Highly variable. MR, FD, for example widely spaced eyes and downslanting palpebral fissures, velopharyngeal insufficiency, conotruncal heart disease
22q11.2 distal microdeletion	1.4–2.1	+	611867	MR, prematurity, prenatal/postnatal growth delay, mild skeletal abnormalities, arched eyebrows, deep set eyes, smooth philtrum, thin upper lip, hypoplastic alae nasi, small pointed chin
Xq28 microduplication	0.4–0.8	–	–	MR, severe hypotonia, progressive lower limb spasticity, absent or very limited speech

[a]Common region.
[b]Potocki–Lupski syndrome.
FD, facial dysmorphisms; LCR, low copy repeat; MC, microcephaly; MR, mental retardation.
From Vissers et al. (*7*).

TABLE 37-3 Dosage-Sensitive Genes Causing ID and Identified by Deletion and/or Duplication Strategies

Syndrome	Chromosome	Size	Gene(s) Involved	MIM	Reference
Cystinuria with mitochondrial disease	del(2)(p16)	179 kb	*SLC3A1, PPM1B, KIAA0436*	606407	Parvari et al.
Adrenal hyperplasia with hypermobility	del(6)(p21)	33 kb	*TNBX, CYP21A*	–	Koppens et al.
CHARGE syndrome	del(8)(q12)	2300 kb	*CHD7*	214800	Vissers et al.
Oto–facial–cervical syndrome	del(8)(q13.3)	316 kb	*EYA1*	166780	Rickard et al.
9q subtelomeric deletion syndrome	del(9)(q34)	Diverse	*EHMT1*	610253	Kleefstra et al.
Potocki–Shaffer syndrome	del(11)(p11.2)	2100 kb	*EXT2, ALX4*	601224	Potocki et al.
Infantile hyperinsulinism enteropathy and deafness	del(11)(p15p14)	122 kb	*USH1C, ABCC8, KCNJ11*	606528	Bitner-Glindzicz et al.
12q14 microdeletion syndrome	del(12)(q14)	3440 kb	*LEMD3, HMGA2, GRIP1*	–	Menten et al.
Peters Plus syndrome	del(13)q12.3q13.1)	1500 kb	*B3GALTL*	261540	Lesnik Oberstein et al.
Tuberous sclerosis polycystic kidney disease	del(16)(p13)	87 kb	*TSC2, PKD1*	173900	Brook-Carter et al.
Potocki–Lupski syndrome	dup(17) (p11.2p11.2)	3700 kb	*RAI1*	610883	Potcoki et al.
Alport leiomyomatosis	del(X)(q22.3)	133 kb	*COL4A5, COL4A6*	301050	Zhou et al.
MECP2 duplication syndrome	dup(X)(q28)	Variable	*MECP2*	–	Van Esch et al.

From Vissers et al. (*7*).

TABLE 37-4 Monogenic Causes of Intellectual Disability, by Mechanism

Gene	Locus	Disorder/Phenotype	Function of Encoded Protein; Subcellular Localization[a]
Genes Required for Neurogenesis			
Microcephalin	MCPH1/8p22-pter	Microcephaly vera	Cell cycle control and DNA repair
CDK5RAP2	MCPH3/q34	Microchephaly vera	Mitotic spindle function in embryonic neuroblasts
ASPM	MCPH5/1q31	Microchephaly vera	Formation of mitotic spindle during mitosis and meiosis
CENPJ	13q12.2	Microchephaly vera	Localization to the spindle poles of mitotic cells
Genes Required for Neuronal Migration			
LIS1	17p13.3	Miller Dieker syndrome: type 1 lissencephaly, pachygyria, subcortical band heterotopia (double cortex)	Interacts with dynein and plays a role in several function, including nuclear migration and differentiation
DCX/Dbcn	Xq22.3	Type 1 lissencephaly, pachygyria, subcortical band heterotopia (double cortex)	Microtubule-associated protein (MAP)
RELN	7q22	Lissencephaly with cerebellar hypoplasia	Extracellular matrix (ECM) molecule, reelin pathway
VLDLR	9p24	Lissencephaly with cerebellar hypoplasia	Low-density lipoprotein receptor, reelin pathway
POMT1	9q34	Walker–Warburg syndrome (also known as HARD syndrome[b])	Protein *o*-mannosyltransferase 1 (glycosylation of alpha-dystroglycan)
POMT2	14q24.3	Walker–Warburg syndrome	Protein *o*-mannosyltransferase 2 (glycosylation of alpha-dystroglycan)
POMGnT1	1p34	Muscle–eye–brain disease (MEB)	Protein *o*-mannose beta-1,2-n-acetylglucosaminyltransferase
Fukutin	9q31	Fukuyama congenital muscular dystrophy (FCMD) with type 2 lissencephaly	Homology with glycoprotein-modifying enzymes (no biochemical activity has been reported).
FLNA	Xq28	Bilateral periventricular nodular heterotopia (BPNH)	Filamin-1 (actin cross-linking phosphoprotein)
Genes Required for Cellular Processes Involved in Neuronal and Synaptic Functions			
FMR1	Xq27	FXS (Facial anomalies with macro-orchidism)	mRNA-binding protein, role in mRNA translation; potential regulation by RhoGTPase pathways; postsynaptic localization
FGD1	Xp11.2	Aarskog–Scott syndrome (Facial, digital and genital anomalies)	RhoGEF protein (GTP exchange factor), activate Rac1 and Cdc42
PAK3	Xq21.3	Nonsyndromic XLMR	P21-activated kinase 3; effector of Rac1 and Cdc42
ARHGEF6	Xq26	Nonsyndromic XLMR	RhoGEF protein, integrin-mediated activation of Rac1 and Cdc42
OPHN1	Xq12	MR with cerebella and vermis hypoplasia	RhoGAP protein (negative control of RhoGTPases; stimulates GTPase activity of RhoA, Rac1 and Cdc42; pre- and post synaptic localization
TM4SF2	Xq11	Nonsyndromic XLMR	Member of the tetraspanin family, integrin-mediated RhoGTPase pathway regulation
NLGN4	Xp22.3	Nonsyndromic XLMR, autism, Asperger syndrome	Member of the neuroligin family, role in synapse formation and activity; post synaptic localization
DLG3	Xq13.1	Nonsyndromic XLMR	Protein involved in postsynaptic density structures; postsynaptic localization
GDI_1	Xq28	Nonsyndromic XLMR	Regulation of Rab4 and Rab5 activity, and of synaptic vesicle recycling; pre- and post synaptic localization
IL1RAPL	Xp22.1	Nonsyndromic XLMR	Potential involvement in exocytosis and ion channel activity
Transcription Signaling Cascade, Remodeling and Transcription Factors			
NF1	17q11	Neurofibromatosis type 1 (NF1); MR is present in 50% of NF1 cases	RasGAP function, involved in Ras/ERK/MAPK signaling transcription cascade; postsynaptic protein
RSK2	Xp22.2	Coffin–Lowry syndrome (facial and skeletal anomalies)	Serine–threonine protein kinase, phosphorylates CREB, involved in Ras/ERK/MAPK signaling cascade, present in the postsynaptic compartment
CDKL5	Xp22.2	Rett-like syndrome with infantile spasms	Serine–threonine kinase (STK9), interacts with MECP2, potential implication in chromatin remodeling
CBP	16p13.3	Rubinstein–Taybi syndrome (mental retardation, broad thumbs and toes, dysmorphic face)	CREB (cAMP response element-binding protein 1) binding protein; chromatin-remodeling factor involved in Ras/ERK/MAPK signaling cascade
EP300	22q13.1	Rubinstein–Taybi syndrome	Transcriptional coactivator similar to CBP, with potent histone acetyl transferase: chromatin-remodeling factor

TABLE 37-4 Monogenic Causes of Intellectual Disability, by Mechanism—*Cont'd*

Gene	Locus	Disorder/Phenotype	Function of Encoded Protein; Subcellular Localization[a]
XNP	Xq13	Large spectrum of phenotypes including ATRX syndrome (microcephaly, facial dysmorphic face, skeletal anomalies and alpha-thalessemia)	Homology with DNA helicases of the SNF2/SWI2 family, chromatin-remodeling factor, regulation of gene expression
MECP2	Xq28	Rett syndrome (female-specific syndrome) and other phenotypes including nonsyndromic MR	Methy-CpG-binding protein 2; chromatin-remodeling factor, involved in a transcriptional silencer complex
DNMT3B	20q11.2	ICF syndrome: immune deficiency associated with centromeric instability, facial dysmorphy and MR	DNA methyltransferase 3B, involved in chromatin remodeling
ARX	Xp22.1	Large spectrum of MR phenotypes: XLAG (X-linked lissencephaly and abnormal genitalia); West syndrome, Partington syndrome; nonsyndromic MR	Transcription factor of the aristaless homeoprotein-related proteins family
JARID1C	Xp11.2	Spectrum of phenotypes: MR with microcephaly, short stature, epilepsy, facial anomalies and nonsyndromic MR	Transcription factor and chromatin remodeling
FMR2	Xq28	Nonsyndromic MR	Potential transcription factor
SOX3	Xq27	Isolated GH deficiency, short stature and MR	SRY-BOX 3: transcription factor
PHF8	Xp11.2	MR with cleft lip or palate	PHD zinc-finger protein, potential role in transcription
ZNF41	Xp11.2	Nonsyndromic MR	Potential transcription factor
GTF2I/ GTF2RD1	7q11.23	Williams syndrome	Transcription factors, potential regulator of c-Fos and immediate-early gene expression
PHF6	Xq26	Börjeson–Forssman–Lehmann syndrome (hypogonadism, obesity, facial anomalies, epilepsy)	Homeodomain-like transcription factor
Other Genes Involved in MR			
RPSS12	4q24	Nonsyndromic ARMR	Member of the trypsin-like serine protease family, enriched in the presynaptic compartment
CRBN	3p25	Nonsyndromic ARMR	ATP-dependent protease; regulation of mitochondrial energy metabolism
CC2D1A	19p13	Nonsyndromic ARMR	Unknown function, protein contains C2 and DM14 domains
FTSJ1	Xq11.2	Nonsyndromic XLMR	Role in tRNA modification and mRNA translation
PQBP1	Xq11.2	Large spectrum of MR phenotypes including nonsyndromic MR	Polyglutamine-binding protein, potentially involved in pre-mRNA splicing
FACL4	Xq22.3	Nonsyndromic XLMR	Fatty-acid synthase-CoA ligase 4; possible role in membrane synthesis and/or recycling
SLC6A8	Xq28	Creatine deficiency syndrome (MR with epilepsy and dysmorphic features) and nonsyndromic MR	Creatine transporter, role in homeostasis of creatine in the brain
OCRL1	Xq25	Lowe syndrome (MR, bilateral cataract and renal Fanconi syndrome)	Inositolpolyphosphate 5-phosphatase (central domain) and RHoGAP-like C-terminal domain
AGTR2	Xq24	Nonsyndromic XLMR	Angiotensin II receptor type 2, signaling pathway
SLC16A2	Xq13.2	Severe syndromic form MR with abnormal levels of thyroid hormones	Monocarbohydrate transporter, T3 transporter
SMS	Xp22.1	Snyder–Robinson syndrome (macrocephaly, scoliosis, dysmorphic features)	Sperimin synthase, CNS development/function (neuron excitability)
UBE3A	15q11	Angelman syndrome	Ubiquitin–protein ligase E3A; protein degradation (proteasome): CNS development/function (neuron differentiation)

[a]Subcellular localization is indicated mainly for protein shown to be present in the pre- and/or postsynaptic compartments.

[b]HARD syndrome includes hydrocephalus (H), agyria (A), retinal dysplasia (RD), with or without encephalocele, often associated with congenital muscular dystrophies.

The table does not represent an exhaustive list of genes involved in MR disorders. For additional genes, see the review by Inlow and Restifo, and online resources: http://xlmr.interfree.it/home.htm and http://www.ggc.org/xlmr_update.htm.

TABLE 37-5 X-Chromosome Genes Causing Both ID and ASDs

Gene	OMIM	Phenotype or syndrome
MID1	300502	Opitz syndrome
NLGN4X	300427	
AP1S2	300629	Brain calcifications
NHS	300457	Nance-Horan syndrome
CDKL5	300203	Rett-like infantile spasms
PTCHD1	300828	
ARX	300382	XLAG
DMD	300377	Duchenne/Becker dystrophy
OTC	300461	Ornithine transcarbamylase deficiency
CASK	300172	Microcephaly, pontine, cerebellar hypoplasia
IL1RAPL1	300206	MRX 21
ZNF674	300573	
SYN1	313440	Epilepsy
ZNF81	314998	
FTSJ1	300499	
PQBP1	300463	Renpenning syndrome
NDP	300658	Norrie disease
CACNA1F	300110	Severe congenital stationary night blindess
SMC1A	300040	Cornelia de Lange syndrome
PHF8	300560	Siderius–Hamel syndrome
JARID1C	314690	Microcephaly, spasticity, epilepsy
IQSEC2	300522	
FGD1	300546	Aarskog s.
OPHN1	300127	Cerebellar Hypoplasia
MED12	300188	FG/Opitz–Kaveggia syndrome
NLGN3	300336	
KIAA2022	300524	Progressive quadriplegia
ATRX	300032	ATRX syndrome
PCDH19	300460	Female-limited epileptic encephalopathy
DCX	300121	Lissencephaly
ACSL4	300157	
AGTR2	300034	
UPF3B	300298	
LAMP2	309060	Danon disease
GRIA3	305915	
OCRL	300535	Lowe syndrome
PHF6	300414	Borjeson–Forssman–Lehmann syndrome
ARHGEF6	300267	
SLC9A6	300231	Christianson syndrome
FMR1	309550	Fragile X syndrome
AFF2	300806	
SLC6A8	300036	Creatine Transporter def.
MECP2	300005	Rett syndrome
L1CAM	308840	MASA syndrome
RAB39B	300774	Epilepsy, Macrocephaly

If phenotype/syndrome not indicated, there are only "non-syndromic" families noted to date.
Source: Modified from Betancur, C. *Brain Res.* **2011**, *1380*, 42–77.

TABLE 37-6 Routine Metabolic Screening Tests in Patients with ID or ASDs

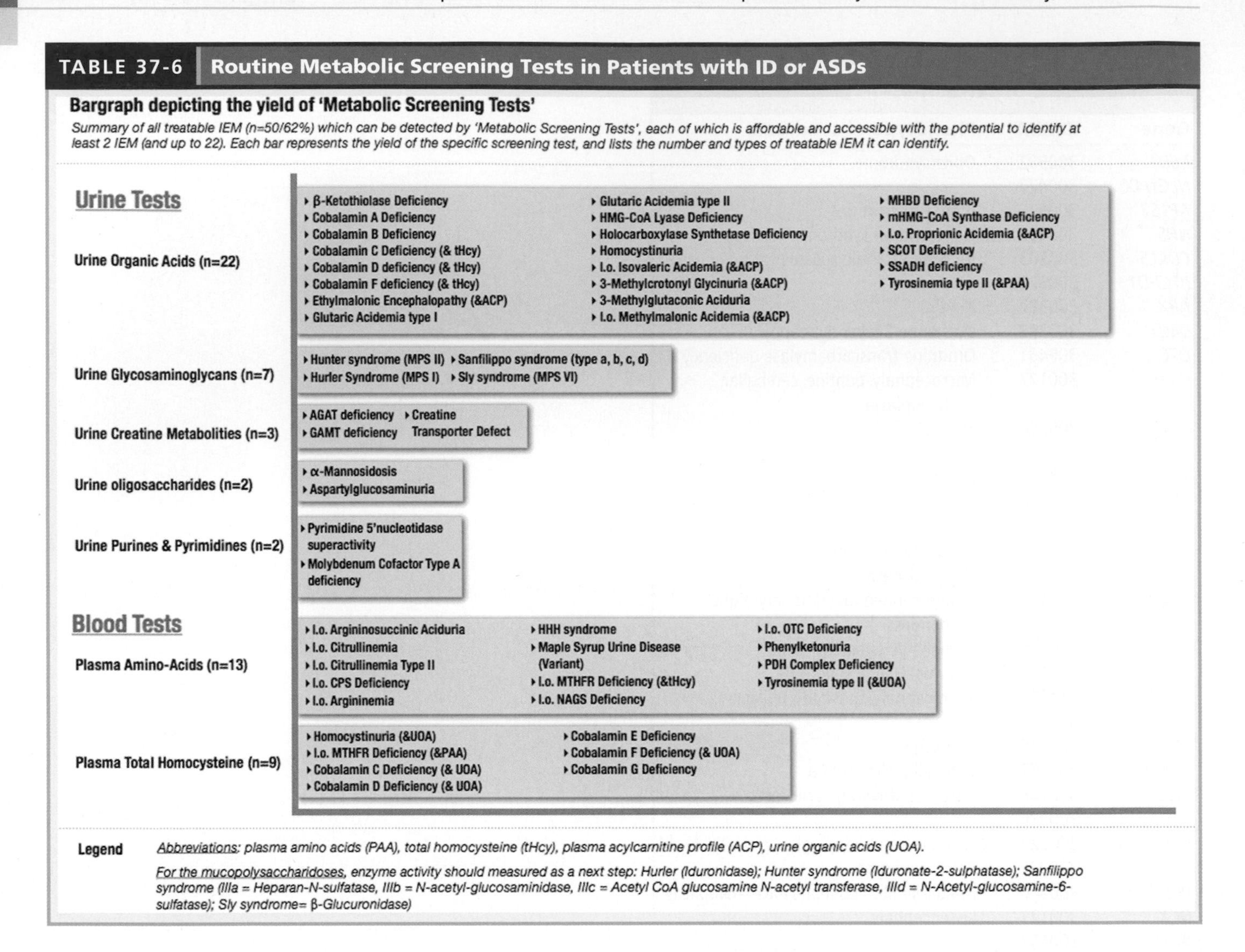

FURTHER READING

1. Michelson, D. J.; Shevell, M. I.; Sherr, E. H.; Moeschler, J. B.; Gropman, A. L.; Ashwal, S. Evidence Report: Genetic and Metabolic Testing on Children with Global Developmental Delay: Report of the Quality Standards Subcommittee of the American Academy of Neurology and the Practice Committee of the Child Neurology Society. *Neurology* **2011**, *77* (17), 1629–1635.
2. Moeschler, J. B. Medical Genetics Diagnostic Evaluation of the Child with Global Developmental Delay or Intellectual Disability. *Curr. Opin. Neurol.* **2008**, *21* (2), 117–122.
3. Schaefer, G. B.; Mendelsohn, N. J. Clinical Genetics Evaluation in Identifying the Etiology of Autism Spectrum Disorders. *Genet. Med.* **2008**, *10* (4), 301–305.
4. Schalock, R. L.; Luckasson, R. A.; Shogren, K. A.; Borthwick-Duffy, S.; Bradley, V.; Buntinx, W. H.; Coulter, D. L.; Craig, E. M.; Gomez, S. C.; Lachapelle, Y., et al. The Renaming of Mental Retardation: Understanding the Change to the Term Intellectual Disability. *Intellect. Dev. Disabil.* **2007**, *45* (2), 116–124.
5. Stevenson, R. E.; Schwartz, C. E. X-Linked Intellectual Disability: Unique Vulnerability of the Male Genome. *Dev. Disabil. Res. Rev.* **2009**, *15* (4), 361–368.
6. van Bokhoven, H. Genetic and Epigenetic Networks in Intellectual Disabilities. *Annu. Rev. Genet.* **2010**.
7. Vissers, L. E.; de Vries, B. B.; Veltman, J. A. Genomic Microarrays in Mental Retardation: From Copy Number Variation to Gene, from Research to Diagnosis. *J. Med. Genet.* **2010**, *47* (5), 289–297.

CHAPTER

38

Abnormal Body Size and Proportion

John M Graham

Division of Clinical Genetics and Dysmorphology, Cedars Sinai Medical Center, Los Angeles, CA, USA

Deepika D'Cunha Burkardt

Cedars Sinai Medical Center, Los Angeles, CA, USA

David L Rimoin

Medical Genetics Institute, Cedars Sinai Medical Center, Los Angeles, CA, USA

Normal stature varies widely among ethnic groups, and also varies within each ethnic group, approximating a normal distribution. Adult height is a classic polygenic trait, influenced by genetic variants in at least 180 loci (*1*). Short stature and tall stature are terms relative to a person's ethnic and familial background, as well as nutritional or psychological factors, or the presence of a chronic disease state. Before beginning a complex diagnostic evaluation, or contemplating growth-promoting or growth-limiting therapy, one must be able to differentiate between pathologic short or tall stature and normal variants at each end of the normal range utilizing standardized growth curves. One must also rule out constitutional delay or acceleration of maturation by correlating bone age and height. This normal variation in the timing of growth accounts for a large proportion of children referred for variations in growth. If possible, it is essential that a specific diagnosis be made because there are literally hundreds of causes of short stature that have differing prognoses, complications, and responses to treatments.

The first step in the clinical evaluation of short stature is to determine whether the body habitus is proportionate or disproportionate (Figure 38-1), best done with the use of anthropometric measurements such as sitting height, upper/lower segment ratio, and/or arm span. Children with disproportionate short stature usually have a skeletal dysplasia, while those with proportionate short stature may have a more generalized disorder (i.e. malnutrition, chronic disease, psychosocial dwarfism, endocrine disorder, genetic syndrome, or chromosomal or teratogenic disorder. Although there are exceptions to this rule as disproportionate, dwarfism may occur in cases of severe cretinism, while proportionate shortening may occur in persons with osteogenesis imperfecta.

Once a person with short stature is found to be proportionate, it is helpful to determine whether the growth deficiency was of prenatal or postnatal onset. Prenatal-onset growth deficiency usually implicates a fetal environmental insult or a generalized cellular genetic defect that may limit cellular mitosis. Late fetal insults are more likely to demonstrate catch-up growth after birth when compared with environmental factors, which have an early impact in fetal life (*2*). On the other hand, postnatal onset of proportionate growth deficiency usually implicates a postnatal environmental insult, such as infection, chronic disease, malnutrition, medication, or an endocrine, psychological, or malabsorption disorder.

Disproportionate body habitus is likely due to a skeletal dysplasia, a heterogeneous group of heritable disorders affecting skeletal connective tissues in which the predominant clinical feature is dwarfism. The osteochondrodysplasias have been divided into 37 groups of disorders and 3 groups of dysostoses based on clinical and radiographic features (Chapter 157) (*3*). Furthermore, each of these disorders is associated with a variety of skeletal and nonskeletal complications, so that an accurate diagnosis will help the clinician to develop a realistic treatment plan. These osteochondrodysplasias and dysostoses have also been classified using molecular pathogenetic and developmental approaches based on the structure and function of the causative genes and proteins; these classification systems are discussed in detail by Warman et al. (*3*).

The array of pathologic conditions that can result in overgrowth is more restricted than the numerous conditions that have a negative impact on growth. There are a number of growth factors that function at various times during development and impact the growth of specific tissues in either a generalized or localized way. Similarly, other factors function to suppress growth. Thus, overgrowth disorders result in either localized or generalized effects on growth. Most prenatal overgrowth disorders persist after birth, and, in some cases, mental deficiency, risk for tumor development, or both, are important associated features. The risk for tumor development usually involves embryonal neoplasms, perhaps related to overexpression of growth factors or lack of suppressing factors. Overgrowth conditions can be subdivided between normal variants, such as familial tall stature or familial rapid maturation, prenatal-onset growth excess, or postnatal-onset growth excess. However, primary growth excess results from intrinsic cellular hyperplasia, while secondary growth excess is due to humorally mediated factors outside the skeletal system. Among pathologic overgrowth disorders, most prenatal-onset growth excess is of the primary type, as is seen with Beckwith–Wiedemann syndrome or Sotos syndrome. Secondary growth excess of prenatal onset is usually not syndromic. Postnatal-onset growth excess is usually secondary, involving overproduction of estrogens or androgens (precocious puberty), overproduction of growth harmone (acromegaly), or overproduction of thyroid hormone. A few primary excess growth disorders manifest the majority of their overgrowth in the postnatal period (e.g. XYY syndrome, XXY syndrome, fragile X syndrome, and Marfan syndrome).

Advances in diagnostic technology have facilitated the diagnosis of previously unknown or unclassified disorders of growth. The use of more advanced molecular genetic and cytogenetic techniques has augmented the clinical diagnosis of known syndromes and has helped define clinically recognizable syndromes based on gene mutation analysis and copy number variants (CNVs) that were previously undetectable using standard cytogenetic techniques. For example, a submicroscopic deletion of 1q24q25 can result in primordial short stature with severe microcephaly possibly due to deletion of the *CENPL* gene, a centrosomal gene in the deleted region (*4*) and deletion of 12q13.3q15 can result in short stature possibly secondary to the deletion of *HMGA2* (*5*). Among genetic causes of syndromic overgrowth, chromosomal deletions and duplications such 898 as dup(4)(p16.3), dup(15)(q26qter), del(9)(q22.32q22.33), del(22)(q13), and del(5)(q35) have been identified, as well as a number of other rare CNVs, emphasizing the usefulness of this technique in unknown overgrowth disorders (*6*).

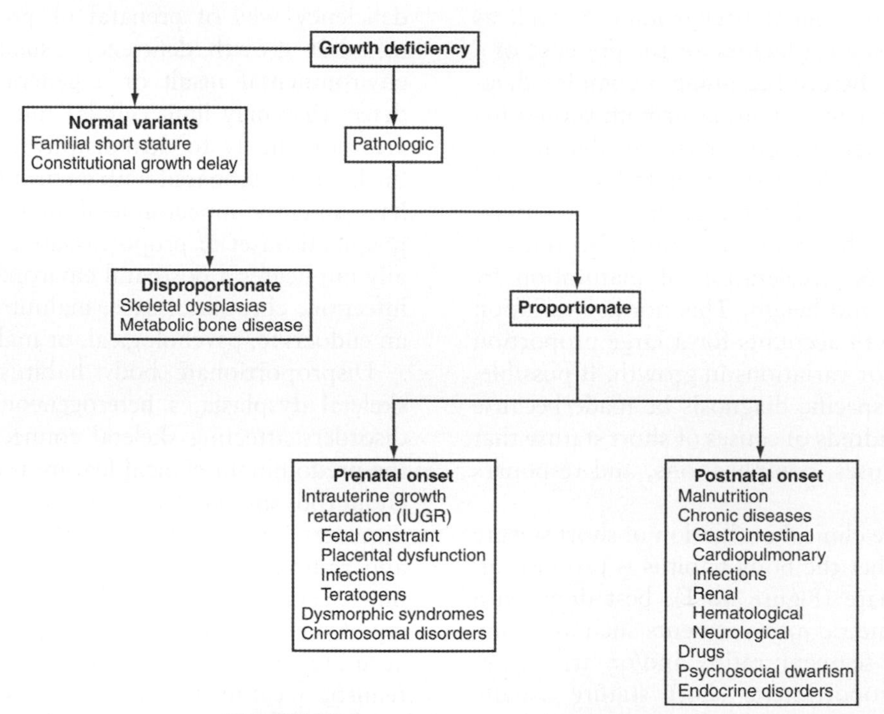

FIGURE 38-1 Classification of short stature

TABLE 38-1 Prenatal-Onset Growth Deficiency Disorders

Disorder	Key Clinical Findings
Teratogenic Disorders	
Fetal alcohol effects	Microcephaly, thin smooth philtrum, short palpebral fissures
Fetal hydantoin effects	Ocular hypertelorism, hypoplastic nails, short nose
Fetal trimethadione effects	Arched eyebrows, cupped helix, heart defects
Fetal warfarin effects	Hypoplastic nose, stippled epiphyses, short limbs
Fetal rubella effects	Deafness, cataracts, patent ductus arteriosus
Fetal varicella effects	Cicatricial skin defects, limb hypoplasia, mental deficiency, seizures
Maternal phenylketonuria effects	Microcephaly, cardiac defects, mental deficiency
Chromosome Abnormalities	
Trisomy 18	Clenched hands, short sternum, low arch dermatoglyphic pattern
Trisomy 13	Scalp defects, polydactyly, holoprosencephaly facies
Triploidy	3-4 Syndactyly, dysplastic calvaria, cystic placenta
4p syndrome	Ocular hypertelorism, hypospadias, preauricular pits
5p syndrome	Cat-like cry, microcephaly, down-slanted palpebral fissures
13q syndrome	Central nervous system defects, short thumbs, anal anomalies
18p syndrome	Ptosis, protruding ears, mental deficiency
18q syndrome	Prominent antihelix, midface hypoplasia, long palms
45,X (Turner) syndrome	Lymphedema of hands and feet, webbed neck, broad chest with wide-set nipples

Proportionate Syndromes

Disorder	Key Clinical Findings	Inheritance	Gene Testing
Brachmann–de Lange (122470)	Microcephaly, synophrys, thin downturned upper lip, small widely spaced teeth, micromelia, hirsutism, autistic tendencies, MR IQ (30–102), occasionally with cardiac septal defects, hearing loss, myopia, gastrointestinal dysfunction, hypoplastic genitalia/cryptorchidism (*1*)	Sporadic, autosomal dominant X-linked (fewer than 1% pts have an affected parent)	3 genes encoding components of the Cohesin Complex: *NIPBL* (Nipped B-like) at 5p13.1 (*2*) *SMC1A* (segregation of mitotic chromosomes 1) gene at Xp11.22 (*3*) *SMC3* at 10q25
Rubinstein–Taybi (180849)	Normal prenatal growth with slowing of growth (height, weight and head circumference) in first few months of life, broad thumbs and halluces, down-slanted palpebral fissures, "Beaked nose" with prominent nasal septum, high arched palate, grimacing smile and talon cusps, moderate to severe mental retardation (IQ from 25 to 79), childhood/adolescent obesity. Variable features: cataracts, colobomas, congenital renal or cardiac defects, cryptorchidism (*4*)	Sporadic, autosomal dominant	*CREBBP* (Creb-binding protein) 16p13.3 (*5*) *EP300* at 22q13 (*6*)
Russell–Silver (180860)	IUGR with persistent postnatal growth deficiency, triangular face, normal-sized cranium, incurved (clinodactyly) fifth finger, asymetric limb length (may lead to decreased growth on the affected side with hemi-hypertrophy), risk of motor and cognitive developmental delay with learning disabilities (*7*)	Sporadic heterogeneous	10%—maternal disomy for 7p11.2 11p15 region: *H19*—maternal (*8*) *IGF2* gene (*9*)
Dubowitz (223370)	IUGR with post natal growth deficiency, microcephaly with a high or sloping forehead and a broad and flat nasal bridge, flat or shallow supraorbital ridges, ptosis and blepharophimosis, with scarce lateral eyebrows, eczema, normal intelligence to mild to moderate MR, behavioral problems including hyperactivity (ADHD). Clinodactyly of the fifth finger, and cutaneous syndactyly of the 2nd and 3rd toes. Genitourinary abnormalities may include hypospadias and cryptorchidism (*10*)	Autosomal recessive	*NSUN2* at 5p15.31

Continued

TABLE 38-1 Prenatal-Onset Growth Deficiency Disorders—*Cont'd*

Disorder	Key Clinical Findings	Inheritance	Gene Testing
Bloom (210900)	Microcephaly, malar erythema/sun sensitivity/telangiectasia, malar hypoplasia, Gastrointestinal Reflux (possibly contributing to infections of lung, middle ear and upper respiratory tract), sparse subcutaneous fat through infancy and early childhood, normal intelligence with poorly defined learning disability, early suceptibility to medical complications such as obstructive pulmonary disease, diabetes mellitus, and various cancers (*11*)	Autosomal recessive	*BLM* (DNA Helicase) at 15q26.1 (*12*) (quadriradial (Qr) in cultured blood lymphocytes and/or increased sister chromatid exchanges (SCEs) in any type of cultured cells)
Fanconi pancytopenia (227650)	Radial hypoplasia, hypoplastic thumbs, hyperpigmentation, pancytopenia (progressive bone marrow failure), increased risk of malignancy-acute myelogenous leukemia or myelodysplastic syndrome and solid tumors) (*13*). May have malformations of the kidneys/urinary tract, heart, GI system, oral cavity, CNS, ears (including hearing loss), developmental delay	Autosomal recessive (*14*) Except FANCB mutations X-inheritance	Diagnostic test is detection of DNA aberations when exposed to dipoxybutane (DEB) or mitomycin C (MMC) 15 genes in the FA complementation group (*15*) (*FANCA*—16q24.3, *FANCB*—Xp22.31, *FANCC*—9q22.3, *FANCD1* [*BRCA2*]—13q12.3, *FANCD2*—3p25.3, *FANCE*—6p22-p21, *FANCF*—11p15, *FANCG*—9q13, *FANCI*—15q25-q26, *FANCJ* [*BRIP1*]—17q22, *FANCL*—2q16.1, *FANCM*—14q21.3, *FANCN* [*PALB2*]—16p12, *FANCO* [*RAD51C*]—17q22, and *FANCP* [*SLX4*]—16p13.3
De Sanctis–Cacchione (278800)	Xeroderma pigmentosum, hypogonadism, microcephaly	Autosomal recessive	*ERCC6* (excision repair cross-complementing) gene at 10q11
Johanson–Blizzard (243800)	Hypoplastic alae nasi, microcephaly, midline scalp defects, sensorineural deafness, exocrine pancreatic insufficiency, nasolacrimal system malformations, absent permanent teeth, hypothyroidism (*16*)	Autosomal recessive	*UBR1 (ubiquitin protein ligase E3 component N-recognin 1)* gene at 15q15-q21.1
Donohue leprechaunism (246200)	Severe IUGR adipose deficiency, thick lips, islet cell hyperplasia results in hyperinsulinemia	Autosomal recessive	*INSR* (insulin receptor) gene at 19p13.2
Seckel syndrome (210600, 606744, 608664, 61376)	IUGR with microcephaly, prominent nose, micrognathia	Autosomal recessive, genetically heterogeneous	SCKL1—*ATR* (Ataxia-Telangiectasia and RAD3-related) at 3q22.1q24 SCKL2—18p11.31-q11.2 (*17*) SCKL3—14q21-q22 (*18*) SCKL4—*CENPJ*-13q12.2 (*19*)
Hallermann–Streiff (234100)	Microphthalmia, small pinched nose, hypotrichosis, dental anomalies, cutaneous atrophy	Sporadic	
Smith–Lemli–Opitz (270400)	Deficiency of enzyme 7-dehydrocholesterol reductase. Prenatal onset growth deficiency with post-natal persistence secondary to abnormal cholesterol metabolism. Microcephaly, ptosis, cleft palate, syndactyly of 2nd and 3rd toes, post-axial polydactyly, hypospadias (*20*)	Autosomal recessive	*DHCR7* at 11q12-q13 (*21*) (diagnostic test is elevation of serum 7DHC)
Williams (130160)	Prominent lips, periorbital fullness, supravalvular aortic stenosis (or other cardiovascular disease), mild MR, specific cognitive profile (overly friendly), endocrine abnormalities (*22*)	Sporadic, autosomal dominant	Genes at 7q11.2 *ELN* gene *LIMK1* *GTF2I* *MAGI2* (*23*) *HIP1* *YWHAG*

TABLE 38-1 Prenatal-Onset Growth Deficiency Disorders—*Cont'd*

Disorder	Key Clinical Findings	Inheritance	Gene Testing
Noonan (163950)	Webbed neck, pectus excavatum, pulmonary stenosis, hypertelorism, downward slanting palpebral fissures, motor delay	Autosomal dominant	*PTPN11* gene 12q24.1. Also *SOS1*, *KRAS*, *BRAF*, *SHOC*
Aarskog (100050)	Ocular hypertelorism, optic nerve hypoplasia, retinal vessel tortuosity, deficient ocular elevation, hyperopia, and anisometropia (*24*) brachydactyly, shawl scrotum,	X-linked	*FGD1* gene at Xp11.2 (*25*)
Robinow (180700)	Broad forehead with flat facial profile, short forearms, hypoplastic genitalia	Autosomal dominant Autosomal recessive	*ROR2* on 9q22
Opitz (300000)	Ocular hypertelorism, hypospadias, swallowing difficulties secondary to abnormalities of the trachea/esophagus and larynx; also associated with cardiac anomalies, imperforate anus, and developmental delay; anteverted nares and posterior pharyngeal cleft were seen only in the X-linked form (*26*)	Autosomal dominant and X-linked	*MID1* on Xp22
Coffin–Siris (135900)	Hypoplastic fifth digits and nails, coarse facies, hirsutism with sparse scalp hair, psychomotor delay/MR (*27*)	Autosomal dominant	*ARID1B* at 6q25 and other *SWI/SNF* subunit genes (*27*)

Source:
(*1*) Deardorff, M. A.; Clark, D. M.; Krantz, I. D. Cornelia de Lange Syndrome. In: Pagon, R. A., Bird, T. D., Dolan, C. R., Stephens, K., Eds.; GeneReviews [Internet]. University of Washington: Seattle, WA: 1993. Available from http://www.ncbi.nlm.nih.gov/bookshelf/br.fcgi?book=gene&part=cdls.
(*2*) Rohatgi, S.; Clark, D.; Kline, A. D.; Jackson, L. G.; Pie, J.; Siu, V.; Ramos, F. J.; Krantz, I. D.; Deardorff, M. A. *Am. J. Med. Genet.* **2010**, *152A*, 1641–1653.
(*3*) Musio, A.; Selicorni, A.; Focarelli, M. L.; Gervasini, C.; Milani, D.; Russo, S.; Vezzoni, P.; Larizza, L. *Nat. Genet.* **2006**, *38*, 528–530.
(*4*) Stevens, C. A. Rubinstein–Taybi Syndrome. In: Pagon, R. A., Bird, T. D., Dolan, C. R., Stephens, K., Eds.; GeneReviews [Internet]. University of Washington: Seattle, WA: 1993. Available from http://www.ncbi.nlm.nih.gov/bookshelf/br.fcgi?book=gene&part=rsts.
(*5*) Chiang, P. W.; Lee, N. C.; Chien, N.; Hwu, W. L.; Spector, E.; Tsai, A. C. *Am. J. Med. Genet. A.* **2009**, *149A* (7), 1463–1467.
(*6*) Bartholdi, D.; Roelfsema, J. H.; Papadia, F.; Breuning, M. H.; Niedrist, D.; Hennekam, R. C.; Schinzel, A.; Peters, D. J. M. *J. Med. Genet.* **2007**, *44*, 327–333.
(*7*) Saal, H. M. Russell–Silver Syndrome. In: Pagon, R. A., Bird, T. D., Dolan, C. R., Stephens, K., Eds.; GeneReviews [Internet]. University of Washington: Seattle, WA: 1993. Available from http://www.ncbi.nlm.nih.gov/bookshelf/br.fcgi?book=gene&part=rss.
(*8*) Bartholdi, D.; Krajewska-Walasek, M.; Ounap, K.; Gaspar, H.; Chrzanowska, K. H.; Ilyana, H.; Kayserili, H.; Lurie, I. W.; Schinzel, A.; Baumer, A. *J. Med. Genet.* **2009**, *46*, 192–197.
(*9*) Nativio, R.; Sparago, A.; Ito, Y.; Weksberg, R.; Riccio, A.; Murrell, A. *Hum. Mol. Genet.* **2011**, *20* (7), 1363–1374.
(*10*) Tsukahara, M.; Opitz, J. M. *Am. J. Med. Genet.* **1996**, *63* (1), 277–289.
(*11*) Wang, Y.; Smith, K.; Waldman, B. C.; Waldman, A. S. *DNA Repair* **2011**, *10* (4), 416–426.
(*12*) Krejci, L.; Van Komen, S.; Li, Y.; Villemain, J.; Reddy, M. S.; Klein, H.; Ellenberger, T.; Sung, P. *Nature* **2003**, *423*, 305–309.
(*13*) Cioc, A. M.; Wagner, J. E.; MacMillan, M. L.; DeFor, T.; Hirsch, B. *Am. J. Clin. Pathol.* **2010**, *133* (1), 92–100.
(*14*) Collins, N. B.; Wilson, J. B.; Bush, T.; Thomashevski, A.; Roberts, K. J.; Jones, N. J.; Kupfer, G. M. *Blood* **2009**, *113* (10), 2181–2190.
(*15*) Chandra, S.; Levran, O.; Jurickova, I.; Maas, C.; Kapur, R.; Schindler, D.; Henry, R.; Milton, K.; Batish, S. D.; Cancelas, J. A., et al. *Mol. Ther.* **2005**, *12*, 976–984.
(*16*) Cheung, J. C.; Thomson, H.; Buncic, J. R.; Héon, E.; Levin, A. V. *J. AAPOS* **2009**,*13* (5), 512–514.
(*17*) Borglum, A. D.; Balslev, T.; Haagerup, A.; Birkebaek, N.; Binderup, H.; Kruse, T. A.; Hertz, J. M. *Europ. J. Hum. Genet.* **2001**, *9*, 753–757.
(*18*) Kilinc, M. O.; Ninis, V. N.; Ugur, S. A.; Tuysuz, B.; Seven, M.; Balci, S.; Goodship, J.; Tolun, A. *Europ. J. Hum. Genet.* **2003**, *11*, 851–857.
(*19*) Al-Dosari, M. S.; Shaheen, R.; Colak, D.; Alkuraya, F. S. *J. Med. Genet.* **2010**, *47*, 411–414.
(*20*) Porter, F. D. *Eur. J. Hum. Genet.* **2008**, *16* (5), 535–541.
(*21*) Witsch-Baumgartner, M.; Loffler, J.; Utermann, G. *Hum. Mutat.* **2001**, *17*, 172–182.
(*22*) Ramocki, M. B.; Bartnik, M.; Szafranski, P.; Kolodziejska, K. E.; Xia, Z.; Bravo, J.; Miller, G. S.; Rodriguez, D. L.; Williams, C. A.; Bader, P. I.; Szczepanik, E.; Mazurczak, T., et al. *Am. J. Hum. Genet.* **2010**, *87*, 857–865.
(*23*) Marshall, C. R.; Young, E. J.; Pani, A. M.; Freckmann, M.-L.; Lacassie, Y.; Howald, C.; Fitzgerald, K. K.; Peippo, M.; Morris, C. A.; Shane, K., et al. *Am. J. Hum. Genet.* **2008**, *83*, 106–111.
(*24*) Taub, M. B., Stanton, A. *Optometry* **2008**, *79* (7), 371–377.
(*25*) Orrico, A., Galli, L.; Cavaliere, M. L.; Garavelli, L.; Fryns, J. P.; Crushell, E.; Rinaldi, M. M.; Medeira, A.; Sorrentino, V. *Eur. J. Hum. Genet.* **2004**, *12* (1), 16–23.
(*26*) Robin, N. H.; Opitz, J. M.; Muenke, M. *Am. J. Med. Genet.* **1996**, *62*, 305–317.
(*27*) Coulibaly, B.; Sigaudy, S.; Girard, N.; Popovici, C.; Missirian, C.; Heckenroth, H.; Tasei, A. M.; Fernandez, C. *Eur. J. Med. Genet.* **2010**, *53* (5), 318–321.

TABLE 38-2 Prenatal-Onset Overgrowth Syndromes

Syndrome	Key Clinical Findings	Inheritance	Gene Testing
Proportionate Syndromes			
Beckwith–Wiedermann (130650)	Macroglossia, infraorbital creases, earlobe creases and pits, abdominal wall defects, neonatal hypoglycemia, visceromegaly, risk for abdominal neoplasms, hemihypertrophy, polyhydramnios, large placenta (*1*)	Sporadic, autosomal dominant	*H19* *LIT1* *CDKN1C*/p57 (*KIP2*)
Perlman (267000)	Hypotonia, mental retardation, serration of upper alveolar ridge, nephromegaly, bilateral cortical hamartomas, nephroblastomatosis (*2*)	Autosomal recessive	
Sotos (117550)	Macrocephaly, dolichocephaly, down-slanted palpebral fissures, hypertelorism, prognathism, high narrow palate, premature eruption of teeth, large hands and feet, kyphoscoliosis, mental deficiency (*3*)	Sporadic, autosomal dominant	*NSD1* at 5q35
Weaver (277590)	Mental retardation, hypertonia, hoarse voice, macrocephaly, round face, ocular hypertelorism, down-slanted palpebral fissures, long philtrum, large ears, micrognathia, camptodactyly, thin deep-set nails, prominent fingertip pads (*4*)	Sporadic, autosomal dominant	*EZH2* at 7q35
Marshall–Smith (602535)	Accelerated linear growth and skeletal maturation, postnatal failure to thrive, hypotonia, developmental delay, structural brain anomalies, respiratory tract anomalies, recurrent pneumonia, pulmonary hypertension, dolichocephaly, coarse eyebrows, shallow orbits, blue sclera, upturned nose, low nasal bridge, small mandibular ramus, hypertrichosis, umbilical hernia, choanal atresia, omphalocele (*6–8*)	Sporadic, autosomal dominant	*NFIX* at 19p13.13
Bannayan–Riley–Ruvalcaba (153480)	Delayed gross motor development, hypotonia, speech delay, mental deficiency, macrocephaly, prominent Schwalbe rings, prominent corneal nerves, pseudopapilledema, mesodermal hamartomas, pigmented penile macules, lipid storage myopathy (*9*)	Autosomal dominant	*PTEN* 10q23.31
Simpson–Golabi–Behmel (300209)/(312870)	Macrocephaly, ocular hypertelorism, short broad nose, large mouth, macroglossia, variable mental retardation, hypotonia, postaxial polydactyly of hands, nail hypoplasia, partial cutaneous syndactyly, cryptorchidism, supernumerary nipples, cardiac defects, gastrointestinal defects, large cystic kidneys (*10*)	X-linked recessive	Type 1 *CXORF5* Xp22.3-p21.2 Type 2 *GPC3* Xq26
Elejalde (256710)	Craniosynostosis, gross edema, short limbs, postaxial polydactyly, redundant neck skin, cystic renal dysplasia, congenital heart defect, spleen anomaly, micromelia (*11*)	Autosomal recessive	
Cantu (239850)	Congenital hypertrichosis, neonatal macrosomia, macrocephaly, coarse facial features, broad nasal bridge, epicanthal folds, wide mouth, thick lips, and distinctive osteochondrodysplasia (*12*)	Autosomal dominant	*ABCC9* at 12p2.1
19q13.13 deletion (613638)	Proportional overgrowth, macrocephaly, frontal bossing, downslanting palpebral fissures with optic/opthalmologic abnormalities like strabismus and/or optic nerve hypoplasia or atrophy, absent corpus callosum, autism (*13*)	Sporadic	*NFIX* deletion
Disproportionate Syndromes			
Proteus (176920)	Regional overgrowth of hands and/or feet, asymmetry of limbs, plantar hyperplasia, hemangiomas, lipomas, lymphangiomas, varicosities, verucous epidermal nevi, macrocephaly, cranial hyperostoses, long bone overgrowth, variable moderate mental deficiency (*14, 15*)	Sporadic	Mosaic *AKT1*

TABLE 38-2 Prenatal-Onset Overgrowth Syndromes—*Cont'd*

Syndrome	Key Clinical Findings	Inheritance	Gene Testing
Hemihyperplasia, lipomatosis, CLOVE		Sporadic	Mosaic *PIK3CA*
Megalencephaly capillary malformation syndrome (602501)	Megalencephaly, prenatal overgrowth, thickened subcutaneous tissue; over half of patients have acquired cerebellar tonsillar herniation (vs congenital Chiari I malformation) with dilated dural sinuses and ventriculomegaly secondary to rapid brain growth and progressive crowding of the posterior fossa (*17–19*)	Sporadic	Mosaic *PIK3CA*
Klippel–Trenaunay syndrome	Cutaneous capillary malformations of a limb, venous malformations, and accompanying segmental hypertrophy of soft tissue and/or bone (*20*)	Sporadic	

Source:

(*1*) Shuman, C.; Beckwith, J. B.; Smith, A. C.; Weksberg, R. Beckwith–Wiedemann Syndrome. In: Pagon, R. A., Bird, T. D., Dolan, C. R., Stephens, K., Eds.; *GeneReviews* [Internet]. University of Washington: Seattle, WA: 1993. Available from http://www.ncbi.nlm.nih.gov/bookshelf/br.fcgi?book=gene&part=bws.
(*2*) Alessandri, J. L.; Cuillier, F.; Ramful, D.; Ernould, S.; Robin, S.; de Napoli-Cocci, S.; Rivière, J. P.; Rossignol, *S. Am. J. Med. Genet. A*. **2008**, *146A* (19), 2532–2537.
(*3*) Tatton-Brown, K.; Cole, T. R. P.; Rahman, N.; 2004 In: Pagon, R. A., Bird, T. D., Dolan, C. R., Stephens, K., Eds.; *GeneReviews* [Internet]. University of Washington: Seattle, WA: 1993. Available from http://www.ncbi.nlm.nih.gov/bookshelf/br.fcgi?book=gene&part=sotos.
(*4*) Hennekam, R. C. M.; Krantz, I.; Allanson, J. Eds. Gorlin's Syndromes of the Head and Neck, 5th ed.; Oxford University Press: New York, 2010.
(*5*) Douglas, J.; Hanks, S.; Temple, I. K.; Davies, S.; Murray, A.; Upadhyaya, M.; Tomkins, S.; Hughes, H. E.; Cole, T. R. P.; Rahman, N. *Am. J. Hum. Genet*. **2003**, *72*, 132–143.
(*6*) Adam, M. P.; Hennekam, R. C.; Keppen, L. D.; Bull, M. J.; Clericuzio, C. L.; Burke, L. W.; Ormond, K. E.; Hoyme, E. H. *Am. J. Med. Genet. A*. **2005**, *137* (2), 117–124.
(*7*) Shaw, A. C., van Balkom, I. D.; Bauer, M.; Cole, T. R.; Delrue, M. A.; Van Haeringen, A.; Holmberg, E.; Knight, S. J.; Mortier, G.; Nampoothiri, S., et al. *Am. J. Med. Genet. A*. **2010**, *152A*, 11, 2714–2726.
(*8*) Butler, M. G. *Am. J. Med. Genet*. **2004**, *126A*, 329–330.
(*9*) Zhou, X. P., Waite, K. A.; Pilarski, R.; Hampel, H.; Fernandez, M. J.; Bos, C.; Dasouki, M.; Feldman, G. L.; Greenberg, L. A.; Ivanovich, J., et al. *Am. J. Hum. Genet*. **2003**, *73*, 404–411.
(*10*) Li, C. C.; McDonald, S. D. *Fetal Diagn. Ther.* **2009**, *25* (2), 211–215.
(*11*) Phadke, S. R.; Aggarwal, S.; Kumari, N. *Clin. Dysmorphol.* **2011**, *20* (2), 98–101.
(*12*) Bregje, W. M.; Gilissen, C.; Grange, D. K., et al. *Am. J. Hum.Genet.* **2012**, *90*, 1094–1101.
(*13*) Malan, V.; Rajan, D.; Thomas, S.; Shaw, A. C.; Louis Dit Picard, H.; Layet, V.; Till, M.; van Haeringen, A.; Mortier, G.; Nampoothiri, S., et al. *Am. J. Hum. Genet*. **2010**, *87* (2), 189–198.
(*14*) Turner, J. T.; Cohen, M. M., Jr.; Biesecker, L. G. *Am. J. Med. Genet.* **2004**, *130A*, 111–122.
(*15*) Biesecker, *L. Eur. J. Hum. Genet*. **2006**, *14* (11), 1151–1157.
(*16*) Heilstedt, H. A.; Bacino, C. A. *BMC Med. Genet.* **2004**, *5* (1).
(*17*) Papetti, L.; Tarani, L.; Nicita, F.; Ruggieri, M.; Mattiucci, C.; Mancini, F.; Ursitti, F.; Spalice, A. *Brain Dev*. **2011**, **Feb 25**.
(*18*) Gonzalez, M. E.; Burk, C. J.; Barbouth, D. S.; Connelly, E. A. *Pediatr. Dermatol.* **2009**, *26* (3), 342–346.
(*19*) Conway, R. L.; Pressman, B. D.; Dobyns, W. B.; Danielpour, M.; Lee, J.; Sanchez-Lara, P. A.; Butler, M. G.; Zackai, E.; Campbell, L.; Saitta, S. C., et al. *Am. J. Med. Genet.* **2007**, *143A*, 2981–3008.
(*20*) Samimi, M.; Lorette, G. *Presse Med*. **2010**, *39* (4), 487–494.

SUMMARY REFERENCES

1. Lango Allen, H.; Estrada, K.; Lettre, G., et al. Hundreds of Variants Clustered in Genomic Loci and Biological Pathways Affect Human Height. *Nature* **2010**, *467* (7317), 832–838.
2. Graham, J. M., Ed. *Smith's Recognizable Patterns of Human Deformation*, 3rd ed.; WB Saunders: Philadelphia, 2007.
3. Warman, M. L.; Cormier-Daire, V.; Hall, C.; Krakow, D.; Lachman, R.; LeMerrer, M.; Mortier, G.; Mundlos, S.; Nishimura, G.; Rimoin, D. L., et al. Nosology and Classification of Genetic Skeletal Disorders: 2010 Revision. *Am. J. Med. Genet. A* **2011**, *9999*, 1–26.
4. Burkardt, D. D.; Rosenfeld, J. A.; Helgeson, M.; Angel, B.; Banks, V.; Smith, W.; Gripp, K.W.; Moline, J.; Moran, R.; Niyazov, D.M., et al. Distinctive Phenotype in 9 Patients with Deletion of Chromosome 1q24-q25. *AJMG* (in press).
5. Lynch, S. A.; Foulds, N.; Thuresson, A. C.; Collins, A. L.; Annerén, G.; Hedberg, B. O.; Delaney, C. A.; Iremonger, J.; Murray, C. M.; Crolla, J. A., et al. The 12q14 Microdeletion Syndrome: Six New Cases Confirming the Role of HMGA2 in Growth. *Eur. J. Hum. Genet.* **2011 May,** *19* (5), 534–539.
6. Malan, V.; Chevallier, S.; Soler, G.; Coubes, C.; Lacombe, D.; Pasquier, L.; Soulier, J.; Morichon-Delvallez, N.; Turleau, C.; Munnich, A., et al. Array-Based Comparative Genomic Hybridization Identifies a High Frequency of Copy Number Variations in Patients with Syndromic Overgrowth. *Eur. J. Hum. Genet.* **2010 Feb,** *18* (2), 227–232, Epub 2009 Oct 21.

CHAPTER

39

Susceptibility and Response to Infection

Michael F Murray

There is an ongoing struggle for the genetic upper hand that is taking place in each of us; the struggle pits your genomic repertoire against the genomic repertoire of millions of microbial organisms. There is a growing list of human genetic variants that either improve or diminishe an individual's chances in these struggles. The struggle can take place at the human barrier surfaces, in the interstitial spaces between cells, or in the cells themselves. In every instance of human infection three forces intersect: the individual patient's genome, the microbial genome(s), and environmental factors.

This chapter does not review the classic immunodeficiency states but rather the host variations that influence the more subtle susceptibilities and resistance associations. As such there is attention to the classic concept of heterozygous advantage and homozygous disadvantage and some new observations related to that concept as well as new observations in the areas of genome-wide association studies (GWAS) and copy number effects on infectious disease susceptibility. The rare instances of nearly absolute resistance to a given infection, as well as the more common scenarios of host genetic variability leading to differences in the rate of clearance of an infection or the consequences of the infection, are reviewed.

New knowledge of variable host resistance to infection is now coming from multiple sources including GWAS. The infections examined include human immunodeficiency virus 1 (HIV1), hepatitis B virus (HBV), hepatitis C virus (HCV), *Neisseria meningitides*, *Mycobacterium tuberculosis*, *Mycobacterium lepra*, and bovine spongiform encephalopathy prions. As in other areas of genetics and genomics, these studies have both confirmed previous observations as well as given new insights into genetic pathways that may not have been considered in a candidate gene approach to research in this area.

New genomic tools will allow ongoing interrogation of the details of host resistance and susceptibility. This will potentially allow for confirmation of long standing speculations, such as the overlap between mycobacterial disease and inflammatory bowel disease. They will also continue to build on genomic and genetic principles such as heterozygous advantage to explain findings where carrier frequency of mutations is increased and has a link to decreased susceptibly to disease, such as that demonstrated by Genovese and colleagues with regard to APOL1 variants, resistance to trypanosomes, and risk for renal disease. Finally, they will also allow us to better understand additional mechanisms of host response, such as non-coding RNAs from both the host and the microbes, and how variations in these genes and their biological targets can alter the clinical course of human infections in individual patients.

By far the largest set of associations is between microbes and the *HLA* genes, but a growing number of associations exist outside of this gene group. Tables 39-1 and 39-2 lay out some of those important observations in both categories.

The hopes and speculations of the mid twentieth century to abolish human infections have not been realized and will not likely ever be realized. Microbial infection will continue to be a driving force of natural selection in humans. A detailed understanding of host genetic susceptibility and response to infection will facilitate improvements in existing therapies and the development of new therapies.

TABLE 39-1 HLA Class Ia and Class II Associated Diseases

Microbe	Type of Organism	Disease(s)	Class
Human T-lymphotropic virus-1 (HTLV-1)	RNA Virus	Myelopathy	Ia
Human immunodeficiency virus-1	RNA virus	AIDS	Ia and II
Hepatitis C virus	RNA virus	Viremia	II
		Mixed cryoglobulinemia	Ia and II
Dengue virus	RNA virus	Hemorrhagic fever Shock	Ia
Puumala hantavirus	RNA virus	Renal disease	Ia and II
Ross River virus	RNA virus	Arthritis	II
Measles	RNA virus	Subacute sclerosing panencephalitis	Ia
Hepatitis B virus	DNA virus	Viremia	II
Herpes simplex virus 2 (HSV-2)	DNA virus	Genital ulcers	Ia
Epstein–Barr virus	DNA virus	Nasopharyngeal cancer	Ia and II
Cytomegalovirus	DNA virus	Retinitis	Ia and II
Human herpes virus 8 (HHV-8)	DNA virus	Kaposi sarcoma	Ia
Human papilloma virus	DNA virus	Cervical cancer	Ia and II
Mycobacterium tuberculosis	Bacteria	Tuberculosis	II
Mycobacterium leprae	Bacteria	Leprosy	II
Mycobacterium avium	Bacteria	Bacteremia	II
		Pneumonitis	Ia and II
Chlamydia trachomatis	Bacteria	Trachoma	Ia
		Fallopian tube scarring	II
Haemophilus influenzae	Bacteria	Meningitis and bacteremia	Ia
Helicobacter pylori	Bacteria	Atrophic gastritis	II
Borrelia burgdorferi	Bacteria	Lyme arthritis	II
Streptococcus pyogenes	Bacteria	Rheumatic fever	II
Plasmodium falciparum	Protozoa	Malaria	Ia and II
Trypanosoma cruzi	Protozoa	Chagas disease	Ia
Toxoplasma gondi	Protozoa	Congenital hydrocephalus	II
Leishmania spp.	Protozoa	Mucocutaneous ulceration	II
Giardia lamblia	Protozoa	Gastroenteritis	II
Schistosomiasis spp.	Helminth	Hepatic fibrosis	II
Echinococcus spp.	Helminth	Liver abscess	II
Brugia spp.	Helminth	Elephantiasis	Ia and II
Onchocerca volvulus	Helminth	River blindness	II
Paracoccidiodes spp.	Fungus	Mycosis	Ia
Fonsecaea pedrosoi	Fungus	Chromoblastomycosis	Ia
Trichosporon spp.	Fungus	Hypersensitivity pneumonitis	II
Cryptococcus neoformans	Fungus	Meningitis	Ia

TABLE 39-2 Non-HLA Microbe–Gene Susceptibility Associations

Microbe	Gene/Gene Product	Associated Susceptibility
HIV-1	FUT 2	Heterosexually acquired infection
	CCR5	Infection and progression to AIDS
	CCR2	Progression to AIDS
	CX_3CR1	Progression to AIDS
	NRAMP1	Progression to AIDS
	SDF1-3′A	Progression to AIDS
Parvovirus B19	PK synthase	Infection
Hepatitis B virus	VDR	Persistent infection
Hepatitis C virus	IL28B	Persistent infection treatment response
H. influenzae	FUT 2	Invasive infection
Streptococcus pneumoniae	FUT 2	Invasive infection
Neisseria meningitidis	FUT 2	Invasive infection
Pseudomonas aeruginosa	CFTR	Pulmonary infection
Salmonella spp.	CFTR	Gastrointestinal infection
	IFNGR	Systemic infection
Vibrio cholerae	Blood group O	Gastrointestinal infection
Escherichia coli	GALT	Bacteremia
H. pylori	Lewis Ag	Gastric infection
	IL 1β	Gastric carcinoma
Mycobacterium spp.	VDR	Symptomatic infection
	IFNGR	Systemic infection
	IL 12RB1	Systemic infection
	NRAMP 1	Symptomatic infection
	IL1β	Pleural tuberculosis
	TNF2	Symptomatic infection
	MBP	Pulmonary tuberculosis
P. falciparum	Blood group A	Severe infection
	αHgb	Severe infection
	βHgb	Severe infection
	G6PD	Severe infection
	TNF2	Cerebral malaria
Plasmodium vivax	DARC	Infection
Leishmania spp.	TNF2	Mucocutaneous infection

FURTHER READING

1. Brown, K. E.; Hibbs, J. R.; Gallinella, G., et al. Resistance to Parvovirus B19 Infection due to Lack of Virus Receptor (Erythrocyte P Antigen). *N. Engl. J. Med.* **1994,** *330,* 1192–1196.
2. Centers for Disease Control and Prevention (CDC) Fatal Laboratory-Acquired Infection with an Attenuated Yersinia Pestis Strain—Chicago, Illinois, 2009. *MMWR Morb. Mortal. Wkly. Rep.* **2011,** *60* (7), 201–205.
3. Genovese, G.; Friedman, D. J.; Ross, M. D., et al. Association of Trypanolytic ApoL1 Variants with Kidney Disease in African Americans. *Science* **2010,** *329* (5993), 841–845.
4. Gonzalez, E.; Kulkarni, H.; Bolivar, H., et al. The Influence of CCL3L1 Gene-Containing Segmental Duplications on HIV-1/AIDS Susceptibility. *Science* **2005,** *307,* 1434–1440.
5. Haldane, J. B. S. Disease and Evolution. *Ric. Sci.* **1949,** *19* (Suppl.), 68.
6. Hindorff, L. A.; Junkins, H. A.; Hall, P. N.; Mehta, J. P.; Manolio, T. A. A Catalog of Published Genome-Wide Association Studies. <www.genome.gov/gwastudies> 2011 (accessed 30 May 2011).

CHAPTER

40

Transplantation Genetics

Steven Ringquist, Ying Lu, and Massimo Trucco

Division of Immunogenetics, Department of Pediatrics, John G Rangos Sr. Research Centre, Children's Hospital of Pittsburgh of UPMC, Pittsburgh, PA, USA

Gaia Bellone

Department of Statistics, Carnegie Mellon University, Pittsburgh, PA, USA

The idea of replacing a diseased, or damaged part of the body of an individual, the recipient, with the same part provided by another healthy individual, the donor, has always been present in the human mind. Mythology is full of examples of chimeras—imaginary animals composed of parts coming from individuals of different species. We had to wait, however, until the beginning of the twentieth century to find evidence of a fresh new look to the transplantation problem. It was, in fact, in 1912 when George Schöne concluded that the "laws of transplantation" were defined as (1) autografts generally succeed; (2) allografts usually fail; and (3) xenografts invariably fail. Twenty-five years later Peter Gorer, an expert on tumor resistance in mice, defined the scientific basis of this postulated blood relationship (*1*). Gorer was interested in explaining why tumors that could often be transplanted from one mouse to another failed to grow when transplanted in another strain of recipient mice. The permissive trait was attributed to two independent genetic loci. Using congenic strains of mice, Gorer and Snell were able to determine that engrafted tumors were most likely able to "take" when the recipient mouse was of the same strain as the donor (*1,2*). More importantly, they also realized that survival of donor tissue was not limited to tumors but applied to any graft. Snell and Gorer identified a number of genetically independent loci controlling these immune responses, which they designated H for histocompatibility (i.e. compatibility between different tissues), followed by an arabic numeral in the order of discovery. The locus designated H-2 appeared to be more immunogenic than the others and corresponded to Gorer's original type II antigen. This locus became known with time as the murine major histocompatibility gene complex (MHC). MHC (in humans known as the human leukocyte antigen, HLA) can be considered the molecular embodiment of Schöne's laws of transplantation.

The historical interpretation of the genetics of histocompatibility was based on experiments in which the histocompatibility antigens were only seen as the factors limiting tissue transplantation by promoting graft rejection. Later on, Pamela Bjorkman and colleagues discerned the physiologic function of the MHC molecules when they solved the X-ray crystallography structure of HLA-A2 and recognized the presence of a heterogenic population of bound peptides to what is now referred to as Bjorkman's groove (*3*). The function of this site, the so-called peptide binding groove, was clearly determined once the interpretation of the crystal structure of the HLA molecule was completed.

The HLA molecules constitute the most highly polymorphic genetic system in humans (*4,5*). A stable, inherited polymorphism gives rise to alternative forms of the protein, the alleles. Nearly all the HLA molecules have various alleles with HLA-B being the most polymorphic with more than 1800 reported alleles (*4*). The molecular basis for the HLA polymorphism resides in nucleotide sequence differences present in the coding regions of the *HLA* genes. The polymorphism is clustered into discrete hypervariable regions, which become more evident when the amino acid sequences of these alleles are aligned. These nucleic acid sequence differences account for amino acid differences among the various HLA molecules. Although most HLA alleles occur in all ethnic groups, they may vary in frequency from group to group (*6*).

HLA molecules are grouped into two classes based on their structure, function, and cellular expression. Class I (HLA-A, -B, and -C) molecules are characterized by a 43-kDa α-chain with three domains (α_1, α_2, and α_3) that dimerizes with the β_2 microglobulin, a short (12 kDa) protein, which provides the fourth domain necessary to stabilize the HLA molecule (*3*). Class I molecules are coexpressed on nearly all nucleated cells and platelets and are anchored via the transmembrane portion and the short cytoplasmic tail of the α-chain only. Class II molecules (HLA-DR, -DQ, and -DP) consist of α- (34 kDa) and β- (29 kDa) chains, each with two domains. Class II molecules are anchored by both chains, each of which can communicate across the membrane with cytoplasmic structures. These molecules are expressed primarily on B lymphocytes, macrophages, dendritic cells, activated T cells, and endothelial cells.

In the 1980s, molecular biology technologies, such as gene cloning and sequencing, became simple to perform. This facilitated the isolation and characterization of the HLA genes at different loci. By comparing sequence variations from person to person, the nature and locations of the polymorphism of the HLA complex was revealed. During the past 10 years, substantial progress has been made in converting the protocols for molecular HLA typing to microsphere, microarray, and next-generation sequencing technologies. For example, microsphere technology has been widely adopted by the HLA typing community. The method exploits the use of color-coded beads developed in order to advance DNA hybridization-based typing methods (*7*). Assignment of HLA genotype is based on the hybridization reaction pattern compared with the pattern associated with published *HLA* gene sequences. The HLA loci are amplified from genomic DNA using sequence-specific oligonucleotide primer mixes optimized for PCR amplification of particular HLA alleles and have been used with dried blood spots as well as DNA isolated from blood draws and buccal swabs.

More recently, the international MHC and Autoimmunity Genetics Network (IMAGEN) has exploited parallel genotyping of 1472 SNPs using microarrays to determine complete HLA genotype from more than 10,000 subjects (*8*). The accuracy of the method for determining 2- and 4-digit resolution at HLA loci has been estimated to be as great as 95% (*9*). Oligonucleotide array–based strategies can provide high-throughput genotyping by virtue of parallel analysis of multiple genetic regions (*8,10*). Additional advances in HLA genotyping methods have occurred as a result of technologies developed for the human genome project. For example, next-generation sequencing technology has advanced substantially (*11*) and a number of publications have reported achieving high resolution typing of HLA loci using genomic DNA isolated from human cell lines, blood samples, as well as from complementaryDNA created from isolations of mRNA (*12–15*). Along with translating the methodology from the research-based setting to the clinical HLA typing laboratory, it is critical to develop computational methods for analyzing the data (*15*). It seems likely that some of the alignment and assembly methods developed for the human genome project (*16*) along with advanced statistical methodology for sorting data obtained from heterozygous subjects will be required in order to fully achieve the goal of using next-generation sequencing for routine HLA typing (*17,18*). Given the ability of the methodology to generate high resolution HLA typing from thousands of samples in a single sequencing run and at low cost, these final hurdles will no doubt be overcome in the near future.

Tissue typing for transplantation refers to determination of the HLA genotypes of both the potential donor(s) and the recipient. Finding the best donor for a kidney transplant generally means finding a six-antigen match by looking at each of the two alleles at HLA-A, -B, and -DR. For bone marrow transplant, however, HLA-C, -DQ, and -DP are also typed and often considered (*19*). A possible criticism to DNA typing is that it gives "too much" information. Although it has been hypothesized that even one nucleotide difference, and hence one amino acid disparity, can result in increased risk of alloreactivity in vivo (*1*), many nucleotide substitutions do not cause amino acid changes. Given the huge number of detectable alleles at the various loci, it will be important to determine which mismatches are allowable and which ones are instead nonpermissive for a positive clinical outcome. A number of studies have been done to determine the clinical impact of HLA matching in unrelated bone marrow transplants. Many reports have shown that the incidence of both acute graft versus host disease (GVHD) and rejection are significantly higher among patients transplanted with phenotypically matched bone marrow cells from unrelated donors compared with cells from genotypically identical siblings, in particular when serologic techniques were used for HLA typing (*1*).

The medical impact of the HLA loci is substantial. Inheritance of select alleles affects an individual's ability to respond to infectious agents but can also affect individual susceptibility to autoimmune disease (*1,2,5*). In transplantation, medicine genetic matching of HLA alleles between tissue recipient and donor is essential for successful outcome. Data clearly show that HLA matching does not and cannot influence solid organ distribution significantly. On average, only about 30% of candidates for bone marrow transplantation will have an HLA-identical sibling or other suitable related donor. Unfortunately, those patients with uncommon HLA phenotypes will not survive a lengthy search for an unrelated donor. It is anticipated that improvement will be seen now that typing by DNA-based methods for HLA typing are in place. The next major effort for the future will be to identify which minor differences between variants can be tolerated with little or no risk of rejection or GVHD, or which tolerogenic protocol can offer safe but effective means to allow transplants across HLA boundaries or even across different species without the need for massive immunosuppression.

compared to the general population. Specific mutations (Y165C and G382D) in the *MUTYH* gene are quite frequently detected in affected individuals (*4*). Thus, MUTYH-associated polyposis is an autosomal recessive colon cancer disorder with at most a minor cancer phenotype in heterozygous carriers. There is also little evidence for a significantly increased risk of other tumor types even though the MUTYH protein is expressed ubiquitously.

42.3 DISORDERS OF MISMATCH REPAIR: LYNCH SYNDROME OR HEREDITARY NON-POLYPOSIS COLON CANCER AND MISMATCH REPAIR DEFICIENCY SYNDROME (AUTOSOMAL RECESSIVE TURCOT SYNDROME)

Lynch Syndrome or hereditary non-polyposis colon cancer is associated with susceptibility to adult onset cancers including colorectal, endometrial and ovarian resulting from inheritance of a heterozygous mutation in one of four different mismatch repair genes *MSH2*, *MLH1*, *MSH6*, and *PMS2* (*5*). These proteins function in heterodimers to recognize and repair mismatches caused by replication errors. Hereditary non-polyposis colon cancer (HNPCC) results from inheriting a single deleterious mutation followed by second somatic inactivating event in the same gene either through mutation, loss, or epigenetic silencing. The lifetime incidence of colon and endometrial cancers is estimated to be 67–70%, with age of onset beginning in young adulthood and clustering in the fifth to sixth decades.

Tumors from patients with HNPCC display an unusual DNA pattern, termed microsatellite instability or replication errors positive (RER+), characterized by changes (both increases and decreases) in the length of repetitive sequences spread throughout the genome due to mismatch repair deficiency in the tumor cell. This represents one of the few examples where the molecular diagnostic test relies on a direct assessment of the localized repair deficiency in the tumor environment. In addition to preventing replication errors, these protein complexes also inhibit recombination between heterologous sequences diminishing deletions and duplications.

The autosomal recessive form of Turcot syndrome is also designated "mismatch repair deficiency syndrome" (MMR-D) (*6*). MMR-D syndrome describes children who inherit homozygous or compound heterozygous mutations in the same MMR genes *MSH2*, *MLH1*, *MSH6* and *PMS2* as those associated with HNPCC. These children have a profound clinical phenotype resulting from mismatch repair deficiency in all cells of the body including café-au-lait spots, axillary freckling and a strikingly high incidence of childhood cancers including brain tumors (gliomas), hematopoietic cancers and Lynch syndrome-associated tumors.

42.4 DISORDERS ASSOCIATED WITH DOUBLE STRAND BREAK RECOGNITION AND REPAIR: ATAXIA TELANGIECTASIA AND RELATED CONDITIONS

Children with ataxia telangiectasia (AT) demonstrate truncal ataxia during early years of childhood, which is progressive, ocular telangiectasias, immunodeficiency and significant cancer risk (*7*). AT is an autosomal recessive disorder due to inheritance of biallelic inactivating mutations of the *ATM DNA* damage response gene. Women heterozygous for *ATM* mutations have an approximately twofold increase in risk of developing breast cancer during adulthood with no other clinical features of AT (*7*).

The cellular phenotype associated with AT is increased killing of lymphocytes (or immortalized lymphoblastoid cells) in response to ionizing radiation exposure. ATM phosphorylates a very large network of regulatory proteins that mediate recognition and repair of DNA lesions including p53 and BRCA1 (*8*). Patients with AT and other defective double strand break repair (DSBR) disorders require substantially modified treatment regimens due to clinically significant cytotoxicity in response to chemotherapy and radiation treatments. The link between ATM deficits and neurologic disability is much more complex and thought to potentially result from excess apoptosis in neurons that accumulate DNA damage or perhaps accumulation of reactive oxygen species mediated damage in neurons (*3*).

There are a large number of proteins required for DSBR and the deficiency of which result in a series of related disorders. The recombination activating genes, *RAG1* and *RAG2*, play an important role in initiation of VDJ recombination. Patients with two truncating mutations, e.g. nonsense or frameshit mutations in *RAG1* or *RAG2*, demonstrate classic severe combined immunodeficiency (SCID) impacting both T and B lymphocytes. Patients who inherit at least one missense allele, encoding a hypomorphic form of the RAG protein that generates greater than 1% enzymatic activity, have a more complex phenotype referred to as Omenn syndrome (*9*). Classic Omenn syndrome includes erythroderma, lymphadenopathy, splenomegaly, granuloma formation in multiple organs and infectious complications. Radiation sensitive-SCID results from inheritance of mutations in the *DCLRE1C* gene, which encodes the Artemis protein required for cleavage of hairpin junctions during VDJ recombination.

Children with hypomorphic mutations in the essential gene that encodes DNA Ligase IV (utilized during non-homologous end joining (NHEJ)) demonstrate a varying phenotype including microcephaly, immunodeficiency and increased risk of lymphoma/leukemia (*10*). However, cells from these patients demonstrate normal DNA damage checkpoint signaling and lack the ionizing radiation sensitivity seen in patients with AT.

AT-like disorder (ATLD) and Nijmegen breakage syndrome (NBS) are two disorders that result from mutations in *MRE11* and *NBN*, respectively, encoding subunits of the MRN (MRE11, Rad50, NBN) heterotrimeric complex central to NHEJ repair (*9,10*). Despite the fact that MRE11 and NBN function in the same complex, they are very distinct disorders clinically. Patients with ATLD demonstrate a similar but milder neurodegeneration phenotype as patients with AT. Patients with NBS demonstrate microcephaly, developmental delay, dysmorphic features, reduced levels of immunoglobulins, radiation sensitivity and increased risk of lymphoma.

The *ATR* gene encodes the AT- and Rad3-related (ATR) protein which is a related PI3 kinase to ATM (*11*). *ATR* encodes an essential function, and thus patients with the ATR-related disorder, Seckel syndrome, are found to carry hypomorphic alleles of the *ATR* gene and demonstrate severe growth defects with intrauterine growth retardation, small stature and marked microcephaly (*10*). ATR protein plays a crucial role in the response of cells to single-stranded DNA that is generated upon replication fork stalling and in repairing templates that contain bulky adducts. Patients with Seckel syndrome and *ATR* mutations do not demonstrate significant ionizing radiation sensitivity and cancer predisposition.

42.5 CROSSLINK REPAIR AND HOMOLOGOUS RECOMBINATION DEFECTS: BREAST–OVARIAN CANCER AND FANCONI ANEMIA

Extensive research on the genetic basis of hereditary breast and ovarian cancer led to the identification of the *BRCA1* and *BRCA2* cancer susceptibility genes (*12*). Both *BRCA1* and *BRCA2* encode proteins which act through large protein complexes to respond to double-strand DNA breaks and mediate homologous recombination. Rare forms of Fanconi anemia (FA subtype D1) resulted from biallelic inheritance of *BRCA2* mutations. There is a complex overlap between interstrand crosslink repair (hallmark of Fanconi phenotype) and homologous recombination mediated repair (BRCA1, BRCA2 and associated proteins) (*13*).

FA is an autosomal or X-linked recessive condition associated with bone marrow failure, congenital anomalies, and a substantially increased risk of malignancy, particularly acute myelogenous leukemia in children and young adults (*13*). The cellular phenotype of Fanconi cells includes a profound sensitivity to crosslinking agents such as diepoxybutane and mitomycin C including the production of breaks and abnormal karyotypic figures. This sensitivity was used to identify at least 15 different FA complementation groups (corresponding to 15 different *FANC* genes). The FANC proteins function in DNA repair through participation in a number of different complexes. FANC A, C, B, E, F, G and L proteins make up the FA core complex which is thought to be function in the first portion of the FA pathway. When activated by DNA damage or replication fork stalling the FA complex proteins result in monoubiquitination of two proteins FANCD2 and FANCI, which together form a heterodimer. An additional upstream component is FANCM, which is found complexed with FAAP24 although the specific clinical phenotype of carrying mutations in *FANCM* or *FAAP24* is not clear.

In normal cells, once FANCD2 and FANCI are monoubiqutinated, this sends a signal which is transduced to the DNA repair complexes that include FANC-D1/BRCA2, FANCJ/BRIP1, FANC-N/PALB2 and FANCP/Rad51C (*14*). These latter proteins play a direct role in repair pathways that utilize homologous recombination. Patients in these FA complementation groups often have a particularly severe cancer phenotype with the vast majority being diagnosed with malignancies (solid tumors, brain tumors and leukemias) by the age of five (*15*). In contrast, analyses of families with heterozygous carriers of *PALB2* and *BRIP1/FANCJ* mutations demonstrated a two- to fourfold relative risk of breast cancer compared to the general population (*14*). The most recently discovered Fanconi gene *FANCP* encodes the Rad51C homologous recombination protein associated with significant childhood and young adult cancer risk (*14*). Heterozygous *RAD51C* carriers have a breast and ovarian cancer risk close to that of *BRCA2* heterozygous carriers (*16*). Thus, these distal members of the FA pathway play an important role in DNA repair that is mediated through homologous recombination. Mutations in these genes result in increased cancer risk in adult heterozygous carriers and a higher risk of solid tumors in children who are biallelic mutation carriers.

42.6 DISORDERS ASSOCIATED WITH RECQ HELICASE DEFICIENCY: BLOOM, WERNER AND ROTHMUND–THOMSON SYNDROMES

There are three autosomal recessive disorders that demonstrate distinct clinical features but share predisposition to malignancy and are the result of mutations in the same family of helicases (Table 42-1). The three disorders, Bloom, Werner and Rothmund–Thomson syndromes, are grouped together because they all result from biallelic mutations in members of the RecQ helicase family, *BLM*, *WRN* and *RECQL4*, respectively (*17*).

Children with Bloom syndrome are very small at birth and have very significant lifelong short stature, have a photosensitive rash, particularly prominent on the face, immunodeficiency and predisposition to a variety of malignancies including leukemia/lymphomas and solid tumors (*17*). The primary diagnostic test is a cytogenetic assay that demonstrates increased sister chromatid exchange (SCE) events as the result of a hyperrecombination phenotype. BLM plays an important role in resolving stalled replication forks and modulating recombination

TABLE 42-1 RecQ Helicase Disorders

Disease	Gene/Location	Cellular Phenotype and Proposed RECQ Function	Cancer Predisposition
Bloom syndrome	*BLM*/15q26.1	Increased sister chromatid exchange Repair of stalled replication forks	Leukemia/lymphoma and solid tumors
Werner syndrome	*WRN*/8p11	Premature senescence Maintenance of telomeres	Soft tissue sarcomas and skin cancers
Rothmund–Thomson syndrome	*RECQL4*/8q24.3[a]	Minor excess cytotoxicity to S phase agents Replication initiation	Osteosarcoma and skin cancers

[a]Only approximately 60% of patients with Rothmund–Thomson syndrome carry mutations in the *RECQL4* gene. The risk of osteosarcoma is associated with positive *RECQL4* mutation status (Wang, L. L.; Gannavarapu, A.; Kozinetz, C. A.; et al. *J. Natl. Cancer Inst.* **2003**, *95*, 669–674).

between sister chromatids. BLM helicase also interacts directly with many of the other homologous recombination proteins such as BRCA1 (*18*).

Werner syndrome is characterized by premature aging (including early-onset atherosclerosis, diabetes and cataracts beginning in the second decade) with increased incidence of soft tissue sarcomas (*17*). The premature aging is manifested at a cellular level as early senescence in fibroblasts from these patients. The WRN helicase encodes both a RecQ helicase module and 3′–5′ nuclease activity. Current data suggest that WRN plays a role in the replication of the telomeric lagging strand, specifically in preventing telomere SCEs (*17*).

The third disorder in this group, Rothmund–Thomson syndrome, is characterized by a rash termed "poikiloderma" which begins in infancy, skeletal dysplasias including radial ray abnormalities, cataracts and increased risk of osteosarcoma (*19*). There is not a clear cellular phenotype that can be used as a diagnostic test. Studies of the RecQ4 helicase across species demonstrate that RecQ4 helicases appear to have two fundamental roles in the initiation of DNA replication: (1) an essential function carried out by an N-terminal region of ReclQ4 that is homologous to the SLD2 family of proteins and (2) functions associated with the intrinsic helicase domain (*20*).

REFERENCES

1. Karalis, A.; Tischkowitz, M.; Millington, G. W. Dermatological Manifestations of Inherited Cancer Syndromes in Children. *Br. J. Dermatol.* **2011**, *164*, 245–256.
2. Cleaver, J. E.; Lam, E. T.; Revet, I. Disorders of Nucleotide Excision Repair: The Genetic and Molecular Basis of Heterogeneity. *Nat. Rev. Genet.* **2009**, *10*, 756–768.
3. Rao, K. S. Mechanisms of Disease: DNA Repair Defects and Neurological Disease. *Nat. Clin. Pract. Neurol.* **2007**, *3*, 162–172.
4. Cheadle, J. P.; Sampson, J. R. MUTYH-Associated Polyposis—From Defect in Base Excision Repair to Clinical Genetic Testing. *DNA Repair (Amst.)* **2007**, *6*, 274–279.
5. Manceau, G.; Karoui, M.; Charachon, A.; Delchier, J. C.; Sobhani, I. HNPCC (Hereditary Non-polyposis Colorectal Cancer) or Lynch Syndrome: A Syndrome Related to a Failure of DNA Repair System. *Bull. Cancer* **2011**, *98*, 323–336.
6. Wimmer, K.; Kratz, C. P. Constitutional Mismatch Repair-Deficiency Syndrome. *Haematologica* **2010**, *95*, 699–701.
7. Lavin, M. F. Ataxia-Telangiectasia: From a Rare Disorder to a Paradigm for Cell Signalling and Cancer. *Nat. Rev. Mol. Cell Biol.* **2008**, *9*, 759–769.
8. Derheimer, F. A.; Kastan, M. B. Multiple Roles of ATM in Monitoring and Maintaining DNA Integrity. *FEBS Lett.* **2010**, *584*, 3675–3681.
9. McKinnon, P. J.; Caldecott, K. W. DNA Strand Break Repair and Human Genetic Disease. *Annu. Rev. Genomics Hum. Genet.* **2007**, *8*, 37–55.
10. O'Driscoll, M.; Gennery, A. R.; Seidel, J.; Concannon, P.; Jeggo, P. A. An Overview of Three New Disorders Associated with Genetic Instability: LIG4 Syndrome, RS-SCID and ATR-Seckel Syndrome. *DNA Repair (Amst.)* **2004**, *3*, 1227–1235.
11. Smith, J.; Tho, L. M.; Xu, N.; Gillespie, D. A. The ATM-Chk2 and ATR-Chk1 Pathways in DNA Damage Signaling and Cancer. *Adv. Cancer Res.* **2010**, *108*, 73–112.
12. Walsh, T.; King, M. C. Ten Genes for Inherited Breast Cancer. *Cancer Cell* **2007**, *11*, 103–105.
13. Kee, Y.; D'Andrea, A. D. Expanded Roles of the Fanconi Anemia Pathway in Preserving Genomic Stability. *Genes Dev.* **2010**, *24*, 1680–1694.
14. Levy-Lahad, E. Fanconi Anemia and Breast Cancer Susceptibility Meet Again. *Nat. Genet.* **2010**, *42*, 368–369.
15. Alter, B. P.; Rosenberg, P. S.; Brody, L. C. Clinical and Molecular Features Associated with Biallelic Mutations in FANCD1/BRCA2. *J. Med. Genet.* **2007**, *44*, 1–9.
16. Meindl, A.; Hellebrand, H.; Wiek, C., et al. Germline Mutations in Breast and Ovarian Cancer Pedigrees Establish RAD51C as a Human Cancer Susceptibility Gene. *Nat. Genet.* **2010**, *42*, 410–414.
17. Chu, W. K.; Hickson, I. D. RecQ Helicases: Multifunctional Genome Caretakers. *Nat. Rev. Cancer* **2009**, *9*, 644–654.
18. Tikoo, S.; Sengupta, S. Time to Bloom. *Genome Integr.* **2010**, *1*, 14.
19. Siitonen, H. A.; Sotkasiira, J.; Biervliet, M., et al. The Mutation Spectrum in RECQL4 Diseases. *Eur. J. Hum. Genet.* **2009**, *17*, 151–158.
20. Capp, C.; Wu, J.; Hsieh, T. S. RecQ4: The Second Replicative Helicase? *Crit. Rev. Biochem. Mol. Biol.* **2010**, *45*, 233–242.

Applications to Specific Disorders

Chromosomal Disorders, 153
Cardiovascular Disorders, 167
Respiratory Disorders, 223
Renal Disorders, 239
Gastrointestinal Disorders, 265
Hematologic Disorders, 281
Endocrinologic Disorders, 319
Metabolic Disorders, 363
Mental and Behavioral Disorders, 413
Neurologic Disorders, 433
Neuromuscular Disorders, 471
Ophthalmologic Disorders, 491
Deafness, 515
Craniofacial Disorders, 521
Dermatologic Disorders, 539
Connective Tissue Disorders, 565
Skeletal Disorders, 585

CHAPTER

44

Sex Chromosome Abnormalities

Claus H Gravholt

Department of Endocrinology and Internal Medicine and Medical Research Laboratories;
Department of Molecular Medicine, Aarhus University Hospital, Aarhus C,
Aarhus, Denmark

Sex chromosome abnormalities occur rather frequently and Turner syndrome (TS) is seen in about 50 per 100,000 females, Klinefelter syndrome (KS) in 150 per 100,000 males, 47,XXX in 85 per 100,000 females, and 47,XYY in 100 per 100,000 males with 47,XYY. Patients with sex chromosome abnormalities are frequently seen by many parties in the health care system. A significant delay in diagnosis or even non-diagnosis exists for all these four syndromes. It is estimated that only about 65% of TS, 25% of KS, 12% of 47,XXX, and 14% of 47,XYY patients are diagnosed. The prevalence of prenatally detected cases with sex chromosome trisomies is also low. Thus, current clinical literature is based on diagnosed individuals and may, therefore, be biased by selection, especially, for example, if non-diagnosed individuals are less stigmatized or perhaps even more stigmatized suffering an early demise and thus remain undiagnosed.

Mortality is increased in all major sex chromosome abnormalities with hazard rates ranging from 1.4 to 1.6 for KS, 2.1–2.5 for 47,XXX, 1.9–3.6 for 47,XYY, and 2.9–4.2 for TS. Rates of morbidity have been studied in TS and KS and have been found to be increased. In TS, this increase is closely linked to a range of conditions known from clinical studies to be more prevalent. Deficiency of the product of *SHOX* explains some of the phenotypic characteristics in TS, principally short stature, which can be improved with growth hormone therapy. Puberty has to be induced in most cases, and female sex hormone replacement therapy should continue during adult years. Morbidity and mortality is increased, especially due to the risk of dissection of the aorta and other cardiovascular diseases, as well as the risk of type 2 diabetes, hypertension, osteoporosis, and thyroid disease. Despite the heavy burden of morbidity, most females with TS rate themselves as having a high quality of life.

In KS, there is an increased risk of being admitted to hospital for virtually any disease, including psychiatric disorders and congenital malformations, which is difficult to attribute specifically to the chromosome aberrations or the known consequences of the condition, specifically hypogonadism, infertility, and cognitive problems. The gene–dose effect may be a plausible cause of the congenital malformations and testicular failure and may also account for the increased risk of delayed speech, learning difficulties, and psychiatric diseases. Disturbed lateralization of the brain hemispheres has been associated with aneuploid number of X-chromosomes and increased risk of dyslexia, disturbed verbal execution, and schizophrenia. The "prototypical" man suffering from KS has traditionally been described as tall, with narrow shoulders, broad hips, sparse body hair, gynecomastia, small testicles, androgen deficiency, azoospermia, and decreased verbal intelligence. Most adult patients with KS are diagnosed because of infertility or hypogonadism, whereas boys with KS are more likely to be diagnosed because of tall stature, learning disabilities, or gynecomastia. In adulthood, KS is associated with hypogonadism, type 2 diabetes, osteoporosis with fractures, infertility, psychiatric diseases, and certain cancers, especially cancer of the breast, mediastinal cancers, and possibly others. A significant reduction in life span is the consequence. Many men with KS report sexual problems and reduced quality of life.

There are no large studies describing the clinical phenotype of 47,XXX and 47,XYY syndromes. No specific features characterize these aneuploidies, although the average height is above normal in most individuals due to the presence of an extra *SHOX* gene. Many remain undiagnosed, although a disturbance of the neurocognitive profile is present in most, with slight reductions in IQ, especially verbal performance.

A number of other sex chromosome abnormalities are seen but with a much lower frequency. Persons with additional sex chromosomes, either X or Y chromosomes, are generally more severely affected than the trisomies. Mosaicism is frequently seen and the phenotype can often be difficult to discern based on the karyotype. Thus, care needs to be based on clinical phenotype and the related medical problems.

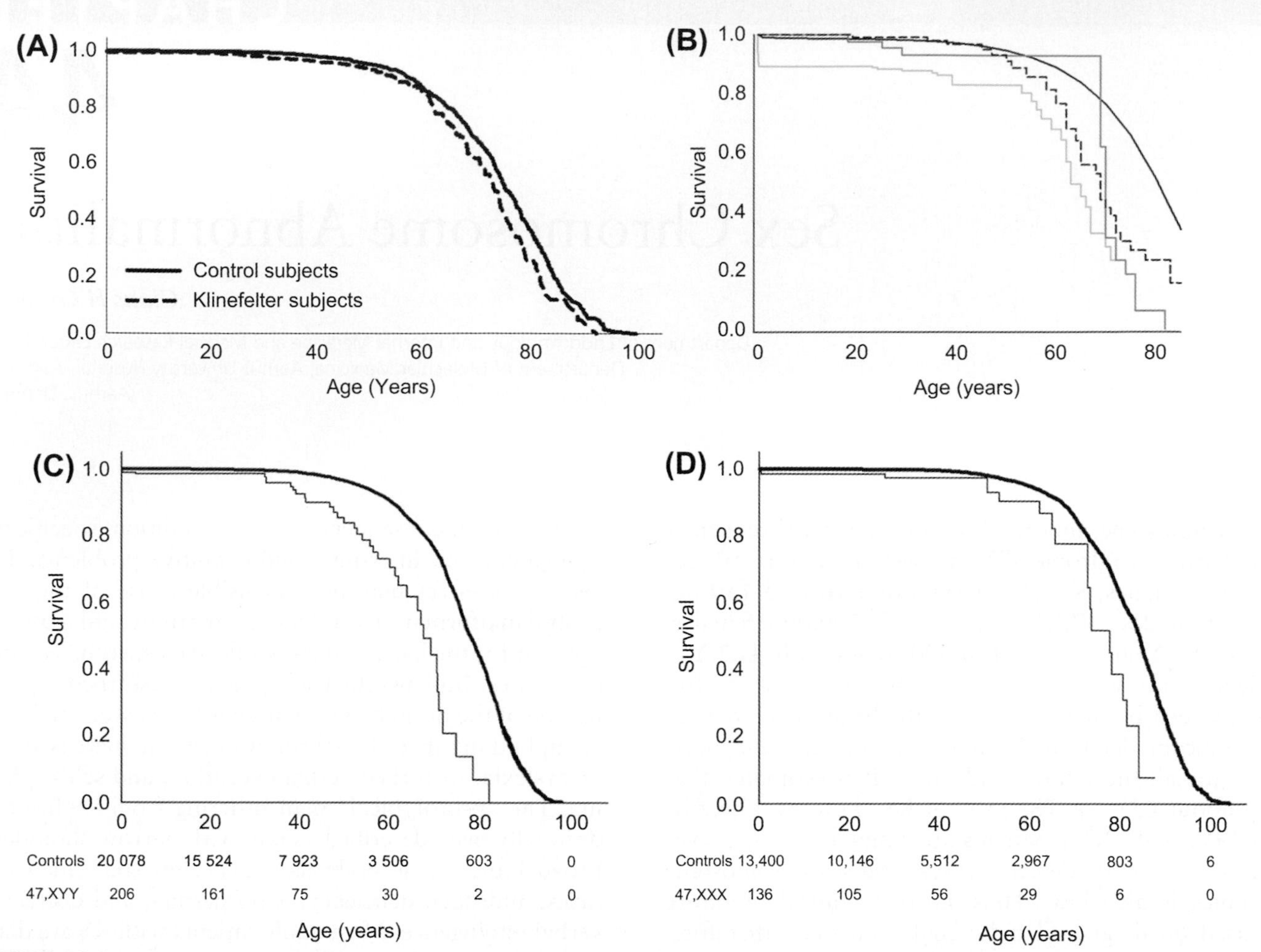

FIGURE 44-1 A. Survival of KS subjects vs. control subjects (hazard ratio, 1.40; 95% CI, 1.13–1.74; *P* _ 0.002). KS subjects lost 2.1 year (95% CI, 0.3–3.9 year; median survival) compared with control subjects. *(From Bojesen, A.; Juul, S.; Birkebaek, N.; Gravholt, C. H. Increased Mortality in Klinefelter Syndrome. J. Clin. Endocrinol. Metab. 2004, 89, 3830–3834.)* B. Kaplan–Meier plots of cumulated mortality in the general population (*black line*), females with 45, X (*light gray line*), females with an isochromosome Xq (*dark gray line*), and females with all other karyotypes associated with TS (*dashed black line*). *(From Stochholm, K.; Juul, S.; Juel, K.; Naeraa, R. W.; Gravholt, C. H. Prevalence, Incidence, Diagnostic Delay, and Mortality in Turner Syndrome. J. Clin. Endocrinol. Metab. 2006, 91, 3897–3902.)* C. Kaplan–Meier survival graphs in *47,XYY* compared to an age-matched male background population. *Solid line* controls, and *thin line* persons. Survival is significantly lower in *47,XYY* persons, log-rank $p<0.0001$. Number of persons and controls are indicated below the figure. *(From Stochholm, K.; Juul, S.; Gravholt, C. H. Diagnosis and Mortality in 47,XYY Persons: A Registry Study. Orphanet J. Rare. Dis. 2010, 5, 15.)* D. Survival graph of all women with a 47,XXX karyotype (*thin line*) and their age and gender-matched control (*solid line*). Log-rank: $p<0.0001$. Numbers of 47,XXX persons and controls are indicated below the figure. *(From Stochholm, K.; Juul, S.; Gravholt, C. H.; Mortality and Incidence in Women with 47,XXX and Variants. Am. J. Med. Genet. A. 2010, 152A, 367–372.)*

BOX 44-1 Suggested Clinical Out-Patient Program for Patients with TS

Baseline
Karyotype
Renal and pelvic ultrasound
Cardiac and aortic magnetic resonance imaging
Echocardiography
Thyroid status and antibodies
Celiac screen
Gonadotropins
Renal and liver function
Bone densitometry (DEXA scan)

Annually
Physical examination, including blood pressure and heart auscultation
Thyroid function
Body composition status (goal of BMI < 25), including physical exercise and diet instruction
Fasting lipids
Fasting blood glucose
Renal and liver function

Every 3–5 years
Echocardiography, cardiac and aortic magnetic resonance imaging
Bone densitometry (DEXA scan)
Audiogram
Celiac screen
Thyroid antibodies (thyroid peroxidase)

BOX 44-2 Abnormalities Associated with KS and Their Tentative Frequencies

Feature	Frequency (%)
Infertility (adults) (*1*)	>99
Small testes (<4ml) (*1*)	>95
Increased gonadothropin levels (*1*)	>95
Azoospermia (adults) (*1*)	>95
Learning disabilities (children) (*2*)	>75
Decreased testosterone levels (*1*)	63–85
Decreased facial hair (adults) (*1*)	60–80
Decreased pubic hair (adults) (*1*)	30–60
Gynecomastia (children) (*2,3*)	38–75
Delay of speech development (children) (*2*)	40–?
Increased height (prepubertal) (*2*)	30–?
Abdominal adiposity (adults) (*4*)	~50
Metabolic syndrome (adults) (*4*)	46
Osteopenia (adults) (*5*)	~40
Type 2 diabetes (adults) (*4*)	10–39
Crypthorcism (*2,6*)	27–37
Decreased penile size (children) (*2*)	10–25
Psychiatric disturbances (children) (*2*)	25
Congenital malformations cleft palate, inguinal hernia (*7*)	~18
Osteoporosis (adults) (*5*)	10
Mitral valve prolapse (adults) (*8*)	?–55
Breast cancer (adults) (*9,10*)	Increased risk (~50-fold)
Mediastinal cancers (children) (*11*)	increased risk (~500-fold)
Fractures (*12,13*)	Increased risk (2- to 40-fold)

BOX 44-3 Proposed Assessment and Follow-Up Program for Patients with KS

At Baseline

Confirmation of karyotype, if necessary
Sex hormones: testosterone, estrogen, SHBG, FSH, and LH
Fasting glucose and lipids
Thyroid status, hemoglobin, hematocrit, prostata specific antigen (PSA)
Physical examination including BP, height, weight, waist, testes, breasts, and varicose veins
Bone desitometry (DEXA scan) and vitamin D status, plasma calcium
Information about the syndrome
Initiation of androgen treatment (injections, transdermally, or oral)
Questions about well-being, physical activity, energy, sexual activity, and libido
Echocardiography
Fertility issues

Annual (every 3 months initially)

Physical examination including BP, height, weight, waist, testes, breasts, and varices
Sex hormones: testosterone, estrogen, SHBG, FSH, and LH (nadir values)
Fasting glucose and lipids
Thyroid status, hemoglobin
Questions about well-being, physical activity, energy, sexual activity, and libido

Every 2-5-10 years

Bone densitometry (DXA scan) and vitamin D status, plasma calcium, PSA

REFERENCES

1. Smyth, C. M.; Bremner, W. J. Klinefelter Syndrome. *Arch. Intern. Med.* **1998 June 22,** *158* (12), 1309–1314.
2. Ratcliffe, S. Long-Term Outcome in Children of Sex Chromosome Abnormalities. *Arch. Dis. Child* **1999 February,** *80* (2), 192–195.
3. Salbenblatt, J. A.; Bender, B. G.; Puck, M. H.; Robinson, A.; Faiman, C.; Winter, J. S. Pituitary-Gonadal Function in Klinefelter Syndrome before and during Puberty. *Pediatr. Res.* **1985 January,** *19* (1), 82–86.
4. Bojesen, A.; Kristensen, K.; Birkebaek, N. H.; Fedder, J.; Mosekilde, L.; Bennett, P., et al. The Metabolic Syndrome Is Frequent in Klinefelter's Syndrome and Is Associated with Abdominal Obesity and Hypogonadism. *Diabetes Care* **2006 July,** *29* (7), 1591–1598.
5. van den Bergh, J. P.; Hermus, A. R.; Spruyt, A. I.; Sweep, C. G.; Corstens, F. H.; Smals, A. G. Bone Mineral Density and Quantitative Ultrasound Parameters in Patients with Klinefelter's Syndrome after Long-Term Testosterone Substitution. *Osteoporos. Int.* **2001,** *12* (1), 55–62.
6. Lanfranco, F.; Kamischke, A.; Zitzmann, M.; Nieschlag, E. Klinefelter's Syndrome. *Lancet* **2004 July 17,** *364* (9430), 273–283.
7. Stewart, D. A.; Netley, C. T.; Park, E. Summary of Clinical Findings of Children with 47, XXY, 47, XYY, and 47, XXX Karyotypes. *Birth Defects Orig. Artic. Ser.* **1982,** *18* (4), 1–5.
8. Fricke, G. R.; Mattern, H. J.; Schweikert, H. U.; Schwanitz, G. Klinefelter's Syndrome and Mitral Valve Prolapse. An Echocardiographic Study in Twenty-Two Patients. *Biomed. Pharmacother.* **1984,** *38* (2), 88–97.
9. Hultborn, R.; Hanson, C.; Kopf, I.; Verbiene, I.; Warnhammar, E.; Weimarck, A. Prevalence of Klinefelter's Syndrome in Male Breast Cancer Patients. *Anticancer Res.* **1997 November,** *17* (6D), 4293–4297.
10. Swerdlow, A. J.; Schoemaker, M. J.; Higgins, C. D.; Wright, A. F.; Jacobs, P. A. Cancer Incidence and Mortality in Men with Klinefelter Syndrome: A Cohort Study. *J. Natl. Cancer Inst.* **2005 August 17,** *97* (16), 1204–1210.
11. Hasle, H.; Mellemgaard, A.; Nielsen, J.; Hansen, J. Cancer Incidence in Men with Klinefelter Syndrome. *Br. J. Cancer* **1995 February,** *71* (2), 416–420.
12. Bojesen, A.; Juul, S.; Birkebaek, N.; Gravholt, C. H. Increased Mortality in Klinefelter Syndrome. *J. Clin. Endocrinol. Metab.* **2004 August,** *89* (8), 3830–3834.
13. Swerdlow, A. J.; Higgins, C. D.; Schoemaker, M. J.; Wright, A. F.; Jacobs, P. A. Mortality in Patients with Klinefelter Syndrome in Britain: A Cohort Study. *J. Clin. Endocrinol. Metab.* **2005 October 4.**

FURTHER READING

1. Bojesen, A.; Juul, S.; Gravholt, C. H. Prenatal and Postnatal Prevalence of Klinefelter Syndrome: A National Registry Study. *J. Clin. Endocrinol. Metab.* **2003,** *88* (2), 622–626.
2. Stochholm, K.; Juul, S.; Juel, K.; Naeraa, R. W.; Gravholt, C. H. Prevalence, Incidence, Diagnostic Delay, and Mortality in Turner Syndrome. *J. Clin. Endocrinol. Metab.* **2006,** *91*, 3897–3902.
3. Stochholm, K.; Juul, S.; Gravholt, C. H. Mortality and Incidence in Women with 47, XXX and Variants. *Am. J. Med. Genet. A* **2010,** *152A* (2), 367–372.
4. Stochholm, K.; Juul, S.; Gravholt, C. H. Diagnosis and Mortality in 47, XYY Persons: A Registry Study. *Orphanet J. Rare Dis.* **2010,** *5*, 15.
5. Schoemaker, M. J.; Swerdlow, A. J.; Higgins, C. D.; Wright, A. F.; Jacobs, P. A. Mortality in Women with Turner Syndrome in Great Britain: A National Cohort Study. *J. Clin. Endocrinol. Metab.* **2008,** *93* (12), 4735–4742.
6. Swerdlow, A. J.; Schoemaker, M. J.; Higgins, C. D.; Wright, A. F.; Jacobs, P. A. Mortality and Cancer Incidence in Women with Extra X Chromosomes: A Cohort Study in Britain. *Hum. Genet.* **2005,** 1–6.
7. Marchini, A.; Rappold, G.; Schneider, K. U. SHOX at a Glance: From Gene to Protein. *Arch. Physiol. Biochem.* **2007,** *113* (3), 116–123.
8. Bondy, C. A. Aortic Dissection in Turner Syndrome. *Curr. Opin. Cardiol.* **2008,** *23* (6), 519–526.
9. Bondy, C. A. Care of Girls and Women with Turner Syndrome: A Guideline of the Turner Syndrome Study Group. *J. Clin. Endocrinol. Metab.* **2007,** *92* (1), 10–25.
10. Boada, R.; Janusz, J.; Hutaff-Lee, C.; Tartaglia, N. The Cognitive Phenotype in Klinefelter Syndrome: A Review of the Literature Including Genetic and Hormonal Factors. *Dev. Disabil. Res. Rev.* **2009,** *15* (4), 284–294.
11. Lanfranco, F.; Kamischke, A.; Zitzmann, M.; Nieschlag, E. Klinefelter's Syndrome. *Lancet* **2004,** *364* (9430), 273–283.
12. Otter, M.; Schrander-Stumpel, C. T.; Curfs, L. M. Triple X Syndrome: A Review of the Literature. *Eur. J. Hum. Genet.* **2009.**
13. Ratcliffe, S. Long-Term Outcome in Children of Sex Chromosome Abnormalities. *Arch. Dis. Child.* **1999,** *80* (2), 192–195.

CHAPTER

45

Deletions and Other Structural Abnormalities of the Autosomes

Nancy B Spinner, Laura K Conlin, and Surabhi Mulchandani

Department of Pathology and Laboratory Medicine at the Children's Hospital of Philadelphia and the Perelman School of Medicine at the University of Pennsylvania, Philadelphia, PA, USA

Beverly S Emanuel

Department of Pediatrics, Perelman School of Medicine, University of Pennsylvania, Division of Human Genetics, The Children's Hospital of Philadelphia, Abramson Research Center, Philadelphia, PA, USA

Deletion or duplication of many regions of the human genome is associated with clinical abnormalities. A significant proportion of the 23,000 genes in the human genome is dosage sensitive, and in this chapter, we discuss many of the regions whose deletion or duplication is associated with a clinical phenotype. We provide updates on some of the well-recognized syndromes that have been described over the past 50 years. We also discuss a number of newer syndromes, which are still being refined. These newer syndromes were identified following the introduction of array-based techniques that allow identification of genomic alterations that are not detectable using standard cytogenetic analysis. This chapter also reviews translocations (mechanisms and consequences) as well as uniparental disomy (UPD).

While the introduction of array-based diagnostic techniques into the clinical laboratory has greatly improved the diagnosis of clinically significant deletions and duplications, it has also revealed clinically benign changes. The differentiation of clinically pathogenic changes from benign changes is now a crucial element of cytogenomic diagnostics. The clinically significant genomic alterations discussed here fall into two categories, recurrent and non-recurrent. Analysis of DNA sequence in and around clinically identified deletions and duplications has revealed that there is a class of abnormalities that occurs by the underlying genomic architecture. The presence of low copy repeats (LCRs) predisposes intervening regions to result in deletions and duplications. These recurrent deletions and duplications have similar break points in many individuals. Among the recurrent, genomic disorders are some of the classic microdeletion syndromes, as well as some of the newly described disorders. Deletions are the most common genomic alteration associated with clinical phenotypes, accounting for more than 50% of genomic alterations identified by chromosome microarray (CMA). Duplications account for about 27% of alterations identified by CMA.

Historically, most genomic disorders (recurrent and non-recurrent) were recognized based on clinical presentation and were subsequently found to be caused by deletion or duplication of a region of the genome. Deletion syndromes that were initially defined by the clinical phenotype and have been shown to occur recurrently include Williams syndrome (deletions of 7q11.23), Prader–Willi or Angelman syndromes (deletions of 15q11q13), Smith–Magenis syndrome (deletions of 17p11.2), and the 22q11.2 deletion syndrome. In contrast, the Wolf–Hirschhorn (4p deletions), Cri du Chat (5p deletions), Langer–Giedion (8q24 deletions), Miller–Dieker syndrome (deletion 17p13), 18p and 18q deletion syndromes are not mediated by repeats, and the break points are variable.

Since the introduction of CMA testing for genomic deletions and duplications, several new syndromes have been identified. These newer syndromes were recognized first as a deletion or duplication, with subsequent analysis of patient phenotypes, and the clinical criteria for these syndromes is just being recognized. Many of these syndromes are recurrent, mediated by LCRs flanking dosage-sensitive genes. Among the newer microdeletion syndromes, several recurrent disorders have been identified and these include deletions of 1q21.1, 8p23.1, 15q13.3, 16p11.2, 17q12, and 17q21.3. The 17q21.3 microdeletion syndrome was the first new disorder identified using CMA, and the features of this syndrome are consistent between patients, such that the phenotype can now be clinically recognized. However, many of these recently identified microdeletion disorders have proven challenging in that the deletions are often inherited from a mildly affected or unaffected parent, suggesting that these deletions are characterized by variable expressivity and/or low penetrance. Included in this group with unclear phenotypes are the 1q21.1 and 15q13.3 deletions. We summarize the available data to provide a guide to interpretation of the findings for these abnormalities.

Clinically significant duplications include several that are associated with supernumerary chromosomes, such as Pallister–Killian syndrome (associated with tetrasomy for the short arm of chromosome 12), the isodicentric 15 syndrome (associated with tetrasomy for proximal 15q), the isochromosome 18p (tetrasomy 18p), and the Cat eye syndrome (tetrasomy for proximal 22q). Isodicentric 15 sydrome and Cat eye syndrome are associated with recurrent repeats, which mediate their formation. Several of the clinically significant microduplication syndromes are the reciprocal of the microdeletion syndromes, as these also arise by non-allelic homologous recombination (NAHR) between LCRs. The reciprocal microduplication syndromes discussed in this chapter include 7q11.23, 8p23.1, 16p11.2, 17p11.2 (Potocki–Lupski syndrome), and 22q11.2. In general, the phenotypes associated with duplications are milder than those associated with the reciprocal deletions.

The goals of this chapter are to introduce the reader to the general concepts common to autosomal deletions, duplications, translocations, and UPD, and provide details on some of the more common, or mechanistically illustrative, disorders.

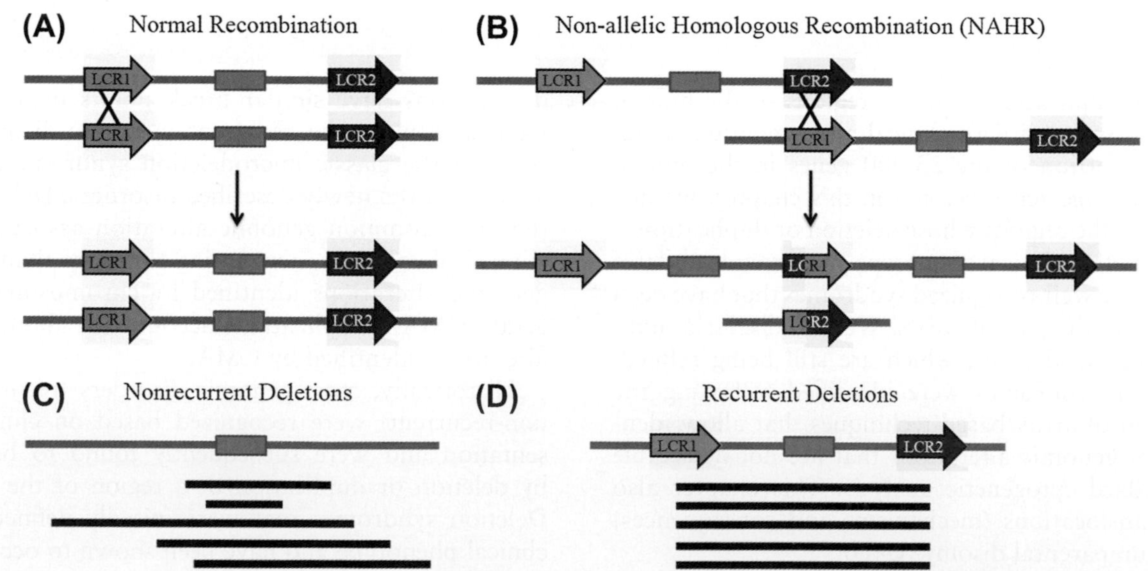

FIGURE 45-1 (A) Diagram demonstrates a genomic section with a gene (rectangle) flanked by low copy repeats (LCR1,2) undergoing normal recombination, which maintains the content of the gene with no deletions or duplications. (B) Demonstration of mispairing of the homologous repeats (LCR1,2) resulting in a deletion and a duplication. Following a recombination event within the mispaired repeats, one chromatid has a duplication of the gene, while the other chromatid is deleted for the gene. (C) Demonstration of random break points associated with deletions that are not mediated by low copy repeats. (D) Demonstration of recurrent break points resulting from non-allelic homologous recombination (NAHR) between repeats LCR1 and LCR2, which flank a gene.

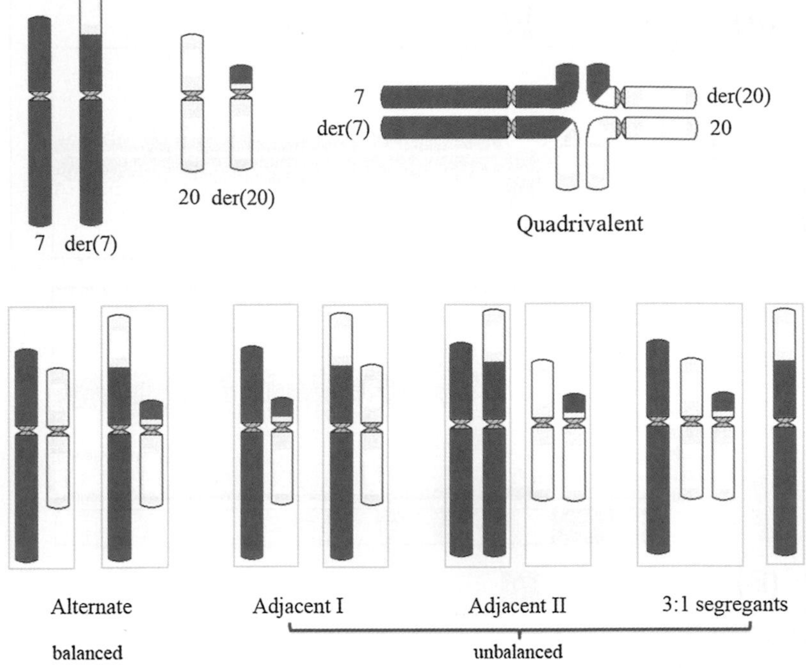

FIGURE 45-2 Pachytene diagram of a hypothetical translocation between chromosomes 7p and 20p is shown. The normal 7 and 20 and derivative 7 and 20 all line up together, with several possible segregation outcomes. If alternate centromeres move into the same gamete, then the resulting fetus would either be normal with respect to chromosomes 7 and 20, or have a balanced rearrangement, similar to the parent. However, if adjacent centromeres move into the same gamete, several types of unbalanced gametes will form. The normal 7 paired with the der(20) will result in monosomy for part of 20p, and trisomy for part of 7p; or if the normal 20 segregates with the der(7), partial trisomy 20p and monosomy 7p results. These latter alternatives, with proper segregation of the different centromeres, but partial monosomy and trisomy for distal segments, are considered Adjacent I segregants. Adjacent 2 segregation occurs when the chromosome 7 and der(7) or 20 and der(20) segregate together, resulting in more profound partial trisomy or monosomy, including the centromeric area. Three to one segregation occurs when three centromeres move into one gamete, with only one chromosome in another. This is a common observation in the recurrent translocation 11;22.

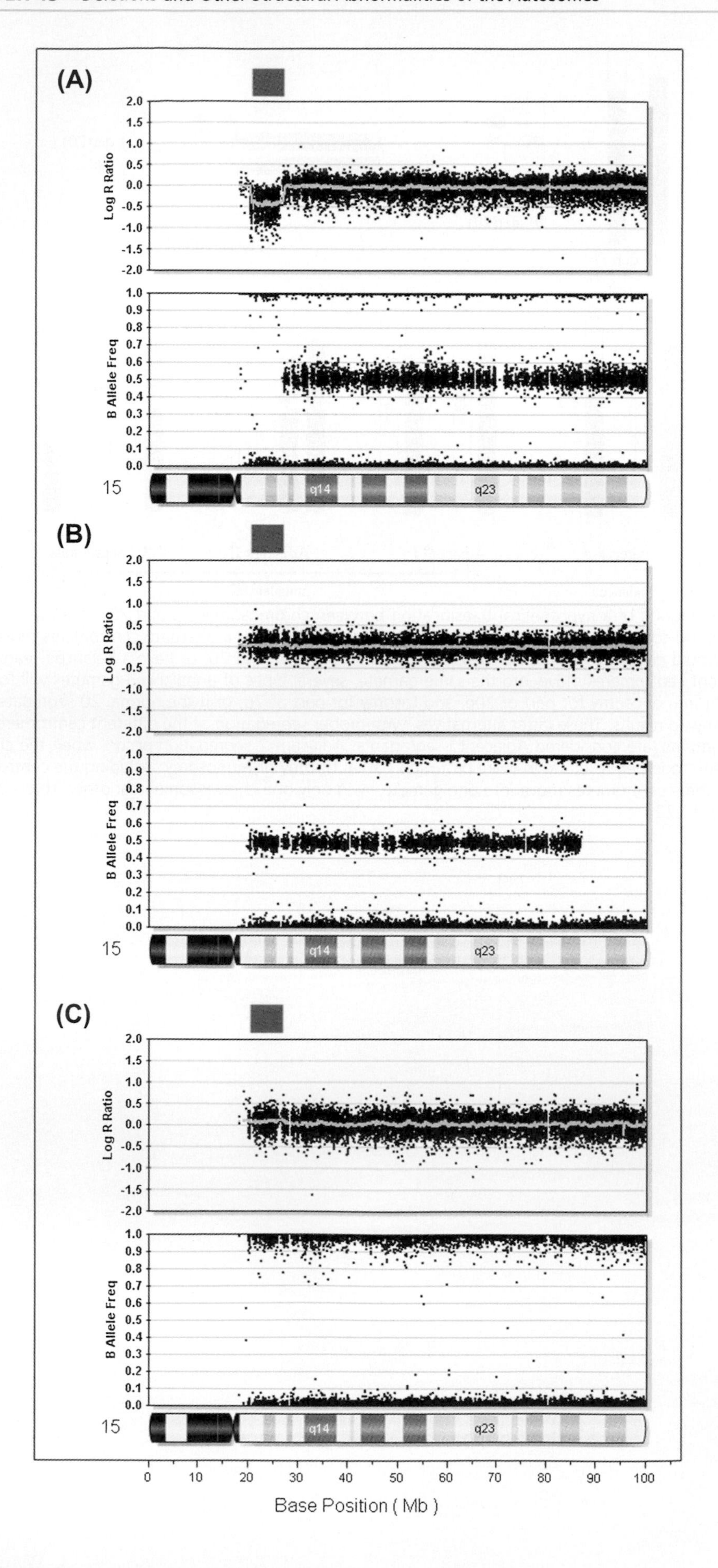
(A)
Log R Ratio
B Allele Freq
15
q14
q23
(B)
Log R Ratio
B Allele Freq
15
q14
q23
(C)
Log R Ratio
B Allele Freq
15
q14
q23
Base Position (Mb)

FIGURE 45-3 Data from the utilization of a genome-wide SNP array demonstrating diagnosis of genomic abnormalities. In each panel, the data are presented using two transformations, the $\log_2 R$ ratio (top) and B allele frequency (bottom). For each panel, each SNP is represented by a dot, and the data are displayed along the length of the chromosome, with the ideogram underneath the B allele frequency plot. The $\log_2 R$ ratio indicates the intensity of the signal at each SNP, compared to a set of "control" individuals, who presumably have two copies at that locus. Therefore if the subject has two copies, the SNPs should plot out at 0 [$\log_2(2/2) = \log_2(1) = 0$]. Deletions are seen as negative values, while duplications are seen as positive values. The genotype of each SNP is also determined, and transformed into a B allele frequency. Each SNP is bi-allelic, with each allele being either an "A" or "B." The B allele frequencies are then determined by the number of B alleles present. A genotype of AA has a B allele frequency of 0%, AB has a B allele frequency of 50%, and BB has a B allele frequency of 100%. Panel A shows a deletion of chromosome 15q11q13. Note the decreased $\log_2 R$ ratio where the intensity drops, as well as the lack of heterozygotes shown in the B allele frequency track. This is a deletion associated with Prader–Willi or Angelman syndrome. Panel B demonstrates a normal $\log_2 R$ ratio across all of chromosome 15, with a region of homozygosity toward the q terminus. This patient has uniparental disomy for chromosome 15, with heterodisomy near the centromere (presence of AB genotypes), and isodisomy near the telomere (absence of AB genotypes). This transition of hetero/isodisomy marks a meiotic crossover. Panel C demonstrates a patient with uniparental isodisomy for chromosome 15. Note the normal $\log_2 R$ ratio, but complete lack of heterozygosity across the entire chromosome. The critical region for Prader–Willi or Angelman syndrome is noted by a dark gray rectangle above the array plots. *(From Conlin, L. K.; Thiel, B. D.; Bonnemann, C. G.; et al.* Hum. Mol. Genet. *2010,* 19, *1263–1275.)*

FURTHER READING

1. Conlin, L. K.; Thiel, B. D.; Bonnemann, C. G., et al. Mechanisms of Mosaicism, Chimerism and Uniparental Disomy Identified by Single Nucleotide Polymorphism Array Analysis. *Hum. Mol. Genet.* **2010**, *19*, 1263–1275.
2. Cooper, G. M.; Mefford, H. C. Detection of Copy Number Variation using SNP Genotyping. *Methods Mol. Biol.* **2011**, *767*, 243–252.
3. Firth, H. V.; Richards, S. M.; Bevan, A. P., et al. DECIPHER: Database of Chromosomal Imbalance and Phenotype in Humans Using Ensembl Resources. *Am. J. Hum. Genet.* **2009**, *84*, 524–533.
4. Miller, D. T.; Adam, M. O.; Aradhya, S., et al. Consensus Statement: Chromosomal Microarray Is a First-Tier Clinical Diagnostic Test for Individuals with Developmental Disabilities or Congenital Anomalies. *Am. J. Hum. Genet.* **2010**, *86*, 749–764.
5. Schaaf, C. P.; Wiszniewska, J.; Beaudet, A. L. Copy Number and SNP Arrays in Clinical Diagnostics. *Annu. Rev. Genomics Hum. Genet.* **2010**, Sep 29 [Epub ahead of print].
6. Stankiewicz, P.; Lupski, J. R. Structural Variation in the Human Genome and its Role in Disease. *Annu. Rev. Med.* **2010**, *61*, 437–455.
7. Warburton, D. De Novo Balanced Chromosome Rearrangements and Extra Marker Chromosomes Identified at Prenatal Diagnosis: Clinical Significance and Distribution of Breakpoints. *Am. J. Hum. Genet.* **1991**, *49*, 995–1013.

Cardiovascular Disorders

Congenital Heart Defects, 169
Inherited Cardiomyopathies, 175
Heritable and Idiopathic Forms of Pulmonary Arterial Hypertension, 181
Hereditary Hemorrhagic Telangiectasia (Osler – Weber – Rendu Syndrome), 184
Hereditary Disorders of the Lymphatic System and Varicose Veins, 192
The Genetics of Cardiac Electrophysiology in Humans, 196
Genetics of Blood Pressure Regulation, 201
Preeclampsia, 206
Common Genetic Determinants of Coagulation and Fibrinolysis, 209
Genetics of Atherosclerotic Cardiovascular Disease, 213
Disorders of the Venous System, 215
Capillary Malformation/ Arteriovenous Malformation, 219

CHAPTER

46

Congenital Heart Defects

Rocio Moran
Department of Pediatric Genetics, Cleveland Clinic Lerner College of Medicine, Cleveland, OH, USA

Nathaniel H Robin
Department of Genetics and Pediatrics, University of Alabama at Birmingham, Birmingham, AL, USA

Congenital heart defects have always had special place among birth defects for clinical geneticists. They are both common and serious with an estimated prevalence of approximately 8 in 1000 births and are the most common cause of birth defect-related death (*1*). They are often the first anomaly to present in a child and, as many are life threatening, they also typically command the greatest attention. For these reasons, they are the defining birth defect for many genetic syndromes. For example, supravalvular aortic stenosis is linked with Williams syndrome, pulmonary stenosis with Noonan syndrome, and coarctation of the aorta with Turner syndrome. However, only one-third of heart defects present as one component of a genetic syndrome, with the remainder occurring as an isolated finding (*2*).

It is the primary role of the medical geneticist to determine if the congenital heart defect is an isolated finding, or represents one manifestation of an underlying syndrome. A syndrome is typically suspected when the heart lesion is associated with findings identified on the medical, prenatal, or family history, or on physical examination, especially noncardiac major and minor anomalies. When a syndrome is suspected, confirmatory testing may be ordered, be it chromosome analysis, fluorescence in situ hybridization, or analysis of single gene(s) for the syndrome of interest. For other cases with additional findings that are not in a recognizable pattern, array comparative genome hybridization (aCGH) should be considered as a comprehensive genetic "screening test." For those patients with isolated defects, our understanding of the genetic basis of congenital heart defects remains limited. That is because congenital heart defects are not variable manifestations of a single developmental aberration. Rather, they are an etiologically heterogeneous collection of malformations, with overlapping genetic and environmental factors that contribute to disease presentation. Further hindering genetic discovery is the fact that it is rare for isolated heart defects to present with traditional Mendelian inheritance, so most of the progress in this area has been made using animal models (*3*). However, a small but growing list of single genes may be analyzed to confirm the clinical suspicion. For those that remain without a diagnosis in both suspected syndromic and isolated congenital heart defects, follow up should be arranged.

Making an accurate genetic diagnosis for a patient with a congenital heart defect has several clear benefits including the ability to provide accurate counseling on prognosis, precise causality, and recurrence risk for the patient and the family (*4*). While the same benefits exist when it is determined that the heart lesion is isolated, there is unfortunately simply less information. Recurrence risk counseling, for example, remains largely unchanged and is based on empirical data (*5*).

Advances in cardiac imaging, cardiac catheterization, electrophysiology, and improving surgical techniques have resulted in the exponential growth of the population of adult patients with congenital heart defects, both syndromic and isolated. As this population transitions from the care of the pediatric provider to the adult provider, it is important for members of the medical team to appreciate the specific needs unique to this population of patients. Long-term follow-up of patients with congenital heart defects has identified additional complications that may represent a continuum of the primary defect and not necessarily acquired as the result of a previous repair. In addition, studies on adult patients with more common genetic syndromes are emerging and personalization of their care to their unique medical needs, depending on specific diagnosis, is necessary to prevent significant morbidity and mortality (*6*).

This chapter will provide an overview of congenital heart defects from a perspective most relevant to the clinical geneticist. Classic cardiac embryology will be reviewed along with the specific genes that, when abnormal, are implicated in the development of a specific heart lesion. We will also review the common genetic syndromes that have a cardiac defect as a major manifestation, categorized by etiology: chromosomal/aneuploidy syndromes, segmental chromosomal deletion/duplication syndromes, single gene mutations, and teratogenic. Also, extensive tables are provided that group syndromes by the type of heart defect. Lastly, recurrence

TABLE 46-1 Nonsyndromic CHD

Heart Defect	Gene	Gene Function	OMIM
Atrioventricular canal	*CRELD1*	Member of a subfamily of epidermal growth factor-related protein	606217
	GDF1	Member of the bone morphogenetic protein (BMP) family and the TGF-beta superfamily. The members of this family are regulators of cell growth and differentiation in both embryonic and adult tissues. Studies suggest that this protein is involved in the establishment of left–right asymmetry in early embryogenesis	600309
	GJA1	Member of the connexin gene family. Major protein of gap junctions in the heart that are thought to have a crucial role in the synchronized contraction of the heart and in embryonic development	121014
Hypoplastic left heart	*GJA1*	Member of the connexin gene family. Major protein of gap junctions in the heart that are thought to have a crucial role in the synchronized contraction of the heart and in embryonic development	121014 241550
	HAND1	Part of a family of basic helix-loop-helix transcription factors. One of two HAND proteins asymmetrically expressed in the developing ventricular chambers. Working in a complementary fashion, they function in the formation of the right ventricle and aortic arch arteries	241550 602406
Septal defects	*NKX2.5*	Functions in early determination of the heart field and may play a role in formation of the cardiac conduction system	108900 600584
	CITED2	Inhibits transactivation of HIF1A-induced genes by competing with binding of HIF1a to p300-CH1	602937
	GATA4	Family of zinc-finger transcription factors. Regulates genes involved in embryogenesis and in myocardial differentiation and function	607941
	HAND1	Part of a family of basic helix-loop-helix transcription factors. One of two HAND proteins asymmetrically expressed in the developing ventricular chambers. Working in a complementary fashion, they function in the formation of the right ventricle and aortic arch arteries	241550 602406
	HAND2	Part of a family of basic helix-loop-helix transcription factors. One of two HAND proteins asymmetrically expressed in the developing ventricular chambers. Working in a complementary fashion, they function in the formation of the right ventricle and aortic arch arteries	602407
	TBX20	Interacts physically, functionally, and genetically with other cardiac transcription factors, including NKX2-5, GATA4, and TBX5	611363
	ACTC1	Actins are highly conserved proteins that are involved in various types of cell motility and is major constituent of the contractile apparatus of the heart	612794
	MYH6	Encodes the alpha heavy chain subunit of cardiac myosin	160710
	THRAP2	Putative member of the TRAP family involved in early embryonic patterning expressed in brain, heart, skeletal muscle, kidney, placenta, and peripheral blood leukocytes	
Double outlet right ventricle	*CFC1*	Member of the epidermal growth factor (EGF)-Cripto, Frl-1, and Cryptic (CFC) family. Plays key role in intercellular signaling pathways during vertebrate embryogenesis including left–right patterning in the heart	605194
	GDF1	Member of the BMP family and the TGF-beta superfamily. The members of this family are regulators of cell growth and differentiation in both embryonic and adult tissues. Studies suggest that this protein is involved in the establishment of left–right asymmetry in early embryogenesis	217095
	NKX2.5	Functions in early determination of the heart field and may play a role in formation of the cardiac conduction system	600584
Tetralogy of Fallot	*GATA4*	Member of the GATA family of zinc-finger transcription factors. Thought to regulate genes involved in embryogenesis and in myocardial differentiation and function	187500 600576
	JAG1	Ligand for the receptor notch 1	187500 601920
	NKX2.5	Functions in early determination of the heart field and may play a role in formation of the cardiac conduction system	600584
	ZFPM2/FOG2	Member of the FOG family of transcription factors. Modulates the activity of GATA family proteins	187500 603693
	GDF1	Member of the BMP family and the TGF-beta superfamily. The members of this family are regulators of cell growth and differentiation in both embryonic and adult tissues. Studies suggest that this protein is involved in the establishment of left–right asymmetry in early embryogenesis	

TABLE 46-1 Nonsyndromic CHD—*cont'd*

Heart Defect	Gene	Gene Function	OMIM
Bicuspid aortic valve	*NOTCH1*	Member of the Notch family of proteins which function as a receptor for membrane bound ligands, and may play multiple roles during development	109730
Supravalvular aortic stenosis	*ELN*	Protein product is one of the two components of elastic fibers	130160 185500
Transposition of the great arteries	*CFC1*	Member of the EGF-CFC family. Plays key role in intercellular signaling pathways during vertebrate embryogenesis including left–right patterning in the heart	217095 605194
	GDF1	Member of the BMP family and the TGF-beta superfamily. The members of this family are regulators of cell growth and differentiation in both embryonic and adult tissues. Studies suggest that this protein is involved in the establishment of left–right asymmetry in early embryogenesis	600309
	THRAP2	Putative member of the TRAP family involved in early embryonic patterning expressed in brain, heart, skeletal muscle, kidney, placenta, and peripheral blood leukocytes	608771 608808
Aortic coarctation	*THRAP2*	Putative member of the TRAP family involved in early embryonic patterning expressed in brain, heart, skeletal muscle, kidney, placenta, and peripheral blood leukocytes	608771 608808

TABLE 46-2 Some Common Teratogens

Teratogenic Influence	Risk of Heart Defect (%)	Most Common Type
Maternal rubella	35	PDA, peripheral pulmonary arterystenosis, septal defects
Maternal diabetes	3–5	VSD, coarctation, TGA
Maternal phenylketonuria	25–50	Tetralogy of Fallot
Systemic lupus erythematosus	20–40	Complete heart block
Maternal alcohol abuse	25–30	Septal defects
Hydantoin	2–3	Pulmonary and aortic stenosis, coarctation of aorta, PDA
Trimethadione	15–30	Tetralogy of Fallot, TGA, hypoplastic left heart
Thalidomide	<5	Tetralogy of Fallot, septal defect struncus arteriosus
Lithium		Ebstein anomaly, tricuspid atresia ASD
Retinoic acid	10–20	Conotruncal heart defects
Cocaine	5	Excess situs disturbance

TABLE 46-3 Common Management Issues for Adults with Genetics Syndromes and Cardiovascular Abnormalities: General Information for Primary Care Providers[a]

Syndrome	Heart Defect	Complications (postoperative unless specified)	Management
Chromosome Abnormality			
22q11.2 deletion	Interrupted aortic arch type B	Left ventricular outflow tract obstruction. Residual coarctation	Depending on degree of obstruction, balloon dilation or surgical correction may be needed.
	Truncus arteriosus	Right ventricular outflow tract obstruction or insufficiency. Truncal valve regurgitation and stenosis.	Depending on the degree of right ventricular, or truncal obstruction or regurgitation, surgery may be required, e.g. valve repair or replacement.
	Tetralogy of Fallot	Right ventricular outflow tract obstruction, including pulmonary artery obstruction, pulmonary regurgitation, right ventricular dilation.	Depending on degree of obstruction or regurgitation, surgery may be required.

Continued

TABLE 46-3 Common Management Issues for Adults with Genetics Syndromes and Cardiovascular Abnormalities: General Information for Primary Care Providers[a]—*cont'd*

Syndrome	Heart Defect	Complications (postoperative unless specified)	Management
Down syndrome	Complete AVC[b]	Common atrioventricular valve regurgitation or stenosis.	Depending on degree of regurgitation or stenosis, surgery may be required.
		Residual shunt at the site of the septal defect closure.	Depending on size of shunt or residual lesion, surgery may be required.
	Arrhythmia	Postoperative bradycardia, syncope	Monitor all cases. May need pacemaker.
	Mitral valve prolapse, less frequently aortic and tricuspid prolapse.	Usually asymptomatic. Mitral, aortic and tricuspid regurgitation may develop	Auscultation in all individuals;obtain echocardiogram if abnormal.
	Coronary artery disease	Not known to have increased risk of angina or myocardial infarction.	Routine population screening.
Turner Syndrome	Bicuspid aortic valve	Can progress to aortic stenosis. As in general population, there is associated risk of aortic dilatation, aneurysm and rupture.	Monitor all cases with meticulous imaging of aortic root using echocardiography or MRI.
	Coarctation	Residual obstruction at the site of coarctation repair may lead to hyper-tension. Some mild coarctation may go undetected until adulthood. Repaired or unrepaired, associated with risk of aortic dilatation, dissection, and rupture.	Depending on severity of obstruction, may require anti-hypertensive medication, interventional catheterization or surgical correction. Aorta requires meticulous imaging using echocardiograph,preferably MRI.
	Hypertension	Can be mild to severe.	Seek underlying cause, e.g. unrepaired coarctation. Treat aggressively when hyper-tension is essential since it is a risk factor for aortic dissection.
	Aortic root dilation	Aortic dissection and rupture area associated with dilatation, although the natural history from dilatation to dissection has not been completely delineated.	Baseline imaging of aorta should be obtained in all individuals; at least one MRI examination may be needed to provide optimal imaging. Imaging should be repeated. Beta-blockade can be useful.
	ECG abnormalities	Aside from predisposition to resting tachycardia, not known to have pathologic arrhythmias.	Because of observed cardiac conduction and repolarization abnormalities, avoid drugs that increase risk of QT prolongation.
Williams–Beuren Syndrome	Supravalvar aortic stenosis, focal or diffuse	Residual left ventricular outflow tract obstruction Re-stenosis	Depending on degree of obstruction, surgical correction may be needed.
	Long-segment stenosis/ hypoplasia of thoracic or descending aorta ("coarctation") interventional catheterization	Re-coarctation	Depending on severity of obstruction, may require anti- hypertensive medication,or surgical correction.
	Coronary artery stenosis (due to medial hypertrophy or dysplastic Aortic valve leaflet) catheterization (if these	Coronary insufficiency	Screen coronary patency by echocardiogram (limited sensitivity), CTA, MRA or procedures are indicated). Exercise stress testing is of limited utility due to poor exercise tolerance in most adults with WBS.
	Hypertension Treat with antihypertensive	Can occur at any age	Screen for renovascular cause. medication.
	Pulmonary artery stenoses (PPS, main branch stenoses)	Balloon dilatation or stenting for significant obstruction. Complications such as tears or aneurysms reported.	

TABLE 46-3 Common Management Issues for Adults with Genetics Syndromes and Cardiovascular Abnormalities: General Information for Primary Care Providers[a]—cont'd

Syndrome	Heart Defect	Complications (postoperative unless specified)	Management
Gene Abnormality			
Holt–Oram Syndrome	ASD, secundum-type, VSD, membranous, muscular	Residual shunt, rarely hemodynamic risk. Potential for stroke.	Depending on size of shunt, may need surgical closure.
	ASD, secundum-type, VSD, Membranous, muscular	Eisenmenger syndrome	Medication for pulmonary hypertension.
	Conduction block	Can be progressive; when complete heart block is present, may be associated with atrial fibrillation.	Need regular monitoring with ECG and 24-hour Holter monitoring in those with known conduction disease for development and/or progression. Severe cases may require permanent pacemaker; antiarrhythmics and anticoagulants (for atrial fibrillation) may be required
Marfan Syndrome	Aortic dilatation	Aortic dissection, rupture aortic regurgitation	Prophylactic aortic root repair to prevent dissection (>45–50 mm); valve sparing technique option.
	Aortic dilatation	Progressive dilatation, rupture	Prophylactic grafting of any portion of the aorta when the risks of surgery are less than the risks of medical management, or in the case of end-organ ischemia, persistent pain
	Mitral valve prolapse	Mitral regurgitation, heart failure	Mitral valve repair or replacement.
TGF-B receptor syndromes	Aortic dilatation	Aortic dissection, rupture	Prophylactic aortic root repair to prevention dissection, but using aortic diameter criteria less than what is used for Marfan syndrome (>40 mm instead of 50mm in the adult)
	Aneuryszm of other major arteries	Arterial dissection, rupture	Prophylactic repair
Hereditary Hemorrhagic Telangectasia	Pulmonary arterio-venous malformations	Cyanosis; paradoxic embolization leading to stroke or brain abscess	Therapeutic or prophylactic embolization of the AVM with coils when the diameter of the feeding artery is 1 mm or greater
	Hepatic arterio-venous malformations	Cirrhosis; hepatic encephalopathy	Medical management of liver dysfunction
Noonan Syndrome	Pulmonic stenosis	Can present in adulthood for the first time; those with previous surgery or balloon angioplasty in childhood can have residual PS and/or pulmonary insufficiency	Options for severe pulmonary stenosis include balloon valvuplasty or valve replacement; severe pulmonary insufficiency often requires valve replacement
	ASD, secundum-type	Residual shunt after repair in childhood; some adults have undetected large defect and fatigue symptoms; in combination with pulmonary stenosis may have risk for stroke	Depending on size of shunt may require surgical closure or closure by a septal occluder device
	Hypertrophic cardiomyopathy	Right ventricular outflow obstruction; can coexist with a CHD such as pulmonary stenosis; risk of arrhythmias	Treatment considerations include surgical myomectomy and beta blockers; careful monitoring for arrhythmias may also include treatment such as amiodarone, pacemaker defibrillator

ASD, atrial septal defect; AVC, atrioventricular canal defect; CAVC, complete atrioventricular canal; CHD, congenital heart defects; CTA, CT angiography; ECG, electrocardiographic; VSD, ventricular septal defect.

[a]Guidelines provide general information for primary care providers, but individuals require specific management by cardiologists. In the interest of brevity, each cardiac abnormality for each disorder was not listed.

[b]ASD, VSD and tetralogy of Fallot were not discussed. Antibiotic prophylaxis for subacute bacterial endocarditis should be offered in appropriate doses per standard practice.

Lin, A. E.; Basson, C. T., et al. Adults with Genetic Syndromes and Cardiovascular Abnormalities: Clinical History and Management. *Genet. Med.* **2008**, *10* (7), 469–494.

risk counseling will be reviewed, for both syndromic and nonsyndromic lesions.

REFERENCES

1. Hoffman, J. I.; Kaplan, S. The Incidence of Congenital Heart Disease. *J. Am. Coll. Cardiol.* **2002,** *39* (12), 1890–1900.
2. Ferencz, C., et al. Congenital Cardiovascular Malformations: Questions on Inheritance. Baltimore–Washington Infant Study Group. *J. Am. Coll. Cardiol.* **1989,** *14* (3), 756–763.
3. Srivastava, D. Making or Breaking the Heart: From Lineage Determination to Morphogenesis. *Cell* **2006,** *126* (6), 1037–1048.
4. Robin, N. H. It Does Matter: The Importance of Making the Diagnosis of a Genetic Syndrome. *Curr. Opin. Pediatr.* **2006,** *18* (6), 595–597.
5. Shieh, J. T.; Srivastava, D. Heart Malformation: What Are the Chances It Could Happen Again?. *Circulation* **2009,** *120* (4), 269–271.
6. Lin, A. E., et al. Adults with Genetic Syndromes and Cardiovascular Abnormalities: Clinical History and Management. *Genet. Med.* **2008,** *10* (7), 469–494.

CHAPTER 47

Inherited Cardiomyopathies

Polakit Teekakirikul

Department of Genetics, Harvard Medical School, Boston, MA, USA

Carolyn Y Ho

Cardiovascular Genetics Center, Cardiovascular Division, Brigham and Women's Hospital, Harvard Medical School, Boston, MA, USA

Christine E Seidman

Department of Genetics, Harvard Medical School; Cardiovascular Genetics Center, Cardiovascular Division, Brigham and Women's Hospital; Howard Hughes Medical Institute, Boston, MA, USA

Cardiomyopathies are disorders of the myocardium that culminate in remodeling the heart, leading to hypertrophy, enlargement, and increased susceptibility to arrhythmias. Inherited and de novo gene mutations are the most prevalent causes for primary cardiomyopathies. Elucidating the molecular mechanisms leading from genetic mutation to the clinical expression of these disorders profoundly impacts our understanding of basic myocyte biology and the practical approach to managing the diseases.

47.1 HYPERTROPHIC CARDIOMYOPATHY

Hypertrophic cardiomyopathy (HCM) is defined by unexplained left ventricular hypertrophy (LVH) in a nondilated ventricle without underlying systemic predisposition (such as hypertension or valvular heart disease) (Figure 47-1).

Unexplained LVH occurs in 1 of 500 young individuals, with up to 60–75% caused by mutations in one of the genes encoding the protein constituents of the cardiac sarcomere (Figure 47-2). HCM causes several patterns of LVH, including asymmetrical LVH at interventricular septum, concentric hypertrophy, and apical hypertrophy. The histopathologic hallmarks are myocyte enlargement, myocyte disarray, and myocardial fibrosis, contributing to diastolic dysfunction. HCM is inherited in an autosomal dominant pattern with age-dependent penetrance. Mutations in sarcomere proteins may alter actin–myosin interaction, intracellular calcium cycling, pathways involved in myocardial fibrosis, force generation, or force transmission (*1,2*). Gene mutations have also been described in non-sarcomere proteins in metabolic pathways (e.g. PRKAG2, LAMP2) that produce a similar appearance of LVH resulting in phenocopies of HCM (*3,4*) (Figure 47-3).

Clinical manifestations of HCM are remarkably diverse. Patients typically present in adolescence, and the clinical spectrum spans from no or minor symptoms to progressive heart failure and sudden cardiac death. Shortness of breath, particularly on exertion, is the most common symptom of HCM.

The presence of unexplained LVH diagnosed by noninvasive cardiac imaging typically leads to the diagnosis of HCM. Clinically available genetic testing also allows for precise and age-independent identification of individuals at risk for developing HCM and should be incorporated into the diagnosis of HCM (Figure 47-4) (see "GeneTests.org" Web site). Treatment of disease manifestations and assessment of the risk for sudden death are two major aspects of HCM management. Medical therapy typically includes beta-adrenergic and calcium channel antagonists. For medically refractory patients with symptomatic obstruction, ethanol septal ablation, surgical myectomy, and cardiac transplantation may be necessary. As with all heritable disease, family screening is also an important component of management.

47.2 DILATED CARDIOMYOPATHY

Dilated cardiomyopathy (DCM) is characterized by an increased left ventricular chamber dimension and diminished contractile function (Figure 47-1). Histopathologic changes may be subtle and nonspecific with minor myocyte hypertrophy, degeneration, and interstitial fibrosis. Idiopathic DCM occurs in 1 in 2500 individuals (*5*). The prevalence of genetic etiologies is not clearly established, but likely accounts for at least 30–40% of disease.

Characteristics symptoms of DCM include exertional dyspnea, fatigue, orthopnea, and lower extremity edema. The age of onset can vary widely. Familial DCM may be associated with additional cardiac (e.g. cardiac conduction system disease) as well as extra-cardiac manifestations. The genetic basis of DCM is highly variable. Most DCM mutations (~85%) are transmitted as autosomal dominant pattern, but other modes of transmission have also been reported. DCM mutations have been found in a broad spectrum of genes encoding proteins that participate in a wide range of myocyte function, including alterations of force generation and transmission (e.g. sarcomere and nuclear intermediate filament), alterations of energy production and regulation (mitochondria), and alterations of intracellular calcium handling (Figure 47-5) (*2*). The prevalence of mutations in each individual gene is quite low, making clinical genetic testing challenging.

The diagnosis of DCM is based on clinical evaluation and the finding of dilated cardiac chambers with only modest hypertrophy. The management of DCM follows management strategies of systolic dysfunctional heart failure, including neurohormonal inhibition and diuresis as well as advanced surgical and mechanical therapies in medically refractory patients.

47.3 ARRHYTHMOGENIC RIGHT VENTRICULAR DYSPLASIA

Arrhythmogenic right ventricular dysplasia is characterized by necrotic and/or apoptotic myocyte loss and fibrofatty replacement of the myocardium with progressive dysfunction, electric instabilities, and sudden death.

The prevalence is estimated to be 1 in 5000 individuals. Thirty percent of cases are familial, typically with autosomal dominant trait. Dominant mutations have been identified in five genes encoding protein components of the desmosomes: plakoglobin (JUP), desmoplakin (DSP), desmocollin-2 (DSC2), desmoglein-2 (DSG2), and plakophilin-2 (PKP2) (*2,6,7*). A rare autosomal recessive is present in Naxos syndrome and Carvajal syndrome with associated features of palmoplantar keratosis and woolly hair. Less common mutations have been reported in the cardiac ryanodine receptor (RYR2) (*8*), the untranslated regions of TGFβ3 and transmembrane protein 43 (TMEM43) (*9*).

Electrocardiography, echocardiography, and cardiac magnetic resonance imaging (MRI) studies are typically used to assess right ventricular size and function. Patients at risk for malignant arrhythmias and sudden death should be considered for prophylactic implantable cardioverter defibrillator placement. Standard medical therapy with ACE inhibition, beta-adrenergic blockade, and diuresis is indicated for symptomatic heart failure. Clinical screening of first-degree relatives is also recommended (*10*).

47.4 VENTRICULAR NONCOMPACTION

Noncompaction of the ventricular myocardium is thought to be caused by failure of the spongiform myocardium to fully compact during embryonic development, resulting in a trabeculated myocardium. There are no pathognomonic histologic findings, although increased interstitial fibrosis and necrotic myocytes have been observed.

The estimated prevalence is 0.014% of patients referred for echocardiography. Patients typically present in childhood or early adulthood. The clinical course may be benign, but some patients experience heart failure, dysrhythmias, and an increased risk of thromboembolic events. There is also an association with neuromuscular disorders and facial dysmorphism in children. To date, ventricular noncompaction has been reported in several patients with mitochondrial disorders (*11*). In addition, mutations in genes encoding tafazzins (G4.5), LIM-binding protein (LDB3), Lamin A/C (LMNA), α-dystrobrevin (DTNA), NKX2.5, and sarcomere proteins (ACTC1, MYH7, MYBPC3, and TNNT2) have been observed (*11*).

Ventricular noncompaction is diagnosed based on echocardiographic and/or cardiac MRI findings of multiple ventricular trabeculations with deep intertrabecular recesses. Management consists of standard medical therapy for symptomatic heart failure.

47.5 CONCLUSION

Unraveling the molecular and genetic basis of inherited cardiomyopathies provides important insight into their fundamental pathophysiology. The study of gene mutations associated with structural heart disease has been instrumental in the identification of key pathways involved in the development of abnormal cardiac morphology, including deficits in force transmission and generation, energy production and regulation, and intracellular calcium handling (Figure 47-6). Knowledge gleaned from these basic science discoveries will ultimately advance opportunities for personalized approaches to diagnosis, intervention, and disease prevention.

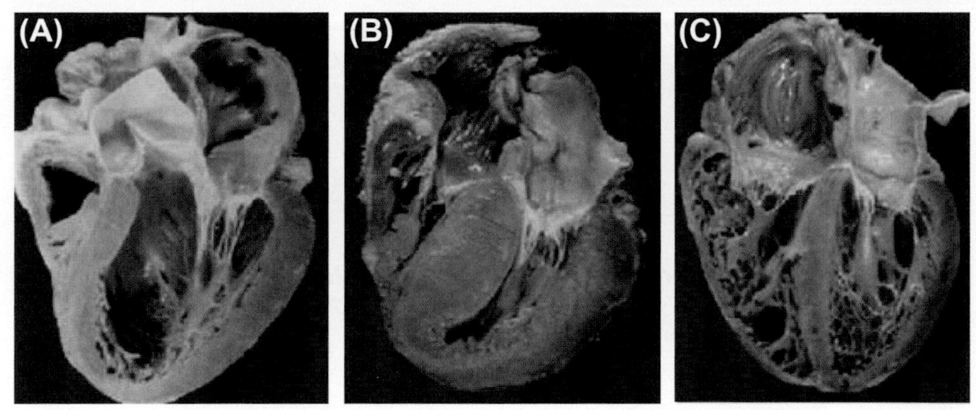

FIGURE 47-1 A comparison of normal cardiac anatomy and cardiac remodeling. A. Normal myocardium with left ventricular (LV) wall thickness ≤ 11 mm and normal LV volume. B. Hypertrophic cardiomyopathy causes increased LV wall thickness without cavity dilation. C. Dilated cardiomyopathy causes enlargement of LV and occasionally right ventricular cavity size. *(Reproduced with permission from Ahmad, F.; Seidman, J. G.; Seidman, C. E. The Genetic Basis for Cardiac Remodeling.* Ann. Rev. Genomics Hum. Genet. ***2005****,* 6, *185–216.)*

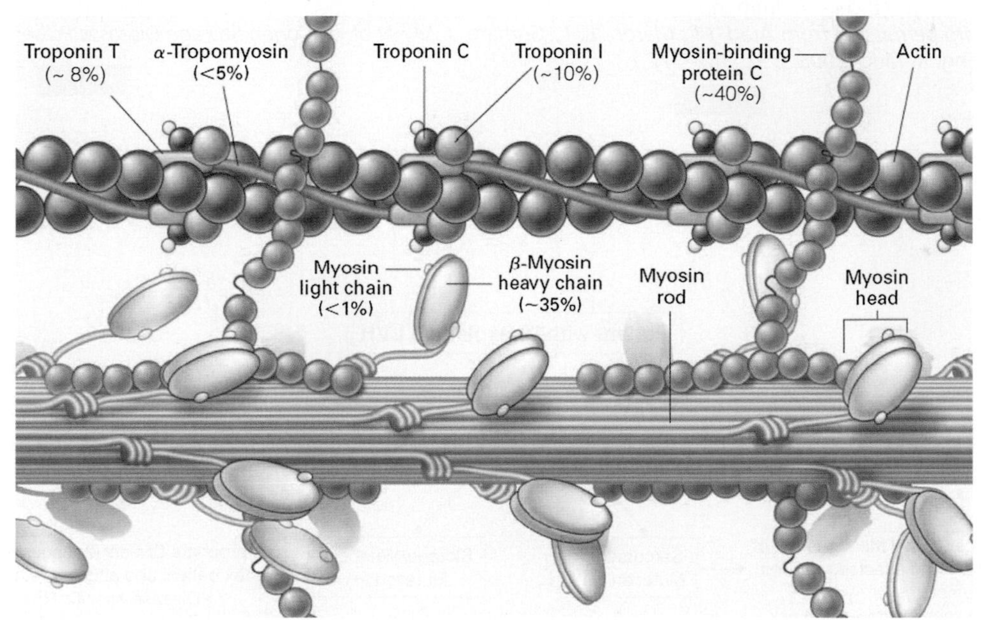

FIGURE 47-2 Mutations in sarcomere protein genes are the genetic etiology of hypertrophic cardiomyopathy as well as some forms of dilated cardiomyopathy. *(Modified and reproduced with permission from Spirito, P.; Seidman, C. E.; McKenna, W. J., et al. The Management of Hypertrophic Cardiomyopathy.* N. Engl. J. Med. ***1997****,* 336, *775–785.)*

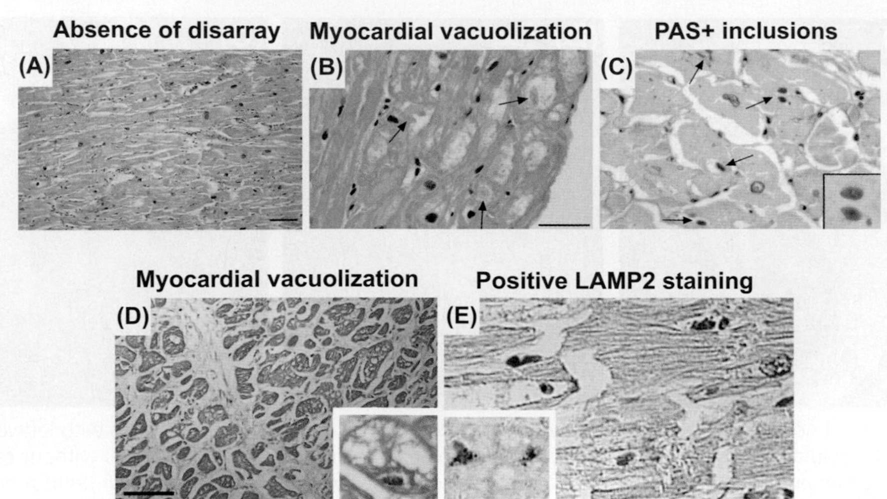

FIGURE 47-3 Histopathology of glycogen storage cardiomyopathies. A and B. PRKAG2 mutations cause non-membrane bound vacuoles (arrows) that stain for glycogen and amylopectin in myocytes. There is less fibrosis than in HCM and no myocyte disarray. C. Homogenous inclusions within vacuoles (arrows and inset) stained positive with PAS are mostly diastase-resistant. *(Reproduced from Arad, M.; Benson, D. W.; Perez-Atayde, A. R., et al. Constitutively Active AMP Kinase Mutations Cause Glycogen Storage Disease Mimicking Hypertrophic Cardiomyopathy.* J. Clin. Invest. ***2002***, 109, *357–362, with permission from The American Society for Clinical Investigation.)* D. LAMP2 mutations cause myocyte enlargement with prominent pleomorphic nuclei, and numerous cytoplasmic vacuoles. A vacuolated myocyte with a "spider cell" (inset) resembles rhabdomyoma cells. E. Immunohistochemical staining with LAMP2-specific antibodies exhibit positive (red) staining within vacuoles. *(Adapted with permission from Arad, M.; Maron, B. J.; Gorham, J. M., et al. Glycogen Storage Diseases Presenting as Hypertrophic Cardiomyopathy.* N. Engl. J. Med. ***2005***, 352, *362–372.)*

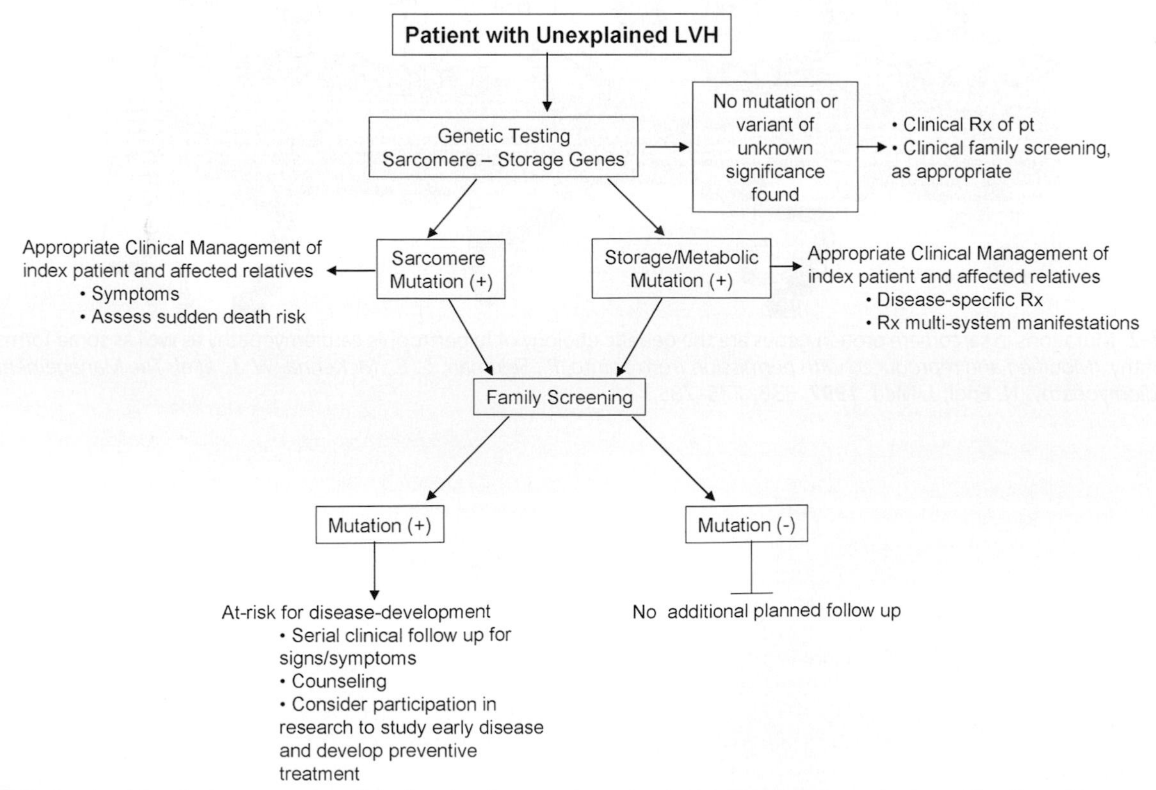

FIGURE 47-4 An algorithm for incorporating clinical molecular genetic testing in the approach to patients and families with unexplained LVH. *(Reproduced with permission from Ho, C. Y. Is Genotype Clinically Useful in Predicting Prognosis in Hypertrophic Cardiomyopathy?* Circulation ***2010***, 122, *2430–2440.)*

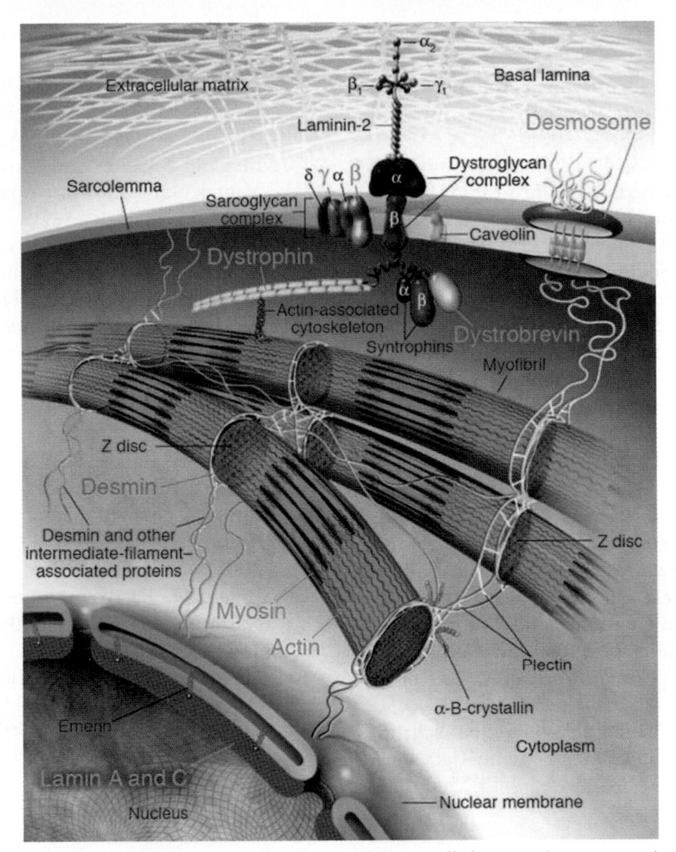

FIGURE 47-5 Mutations in components of the myocyte cytoskeleton and extracellular matrix are associated with dilated cardiomyopathy and heart failure. Contractile force generated by the sarcomere is propagated through the actin cytoskeleton and dystrophin into the dystrophin-associated glycoprotein complex (α- and β-dystroglycans, α-, β-, γ-, and δ-sarcoglycans, caveolin-3, syntrophin, and dystrobrevin). Protein elements of the desmosome, plakoglobin, desmoplakin, and plakophilin-2 provide functional and structural contacts among neighboring cells. Intermediate filament proteins, including desmin, link this network to the nuclear membrane, where lamin A/C is located. *(Reproduced from Morita, H.; Seidman, J.; Seidman, C.E. Genetic Causes of Human Heart Failure.* J. Clin. Invest. ***2005****,* 115, *518–526, with permission from The American Society for Clinical Investigation.)*

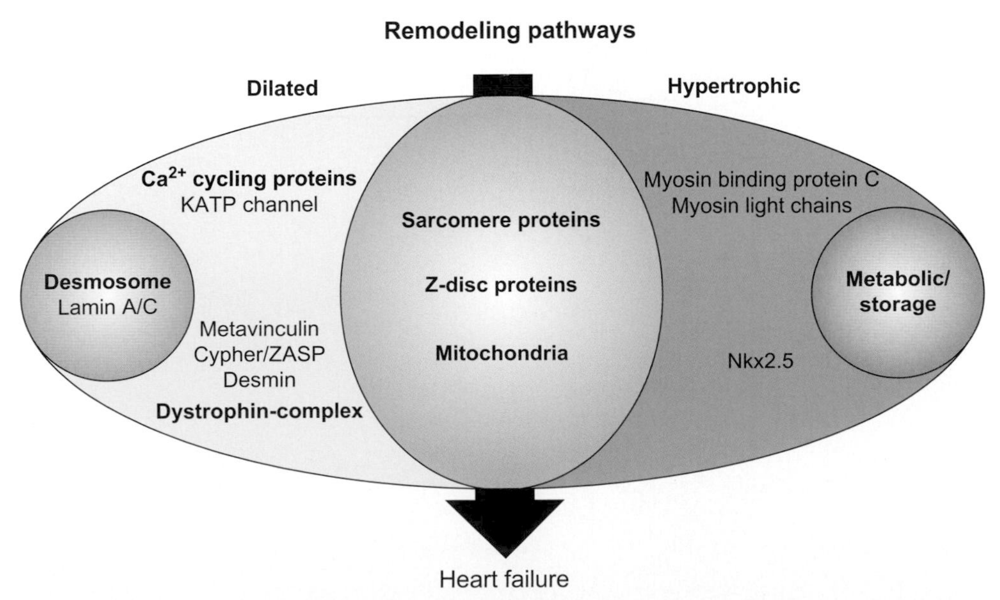

FIGURE 47-6 Human gene mutations can cause cardiac hypertrophy (blue), dilation (yellow), or both (green). Pathways that lead to pathologic remodeling of the heart involve force generation and transmission (sarcomere proteins and cytoskeletal elements), energy production and regulation [metabolic/glycogen storage proteins (pink) and mitochondrial proteins], and intracellular calcium cycling (sarcomere proteins, Ca2+ cycling proteins). *(Reproduced with permission from Morita, H; Seidman, J; Seidman, C.E. Genetic Causes of Human Heart Failure.* J. Clin. Invest. ***2005****,* 115, *518–526.)*

REFERENCES

1. Seidman, C. E.; Seidman, J. G. Identifying Sarcomere Gene Mutations in Hypertrophic Cardiomyopathy: A Personal History. *Circ. Res.* **2011,** *108* (6), 743–750.
2. Morita, H.; Seidman, J.; Seidman, C. E. Genetic Causes of Human Heart Failure. *J. Clin. Invest.* **2005,** *115* (3), 518–526.
3. Arad, M.; Maron, B. J.; Gorham, J. M.; Johnson, W. H., Jr.; Saul, J. P.; Perez-Atayde, A. R.; Spirito, P.; Wright, G. B.; Kanter, R. J.; Seidman, C. E.; Seidman, J. G. Glycogen Storage Diseases Presenting as Hypertrophic Cardiomyopathy. *N. Engl. J. Med.* **2005,** *352* (4), 362–372.
4. Maron, B. J.; Roberts, W. C.; Arad, M.; Haas, T. S.; Spirito, P.; Wright, G. B.; Almquist, A. K.; Baffa, J. M.; Saul, J. P.; Ho, C. Y., et al. Clinical Outcome and Phenotypic Expression in LAMP2 Cardiomyopathy. *JAMA* **2009,** *301* (12), 1253–1259.
5. Maron, B. J.; Towbin, J. A.; Thiene, G.; Antzelevitch, C.; Corrado, D.; Arnett, D.; Moss, A. J.; Seidman, C. E.; Young, J. B. Contemporary Definitions and Classification of the Cardiomyopathies: An American Heart Association Scientific Statement from the Council on Clinical Cardiology, Heart Failure and Transplantation Committee; Quality of Care and Outcomes Research and Functional Genomics and Translational Biology Interdisciplinary Working Groups; and Council on Epidemiology and Prevention. *Circulation* **2006,** *113* (14), 1807–1816.
6. Syrris, P.; Ward, D.; Evans, A.; Asimaki, A.; Gandjbakhch, E.; Sen-Chowdhry, S.; McKenna, W. J. Arrhythmogenic Right Ventricular Dysplasia/Cardiomyopathy Associated with Mutations in the Desmosomal Gene Desmocollin-2. *Am. J. Hum. Genet.* **2006,** *79* (5), 978–984.
7. Pilichou, K.; Nava, A.; Basso, C.; Beffagna, G.; Bauce, B.; Lorenzon, A.; Frigo, G.; Vettori, A.; Valente, M.; Towbin, J.; Thiene, G.; Danieli, G. A.; Rampazzo, A. Mutations in Desmoglein-2 Gene Are Associated with Arrhythmogenic Right Ventricular Cardiomyopathy. *Circulation* **2006,** *113* (9), 1171–1179.
8. Tiso, N.; Stephan, D. A.; Nava, A.; Bagattin, A.; Devaney, J. M.; Stanchi, F.; Larderet, G.; Brahmbhatt, B.; Brown, K.; Bauce, B.; Muriago, M.; Basso, C.; Thiene, G.; Danieli, G. A.; Rampazzo, A. Identification of Mutations in the Cardiac Ryanodine Receptor Gene in Families Affected with Arrhythmogenic Right Ventricular Cardiomyopathy Type 2 (ARVD2). *Hum. Mol. Genet.* **2001,** *10* (3), 189–194.
9. Beffagna, G.; Occhi, G.; Nava, A.; Vitiello, L.; Ditadi, A.; Basso, C.; Bauce, B.; Carraro, G.; Thiene, G.; Towbin, J. A.; Danieli, G. A.; Rampazzo, A. Regulatory Mutations in Transforming Growth Factor-Beta3 Gene Cause Arrhythmogenic Right Ventricular Cardiomyopathy Type 1. *Cardiovasc. Res.* **2005,** *65* (2), 366–373.
10. Judge, D. P. Arrhythmogenic Right Ventricular Dysplasia/Cardiomyopathy: A Family Affair. *Circulation* **2011,** *123* (23), 2661–2663.
11. Finsterer, J. Cardiogenetics, Neurogenetics, and Pathogenetics of Left Ventricular Hypertrabeculation/Noncompaction. *Pediatr. Cardiol.* **2009,** *30* (5), 659–681.

CHAPTER

48

Heritable and Idiopathic Forms of Pulmonary Arterial Hypertension

Eric D Austin

Vanderbilt University Medical Center, Department of Pediatrics, Division of Allergy, Pulmonary, and Immunology Medicine, Nashville, TN, USA

John H Newman and James E Loyd

Vanderbilt University Medical Center, Department of Medicine, Division of Allergy, Pulmonary & Critical Care Medicine, Nashville, TN, USA

John A Phillips

Vanderbilt University Medical Center, Department of Pediatrics, Division of Medical Genetics and Genomic Medicine, Nashville, TN, USA

Pulmonary arterial hypertension (PAH) is characterized by widespread obstruction or obliteration of the smallest pulmonary arteries. These arteries are obstructed by changes in all the layers of the vascular wall, including smooth muscle hypertrophy in the vascular media, obstruction of the lumen by concentric intimal fibrosis and microthrombi, and changes in the surrounding adventitia as well. When a sufficient number of vessels are occluded, the resistance to blood flow through the lungs increases. In order to maintain adequate pulmonary blood flow, the right ventricle compensates by generating higher pressure. The early phase of disease may manifest few or no symptoms until the right ventricle can no longer compensate for the increased resistance, and then progressive heart failure ensues. Initial symptoms include dyspnea, fatigue, chest pain, palpitation, syncope, or edema.

PAH was previously known as primary pulmonary hypertension (PPH), although this label no longer exists. Several classification schemes have been used to characterize PAH since its initial descriptions in the 1950s as PPH. The currently accepted system was established at the most recent World Symposium on Pulmonary Hypertension in 2008. Group 1 pulmonary hypertension, known as PAH, includes PAH that occurs both in families and as a sporadic disease. In recognition of genetic advances in understanding, including the discovery that mutations in select genes occur among patients without family history of PAH previously felt to have sporadic disease, patients with PAH are labeled heritable (HPAH) if they possess a family history of PAH and/or a detectable mutation in a PAH-associated gene. Those subjects who would otherwise meet criteria for what is classically considered PPH but lack family history or detectable mutation are considered to have the sporadic disease idiopathic PAH (IPAH).

It is likely that as our genetic understanding evolves and detection methods improve that the distinction between HPAH and IPAH will further narrow.

Individuals who have HPAH have identical symptoms, signs, and clinical course as those with IPAH (*1*). The discovery of mutations in the gene bone morphogenetic protein (BMP) receptor type 2 (*BMPR2*) as the major genetic association provided the first real understanding of the central pathogenesis and provided a basis for genetic testing, as well as the potential for new and innovative future therapies to address the basic pathogenesis at the molecular level. While to date the exact mechanism by which *BMPR2* gene mutations increase the susceptibility to PAH has not been elucidated, several features of HPAH related to *BMPR2* mutations are noteworthy. These include reduced penetrance, variable expressivity, gender dimorphism, and possible genetic anticipation. These features suggest that additional genetic and non-genetic factors modify PAH in the setting of a *BMPR2* mutation, and perhaps in other types of PAH.

PAH affects individuals of all ages, including the very young and the elderly. While females are more than twice as likely to be affected as males, whether or not one gender has more severe disease is an area of current debate. In HPAH, mortality occurs at all ages regardless of gender. The clinical characteristics and natural history of untreated PAH (then known as PPH) were reported in a multicenter study (*1*), which was conducted before the identification of the first widely effective therapies. This study included 194 affected individuals from 32 US centers where other causes of pulmonary hypertension were rigorously excluded. The mean age at diagnosis was 36 years, and while the clinical course varies, most untreated individuals gradually deteriorated, with overall mean survival of only 2.8 years. Preceding the availability of current therapies, the clinical functional capacity correlated very closely with survival. Patients who were New York Heart Association class IV had a mean survival of 6 months, whereas those who were class III had a mean survival of 2.5 years, and those who were class I or II had a mean survival of 5 years. Family history was positive for only 6% of cases in that initial report.

There are currently several classes of PAH-specific medications for therapy. While calcium channel blockers are an option for some patients, most require therapy with one or more of the following: prostaglandin analogs, endothelin receptor antagonists, and phosphodiesterase-5 inhibitors. In the current treatment era, quality of life, capacity for activity, and survival appear to have improved, although to what extent is somewhat unclear at this time. Regardless, when possible, PAH patients should be managed by physicians experienced in the diagnosis and management of this challenging disease.

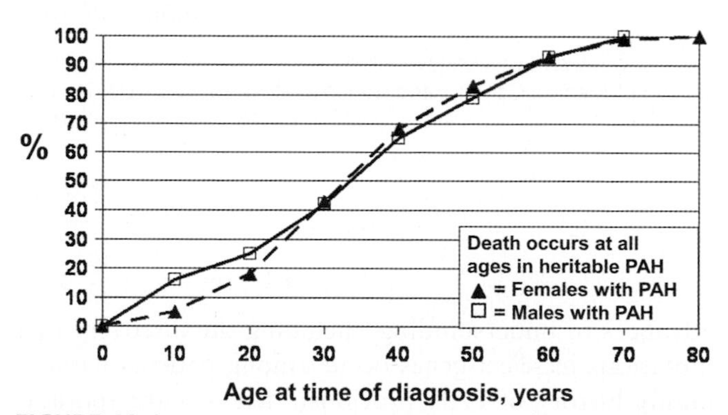

FIGURE 48-1 Cumulative mortality of patients with heritable PAH (HPAH) in families. Circles represent females, squares represent males.

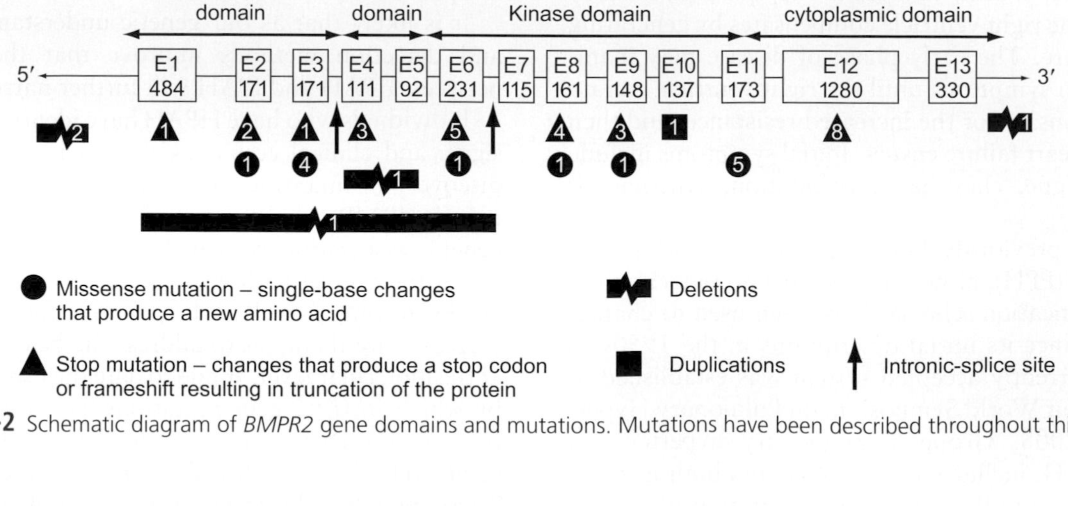

BMPR2 MUTATIONS

Ligand-binding extracellular domain | Trans-membrane domain | Kinase domain | C-terminal cytoplasmic domain

5′ E1 484 | E2 171 | E3 171 | E4 111 | E5 92 | E6 231 | E7 115 | E8 161 | E9 148 | E10 137 | E11 173 | E12 1280 | E13 330 3′

Missense mutation – single-base changes that produce a new amino acid

Stop mutation – changes that produce a stop codon or frameshift resulting in truncation of the protein

Deletions

Duplications

Intronic-splice site

FIGURE 48-2 Schematic diagram of *BMPR2* gene domains and mutations. Mutations have been described throughout this large gene.

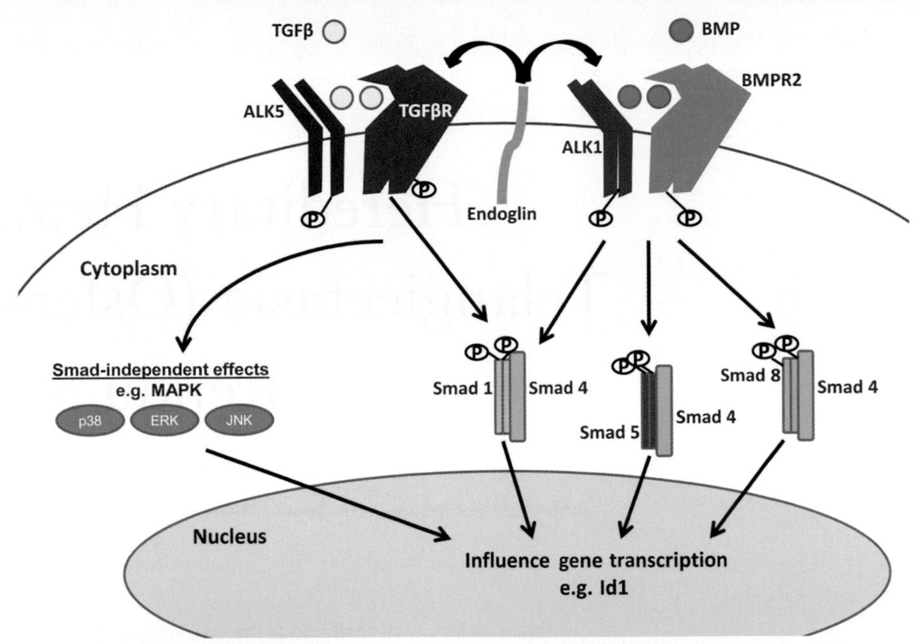

FIGURE 48-3 Schematic diagram of BMP signaling.

REFERENCE

1. Rich, S., et al. Primary Pulmonary Hypertension. A National Prospective Study. *Ann. Intern. Med.* **1987**, *107* (2), 216–223.

FURTHER READING

1. Dresdale, D. T.; Schultz, M.; Michtom, R. J. Primary Pulmonary Hypertension. I. Clinical and Hemodynamic Study. *Am. J. Med.* **1951**, *11* (6), 686–705.
2. Simonneau, G., et al. Updated Clinical Classification of Pulmonary Hypertension. *J. Am. Coll. Cardiol.* **2009**, *54* (1 Suppl), S43–S54.
3. Runo, J. R.; Loyd, J. E. Primary Pulmonary Hypertension. *Lancet* **2003**, *361* (9368), 1533–1544.
4. Farber, H. W.; Loscalzo, J. Pulmonary Arterial Hypertension. *N. Engl. J. Med.* **2004**, *351* (16), 1655–1665.
5. Rich, S., et al. Primary Pulmonary Hypertension. A National Prospective Study. *Ann. Intern. Med.* **1987**, *107* (2), 216–223.
6. Machado, R. D., et al. Genetics and Genomics of Pulmonary Arterial Hypertension. *J. Am. Coll. Cardiol.* **2009**, *54* (1 Suppl), S32–S42.
7. Barst, R. J., et al. A Comparison of Continuous Intravenous Epoprostenol (prostacyclin) with Conventional Therapy for Primary Pulmonary Hypertension. The Primary Pulmonary Hypertension Study Group. *N. Engl. J. Med.* **1996**, *334* (5), 296–302.
8. Badesch, D. B., et al. Pulmonary Arterial Hypertension: Baseline Characteristics from the REVEAL Registry. *Chest* **2010**, *137* (2), 376–387.
9. McLaughlin, V. V., et al. ACCF/AHA 2009 Expert Consensus Document on Pulmonary Hypertension a Report of the American College of Cardiology Foundation Task Force on Expert Consensus Documents and the American Heart Association Developed in Collaboration with the American College of Chest Physicians; American Thoracic Society, Inc.; and the Pulmonary Hypertension Association. *J. Am. Coll. Cardiol.* **2009**, *53* (17), 1573–1619.

CHAPTER

49

Hereditary Hemorrhagic Telangiectasia (Osler–Weber–Rendu Syndrome)

Alan E Guttmacher, Douglas A Marchuk and Scott O Trerotola

Center for the Integration of Genetic Healthcare Technologies, University of Pennsylvania, School of Medicine, Philadelphia, PA, USA

Reed E Pyeritz

Perelman School of Medicine at the University of Pennsylvania, Philadelphia, PA, USA

49.1 INTRODUCTION

Various features of hereditary hemorrhagic telangiectasia (HHT) were first described in the nineteenth century, and especially epistaxis was recognized as being familial. Around the turn of the twentieth century, Osler, Weber, and Rendu independently associated multiple phenotypic features into what came to be known for a time as Osler–Weber–Rendu syndrome. The condition is characterized by mucocutaneous telangiectases, epistaxis (due to telangiectases in the nasal mucosa), visceral arteriovenous malformations (AVMs, primarily in brain, lung, gastrointestinal tract and liver), and autosomal dominant inheritance. Two of the genes, which when mutated account for over two-thirds of cases, were discovered in the 1990s. The pathogenesis remains poorly understood, but involves disruption of the transforming growth factor-beta (TGF-β) and bone morphogenic protein (BMP) signaling pathway in endothelial cells. Early diagnosis, aggressive management of bleeding, and prophylactic treatment of visceral AVMs greatly improve morbidity and mortality.

49.2 PREVALENCE

Population-based surveys suggest a prevalence of at least one per 5000–8000. However, the condition is underdiagnosed. Several islands have markedly increased prevalence because of founder effect.

49.3 PHENOTYPE AND NATURAL HISTORY

The fundamental problem is in the development of blood vessels, especially the connections between arteries and veins. Problems start in the embryo, but they often do not become clinically important until adulthood. If the tiniest connections, the capillaries, are involved, then telangiectases develop. These 1–2 mm enlargements appear dark red on the skin, the lips, the tongue, and the mucosa of the nose and the bowel. Because the telangiectases are close to the surface and have very thin walls, they bleed easily. When the abnormal connections are larger, they are termed AVMs and are primarily found in the brain, lung, and liver. An AVM causes problems because capillaries are bypassed; the resulting shunt places strain on the heart, because of the increased blood flow demands. In the lungs, an AVM provides a direct connection between bacteria and blood clots in veins to travel directly to the arterial circulation without being filtered by lung capillaries; the result can be a stroke or cerebral abscess if the final resting place is the brain, or ischemic damage to an end organ such as the kidney.

The condition is often misdiagnosed, both because the features vary considerably among those affected (even relatives), and features accumulate over the lifespan of patients. Both marked inter- and intra-familial variability characterize HHT. Inter-familial variability is due, in part, to different mutations in different genes. Intra-familial variability is typical of many dominant disorders.

In HHT, punctuate telangiectases can appear anywhere on the skin or mucus membranes, but are most common on the fingers, palms, face, lips, buccal cheek, and tongue (Figure 49-1).

Telangiectases may be present in children, but they typically appear later and increase with age.

Often the first sign of HHT is recurrent bleeding from the nose (epistaxis). Because epistaxis occurs so frequently in the general population, until the episodes become frequent or severe, the patient or parent may not seek medical attention. Even then, the primary care or emergency physician usually treats or refers to the rhinologist but does not delve into the cause. The family history of a child that reveals severe epistaxis in close relatives can be the most important piece of evidence that a familial disorder should be entertained, and HHT is at the top of a very short list.

Bleeding from mucosal telangiectases in any portion of the gastrointestinal (GI) track can lead to chronic anemia. At the most proximal end, lesions on the lips, tongue, and oral mucosa are of cosmetic concern primarily, but bleeding can be annoying and embarrassing. Bleeding from the stomach, small intestine, and colon becomes more common with age; rarely this is a problem in children or adolescents.

One uncommon form of HHT is associated with juvenile polyposis and is due to mutations in *SMAD4*. Juvenile polyps can bleed or obstruct but are most worrisome because of their susceptibility to malignancy.

Three forms of developmental lesions and one acquired form are common in the brain and spinal column in HHT. Telangiectases, venous malformations, and cerebral arteriovenous malformations (CAVMs) are often clinically silent.

The major acquired lesion is brain abscess, which is thought due to paradoxic embolization of bacteria through a pulmonary arteriovenous malformation (PAVM).

The single clinical finding that is most likely to prompt consideration of the diagnosis of HHT is the pulmonary AVM (PAVM; Figure 49-2A). PAVMs should be looked for carefully in any person with, or at risk for, HHT, including infants, because considerable morbidity results from these lesions. A PAVM of any size can serve as a conduit for bacteria, air bubbles, or clots from the venous circulation to be transmitted directly to the systemic arterial circulation. The risk of cerebral infarction due to embolization is greater in patients with multiple PAVMs. The patient may present with progressively worsening dyspnea on exertion, cyanosis, and clubbing. Chronic right-to-left shunting can also lead to pulmonary hypertension and failure of the right side of the heart.

Infants and children should be screened for PAVMs, and occasionally, they have one or more of clinical importance. Typically, however, PAVMs emerge in adolescence and young adulthood. A lesion has a tendency to expand over time, and new ones emerge, which emphasizes the need for lifelong assessment. Some young adults will have no evidence of an intrapulmonary shunt, and are unlikely to develop any.

A small fraction of patients with HHT have pulmonary hypertension unrelated to their degree of shunting. These patients typically have a mutation in *ALK1*. Some families with primary pulmonary hypertension but without signs of HHT have mutations in *ALK1* (OMIM 178600; see Chapter 48).

Four types of hepatic vascular lesions occur in HHT: (1) telangiectases; (2) direct communication of hepatic arteries to hepatic veins; (3) direct communication of hepatic arteries to portal veins; and (4) portal vein–to–hepatic vein connections. Communication between the portal circulation and the systemic venous circulation leads to hepatic encephalopathy, ascites, and an increase in GI hemorrhage.

49.4 GENETICS

The familial nature of HHT was recognized from the earliest descriptions of the phenotype. Inheritance is autosomal dominant with virtually complete penetrance if careful phenotypic assessment is performed. However, considerable intra-familial variability occurs.

Mutations in three genes are known to cause HHT and 87% of probands will have a mutation in one of them. Endoglin (*ENG*), ALK1 (*ACVRL1*), and SMAD4 (*SMAD4*) all function in the TGFβ-BMP signaling pathway. Mutations in *SMAD4* also cause juvenile polyposis. At least two other causative loci exist.

49.5 PATHOGENESIS

A reduction to 50% of functional endoglin or ALK1 levels is compatible with development of a normal vascular system in utero, because there is no evidence of increased miscarriage in HHT families. Yet mutation carriers are at nearly 100% risk of developing the vascular lesions observed in HHT. The vascular lesions in HHT are localized to discrete regions within specific organs in the affected tissue, with little evidence of pathology outside the lesions themselves. This suggests that some genetic, physiologic, or mechanical event initiates the formation of each vascular lesion. The pathobiology of the disease may be related to remodeling of the vascular endothelium following an unknown initiating event. TGFβ1 mediates vascular remodeling through effects on extracellular matrix production by endothelial cells, stromal interstitial cells, smooth muscle cells, and pericytes.

49.6 DIAGNOSIS

Until very recently, the diagnosis of HHT could only be made on clinical grounds. However, with the clinical availability of DNA-based genetic testing, this has changed. In some situations when signs, symptoms, and/or family history raise the possibility of HHT, genetic testing can now provide a definitive answer. However, because of technical challenges, including that HHT can be due to a mutation in more than one gene, and that most families tested to date have had unique mutations, genetic testing is not always currently simple or definitive. Genetic testing is particularly helpful when the "familial" mutation has been identified; in such cases, presence or absence of the mutation, and thus of affected status, in other family members can usually be readily determined. In other situations, clinical grounds often remain the mainstay for making the diagnosis.

The classic findings in HHT are the quadrad of recurrent epistaxis, multiple telangiectases, visceral AVMs, and a positive family history. However, clinical diagnosis can be challenging, especially because many patients with HHT do not show evidence of all this quadrad and recurrent epistaxis and multiple telangiectases are each common findings in the general population.

49.7 MANAGEMENT

49.7.1 Mucocutaneous Telangiectases

Lesions on the tongue, lips, and fingers may bleed when traumatized. Photocoagulation with a laser can be effective in stopping acute bleeding. Smaller lesions can be eliminated, and this is of some cosmetic utility around the face.

49.7.2 Epistaxis

Often the most frustrating and debilitating feature of HHT is recurrent bleeding from the nose. Bleeding often starts spontaneously, even during sleep, and can be profuse. The unpredictability of epistaxis and the difficulty in stemming the acute hemorrhage may severely impact a patient's occupation, social interactions, and activities. Some patients achieve long-term improvement from periodic photocoagulation. However, those who continue to bleed enough to require transfusion of packed erythrocytes on a regular basis need to be considered for septal dermoplasty.

Prevention of epistaxis, although imperfect, is important to practice. Several agents have been reported to be effective in reducing severe epistaxis in individual patients or small, nonradomized series. The anti-VEGF drug, bevacizumab, given topically or systemically, holds promise but can be associated with severe adverse effects. Thalidomide and tranexamic acid also hold promise.

Drugs that alter platelet function, such as aspirin and nonsteroidal anti-inflammatory agents, should be avoided.

49.7.3 Central Nervous System

The developmental lesions are usually asymptomatic. Large venous malformations and CAVMs, in the absence of bleeding, may cause problems because of their size or location. Whether any particular lesion needs to be treated by radiation, surgery, or vascular occlusion needs to be considered on the basis of the patient's symptoms and age.

Brain abscess, due to paradoxic embolization of bacteria through a PAVM, needs to be treated aggressively.

49.7.4 Lung

Plain chest radiography, pulse oximetry, arterial blood gases, or a combination thereof are insensitive for detecting all but major right-to-left shunts in the lung. The contrast echocardiogram, in which agitated saline is injected into an antecubital vein as the four-chambered view of the heart is observed, is a useful screening tool for PAVMs. Delayed appearance of more than a few bubbles in the left side of the heart requires follow-up imaging.

The standard protocol to address the size, number, and location of PAVMs is a high-resolution CT scan of the lung before and after injection of radio-opaque contrast.

Until the early 1980s, patients with a PAVM of any cause were not treated unless they had recurrent hemoptysis, severe arterial desaturation, or high-output heart failure. Then, the segment or segments of lung with the offending or largest PAVMs were resected. Unfortunately, because PAVMs in HHT can progress over time, this surgical approach could be used only a limited number of times. Lesions are now treated by catheterization with insertion of an occlusive device or wire coils that invoke a thrombus in the feeding artery. Now, under fluoroscopic guidance, four to six or more PAVMs can be treated in one outpatient session (Figure 49-3).

PAVMs with feeding arteries of greater than 2 mm diameter should be untreated. After all PAVMs are occluded, the patient should have a repeat CT scan in 6 months to document that all remain occluded (i.e. that no recanalization has occurred).

Major complications related to PAVMs can occur during pregnancy, due to either pulmonary hemorrhage, increased right-to-left shunting, or cerebrovascular events. Any patient diagnosed with HHT during pregnancy should undergo high-resolution CT of the lung with contrast. Embolotherapy is indicated during pregnancy for PAVMs that can be reached by an intervention catheter.

49.7.5 Liver

Vascular lesions in the liver are much less amenable to treatment than are PAVMs. Occlusion of the feeding vessel or vessels, although technically feasible, carries an unacceptable risk of infarction of substantial regions of liver parenchyma. There also is an increased risk of infection associated with endoscopic retrograde cannulation of the pancreas, so ERCP is contraindicated in HHT.

Patients who have hepatic encephalopathy can be treated with nitrogen restriction and lactulose. Those with high-output cardiac failure respond initially to diuretics. The therapy of last resort is liver transplantation.

49.8 GASTROINTESTINAL

Hemorrhage from the GI tract can be one of the most debilitating and frustrating problems of HHT, especially in older people. Upper and lower endoscopy typically shows multiple mucosal telangiectases from the oral cavity to the duodenum and from the rectum to the ileocecal valve. The number of telangiectases roughly correlates with the anemia and transfusion requirement. Any lesion that appears to be bleeding can be cauterized. However, there is no point in attempting to treat every telangiectasis, especially if repeated endoscopies are contemplated. Capsule endoscopy confirms that telangiectases exist throughout the jejunum and ileum; but push enterostomy or major surgery is rarely required. As with refractory epistaxis, hormonal (estrogen/progesterone in women without contraindications, and danazol in men) and antifibrinolytic therapy may have a role in chronic GI bleeding. Similarly, experimental agents such as thalidomide and bevacizumab are worth considering.

49.9 ANEMIA

Unless a contraindication exists, virtually all adults with HHT should take supplemental iron. The amount and form of iron depends on the severity of the anemia. Additional cofactors, folate, and B_{12} should be considered. Some patients require episodic or even regular transfusions of packed erythrocytes.

49.10 COUNSELING

The same issues that arise in most serious autosomal dominant disorders occur in HHT. The recent availability of molecular diagnosis offers promise of presymptomatic and prenatal diagnosis. However, complexities often arise due to intergenic heterogeneity, “negative” results, and DNA sequence variants of uncertain pathogenic importance.

The Hereditary Hemorrhagic Telangiectasia Foundation International (www.hht.org) plays an important role in several areas. The organization facilitates the establishment of centers at academic hospitals that provide comprehensive medical and counseling services for patients and families with HHT. There are now 29 centers worldwide, including 15 in North America. The Foundation also sponsors national conferences for patient’s education and support, and a biennial international research symposium.

49.11 LIFE EXPECTANCY

Infants and children with HHT and unusually severe vascular malformations in the CNS or lung can die. Many adults with HHT live into their eighth and ninth decade despite severe anemia from epistaxis and GI bleeding. A decade ago, a mortality analysis of patients in one county in Denmark showed that patients younger than age 60 died at twice the population average. A more recent study from a HHT center in Italy confirmed an increased peak in mortality rate in patients under the age of 50 years, and somewhat higher mortality in those aged 60–79. Life expectancy was not related to gender or to genotype.

ACKNOWLEDGMENTS

This revision was accomplished during a time of support for REP by the National Heart, Lung and Blood Institute (GenTAC) and the National Human Genome Research Center (Center of Excellence in ELSI Research P50-HG-004487), both of the U.S. National Institutes of Health, and while in residence at the Brocher Foundation, Hermance, Switzerland. We also thank all of our colleagues who shared their published and unpublished work and commented on sections of the chapter.

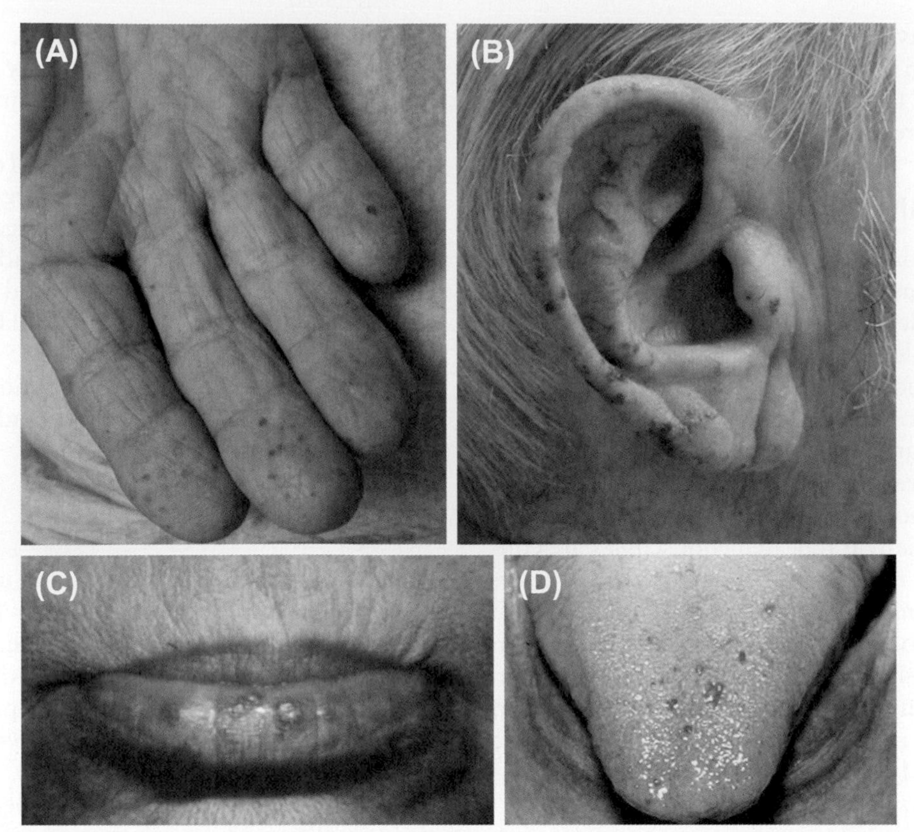

FIGURE 49-1 Dermal and mucocutaneous features of HHT. (A) Digits, (B) ear, (C) lips, and (D) tongue.

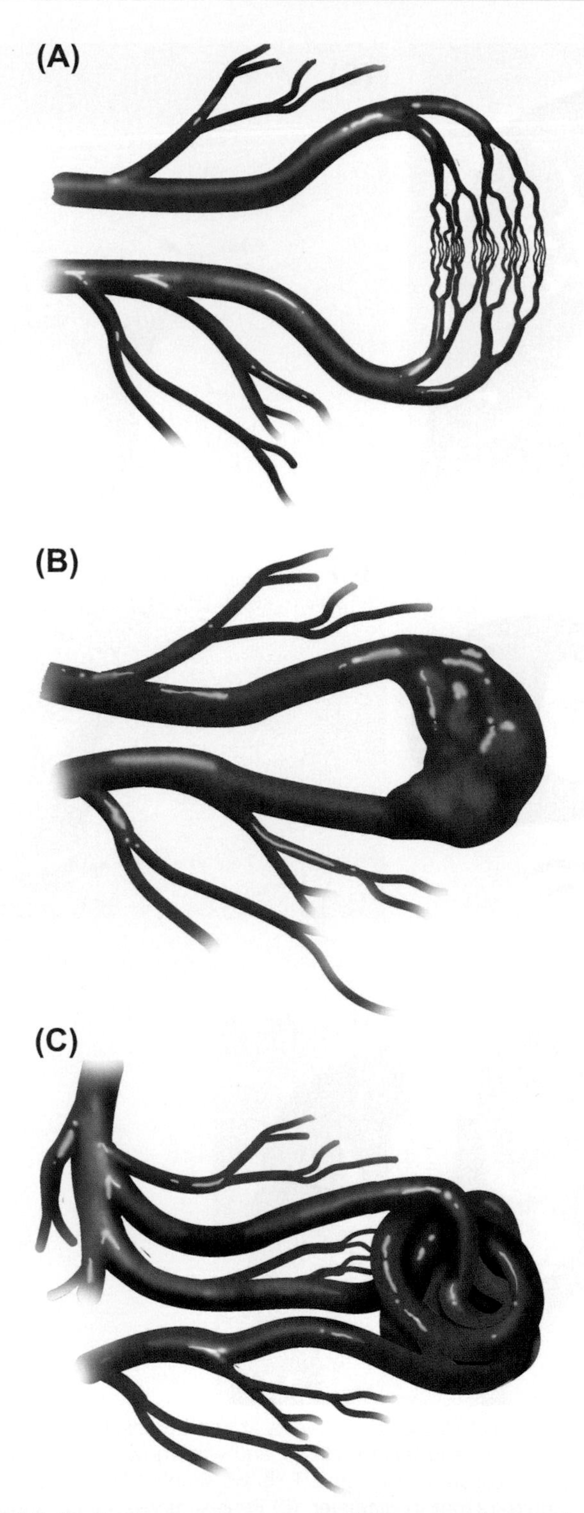

FIGURE 49-2 Filtration function of pulmonary capillary bed and how a PAVM bypasses that function to produce paradoxical emboli can be envisioned from these renderings. (A) Normal capillary bed. *(Reprinted with permission from Trerotola, S. O.; Pyeritz, R. E. PAVM Embolization: An Update. AJR* ***2010****, 195, 837–845.)* (B) A simple PAVM. (C) A complex, large PAVM. *(Reprinted with permission from Trerotola, S. O.; Pyeritz, R. E. PAVM Embolization: An Update. AJR* ***2010****, 195, 837–845.)*

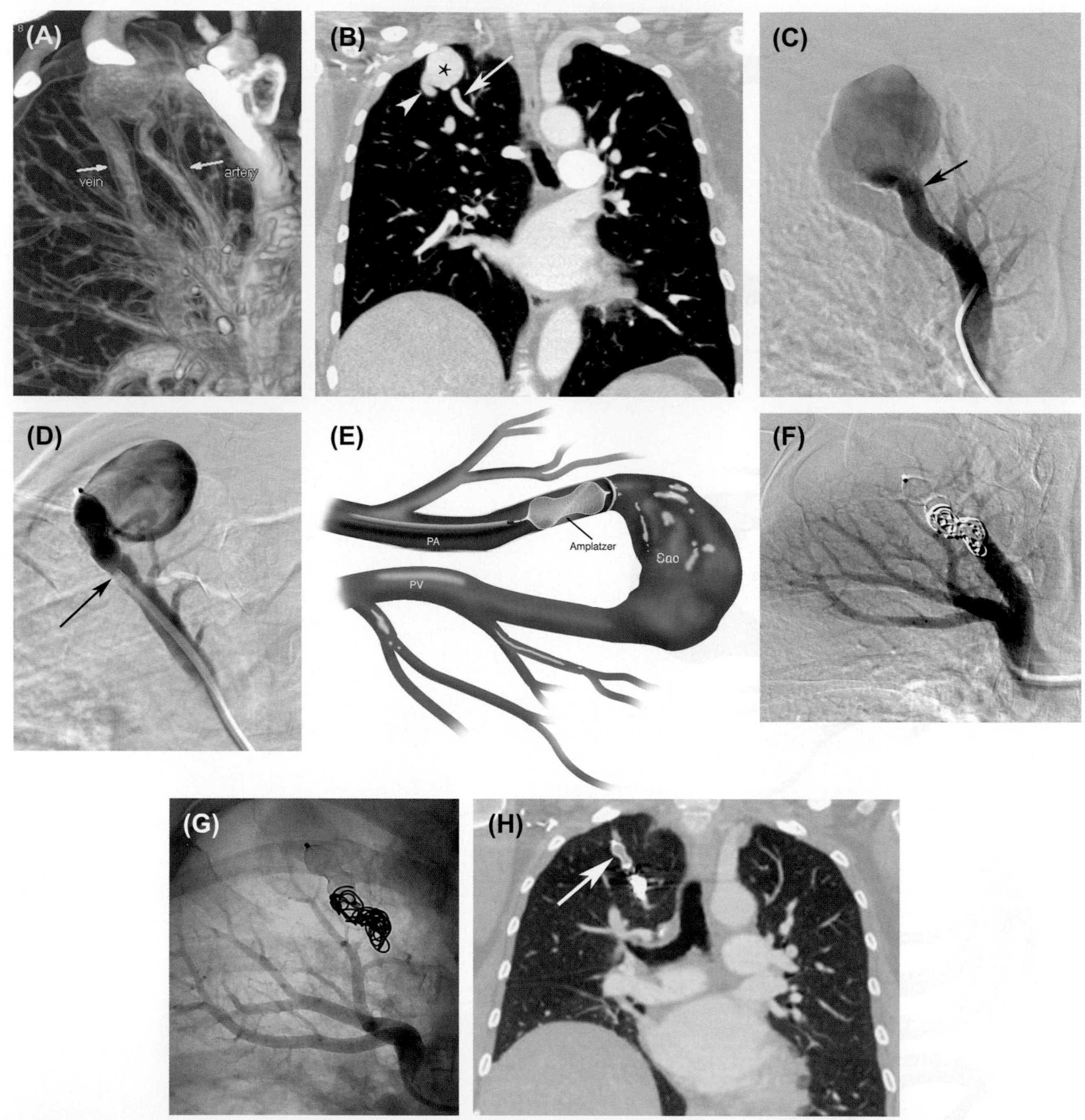

FIGURE 49-3 A 52-year-old woman with known right upper lobe PAVM. Patient was followed for 14 years at another institution and told treatment was not needed. She has obvious HHT with a strong family history, epistaxis, and telangiectases, and was known by her primary care physician to have HHT. She was asymptomatic. (A) Color-enhanced coronal reconstruction from CT shows simple PAVM in the right upper lobe; feeding artery and draining vein are labeled. Feeding artery measured 12 mm in diameter. (B) Representative image showing feeding artery (*arrow*), sac (*asterisk*), and draining vein (*arrowhead*). (C) Image from selective right upper lobe pulmonary angiography confirms simple angioarchitecture; the feeding artery is indicated. (D) Image obtained during contrast injection with 16 mm Amplatzer vascular occluder deployed at the mouth of PAVM sac. Arrow shows occluder still attached to delivery cable. (E) Diagrammatic representation of panel D. F and G. Subtracted (F) and unsubtracted (G) postembolization pulmonary arteriograms after placement of coils in addition to Amplatzer vasclar occluder show occlusion of PAVM. H. Representative image from coronal CT reconstruction 6 months after embolization. Sac has almost completely disappeared. Characteristic appearance of Amplatzer device (*arrow*). Patient reported marked improvement in her exercise tolerance. *(Reprinted with permission from Trerotola, S. O.; Pyeritz, R. E. PAVM Embolization: An Update. AJR* ***2010****, 195, 837–845.)*

FURTHER READING

1. Al-Saleh, S.; Mei-Zahav, M.; Faughnan, M. E., et al. Screening for Pulmonary and Cerebral Arteriovenous Malformations in Children with Hereditary Haemorrhagic Telaniectasia. *Eur. Respir. J.* **2009**, *34,* 875–881.
2. Bayrak-Toydemir, P.; McDonald, J.; Markewitz, B., et al. Genotype-Phenotype Correlation in Hereditary Hemorrhagic Telangiectasia: Mutations and Manifestations. *Am. J. Med. Genet. A.* **2006**, *140,* 463–470.
3. Brinkerhoff, B. T.; Poetker, D. M.; Choong, N. W. Long-Term Therapy with Bevacizumab in Hereditary Hemorrhagic Telangiectasia. *N. Engl. J. Med.* **2011**, *364,* 688–689.
4. Cohen, J. H.; Faughnan, M. E.; Letarte, M., et al. Cost Comparison of Genetic and Clinical Screening in Families with Hereditary Hemorrhagic Telangiectasia. *Am. J. Med. Genet.* **2005**, *137A,* 153–160.
5. Faughnan, M. E.; Palda, V. A.; Garcia-Tsao, G., et al. International Guidelines for the Diagnosis and Management of Hereditary Hemorrhagic Telangiectasia. *J. Med. Genet.* **2011**, *48,* 73–87.
6. Gallione, C.; Aylsworth, A. S.; Beis, J., et al. Overlapping Spectra of SMAD4 Mutations in Juvenile Polyposis (JP) and JP-HHT Syndrome. *Am. J. Med. Genet. A.* **2010**, *152A,* 333–339.
7. Gallione, C. J.; Repetto, G. M.; Legius, E., et al. A Combined Syndrome of Juvenile Polyposis and Hereditary Haemorrhagic Telangiectasia Associated with Mutations in MADH4 (SMAD4). *Lancet* **2004**, *363,* 852–859.
8. Gallione, C. J.; Richards, J. A.; Letteboer, T. G., et al. SMAD4 Mutations Found in Unselected HHT Patients. *J. Med. Genet.* **2006**, *43,* 793–797.
9. Garcia-Tsao, G.; Korzenik, J. R.; Young, L., et al. Liver Disease in Patients with Hereditary Hemorrhagic Telangiectasia. *N. Engl. J. Med.* **2000**, *343,* 931–936.
10. Girerd, B.; Montani, D.; Coulet, F., et al. Clinical Outcomes of Pulmonary Arterial Hypertension in Patients Carrying an ACVRL1 (ALK1) Mutation. *Am. J. Respir. Crit. Care Med.* **2010**, *181,* 851–861.
11. Krings, T.; Ozanne, A.; Chang, S. M., et al. Neurovascular Phenotypes in Hereditary Haemorrhagic Telangiectasia Patients According to Age: Review of 50 Consecutive Patients Aged 1 day-60 Years. *Neuroradiol.* **2005**, *47,* 711–720.
12. McDonald, J. M.; Bayrak-Toydemir; Pyeritz, R. E. Hereditary Hemorrhagic Telangiectasia: An Overview of Diagnosis, Management, and Pathogenesis. *Genet. Med.* **2011**, *13,* 607–616.
13. McDonald, J.; Damjanovich, K.; Millson, A., et al. Molecular Diagnosis in Hereditary Hemorrhagic Telangiectasia: Findings in a Series Tested Simultaneously by Sequencing and Deletion/Duplication Analysis. *Clin. Genet.* **2011**, *79,* 335–344.
14. Ross, D. A.; Nguyen, D. B. Inferior Turbinectomy in Conjunction with Septodermoplasty for Patients with Hereditary Hemorrhagic Telangiectasia. *Laryngoscope* **2004**, *114,* 779–781.
15. Sabbà, C.; Pasculli, G.; Suppressa, P., et al. Life Expectancy in Patients with Hereditary Hemorrhagic Telangiectasia. *QJM* **2006**, *99,* 327–334.
16. Ten Dijke, P.; Goumans, M.-J.; Pardali, E. Endoglin in Angiogenesis and Vascular Diseases. *Angiogenesis* **2008**, *11,* 79–89.
17. Trerotola, S. O.; Pyeritz, R. E. PAVM Embolization: An Update. *AJR* **2010**, *195,* 837–845.
18. van Tuyl, S. A.; Letteboer, T. G.; Rogge-Wolf, C., et al. Assessment of Intestinal Vascular Malformations in Patients with Hereditary Hemorrhagic Teleangiectasia and Anemia. *Eur. J. Gastroenterol. Hepatol.* **2007**, *19,* 153–158.
19. www.HHT.org.
20. McDonald, J.; Pyeritz, R. E. Hereditary Hemorrhagic Telangiectasia. *Gene Reviews*. Updated 19 May 2009. www.ncbi.nlm.nih.gov/books/NBK1351/

CHAPTER

50

Hereditary Disorders of the Lymphatic System and Varicose Veins

Robert E Ferrell

Department of Human Genetics, University of Pittsburgh, Pittsburgh, PA, USA

Reed E Pyeritz

Perelman School of Medicine at the University of Pennsylvania, Philadelphia, PA, USA

Lymphatics are thin-walled vessels that carry a protein-rich fluid from the periphery back to the central circulation. As in the systemic venous system, unidirectional flow is assisted by valves throughout lymphatics of all calibers. Lymphatics draining the legs, pelvis, and abdomen merge into the cisterna chyli, which passes with the aorta through the diaphragm and becomes the thoracic duct. Lymphatics comprising the left jugular, subclavian, and mediastinal trunks usually join the thoracic duct as it loops over the left subclavian artery and drain into the left innominate vein. Lymphatic drainage of the right subclavian, jugular, and mediastinal trunks usually merges into a right lymphatic duct, which opens into the right innominate vein. However, any of the trunks may open independently into the great veins. Arranged periodically along lymphatic trunks are lymph nodes, comprising a stroma packed with cells of the immune system, which serve as filters of the lymphatic fluid.

50.1 DEVELOPMENT OF THE LYMPHATIC SYSTEM

The lymphatic vessels arise from the cardinal vein after the cardiovascular system is established and functional. Lymphatic vessel development commences about embryonic weeks 6–7 in humans, when a distinct subpopulation of endothelial cells in the lateral part of the anterior cardinal veins becomes committed to a lymphatic lineage and sprout laterally to form the vascular sacs. The peripheral lymphatic vasculature is generated by the sprouting of lymphatic vessels from the lymph sacs, followed by merging of the separate lymphatic capillary networks and remodeling and maturation of the lymphatic capillary plexus.

Direct communication between the lymphatic network and the vascular system occurs around 40 days postconception in humans, whereby the lymphatic drainage of the right and left sides of the upper body, which had collected in bilateral jugular lymph sacs, empties into the ipsilateral jugular veins. Failure of delay in completing these communications leads to widespread distortion of lymphatic development termed the *jugular lymphatic obstruction sequence.* The clinical consequences of edema distal to the obstruction will vary depending on whether communication is established and when it occurs. Complete absence of communication is lethal.

Defective lymphatic development has consequences for the cardiovascular system. Peripheral veins tend to be of large caliber, presumably because of increased venous return from edematous tissues. The frequency of left-sided flow defects (e.g. aortic coarctation, bicuspid aortic valve, hypoplastic left heart) is increased, in part due to the space occupied by the distended jugular sacs.

50.2 DISORDERS OF THE LYMPHATIC SYSTEM

Diffuse, acquired blockage of lymphatics, such as by fibrosis, tumor, or infection (e.g. *Microfilaria bancrofti*), usually results in edema of the body parts distal to the blockage. However, if the thoracic duct is blocked, anastamosing channels develop between the lymphatic system and systemic veins, and edema does not persist (Table 50-1).

A prenatal cystic hygroma is often seen in fetal aneuploidy. If the fetal karyotype is normal, a cystic hygroma detected in the first trimester carries a good prognosis; if it resolves by the mid-second trimester, little distortion of the infant is likely to be noted.

Both the Turner syndrome (Chapter 50) and the Noonan syndrome (Chapter 52) have prominent multisystem malformations due to fetal lymphatic obstruction. Failure of the jugular lymphatic–venous communication to occur may be the major reason that 98% of embryos with a 45,X karyotype do not survive to term.

The National Lymphedema Network, Inc. maintains a list of various support groups; while many pertain to acquired lymphedema, issues pertinent to various congenital and hereditary conditions can be found (www.lymphnet.org).

50.2.1 Hereditary Lymphedema I

Hereditary lymphedema 1A (OMIM 153100) is a mainly congenital onset lymphedema, the swelling of the extremities due to failure in the development and/or function of the lymphatics, leading to the accumulation of lymph in the interstitial fluid space. It is highly variable in expression and penetrance, and unaffected obligate heterozygous individuals occur in almost all pedigrees of sufficient size. The age at onset varies from embryonic (as seen on prenatal ultrasound) to middle age. The classic presentation includes congenital edema of the lower half of the body with no other obvious manifestations. The complications of Milroy disease include hypoproteinemia due to intestinal loss of albumin, increased susceptibility to infection in the affected limb, and, in some cases, angiosarcoma. Milroy disease maps to chromosome 5q34-q35 and is caused by mutations in the kinase domains of the vascular endothelial growth factor receptor-3 (*VEGF3*, *FLT4*).

Hereditary lymphedema 1C (OMIM 613480) has an early onset of uncomplicated lymphedema of the legs and hands, and is inherited in an autosomal dominant pattern with variation in age at onset, penetrance, and expressivity. Phenotypically it is indistinguishable from Milroy disease, but is caused by missense mutations in *GJC2* coding for connexin 47.

50.2.2 Hereditary Lymphedema II

Hereditary lymphedema II (OMIM 153200) is a dominantly inherited, pubertal or adult onset lymphedema predominantly affecting the legs, but with involvement of the arms, face, and larynx in some families.

50.2.3 Lymphedema–Distichiasis Syndrome

Lymphedema–distichiasis syndrome (OMIM 153400) is characterized by bilateral lymphedema, usually of the legs, and having a peripubertal age at onset and distichiasis. These patients often come to medical attention because of corneal abrasions caused by the presence of a double row of eyelashes (distichiasis). The disorder was localized to 16q24.3 by linkage in three families, and *FOXC2* was implicated in the study of a patient with a (Y;16) chromosomal translocation.

50.3 MENDELIAN DISORDERS AFFECTING BOTH THE LYMPHATIC AND VENOUS SYSTEM

50.3.1 Klippel–Trenaunay Syndrome

Klippel–Trenaunay syndrome (KTS; OMIM 14900) is defined by cutaneous hemangiomata and hypertrophy of bones and soft tissues (*19*). There is clear overlap with the Parkes Weber syndrome (OMIM 608355), which has hypertrophy due to arteriovenous malformations. Lymphatic abnormalities are restricted to KTS. Very few familial cases of KTS have been described.

50.4 VARICOSE VEINS

This term most commonly refers to increased caliber and tortuosity associated with engorgement of the superficial veins of the leg, with the saphenous being the most prominent. Varicosities may be congenital, but develop in middle age in the vast majority of affected people. As much as 10% of the population of developed countries have varicosities of the legs, and perhaps 10% of them have symptomatic venous insufficiency. Anything that increases pressure and flow in the superficial veins, such as proximal obstruction (pregnant uterus, tumor) or thrombosis of the deep veins, can increase the lumenal diameter. Little dilatation is necessary before the venous valves become incompetent, which in turn increases pressure upstream and causes further dilatation. Chronic dilatation results in vascular remodeling and persistent valvular failure. The clinical results are discomfort, mild edema, susceptibility to venous rupture with minor trauma, and venous stasis, which increases the susceptibility to thrombosis.

Many factors predispose to varicose veins and venous insufficiency of the lower extremity, including age (more common the older the subject), gender (increased in females), number of pregnancies, weight (increased obesity), lifestyle (increased with prolonged standing), and positive family history. The impact of family history is subject to the typical inadequacies (imperfect sensitivity and specificity) of this method. Certain heritable disorders of connective tissue, especially Ehlers–Danlos of the classical and vascular types and the Marfan syndrome, predispose to varicose veins. However, the vast majority of venous varicosities are not simply inherited. Multifactorial inheritance was suggested by studies of large families. Another study of 134 relatively small families found that when neither parent was affected, the risk that a child had varicosities was 20%. When one parent was affected, the risk to a son was 25% and to a daughter was 62%. When both parents were affected, the risk to any child was 90%. Having an affected mother may confer greater risk than if father is the one parent involved. Ongoing efforts to define the genes involved in pathogenesis have revealed a few candidates involved in smooth muscle and endothelial structure and function.

50.5 GENETIC COUNSELING

Genetic diseases that affect the lymphatics are rare and heterogeneous making them problematic from a counseling perspective. Milroy disease, hereditary lymphedema 1C, lymphedema–distichiasis syndrome, and Hennekam syndrome are amenable to molecular diagnosis, but there are no prevalent mutations that can be screened in the general population. In families with known mutations, screening while possible is complicated by the extreme variation occurrence of any feature (because of lack of penetrance), of age of onset of features, and of severity of the disease in carriers of the mutation. Lymphedema can be detected by prenatal ultrasound, but the specificity of ultrasound for the detection of lymphedema is uncertain and the clinical outcome, in cases where an abnormal ultrasound is present, cannot be reliably predicted.

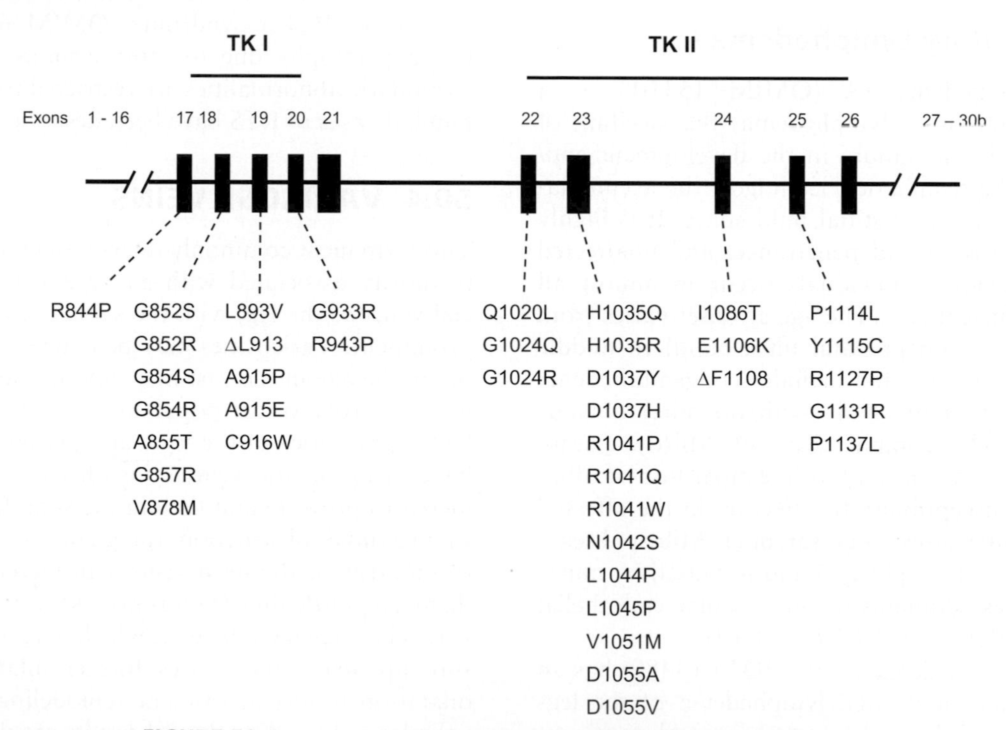

FIGURE 50-1 VEGFR-3 (FLT4) mutations causing hereditary lymphedema.

TABLE 50-1 Mendelian Disorders Affecting the Lymphatics

Disorder	Inheritance	OMIM No.	Locus	Gene	Features
Lymphedema, hereditary, 1A, Milroy disease	AD	153100	5q34q35	*FLT4* (*VEGFR3*)	Congenital onset
Lymphedema, hereditary, 1B	AD	611944	6q16.2—q02.1		Congenital onset, fully expressive at puberty
Lymphedema, hereditary, 1C	AD	613480	1q41—q42	*GJC2*	Congenital (leg), variable onset (arm), pubertal to adult
Lymphedema, hereditary, II	AD	153200			Pubertal or adult onset
Lymphedema–Distichiasis syndrome	AD	153400	16q24.3	*FOXC2*	Bilateral dilated lymphatics; double row of eyelashes
Lymphedema, microcephaly–chorioretinopathy syndrome	AD	152950			Microcephaly frequently with normal intelligence; chorioretinopathy
Hennekam lymphangiectasia–lymphedema syndrome	AR	235510	18q21.32	*CCBE1*	Generalized lymphatic dysplasia; intestinal lymphangiectasia
Hypotrichosis lymphedema–telangiectasia syndrome	AR/AD	607823	20q13.33	*SOX18*	Autosomal dominant and autosomal recessive inheritance reported
Cholestasis–lymphedema syndrome	AR	124900	15q		Aagenaes syndrome, jaundice, and lymphatic hyperplasia
Yellow nail syndrome	AR	153300			With frequent respiratory tract involvement
Choanal atresia and lymphedema	AR	613611	1q32	*PTPN14*	Listed under OMIM 603155. May be the same disorder
Ectodermal dysplasia, anhidrotic, with immunodeficiency, osteopetrosis, and lymphedema	X	300301	Xq28		Possible mutation in NEMO

FURTHER READING

1. Alders, M.; Hogan, B. M.; Gjini, E., et al. Mutations in CCBE1 Cause Generalized Lymph Vessel Dysplasia in Humans. *Nat. Genet.* **2009**, *41*, 1272–1274.
2. Alitalo, K.; Tammela, T.; Petrova, T. V. Lymphangiogenesis in Development and Human Disease. *Nature* **2005**, *438*, 946–953.
3. Au, A. C.; Hernandez, P. A.; Lieber, E., et al. Protein Tyrosine Phosphatase PTPN14 Is a Regulator of Lymphatic Function and Choanal Development in Humans. *Am. J. Hum. Genet.* **2010**, *87*, 436–444.
4. Cohen, M. M., Jr. Klippel-Trenaunay Syndrome. *Am. J. Med. Genet.* **2000**, *93*, 171–175.
5. Ghaderian, S. M.; Lindsey, N. J.; Graham, A. M., et al. Pathogenic Mechanisms in Varicose Vein Disease: The Role of Hypoxia and Inflammation. *Pathology* **2010**, *42*, 446–453.
6. Ghalamkarpour, A.; Debauche, C.; Haan, E., et al. Sporadic in Utero Generalized Edema Caused by Mutations in the Lymphangiogenic Genes VEGFR3 and FOXC2. *J. Pediatr.* **2009**, *155*, 90–93.
7. Irrthum, A.; Devriendt, K.; Chitayat, D., et al. Mutations in the Transcription Factor Gene SOX18 Underlie Recessive and Dominant forms of Hypotrichosis-Lymphedema-Telangiectasia. *Am. J. Hum. Genet.* **2003**, *72*, 1470–1478.
8. Jurisic, G.; Detmar, M. Lymphatic Endothelium in Health and Disease. *Cell Tissue Res.* **2009**, *335*, 97–108.
9. Karkkainen, M. J.; Ferrell, R. E.; Lawrence, E. C., et al. Missense Mutations Interfere with VEGFR-3 Signalling in Primary Lymphoedema. *Nat. Genet.* **2000**, *25*, 153–159.
10. Karpanen, T.; Alitalo, K. Molecular Biology and Pathology of Lymphangiogenesis. *Annu. Rev. Pathol.* **2008**, *3*, 367–397.
11. Mellor, R. H.; Brice, G.; Stanton, A. W.B., et al. Mutations in *FOXC2* Are Strongly Associated with Primary Valve Failure in Veins of the Lower Limb. *Circulation* **2007**, *115*, 1912–1920.
12. Oliver, G. Lymphatic Vasculature Development. *Nat. Rev. Immunol.* **2004**, *4*, 35–45.
13. Tammela, T.; Alitalo, K. Lymphangiogenesis: Molecular Mechanisms and Future Promise. *Cell* **2010**, *140*, 460–476.

CHAPTER

51

The Genetics of Cardiac Electrophysiology in Humans

Reed E Pyeritz

Perelman School of Medicine at the University of Pennsylvania, Philadelphia, PA, USA

51.1 INTRODUCTION

Effective circulation of blood depends on rhythmic and forceful contraction of the heart. This occurs about 2.5 billion times during the normal human lifespan. The heart, composed primarily of cardiomyocytes, contracts synchronously in response to the generation and transmission of an electrical impulse that originates in the sinoatrial node. This initial signal activates the atria and is propagated to the atrioventricular (AV) node. While the signal transits the AV node, the atria contract and fill the relaxed ventricles. The electrical impulse exits the AV node into the fast-conducting AV bundle, the bundle branches and ultimately the Purkinje fibers, which initiate ventricular contraction. Proper orchestration of excitation–contraction coupling requires heterogeneity in cardiomyocytes throughout the myocardium, which is a finely tuned developmental process. Various transcription factors, microRNAs, and epigenetic signals control the expression of ion channel and gap junction genes such that gradients of function exist, both along the axes of the heart and through the myocardial wall (Table 51-1). Disruptions of development, whether from mutation, toxin, or other environmental insult, can have a profound effect on electrophysiology. For example, a layer of connective tissue must insulate the atria from the ventricles, with the AV conduction axis as the only electrical pathway. A common developmental defect is an accessory conduction pathway within the AV canal that predisposes to Wolf–Parkinson–White (WPW) syndrome. Most cases of WPW are sporadic with no clear explanation for the accessory pathway. However, around 5% of people with WPW have a near relative with accessory conduction, and mutations in *PRKAG2* cause WPW in a few families. Another congenital defect of AV conduction is complete heart block, which may require a pacemaker. As additional genes intrinsic to this developmental pathway are defined, further candidate genes for congenital and early onset conduction disorders will be identified.

Extremes of cardiac cycling—bradycardia and tachycardia—are detrimental to the circulation. Thus, the heart rate is subject to physiologic homeostasis in the sense of Claude Bernard and Walter Cannon. For example, cardiac repolarization requires control of inward and outward currents of multiple ions. The length of the action potential of any given cardiomyocyte is a crucial determinant of the risk of dysarrhythmogenesis. Expression of cardiac ion channels fluctuates under the control of a circadian oscillator; defects in circadian rhythmicity predispose to ventricular arrhythmias in mice. This type of mechanism might explain the well-documented diurnal variation in human arrhythmogenesis.

Life-threatening cardiac dysrhythmia is usually associated with overt cardiac pathology that is evident on gross or microscopic examination. Damage due to ischemia is the most common cause for acute and chronic disorders of conduction, rhythm, or both. Other common structural causes are most congenital heart defects and the wide variety of inherited cardiomyopathies. Inflammation (e.g. myocarditis), toxins (e.g. alcohol), and drugs can also disrupt electrophysiology. However, about 10% of hearts of people who die suddenly have no evident pathologic changes. Ventricular tachyarrhythmias, such as ventricular tachycardia (VT) and ventricular fibrillation (VF), are the major cause of sudden cardiac death (SCD) or events that are interpreted as near death such as unexpected syncope. Such events in people who do not have structural heart disease increasingly fall into one of the recognized forms of familial dysrhythmia or conduction defects. Thus far, these phenotypes have coalesced around long and short QT syndromes, Brugada syndrome, and catecholaminergic polymorphic ventricular tachycardia (CPVT). Recent estimates have about a third of young people who die suddenly having a mutation in an ion channel gene. Similarly, about 10% of cases of sudden infant death syndrome are due to channelopathies. Both of these estimates are likely to rise as additional genes are discovered.

In terms of defining the genetic contribution to cardiac cycling and conduction, a first and successful approach has been to identify genes causing the Mendelian phenotypes. This has led to the confirmation of the importance of some channels and receptors implicated in the action potential. A second approach has been to study genes that may contribute to the variance of electrophysiology due to their well-known effect on the cardiovascular system. A third approach has been genome-wide association studies. Heart rate was associated with markers near *MYH6*, PR interval was associated with *SCN10A*, *TBX5*, *CAV1*, and *ARHGAP24*, and QRS duration was associated with *TBX5*, *SCN10A*, 6p21, and 10q21. A fourth approach, which will become feasible in the near future, involves sequencing the entire whole exome syndrome or whole genome syndrome of individuals with marked perturbation in electrophysiology, and comparing genetic variation between them and closely related individuals with normal phenotypes. The loci that have been associated with cardiac conduction are shown in Table 51-1.

Common to all of the Mendelian forms of conduction disturbance and dysrhythmia are the features of variable expression and genetic heterogeneity, both at the intragenic and at the intragenic levels. Many of the Mendelian forms are autosomal dominant, and because reproductive fitness is little affected, few cases are de novo.

51.2 FAMILIAL FORMS OF CONDUCTION DISTURBANCE AND DYSRHYTHMIA

Many pleiotropic syndromes caused by mutations in single genes have conduction disturbance as a feature. A few of these are listed in Table 51-2.

A number of Mendelian conditions are associated predominantly or only with persistent or episodic dysrhythmia (Table 51-3). The long QT syndrome (LQTS) is relatively common (1/3000–5000), characterized by a corrected QT interval (QTc) of >460 ms, and a predisposition to a polymorphic ventricular tachycardia, *torsades de pointes*. The initial presentation is often syncope, but may be SCD. LQTS is due to mutation in any of more than a dozen genes that impact ion transport across membranes (Table 51-4). A minority of cases is sporadic due to new mutations in one of these genes. Thus, genetic counseling and screening of relatives are important. The major predictors of clinical outcome include male sex (which predisposes to symptoms during childhood), early occurrence of syncope, late onset of institution of β-adrenergic blockage, and length of the corrected QT interval. While some genotype–phenotype correlation with syncope and sudden death exists, a family history of sudden death in a sib is not especially predictive. Institution of appropriate medical therapy (generally β-adrenergic blockade) and an implantable cardiac defibrillator (ICD) are effective. A number of drugs are associated with QT prolongation and ventricular tachycardia. In some of these individuals, mutations in the same genes causing congenital LQTS have been identified.

The short QT syndrome is associated with a QTc < 300 ms, palpitations, syncope, atrial fibrillation, and SCD. Gain-of-function mutations in three potassium channel genes, *KCNH2, KCNQ1*, and *KCNJ2*, have thus far been found. Additional loci need to be identified.

The Brugada syndrome is characterized by an electrocardiographic pattern of right bundle branch block and ST elevation in the right precordial leads associated with a susceptibility to VT and SCD. A striking degree of ethnic, gender, and geographic variability exists. Brugada syndrome is more prevalent in Asia and males are more likely to be symptomatic. While a number of genes have been identified as causal, and a surprising number of cases seem to be de novo, about 75% of cases test negative on routine clinical molecular panels.

In CPVT, the baseline ECG is unremarkable, but episodic polymorphic VT induced by adrenergic stress leads to syncope or SCD. The fundamental defect involves abnormal calcium processing. Most patients have an autosomal dominant condition due to mutations in the cardiac ryanodine receptor (*RYR2*). An uncommon recessive form due to mutations in the cardiac calsequestrin gene (*CASQ2*) occurs. This leaves about one-half of patients without a known genetic mutation. Men with mutations in *RYR2* are more likely to be symptomatic and display bradycardia. Exercise testing may identify some relatives at risk. People with a known mutation should be treated with β-adrenergic blockade, and if symptoms recur, with an ICD.

In idiopathic ventricular fibrillation, the baseline EHG may be normal; specifically, the QTc is not prolonged. The pathophysiology is unclear, but two genes have been identified as causative for this autosomal dominant condition.

51.3 PERSONALIZED MANAGEMENT OF DISORDERS OF CARDIAC ELECTROPHYSIOLOGY

Understanding the molecular basis of the Mendelian disorders of cardiac conduction and rhythm may guide effective treatment. For example, patients with LQTS3 have a mutation in the sodium channel and might benefit from the late sodium current blocker ranolazine, which shortens the QTc. Careful long-term follow-up of well-characterized patients and families may reveal clinical predictors of outcome and may suggest how aggressive management should be.

In health care systems that do not foster close follow-up, especially in developing areas of the world, achieving adequate control of the electrophysiological disturbance is infrequent. Adverse effects of medications and costs are contributors to non-adherence. One of the goals of determining genetic variations that contribute to electrophysiological disturbance is to identify new targets for pharmacologic therapies. This is one way in which management could be "personalized."

51.4 CONCLUSIONS

Considerable progress has been achieved in the past 5 years in defining genetic contributors to both familial forms of conduction disturbances and dysrhythmias, and to some degree on normal cardiac electrophysiology. Even if another common, potential cause is present, such as ischemia, it is incumbent on the clinician to determine if the patient has a primary cause of their condition. This may lead to both directed and effective therapy and the identification of relatives at risk. Once a primary cause is eliminated, then the patient must be managed empirically.

TABLE 51-1 Some Loci Associated with Normal and Abnormal Cardiac Cycle Electrophysiology

Locus	OMIM #Function	Developmental Role
TBX2	600747 Transcription factor	AV canal development
TBX3	601621 Transcription factor	Conduction system development
TBX5	601620 Transcription factor	Conduction and chamber development
IRX5	606195 Transcription factor	Transmural gradient of K-channels
NKX2-5	600584 Transcription factor	Multifunctional cardiac co-factor
GJC3	611925 Gap junction	Establishes slow conduction
HCN4	605206 K channel	Activation

Source: Adapted from Postma, A.V., et al. *Cardiovasc. Res.* **2011**, *91*, 243–251.

TABLE 51-2 Some Mendelian Conditions Associated with Disturbances of Cardiac Conduction

Condition	Phenotype	OMIM No.	Locus	Gene
Disorders primarily of electrophysiology				
Brugada syndrome	SSS, AVB	600163	3p22.2	*SCN5A*
		600235	19q13.12	*SCN1B*
Andersen–Tawil syndrome	AVB, BBB	600681	17q24.3	*KCNJ2*
CPVT	AVB	180902	1q43	*RYR2*
CPVT	SB	114251	1p13.1	*CASQ2*
	AVB, BBB	600584	5q35.1	*NKX2-5*
Pleiotropic syndromes				
Emery–Dreifuss muscular dystrophy	AVB	150330	1q22	*LMNA*
Fabry disease	Short PR	301500	Xq22	*GLA*
Glycogen storage disease	WPW, AVB	602743	7q36.1	*PRKAG2*
Hereditary amyloidosis	Heart block	105210	18q12.1	*TTR*
Holt–Oram syndrome	SB, AVB, BBB	142900	12q24.21	*TBX5*
Steinert's disease	AVB, IVCB	605377	19q13.32	*DMPK* 3′UTR
Ulnar-mammary syndrome		600747	17q23.2	*TBX2*

AVB, atrioventricular block; BBB, bundle branch block; IVCB, intraventricular conduction block; SB, sinus bradycardia; SSS, sick sinus syndrome; WPW, Wolf–Parkinson–White syndrome.

TABLE 51-3 Some Mendelian Conditions Associated with Dysrhythmia

Condition	Phenotype	OMIM No.	Locus	Gene
Brugada syndrome	ST-segment elevation in V1-V3, RBBB, VT	600163	3p22.2	*SCN5A*
		611777	3p22.3	*GPD1L*
		611875	12p13.33	*CACNA1C*
		611876	10p12.33	*CACNB2*
		600235	19q13.12	*SCN1B*
		613119	11q13.4	*KCNE3*
		613120	11q24.1	*SCN3B*
		613123	15q24.1	*HCN4*
Catecholaminergic VT	Episodic polymorphic VT Bradycardia	604772	1q43	*RYR2*
		611938	1p13.1	*CASQ2*
		614021	7p22-p14	*CPVT3*
Idiopathic VF	Normal QTc, VF	603829	3p22.2	*SCN5A*
		612956	7q36.2	*DPP6*
Short QT syndrome	QTc < 350 ms	152427	7q35	*HCNH2*
		607542	11p15	*KCNQ1*
		600681	17q23.1-q24.2	*KCNJ2*
		114205	12p13.3	*CACNA1C*
		600003	10p12.33	*CACNB2*

TABLE 51-4 The Long QT Syndromes

Location	Phenotype	Phenotype OMIM #	Gene/Locus	Gene/Locus OMIM #
3p25.3	Long QT syndrome-9	611818	*CAV3, LGMD1C, LQT9*	601253
3p22.2	Long QT syndrome-3	603830	*SCN5A, LQT3, VF1, HB1, SSS1, CMD1E, CDCD2*	600163
4q25-q26	Cardiac arrhythmia, ankyrin-B-related	600919	*ANK2, LQT4*	106410
4q25-q26	Long QT syndrome-4	600919	*ANK2*, LQT4	106410
7q21.2	Long QT syndrome-11	611820	*AKAP9, YOTIAO, AKAP450*	604001
7q36.1	Long QT syndrome-2	613688	*KCNH2, LQT2, HERG, SQT1*	152427
7q36.1	{Long QT syndrome-2, acquired susceptibility to}	613688	*KCNH2, LQT2, HERG, SQT1*	152427
11p15.5-p15.4	{Long QT syndrome 1, acquired susceptibility to}	192500	*KCNQ1, KCNA9, LQT1, KVLQT1, ATFB3, SQT2*	607542
11p15.5-p15.4	Long QT syndrome-1	192500	*KCNQ1, KCNA9, LQT1, KVLQT1, ATFB3, SQT2*	607542
11q23.3	Long QT syndrome-10	611819	*SCN4B*	608256
11q24.3	Long QT syndrome-13	613485	*KCNJ5, GIRK4, KATP1, LQT13*	600734
12p13.33	Timothy syndrome	601005	*CACNA1C, CACNL1A1, CCHL1A1, TS*	114205
12p11.1	{Acquired long QT syndrome reduced susceptibility to}	613688	*ALG10, KCR1*	603313
17q24.3	Long QT syndrome-7	170390	*KCNJ2, HHIRK1, KIR2.1, IRK1, LQT7, SQT3, ATFB9*	600681
20q11.21	Long QT syndrome-12	612955	*SNT1, LQT12*	601017
21q22.11	Long QT syndrome-6	613693	*KCNE2, MIRP1, LQT6, ATFB4*	603796
21q22.12	Long QT syndrome-5	613695	*KCNE1, JLNS, LQT5, JLNS2*	176261

Source: Adapted from Online Mendelian Inheritance in Man, OMIM®. McKusick-Nathans Institute of Genetic Medicine, Johns Hopkins University (Baltimore, MD), 24 March 2012. World Wide Web URL: http://omim.org/

FURTHER READING

1. Ackerman, M. J.; Mohler, P. J. Defining a New Paradigm for Human Arrhythmia Syndromes. *Circ. Res.* **2010,** *107,* 457–465.
2. Barsheshet, A., et al. Genetics of Sudden Cardiac Death. *Curr. Cardiol. Rep.* **2011,** *13,* 364–376.
3. Chockalingam, P.; Wilde, A. A. M. Loss-of-Function Sodium Channel Mutations in Infancy: A Pattern Unfolds. *Circulation* **2012,** *125,* 6–8.
4. Chopra, N.; Knollmann, B. C. Genetics of Sudden Cardiac Death Syndromes. *Curr. Opin. Cardiol.* **2011,** *26,* 196–203.
5. Jeyaraj, D., et al. Circadian Rhythms Govern Cardiac Repolarization and Arrhythmogenesis. *Nature* **2012,** *483,* 96–101.
6. Kannankeril, P., et al. Drug-Induced Long QT Syndrome. *Pharmacol. Rev.* **2010,** *62,* 760–781.
7. Kanter, R. J., et al. Brugada-Like Syndrome in Infancy Presenting with Rapid Ventricular Tachycardia and Intraventricular Conduction Delay. *Circulation* **2012,** *125,* 14–22.
8. Kato, T., et al. Connexins and Atrial Fibrillation: Filling in the Gaps. *Circulation* **2012,** *125,* 203–206.
9. Liu, X., et al. Common Variants for Atrial Fibrillation: Results from Genome-Wide Association Studies. *Hum. Genet.* **2012,** *131,* 33–39.
10. Nunn, L. M.; Lambiase, P. D. Genetics and Cardiovascular Disease—Causes and Prevention of Unexpected Sudden Adult Death: The Role of the SADS Clinic. *Heart* **2011,** *97,* 1122–1127.
11. Park, D. S.; Fishman, G. I. The Cardiac Conduction System. *Circulation* **2011,** *123,* 904–915.
12. Postma, A. V., et al. Developmental Aspects of Cardiac Arrhythmogenesis. *Cardiovasc. Res.* **2011,** *91,* 243–251.
13. Priori, S. G.; Chen, S. R. Q. Inherited Dysfunction of Sarcoplasmic Reticulum Ca^{2+} Handling and Arrhythmogenesis. *Circ. Res.* **2011,** *108,* 871–883.
14. Roden, D. M. Personalized Medicine and the Genotype-Phenotype Dilemma. *J. Interv. Card. Electrophysiol.* **2011,** 3117–3123.
15. Saffitz, J. E. Arrhythmogenic Cardiomyopathy: Advances in Diagnosis and Disease Pathogenesis. *Circulation* **2011,** *124,* e390–e392.
16. Wang, Z. The Role of MicroRNA in Cardiac Excitability. *J. Cardiovasc. Pharmacol.* **2010,** *56,* 460–470.
17. Wilde, A. A. M.; Brugada, R. Phenotypical Manifestations of Mutations in the Genes Encoding Subunits of the Cardiac Sodium Channel. *Circ. Res.* **2011,** *108,* 884–897.
18. Xiao, J., et al. The Genetics of Atrial Fibrillation: From the Bench to the Bedside. *Annu. Rev. Genomics Human. Genet.* **2011,** *12,* 73–96.

CHAPTER

52

Genetics of Blood Pressure Regulation

Frank S Ong, Kenneth E Bernstein, and Jerome I Rotter

Cedars-Sinai Medical Center, Los Angeles, CA, USA

52.1 INTRODUCTION

Human blood pressure represents both a classic complex trait and a quantitative trait, and many factors interact to regulate it. Extremes of blood pressure, high or low, are deleterious. Thus, systolic, diastolic and mean blood pressures are subject to physiologic homeostasis in the sense of Claude Bernard and Walter Cannon. From a population perspective, elevation of blood pressure, especially essential hypertension, is a very common disorder and has attracted the most attention, both clinically and experimentally. Over one billion people in the world have systemic hypertension. Epidemiologic studies demonstrate that chronic hypertension is a major risk factor for stroke, coronary artery disease, peripheral vascular disease, heart failure, atrial fibrillation, and aortic dissection. Diseases associated with hypertension are the most common causes of preventable death in the world. Because hypertension can be asymptomatic for years, the condition is often not diagnosed until severe end-organ damage has occurred.

The blood pressure of an individual is determined by both genetic predisposition and environmental factors. The heritable component of blood pressure has been documented in studies of both families and twins. Evidence suggests that approximately 30% of the variance of blood pressure is attributable to genetic heritability and 50% to environmental influences. The unimodal distribution of blood pressure within each age group and in each sex strongly suggests that multiple loci are involved. Many investigators have attempted to identify the genes involved in blood pressure regulation, their respective importance in determining blood pressure level, and sometimes their interaction with other genes and environmental components. Different strategies have been undertaken (Figure 52-1). A first and successful approach has been to identify genes causing some rare Mendelian forms of hypertension. This has led to the confirmation of the importance of some enzymes, channels and receptors implicated in sodium handling in the regulation of blood pressure. A second approach has been to study genes that may contribute to the variance of blood pressure due to their well-known effect on the cardiovascular system. The genes of the renin–angiotensin–aldosterone system are a good illustration of such a "candidate gene" approach, since this system is involved in the control of blood pressure and in the pathogenesis of several forms of experimental and human hypertension. A third approach has been genome-wide association studies (GWAS). These studies initially employed small samples and led to spurious results. Several more recent studies involving tens of thousands of hypertensive subjects and controls have identified more than a dozen loci that are associated with elevations of blood pressure. The latest meta-analyses combine these large data sets to increase the power of the observations. A fourth approach, which will become feasible in the near future, involves sequencing the entire whole exome sequencing (WES) or whole genome sequencing (WGS) of individuals with marked perturbation in blood pressure, and comparing genetic variation between them and closely related individuals with normal blood pressure.

The loci that have been associated with blood pressure regulation or essential hypertension are shown in Table 52-1.

52.2 MENDELIAN AND OTHER FORMS OF HYPERTENSION

Hypertension is generally defined as a persistent blood pressure above 140/90 mmHg. The prevalence of hypertension increases with age, and by age 90 years, about 90% of men and women will be so diagnosed.

Primary hypertension encompasses those conditions with a single major cause. Stenosis of the renal artery produces "reno-vascular" hypertension. Chronic renal disease, aortic coarctation, primary aldosteronism, Cushing syndrome, and pheochromocytoma are also causative, and may have a number of single-gene etiologies. Certain drugs, notably oral contraceptives, and pregnancy (pre-eclampsia) are other causes to consider.

Many syndromes caused by mutations in single genes have hypertension as a feature. A few of these are listed in Table 52-2. Psuedohypoaldosteronism type 2 is characterized by early onset hyperkalemia and hypertension, low renin, and inappropriately normal aldosterone and is genetically heterogeneous. In most cases, the metabolic and blood pressure abnormalities respond to thiazide diuretics.

52.3 MENDELIAN AND OTHER FORMS OF HYPOTENSION

A number of Mendelian conditions are associated with persistent or episodic hypotension (Table 52-3). Several themes are common. A variety of progressive neurologic conditions, including Parkinsonism, neuropathic amyloidosis, and hereditary polyneuropathies, have either persistent or orthostatic hypotension as a clinical feature. These conditions emphasize the importance of the autonomic nervous system in maintaining vascular tone. Additionally, disorders of corticosteroid and mineralocorticoid homeostasis lead to hypotension. Psuedohypoaldosteronism type 1 is characterized by renal resistance to aldosterone and is genetically heterogeneous.

In theory, GWAS of individuals with Mendelian or polygenic hypotension could reveal genes associated with low blood pressure. However, the relative paucity of hypotensive individuals, who are not in shock from blood loss or sepsis, precludes GWAS. However, such individuals might be informative in WES or WGS, especially when compared to their normotensive relatives.

52.4 ESSENTIAL HYPERTENSION

Once all so-called primary causes of hypertension (e.g. Mendelian, drug-induced, neuroendocrine and channelopathy disorders) are excluded, the remaining large proportion of individuals with consistently elevated blood pressure are said to have essential hypertension. This condition was first described in 1877, and recognized 50 years ago by Pickering and colleagues and by McKusick to be a complex trait in which multiple genes participate. Determining which genes have a role has been frustrating. Despite high heritability of various measures—31–34% for single-measure systolic and diastolic pressure, to 56–57% for repeated measures, to 63–68% for ambulatory blood pressure readings—no genetic variant of large effect has been found in population studies. The genes involved in the Mendelian forms of hyper- and hypotension, which have served as candidates in numerous linkage and association studies, contribute little or none of the heritability of essential hypertension. Other candidate genes have been unrevealing as well. GWAS affords an unbiased search for genetic variants across the entire genome and several dozen loci have been identified with high statistical validity (Table 52-1). None of these loci, however, accounts for a high relative risk. Moreover, few of the SNPs occur within genes, although some are near loci that may suggest new candidate genes involved in blood pressure regulation.

52.5 PERSONALIZED MANAGEMENT OF DYSREGULATED BLOOD PRESSURE

Understanding the molecular basis of one of the uncommon Mendelian forms of hypertension may guide effective treatment. For example, in glucocorticoid suppressible hypertension, suppression of ACTH with dexamethasone can normalize blood pressure. Similarly, in Liddle syndrome, with gain-of-function mutations in the gene encoding the β subunit of the epithelial amiloride-sensitive sodium channel (*SCNN1B*), triamterene that specifically blocks the distal renal epithelial sodium channel reduces salt retention and lowers blood pressure.

However, the vast majority of hypertensive individuals are treated empirically. Initially, weight reduction, healthy dietary habits, restriction of dietary sodium, enhanced physical activity, and moderation of alcohol consumption are recommended. If these approaches are unsuccessful, then pharmacologic therapy is prescribed. This requires considerable trial and error, both as to choice of medications and their dosages. In health care systems that do not foster close follow-up, especially in developing areas of the world, achieving a blood pressure of less than 140/90 occurs in less than one-third of cases. Adverse effects of medications and costs are contributors to non-adherence. One of the goals of determining genetic variations that contribute to elevated blood pressure is to identify new targets for pharmacologic therapies. This is one way in which antihypertensive management could be "personalized."

52.6 CONCLUSIONS

Considerable progress has been achieved in the past 5 years in defining genetic contributors to both Mendelian and complex forms of blood pressure dysregulation, and to some degree on blood pressure homeostasis. It is incumbent on the clinician to determine first if the hypertensive or hypotensive patient has a primary cause of their condition. This may lead to both directed and effective therapy and the identification of relatives at risk. Once a primary cause is eliminated, then the patient must be managed empirically. Genetic loci thus far identified for essential hypertension each contributes a relatively modest relative risk, and have not yet led to targeted therapies.

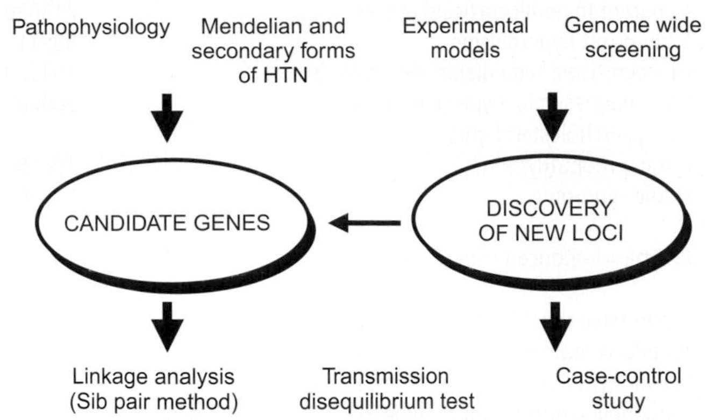

FIGURE 52-1 Main strategies used for studying the genetics of arterial hypertension. (*Revised by Reed E. Pyeritz. April 2006.*)

TABLE 52-1 Some Loci Associated with Abnormal Regulation of Blood Pressure or Hypertension Determined by Large-Scale GWAS

Ethnicity	Phenotypes	Locus	*P* Value	Nearest Gene
Asian	SBP	12-88,584,717	9.1×10^{-7}	*ATPB1*
Asian	DBP	12-88,584,719	1.2×10^{-6}	
Americans	SBP	2-190,446,083	2.1×10^{-11}	*PMS1*
African descent		6-16,031,005	3.4×10^{-9}	*MYLIP*
		8-102,026,053	1.6×10^{-8}	*YWHAZ*
		11-9,388,666	4.8×10^{-8}	*IPO7*
		14-91,877,083	1.5×10^{-8}	*SLC24A4*
European	SBP	5-162,604,350	3.5×10^{-7}	*CCNG1*
European	HTN	12-24,872,878	7.4×10^{-6}	*BCAT1*
European	SBP	1-11,785,365	1×10^{-5}	*MTHFR*
Mixed	SBP	10-104,836,168	3×10^{-7}	*NT5C2*
	SBP	17-40,563,647	4×10^{-6}	*PLCD3*
	DBP	4-81,403,365	7×10^{-9}	*FGF5*
	DBP	10-63,194,597	3×10^{-6}	C10orf107
	DBP	12-110,492,139	1×10^{-7}	*ATXN2*
	DBP	15-72,865,396	6×10^{-8}	*CSK*
	DBP	17-44,795,465	5×10^{-6}	*ZNF652*
European	SBP	10-104,584,497	2×10^{-6}	*CYP17A1*
	SBP	11-16,858,844	5.8×10^{-7}	*PLEKH7*
	SBP	12-88,537,220	3×10^{-11}	*ATP2B1*
	DBP, HTN	12-88,533,090	3.7×10^{-7}	*SH2B3*
	SBP, DBP	12-110,368,991	5.7×10^{-7}	*ULK4*
	DBP	10-18,748,804	8.7×10^{-7}	*CACNB2*
	DBP	15-72,912,698	8.1×10^{-7}	*CPLX3*

Phenotypes: DBP, diastolic blood pressure; SBP, systolic blood pressure; HTN, hypertension.
Locus: chromosome-position.
Source: Adapted from Ehret, G. B. *Curr. Hypertens. Rep.* **2010**, *12*, 17–25.

TABLE 52-2 Some Mendelian and Mitochondrial Conditions Associated with Hypertension

Phenotype	OMIM No.	Locus	Gene
Apparent mineralocorticoid excess	218030	16q22.1	*HSD11B2*
Early-onset hypertension	605115	4q31.23	*NR3C2*
Glucocorticoid remediable aldosteronism	103900	8q21	*CYP11B1*
Hypomagnesemia, hypertension, hypercholesterolemia	500005	Mitochondrial	*tRNA Ile*
IgA nephropathy	106150	1q42.2	*AGT*
Liddle syndrome	177200	16p12.2	*SCNN1G*
		16p12.2	*SCNN1B*
Pregnancy-induced hypertension	189800	7q36.1	*NOS3*
		2p13	
Progressive renal failure	161900	1q21	
Pseudohypoaldosteronism, type 2A	145260	1q31-q42	
Pseudohypoaldosteronism, type 2B	614491	17q21.31	*WNK4*
Pseudohypoaldosteronism, type 2C	614492	12p13.33	*WNK1*
Pseudohypoaldosteronism, type 2D	614495	5q31.2	*KLHL3*
Susceptibility to hypertension	145500	1p36.12	*ECE1*
		1q42.2	*AGT*
		3q14	*AGTR1*
		7q36.1	*NOS3*
		12p13.31	*GNB3*
		12p12.2-p12.1	*HYT4*
		15q	*HYT2*
		17q	*HYT1*
		20q13.13	*PTGIS*
Williams–Beuren syndrome	194050	7q11.23	

TABLE 52-3 Some Mendelian Conditions Associated with Hypotension

Phenotype	OMIM No.	Locus	Gene
Achasia–Addisonianism–Alacrima syndrome	231550	12q13.13	
Aromatic L-amino acid decarboxylase def.	608643	7p12.1	*DDC*
Corticosteroid binding globulin deficiency	611489	14q32.13	*SERPINA6*
Corticosterone methyloxidase def, type II	610600	8q24.3	*CYP11B2*
Demyelinating leukodystrophy	169500	5q23.2	*LMNB1*
Fabry disease	301500	Xq22	*GLA*
Hereditary amyloidosis	105210	18q12.1	*TTR*
Hereditary sensory and autonomic neuropathy	223900	9q13.3	*IKBKAP*
Multiple system atrophy	223360	4q22.1	*SNCA*
	223360	9q34.2	*DBH*
Orthostatic hypotensive disorder	143850	18q	
Pseudohypoaldosteronism, type 1	264350	12p13.31	*SCNN1A*
		16p12.2	*SCNN1G*
		16p12.2	*SCNN1B*
	177735	4q31.23	*NR3C2*

FURTHER READING

1. Butler, M. G. Genetics of Hypertension. Current Status. *J. Med. Liban.* **2010 Jul–Sep,** *58* (3), 175–178, PMID: 21462849.
2. Coffman, T. M. Under Pressure: The Search for the Essential Mechanisms of Hypertension. *Nat. Med.* **2011 Nov 7,** *17* (11), 1402–1409, PMID: 22064430.
3. Doris, P. A. The Genetics of Blood Pressure and Hypertension: The Role of Rare Variation. *Cardiovasc. Ther.* **2011 Feb,** *29* (1), 37–45. 10.1111/j.1755-5922.2010.00246.x, Epub 2010 Dec 6. PMID: 21129164.
4. Ehret, G. B. Genome-Wide Association Studies: Contribution of Genomics to Understanding Blood Pressure and Essential Hypertension. *Curr. Hypertens. Rep.* **2010 Feb,** *12* (1), 17–25, PMID: 20425154.
5. Evaluation of Genomic Applications in Practice and Prevention (EGAPP) Working Group Recommendations from the EGAPP Working Group: Genomic Profiling to Assess Cardiovascular Risk to Improve Cardiovascular Health. *Genet. Med.* **2010 Dec,** *12* (12), 839–843, PMID: 21042222.
6. Franceschini, N.; Reiner, A. P.; Heiss, G. Recent Findings in the Genetics of Blood Pressure and Hypertension Traits. *Am. J. Hypertens.* **2011 Apr,** *24* (4), 392–400, Epub 2010 Oct 14. PMID: 20948529.
7. Fung, M. M.; Zhang, K.; Zhang, L.; Rao, F.; O'Connor, D. T. Contemporary Approaches to Genetic Influences on Hypertension. *Curr. Opin. Nephrol. Hypertens.* **2011 Jan,** *20* (1), 23–30, PMID: 21045684.
8. Geller, D. S. Clinical Evaluation of Mendelian Hypertensive and Hypotensive Disorders. *Semin. Nephrol.* **2010 Jul,** *30* (4), 387–394, PMID: 20807611.
9. Holtzclaw, J. D.; Grimm, P. R.; Sansom, S. C. Role of BK Channels in Hypertension and Potassium Secretion. *Curr. Opin. Nephrol. Hypertens.* **2011 Sep,** *20* (5), 512–517, PMID: 21670674.
10. Konoshita, T. Genomic Disease Outcome Consortium (G-DOC) Study Investigators. Do Genetic Variants of the Renin-Angiotensin System Predict Blood Pressure Response to Renin-Angiotensin System-Blocking Drugs? A Systematic Review of Pharmacogenomics in the Renin-Angiotensin System. *Curr. Hypertens. Rep.* **2011 Oct,** *13* (5), 356–361, PMID: 21562941.
11. Kraja, A. T.; Hunt, S. C.; Rao, D. C.; Dávila-Román, V. G.; Arnett, D. K.; Province, M. A. Genetics of Hypertension and Cardiovascular Disease and Their Interconnected Pathways: Lessons from Large Studies. *Curr. Hypertens. Rep.* **2011 Feb,** *13* (1), 46–54, PMID: 21128019.
12. Rafiq, S.; Anand, S.; Roberts, R. Genome-Wide Association Studies of Hypertension: Have They Been Fruitful? *J. Cardiovasc. Transl. Res.* **2010 Jun,** *3* (3), 189–196, Epub 2010 Mar 30. PMID: 20560039.
13. Sanada, H.; Jones, J. E.; Jose, P. A. Genetics of Salt-Sensitive Hypertension. *Curr. Hypertens. Rep.* **2011 Feb,** *13* (1), 55–66, PMID: 21058046.
14. Sander, G. E.; Giles, T. D. Resistant Hypertension: Concepts and Approach to Management. *Curr. Hypertens. Rep.* **2011 Oct,** *13* (5), 347–355, PMID: 21922182.

CHAPTER

53

Preeclampsia

Anthony R Gregg

Department of Obstetrics and Gynecology, University of Florida College of Medicine, Gainesville, FL, USA

There are currently four disorders of pregnancy that include hypertension as a defining feature. These are chronic hypertension, gestational hypertension, preeclampsia, and chronic hypertension with superimposed preeclampsia. Preeclampsia is distinguished from the other three by the gestational age of onset of the hypertension. In preeclampsia, this occurs after 20 weeks gestation and is seen in concert with proteinuria. Preeclampsia is a quantitative trait disorder. The blood pressure threshold used to define this condition is 140 mm of mercury systolic or 90 mm of mercury diastolic. The degree of proteinuria is most often defined as 300 mg in 24 h after a 24-hour urine collection. However, a urine dipstick value of 1+, which correlates with 30 mg/dl, or a protein/creatinine ratio of 30 mg per mM, has also been considered in this definition.

Since preeclampsia is a quantitative trait disorder, one would naturally conclude that its molecular basis would be somewhat elusive. This has proved true; however, it has long been known that preeclampsia is a heritable condition. In the past, preeclampsia was considered non-recurrent as long as one's partner remained constant. Recurrence risk is now much more complicated than this dictum would suggest. Preeclampsia has a recurrence risk of about 18–55% (Table 53-1). Factors that influence the recurrence risk include the gestational age of onset, maternal age, ethnicity, severity of the condition during the index pregnancy, and the presence of comorbid conditions (e.g. thrombophilia, autoimmune condition, underlying renal disease, and chronic hypertension). As might be expected for quantitative trait conditions, a single gene has not been identified as causative for this condition.

Clues to the many genes that might contribute to this condition surround the observation that removal of the placenta (delivery of the fetus) leads to a cure within days to weeks. Investigations have elucidated abnormalities in placentation when preeclampsia is present. In pregnancies not destined to end in preeclampsia, villi that anchor the placenta are capable of invading the decidua for attachment. This process requires a shift in the immune system from a Th-1 pro-inflammatory to a Th-2 anti-inflammatory milieu within the decidua (Figure 53-1). Evidence suggests that the Fas/FasL pathway plays a critical role mediated by cytokines. Although this pathway was first identified in the early steps of apoptosis in activated T lymphocytes, it can also drive apoptosis in cytotrophoblast. The receptor Fas is present not only on the surface of T lymphocytes but it is also present in a soluble form within serum. The soluble form is capable of binding FasL preventing it from initiating apoptosis through its interaction with membrane associated Fas. Perturbations in the Fas/FasL pathway have been implicated in abnormal placentation.

Invasion of the spiral arteries by extra villous cytotrophoblast is also a critical process in normal placentation (Figure 53-2A). These cells are shed from anchoring villi and migrate through the decidua to plug nearby spiral arteries (Figure 53-2A). This migratory process requires that anchoring villi escape recognition by natural killer cells. Furthermore, it seems that extra villous cytotrophoblast cells go through a maturation process that requires a vital oxygen milieu. Once spiral arteries are plugged, they recanalize through a process called "pseudovasculogenesis." In this process, vascular smooth muscle and endothelial cells lining the spiral arteries are replaced by cytotrophoblast (Figure 53-2B). Biopsies of the placental bed from patients with preeclampsia show evidence of incomplete or shallow spiral artery invasion by cytotrophoblast cells. In some cases, the process of pseudovasculogenesis is incomplete leading to the pathognomonoic histologic feature of the placenta in preeclampsia, "acute atherosis."

The multifaceted process of cytotrophoblast invasion has led to experimental lines of evidence that implicate specific genes and the molecules they produce. Hif-1, a transcription factor regulates genes important in trophoblast differentiation and migration. Not a complete leap of faith, low oxygen tension at the tissue level leading to alteration in the Hif-1 pathway may be linked to the concept that patients with inherited or acquired thrombophilias are at an increased risk of developing preeclampsia (Figure 53-3). HLA-G expressed on the surface of cytotrophoblast cells is capable of preventing natural killer cell lysis through specific killing inhibitory receptors. It is conceivable that mutations resulting in conformational changes in the HLA-G molecule on cytotrophoblast cells (from a genetic perspective a fetal tissue) as a result of a paternal or maternal inheritance leads to impaired passage of cytotrophoblast through the decidua.

Vascular endothelial growth factor (VEGF) and placental growth factor (PlGF) are angiogenic factors. VEGF has been shown capable of transforming stem cells to endothelial cells. These growth factors are thought to be important in the process of "pseudovasculogenesis." A soluble protein, sFlt1, circulates in serum and like VEGF receptor is capable of binding to the angiogenic factors VEGF and PlGF. A link between increased sFlt1 and reduced angiogenic factors has been identified in patients with preeclampsia. Circulating angiogenic factors hold some promise as future biomarkers of preeclampsia.

A description of the many processes necessary for normal placentation is meant to highlight the many possible genetic contributions to abnormal placentation and its hallmark condition, preeclampsia. Whether perturbations in a single pathway or multiple pathways are required is still to be determined. A description of these pathways is incomplete without pointing out that genome-wide scans and studies of molecules that mediate vascular smooth muscle responsiveness (e.g. nitric oxide and serotonin) have also provided insights into the preeclampsia phenotype. Finally, animal models offer hope to a more complete understanding of the pathways involved in this condition, one which contributes to the burden of maternal and neonatal morbidity and mortality.

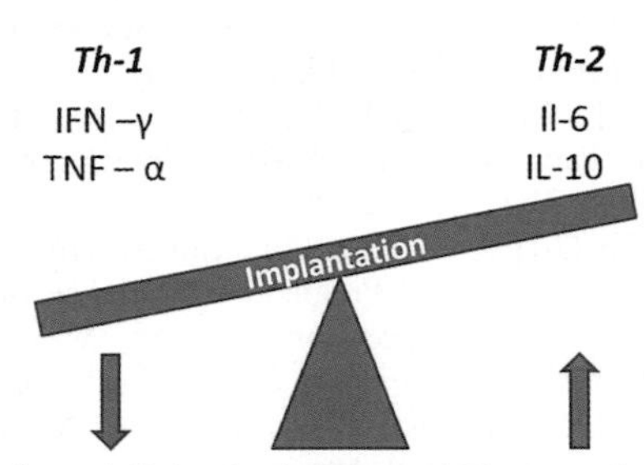

FIGURE 53-1 A shift in the balance of Th-1 pro-inflammatory to Th-2 anti-inflammatory cytokines is implicated in trophoblast invasion of the decidua.

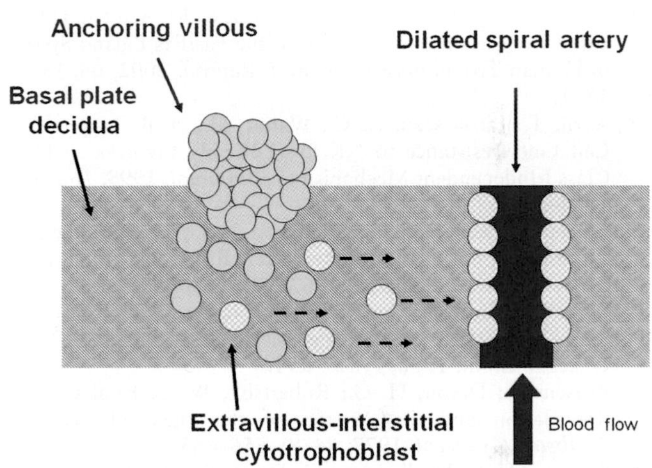

FIGURE 53-2B After plugging nearby spiral arteries, the vessels re-canalize through a process called "pseudovasculogenesis." Cytotrophoblast replaces the endothelium and vascular smooth muscle resulting in a dilated spiral artery delivering blood under low pressure to the already attached placenta.

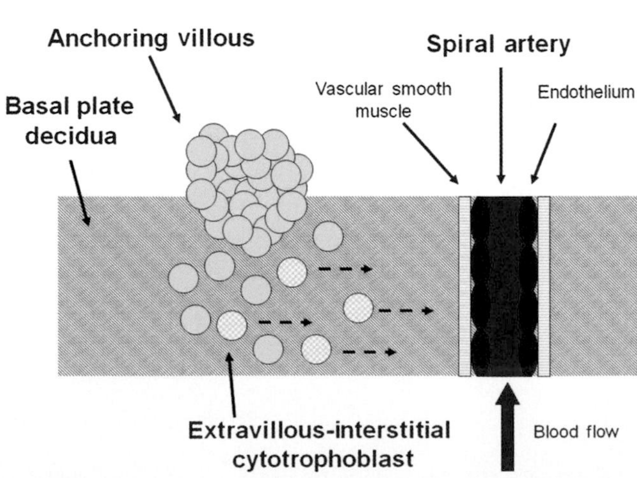

FIGURE 53-2A In normal placentation, extravillous cytotrophoblast cells migrate through the decidua and invade nearby spiral arteries.

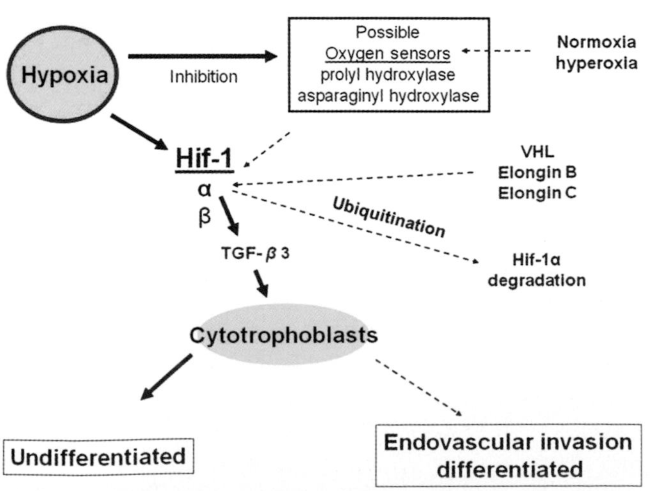

FIGURE 53-3 A hypoxic mileu in vitro results in poorly differentiated cytotrophoblast through the actions of Hif1 and TGFβ3 signaling. Under normoxic or hyperoxic conditions, Hif1 is degraded and cytotrophoblast is differentiated and demonstrated a phenotype consistent with endovascular invasion.

TABLE 53-1 The Recurrence Risk of Severe Preeclampsia or the Risk of Chronic Hypertension When the Index Pregnancy was Complicated by Severe Preeclampsia

Index Pregnancy	Complicated by Early-Onset, Severe Preeclampsia (HELLP) Syndrome, or Eclampsia (*28,66,136*)	Early-Onset Preeclampsia (Delivery <34 Weeks) (*89*)	Eclampsia or Preeclampsia (*136*)
Future Pregnancy or Condition			
Overall recurrence	17.9–55%		
Complication requiring delivery <34 weeks		17%	
Chronic Hypertension with 10 years			4.4%
Chronic Hypertension greater than 10 years later			51%

FURTHER READING

1. Aschkenazi, S.; Straszewski, S.; Verwer, K. M. A., et al. Differential Regulation and Function of the Fas/Fas Ligand System in Human Trophoblast Cells. *Biol. Reprod.* **2002,** *66,* 1853–1861.
2. Avril, T.; Jarousseau, A. C.; Watier, H., et al. Trophoblast Cell Line Resistance to NK Lysis Mainly Involves an HLA Class I-Independent Mechanism. *J. Immunol.* **1999,** *162* (10), 5902–5909.
3. Brosens, I.; Robertson, W. B.; Dixon, H. G. The Physiological Response of the Vessels of the Placental Bed to Normal Pregnancy. *J. Pathol. Bacteriol.* **1967,** *93* (2), 569–579.
4. Brosens, I. A.; Robertson, W. B.; Dixon, H. G. The Role of the Spiral Arteries in the Pathogenesis of Preeclampsia. *Obstet. Gynecol. Annu.* **1972,** *1,* 177–191.
5. Brosens, I.; Dixon, H. G.; Robertson, W. B. Fetal Growth Retardation and the Arteries of the Placental Bed. *Br. J. Obstet. Gynaecol.* **1977,** *84* (9), 656–663.
6. Brown, M. A.; Lindheimer, M. D.; de Swiet, M., et al. The Classification and Diagnosis of the Hypertensive Disorders of Pregnancy: Statement from the International Society for the Study of Hypertension in Pregnancy (ISSHP). *Hypertens. Pregnancy* **2001,** *20* (1), ix–xiv.
7. Gregg, A. R. Hypertension in Pregnancy. *Obstet. Gynecol. Clin. North. Am.* **2004,** *31* (2), 223–241.
8. King, A.; Hiby, S. E.; Gardner, L., et al. Recognition of Trophoblast HLA Class I Molecules by Decidual NK Cell Receptors—A Review. *Placenta* **2000,** *21* (Suppl A), S81–S85.
9. Kolarz, D. D.; Leszczynska-Gorzelak, B.; Rolinski, J.; Oleszczuk, J. The Expression and Concentrations of Fas/APO-1 (CD95) Antigen in Patients with Severe Pre-Eclampsia. *J. Reprod. Immunol.* **2001,** *49,* 153–164.
10. Levine, R. J.; Maynard, S. E.; Qian, C., et al. Circulating Angiogenic Factors and the Risk of Preeclampsia. *N. Engl. J. Med.* **2004,** *350* (7), 672–683.
11. Kovanci, E.; Gregg, A. R. A Quantitative Analysis of Blood Pressure Across Normal Pregnancy: Evidence of a Dynamic Change in Gene Expression Among Mexican-Hispanic Patients. *Hypertens. Pregnancy* **2010,** *29* (2), 236–247.
12. Neale, D.; Demasio, K.; Illuzi, T., et al. Maternal Serum of Women with Pre-Eclampsia Redcues Trophoblast Cell Viability: Evidence for an Increased Sensitivity to Fas-Mediated Apoptopsis. *J. Matern. Fetal Neonatal Med.* **2003,** *13,* 39–44.
13. Report of the National High Blood Pressure Education Program Working Group on High Blood Pressure in Pregnancy. *Am. J. Obstet. Gynecol.* **2000,** *183*(1), S1–S22.

CHAPTER

54

Common Genetic Determinants of Coagulation and Fibrinolysis

Angela M Carter, Kristina F Standeven, and Peter J Grant

Division of Cardiovascular & Diabetes Research, The LIGHT Labs, University of Leeds, Leeds, UK

The hemostatic system maintains a delicate balance between thrombus formation and clot lysis to prevent blood loss and maintain vascular patency. Activation of coagulation and platelets leads to thrombin generation, platelet deposition and fibrin formation, while naturally occurring anticoagulants limit thrombin generation and the fibrinolytic system is involved in clot remodeling and lysis. Perturbation of these processes gives rise to bleeding disorders or to thrombosis. The contribution of genetic factors to the pathogenesis of venous thrombosis is indicated by a strong family history in individuals with premature thrombosis and causal mutations in a variety of genes encoding hemostatic factors have been identified. The development of arterial thrombosis is more complex; however, an important role for thrombosis is supported by familial clustering and the identification of variants in candidate genes encoding hemostatic factors that are associated with arterial thrombosis. Genome-wide association studies (GWAS) have successfully identified novel genetic loci associated with a number of intermediate hemostatic phenotypes. However, although gene variants associated with cardiovascular disease have also been identified through GWAS approaches, associations of disease-associated SNPs with hemostatic intermediate phenotypes have not been consistently demonstrated to date.

54.1 THE COAGULATION CASCADE

54.1.1 Factor VIII and von Willebrand Factor

Evidence of a genetic contribution to FVIII and von Willebrand factor (vWF) originated from associations between FVIII, vWF, and ABO blood group and clustering of FVIII and vWF levels in families. Common polymorphisms of the vWF gene (*VWF*) have been identified, including four within the 5′ gene regulatory region. A recent GWAS identified SNPs associated with FVIII localized to 5, and vWF to 8, chromosomal regions (five of which coincided with those for FVIII).

54.1.2 Factor VII

Common polymorphisms in the FVII gene include a 10-bp insertion and Arg353Gln which account for 20% of variance in plasma factor VII. Gln353 is associated with lower plasma FVII, being less efficiently secreted in vitro. GWAS of FVII identified 305 SNPs clustered in five distinct chromosomal regions, including 13q34 (FVII structural gene).

54.1.3 Factor V

Activated FV (FVa) is a cofactor for conversion of prothrombin to thrombin, whereas zymogen factor V is a cofactor with protein S for activated protein C (APC)-mediated inhibition of FVIIIa. APC resistance is associated with venous thromboembolism. A missense polymorphism of the factor V gene (Arg506Gln, Factor V Leiden) accounts for most APC. The Gln506 variant of FVa shows reduced inactivation by APC and is a less effective cofactor for inactivation of FVIIIa, leading to a hypercoagulable state. The Gln506 allele increases risk for venous thrombosis between three- and eightfold in heterozygotes and 50- to 80-fold in homozygotes.

54.1.4 Prothrombin

G20210A in the 3′-untranslated region and A19911G in intron M of the gene associate with levels of prothrombin influence prothrombin messenger RNA splicing and relate to plasma prothrombin levels. Both polymorphisms are associated with venous thrombosis.

54.1.5 Fibrinogen

Fibrinogen β-chain polymorphisms (-1420 G/A, -993 C/T, -455 G/A, -148 C/T, and Arg448Lys) consistently associate with increased fibrinogen levels. Both polymorphisms in coding regions (α-Thr312Ala, β-Arg448Lys)

influence fibrin properties and α-Thr312Ala shows associations with arterial and venous disease.

54.1.6 Factor XIII

Common polymorphisms in the FXIII gene include Val34Leu and Pro564Leu, which significantly associate with FXIII cross-linking activity. Joint linkage and association analysis indicate that Val34Leu is the dominant genetic determinant of FXIII activity. The Leu34 allele results in enhanced thrombin activation and altered fibrin structure/function. Inverse associations of Leu34 with MI and venous thromboembolism are supported by meta-analyses.

54.2 NATURAL ANTICOAGULANTS

Thrombomodulin is a thrombin receptor that converts thrombin into an anticoagulant capable of activating protein C, while down-regulating fibrinolysis and inflammation through activation of thrombin-activatable fibrinolysis inhibitor. Associations between polymorphisms of the thrombomodulin gene and thrombosis have been contradictory.

Protein C exerts an anticoagulant action by inactivating factors V and VIII, and protein C deficiency is associated with venous thromboembolism. Rare mutations in the gene encoding protein C cause familial venous thrombosis. A GWAS identified 504 SNPs associated with protein C, localized to four main chromosomal regions. The SNP most strongly associated with protein C was rs867186 in 20q11 in the gene encoding the endothelial protein C receptor, which accounted for ~10% of variance.

Protein S deficiency is associated with venous thrombosis and mutations give rise to rare forms of familial venous thrombosis. Common polymorphisms of the protein S gene include Pro626 and the 2698 C/A polymorphisms which associate with plasma levels of protein S antigen.

Antithrombin inhibits thrombin and factor Xa, and deficiency is a risk factor for venous thrombosis. Gene mutations account for familial cases of venous thrombosis and a common SNP in intron 1 of the antithrombin gene has been associated with venous thrombosis.

54.2.1 Tissue Factor Pathway Inhibitor

Reduced tissue factor pathway inhibitor (TFPI) levels associate with arterial and venous thrombosis and polymorphisms associate with plasma TFPI; -287 T/C and -33 T/C influence promoter activity by binding of nuclear proteins and -33 T/C associates with venous thrombosis.

54.3 THE FIBRINOLYTIC CASCADE

Tissue-type plasminogen activator (tPA) activates fibrinolysis by converting plasminogen to plasmin. Elevated levels associate with risk of myocardial infarction which may reflect associations with plasma PAI-1 or other vascular risk factors. None of the polymorphisms in the tPA gene associate with basal tPA levels. Allele dose-dependent vascular tPA release in a perfused forearm model is seen with the Alu insertion/deletion, -7351 C/T, 20099 T/C, and 27445 T/A.

Plasminogen activator inhibitor-1 (PAI-1) is the major circulating inhibitor of tPA. Elevated PAI-1 associates with cardiovascular disease and clusters with insulin resistance. Plasma PAI-1 consistently associates with -675 4 G/5 G, with 25% higher levels in those homozygous for the 4 G allele. Meta-analyses support associations of the 4 G allele with MI and venous thrombosis.

54.4 PLATELET FUNCTION AND PLATELET GLYCOPROTEIN RECEPTOR VARIANTS

GPIb-V-IX mediates initial adhesion of platelets under high shear stress to vWF. The $\alpha_2\beta_1$ and GPVI receptors mediate platelet interactions with collagen. The $\alpha_{IIb}\beta_3$ receptor binds fibrinogen to mediate platelet aggregation. Polymorphisms in the GPIbα gene influence binding to vWF, platelet plug formation under high shear rates and receptor density. Polymorphisms in $\alpha_2\beta_1$ (807 C/T and 873 G/A) are associated with lower $\alpha_2\beta_1$ receptor density. $\alpha_{IIb}\beta_3$ Pro33 is associated with decreased activation and enhanced fibrinogen binding and clot retraction. Studies have identified inconsistent associations between Leu33Pro and cardiovascular disease. GWAS of platelet function identified SNPs localized to regions associated with ADP-induced, epinephrine-induced and collagen-induced aggregation. The genes identified encoded platelet receptors or intracellular signaling pathways.

54.5 GWAS FOR VENOUS AND ARTERIAL THROMBOSIS

54.5.1 Venous Thrombosis

A number of SNPs related to hemostasis have been consistently associated with DVT, including SNPs associated with plasma factor XI and ABO blood group.

54.5.2 Arterial Thrombosis

GWAS analyses for CVD replicated a susceptibility locus at 9p21.3. Significant associations occurred between SNPs in 9p21.3 and measures of platelet aggregation in response to collagen and epinephrine. A further study identified the locus most strongly associated with myocardial infarction as *ABO*.

54.6 CONCLUSIONS

The fluid (coagulation/fibrinolysis) and cellular (platelet) aspects of thrombosis have relatively high degrees of heritability with multiple genotypes in coding genes that relate to levels of coagulation proteins or to platelet function. The GWAS studies reported to date support contributions of many genetic variants, with minor individual effect sizes, to variance in hemostatic phenotypes and risk for thrombotic disease. However, additively the significant SNPs account for only a minor proportion of variance (or risk) despite the strong genetic component estimated for many phenotypes in twin and family studies. Population stratification and poorly defined phenotypes may partly account for this missing heritable component. However, it is likely that new approaches for data analysis will enable evaluation of the combined effects of multiple genes and environmental factors to reveal additional genetic factors which modestly contribute to variance in intermediate and disease phenotypes. The impact of these discoveries on clinical practice remains uncertain.

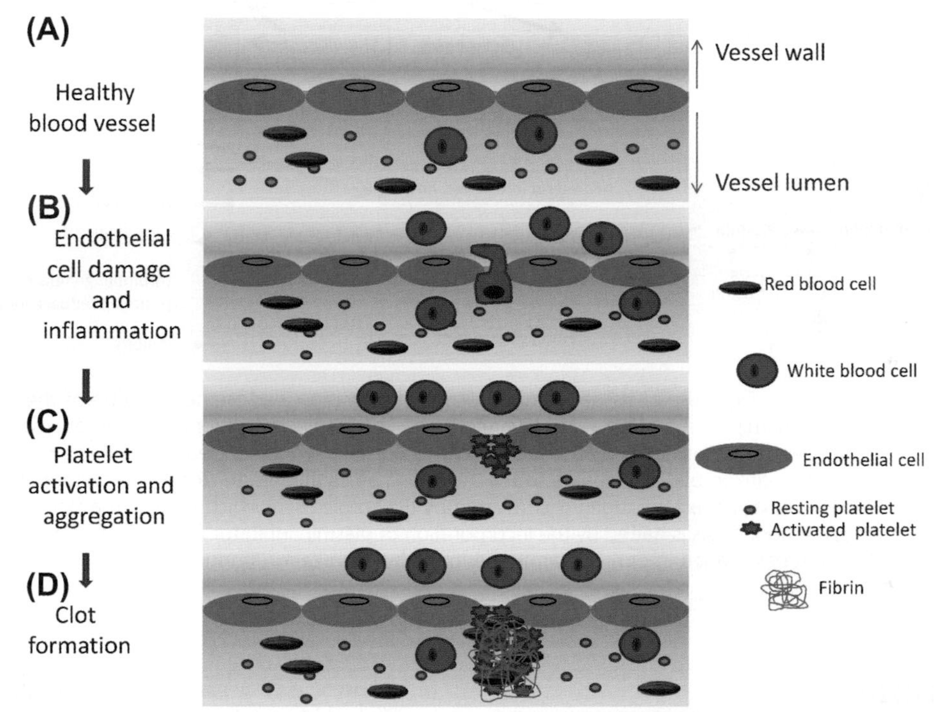

FIGURE 54-1 Dynamic process of clot formation. A. Clotting is initiated upon endothelial cell injury. The normal vessel endothelium is characterized by an anti-inflammatory anti-thrombotic phenotype that maintains vascular patency. B. Pro-inflammatory cells such as monocytes migrate from the vessel lumen into the vessel wall and release cytokines which further promote inflammatory cell transmigration. C. Exposure of sub-endothelial collagen and von Willebrand factor lead to platelet adhesion activation and aggregation forming a platelet plug at the site of injury. D. Exposure of tissue factor at the site of injury initiates activation of the coagulation cascade (see Figure 54.2) which results in thrombin generation and the formation of a fibrin mesh that traps circulatory blood cells such as platelets and erythrocytes to enable consolidation of the thrombus.

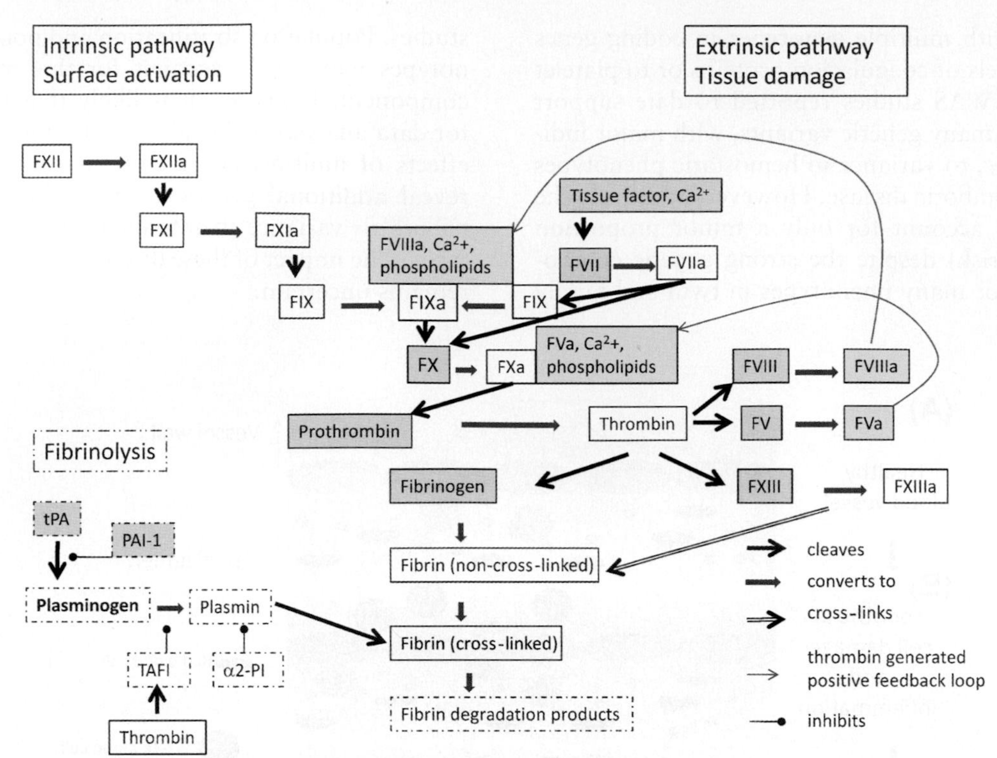

FIGURE 54-2 The coagulation cascade. Exposure of tissue factor and contact activation lead to the initiation of the clotting cascade, which culminates in the conversion of prothrombin to thrombin. A positive feedback loop results in further activation of cofactors V and VIII, leading to a burst in thrombin activity sufficient to ensure effective fibrinogen conversion. Fibrinogen is cleaved by thrombin to give rise to polymerized fibrin, which in turn becomes cross-linked by FXIII, leading to the formation of a stable fibrin clot. Fibrin formation is counteracted by the activation of the fibrinolytic cascade, in which plasminogen is converted by tPA to plasmin, which digests the fibrin clot, giving rise to fibrin degradation products. The process of fibrinolysis is halted by the incorporation of plasmin inhibitor (α2-antiplasmin) and TAFI into the clot (the latter being activated by thrombin). Areas shaded in pink are components of the coagulation/fibrinolysis system. Areas shaded in blue denote cofactor activity.

FURTHER READING

1. Smith, N. L.; Chen, M. H.; Dehghan, A., et al. Novel Associations of Multiple Genetic Loci with Plasma Levels of Factor VII, Factor VIII, and von Willebrand Factor: The CHARGE (Cohorts for Heart and Aging Research in Genome Epidemiology) Consortium. *Circulation* **2010,** *121,* 1382–1392.
2. Gohil, R.; Peck, G.; Sharma, P. The Genetics of Venous Thromboembolism. A Meta-analysis Involving Approximately 120,000 Cases and 180,000 Controls. *Thromb. Haemost.* **2009,** *102,* 360–370.
3. Standeven, K. F.; Uitte de, W. S.; Carter, A. M., et al. Heritability of Clot Formation. *Semin. Thromb. Hemost.* **2009,** *35,* 458–467.
4. Tang, W.; Basu, S.; Kong, X., et al. Genome-Wide Association Study Identifies Novel Loci for Plasma Levels of Protein C: The ARIC Study. *Blood* **2010.**
5. Burr, D.; Doss, H.; Cooke, G. E., et al. A Meta-analysis of Studies on the Association of the Platelet PlA Polymorphism of Glycoprotein IIIa and Risk of Coronary Heart Disease. *Stat. Med.* **2003,** *22,* 1741–1760.
6. Johnson, A. D.; Yanek, L. R.; Chen, M. H., et al. Genome-Wide Meta-analyses Identifies Seven Loci Associated with Platelet Aggregation in Response to Agonists. *Nat. Genet.* **2010,** *42,* 608–613.
7. Bezemer, I. D.; Bare, L. A.; Doggen, C. J., et al. Gene Variants Associated with Deep Vein Thrombosis. *JAMA* **2008,** *299,* 1306–1314.
8. Tregouet, D. A.; Heath, S.; Saut, N., et al. Common Susceptibility Alleles Are Unlikely to Contribute as Strongly as the FV and ABO Loci to VTE Risk: Results from a GWAS Approach. *Blood* **2009,** *113,* 5298–5303.
9. Musunuru, K.; Post, W. S.; Herzog, W., et al. Association of Single Nucleotide Polymorphisms on Chromosome 9p21.3 with Platelet Reactivity: A Potential Mechanism for Increased Vascular Disease. *Circ. Cardiovasc. Genet.* **2010,** *3,* 445–453.
10. Reilly, M. P.; Li, M.; He, J., et al. Identification of ADAMTS7 as a Novel Locus for Coronary Atherosclerosis and Association of ABO with Myocardial Infarction in the Presence of Coronary Atherosclerosis: Two Genome-Wide Association Studies. *Lancet* **2011,** *377,* 383–392.

CHAPTER

55

Genetics of Atherosclerotic Cardiovascular Disease

Atif N Qasim

Division of Cardiology, University of California, San Francisco, CA, USA

Muredach P Reilly

Cardiovascular Institute, Perelman School of Medicine, University of Pennsylvania, Philadelphia, PA, USA

Estimates of the heritability for coronary heart disease (CHD), a major cause of worldwide morbidity and mortality, range from 30% to 60%. Over the last 5 years there has been intense focus on discovery of novel CHD loci and genes through genome-wide association studies (GWAS), a technique that has also advanced our understanding of previously identified candidate genes. This chapter reviews recent advances and highlights continued efforts that will drive continued rapid evolution of the genomics of atherosclerosis and CHD.

Several major challenges have hindered the genetic study of CHD. These primarily include (1) the presence of numerous heterogeneous CHD phenotypes (myocardial infarction(MI), chronic stable angina, coronary artery spasm, and so forth) and use of surrogate phenotypes (coronary artery calcium (CAC), carotid IMT, and so forth), (2) complex non-Mendelian inheritance in most cases of CHD with complex gene–gene and gene–environment underpinning, and (3) failure of mouse models to translate into accurate depictions of disease in humans. Solutions to these problems have come through the use of large-scale case-control association studies with more accurately defined phenotypes, and an eye toward focusing mouse models on genes already validated in large human studies.

Candidate gene studies in the past have provided numerous possibilities for biological association with CHD, but only a few of them have withstood rigorous, contemporary methods for investigating causation, such as Mendelian randomization. *PCSK9* and *LPA* are prime examples of genes that have reproducible associations with CAD and MI and subclinical atherosclerosis traits, whereas CRP appears not to be causative but nonetheless still may be useful clinically. Although family-based linkage studies appear to be losing traction given that recent findings such as *MEF2A* or *ALOX5AP* have yet to replicated, family studies may remain a viable tool for studying extreme traits as exemplified by the story behind the novel lipid locus, *ANGPTL3*.

Without question, GWAS were the breakthrough technology in this field in the last 5 years and to date approximately 30 novel and now well-replicated loci have been associated with CHD, the first and most well-replicated being the 9p21.3 locus. Although 9p21.3 is associated with CHD events across multiple races, CAC, and other non-atherosclerotic vascular wall traits, the biology behind 9p21's association remains largely a mystery. It may be related to the function of two nearby tumor suppressor genes, *CDKN2A/B* or a predicted noncoding RNA named ANRIL, or via the impact on inflammatory pathway as evidenced by the enrichment of enhancers within the 9p21 region. Findings for other loci also have important emerging biology. For example, genetic variation at the 1p13 locus (associated with both LDL-C and CHD) was found to modulate liver expression of the SORT1 protein, which regulates hepatic VLDL secretion and perhaps also LDL clearance. Mechanisms for involvement of genes at other noteworthy loci including *ABO* at 9q34.2, *ADAMTS7* at 15q25.1, *TCF21* at 6q23.2, and numerous others covered in detail in this chapter are still being deciphered. Interestingly, findings from GWAS of specific CVD risk factors do not overlap as much as we might have expected with those of CHD, albeit more so for genome-wide association (GWA) of plasma lipids than others. This suggests GWA findings may indeed act via novel biological mechanisms.

What does this mean for patient care? Attempts to incorporate the most striking finding from GWAS, the 9p21 locus, into risk predication models have been disappointing. Other efforts using more complex genetic scores from existing data have also been underwhelming. This is likely because we have only begun to scratch the surface in explaining the heritability of CHD. Up to now, the GWAS approach has focused on less than 1%

of the human genome and was biased toward common less-impactful variation. In the coming decade, we will also begin to finally identify the rare, high-impact causal alleles in the existing genomic regions as well as novel CHD loci via whole genome and exome sequencing. From these, there should be exponential advances in risk prediction and personalized preventive medicine, and thus the next few years have the potential to fundamentally change our current model of human atherosclerosis.

FURTHER READING

1. Marenberg, M. E.; Risch, N.; Berkman, L. F.; Floderus, B.; de Faire, U. Genetic Susceptibility to Death from Coronary Heart Disease in a Study of Twins. *N. Engl. J. Med.* **1994,** *330* (15), 1041–1046.
2. Benn, M.; Nordestgaard, B. G.; Grande, P.; Schnohr, P.; Tybjaerg-Hansen, A. PCSK9 R46L, Low-Density Lipoprotein Cholesterol Levels, and Risk of Ischemic Heart Disease: 3 Independent Studies and Meta-Analyses. *J. Am. Coll. Cardiol.* **2010,** *55* (25), 2833–2842.
3. Clarke, R.; Peden, J. F.; Hopewell, J. C.; Kyriakou, T.; Goel, A.; Heath, S. C.; Parish, S.; Barlera, S.; Franzosi, M. G.; Rust, S., et al. Genetic Variants Associated with Lp(a) Lipoprotein Level and Coronary Disease. *N. Engl. J. Med.* **2009,** *361* (26), 2518–2528.
4. Elliott, P.; Chambers, J. C.; Zhang, W.; Clarke, R.; Hopewell, J. C.; Peden, J. F.; Erdmann, J.; Braund, P.; Engert, J. C.; Bennett, D., et al. Genetic Loci Associated with C-Reactive Protein Levels and Risk of Coronary Heart Disease. *JAMA* **2009,** *302* (1), 37–48.
5. Ridker, P. M.; Danielson, E.; Fonseca, F. A.; Genest, J.; Gotto, A. M., Jr.; Kastelein, J. J.; Koenig, W.; Libby, P.; Lorenzatti, A. J.; MacFadyen, J. G., et al. Rosuvastatin to Prevent Vascular Events in Men and Women with Elevated C-Reactive Protein. *N. Engl. J. Med.* **2008,** *359* (21), 2195–2207.
6. Musunuru, K.; Pirruccello, J. P.; Do, R.; Peloso, G. M.; Guiducci, C.; Sougnez, C.; Garimella, K. V.; Fisher, S.; Abreu, J.; Barry, A. J., et al. Exome Sequencing, ANGPTL3 Mutations, and Familial Combined Hypolipidemia. *N. Engl. J. Med.* **2010,** *363* (23), 2220–2227.
7. Schunkert, H.; Konig, I. R.; Kathiresan, S.; Reilly, M. P.; Assimes, T. L.; Holm, H.; Preuss, M.; Stewart, A. F.; Barbalic, M.; Gieger, C., et al. Large-Scale Association Analysis Identifies 13 New Susceptibility Loci for Coronary Artery Disease. *Nat. Genet.* **2011,** *43* (4), 333–338.
8. Butterworth, A. S.; Braund, P. S.; Farrall, M.; Hardwick, R. J.; Saleheen, D.; Peden, J. F.; Soranzo, N.; Chambers, J. C.; Sivapalaratnam, S.; Kleber, M. E., et al. Large-Scale Gene-Centric Analysis Identifies Novel Variants for Coronary Artery Disease. *PLoS Genet.* In press.
9. Musunuru, K.; Strong, A.; Frank-Kamenetsky, M.; Lee, N. E.; Ahfeldt, T.; Sachs, K. V.; Li, X.; Li, H.; Kuperwasser, N.; Ruda, V. M., et al. From Noncoding Variant to Phenotype via SORT1 at the 1p13 Cholesterol Locus. *Nature* **2010,** *466* (7307), 714–749.
10. Reilly, M. P.; Li, M.; He, J.; Ferguson, J. F.; Stylianou, I. M.; Mehta, N. N.; Burnett, M. S.; Devaney, J. M.; Knouff, C. W.; Thompson, J. R., et al. Identification of ADAMTS7 as a Novel Locus for Coronary Atherosclerosis and Association of ABO with Myocardial Infarction in the Presence of Coronary Atherosclerosis: Two Genome-Wide Association Studies. *Lancet* **2011,** *377* (9763), 383–392.

CHAPTER 56

Disorders of the Venous System

Pascal Brouillard and Nisha Limaye

Laboratory of Human Molecular Genetics, de Duve Institute, Université Catholique de Louvain, Brussels, Belgium

Laurence M Boon

Laboratory of Human Molecular Genetics, de Duve Institute; Center for Vascular Anomalies, Cliniques Universitaires St-Luc, Université catholique de Louvain, Brussels, Belgium

Miikka Vikkula

Laboratory of Human Molecular Genetics, de Duve Institute, and Walloon Excellence in Life Sciences and Biotechnology (WELBIO), Université catholique de Louvain, Brussels, Belgium

56.1 INTRODUCTION

Vascular anomalies are localized defects that occur during vascular development. They are subdivided into vascular tumors (mainly hemangiomas) and vascular malformations, which in turn are subcategorized according to the type(s) of vessel(s) altered. Venous malformations (VMs), the focus of this chapter, are the most frequently seen in interdisciplinary vascular anomaly centers due to the pain, aesthetic issues and functional difficulties associated with their size and localization in different organ systems.

56.2 GLOMUVENOUS MALFORMATION

The most frequently inherited VM, glomuvenous malformation (GVM), accounts for roughly 5% of VMs (*4*), and is often improperly called "glomangioma" or "(multiple) glomus tumor," although it is not a tumor and never becomes malignant. GVM is characterized by the presence of a variable number of rounded "glomus cells" around distended venous channels (*13,14*), but is a distinct entity from paraganglioma and (subungual) solitary glomus tumor, both of which also have rounded cells, and are sometimes referred to as "glomus tumor."

GVM segregates as an autosomal dominant disease, with incomplete penetrance and variable expressivity. Clinical distinction of GVM from VMs and mucocutaneous VMs (VMCM) can be difficult in patients with few small lesions and without familial history of the disease, but a series of clinical criteria and a few biomarkers have been defined. GVM is usually raised, nodular, present at birth, and slowly expands during childhood. It is often multifocal and hyperkeratotic (Figure 56-1). Its color varies from pink to purplish-dark blue. Plaque-like GVM is flat and purple, and usually darkens with time. GVM is usually located on the extremities, involves skin and subcutis, and is soft and often painful on palpation. It cannot be completely emptied by compression. Contrary to VM, GVM is rarely encountered in mucosal tissue. Currently, the best therapy is surgical resection of the entire GVM, which is curative, and is facilitated by the non-infiltration of GVM in underlying tissues. Alternatively, sclerotherapy can be used.

GVMs are caused by loss-of-function mutations in a gene named glomulin (*GLMN*), the function of which is still unknown. So far, 38 distinct mutations have been reported in 154 families. Fifteen of the mutations account for almost 85% of the pedigrees with a *GLMN* mutation. The most frequent mutation is *c.157_161del* (aka *157delAAGAA*), present in 68 kindreds (44.2%).

There is no correlation between the position of a mutation in the *GLMN* gene and the characteristics of the disorder, such as the number of glomus cells, the extent or number of lesions. Expressivity is variable from patient to patient even with the same mutation, suggesting a paradominant mode of inheritance. GVM has an age-dependent variation in penetrance, which reaches its maximum (92.7%) by 20 years of age. Unaffected mutation carriers have been identified in GVM families and affected individuals develop new, albeit often small, lesions with time. This multifocality is explained by the occurrence of somatic second-hit mutations in glomulin. In GVMs, the mural glomus cells are round or polygonal, instead of elongated like the normal vascular smooth muscle cells. They stain positively for smooth muscle α-actin and vimentin, whereas they are negative for desmin, von Willebrand factor and S-100 neuronal marker. During murine development, glomulin RNA was first detected at E10.5 dpc in cardiac outflow tracts and later, strong expression was seen in vascular smooth muscle cells (vSMCs). Glomus cells do not express late markers of vSMC differentiation, glomulin and smoothelin-b, whereas two earlier markers, smooth muscle myosin heavy chain and h-caldesmon, are detected. Thus, it seems that glomus cells deviate in the differentiation process due to lack of glomulin.

Glomulin has no known motif or conserved domain. It may act in both the transforming growth factor beta (TGFβ) and hepatocyte growth factor (HGF) pathways, which are crucial for vSMC differentiation (Figure 56-2). In vitro, glomulin, encoded by *GLMN*, interacts with FKBP12, which in turn can inhibit TGFβ receptors as well as mTOR signaling. Complete loss of glomulin in GVMs could thus result in inhibition of both. Glomulin also interacts with the intracellular part of c-Met, the receptor for HGF, which mediates vascular SMC migration. Upon HGF binding, glomulin is tyrosine-phosphorylated, released, and induces phosphorylation of p70S6-kinase, thereby controlling protein synthesis. Glomulin was also reported to interact with Cul7, with which it forms a SCF-like complex. These complexes are E3-ubiquitin-ligases determining the specificity for the substrate to ubiquitinate. Thus, glomulin-Cul7-containing complex may also regulate protein degradation in vSMC, and affect their differentiation.

56.3 INHERITED VENOUS MALFORMATION

A second, less common inherited VM is the autosomal dominant mucocutaneous VM (VMCM), which accounts for another 1–2% of all VMs. It is characterized by multiple small, compressible blue lesions on the skin and mucosa, commonly located in the cervico-facial region and the limbs, and less often on the trunk (Figure 56-1). Histologically, distended venous channels are lined by a single endothelial cell layer surrounded by sparse, irregularly distributed vascular smooth muscle cells. VMCM is caused by mutations in *TEK*, located on chromosome 9p21 which encodes an endothelial cell tyrosine kinase receptor that binds the angiopoietins ANGPT 1, 2, and 4 (Angpt3 in mice). The resulting modulation of downstream signaling molecules, including PI3K/Akt and MAPK, is critical to endothelial cell survival and function (Figure 56-2). Eight intracellular mutations have been identified, which cause widely variable ligand-independent hyperphosphorylation of the receptor in vitro. The most commonly identified change R849W (10/17 families reported) was shown to be accompanied by a somatic loss-of-function of the wild-type receptor in one lesion, locally abolishing some putative protective/competitive effect of the latter. As in the case of GVM, this likely explains why the germline mutations do not cause generalized vascular abnormalities despite being ubiquitous, instead resulting in highly focal malformations only where their effects are compounded by an additional somatic hit.

56.4 SPORADIC VENOUS MALFORMATION

By far, the most frequent (95%) of the VMs are those that occur sporadically, and therefore unpredictably. Histologically similar to VMCMs, sporadic VMs are usually much larger, single lesions that affect the skin and mucosa (Figure 56-1), but can also infiltrate underlying tissues in various organ systems. They can cause significant morbidity due to their size, localization, or expansion, making them the malformations most frequently treated at centers specializing in vascular anomalies. In about 42% of patients, localized intravascular coagulopathy, characterized by elevated levels of D-dimers, is observed, and correlates with the size and depth of lesions and the presence of phleboliths. While compression, sclerotherapy and surgical removal are currently employed to treat VM, they can be problematic and ineffective depending on the nature and location of the lesion, and can lead to regrowth.

At least 50% of common sporadic VMs are caused by somatic mutations in TIE2/*TEK*. As with VMCM, these intracellular changes cause ligand-independent hyperphosphorylation of the receptor in vitro. Among the changes is the frequent L914F (69% of mutation-positive samples), which has never been observed in germ line among VMCM families, suggesting it is lethal when ubiquitous. The somatic changes also include a series of double-mutations in cis (on the same allele), which seem to be enriched in rare, multifocal sporadic VM. Among the constituent single mutations are three that cause premature truncation in the C-terminal tail domain, yet cause receptor phosphorylation, likely by destroying its customary inhibitory fold.

The mechanism by which chronic TIE2 activation causes VMs is unknown. R849W has been shown to have a Shc and pAkt-dependent pro-survival effect on endothelial cells (Figure 56-2). It does not seem to influence their proliferation or migration in vitro, but does lead to the formation of more unstable tubes as compared to cells bearing the wild-type receptor. It causes increased basal as well as LPS-induced phosphorylation and activation of STAT1, an inflammatory mediator. Inherited R849W and somatic L914F have differential but overlapping effects on the subcellular localization of the receptor, as well as its translocation in response to ligand. The lack of any clear correlation between the patient phenotype and the strength of receptor phosphorylation suggests that pathogenesis results, not merely from a quantitative increase, but also qualitative abnormalities in signaling downstream of TIE2, potentially due to its localization in compartments with inappropriate interacting molecules.

56.5 HYPERKERATOTIC CUTANEOUS CAPILLAROVENOUS MALFORMATION

Multiple cutaneous VMs and capillaro-VMs sometimes occur in association with cerebral cavernous malformation (CCM), described in detail in Chapter 57

Hyperkeratotic cutaneous capillary-VMs (Figure 56-1) are exclusively associated with mutations in *CCM1* (KRIT1). Nodular VMs are associated with mutations in *CCM3* (PDCD10), and sometimes *CCM1* (*22,24*). Histologically, they are similar to the cerebral lesions of CCM, and consist of several thin-walled channels packed together and surrounded by a fibrous layer containing smooth muscle cells. Mutations in *CCM2* (malcavernin/MGC4607), the third gene known to cause CCM, are rarely if ever associated with VMs.

56.6 OTHER VM-ASSOCIATED SYNDROMES

VMs also arise in the context of sporadic syndromes such as blue rubber bleb naevus (BRBN) and Klippel–Trenaunay, the etiology of which is unknown. BRBN is characterized by multiple small, rubbery cutaneous VMs, often on the plantar surfaces (Figure 56-1), associated with gastrointestinal VMs. Klippel–Trenaunay syndrome, is characterized by capillary-lymphatico-VMs (CLVMs) on a hypertrophic extremity, most often a lower limb. Abnormal capillaro-venous channels in the form of spindle cell hemangioendotheliomas are seen in Maffucci syndrome (Figure 56-1), which also involves multiple enchondromas, benign tumors of the cartilage, with a high rate of malignant transformation, most often to chondrosarcoma. Genome-wide single nucleotide polymorphism mapping arrays have revealed several somatic deletions and losses of heterozygosity in the tumors, but none in common to all or most samples.

56.7 CONCLUSION

The discovery that loss of glomulin function or aberrant TIE2 activation cause more than 50% of all VMs makes known or novel inhibitors of the related pathways potentially useful therapeutic candidates. Further insights into the pathways dysregulated by these molecules and other that cause venous disorders could reveal additional suitable targets for these anomalies. The availability of good animal models of (glomu) VMs, currently lacking, would allow for the testing of potential therapies for their safety and efficacy.

ACKNOWLEDGMENTS

The authors' studies are partially funded by the Interuniversity Attraction Poles initiated by the Belgian Federal Science Policy, network 6/05; Concerted Research Actions (A.R.C.)—Convention No. 07/12-005 of the Belgian French Community Ministry; the National Institute of Health, Program Project P01 AR048564; the F.R.S.-FNRS (Fonds de la Recherche Scientifique); and la Communauté française de Wallonie-Bruxelles et de la Lotterie nationale, Belgium (to MV). NL is a "Chercheur Qualifié" of the Fonds de la Recherche Scientifique-FNRS.

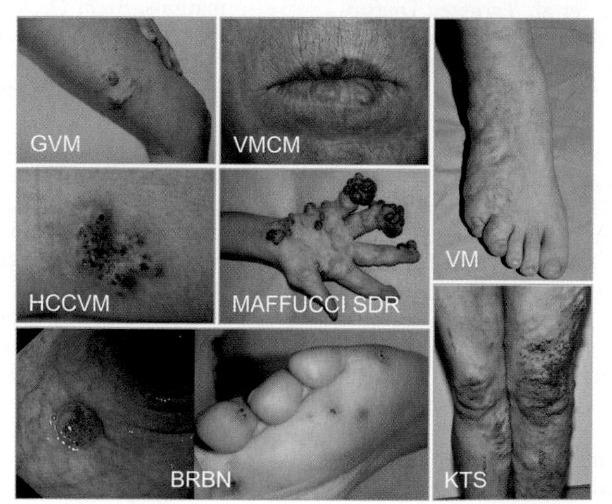

FIGURE 56-1 Venous anomalies: GVM on leg; multiple cutaneous and mucosal venous malformations (VMCMs) on lips; sporadic VM on foot; HCCVM on arm; multiple spindle cell hemangioendotheliomas and enchondromas on hand of patient with Maffucci syndrome; intestinal and cutaneous VMs of patient with Blue Rubber Bleb Nevus syndrome (BRBN); capillaro-lymphatico-venous malformation on legs of patient with Klippel–Trenaunay syndrome (left side more affected).

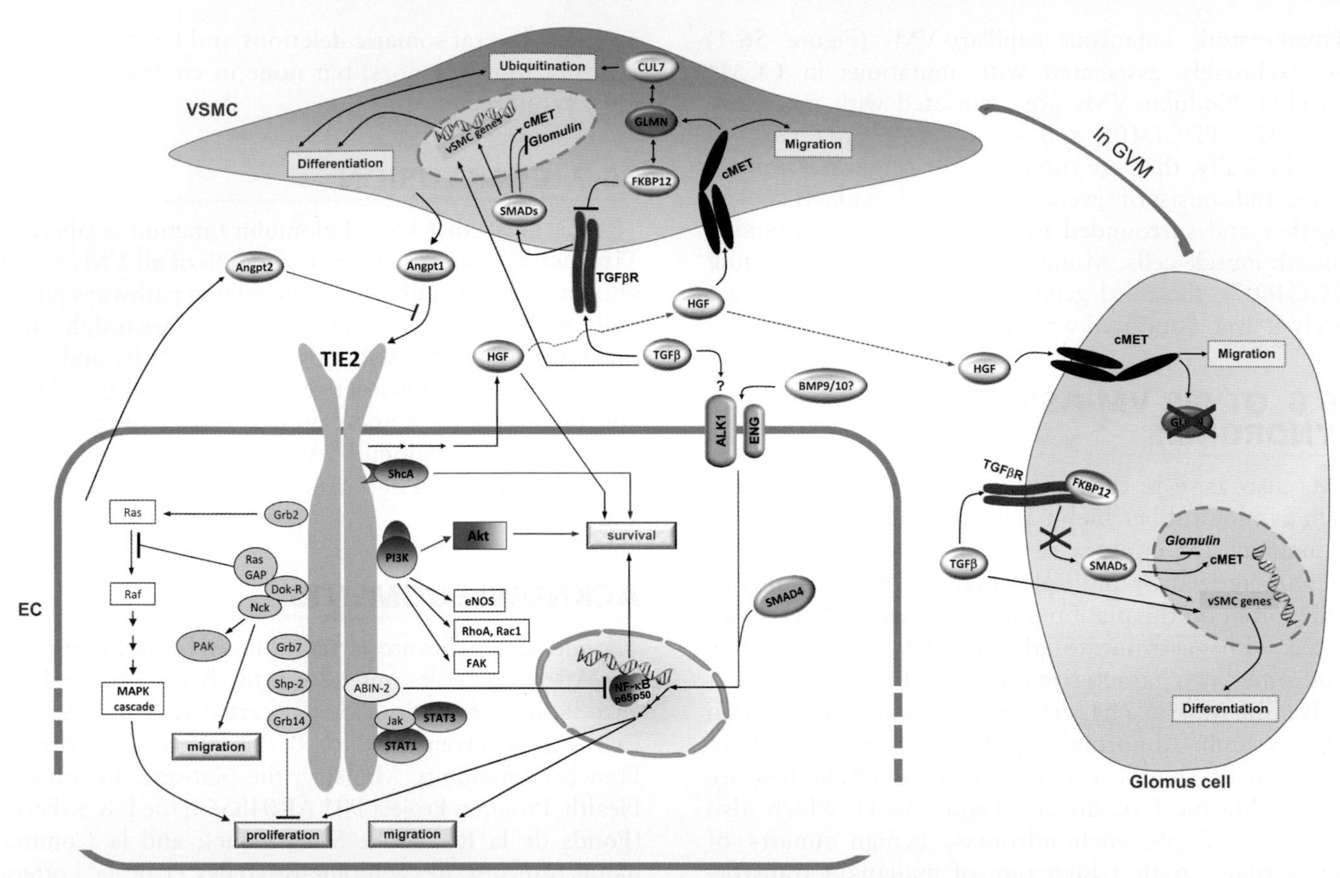

FIGURE 56-2 Signaling pathways involving glomulin (GLMN) and TIE-2 in normal and pathological endothelial and vSMCs. Lack of GLMN signaling results in aberrant differentiation towards glomus cells. Pathways chronically activated by *TIE-2* mutations highlighted in orange; with the exception of STAT1, all are also activated by wild-type receptor, upon stimulation.

FURTHER READING

1. Arai, T.; Kasper, J. S.; Skaar, J. R., et al. Targeted Disruption of *p185/Cul7* Gene Results in Abnormal Vascular Morphogenesis. *Proc. Natl. Acad. Sci. U.S.A.* **2003**, *100* (17), 9855–9860.
2. Boon, L. M.; Mulliken, J. B.; Enjolras, O.; Vikkula, M. Glomuvenous Malformation (Glomangioma) and Venous Malformation: Distinct Clinicopathologic and Genetic Entities. *Arch. Dermatol.* **2004**, *140* (8), 971–976.
3. Boon, L. M.; Vikkula, M. Vascular Anomalies. In *Fitzpatrick's Dermatology in General Medicine*, 8th ed., Wolff, K., Goldsmith, L. A., Katz, S. I., et al. Eds.; McGraw-Hill Professional Publishing Vol. 2, in press.
4. Brouillard, P.; Boon, L. M.; Mulliken, J. B., et al. Mutations in a Novel Factor, Glomulin, Are Responsible for Glomuvenous Malformations ("Glomangiomas"). *Am. J. Hum. Genet.* **2002**, *70* (4), 866–874.
5. Brouillard, P.; Boon, L. M.; Mulliken, J. B.; et al. Genotype and Phenotype of 153 Families with a Mutation in GLMN. *Mol. Syndromol.* in press.
6. Brouillard, P.; Enjolras, O.; Boon, L. M.; Vikkula, M. In *Glomulin and Glomuvenous Malformation. Inborn Errors of Development*, 2nd ed.; Epstein, C. J.; Erickson, R. P.; Wynshaw-Boris, A., Eds.; Oxford University Press: New York, 2008; pp 1561–1565.
7. Brouillard, P.; Ghassibe, M.; Penington, A., et al. Four Common Glomulin Mutations Cause Two Thirds of Glomuvenous Malformations ("Familial Glomangiomas"): Evidence for a Founder Effect. *J. Med. Genet.* **2005**, *42* (2), e13.
8. Brouillard, P.; Vikkula, M. Genetic Causes of Vascular Malformations. *Hum. Mol. Genet.* **2007**, *16* (Spec No. 2), R140–R149.
9. Dompmartin, A.; Acher, A.; Thibon, P., et al. Association of Localized Intravascular Coagulopathy with Venous Malformations. *Arch. Dermatol.* **2008**, *144* (7), 873–877.
10. Dompmartin, A.; Ballieux, F.; Thibon, P., et al. Elevated D-Dimer Level in the Differential Diagnosis of Venous Malformations. *Arch. Dermatol.* **2009**, *145* (11), 1239–1244.
11. Limaye, N.; Boon, L. M.; Vikkula, M. From Germline Towards Somatic Mutations in the Pathophysiology of Vascular Anomalies. *Hum. Mol. Genet.* **2009**, *18* (R1), R65–R74.
12. Limaye, N.; Wouters, V.; Uebelhoer, M., et al. Somatic Mutations in Angiopoietin Receptor Gene *TEK* Cause Solitary and Multiple Sporadic Venous Malformations. *Nat. Genet.* **2009**, *41* (1), 118–124.
13. McIntyre, B. A.; Brouillard, P.; Aerts, V.; et al Glomulin Is Predominantly Expressed in Vascular Smooth Muscle Cells in the Embryonic and Adult Mouse. *Gene Expr. Patterns* **2004**, *4* (3), 351–358.
14. Sirvente, J.; Enjolras, O.; Wassef, M., et al. Frequency and Phenotypes of Cutaneous Vascular Malformations in a Consecutive Series of 417 Patients with Familial Cerebral Cavernous Malformations. *J. Eur. Acad. Dermatol. Venereol.* **2009**, *23* (9), 1066–1072.
15. Soblet, J.; Limaye, N.; Uebelhoer, M.; et al. Variable Somatic TIE2 Mutations in Half of Sporadic Venous Malformations. *Mol. Syndromol.* in press.
16. Toll, A.; Parera, E.; Gimenez-Arnau, A. M., et al. Cutaneous Venous Malformations in Familial Cerebral Cavernomatosis Caused by *KRIT1* Gene Mutations. *Dermatology* **2009**, *218* (4), 307–313.
17. Wouters, V.; Limaye, N.; Uebelhoer, M., et al. Hereditary Cutaneomucosal Venous Malformations Are Caused by TIE2 Mutations with Widely Variable Hyper-Phosphorylating Effects. *Eur. J. Hum. Genet.* **2010**, *18* (4), 414–420.

CHAPTER

57

Capillary Malformation/ Arteriovenous Malformation

Nicole Revencu

Center for Human Genetics, Cliniques universitaires St Luc and Université catholique de Louvain; and Laboratory of Human Molecular Genetics, de Duve Institute, Université catholique de Louvain, Brussels, Belgium

Laurence M Boon

Center for Vascular Anomalies, and Division of Plastic Surgery, Cliniques universitaires St Luc and Université catholique de Louvain; and Laboratory of Human Molecular Genetics, de Duve Institute, Université catholique de Louvain, Brussels, Belgium

Miikka Vikkula

Walloon Excellence in Lifesciences and Biotechnology (WELBIO), and Laboratory of Human Molecular Genetics, de Duve Institute, Université catholique de Louvain, Brussels, Belgium

57.1 INTRODUCTION

Vascular malformations are localized defects of vascular morphogenesis, classified into capillary, venous, arterial, lymphatic, or combined (e.g. arteriovenous malformation (AVM), capillary-venous malformation). Capillary malformation (CM) or "port-wine stain" is a characteristic macular stain, which affects 0.3% of newborns. It is usually an isolated, sporadic, solitary flat lesion present at birth, and grows proportionately with the child and persists throughout life. CM is a slow-flow, homogenous lesion of variable size, with geographic borders, located mainly on the skin, sometimes on the mucosa, as a red macula that darkens and often thickens with age. CMs are asymptomatic, but can generate important psychosocial distress. Often they do not require any treatment. When necessary for unsightly reasons, laser, mainly pulsed dye, is the first line treatment.

Particular attention has to be paid to CM located on the territory of the first branch of the trigeminal nerve (V1), which in 10% of cases is part of Sturge–Weber syndrome (OMIM 185300). This is a sporadic, severe neurocutaneous disorder, with an estimated incidence at 1/50,000 and unknown etiology. It manifests as a unilateral CM present at birth and located on the forehead and the upper eyelid (V1), ipsilateral leptomeningeal capillary-venous anomaly and ocular involvement. The major associated risks are seizures, most often before age 2 years, with a risk of controlateral neurologic deficit and learning difficulties and glaucoma. Management of epilepsy and glaucoma is an emergency.

CMs have to be differentiated from common *nevus flammeus*, present in up to 50% of newborns and located on the back of the neck, forehead, eyelids, or upper lip. Hypertrophic CM located on a limb has to be differentiated from Klippel–Trenaunay syndrome (capillaro-veno-lymphatic malformation, OMIM 149000, Chapter 50) and Parkes Weber syndrome (PKWS; capillary blush on an extremity, arteriovenous microfistulas and hypertrophy, OMIM 608355, see below CM-AVM). Sometimes, a capillary blush can be observed on top of an AVM and can be misinterpreted as a CM.

Even if most CMs are sporadic, some are familial, as those seen in CM-AVM, cerebral cavernous malformation (CCM), or hereditary hemorrhagic telangiectasia (HHT). These three conditions have an autosomal dominant transmission. The first two conditions are discussed here, whereas HHT in Chapter 49.

57.2 CAPILLARY MALFORMATION—ARTERIOVENOUS MALFORMATION

Linkage studies in families with inherited CMs allowed the identification of a susceptibility locus on 5q and subsequently the recognition of *RASA1* on 5q14.1 (OMIM 139150) as the mutated gene. The condition is characterized by multifocal CMs and fast-flow vascular malformations and was called CM-AVM (CM-AVM; OMIM 608354). About 100 families with heterozygous *RASA1* mutations have been identified until now. The penetrance is high, more than 95%, and de novo occurrence is around 20%. The expressivity is variable even within the same family and the prevalence at least 1/100,000. All the affected people have CMs, but in contrast to classical CM, CMs of CM-AVM are often multifocal and small, less than 1cm, round or oval, randomly distributed and frequently surrounded by a pale halo. New lesions can appear during childhood.

In addition, about 30% of the individuals with *RASA1* mutation have fast-flow vascular malformations with in or out of the central nervous system (CNS), such as AVM or arteriovenous fistula (AVF) or PKWS. In contrast to hereditary hemorrhagic telangiectasia, no lung or liver AVMs/AVFs have been observed. Lesions in the CNS are usually symptomatic early in life. PKWS is characterized by a capillary blush on an extremity, multiple arteriovenous microfistulas and bony and soft tissue hypertrophy (OMIM 608355).

CMs are harmless lesions; they do not respond well to pulsed dye laser. Fast-flow lesions in the CNS seem to manifest early in life. Nevertheless, a regular brain MRI is recommended until more data are available. These lesions as well as extra-cranial fast-flow lesions require a multidisciplinary approach, as isolated AVM and AVF. Patients with PKWS should be treated conservatively; epiphysiodesis for leg length discrepancy should be avoided when possible as it can aggravate the fast-flow vascular malformation.

Most of the *RASA1* mutations lead to premature termination, which suggests loss-of-function. The pathophysiology of CM-AVM is unknown, but the number of lesions and their localized nature could indicate that a somatic second-hit is required for the lesions to develop, as previously reported in other multifocal vascular malformations (1, 2, 10, 11). *RASA1* encodes p120RASGAP, a multidomain cytoplasmic protein that acts as a negative regulator of the RAS-signaling pathway. *RASA1* is an NF1 homolog, and thus CM-AVM is part of the RAS/MAPK-related disorders, e.g. neurofibromatosis type I, Legius, Noonan, LEOPARD, Costello and cardio-facio-cutaneous syndromes.

57.2.1 Cerebral Cavernous Malformation

CCM (OMIM 116860) is a vascular anomaly located in the CNS: intracranial or intra-spinal. It can be associated with retinal or cutaneous lesions. CCM consists of clusters of dilated capillary-like vessels in a dense collagenous matrix, without normal vascular supporting cells and with abnormal blood–brain barrier. Based on cerebral magnetic resonance and autopsy studies the prevalence is considered to be 1/200–1/1000.

Sporadic and familial forms have been described; usually one lesion is observed in sporadic patients and multifocal lesions in familial patients by gradient-echo MRI. Linkage studies followed by sequencing of positional candidate genes or loss-of-heterozygosity mapping identified three *CCM* genes: *CCM1* (7q; *KRIT1*), *CCM2* (7p;*MGC4607*) and *CCM3* (3q;*PDCD10*). A heterozygous mutation is identified in about 95% of patients with multifocal CCMs and positive family history, but only in about half of the patients with multifocal CCMs and unaffected parents. No *CCM* germ line mutation is expected in sporadic cases with one cerebral lesion on gradient-echo MRI.

About 60% of the mutation carriers are symptomatic; associated symptoms are mainly seizures and cerebral hemorrhages, but also headaches and focal neurological deficit, irrespective of the gene involved. The radiological penetrance is much higher but not complete, even with the very sensitive gradient-echo MRI sequences and this observation is important for genetic counseling. Symptoms do not seem to correlate with the number of lesions, but rather with their location. The most severe outcome is associated with those in brainstem and basal ganglia. Current clinical guidelines for symptomatic patients recommend medical treatment for seizures and surgical removal in case of hemorrhage, focal neurological deficit, or lesions associated with intractable epilepsy.

More than 150 distinct mutations in the three *CCM* genes have been reported until now and most of them suggest loss-of-function and haploinsufficiency. To explain the localized nature of the malformation and the number of lesions (one in sporadic versus multifocal in familial CCM), a double-hit mechanism with bi-allelic loss was proposed and recently confirmed by genetic and immunohistochemistry studies.

In parallel with the genetic studies, many biochemical and in vivo studies have been conducted in an effort to unravel the disease mechanism. Yet, many pieces from the puzzle are still missing. The three *CCM* genes are conserved across species. *CCM1* encodes KRIT1 (Krev interaction trapped 1; OMIM 604214), a scaffold protein, which contains several domains involved in protein–protein interaction. *CCM2 (MGC4607*; OMIM 607929) encodes malcavernin, a scaffold protein containing a phosphotyrosine binding domain, similar to that of ICAP1α, one of the KRIT1 partners. *CCM3* (*PDCD10*; OMIM 603285) encodes PDCD10, a protein with no known functional domain. It has been shown that the three CCM proteins interact (CCM2 is the linker between CCM1 and CCM3), but they also have specific partners. These proteins seem to play a role in cell–cell junctions, cell shape and polarity and possibly cell-extracellular matrix adhesion.

Besides the disorders described in this chapter, CMs are observed as part of many other rare entities/syndromes of unknown etiology, such as macrocephaly-CM (OMIM 602501; previously described as M-CMTC macrocephaly-cutis marmorata telangiectatica congenita), diffuse CM with overgrowth, or Wyburn-Mason syndrome, etc.

ACKNOWLEDGMENTS

The authors' studies are partially funded by the Interuniversity Attraction Poles initiated by the Belgian Federal Science Policy, network 6/05; Concerted Research Actions (A.R.C.)—Convention No. 07/12-005 of the Belgian French Community Ministry; the F.R.S.-FNRS (Fonds de la Recherche Scientifique); and la Communauté Française de Wallonie-Bruxelles and Lotterie nationale, Belgium (to MV).

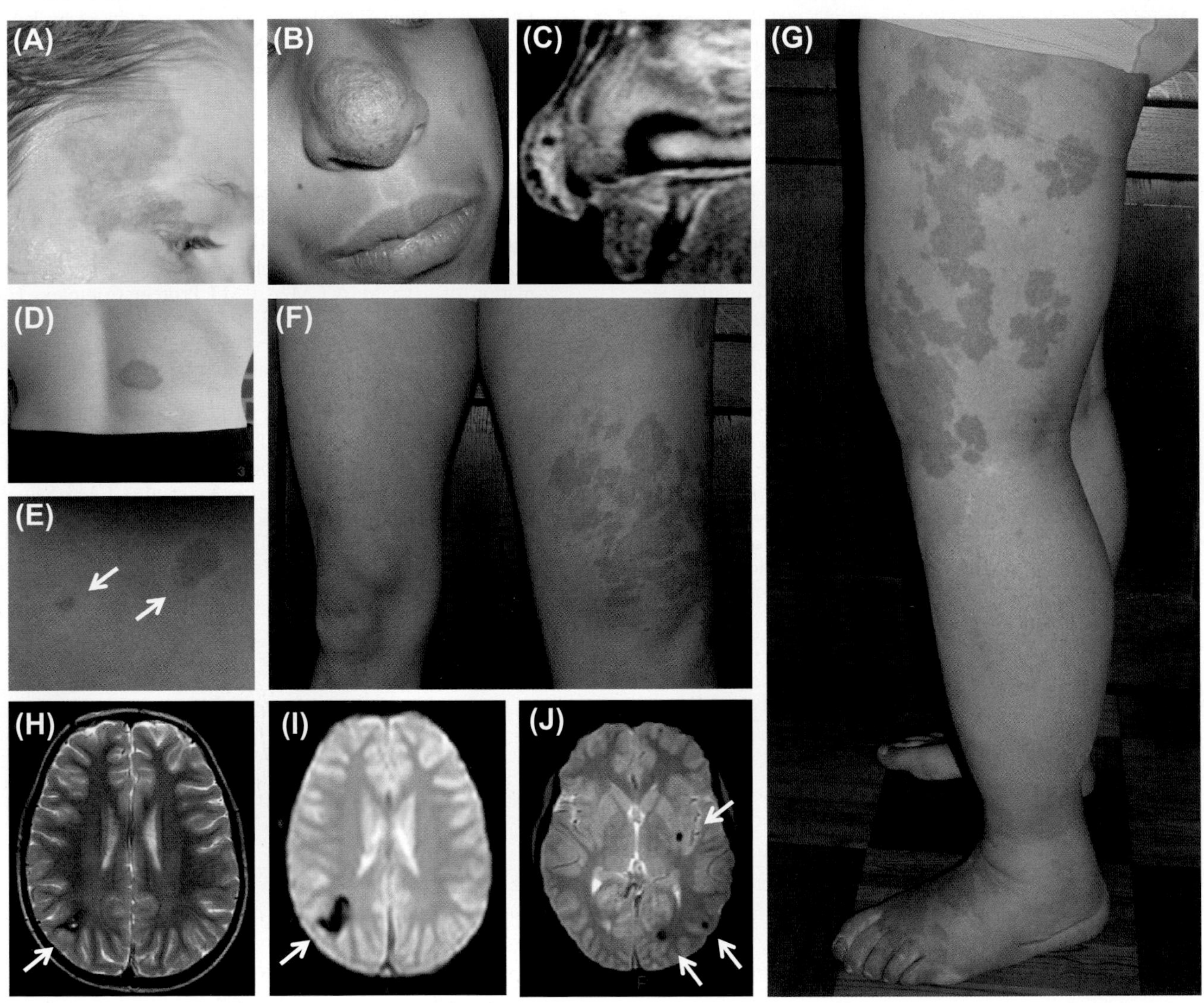

FIGURE 57-1 Disorders of the capillaries. (A) sporadic CM; (B–G) phenotypic variation in CM-AVM: (B) nose AVM and small CM on the left cheek; (C) T1-weighted MRI-enhanced sequence of the AVM shown in (B); (D, E) CM of CM-AVM—small multifocal CM with halo (arrows); (F, G) Parkes Weber syndrome in the lower extremity: patchy CM and hypertrophy, (H, I) sporadic CCM lesion (arrow) in the parietal lobe: axial T2-weighted MRI sequence (H) and axial T2-weighted gradient-echo sequence (I), (J) multiple supratentorial CCMs (arrows)—axial T2-weighted gradient-echo sequence.

FURTHER READING

1. Akers, A. L.; Johnson, E.; Steinberg, G. K., et al. Biallelic Somatic and Germline Mutations in Cerebral Cavernous Malformations (CCMs): Evidence for a Two-Hit Mechanism of CCM Pathogenesis. *Hum. Mol. Genet.* **2009,** *18* (5), 919–930.
2. Carr, C. W.; Zimmerman, H. H.; Martin, C. L., et al. 5q14.3 Neurocutaneous Syndrome: A Novel Continguous Gene Syndrome Caused by Simultaneous Deletion of RASA1 and MEF2C. *Am. J. Med. Genet. Part A* in press.
3. Denier, C.; Labauge, P.; Bergametti, F., et al. Genotype–Phenotype Correlations in Cerebral Cavernous Malformations Patients. *Ann. Neurol.* **2006,** *60* (5), 550–556.
4. Engels, H.; Wohlleber, E.; Zink, A., et al. A Novel Microdeletion Syndrome Involving 5q14.3-q15: Clinical and Molecular Cytogenetic Characterization of Three Patients. *Eur. J. Hum. Genet.* **2009,** *17* (12), 1592–1599.
5. Faurobert, E.; Albiges-Rizo, C. Recent Insights into Cerebral Cavernous Malformations: A Complex Jigsaw Puzzle Under Construction. *FEBS J.* **2010,** *277* (5), 1084–1096.
6. Hilder, T. L.; Malone, M. H.; Bencharit, S., et al. Proteomic Identification of the Cerebral Cavernous Malformation Signaling Complex. *J. Proteome Res.* **2007,** *6* (11), 4343–4355.
7. Labauge, P.; Denier, C.; Bergametti, F.; Tournier-Lasserve, E. Genetics of Cavernous Angiomas. *Lancet Neurol.* **2007,** *6* (3), 237–244.
8. Le Meur, N.; Holder-Espinasse, M.; Jaillard, S., et al. MEF2C Haploinsufficiency Caused by Either Microdeletion of the 5q14.3 Region or Mutation Is Responsible for Severe Mental Retardation with Stereotypic Movements, Epilepsy and/or Cerebral Malformations. *J. Med. Genet.* **2010,** *47* (1), 22–29.
9. Limaye, N.; Boon, L. M.; Vikkula, M. From Germline Towards Somatic Mutations in the Pathophysiology of Vascular Anomalies. *Hum. Mol. Genet.* **2009,** *18* (R1), R65–R74.
10. Limaye, N.; Wouters, V.; Uebelhoer, M., et al. Somatic Mutations in Angiopoietin Receptor Gene *TEK* Cause Solitary and Multiple Sporadic Venous Malformations. *Nat. Genet.* **2009,** *41* (1), 118–124.
11. Pagenstecher, A.; Stahl, S.; Sure, U.; Felbor, U. A Two-Hit Mechanism Causes Cerebral Cavernous Malformations: Complete Inactivation of CCM1, CCM2 or CCM3 in Affected Endothelial Cells. *Hum. Mol. Genet.* **2009,** *18* (5), 911–918.
12. Revencu, N.; Boon, L. M.; Mulliken, J. B., et al. Parkes Weber Syndrome, Vein of Galen Aneurysmal Malformation, and Other Fast-Flow Vascular Anomalies are Caused by RASA1 Mutations. *Hum. Mutat.* **2008,** *29* (7), 959–965.
13. Revencu, N.; Vikkula, M. Cerebral Cavernous Malformation: New Molecular and Clinical Insights. *J. Med. Genet.* **2006,** *43* (9), 716–721.
14. Riant, F.; Bergametti, F.; Ayrignac, X., et al. Recent Insights into Cerebral Cavernous Malformations: The Molecular Genetics of CCM. *FEBS J.* **2010,** *277* (5), 1070–1075.
15. Stahl, S.; Gaetzner, S.; Voss, K., et al. Novel CCM1, CCM2, and CCM3 Mutations in Patients with Cerebral Cavernous Malformations: In-Frame Deletion in CCM2 Prevents Formation of a CCM1/CCM2/CCM3 Protein Complex. *Hum. Mutat.* **2008,** *29* (5), 709–717.
16. Thiex, R.; Mulliken, J. B.; Revencu, N., et al. A Novel Association between RASA1 Mutations and Spinal Arteriovenous Anomalies. *Am. J. Neuroradiol.* **2010,** *31* (4), 775–779.
17. Voss, K.; Stahl, S.; Schleider, E., et al. CCM3 Interacts with CCM2 Indicating Common Pathogenesis for Cerebral Cavernous Malformations. *Neurogenetics* **2007,** *8* (4), 249–256.
18. Zheng, X.; Xu, C.; Di Lorenzo, A., et al. CCM3 Signaling through Sterile 20-Like Kinases Plays an Essential Role During Zebrafish Cardiovascular Development and Cerebral Cavernous Malformations. *J. Clin. Invest.* **2010,** *120* (8), 2795–2804.

Respiratory Disorders

Cystic Fibrosis, 225
Genetic Underpinnings of Asthma and Related Traits, 229
Hereditary Pulmonary Emphysema, 234
Interstitial and Restrictive Pulmonary Disorders, 237

CHAPTER

58

Cystic Fibrosis

Garry R Cutting

McKusick-Nathans Institute of Genetic Medicine, Johns Hopkins University School of Medicine, Baltimore, MD, USA

Cystic fibrosis (CF) is a single-gene recessive disorder that affects ~70,000 individuals worldwide. Median survival for CF patients has progressively increased to 38 years for an affected child born in 2009. Patients with CF manifest disease in the lungs, pancreas, intestine, male reproductive tract, and sweat gland. CF transmembrane conductance regulator (CFTR), the dysfunctional protein in CF patients, conducts chloride across the apical membranes of polarized epithelia. Loss of CFTR function affects the transport of chloride, sodium, and water across epithelial tissues leading to inadequate hydration of mucous secretions of CF patients. Obstruction of luminal space follows and recurrent cycles of inflammation and fibrosis ultimately destroys affected organs. Obstruction of the exocrine pancreas causes intestinal malabsorption and an abnormal nutritional status in most CF patients. Obstructive lung disease is the cause of death in almost 90% of patients. A minor fraction of CF patients (~10%) manifest disease in a subset of the aforementioned organ systems and are termed non-classic CF.

Over 1800 disease-associated mutations have been reported in the CFTR gene, although one mutation, F508del, accounts for approximately 70% of CF alleles worldwide. The F508del mutation causes misfolding of CFTR that leads to intracellular degradation and loss of functional product. The functional consequences of only a minor fraction of the remaining mutations have been evaluated. The commonness of F508del homozygosity among CF patients (~50%) creates a reference population for analyzing the phenotypic consequences of variability in CFTR genotype. CFTR genotype correlates with the severity of pancreatic exocrine disease, somewhat with sweat chloride concentration but poorly with variation in nutritional status and lung disease. Both environmental and genetic modifiers that influence severity of lung disease and risk for complications such as diabetes, intestinal obstruction, and liver disease have been identified. There are rare cases of genetic heterogeneity in cases of non-classic CF.

Diagnosis of CF is based on clinical features and demonstration of elevated concentrations of chloride in sweat. Mutation analysis of CFTR and evaluation of ion transport in nasal epithelium can aid in diagnosis. Mutations in CFTR can cause congenital bilateral absence of the vas deferens (obstructive male infertility) or pancreatitis, and can be a risk factor for bronchiectasis and chronic rhinosinusitis. Newborn screening for CF is now widespread in North America, Europe, and Australasia. Population screening for CF carriers has been in effect in the United States and regions of Europe for almost a decade. Symptomatic treatment is currently the mainstay of CF therapy although molecular-based therapies aimed at augmenting function of defective CFTR are on the horizon.

TABLE 58-1 Phenotypes Associated with Mutations in the CFTR Gene

	Classic CF	Non-classic	CBAVD
Chronic pulmonary disease	+	+	+/−
Pancreatic exocrine disease	+	+/−	−
Elevated sweat chloride (>60 mM)	+	+/−	+/−
Male infertility	+	+/−	+

TABLE 58-2 Signs and Symptoms Suggesting a CF Diagnosis[a]

	Percent
Respiratory symptoms	51.6
Failure to thrive/malnutrition	43.2
Steatorrhea/abnormal stools	34.8
Meconium ileus/intestinal obstruction	20.6
Family history	17.1
Electrolyte imbalance	5.3
Rectal prolapsed	3.8
Neonatal screening	3.3
Nasal polyps/sinus disease	2.7
Genotype	2.3
Prenatal diagnosis	1.3
Liver problems	1.1
Edema/hypoproteinema/hypoalbuminemia	0.4
Other	2.0
Unknown	2.4

[a]Patients may present with one or more sign or symptom.
Source: Cystic Fibrosis Foundation, Cystic Fibrosis Foundation Patient Registry Annual Data Report 1999, Sep 1.

TABLE 58-3 Phenotypic Features Consistent with a Diagnosis of CF

1. Chronic sinopulmonary disease manifested by
   a. Persistent colonization/infection with typical CF pathogens including *Staphylococcus aureus*, nontypeable *Haemophilus influenzae*, mucoid and nonmucoid *Pseudomonas aeruginosa*, and *Burkholderia cepacia*
   b. Chronic cough and sputum production
   c. Persistent chest radiograph abnormalities (e.g. bronchiectasis, atelectasis, infiltrates, hyperinflation)
   d. Airway obstruction manifested by wheezing and air trapping
   e. Nasal polyps: radiographic or computed tomographic abnormalities of the paranasal sinuses
   f. Digital clubbing
2. Gastrointestinal and nutritional abnormalities including
   a. Intestinal: meconium ileus, distal intestinal obstruction syndrome, rectal prolapse
   b. Pancreatic: pancreatic insufficiency, recurrent pancreatitis
   c. Hepatic: chronic hepatic disease manifested by clinical or histologic evidence of focal biliary cirrhosis or multilobular cirrhosis
   d. Nutritional: failure to thrive (protein–calorie malnutrition), hypoproteinemia and edema, complications secondary to fat-soluble vitamin deficiency
3. Salt loss syndromes: acute salt depletion, chronic metabolic alkalosis
4. Male urogenital abnormalities resulting in obstructive azoospermia (CBAVD)

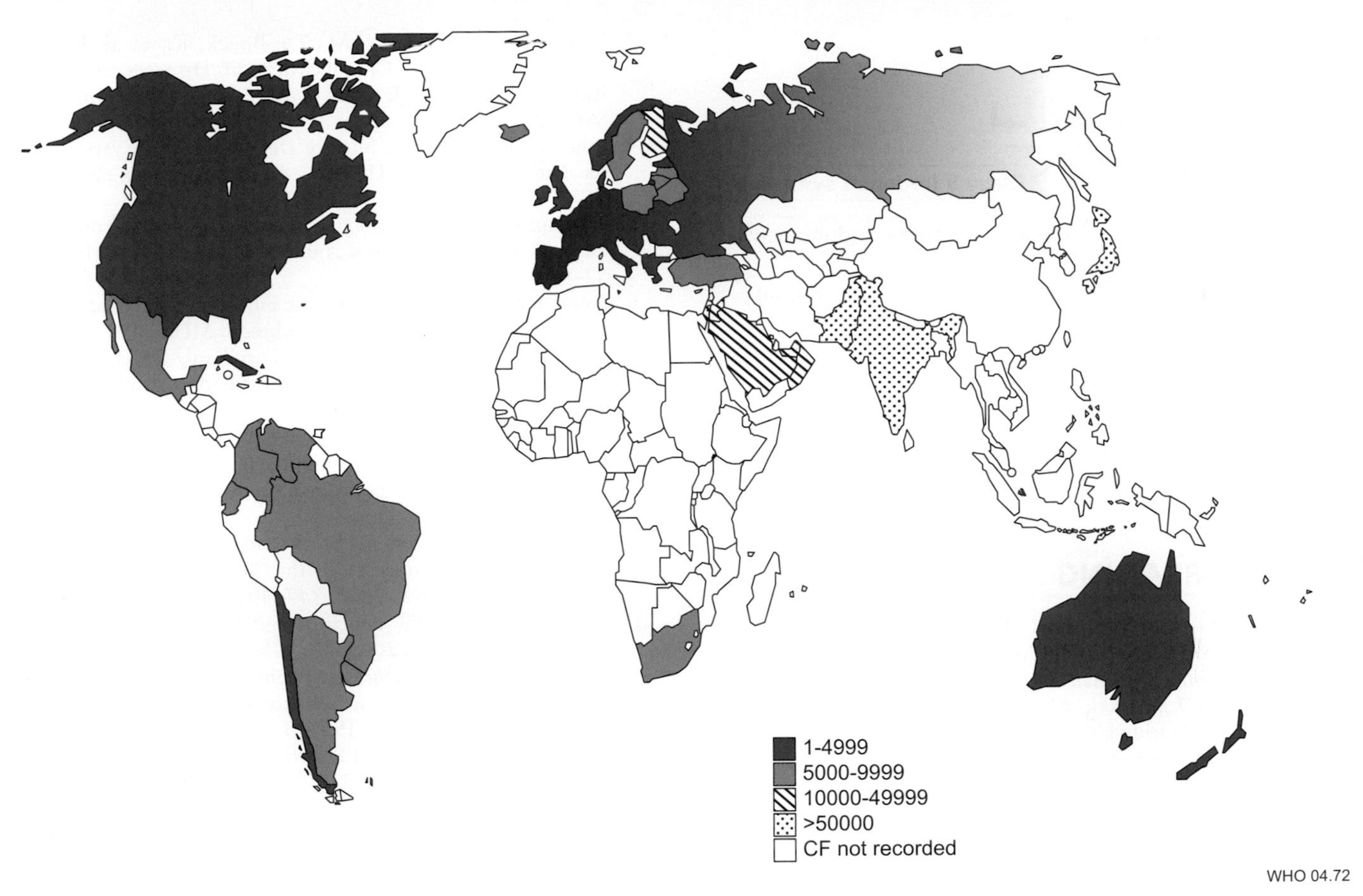

FIGURE 58-1 Worldwide Incidence of CF.

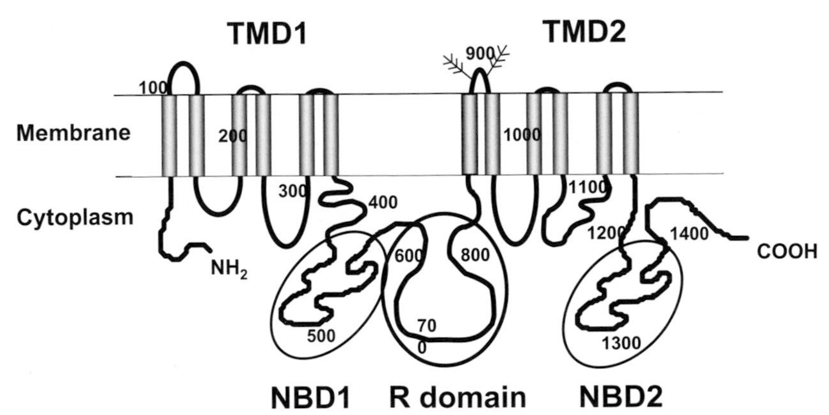

FIGURE 58-2 Structural and Functional Domains of CFTR. CFTR is composed of two transmembrane domains (TMD1 and TMD2) each containing six hydrophobic segments inserted into the cell membrane, two intracellular nucleotide binding domains (NBD1 and NBD2) that interact with ATP, and an intracellular regulatory (R) domain that is phosphorylated by protein kinase A. The fourth extracellular loop has two sites for asparagine-linked glycosylation. Residue numbers are shown.

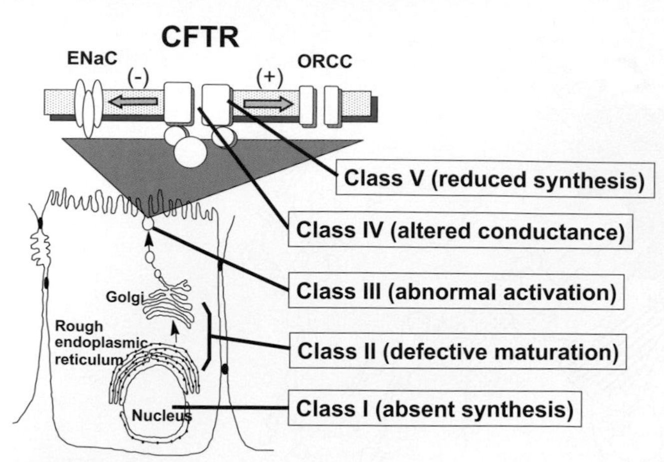

FIGURE 58-3 The Cellular Consequences of Five Classes of Mutations.

FURTHER READING

1. Ashlock, M. A.; Olson, E. R. Therapeutics Development for Cystic Fibrosis: A Successful Model for a Multisystem Genetic Disease. *Annu. Rev. Med.* **2011,** *62,* 107–125.
2. Bobadilla, J. L.; Macek, M.; Fine, J. P.; Farrell, P. M. Cystic Fibrosis: A Worldwide Analysis of CFTR Mutations—Correlation with Incidence Data and Application to Screening. *Hum. Mutat.* **2002,** *19,* 575–606.
3. Bombieri, C.; Claustres, M.; De Boeck, K., et al. Recommendations for the Classification of Diseases as CFTR-Related Disorders. *J. Cyst. Fibros.* **2011,** *10* (Suppl 2), S86–102.
4. Boucher, R. C. Airway Surface Dehydration in Cystic Fibrosis: Pathogenesis and Therapy. *Annu. Rev. Med.* **2007,** *58,* 157–170.
5. Cutting, G. R. Modifier Genes in Mendelian Disorders: The Example of Cystic Fibrosis. *Ann. N. Y. Acad. Sci.* **2010,** *1214,* 57–69.
6. Cystic Fibrosis Foundation Cystic Fibrosis Foundation Patient Registry Annual Data Report: Bethesda, MD, 2009.
7. De Boeck, K.; Wilschanski, M.; Castellani, C., et al. Cystic Fibrosis: Terminology and Diagnostic Algorithms. *Thorax* **2006,** *61,* 627–635.
8. Farrell, P. M.; Rosenstein, B. J.; White, T. B., et al. Guidelines for Diagnosis of Cystic Fibrosis in Newborns Through Older Adults: Cystic Fibrosis Foundation Consensus Report. *J. Pediatr.* **2008,** *153,* S4–S14.
9. Groman, J. D.; Meyer, M. E.; Wilmott, R. W., et al. Variant Cystic Fibrosis Phenotypes in the Absence of CFTR Mutations. *N. Engl. J. Med.* **2002,** *347,* 401–407.
10. Riordan, J. R. CFTR Function and Prospects for Therapy. *Annu. Rev. Biochem.* **2008,** *77,* 701–726.
11. The Cystic Fibrosis Genotype-Phenotype Consortium Correlation between Genotype and Phenotype in Patients with Cystic Fibrosis. *N. Engl. J. Med.* **1993,** *329,* 1308.
12. Zielenski, J. Genotype and Phenotype in Cystic Fibrosis. *Respiration* **2000,** *67,* 117–133.

CHAPTER

59

Genetic Underpinnings of Asthma and Related Traits

Hakon Hakonarson

Department of Pediatrics, University of Pennsylvania, School of Medicine, PA, USA

Michael E March and Patrick M A Sleiman

Center for Applied Genomics, Children's Hospital of Philadelphia, PA, USA

Asthma is a chronic inflammatory condition of the lungs, characterized by excessive responsiveness of the lungs to stimuli, in the forms of infections, allergens, and environmental irritants. Although it is usually recognized through the acute episodes of asthma attacks, including wheezing and sometimes irreversible declines in lung function, asthma has an important immune system component. There is a clear connection between asthma and atopy, with an estimated 50% of asthma cases attributable to specific allergies. In the United States, it is estimated that at least 22.9 million Americans suffer asthma. Asthma is the leading chronic illness in US children, with 6.8 million affected in 2006. Global rates of asthma range from 1 to 18% of the population from different countries. Prevalence is rising in locations where rates were previously low, and variation in rates from country to country appears to be diminishing.

Twin studies have shown that there is a genetic element to asthma susceptibility, with heritability of the condition estimated at between 0.36 and 0.77. Although researchers have successfully identified the genetic causes of nearly 3000 single-gene disorders, it has been comparatively difficult to identify the genetic basis of complex genetic disorders, such as asthma. Three study designs are routinely employed to investigate genetic contributions in complex diseases: candidate-gene association studies, genome-wide linkage studies, and genome-wide association studies (GWAS).

In a candidate gene association study, a particular gene (or set of genes) is selected for study based on its biological plausibility or suspected role in the phenotype of interest. More than 600 candidate genes have been associated with asthma and related phenotypes in over 1000 publications, although relatively few of these genes have been successfully replicated and many have been examined and failed replication. Associated candidate genes include molecules with known immunological functions, genes involved in lung function and development, and molecules involved in epithelial barrier function, tissue responses to environmental exposures, and innate immunity. Genes that have been identified in five or more candidate gene studies as having a positive association with asthma, or asthma related phenotypes, are listed in Table 59-1.

Genome-wide linkage study design focuses on families affected by the disease of interest. With less genetic recombination occurring between closely related individuals, it is possible to screen the entire genome with a panel of relatively few, evenly spaced markers, searching for variants that are either unique to or over-represented in affected individuals. The list of genes that have been identified through positional cloning following linkage studies is relatively small, and is summarized in Table 59-2. *GPRA* and *ADAM33* are of particular interest, given their likely involvement in lung function.

In GWAS, hundreds of thousands of single nucleotide polymorphisms (SNPs) are compared across the entire genome between cases and controls. Numerous genes with known functions in immunology have been associated with asthma or related phenotypes through GWAS, although multiple genes with no obvious connection to asthma have also been identified (*DENND1B*, *ORMDL3/GSDMB*).

A summary of loci identified in GWAS targeting asthma or related conditions is presented in Table 59-3.

The numerous genome-wide linkage studies, candidate gene association studies, and GWAS performed on asthma and asthma-related phenotypes have resulted in an increasingly large list of genes implicated in asthma susceptibility and pathogenesis. This list can be categorized into four broad functional groups, from which several themes have emerged, as summarized in Figure 59-1.

T_H2 –cell–mediated adaptive immune responses have been widely recognized as a crucial component of allergic disease. Genes involved in T_H2 -cell differentiation and function have been extensively studied in asthma candidate gene association studies, and as one might expect, SNPs in many of these genes have been associated with asthma and other allergic phenotypes. Genes involved in the development and pathogenesis of allergy, as well as molecules involved in T_H1 and T_H2-cell differentiation, have been associated with asthma.

A second class of associated genes is involved in detection of pathogens and allergens. These genes include pattern recognition receptors and extracellular receptors, such as *CD14* and the toll-like receptor family. These molecules are important in the identification of foreign pathogens, so their involvement in asthma is unsurprising.

A variety of genes involved in mediating the response to allergic inflammation and oxidant stress on the tissue level appear to be important contributors to asthma susceptibility. A prominent example is *ADAM33*, a disintegrin and metalloprotease expressed in lung fibroblasts and smooth muscle cells, and suspecting in the remodeling that is observed in the asthmatic lung.

Studies of asthma genetics have raised new interest in the body's first line of immune defense, the epithelial barrier, in the pathogenesis of asthma. Mutations in the filaggrin gene (*FLG*) are strongly associated with asthma, both dependent and independent of atopic dermatitis. Other genes involved in recruitment of immune cells to epithelium and the response of the tissue to inflammation have also been implicated.

Large challenges remain in our understanding of the genetic underpinnings of asthma. GWAS allows for the association of unsuspected novel genes with the disease, and many functional studies will need to be performed in order to elucidate how these genes fit into the disease pathobiology. Techniques are still in development to allow the characterization of interactions between multiple susceptibility genes, and between genes and environmental influences. It is likely that considerable breakthroughs in our understanding of the genetics of asthma will come only with the development of tools and study designs to analyze the networks of factors that contribute to complex genetic diseases. Next-generation sequencing will undoubtedly play an important role in this process.

TABLE 59-1 Summary of Well-Replicated Loci Identified Through Candidate Gene Studies

Gene	Chromosomal Locus	Function
Immune Function		
IL10	1q31-q32	Cytokine—immune regulation
CTLA4	2q33	Control/inhibition of T cell responses/immune regulation
IL13	5q31	Induces T_H2 effector functions
IL4	5q31.1	T_H2 differentiation
CD14	5q31.1	Microbe detection—recognizes pathogen Associated molecular patterns
HAVCR1	5q33.2	T cell responses—hepatitis A virus receptor
LTC4S	5q35	Leukotriene synthase—inflammatory mediator
LTA	6p21.3	Inflammatory mediator
TNF	6p21.3	Inflammatory mediator
HLA-DRB1	6p21	Major histocompatibilty complex Class II—antigen presentation
HLA-DQB1	6p21	
HLA-DPB1	6p21	
FCER1B	11q13	Receptor for IgE—Atopy
IL18	11q22.2-q22.3	Inflammation
STAT6	12q13	IL-4 and IL-13 signaling
CMA1	14q11.2	Chymase—mast cell expressed serine protease
IL4R	16p12.1-p12.2	Alpha chain of receptors for IL-4 and IL-13
Barrier Function/Innate Immunity		
FLG	1q21.3	Epithelial integrity and barrier function
SPINK5	5q32	Epithelial serine protease inhibitor
CC16	11q12.3-q13.1	Potential immunoregulatory function—epithelial expression
NOS1	12q24.2-q24.31	Nitric oxide synthase—cellular communication
CCL11	17q21.1-q21.2	Eoxtaxin-1—eosinophil chemoattractant
CCL5	17q11.2-q12	RANTES—chemoattactant for T cells, eosinophils, basophils
Tissue Response		
GSTM1	1p13.3	Detoxification, removal of products of oxidative stress
ADRB2	5q31-q32	Smooth muscle relaxation
GPRA	7p14.3	Regulation of metalloprotease expression, neuronal effects
NAT2	8p22	Detoxification
GSTP1	11q13	Detoxification, removal of products of oxidative stress
ACE	17q23.3	Regulation of inflammation
TBXA2R	19p13.3	Platelet aggregation
TGFB1	19q13.1	Influences cell growth, differentiation, proliferation, apoptosis
ADAM33	20p13	Cell–cell and cell–matrix interactions
GSTT1	22q11.23	Detoxification, removal of products of oxidative stress

Genes identified as asthma susceptibility loci in candidate gene studies. Genes are grouped loosely based on their functions in immunity, epithelial barrier function, or tissue response and remodeling.

TABLE 59-2 Table of Loci Identified Though Linkage Studies and Positional Cloning

Gene	Chromosomal Locus
CYFIP2	5q33.3
DPP10	2q14.1
HLAG	6p21.33
PHF11	13q14.3
GPRA	7p14.3
ADAM33	20p13

TABLE 59-3 Summary of GWAS Loci Referenced in this Review, Including Chromosome Location, the Most Significant SNP Identified, and the End Point of the Study

Reported Gene	Locus	Top SNP	End Point Analyzed
RAD50	5q31.1	rs2244012	Asthma
HLA-DR/DQ	6p21.32	rs3998159	
DENND1B	1q31.3	rs2786098	Asthma
TLE4	9q21.31	rs2378383	Asthma
PDE4D	5q12.1	rs2548659	Asthma
ORMDL3	17q12	rs7216389	Asthma
PDE11A	2q31.2	rs11684634	Asthma
CHI3L1	1q32.1	rs4950928	Asthma/YKL-40 serum levels
FCER1A	1q23.2	rs2251746	Serum IgE levels
STAT6	12q13	rs12368672	
RAD50	5q31.1	rs2040704	
IL1RL1	2q12.1	rs1420101	Blood eosinophil count/asthma
IKZF2	5q31.1	rs12619285	
GATA2	3q21.3	rs4857855	
IL5	2q12.1	rs4143832	
SH2B3	12q24.12	rs3184504	
CHRNA 3/5	15q24	rs8034191	COPD
HHIP	4q31.22	rs13147785	FEV_1/FVC
TNS1	2q35	rs2571445	FEV_1
GSTCD	4q24	rs10516526	FEV_1
HTR4	5q33.1	rs3995090	FEV_1
AGER	6p21.32	rs2070600	FEV_1/FVC
THSD4	15q23	rs12899618	FEV_1/FVC
GPR126	6q24.1	rs3817928	FEV_1/FVC
ADAM19	5q33	rs2277027	FEV_1/FVC
AGER-PPT2	6p21.3	rs2070600	FEV_1/FVC
FAM13A	4q22.1	rs2869967	FEV_1/FVC
PTCH1	9q22.32	rs16909898	FEV_1/FVC
PID1	2q36.3	rs1435867	FEV_1/FVC
HTR4	5q33.1	rs7735184	FEV_1/FVC
INTS12-GSTCD-NPNT	4q24	rs17331332	FEV_1
IL1RL1/IL18R1	2q12.1	rs3771166	Asthma
HLA-DQ	6p21.32	rs9273349	
IL33	9p24.1	rs1342326	
SMAD3	15q22.33	rs744910	
IL2RB	22q12.3	rs2284033	
ORMDL3/GSDMB	17q12	rs2305480	Childhood onset asthma

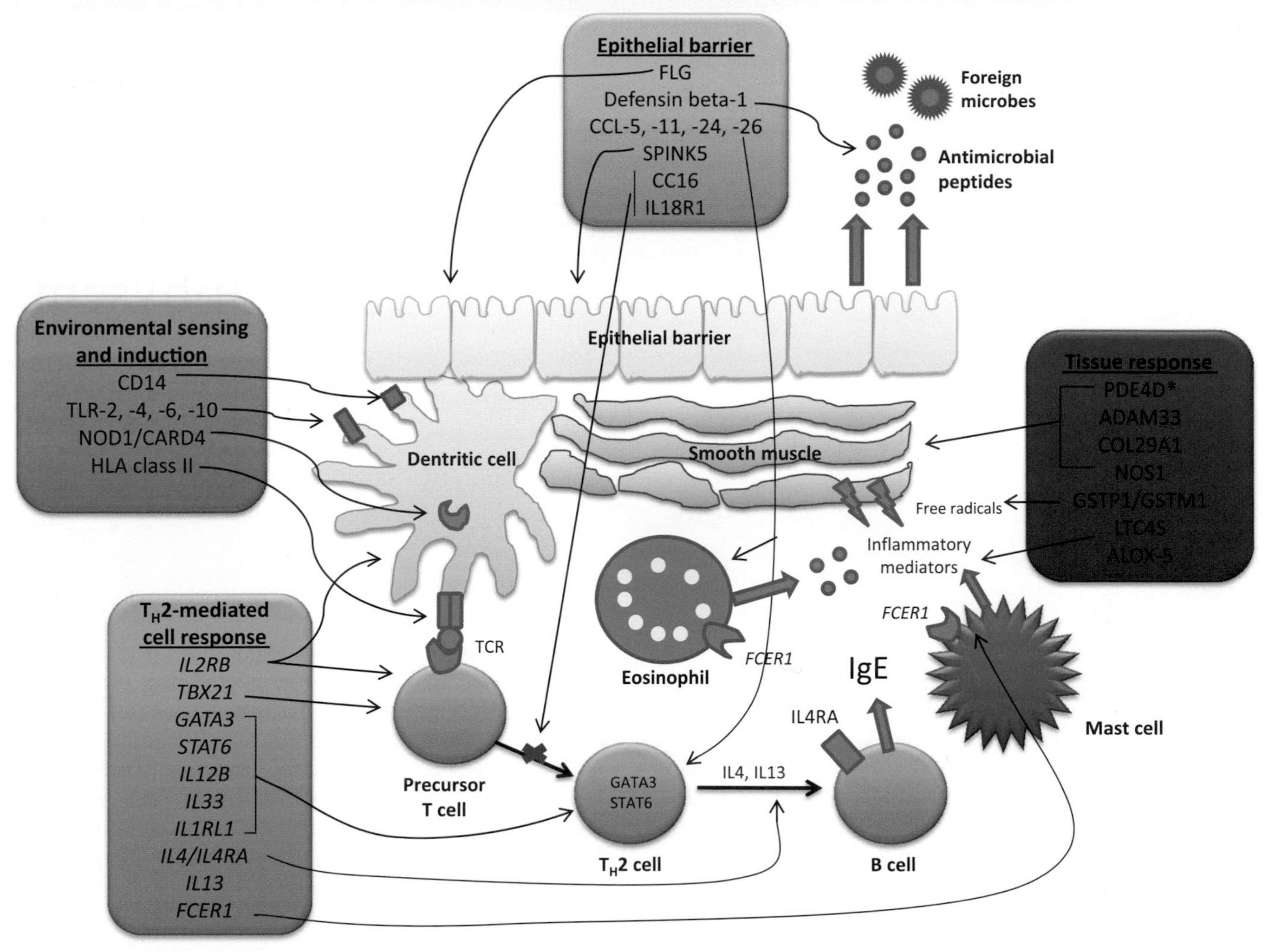

FIGURE 59-1 Genes implicated in asthma susceptibility and pathogenesis.

FURTHER READING

1. Gudbjartsson, D. F.; Bjornsdottir, U. S.; Halapi, E., et al. Sequence Variants Affecting Eosinophil Numbers Associate with Asthma and Myocardial Infarction. *Nat. Genet.* **2009**, *41*, 342–347.
2. Hancock, D. B.; Eijgelsheim, M.; Wilk, J. B., et al. Meta-Analyses of Genome-Wide Association Studies Identify Multiple Loci Associated with Pulmonary Function. *Nat. Genet.* **2010**, *42*, 45–52.
3. Himes, B. E.; Hunninghake, G. M.; Baurley, J. W., et al. Genome-Wide Association Analysis Identifies PDE4D as an Asthma-Susceptibility Gene. *Am. J. Hum. Genet.* **2009**, *84*, 581–593.
4. Martinez, F. D. Gene-Environment Interactions in Asthma and Allergies: A New Paradigm to Understand Disease Causation. *Immunol. Allergy Clin. North Am.* **2005**, *25*, 709–721.
5. Moffatt, F.; Gut, I. G.; Demenais, F., et al. A Large-Scale, Consortium-Based Genomewide Association Study of Asthma. *N. Engl. J. Med.* **2010**, *363*, 1211–1221.
6. Ober, C.; Thompson, E. E. Rethinking Genetic Models of Asthma: The Role of Environmental Modifiers. *Curr. Opin. Immunol.* **2005**, *17*, 670–678.
7. Pearce, N.; Ait-Khaled, N.; Beasley, R., et al. Worldwide Trends in the Prevalence of Asthma Symptoms: Phase III of the International Study of Asthma and Allergies in Childhood (ISAAC). *Thorax* **2007**, *62*, 758–766.
8. Pillai, S. G.; Ge, D.; Zhu, G., et al. A Genome-Wide Association Study in Chronic Obstructive Pulmonary Disease (COPD): Identification of Two Major Susceptibility Loci. *PLoS Genet.* **2009**, *5* (3), e1000421.
9. Sleiman, P. M.; Flory, J.; Imielinski, M., et al. Variants of DENND1B Associated with Asthma in Children. *N. Engl. J. Med.* **2010**, *362*, 36–44.
10. Vercelli, D. Discovering Susceptibility Genes for Asthma and Allergy. *Nat. Rev. Immunol.* **2008**, *8*, 169–182.
11. Wilk, J. B.; Chen, T. H.; Gottlieb, D. J., et al. A Genome-Wide Association Study of Pulmonary Function Measures in the Framingham Heart Study. *PLoS Genet.* **2009**, *5* (3), e1000429.

CHAPTER

60

Hereditary Pulmonary Emphysema

Chad K Oh and Nestor A Molfino

Clinical Development, MedImmune, LLC, Gaithersburg, MD, USA

Emphysema is included in the syndrome of chronic obstructive pulmonary disease (COPD) (*1*). COPD is the fourth leading cause of morbidity and mortality in the United States and is expected to rank third as the cause of death worldwide by 2020 (*2*). It is estimated that there are 10 million individuals in the United States with physician-diagnosed COPD and many more affected individuals who are undiagnosed (*3*).

Although case reports of familial COPD were published in the 1950s, interest in the role of genetic factors in COPD largely began with the discovery of severe alpha 1-antitrypsin (AAT) deficiency by Laurell and Eriksson in 1963 (*4*). AAT deficiency is a proven genetic determinant of COPD (*5*); therefore, we discuss the molecular and population genetics of AAT deficiency in detail. In addition, we review the evidence for genetic factors in non-AAT deficiency COPD, including assessment of risk to relatives, segregation analysis, linkage analysis, and association studies. We also discuss the utility of animal models in identifying the genetic determinants of COPD.

Only a small percentage of COPD patients inherit severe AAT deficiency, and additional genetic factors likely influence the development of COPD (*6*). Further efforts in linkage analysis, association studies, and animal models may lead to the identification of such factors. To achieve a complete understanding of COPD pathophysiology, characterization of the interactions between genetic determinants and cigarette smoking (and potentially other environmental factors) will be required (*7*).

A key component in genetic studies is the identification of a distinct and well-characterized phenotype (*8*). A variety of methods, including pulmonary function test data, chest computed tomography (CT) scan measurements, questionnaire-derived assessments of respiratory symptoms, and development of biochemical markers that relate to the pathophysiology of COPD have been successfully used to determine phenotypes (*9*). Particularly, the development of high-resolution CT (HRCT) chest scans has provided a powerful, noninvasive tool for assessment of the anatomic presence of emphysema (*10*). HRCT scans are much more sensitive than conventional chest X-rays for the detection of emphysema. High-resolution (thin cut) images allow identification and quantification of the extent of emphysema, which was not possible with conventional (thick cut) chest CT scan images.

A variety of biochemical markers have been developed in an effort to reflect the inflammation and tissue destruction of COPD (*11*). Recently, several large-scale clinical studies have been conducted to identify novel biomarkers (*12*). One such study is Evaluation of COPD Longitudinally to Identify Predictive Surrogate End points (ECLIPSE), a 3-year longitudinal study to define the parameters that characterize subgroups and predict disease progression in more than 2000 individuals with COPD (*13*). Other large-scale studies are being conducted, including SubPopulations and InteRmediate Outcome Measures In COPD Study (SPIROMICS) and more recently, the Genetic Epidemiology of COPD (COPDGene) study (*14*).

Genetic studies discussed in this chapter include segregation analysis, linkage analysis, and genetic association studies. A large number of association studies have compared the distribution of variants in candidate genes hypothesized to be involved in the development of COPD in COPD patients and control subjects. Identification of genetic factors influencing the development of COPD unrelated to AAT deficiency could clarify the biochemical mechanisms causing COPD, allow identification of highly susceptible individuals, and lead to new therapeutic interventions for COPD.

TABLE 60-1 Age, Smoking History and Spirometry in First-Degree Relatives of Boston Early-Onset COPD Study Probands Compared to Control Subjects[a]

Group	*n*	FEV_1/FVC (% Pred)	FEV_1 (% Pred)	Age	Pack-Years
Smoking first-degree relatives	112	83.5 ± 16.1[b]	76.1 ± 20.9[b]	45.9 ± 17.3	28.5 ± 26.6
Smoking control subjects	48	94.3 ± 10.3	89.2 ± 14.4	48.6 ± 13.9	22.1 ± 22.1
Nonsmoking first-degree relatives	91	92.7 ± 7.6	93.4 ± 12.9	34.5 ± 18.9	0.00
Nonsmoking control subjects	35	95.5 ± 7.2	93.4 ± 14.2	39.9 ± 18.2	0.00

[a]Values are mean ± SD. No other pairwise comparisons between first-degree relatives and control subjects with similar smoking histories were significant at $P < 0.05$.
[b]Indicates $P < 0.01$ compared to control subjects with similar smoking history.
Adapted from Silverman et al. (*15*).

TABLE 60-2 Linkage Analysis of Quantitative Spirometric Phenotypes in the General Population[a]

Author (Date)	Population (*n*)	FEV_1		FEV_1/FVC	
		Region	Maximum LOD	Region	Maximum LOD
Joost 2002	Framingham Study (*n* = 1578)	4p	1.6	—	—
		6q	2.4	—	—
Wilk (2000)	Family Heart Study (*n* = 2178)	3q	2.0	1p	1.8[b]
				4p	2.6[b]
				8p	1.6[b]
				11q	1.9[b]
				15q	1.6[b]
Malhotra (2003)	CEPH Pedigrees (*n* = 264)			2q	2.0
				5q	2.2[c]

[a]LOD scores above 1.5 are presented.
[b]Based on normalized phenotypic values for FEV_1/FVC.
[c]Corresponds to parametric heterogeneity LOD score.

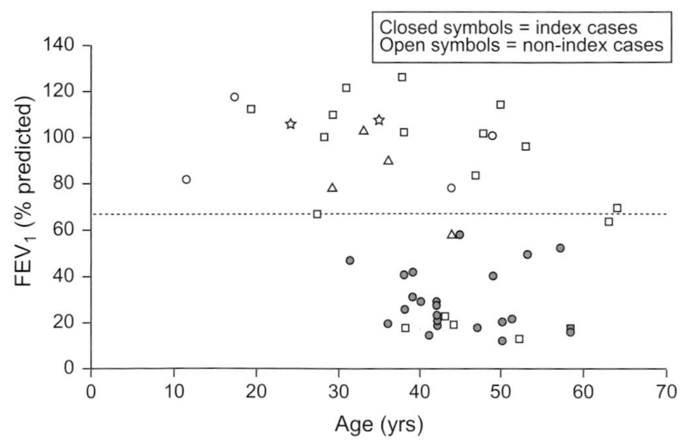

FIGURE 60-1 Effect of ascertainment bias on percent predicted FEV_1 among 52 severely AAT-deficient subjects (PI Z) in the study by Silverman et al. Closed circles represent index PI Z subjects (individuals diagnosed with severe AAT deficiency because they had COPD, and who were the first PI Z subject identified in their family); open symbols represent non-index subjects. Non-index subjects were ascertained by liver disease (open circles), family studies (squares), population screening (triangles), and other pulmonary problems (stars). Marked variability in FEV_1 values is evident for the non-index PI Z subjects. *(From Silverman, E. K.; Pierce, J. A.; Province, M. A.; Rao, D. C.; Campbell, E. J. Variability of Pulmonary Function in Alpha-1-Antitrypsin Deficiency: Clinical Correlates.* Ann. Intern. Med. ***1989,*** 111, *982–991 (16)).*

REFERENCES

1. Pauwels, R. A.; Buist, A. S.; Calverley, P. M.; Jenkins, C. R.; Hurd, S. S. GOLD Scientific Committee Global Strategy for the Diagnosis, Management, and Prevention of Chronic Obstructive Pulmonary Disease. NHLBI/WHO Global Initiative for Chronic Obstructive Lung Disease (GOLD) Workshop Summary. *Am. J. Respir. Crit. Care Med.* **2001,** *163,* 1256–1276.
2. Rabe, K. F.; Hurd, S.; Anzueto, A.; Barnes, P. J.; Buist, S. A.; Calverley, P.; Fukuchi, Y.; Jenkins, C.; Rodriguez-Roisin, R.; van Weel, C.; Zielinski, J. Global Initiative for Chronic Obstructive Lung Disease Global Strategy for the Diagnosis, Management, and Prevention of Chronic Obstructive Pulmonary Disease: GOLD Executive Summary. *Am. J. Respir. Crit. Care Med.* **2007,** *176,* 532–555.
3. Mannino, D. M.; Homa, D. M.; Akinbami, L. J.; Ford, E. S.; Redd, S. C. Chronic Obstructive Pulmonary Disease Surveillance—United States, 1971–2000. *MMWR Surveill. Summ.* **2002,** *51,* 1–16.
4. Hurst, A. Familial Emphysema. *Am. Rev. Respir. Dis.* **1959,** *80,* 179–180.
5. Laurell, C. B.; Eriksson, S. The Electrophoretic a_1-Globulin Pattern of Serum in a_1-Antitrypsin Deficiency. *Scand. J. Clin. Lab. Invest.* **1963,** *15,* 132–140.
6. Lieberman, J.; Winter, B.; Sastre, A. Alpha 1-Antitrypsin Pi-Types in 965 COPD Patients. *Chest* **1986,** *89,* 370–373.
7. Hogg, J. C. Identifying Smokers at Risk for Developing Airway Obstruction. *Chest* **1998,** *114,* 355.
8. Kitaguchi, Y.; Fujimoto, K.; Kubo, K.; Honda, T. Characteristics of COPD Phenotypes Classified According to the Findings of HRCT. *Respir. Med.* **2006,** *100,* 1742–1752.
9. Ferris, B. G. Epidemiology Standardization Project (American Thoracic Society). *Am. Rev. Respir. Dis.* **1978,** *118,* 1–120.
10. Guest, P. J.; Hansell, D. M. High Resolution Computed Tomography (HRCT) in Emphysema Associated with Alpha-1-Antitrypsin Deficiency. *Clin. Radiol.* **1992,** *45,* 260–266.
11. Montuschi, P.; Collins, J. V.; Ciabattoni, G.; Lazzeri, N.; Corradi, M.; Kharitonov, S. A.; Barnes, P. J. Exhaled 8-Isoprostane as an In Vivo Biomarker of Lung Oxidative Stress in Patients with COPD and Healthy Smokers. *Am. J. Respir. Crit. Care Med.* **2000,** *162,* 1175–1177.
12. Vestbo, J.; Anderson, W.; Coxson, H. O.; Crim, C.; Dawber, F.; Edwards, L.; Hagan, G.; Knobil, K.; Lomas, D. A.; MacNee, W.; Silverman, E. K.; Tal-Singer, R. ECLIPSE Investigators Evaluation of COPD Longitudinally to Identify Predictive Surrogate End-Points (ECLIPSE). *Eur. Respir. J.* **2008,** *31,* 869–873.
13. Lomas, D. A.; Silverman, E. K.; Edwards, L. D.; Miller, B. E.; Coxson, H. O.; Tal-Singer, R. Evaluation of COPD Longitudinally to Identify Predictive Surrogate Endpoints (ECLIPSE) Investigators Evaluation of Serum CC-16 as a Biomarker for COPD in the ECLIPSE Cohort. *Thorax* **2008,** *63,* 1058–1063.
14. Regan, E. A.; Hokanson, J. E.; Murphy, J. R.; Make, B.; Lynch, D. A.; Beaty, T. H.; Curran-Everett, D.; Silverman, E. K.; Crapo, J. D. Genetic Epidemiology of COPD (COPDGene) Study Design. *COPD* **2010,** *7,* 32–43.
15. Silverman, E. K.; Chapman, H. A.; Drazen, J. M.; Weiss, S. T.; Rosner, B.; Campbell, E. J.; O'Donnell, W. J.; Reilly, J. J.; Ginns, L.; Mentzer, S.; Wain, J.; Speizer, F. E. Genetic Epidemiology of Severe, Early-Onset Chronic Obstructive Pulmonary Disease. Risk to Relatives for Airflow Obstruction and Chronic Bronchitis. *Am. J. Respir. Crit. Care Med.* **1998,** *157,* 1770–1778.
16. Silverman, E. K.; Pierce, J. A.; Province, M. A.; Rao, D. C.; Campbell, E. J. Variability of Pulmonary Function in Alpha-1-Antitrypsin Deficiency: Clinical Correlates. *Ann. Intern. Med.* **1989,** *111,* 982–991.

CHAPTER

61

Interstitial and Restrictive Pulmonary Disorders

William E Lawson and James E Loyd

Division of Allergy, Pulmonary, and Critical Care Medicine, Vanderbilt University School of Medicine, Department of Veterans Affairs Medical Center, Nashville, TN, USA

Interstitial lung diseases (ILDs) represent a heterogeneous group of lung disorders in which inflammation and/or fibrosis in the alveolar interstitial space leads to pulmonary restriction, hypoxemia, and respiratory failure. Over 150 diseases are associated with ILD, some with known causes including environmental exposures, autoimmune and rheumatologic disease, and underlying genetic syndromes. However, in many cases, the cause remains unknown, comprising a group known as the idiopathic interstitial pneumonias (IIPs). Over the past decade, evidence has emerged that a component of these IIPs has a genetic basis. The IIPs break down into seven categories, of which idiopathic pulmonary fibrosis (IPF) is the most common and most severe. Individuals with IPF present with shortness of breath on exertion, restriction of pulmonary function testing, and radiographic evidence of ILD. Patients typically progress to hypoxemia and respiratory failure, with most patients dying within 5 years of diagnosis. Unfortunately, effective treatments remain elusive.

Familial interstitial pneumonia (FIP) is defined when two or more individuals have an IIP within a single family. In most families, inheritance is autosomal dominant. Studies suggest that familial cases account for ~2–4% of IPF, but this percentage may be higher. FIP is indistinguishable from sporadic IPF in terms of clinical presentation, radiography, histopathology, and survival except that FIP patients tend to present at an earlier age. To date, four genes have been linked to adult FIP [surfactant protein C (*SFTPC*), surfactant protein A2 (*SFTPA2*), telomerase reverse transcriptase (*TERT*), and telomerase RNA component (*TERC*)], but together only account for ~15% of FIP cases. Pediatric ILD has been linked to mutations in *SFTPC* and ATP-binding cassette transporter A3 (*ABCA3*).

SFTPC was the first gene to be linked to FIP (in both children and adults). These mutations lead to aberrant protein products that cause endoplasmic reticulum (ER) stress with activation of the unfolded protein response (UPR) and toxicity to type II alveolar epithelial cells (AECs). Disease-causing mutations in *SFTPA2* cause ER stress in type II AECs also. Interestingly, two groups have shown that ER stress and UPR activation are present in the lungs of subjects with sporadic IPF. Like *SFTPC*, mutations in *ABCA3* also lead to derangements in surfactant protein processing. Two groups linked mutations in *TERT* and *TERC* to FIP cases and revealed that these mutations cause telomere shortening. Subsequent studies demonstrated that telomere shortening is a common finding in sporadic IPF.

Some systemic diseases with prominent ILD likely have important genetic contributions. Systemic sclerosis is a connective tissue disorder characterized by excessive deposition of connective tissue components including collagen in the skin and other tissues, prominent alterations in the microvasculature, and abnormalities in cellular and humoral immunology. In some cases, lung fibrosis can be severe. Multiple studies suggest disease risk associations with specific polymorphisms in selected genes. Sarcoidosis is a multisystem disorder in which granulomas accumulate in the tissues. This process can occur in any organ, but over 90% of individuals present with lung, ocular, and/or skin manifestations. Of genetic studies to date, the most revealing has been the observation of an association with a polymorphism in the butyrophilin-like2 (*BTNL2*) gene, but the mechanism by which this impacts disease is unknown.

Multiple genetic disorders include ILD as a manifestation. Dyskeratosis congenita is a multisystem disorder that has prominent mucocutaneous manifestations, but also causes bone marrow failure and lung fibrosis. Mutations have been described in the gene for dyskerin (*DKC1*), *TERT* (*24*), *TERC*, and other genes involved in telomere maintenance. Hermansky–Pudlak syndrome is a group of autosomal recessive disorders characterized by oculocutaneous albinism and prolonged bleeding from defective platelet function. Eight subtypes each have their own genetic association, two with prominent ILD. Lymphangioleiomyomatosis (LAM) is characterized by infiltration of benign smooth muscle-like cells in lung parenchyma with cyst formation. Patients with tuberous sclerosis complex (TSC) can present with LAM, with TSC caused by mutations in *TSC1* or *TSC2*, encoding hamartin and tuberin, respectively. Sporadic LAM is caused by somatic mutations in *TSC2*. Pulmonary Langerhans cell histiocytosis is a lung disorder in which cells of dendritic lineage proliferate in the lung leading to disease. Familial cases have been reported, but genetic associations are not described. Pulmonary alveolar proteinosis is characterized by accumulation of surfactant rich lipoproteinaceous material in alveoli. The idiopathic form is most common, with patients developing neutralizing antibodies against granulocyte–macrophage colony-stimulating factor (GM-CSF). Congenital cases have been linked to mutations in *SFTPC*, *SFTPB*, and the βc chain of the GM-CSF receptor. Lysinuric protein intolerance (LPI) is caused by autosomal recessive mutations in the heterotrimeric amino acid transporter (HAT) y^+LAT-1 gene *SLC7A*. Up to two-thirds of subjects have ILD on radiographic imaging. Neurofibromatosis 1 is caused by mutations in *NF1*, the gene for neurofibromin. Affected individuals can present with ILD or restrictive pulmonary function because of thoracic abnormalities. Familial hypocalciuric hypercalcemia is caused by inactivating mutations in *CaR*, a calcium-sensing receptor gene, with some patients having severe ILD. Pulmonary alveolar microlithiasis is an autosomal recessive disorder characterized by calcium phosphate microlith accumulation within the alveoli caused by mutations in *SLC34A2*, which encodes for a type IIb sodium phosphate co-transporter. Lipoid proteinosis is characterized by widespread deposition of glycoprotein in tissues including the lung and is caused by mutations in the gene encoding extracellular matrix protein 1 (*ECM1*). Marfan syndrome is caused by mutations in the gene for fibrillin-1 (*FBN-1*). ILD is rare, but restrictive pulmonary function may be present due to thoracic skeletal abnormalities. Some storage disorders may have ILD manifestations. Gaucher's disease is a lysosomal storage disease caused by defective glucosylceramide hydrolysis due to mutations in the gene encoding for β-glucosidase. Niemann–Pick disease is a lipid storage disease in which sphinomyelin and cholesterol accumulate. Different genes are responsible across six variants. Fabry's disease is a lysosomal storage disorder of glycosphingolipid metabolism due to α-galactosidase A deficiency. GM_1 gangliosidosis is a storage disorder caused by β-galactosidase deficiency.

FURTHER READING

1. American Thoracic Society. Idiopathic Pulmonary Fibrosis: Diagnosis and Treatment. International Consensus Statement. American Thoracic Society (ATS), and the European Respiratory Society (ERS). *Am. J. Respir. Crit. Care Med.* **2000**, *161*, 646–664.
2. American Thoracic Society/European Respiratory Society International Multidisciplinary Consensus Classification of the Idiopathic Interstitial Pneumonias. This Joint Statement of the American Thoracic Society (ATS), and the European Respiratory Society (ERS) Was Adopted by the ATS Board of Directors, June 2001 and by the ERS Executive Committee, June 2001. *Am. J. Respir. Crit. Care Med.* **2002**, *165*, 277–304.
3. Armanios, M. Y.; Chen, J. J.; Cogan, J. D.; Alder, J. K.; Ingersoll, R. G.; Markin, C.; Lawson, W. E.; Xie, M.; Vulto, I.; Phillips, J. A., III; Lansdorp, P. M.; Greider, C. W.; Loyd, J. E. Telomerase Mutations in Families with Idiopathic Pulmonary Fibrosis. *N. Engl. J. Med.* **2007**, *356*, 1317–1326.
4. Bullard, J. E.; Wert, S. E.; Whitsett, J. A.; Dean, M.; Nogee, L. M. ABCA3 Mutations Associated with Pediatric Interstitial Lung Disease. *Am. J. Respir. Crit. Care Med.* **2005**.
5. Crino, P. B.; Nathanson, K. L.; Henske, E. P. The Tuberous Sclerosis Complex. *N. Engl. J. Med.* **2006**, *355*, 1345–1356.
6. Iannuzzi, M. C.; Rybicki, B. A.; Teirstein, A. S. Sarcoidosis. *N. Engl. J. Med.* **2007**, *357*, 2153–2165.
7. Lawson, W. E.; Crossno, P. F.; Polosukhin, V. V.; Roldan, J.; Cheng, D. S.; Lane, K. B.; Blackwell, T. R.; Xu, C.; Markin, C.; Ware, L. B.; Miller, G. G.; Loyd, J. E.; Blackwell, T. S. Endoplasmic Reticulum Stress in Alveolar Epithelial Cells Is Prominent in IPF: Association with Altered Surfactant Protein Processing and Herpesvirus Infection. *Am. J. Physiol. Lung Cell. Mol. Physiol.* **2008**, *294*, L1119–L1126.
8. Lee, H. L.; Ryu, J. H.; Wittmer, M. H.; Hartman, T. E.; Lymp, J. F.; Tazelaar, H. D.; Limper, A. H. Familial Idiopathic Pulmonary Fibrosis: Clinical Features and Outcome. *Chest* **2005**, *127*, 2034–2041.
9. Loyd, J. E. Pulmonary Fibrosis in Families. *Am. J. Respir. Cell Mol. Biol.* **2003**, *29*, S47–S50.
10. Maitra, M.; Wang, Y.; Gerard, R. D.; Mendelson, C. R.; Garcia, C. K. Surfactant Protein A2 Mutations Associated with Pulmonary Fibrosis Lead to Protein Instability and Endoplasmic Reticulum Stress. *J. Biol. Chem.* **2010**, *285*, 22103–22113.
11. Steele, M. P.; Speer, M. C.; Loyd, J. E.; Brown, K. K.; Herron, A.; Slifer, S. H.; Burch, L. H.; Wahidi, M. M.; Phillips, J. A., III; Sporn, T. A., et al. Clinical and Pathologic Features of Familial Interstitial Pneumonia. *Am. J. Respir. Crit. Care Med.* **2005**, *172*, 1146–1152.
12. Thomas, A. Q.; Lane, K.; Phillips, J., III; Prince, M.; Markin, C.; Speer, M.; Schwartz, D. A.; Gaddipati, R.; Marney, A.; Johnson, J., et al. Heterozygosity for a Surfactant Protein C Gene Mutation Associated with Usual Interstitial Pneumonitis and Cellular Nonspecific Interstitial Pneumonitis in One Kindred. *Am. J. Respir. Crit. Care Med.* **2002**, *165*, 1322–1328.
13. Tsakiri, K. D.; Cronkhite, J. T.; Kuan, P. J.; Xing, C.; Raghu, G.; Weissler, J. C.; Rosenblatt, R. L.; Shay, J. W.; Garcia, C. K. Adult-Onset Pulmonary Fibrosis Caused by Mutations in Telomerase. *Proc. Natl. Acad. Sci. U.S.A* **2007**, *104*, 7552–7557.
14. Wei, M. L. Hermansky–Pudlak Syndrome: A Disease of Protein Trafficking and Organelle Function. *Pigment Cell Res.* **2006**, *19*, 19–42.

Renal Disorders

Congenital Anomalies of the Kidney and Urinary Tract, 241
Cystic Diseases of the Kidney, 252
Nephrotic Disorders, 255
Renal Tubular Disorders, 257
Cancer of the Kidney and Urogenital Tract, 259

CHAPTER 62

Congenital Anomalies of the Kidney and Urinary Tract

Grace J Noh
Cedars-Sinai Medical Center, Medical Genetics Institute, Los Angeles, CA, USA

Rosemary Thomas-Mohtat
Children's Hospital at Montefiore, Department of Pediatrics, Division of Nephrology, Bronx, NY, USA

Elaine S Kamil
Cedars-Sinai Medical Center, Department of Pediatrics, Division of Nephrology, Los Angeles, CA, USA

In this chapter, the congenital anomalies of the kidney and urinary tract (CAKUT) are discussed. It is divided into the following three main categories.

62.1 ERRORS OF ORGANOGENESIS

Renal agenesis refers to the complete absence of one or both kidneys. At birth, bilateral renal agenesis presents with severe respiratory distress due to pulmonary hypoplasia and morphologic features of Potter syndrome (*1*). Complete absence of the kidney is hypothesized to result from failure of the ureteric bud derived from the Wolffian duct to make contact with the nephrogenic blastema. Bilateral renal agenesis is associated with a grim prognosis (*2*).

Renal hypodysplasia (RHD), encompassing renal aplasia, hypoplasia, and dysplasia, is the second leading cause of chronic renal insufficiency in the pediatric population (*3*). Primary renal hypoplasia is a reduction of renal size by two standard deviations from the mean size for age. The cause of RHD is aberrant interactions between the ureteric bud and metanephric mesenchyme. Studies have discovered mutations in PAX2 and HNF1B in up to 15% of Caucasian and 10% of Japanese children with RHD (*4–9*).

Renal dysplasia results from poorly branched and abnormally differentiated nephrons and collecting ducts, increased stroma and occasionally cysts and metaplastic tissues, such as cartilage (*10*). Clinical symptoms include anuria, oliguria, impairment of urinary concentration and acidification, polyuria with polydipsia, urinary salt-wasting, renal tubular acidosis, hematuria, hypertension, uremia, back pain, and symptoms associated with renal insufficiency. Many dysplastic kidneys involute over time, including before birth (*11*).

Agenesis of the ureters is frequently associated with renal agenesis. Duplication of the ureters is common. Abnormal division during ureteric bud formation is postulated as the mechanism for duplication of the ureters. Renal ultrasound imaging will simplify diagnosis of severe ureteral anomalies among first-degree relatives, but will not identify duplicated ureters that are not dilated.

Congenital segmental urethral agenesis or atresia is more common than complete atresia. Children may present with abdominal distention due to enlarged bladder or urinary ascites, laxity of abdominal musculature, bilateral cryptorchidism, oligohydramnios, failure to pass urine, or evidence of urinary fistulae such as a patent urachus. Duplication of the urethra may be complete, with two separate external meati, or partial, with variable outlet of the accessory urethra. Postnatal management includes urinary diversion or vesicostomy which may be necessary before reconstruction of the urethra.

Posterior urethral valves (PUVs) are most commonly found in males. Valvular obstruction results in incomplete emptying of the bladder, which may become trabeculated with thickened walls. Common presenting symptoms, in those not diagnosed prenatally, are abdominal mass (38–53%), oligohydramnios (32%), failure to grow (18–40%), sepsis or urinary tract infections (20%), or a weak urinary stream (14%). Postnatally, primary ablation of the valves with transurethral fulguration is usually successful. When the anterior urethral lumen is too narrow for endoscopy, a temporary vesicostomy may be necessary.

62.2 ERROR OF MIGRATION AND POSITION

Horseshoe kidneys are fused kidneys that are unable to ascend from the embryonic pelvic position. While most are asymptomatic, the most common clinical symptoms are pain, hematuria, and urinary tract symptoms from obstruction of the ureters. Treatment of hydronephrosis and vesicoureteral reflux (VUR), when indicated, can greatly reduce associated clinical symptomatology and renal damage.

62.3 ERRORS RESULTING IN OBSTRUCTION

The most common anomalies resulting in obstructive kidney(s) are bladder outlet obstruction (e.g. from PUV), hydronephrosis due to reflux or vesicoureteral junction obstruction, pelvicaliectasis due to ureteropelvic junction obstruction, ureteral duplication or strictures, and ureterocele. Clinical symptoms are typical of those caused by obstruction.

Primary VUR is the retrograde flow of urine from the urinary bladder to the kidney due to incompetence at the vesicoureteral junction. A small proportion of these patients progress to end stage renal disease. This progressive condition, known as reflux nephropathy, consists of calyceal clubbing or deformity with overlying corticomedullary scarring. There is genetic heterogeneity in the

TABLE 62-1 Frequency of Urinary Tract Anomalies in Cases with Other Structural Defects

Structural Anomaly/Pattern	Percentage with Urinary Tract Anomalies	Most Common Urinary Tract Defects (in Relative Order of Frequency)
Absent gallbladder	32	Cystic dysplasia, renal agenesis, horseshoe kidneys
Agenesis of the corpus callosum	45–55	Reflux, ureterocele, unilateral renal agenesis, crossed fused renal ectopia, bladder diverticulae
Anencephaly	5–16	Hydronephrosis, horseshoe kidneys, polycystic kidneys, renal agenesis, renal hypoplasia, urethral atresia
Anorectal malformation	26–58[a]	Hydronephrosis, unilateral renal agenesis, cystic dysplasia, reflux, cystic dysplasia, renal ectopia, cloacal exstrophy
Biliary atresia	3	Double ureter, hydronephrosis, renal cysts
Caudal dysplasia	40	Renal agenesis, hypoplasia, cystic dysplasia, horseshoe kidneys, crossed renal ectopia, urachal anomalies
CHARGE	42	Unilateral renal agenesis, hydronephrosis, renal hypoplasia
Diaphragmatic hernia	15–18	Renal agenesis, cystic dysplasia, hydronephrosis, ureteropelvic obstruction
Esophageal atresia and tracheoesophageal fistula	33	Unilateral renal agenesis, horseshoe kidneys, reflux
Gastroschisis	15	Unilateral renal agenesis, horseshoe kidneys
Heart defect	5–39[a]	Duplex collecting system, unilateral renal agenesis, renal ectopia
Lateral body wall defect	50–65	Renal agenesis, urethral atresia, hydronephrosis
Limb reduction defects	9	Renal agenesis, hydronephrosis, cystic dysplasia, horseshoe kidney
MURCS association	28–80	Renal agenesis, renal ectopia
Myelomeningocele	9	Renal agenesis, horseshoe kidney
Omphalocele	11–47	Cloacal exstrophy, horseshoe kidneys, patent urachus
Oral clefts	4	Renal agenesis, horseshoe kidney
Penoscrotal transposition	90	Renal agenesis, cystic dysplasia, ectopia, horseshoe kidneys
Persistent cloaca	83	Cloacal and bladder anomalies
Pulmonary hypoplasia	18–21	Cystic dysplasia, renal agenesis, horseshoe kidney, polycystic kidney, urethral atresia
Single umbilical artery (isolated)	26	Dilated renal pelvis, duplicated renal pelvis, reflux, hydronephrosis, horseshoe kidneys, unilateral renal agenesis
Sirenomelia	100	Renal agenesis, cystic dysplasia, urethral atresia
Supernumerary nipples	4	No specific pattern
Tracheal agenesis	38	Renal agenesis, cystic dysplasia, horseshoe kidney
VACTERL association	82–87	Reflux, unilateral renal agenesis, ureteropelvic junction obstruction, crossed fused ectopia
Vertebral defects	27–46	Unilateral renal agenesis, duplicated ureter, renal ectopia, horseshoe kidney

[a]Higher numbers seen in autopsy series and/or those with ≥3 congenital anomalies.
Source: From Stevenson, R. E. Human Malformations and Related Anomalies, 2nd ed.; Oxford University Press: Oxford, New York, 2006; p. 1167.

TABLE 62-2 Common Chromosomal Disorders Associated with Urinary Tract Anomalies

Chromosome Disorder	Urinary Tract Anomaly	Frequency
2q terminal deletion	Wilms tumor, horseshoe kidney, dysplastic kidney, renal hypoplasia, ureteral stenosis	11%
4p	Renal agenesis or hypoplasia, vesicoureteral reflux (VUR), hydronephrosis	33%
4q partial duplication	Horseshoe kidney, renal hypoplasia, renal duplication, ectasia of distal tubule	Frequent
5p	Horseshoe kidney, renal agenesis, renal duplication, ectasia of distal tubules	Occasional
6q partial duplication	Unilateral renal agenesis, cystic dysplasia	Uncertain[a]
7 trisomy	Cystic dysplasia, enlarged kidneys	Frequent
7q partial deletion (Williams–Beuren)	Renal aplasia/hypoplasia, duplicated kidney, bladder diverticula	18%
8 trisomy, 8 trisomy mosaicism	Cystic dysplasia, enlarged kidneys, small cortical cysts, hydronephrosis, duplication of kidneys/ureters/pelvis	Frequent
9p partial duplication	Horseshoe kidney, hydronephrosis	Frequent
9 trisomy	Hydronephrosis, cystic dysplasia, duplication kidneys and ureters	Frequent
10p partial duplication	Unilateral agenesis, cystic kidney, renal dysplasia	Uncertain
10q partial duplication	Hypoplastic kidney, hydronephrosis	Frequent
11p13 deletion	Wilms tumor	Frequent
13q	Hydronephrosis, vesicoureteral junction obstruction	Uncommon
13 ring	Renal hypoplasia and ectopy, duplication of kidney and ureter, unilateral renal agenesis, polycystic kidney	Frequent
13 trisomy	Hydronephrosis, cystic dysplasia, micropolycystic dysplasia, hydroureter, horseshoe kidney, ureteral duplication, duplication of renal pelvis, small cortical cysts	60–70%
17p partial deletion (Smith–Magenis)	Variable; no characteristic structural abnormality	19%
17p13.3 deletion (Miller–Dieker)	Cystic or pelvic kidney	Uncertain
18q	Horseshoe kidney, unilateral agenesis, hydronephrosis	40%
18 ring	Hydronephrosis, tubular dilation	20%
18 trisomy	Horseshoe kidney, ectopia, ureteral duplication, cortical cysts, exstrophy of the cloaca, hydronephrosis	70%
21q	Unilateral renal agenesis, abnormal kidney shape, dilated calyces	Uncertain[a]
21 trisomy	Renal agenesis, hypoplasia, horseshoe kidney, posterior urethral valves, hydronephrosis, ureteropelvic junction obstruction, VUR, fetal hydronephrosis	3–7%
22q11 deletion	Renal agenesis, dysplasia, multicystic kidneys	36%
22 partial trisomy	Renal agenesis, horseshoe kidney, hydronephrosis	Frequent
22 partial tetrasomy (cat eye)	Unilateral or bilateral renal agenesis, hydronephrosis, supernumerary kidneys	Occasional
45,X (plus other Turner karyotype abnormalities)	Horseshoe kidney, duplication collecting system, abnormal rotation, ureteropelvic junction obstruction, cystic malformation of collecting tubules; cystic, double, and ectopic kidneys, intrarenal vascular changes	60–80%
XXXXY	Hydronephrosis	10%
XXXXX	Renal hypoplasia, dysplasia	Uncertain[a]
XXY (Klinefelter)	Renal cysts, hydronephrosis	Uncertain[a]
Triploidy	Hydronephrosis, renal cysts, polycystic kidneys, cystic dysplasia	Frequent

[a]Reported abnormalities may not be in excess over background risk for all urinary tract anomalies.

TABLE 62-3 Urinary Tract Anomalies Associated with Teratogen Exposures

Teratogen	Major Features	Renal Anomaly
Alcohol	Prenatal onset of growth retardation, microcephaly, hypotelorism, short palpebral fissures, smooth philtrum, variable other structural anomalies, including heart and skeletal anomalies	Small rotated kidneys, horseshoe kidney, renal dysplasia, micromulticystic dysplasia, hydronephrosis
Alkylating agents (busulfan, chlorambucil, cyclophosphamide, mechlorethamine)	Growth retardation; cleft palate; microphthalmia; digit anomalies; cardiac defects; anomalies of larynx, trachea, and esophagus	Agenesis of kidneys, hydronephrosis, hydroureter
Angiotensin-converting enzyme (ACE) inhibitors	Growth retardation, hypocalvaria, pulmonary hypoplasia oligohydramnios, patent ductus arteriosus, limb anomalies, fetal hypotension, increased fetal and neonatal mortality	Glomerulopathy, interstitial nephritis, nephrotic syndrome, progressive renal failure, renal artery stenosis, anuria, neonatal acute kidney injury, renal tubular dysgenesis
Cocaine	Structural anomalies resulting from vascular disruption	Renal and ureteral agenesis, hydronephrosis, prune belly syndrome, hypospadias, ambiguous genitalia
Maternal diabetes	Caudal regression, neural tube defects, congenital heart, and other anomalies	Renal agenesis, ureter anomalies, urethral anomalies, cystic dysplasia
Rubella	Cataracts, microphthalmia, pigmentary retinopathy, growth retardation, heart anomalies (patent ductus arteriosus, peripheral pulmonary stenosis), skeletal anomalies, sensorineural deafness, neurologic impairment, microcephaly	Stenosis of renal artery, polycystic kidney, duplication of the ureters, unilateral renal agenesis
Thalidomide	Limb reduction anomalies, micrognathia, neural tube defects; vertebral, heart, and visceral anomalies	Renal agenesis, obstructive uropathy, abnormal kidney rotation, pelvic kidney, horseshoe kidney
Trimethadione	Distinctive facial features, developmental delay, prenatal and postnatal growth impairment, omphalocele; heart, skeletal, and limb anomalies	Absent kidney and ureter, fetal lobulation of kidneys
Vitamin A congeners	Micrognathia, cleft palate, microphthalmia, midfacial hypoplasia, hearing anomalies, anotia, microtia, neural tube defects	Hypoplastic kidneys, hydronephrosis

TABLE 62-4 Disorders Associated with Renal Dysplasia

Disorder	Major Features	Renal Anomaly	Etiology	Genes
Asplenia with cardiovascular anomalies	Bilateral right or left embryonic primordia, asplenia, polysplenia, complex heart anomalies	Renal dysplasia	AR (208530)	
Bardet–Biedl	Mental retardation, pigmentary retinopathy, polydactyly, obesity, hypogenitalism	Renal dysplasia, cystic tubular disease, focal segmental glomerulosclerosis, bladder detrusor instability	AR (209900)	*BBS 1-15*
Beckwith–Wiedemann	Omphalocele, macroglossia, organomegaly, islet cell hyperplasia, adrenal cytomegaly, embryonic tumors, macrosomia	Nephromegaly, renal medullary dysplasia, Wilms tumor, nephrocalcinosis	AD (130650)	*CDKN1C* imprinting error, uniparental disomy, duplication, inversion or translocation of 11p15.5
Branchio-oto-renal	Mixed hearing loss, Mondini cochlear malformations, pinnae anomalies, branchial cleft fistulas, preauricular pits	Renal dysplasia, agenesis, and ectopy; ureteral anomalies	AD (113650)	*EYA1* *SIX1* *SIX5*
Caudal dysplasia	Sacral agenesis/hypoplasia, lower limb and skeletal anomalies, anal atresia, anomalies of the uterus	Renal dysplasia and agenesis; anomalies of urethra, and bladder	Heterogeneous, maternal diabetes in some patients	
Cerebro-renal-digital	Digital and limb anomalies, brain malformations, other anomalies. See Table 62-6	Renal dysplasia, ectopy, agenesis, ureteral anomalies	Heterogeneous	
CHARGE syndrome	Coloboma, heart anomalies, choanal atresia, mental retardation, genital hypoplasia, ear anomalies, deafness	Cystic renal dysplasia, renal agenesis, ureteric anomalies, fused/ectopic kidneys	AD (214800)	*CHD7*
Chromosome abnormalities	See Table 62-2			
Cloacal exstrophy	Persistent cloaca, exstrophy of cloaca, failure of fusion of genital tubercles, omphalocele, vertebral anomalies, spina bifida cystica, abnormal genital structures	Duplication of urethra; urethral and ureteral anomalies; exstrophy of the bladder; renal dysplasia, agenesis, and ectopy	Heterogeneous	
Cornelia de Lange	Microcephaly, in utero growth retardation, distinct facies, micromelia, oligodactyly, heart and other anomalies, mental retardation	Renal dysplasia, agenesis, hypoplasia	AD (122470)	*NIPBL*
Craniosynostosis-mental retardation-clefting syndrome	Craniosynostosis (coronal), hypertelorism, choroidal coloboma, mild mesomelic shortening of limbs, developmental delay, seizures	Segmental renal dysplasia, cystic dysplasia	AR (218650)	
Cystic hamartoma of lung and kidney	Hamartomatous pulmonary cysts	Medullary dysplasia, cellular mesoblastic nephoma	Unknown	
Denys–Drash syndrome	Male pseudohermaphroditism, ambiguous genitalia, gonadal dysgenesis, gonadoblastoma	Diffuse mesangial sclerosis glomerulopathy, nephroblastoma (Wilms tumor), nephrotic syndrome, focal segmental glomerulosclerosis	AD (194080)	*WT1*

Continued

TABLE 62-4 Disorders Associated with Renal Dysplasia—*Cont'd*

Disorder	Major Features	Renal Anomaly	Etiology	Genes
Digeorge syndrome	Sacral meningocele, hydrocephalus, conotruncal heart anomalies, hypoplasia of the thymus and parathyroid glands, hypocalcemia	Unilateral renal agenesis, dysplasia, hydronephrosis	AD (188400)	22q11.2 deletion
Early amnion rupture	Digital and limb amputations, ring constrictions, facial clefts, body wall defects, brain anomalies	Renal dysplasia, agenesis, and ectopy; ureteric anomalies	Sporadic	
Ectrodactyly-ectodermal dysplasia-clefting (EEC)	Ectrodactyly, ectodermal dysplasia, cleft lip/palate, abnormal hearing, internal genital tract anomalies, anal atresia	Renal agenesis or cystic dysplasia, ureteral and bladder anomalies, duplicated collecting system	AD (129900)	*EEC1* *EEC2* *TP63*
Fanconi pancytopenia	Pancytopenia; anemia; radial aplasia/hypoplasia;microcephaly; short stature; variable eye, ear, and heart anomalies; increased chromosome breakage	Renal agenesis (39%) or dysplasia, duplication of pelvis and/or ureter, ectopic or horseshoe kidney, hydronephrosis	AR (227650)	*FANC(A,C-G,I,J,L-P)*
Fraser syndrome	Cryptophthalmia, cleft lip/palate, genital anomalies, atresia of ear canal, anal atresia, syndactyly	Renal dysplasia or agenesis; ureteric anomalies	AR (219000)	*FRAS1 FREM2*
Genitopalatocardiac	Male pseudohermaphroditism, micrognathia, cleft palate, conotruncal cardiac anomalies, other anomalies	Renal cystic dysplasia	AR (231060)	
Harrod syndrome	Arachnodactyly, hypospadias, cryptorchidism, distinctive facial features, anomalous vasculature, gut malrotation	Cortical microcysts of the kidney, renal dysplasia, ureteral anomalies, VUR	?AR (601095)	
Hemifacial microsomia (Oculoauriculovertebral spectrum, Goldenhar)	Facial asymmetry, epibulbar dermoid, coloboma, anotia, preauricular tags, deafness, vertebral anomalies, heart anomalies, variable brain malformations	Renal cystic dysplasia, agenesis, or ectopia; hydronephrosis and hydroureter; abnormal blood supply to kidney	Heterogeneous AD (164210)	
Hypoparathyroidism, sensorineural deafness, and renal disease	Hypoparathyroidism, sensorineural hearing loss, and renal dysplasia	Renal dysplasia, nephrosis	AD (146255)	*GATA3*
Lenz microphthalmia	Microphthalmia, coloboma; mental retardation; skeletal, dental, genital, and cardiovascular anomalies	Renal agenesis or dysplasia, hydroureters	XLR (309800)	*BCOR*
Limb-body wall complex	Lateral body wall deficiency, limb reduction anomalies, neural tube defects, scoliosis, heart anomalies	Renal agenesis or dysplasia; ureteral and urethral anomalies; fused/ectopic kidneys; bladder exstrophy	Sporadic	
McKusick-Kaufman	Hydrometrocolpos, transverse vaginal membrane, vaginal septum, postaxial polydactyly, cardiac anomalies, hypospadias	Polycystic kidney, vesicovaginal fistula, hydronephrosis (secondary to ureteral compression from hydrometrocolpos)	AR (236700)	*BBS6*

TABLE 62-4 Disorders Associated with Renal Dysplasia—*Cont'd*

Disorder	Major Features	Renal Anomaly	Etiology	Genes
Microcephaly, hiatus hernia and nephrotic syndrome (Galloway–Mowat)	Microcephaly, seizures, psychomotor retardation, eye anomalies	Nephrotic syndrome, microcystic dysplasia, focal glomerulosclerosis, diffuse mesangial sclerosis	AR (251300)	
MURCS association	Müllerian duct aplasia, renal agenesis, absent vagina, uterus, cervicothoracic somitic vertebral anomalies, short stature	Renal agenesis, dysplasia,or ectopia; ureteral and urethral anomalies	Sporadic (601076)	
Multinodular goiter, cystic renal disease, and digital anomalies	Multinodular goiter, triphalangeal thumbs, preaxial polydactyly of feet	Renal dysplasia, polycystic kidney anomalies	AD (138790)	
Neural tube defects	Meningomyelocele, anencephaly, encephalocele, vertebral anomalies, anomalies of schisis association, midline anomalies	Renal agenesis, hypoplasia, dysplasia, or fusion; ureteral anomalies	Heterogeneous multifactorial	
Renal hamartomas nephroblastomatosis, and fetal gigantism (Perlman)	Fetal macrosomia, hypotonia, psychomotor retardation, seration of upper alveolar ridge, can be perinatal lethal	Bilateral renal hamartomas, nephroblastomatosis, Wilms tumor	AR (267000)	
Nephronopthisis 2	Hypertension, pulmonary hypoplasia	Cortical microcysts, tubular atrophy, chronic tubulointerstitial nephritis,enlarged hyperechoic kidneys, absence of corticomedullary differentiation, renal failure by age 3	AR (602088)	*INVS*
Potter oligohydramnios sequence	Clinical features resulting from oligohydramnios	Bilateral renal agenesis, aplasia, hypoplasia, or dysplasia	Heterogeneous	
Renal adysplasia	Internal genital tract anomalies; occasional anomalies of anus, heart, spine, hands, and feet	Renal agenesis, hypoplasia, or dysplasia; ureteral and urethral anomalies	AD (191830)	*RET, UPK3A*
Renal-hepatic-pancreatic	See Table 62–5			
Sacral defect with anterior meningocele (Sirenomelia)	Fusion of lower limbs, sacral agenesis, anal atresia, uterine/vaginal anomalies, cardiac defects	Urethral atresia, ectopic urethra, posterior urethral valves, renal agenesis and dysplasia, ureteral and bladder anomalies	AD (600145)	*VANGL1*
Senior–Loken (Renal-retinal)	Pigmentary retinal dysplasia, occasional hypotonia, seizures, hearing loss, psychomotor retardation	Renal dysplasia, juvenile nephronophthisis, medullary cystic disease	AR (266900)	*NPHP1*
Simpson–Golabi–Behmel	Prenatal and postnatal overgrowth, variable mental function, characteristic facial appearance, post-axial polydactyly, structural anomalies of organ systems	Large cystic dysplastic kidneys, hydronephrosis, duplication of renal pelvis, Wilms tumor	XLR (312870)	

Continued

TABLE 62-4 Disorders Associated with Renal Dysplasia—*Cont'd*

Disorder	Major Features	Renal Anomaly	Etiology	Genes
Smith–Lemli–Opitz	Microcephaly, postaxial polydactyly, ambiguous genitalia, facial dysmorphism; 2–3 toe syndactyly, disorder of cholesterol metabolism	Unilateral renal agenesis orcystic dysplasia, fused kidneys	AR (27400)	*DHCR7*
Teratogen exposures	See Table 62–3			
Tuberous sclerosis	Hypopigmented macules, adenoma sebaceum, retinal and brain tumors or phakomas, mental retardation, seizures	Renal angiomyolipomas (70%), renal epithelial cysts (20–30%), oncocytoma, renal cell carcinoma, polycystic kidney disease	AD (191100)	*TSC1 TSC2*
Urethral obstruction	Deficient abdominal muscle, urinary obstruction/ distension, undescended testes, malrotation of the gut, clubfeet, limb reduction anomalies	Posterior urethral valves, urethral atresia, ureteral duplication, bladder distension, renal dysplasia, hydronephrosis	Heterogeneous (prune belly syndrome)	
Urorectal septum malformation	Müllerian duct anomalies, persistent cloaca, ambiguous genitalia, imperforate anus, other anomalies	Renal agenesis, hypoplasia, and dysplasia; ureteral and urethral anomalies	Sporadic, maternal diabetes in some patients	
VACTERL association	Vertebral defects, anal atresia, cardiac malformation, tracheoesophageal fistula, radial ray hypoplasia/ aplasia, genitourinary and limb anomalies	Renal agenesis, hypoplasia, and dysplasia; ureteric and urethral anomalies	Sporadic (192350)	
Zellweger	Hypotonia, seizures, cirrhosis, peroxisomal enzyme deficiency of dihydroxyacetone phosphate (DHAP) acyltransferase	Renal cortical microcysts, absent renal peroxisomes, hydronephrosis	AR (214100)	*PEX 1-3,5,6,10, 12-14,16,19,26*

AD, autosomal dominant; AR, autosomal recessive; XLR, X-linked recessive.

TABLE 62-5 Disorders Associated with Polycystic and Multicystic Kidneys, Including Disorders with Renal-Hepatic-Pancreatic Dysplasia

Disorder	Major Features	Renal Anomaly	Etiology	Genes
Acrorenal-mandibular	Split hand/foot; genital, vertebral, rib and uterine anomalies	Renal agenesis, ureteral anomalies, cystic kidney, renal dysplasia	AR (200980)	
Asplenia with cardiovascular anomalies	Bilateral right or left embryonic primordia, asplenia, polysplenia, complex heart anomalies, situs inversus	Renal dysplasia, cortical cysts	AR (208530)	
Branchio-oto-renal (Melnick–Fraser)	Mixed hearing loss, Mondini cochlear malformation, pinnae anomalies, branchial cleft fistula, preauricular pits	Cystic dysplastic kidneys, renal agenesis, ectopia, ureteral anomalies	AD (113650)	*EYA1* *SIX1* *SIX5*
Campomelia, Cumming type	Severe shortening and bowing of long bones, cervical lymphocele, vertebral anomalies, cystic dysplasia of liver and pancreas, shortgut, polysplenia, pulmonary hypoplasia, other anomalies	Renal dysplasia, polycystic kidneys	AR (211890)	
Carnitine palmito-yltransferase Deficiency, lethal neonatal	Dysmorphic face, cardiomegaly, respiratory failure, hepatomegaly, long digits, hypotonia, hypoglycemia	Enlarged polycystic kidneys, dysplasic renal parenchyma, lipid accumulation in kidney, especially proximal convoluted tubules, double ureters	AR (608836)	*CPT2*
Fryns	Neonatal lethal; diaphragmatic defects; distal digital hypoplasia; distinct facial features; pulmonary hypoplasia; eye, brain, and other anomalies	Renal dysplasia, cortical cysts	AR (229850)	
Glutaric acidemia, type II	Multiple acyl-CoA dehydrogenase deficiency, cerebral dysplasia, fatty liver, biliary dysgenesis, pancreatic dysplasia, Potter facies	Renal dysplasia, multicystic kidneys	AR (231680)	*ETFA* *ETFB* *ETFDH*
Hutterite cerebroos-teonephrodysplasia	Short stature with mild spondylorhizomelic dysplasia, failure to grow, severe mental retardation, postnatal microcephaly, seizures	Terminal nephrotic syndrome, nephrosis, proteinuria, end stage renal disease	AR (236450)	
Jeune (asphyxiating thoracic dystrophy)	Small, bell-shaped thorax; pulmonary hypoplasia, polydactyly, variable rhizomelic limb shortening, trident pelvis, biliary dysgenesis, pancreatic dysplasia, other anomalies	Renal dysplasia, glomerulonephritis, juvenile nephrophthisis, stenosis of uretero-vesical junction, hydroureters, multicystic kidneys	AR (208500)	
Joubert syndrome	Cerebellar vermic aplasia, ataxia,colobomata, Dandy–Walker malformation	Multicystic kidneys	AR (213300, 243910)	*INPP5E*
Meckel–Gruber	Occipital encephalocele, polydactyly, cleft lip/palate, microphthalmia, small or ambiguous genitalia, brain anomalies, biliary dysgen-esis and pancreatic dysplasia	Renal dysplasia or hypodysplasia, ureteral hypoplasia or aplasia, hypoplastic bladder, urethral agenesis, renal cysts, hydroureter	AR (249000)	*MKS1*
Multinodular goiter, cystic renal disease, and digital anomalies	Multinodular goiter, triphalangeal thumbs, preaxial polydactyly of feet	Renal dysplasia, polycystic kidney	AD (138790)	
Orofaciodigital, type I	Lobulated tongue, median pseudocleft of lip, cleft palate, hypoplastic alae nasi, digital anomalies, mental impairment	Adult-onset polycystic kidneys	XLR (311200)	*CXORF5*

Continued

TABLE 62-5 Disorders Associated with Polycystic and Multicystic Kidneys, Including Disorders with Renal-Hepatic-Pancreatic Dysplasia—*Cont'd*

Disorder	Major Features	Renal Anomaly	Etiology	Genes
Polycystic kidney, autosomal dominant	Hepatic cysts (hepatic fibrosis is unusual), mitral valve prolapse (25%), berry aneurysms of cerebral (10–36%) and abdominal vessels, cysts in pancreas, ovaries, lungs and other, diverticulitis	Cystic disease of renal medulla and cortexin nephrons and collecting tubules	AD (173900)	*PKD1* *PKD2*
Polycystic kidney, autosomal recessive	Hepatic fibrosis, biliary dysgenesis, pancreatic dysplasia	Medullary ductal ectasia, cortical cysts	AR (263200)	*PKHD1*
Polycystic kidney disease, Potter type I, with micro-brachycephaly, hypertelorism and brachymelia	Microbrachycephaly, hypertelorism, brachymelia and distinctive facial features	Polycystic kidney disease, with cortical cysts and medullary ductal ectasia (Potter type I)	AR (263210)	
Senior–Loken (renal dysplasia-retinal aplasia)	Pigmentary retinal dysplasia, hypotonia, seizures, hearing loss, mental retardation	Renal dysplasia, juvenile nephrophthisis, medullary cystic disease	AR (266900)	*NPHP1*
Short rib polydactyly, type II (Majewski)	Median cleft lip; pre/postaxial polydactyly; short ribs and limbs; genital, laryngeal, epiglottic and visceral anomalies, pachygyria, biliary dysgenesis; pancreatic dysplasia; lethal	Renal dysplasia, multicystic kidneys, glomerular and renal tubular cysts	AR (263520)	*NEK1*
Short rib polydactyly, type I (Saldino–Noonan)	Phocomelia; metaphyseal dysplasia; defective ossification of calvarium, vertebrae, pelvis, carpal and tarsal bones; cardiac, GI and GU anomalies; biliary dysgenesis; pancreatic dysplasia; lethal	Renal dysplasia, multicystic kidneys	AR (263530)	
Trisomy 9	Potter sequence, cleft palate, joint immobility, biliary dysgenesis,pancreatic dysplasia	Renal dysplasia, multicystic kidneys	Chromosomal	
Tuberous sclerosis	Hypopigmented macules, adenoma sebaceum, retinal and braintumors, seizures, mental retardation	Renal angiomyolipomas (70%), renal epithelial cysts (20-30%), oncocytoma, renal cell carcinoma, polycystic kidney disease	AD (191100)	*TSC1 TSC2*
Von Hippel–Lindau	Retinal angiomas, cerebellar hemangioblastomas	Renal cysts, renal cell carcinoma (40%), bladder papillomas	AD (193300)	*VHL*
Zellweger syndrome	Death in early infancy, hypotonia, seizures, biliary dysgenesis, peroxisomal enzyme deficiency of dihydrohydroxyacetonephosphate (DHAP) acyltransferase	Kidney with subcortical cysts, hydroureter, multicystic kidney	AR (214100)	*PEX 1-3,5,6,10, 12-14,16, 19,26*

AD, autosomal dominant; AR, autosomal recessive; XLR, X-linked recessive.

pattern of inheritance of VUR. Management is directed toward reducing ongoing damage to the kidneys and preventing recurrent infection through the use of prophylactic antibiotics (*12*). The ongoing debate of whether antibiotic prophylaxis prevents urinary tract infections or prevents renal scarring is being addressed by the RIVUR study (*13*).

REFERENCES

1. Potter, E. L. Bilateral Renal Agenesis. *J. Pediatr.* **1946 Jul,** *29,* 68–76.
2. Klaassen, I.; Neuhaus, T. J.; Mueller-Wiefel, D. E.; Kemper, M. J. Antenatal Oligohydramnios of Renal Origin: Long-Term Outcome. *Nephrol. Dial. Transplant.* **2007 Feb,** *22* (2), 432–439.
3. Warady, B. A.; Chadha, V. Chronic Kidney Disease in Children: The Global Perspective. *Pediatr. Nephrol.* **2007 Dec,** *22* (12), 1999–2009.
4. Heidet, L.; Decramer, S.; Pawtowski, A.; Moriniere, V.; Bandin, F.; Knebelmann, B., et al. Spectrum of HNF1B Mutations in a Large Cohort of Patients Who Harbor Renal Diseases. *Clin. J. Am. Soc. Nephrol.* **2010 Jun,** *5* (6), 1079–1090.
5. Amiel, J.; Audollent, S.; Joly, D.; Dureau, P.; Salomon, R.; Tellier, A. L., et al. PAX2 Mutations in Renal-Coloboma Syndrome: Mutational Hotspot and Germline Mosaicism. *Eur. J. Hum. Genet.* **2000 Nov,** *8* (11), 820–826.
6. Weber, S.; Moriniere, V.; Knuppel, T.; Charbit, M.; Dusek, J.; Ghiggeri, G. M., et al. Prevalence of Mutations in Renal Developmental Genes in Children with Renal Hypodysplasia: Results of the ESCAPE Study. *J. Am. Soc. Nephrol.* **2006 Oct,** *17* (10), 2864–2870.
7. Ulinski, T.; Lescure, S.; Beaufils, S.; Guigonis, V.; Decramer, S.; Morin, D., et al. Renal Phenotypes Related to Hepatocyte Nuclear Factor-1beta (TCF2) Mutations in a Pediatric Cohort. *J. Am. Soc. Nephrol.* **2006 Feb,** *17* (2), 497–503.
8. Nakayama, M.; Nozu, K.; Goto, Y.; Kamei, K.; Ito, S.; Sato, H., et al. HNF1B Alterations Associated with Congenital Anomalies of the Kidney and Urinary Tract. *Pediatr. Nephrol.* **2010 Jun,** *25* (6), 1073–1079.
9. Thomas, R.; Sanna-Cherchi, S.; Warady, B. A.; Furth, S. L.; Kaskel, F. J.; Gharavi, A. G. HNF1B and PAX2 Mutations Are a Common Cause of Renal Hypodysplasia in the CKiD Cohort. *Pediatr. Nephrol.* **2011 Mar 5,** *26* (6), 897–903.
10. Woolf, A. S.; Price, K. L.; Scambler, P. J.; Winyard, P. J. Evolving Concepts in Human Renal Dysplasia. *J. Am. Soc. Nephrol.* **2004 Apr,** *15* (4), 998–1007.
11. Rottenberg, G. T.; Gordon, I.; De Bruyn, R. The Natural History of the Multicystic Dysplastic Kidney in Children. *Br. J. Radiol.* **1997 Apr,** *70* (832), 347–350.
12. Kamil, E. S. Recent Advances in the Understanding and Management of Primary Vesicoureteral Reflux and Reflux Nephropathy. *Curr. Opin. Nephrol. Hypertens.* **2000 Mar,** *9* (2), 139–142.
13. Mathews, R.; Carpenter, M.; Chesney, R.; Hoberman, A.; Keren, R.; Mattoo, T., et al. Controversies in the Management of Vesicoureteral Reflux: The Rationale for the RIVUR Study. *J. Pediatr. Urol.* **2009 Oct,** *5* (5), 336–341.

CHAPTER

63

Cystic Diseases of the Kidney

Angela Sun

Cedars-Sinai Medical Center, Los Angeles, CA, USA

Raymond Y Wang

Division of Metabolic Disorders, Children's Hospital of Orange County, Orange, CA, USA

Dechu P Puliyanda

Cedars-Sinai Medical Center, David Geffen School of Medicine, UCLA, Los Angeles, CA, USA

In this chapter, the major forms of cystic renal disease and their genetic bases are reviewed. For each condition, the genetic aspects, molecular pathogenesis, clinical features, and management are discussed.

Autosomal dominant polycystic kidney disease (ADPKD) is a multi-organ system disorder with onset in the fourth to fifth decades. In addition to kidney cysts, affected individuals develop hepatic cysts, cardiac disease (e.g. mitral valve proplapse, compensatory left ventricular hypertrophy, dilatation of the aortic root, etc.), and intracranial aneurysms. Mutations in *PKD1* account for 85% of cases of ADPKD, and most of the remainder are due to *PKD2* mutations. The vast majority of mutations result in truncation of the polycystin-1 and polycystin-2 proteins. Those with a *PKD1* mutation have a more severe phenotype with a median age of onset of end stage renal disease (ESRD) or death at age 53 years, as compared to 69 years in those with a *PKD2* mutation. Mutations at the 5′-end of the gene also produce more severe disease than those at the 3′-end.

Autosomal recessive polycystic kidney disease (ARPKD) is recognized prenatally or shortly after birth based on the findings of oligohydramnios, nephromegaly, renal cysts, Potter facies, and respiratory insufficiency due to pulmonary hypoplasia. Neonatal mortality is estimated to be 20–50%. Those that survive the neonatal period suffer from growth retardation, systemic hypertension, chronic renal failure, and liver disease, specifically Caroli's disease. ARPKD is caused by mutations in the *PKHD1* gene, and the estimated carrier frequency is 1 in 70. As with *PKD1* and *PKD2*, most mutations in *PKHD1* result in truncation of the protein product, in this case, fibrocystin.

Familial nephronophthisis is the most common genetic cause of renal failure in childhood and early adulthood. There are three variants which are classified according to age of onset: juvenile, infantile, and adolescent nephronophthisis. To date, at least nine genes have been found to cause nephronophthisis in addition to those responsible for the syndromic forms of nephronophthisis, which include Joubert syndrome, Senior–Løken syndrome, Meckel–Gruber syndrome, Cogan-type oculomotor apraxia, and Mainzer–Saldino syndrome, among others. The most common mutation identified in patients with nephronophthisis is a homozygous deletion of *NPHP1*. All forms are autosomal recessive in their mode of inheritance. Details of the individual genes as well as genotype–phenotype correlations are discussed in detail in the chapter.

Medullary cystic kidney disease (MCKD) has a clinical presentation similar to juvenile nephronophthisis with symptoms of polyuria and polydipsia. However, onset is in adulthood, and the mode of inheritance is autosomal dominant. Disease progression is rapid with an average interval of 5 years between time of diagnosis and development of ESRD. The *UMOD* gene is associated with MCKD type 2 and encodes uromodulin, also known as Tamm–Horsfall protein, which plays a role in maintaining the impermeability of the thick ascending loop of Henle. The locus for MCKD type 1 has been mapped to chromosome 16p12, but a causative gene has not yet been identified.

There are many genetic syndromes with renal cysts as a major component. Tuberous sclerosis complex, Bardet–Biedl syndrome (BBS), Jeune asphyxiating thoracic dysplasia, Hajdu–Cheney syndrome, campomelia Cumming type, glutaric acidemia type II, carnitine palmitoyltransferase II deficiency, and Zellweger syndrome are reviewed in the chapter. Other syndromes are listed in Table 63-1.

The mechanisms of renal cystogenesis are complex and intricately intertwined. Polycystin-1, polycystin-2, fibrocystin, and the nephrocystins all localize to the primary cilia of renal epithelial cells. Polycystin-1 functions as a mechanosensor of urine flow and triggers calcium influx through polycystin-2 channels. When this system is disrupted, impaired calcium influx leads to increased cAMP levels and ultimately abnormal cell proliferation and cell dedifferentiation. The Wnt signaling pathway also plays a crucial role in this process. When the calcium signaling cascade is initiated, several proteins, including nephrocystin-2 and the BBS proteins, are activated, and shift the Wnt pathway from its canonical to noncanonical mode, which results in β-catenin degredation by proteasomes. When calcium signaling or ciliary proteins are defective, the canonical Wnt pathway predominates, and β-catenin is able to translocate into the nucleus where it acts as a transcriptional activator, promoting the expression of fetal genes to induce cell proliferation. Abnormal planar cell polarity and cyst formation follow.

Much progress has been made in our understanding of renal cystogenesis. Although questions remain to be answered, our increasing knowledge of the genetic and molecular basis of cystic kidney diseases will enhance insight into the pathogenesis, clinical consequences, and ultimately the treatment of these disorders.

TABLE 63-1

Autosomal recessive inheritance
Meckel syndrome
Kaufman–Mckusick syndrome
Retina-renal dysplasia syndromes
Ivemark syndrome
Fryns syndrome
Autosomal dominant inheritance
Brachio-oto-renal syndrome
von Hippel Lindau syndrome
Townes Brocks syndrome
X-linked
Oro-facial-digital syndrome
Chromosomal
Trisomy 18
Trisomy 13
Trisomy 9
Triploidy
Deletion 3p
Inheritance variable
VATER association
Proteus syndrome
Prune-belly syndrome

FURTHER READING

1. Rossetti, S.; Consugar, M. B.; Chapman, A. B.; Torres, V. E.; Guay-Woodford, L. M.; Grantham, J. J.; Bennett, W. M.; Meyers, C. M.; Walker, D. L.; Bae, K., et al. Comprehensive Molecular Diagnostics in Autosomal Dominant Polycystic Kidney Disease. *J. Am. Soc. Nephrol.* **2007,** *18* (7), 2143–2160.
2. Hateboer, N.; v Dijk, M. A.; Bogdanova, N.; Coto, E.; Saggar-Malik, A. K.; San Millan, J. L.; Torra, R.; Breuning, M.; Ravine, D. Comparison of Phenotypes of Polycystic Kidney Disease Types 1 and 2. European PKD1-PKD2 Study Group. *Lancet* **1999,** *353* (9147), 103–107.
3. Torra, R.; Badenas, C.; Darnell, A.; Nicolau, C.; Volpini, V.; Revert, L.; Estivill, X. Linkage, Clinical Features, and Prognosis of Autosomal Dominant Polycystic Kidney Disease Types 1 and 2. *J. Am. Soc. Nephrol.* **1996,** *7* (10), 2142–2151.
4. Rossetti, S.; Chauveau, D.; Kubly, V.; Slezak, J. M.; Saggar-Malik, A. K.; Pei, Y.; Ong, A. C.; Stewart, F.; Watson, M. L.; Bergstralh, E. J., et al. Association of Mutation Position in Polycystic Kidney Disease 1 (*PKD1*) Gene and Development of a Vascular Phenotype. *Lancet* **2003,** *361* (9376), 2196–2201.
5. Rossetti, S.; Burton, S.; Strmecki, L.; Pond, G. R.; San Millan, J. L.; Zerres, K.; Barratt, T. M.; Ozen, S.; Torres, V. E.; Bergstralh, E. J., et al. The Position of the Polycystic Kidney Disease 1 (*PKD1*) Gene Mutation Correlates with the Severity of Renal Disease. *J. Am. Soc. Nephrol.* **2002,** *13* (5), 1230–1237.
6. Roy, S.; Dillon, M. J.; Trompeter, R. S.; Barratt, T. M. Autosomal Recessive Polycystic Kidney Disease: Long-Term Outcome of Neonatal Survivors. *Pediatr. Nephrol.* **1997,** *11* (3), 302–306.
7. Kaplan, B. S.; Fay, J.; Shah, V.; Dillon, M. J.; Barratt, T. M. Autosomal Recessive Polycystic Kidney Disease. *Pediatr. Nephrol.* **1989,** *3* (1), 43–49.
8. Bergmann, C.; Senderek, J.; Kupper, F.; Schneider, F.; Dornia, C.; Windelen, E.; Eggermann, T.; Rudnik-Schoneborn, S.; Kirfel, J.; Furu, L., et al. PKHD1 Mutations in Autosomal Recessive Polycystic Kidney Disease (ARPKD). *Hum. Mutat.* **2004,** *23* (5), 453–463.
9. Harris, P. C.; Rossetti, S. Molecular Genetics of Autosomal Recessive Polycystic Kidney Disease. *Mol. Genet. Metab.* **2004,** *81* (2), 75–85.
10. Zerres, K.; Rudnik-Schoneborn, S.; Steinkamm, C.; Becker, J.; Mucher, G. Autosomal Recessive Polycystic Kidney Disease. *J. Mol. Med.* **1998,** *76* (5), 303–309.
11. Hildebrandt, F.; Otto, E. Molecular Genetics of Nephronophthisis and Medullary Cystic Kidney Disease. *J. Am. Soc. Nephrol.* **2000,** *11* (9), 1753–1761.
12. Otto, E.; Betz, R.; Rensing, C.; Schatzle, S.; Kuntzen, T.; Vetsi, T.; Imm, A.; Hildebrandt, F. A Deletion Distinct from the Classical Homologous Recombination of Juvenile Nephronophthisis Type 1 (NPH1) Allows Exact Molecular Definition of Deletion Breakpoints. *Hum. Mutat.* **2000,** *16* (3), 211–223.
13. Yoder, B. K.; Hou, X.; Guay-Woodford, L. M. The Polycystic Kidney Disease Proteins, Polycystin-1, Polycystin-2, Polaris, and Cystin, Are Co-Localized in Renal Cilia. *J. Am. Soc. Nephrol.* **2002,** *13* (10), 2508–2516.
14. Nauli, S. M.; Alenghat, F. J.; Luo, Y.; Williams, E.; Vassilev, P.; Li, X.; Elia, A. E.; Lu, W.; Brown, E. M.; Quinn, S. J., et al. Polycystins 1 and 2 Mediate Mechanosensation in the Primary Cilium of Kidney Cells. *Nat. Genet.* **2003,** *33* (2), 129–137.
15. Torres, V. E.; Harris, P. C. Autosomal Dominant Polycystic Kidney Disease: The Last 3 Years. *Kidney Int.* **2009,** *76* (2), 149–168.
16. Deltas, C.; Papagregoriou, G. Cystic Diseases of the Kidney: Molecular Biology and Genetics. *Arch. Pathol. Lab. Med.* **2010,** *134* (4), 569–582.

17. Ong, A. C.; Harris, P. C. Molecular Pathogenesis of ADPKD: The Polycystin Complex Gets Complex. *Kidney Int.* 2005, *67* (4), 1234–1247.
18. Simons, M.; Gloy, J.; Ganner, A.; Bullerkotte, A.; Bashkurov, M.; Kronig, C.; Schermer, B.; Benzing, T.; Cabello, O. A.; Jenny, A., et al. Inversin, the Gene Product Mutated in Nephronophthisis Type II, Functions as a Molecular Switch between Wnt Signaling Pathways. *Nat. Genet.* 2005, *37* (5), 537–543.
19. Germino, G. G. Linking Cilia to Wnts. *Nat. Genet.* 2005, *37* (5), 455–457.
20. Gascue, C.; Katsanis, N.; Badano, J. L. Cystic Diseases of the Kidney: Ciliary Dysfunction and Cystogenic Mechanisms. *Pediatr. Nephrol.* 2010.

CHAPTER 64

Nephrotic Disorders

Hannu Jalanko

Department of Pediatric Nephrology, Children's Hospital, University of Helsinki, Finland

Helena Kääriäinen

National Institute for Health and Welfare, Helsinki, Finland

Nephrotic syndrome (NS) is a clinical entity characterized by heavy proteinuria, hypoproteinemia, and edema. NS is caused by acquired or genetic defects of the kidney filtration barrier located in the glomerular capillary wall. This barrier comprises three layers: fenestrated endothelium, glomerular basement membrane (GBM), and epithelial cell (podocyte) layer with distal foot processes and interposed slit diaphragms (SDs). The barrier is a size- and charge-selective molecular sieve, which effectively prevents leakage of plasma proteins into urine. Mutations in genes encoding for podocyte proteins lead to proteinuria and, to date, 10–20% of sporadic and 30–40% of familial cases of therapy resistant NS (SRNS) are associated with a gene defect.

64.1 NS IN NEWBORNS AND CHILDREN

Nephrin is a podocyte protein and a major component of the SD. Mutations in the nephrin gene (*NPHS1*) causes congenital nephrotic syndrome of the Finnish type (CNF, NPHS1), which is an autosomal recessive disease accounting for about half of cases with congenital NS. Typically, CNF children are born prematurely and develop full-blown NS during the first days of life. The index of placental weight/birth weight is over 25% in practically all newborns with CNF. The children show no extrarenal malformations. CNF has been reported from all over the world and roughly 150 different mutations have been described, spanning the whole gene. Mutations in NPHS1 are quite rare in cases manifesting after the first weeks of life. Disease-causing NPHS1 mutations, however, have been found in individual cases with an age of onset ranging from 6 months to 8 years.

Podocin is a hairpin-like protein of the stomatin family and is exclusively expressed in podocytes, where it is an important component of the SD complex. Podocin is encoded by the *NPHS2* gene composed of eight exons and located on chromosome 1q25–q31. Genetic defects in NPHS2 can lead to NS starting at any age. To date, more than 100 different NPHS2 mutations have been identified in patients with autosomal-recessive SRNS. Most mutations cause a severe disease with an onset in the newborn period or early childhood. The kidney biopsy in most cases reveals focal segmental glomerulosclerosis (FSGS). End stage renal disease (ESRD) develops within a few years. The podocin gene mutations do not cause any extrarenal defects.

Wilms tumor factor 1 (WT1) is a transcription factor of the zinc finger family important for the embryonic development of the kidney and genitalia. WT1 is expressed in podocytes and probably controls the expression of nephrin. WT1 contains 10 exons, of which exons 7–10 encode the 4 zinc fingers of the DNA-binding domain. Heterozygous mutations in exons 8 and 9 of WT1 lead to Denys–Drash syndrome, which is a combination of NS, male pseudohermaphroditism, and Wilms tumor. NS develops in early childhood, and the kidney biopsy shows diffuse mesangial sclerosis (DMS). Point mutations in intron 9 of WT1 lead to Frasier syndrome, which is characterized by NS and pseudohermaphroditism. Proteinuria is detected in childhood, usually between 2 and 6 years of age, and kidney biopsy reveals FSGS. The disorder is associated with gonadoblastomas, but not with Wilms tumor. WT1 mutations have also been observed in patients with isolated NS, showing DMS histology. The incidence of WT1 mutations in SRNS has been evaluated to be about 7%.

Phospholipase CE1 (PLCE1) belongs to the phospholipase family of proteins that catalyze hydrolysis of phosphoinositides, which are involved in a wide spectrum of intracellular functions. In podocytes, PLCE1 probably interacts with the SD and actin cytoskeleton. Mutations in PLCE1 cause early onset NS (age range 1 month to 6 years). The patients have no extrarenal manifestations and kidney histology shows DMS in most patients. In a worldwide cohort of idiopathic DMS patients, truncating mutations in PLCE1 were detected in a third of the families. Thus, PLCE1 is a major gene causing isolated DMS.

Laminin β2 is a component of Laminin-521, specifically expressed in the GBM and at some other sites such as intraocular muscles and neuromuscular synapses. Mutations in the *LAMB2* gene are associated with Pierson syndrome, which is a rare autosomal recessive disorder characterized by early-onset NS with variable ocular (especially microcoria) and neurologic defects. After the original observation, milder variants of the syndrome have been reported with less prominent extrarenal abnormalities The age of the patients at diagnosis is mostly less than 3 months and the age at onset of ESRD varies from newborn period to 16 years. Renal biopsy shows mostly DMS histology. Among patients with isolated NS, LAMB2 mutations are rare.

64.2 NS IN ADOLESCENTS AND ADULTS

Podocin gene mutations are the major cause of genetic form of NS in adolescents and adults. Many of these late-onset cases are compound heterozygotes, with one allele harboring an R229Q mutation and the other allele carrying a disease-causing mutation. Besides NPHS2 mutations, genetic defects in four other podocyte proteins are associated with rare cases of late-onset NS. CD2-associated protein (CD2AP) is localized to the SD of the podocyte, where it links podocin and nehrin to the phosphoinositide 3-OH kinase to form a signaling complex. The role of CD2AP in human NS is still being elucidated, but individuals with homozygous mutations in CD2AP have been reported to present with early or late-onset FSGS. α-actinin-4 (ACTN4) is an actin-bundling protein important for the integrity of the podocyte cytoskeleton. Mutations in ACTN4 are associated with an autosomal dominant form of familial FSGS. By now, only a few ACTN4 missense mutations have been described. INF2 is a member of formins and has the unique ability to accelerate both actin polymerization and depolymerization. Patients with INF2 mutations present in adolescence or adulthood typically with moderate or nephrotic range proteinuria. The disease is progressive, often leading to ESRD. Canonical transient receptor potential C6 ion channel (TRPC6) is a receptor-operated cation channel that contributes to changes in the cytosolic-free Ca^{2+}-concentration and possibly interacts with nephrin. The few published studies, to date, suggest that TRPC6 mutations account for 3–7% of cases with familial adult-onset FSGS.

64.3 NS DIAGNOSTICS AND MANAGEMENT

NS is easily detected in newborns, infants, and children. Edema, hypoproteinemia, hyperlipidemia, and heavy proteinuria are the cardinal signs of NS. The exact etiologic diagnosis behind the NS is much more difficult to unravel. In NS, patients with no response to medical therapy, genetic analysis, and kidney biopsy are indicated. Screening for several genes at the same time is time-consuming but expensive. A stepwise approach is often reasonable. The clinical renal and extrarenal findings, kidney biopsy, and the age of onset of the disease will give a clue for the possible gene involved.

NS caused by genetic defects is not responsive to immunosuppressive medication. Renal transplantation is an established mode of therapy for children with NS not responding to medical treatment.

demonstrate somatic *VHL* gene mutation, methylation or loss, resulting in homozygous *VHL* gene inactivation. The *VHL* tumor suppressor gene has multiple functions (*16*) but the best characterized is the regulation of hypoxia gene response pathways.

66.1.2.2 Familial Non-VHL RCC. Familial non-VHL clear cell RCC (FCRC) is characterized by two or more relatives with clear cell RCC and though the genetic basis for most cases of familial non-VHL clear cell RCC has not been elucidated, some may harbor a constitutional translocation or germ line *VHL*, *FLCN* or *SDHB* mutations (*17,18*). Although familial RCC associated with chromosome 3 translocations is rare, cytogenetic analysis should be performed in FCRC and early-onset/multicentric RCC. However, in chromosome 3 translocation carriers without a personal or family history of RCC (and without disruption of known tumor suppressor genes), the risk of RCC appears to be small (*19*).

66.1.2.3 Hereditary Papillary Renal Carcinoma. Histopathologically, papillary cancer may be divided into two subtypes. Germ line *MET* mutations are associated with Type 1 tumors that are characterized by small basophilic cells with pale cytoplasm and inconspicuous nuclei (*20*) and Type 2 papillary RCC susceptibility may be associated with germ line FH mutations (*21*). While homozygous recessive FH mutations cause FH deficiency, heterozygous mutations predispose to cutaneous and uterine leiomyomatosis (fibroids). A subset of patients with hereditary leiomyomatosis develop Type 2 papillary or collecting duct histology RCC (*21,22*), so this dominantly inherited condition is called hereditary leiomyomatosis and renal cell cancer syndrome. Although only a minority of affected individuals develops RCC, the renal cancers are aggressive and should be removed when detected.

Two inherited RCC syndromes associated with tumors of variable histopathology are BHD syndrome and RCC associated with mutations in the succinate dehydrogenase subunit genes. BHD is a dominantly inherited disorder and is characterized by cutaneous fibrofolliculomas, lipomas, and spontaneous pneumothoraces and probably colonic polyps, and cancers (*23*). Germ line mutations in the succinate dehydrogenase B subunit gene (*SDHB*) were, following the demonstration that *SDHD* gene mutations caused head and neck paraganglioma and pheochromocytoma, also demonstrated to predispose to these tumors (*24*). Subsequently, it was found that, in addition to germ line *SDHB* mutations are associated with a high risk of malignant pheochromocytoma, about 5% of patients with inherited RCC will have an *SDHB* mutation (*18*). A variety of histological subtypes of RCC may be associated with *SDHB* mutations (and less frequently *SDHD*) and the lifetime risk of RCC in *SDHB* mutation carriers was estimated to be about 15% (*25*).

Carcinoma of the renal pelvis accounts for approximately 10% of all malignant renal tumors. Environmental causes include occupational exposure (as for bladder cancer) and prolonged excessive phenacetin ingestion.

Examples of familial ureteric and renal pelvis transitional cell carcinoma are uncommon, but transitional carcinoma of the renal pelvis (together with cancer of the bladder and ureter) can be a feature of hereditary non-polyposis colorectal cancer syndrome (HNPCC, Lynch syndrome).

66.2 TUMORS OF THE BLADDER

Carcinoma of the bladder has an incidence of approximately 30 and 10 per 100,000 in males and females, respectively. Most tumors are transitional cell carcinomas, and a notable feature is the propensity of patients to develop multiple urothelial tumors. Occupational exposure to industrial chemicals, including arylamines, is now firmly established as a predisposing factor but only a minority of patients will have a significant occupational history (*26*). Other agents associated with bladder cancer include tobacco smoking, chronic infection (bacterial, schistosomiasis), and phenacetin exposure. Much interest has focused on the interaction of genetic and environmental factors in bladder cancer susceptibility, in particular the association between disease and polymorphisms in genes that encode enzymes that metabolize potential carcinogens. Evidence for familial transitional cell carcinoma of the upper and lower urinary tract has been reviewed by Mueller and coworkers (*27*) and, in addition to rare families suggestive of dominant inheritance with incomplete penetrance, patients with germ line retinoblastoma mutations may have an increased risk of bladder cancer. Bladder cancer risk is increased in hereditary non-polyposis colorectal cancer syndrome (HNPCC, Lynch syndrome)—particularly in men with an MSH2 mutation (*28*). However familial clustering of bladder cancer is uncommon.

66.3 CARCINOMA OF THE PROSTATE

Prostate cancer is the most common cancer, and the second cause of cancer death, in North American men. Environmental factors such as a high-fat diet and androgen stimulation have been implicated and the role of genetic factors has come under intense scrutiny. Genetic epidemiology studies suggested that the inheritability of prostate cancer was greater than that of breast or colorectal cancers and complex segregation analysis suggested that prostate cancer clustering could be caused by a rare, highly penetrant dominantly inherited predisposition gene (*29*). However, family linkage studies of familial prostate cancer, though initially promising, proved to be significantly more challenging than those for familial breast and colorectal cancer. With the advent of genome-wide association studies, a large number of susceptibility alleles have been identified (*30,31*)—though it is unclear how these findings might impact on clinical practice (*32*).

Following reports that carriers of *BRCA1* and *BRCA2* mutations in familial breast/ovary cancer kindred had an increased risk of prostate cancer, these genes were analyzed as candidate high penetrance prostate cancer susceptibility genes and it was found that ~2% of men with early-onset prostate (>55 years) had a germ line *BRCA2* mutation (*33*). In many cases, there was no family history of breast or ovarian cancer. A male *BRCA2* mutation carrier was estimated to have a 23-fold increased risk of developing prostate cancer by age 55 years and it also appears that male *BRCA2* carriers with prostate cancer have a poorer prognosis than similar non-carriers (*34*).

66.4 TESTICULAR NEOPLASMS

Testicular cancer is the most frequent malignancy in men between 20 and 40 years of age. Most tumors are of germ cell origin (seminoma, teratoma) but approximately 5–10% are sex cord/stromal tumors (e.g. Leydig and Sertoli cell), and gonadoblastoma contains germ cell and stromal elements. Epidemiologic studies have suggested that gonadal hormone drive is a major factor in the development of germ cell tumors. The most important risk factor is cryptorchidism, and this and other conditions (e.g. orchitis, infertility) are associated with reduced testicular function, which would produce increased gonadotropin drive to the testicles. Testicular tumors are bilateral in about 4% of patients. Large families with a high incidence of testicular cancer have been described but are rare (*35*) and the majority of families are relative pairs, usually brothers, suggesting that susceptibility may be caused by genes with small or moderate effects. An increased risk of testicular germ cell tumors has been reported in Down syndrome and Klinefelter syndrome (*36*). Linkage studies in familial testicular cancer were hampered by the paucity of large multigenerational pedigrees and though candidate gene studies identified the Y chromosome *gr/gr* deletion as causing two- to threefold increase in risk, this variant occurs in <3% of males (*37*) but recently, genome-wide association studies have identified significant associations to several loci (*38,39*).

Genetic disorders predisposing to testicular tumors can be classified according to the presence or absence of abnormal sexual differentiation. Disorders of sex development (DSD) are defined as conditions of incomplete or disordered genital or gonadal development leading to discordances between genetic sex, gonadal sex and phenotypic sex and are sub-classified into (a) sex chromosomal DSD, (b) 46XY DSD and (c) 46XX DSD (*40*). DSD patients with hypervirilization are not at risk of germ cell tumors and the risk is variable patients with hypovirilzation and gonadal dysgenesis.

Gonadoblastoma occurs in XY gonadal dysgenesis and in patients with the WAGR and Frasier syndromes (see above) and the risk of germ cell tumors in gonadal dysgenesis is associated with the presence of Y chromosome material and, in particular, the presence of Yq material. The association between the presence of a Y chromosome and gonadoblastoma is also a feature of single gene disorders causes of DSD. XX forms of gonadal dysgenesis are probably not associated with an increased risk for gonadoblastomas, but XY pure gonadal dysgenesis (Swyer's syndrome, which may result from SRY mutations) is frequently complicated by this tumor. Affected individuals are of normal stature and do not have the features of Turner syndrome, but have streak gonads. The incidence of gonadal neoplasia in true hermaphrodites (individuals with both testicular and ovarian tissue) appears to be low, although both ovarian and testicular tumors have been reported. Complete androgen insensitivity syndrome (testicular feminization) is inherited as an X-linked recessive disorder and is associated with an increased risk of testicular malignancy (about 5%), most often seminoma.

Mendelian disorders not associated with abnormal sexual development that predispose to testicular tumors include Peutz–Jegher syndrome, Carney complex and possibly X-linked ichthyosis.

REFERENCES

1. Breslow, N.; Beckwith, J. B.; Ciol, M.; Sharples, K. Age Distribution of Wilms' Tumor: Report from the National Wilms' Tumor Study. *Cancer Res.* **1988**, *48,* 1653–1657.
2. Muto, R.; Yamamori, S.; Ohashi, H.; Osawa, M. Prediction by FISH Analysis of the Occurrence of Wilms' Tumor in Aniridia Patients. *Am. J. Med. Genet.* **2002**, *108,* 285–289.
3. Clericuzio, C.; Hingorani, M.; Crolla, J. A.; van Heyningen, V.; Verloes, A. Clinical Utility Gene Card for WAGR Syndrome. *Eur. J. Hum. Genet.* **2011**, *19* (4).
4. Coppes, M. J.; Huff, V.; Pelletier, J. Denys–Drash Syndrome: Relating a Clinical Disorder to Genetic Alterations in the Tumor Suppressor Gene *WT1*. *J. Pediatr.* **1993**, *123,* 673–678.
5. Mueller, R. F. The Denys–Drash Syndrome. *J. Med. Genet.* **1994**, *31,* 471–477.
6 Little, M.; Wells, C. A Clinical Overview of *WT1* Gene Mutations. *Hum. Mutat.* **1997**, *9,* 209–225.
7. Elliott, M. L.; Maher, E. R. Syndrome of the Month: Beckwith–Wiedemann Syndrome. *J. Med. Genet.* **1994**, *31,* 560–564.
8. Maher, E. R.; Reik, W. Beckwith–Wiedemann Syndrome: Imprinting in Clusters Revisited. *J. Clin. Invest.* **2000**, *105,* 247–252.
9. Maher, E. R.; Brueton, L. A.; Bowdin, S. C.; Luharia, A.; Cooper, W.; Cole, T. R.; Macdonald, F.; Sampson, J. R.; Barratt, C. L.; Reik, W.; Hawkins, M. M. Beckwith–Wiedemann Syndrome and Assisted Reproduction Technology (ART). *J. Med. Genet.* **2003**, *40,* 62–64.
10. Lim, D. H.; Maher, E. R. Genomic Imprinting Syndromes and Cancer. *Adv. Genet.* **2010**, *70,* 145–175.
11. Maher, E. R.; Neumann, H. P.; Richard, S. Von Hippel-–Lindau Disease: A Clinical and Scientific Review. *Eur. J. Hum. Genet.* **2011**, *19,* 617–623.
12. Washecka, R.; Hanna, M. Malignant Renal Tumours in Tuberous Sclerosis. *Urology* **1991**, *37,* 340–343.
13. Hes, F. J.; McKee, S.; Taphoorn, M. J.; Rehal, P.; van Der Luijt, R. B.; McMahon, R.; van Der Smagt, J. J.; Dow, D.; Zewald, R. A.; Whittaker, J., et al. Cryptic Von Hippel–Lindau Disease Germline Mutations in Patients with Haemangioblastoma Only. *J. Med. Genet.* **2000**, *37,* 939–943.

14. Woodward, E. R.; Eng, C.; McMahon, R.; Voutilainen, R.; Affara, N. A.; Ponder, B. A.; Maher, E. R. Genetic Predisposition to Phaeochromocytoma: Analysis of Candidate Genes GDNF, RET and VHL. *Hum. Mol. Genet.* **1997,** *6,* 1051–1056.
15. Latif, F.; Tory, K.; Gnarra, J.; Yao, M.; Duh, F. M.; Orcutt, M. L.; Stackhouse, T.; Kuzmin, I.; Modi, W.; Geil, L., et al. Identification of the Von Hippel–Lindau Disease Tumour Suppressor Gene. *Science* **1993,** *260,* 1317–1320.
16. Kaelin, W. G., Jr. Treatment of Kidney Cancer; Insights Provided by the VHL Tumor-Suppressor Protein. *Cancer* **2009,** *115,* 2262–2272.
17. Woodward, E. R.; Ricketts, C.; Killick, P.; Gad, S.; Morris, M. R.; Kavalier, F.; Hodgson, S. V.; Giraud, S.; Bressac-de Paillerets, B.; Chapman, C.; Escudier, B.; Latif, F.; Richard, S.; Maher, E. R. Familial Non-VHL Clear Cell (Conventional) Renal Cell Carcinoma: Clinical Features, Segregation Analysis, and Mutation Analysis of FLCN. *Clin. Cancer Res.* **2008,** *14* (18), 5925–5930.
18. Ricketts, C.; Woodward, E. R.; Killick, P.; Morris, M. R.; Astuti, D.; Latif, F.; Maher, E. R. Germline SDHB Mutations and Familial Renal Cell Carcinoma. *J. Natl. Cancer Inst.* **2008,** *100,* 1260–1262.
19. Woodward, E. R.; Skytte, A. B.; Cruger, D. G.; Maher, E. R. Population-Based Survey of Cancer Risks in Chromosome 3 Translocation Carriers. *Genes Chromosomes Cancer* **2010,** *49* (1), 52–58.
20. Schmidt, L.; Junker, K.; Nakaigawa, N.; Kinjerski, T.; Weirich, G.; Miller, M.; Lubensky, I.; Neumann, H. P.; Brauch, H.; Decker, J.; Vocke, C.; Brown, J. A.; Jenkins, R.; Richard, S.; Bergerheim, U.; Gerrard, B.; Dean, M.; Linehan, W. M.; Zbar, B. Novel Mutations of the MET Proto-Oncogene in Papillary Renal Carcinomas. *Oncogene* **1999,** *18,* 2343–2350.
21. Tomlinson, I. P.; Alam, N. A.; Rowan, A. J.; Barclay, E.; Jaeger, E. E.; Kelsell, D.; Leigh, I.; Gorman, P.; Lamlum, H.; Rahman, S.; Roylance, R. R.; Olpin, S.; Bevan, S.; Barker, K.; Hearle, N.; Houlston, R. S.; Kiuru, M.; Lehtonen, R.; Karhu, A.; Vilkki, S.; Laiho, P.; Eklund, C.; Vierimaa, O.; Aittomäki, K.; Hietala, M.; Sistonen, P.; Paetau, A.; Salovaara, R.; Herva, R.; Launonen, V.; Aaltonen, L. A. Multiple Leiomyoma Consortium. Germline Mutations in FH Predispose to Dominantly Inherited Uterine Fibroids, Skin Leiomyomata and Papillary Renal Cell Cancer. *Nat. Genet.* **2002,** *30,* 406–410.
22. Toro, J. R.; Nickerson, M. L.; Wei, M. H.; Warren, M. B.; Glenn, G. M.; Turner, M. L.; Stewart, L.; Duray, P.; Tourre, O.; Sharma, N.; Choyke, P.; Stratton, P.; Merino, M.; Walther, M. M.; Linehan, W. M.; Schmidt, L. S.; Zbar, B. Mutations in the Fumarate Hydratase Gene Cause Hereditary Leiomyomatosis and Renal Cell Cancer in Families in North America. *Am. J. Hum. Genet.* **2003,** *73,* 95–106.
23. Menko, F. H.; van Steensel, M. A.; Giraud, S.; Friis-Hansen, L.; Richard, S.; Ungari, S.; Nordenskjöld, M.; Hansen, T. V.; Solly, J.; Maher, E. R. European BHD Consortium Birt–Hogg–Dubé Syndrome: Diagnosis and Management. *Lancet Oncol.* **2009,** *10,* 1199–1206.
24. Astuti, D.; Latif, F.; Dallol, A.; Dahia, P. L.; Douglas, F.; George, E.; Skoldberg, F.; Husebye, E. S.; Eng, C.; Maher, E. R. Gene Mutations in the Succinate Dehydrogenase Subunit SDHB Cause Susceptibility to Familial Pheochromocytoma and to Familial Paraganglioma. *Am. J. Hum. Genet.* **2001,** *69,* 49–54.
25. Ricketts, C. J.; Forman, J. R.; Rattenberry, E.; Bradshaw, N.; Lalloo, F.; Izatt.; Cole, T. R.; Armstrong, R.; Kumar, V. K.; Morrison, P. J.; Atkinson, A. B.; Douglas, F.; Ball, S. G.; Cook, J.; Srirangalingam, U.; Killick, P.; Kirby, G.; Aylwin, S.; Woodward, E. R.; Evans, D. G.; Hodgson, S. V.; Murday, V.; Chew, S. L.; Connell, J. M.; Blundell, T. L.; Macdonald, F.; Maher, E. R. Tumor Risks and Genotype–Phenotype-Proteotype Analysis in 358 Patients with Germline Mutations in SDHB and SDHD. *Hum. Mutat.* **2010,** *31,* 41–51.
26. Shirai, T. Etiology of Bladder Cancer. *Semin. Urol.* **1993,** *3,* 113–126.
27 Mueller, C. M.; Caporaso, N.; Greene, M. H. Familial and Genetic Risk of Transitional Cell Carcinoma of the Urinary Tract. *Urol. Oncol.* **2008,** *26,* 451–464.
28. van der Post, R. S.; Kiemeney, L. A.; Ligtenberg, M. J.; Witjes, J. A.; Hulsbergen-van de Kaa, C. A.; Bodmer, D.; Schaap, L.; Kets, C. M.; van Krieken, J. H.; Hoogerbrugge, N. Risk of Urothelial Bladder Cancer in Lynch Syndrome Is Increased, in Particular among MSH2 Mutation Carriers. *J. Med. Genet.* **2010,** *47,* 464–470.
29. Carter, B. S.; Beaty, T. H.; Steinberg, G. D.; Childs, B.; Walsh, P. C. Mendelian Inheritance of Familial Prostate Cancer. *Proc. Natl. Acad. Sci U.S.A.* **1992,** *89,* 3367–3371.
30. Al Olama, A. A.; Kote-Jarai, Z.; Giles, G. G., et al. Multiple Loci on 8q24 Associated with Prostate Cancer Susceptibility. *Nat. Genet.* **2009,** *41,* 1058–1060.
31. Eeles, R. A.; Kote-Jarai, Z.; Al Olama, A. A.; Giles, G. G.; Guy, M.; Severi, G.; Muir, K., et al. Identification of Seven New Prostate Cancer Susceptibility Loci through a Genome-Wide Association Study. *Nat. Genet.* **2009,** *41,* 1116–1121.
32. Pashayan, N.; Duffy, S. W.; Chowdhury, S.; Dent, T.; Burton, H.; Neal, D. E.; Easton, D. F.; Eeles, R.; Pharoah, P. Polygenic Susceptibility to Prostate and Breast Cancer: Implications for Personalised Screening. *Br. J. Cancer* **2011,** *104,* 1656–1663.
33. Edwards, S. M.; Kote-Jarai, Z.; Meitz, J.; Hamoudi, R.; Hope, Q.; Osin, P.; Jackson, R., et al. Two Percent of Men with Early-Onset Prostate Cancer Harbor Germline Mutations in the *BRCA2* Gene. *Am. J. Hum. Genet.* **2003,** *72,* 1–12.
34. Edwards, S. M.; Evans, D. G.; Hope, Q.; Norman, A. R.; Barbachano, Y.; Bullock, S.; Kote-Jarai, Z., et al. *Br. J. Cancer* **2010,** *103,* 918–924.
35. Goss, P. E.; Bulbul, M. A. Familial Testicular Cancer in Five Members of a Cancer-Prone Kindred. *Cancer* **1990,** *66,* 2044–2046.
36. Gilbert, D.; Rapley, E.; Shipley, J. Testicular Germ Cell Tumours: Predisposition Genes and the Male Germ Cell Niche. *Nat. Rev. Cancer* **2011,** *11,* 278–288.
37. Nathanson, K. L.; Kanetsky, P. A.; Hawes, R.; Vaughn, D. J.; Letrero, R.; Tucker, K.; Friedlander, M.; Phillips, K. A.; Hogg, D.; Jewett, M. A., et al. The Y Deletion gr/gr and Susceptibility to Testicular Germ Cell Tumor. *Am. J. Hum. Genet.* **2005,** *77,* 1034–1043.
38. Rapley, E. A.; Turnbull, C.; Al Olama, A. A.; Dermitzakis, E. T.; Linger, R.; Huddart, R. A.; Renwick, A.; Hughes, D.; Hines, S.; Seal, S., et al. A Genome-Wide Association Study of Testicular Germ Cell Tumor. *Nat. Genet.* **2009,** *41,* 807–810.
39. Turnbull, C.; Rapley, E. A.; Seal, S.; Pernet, D.; Renwick, A.; Hughes, D.; Ricketts, M.; Linger, R.; Nsengimana, J.; Deloukas, P., et al. UK Testicular Cancer Collaboration Variants near DMRT1, TERT and ATF7IP Are Associated with Testicular Germ Cell Cancer. *Nat. Genet.* **2010,** *42,* 604–607.
40. Looijenga, L. H.; Hersmus, R.; Oosterhuis, J. W.; Cools, M.; Drop, S. L.; Wolffenbuttel, K. P. Tumor Risk in Disorders of Sex Development (DSD). *Best Pract. Res. Clin. Endocrinol. Metab.* **2007 Sep,** *21* (3), 480–495.

Gastrointestinal Disorders

Gastrointestinal Tract and Hepatobiliary Duct System, 267
Bile Pigment Metabolism and Its Disorders, 270
Cancer of the Colon and Gastrointestinal Tract, 272

CHAPTER

67

Gastrointestinal Tract and Hepatobiliary Duct System

Eberhard Passarge

Institut für Humangenetik, Universitätsklinikum Essen, Essen; Universitätsklinikum Leipzig, Leipzig, Germany

This chapter focuses on gross defects of the intestinal anatomical structures, genetic disorders of the small and the large intestines, and the biliary tract system with emphasis on two genetic disorders, congenital intestinal aganglionosis (Hirschsprung disease) and a complex multisystem disorder affecting the biliary duct system [Alagille syndrome (AGS)]. The different types of congenital malformations of the gastrointestinal (GI) tract can best be understood on the basis of the underlying normal embryological development. In brief, the cells forming the gut derive from the endoderm for the inner lining, the mucosa, from the ectoderm for the wall, containing muscles and connective tissue, and from the neural crest for the intramural ganglion cells required for peristalsis. Following a brief review of the embryological background, genetically relevant GI disorders and malformations are presented according to the following classification:

(1) Gross defects of the intestinal anatomical structures (atresias of the oesophagus, duodenum, jejunum, and anus, malrotation, duplication, defects of the abdominal wall, Meckel's diverticulum, defects of the diaphragm and partial agenesis of the pancreas)
(2) Congenital intestinal aganglionosis (Hirschsprung disease) with reference to the heterogenous etiology, embryological origin, molecular genetics, empirical risk figures, diagnosis, syndromic forms, and animal models
(3) Defects of the hepatobiliary duct system [biliary atresias, arteriohepatic dysplasia syndrome [AGS]]
(4) Functional disorders (hereditary pancreatitis, infantile hypertrophic pyloric stenosis, achalasia, intestinal pseudoobstruction, and intussusception), including 110 references in the online version of this chapter.

Hirschsprung disease is a complex, heterogeneous disorder, resulting from absence of the intramural intestinal ganglion cells (congenital intestinal aganglionosis) owing to failure of migration from the neural crest during embryonic development. Therefore, Hirschsprung disease is considered as a neural crest disorder (neurocristopathy). Nine genes and three signal pathways have been identified in relation to the aetiology of this group of disorders. In addition, disease-related noncoding regions in the *RET* gene contribute to the causes by interacting with at least one of the other predisposing genes. The disease susceptibility between males and females differs by a factor of 5 (5:1 higher disease frequency in males than in females).

Current knowledge of Hirschsprung disease can be summarized as follows:

(1) Hirschsprung disease is currently the best understood multifactorial, non-Mendelian genetic disorder.
(2) Mutations in the receptor tyrosine kinase gene *RET* located at human chromosome 10q11.1 play a pivotal role in the etiology, in combination with other genes.
(3) A common noncoding variant in intron 1 of the *RET* gene within a conserved enhancer-like DNA sequence markedly influences the susceptibility to congenital intestinal aganglionosis.
(4) This variant reduces enhancer activity in vitro and contributes to the disease risk 20-fold more than other rare alleles, in spite of its low penetrance.
(5) The eight other known genes involved in the genetic etiology account for less than 30% of patients.
(6) Three different signal pathways related to ganglion cell migration and intramural intestinal function are involved in the genetic pathogenesis of congenital intestinal aganglionosis: (i) the RET receptor kinase pathway with the *RET* gene and its ligand GDNF (glial cell line-derived neurotrophic factor), (ii) the endothelin type B receptor pathway with the EDNRB receptor and its ligand EDN3 (endothelin-3), and (iii) the transcription factor SOX10.

(7) A single genetic change in any of the known genes and chromosomal regions predisposing to congenital intestinal aganglionosis is neither sufficient nor necessary to cause the disease.
(8) The genetic effects differ in males and females. The frequency of transmission of predisposing factors differs between males and females, resulting in a 5.7-fold and 2.1-fold increase in susceptibility in males and females, respectively. Empirical risk figures for first-degree relatives are presented in the online version of this chapter.
(9) The predisposing *RET* susceptibility allele differs in world-like frequency. It is virtually absent in Africa (less than 5%) in contrast to Asia (40% in China and Japan) and Europe (25%).
(10) In about 12% of patients, congenital intestinal aganglionosis occurs in syndromic form associated with other clinical manifestations outside the GI tract, usually in a monogenetic inheritance pattern.

AGS (ALGS; Online Mendelian Inheritance in Man (OMIM) 118450/601920) is a multisystem developmental disorder with important hepatic manifestations resulting from a paucity or atresia of intrahepatic bile ducts, which leads to cholestasis and liver insufficiency in early childhood. It progresses to cirrhosis and liver failure in a high proportion of patients. Neonatal jaundice is an early manifestation in many, but not in all patients. AGS is the most common form of familial cholestatic liver disease in childhood, with a frequency of about 1:70,000 live births. The hepatic manifestations are documented by liver biopsy showing a paucity of intrahepatic ducts in 80–100% of biopsies, to some extent depending on the age of the patient when the biopsy is taken. Elevated serum bile acids, conjugated bilirubin, alkaline phosphatase, cholesterol, and γ-glutamyl transferase are typical laboratory findings. Two forms, ALGS1 (OMIM 118450) and ALGS2 (OMIM 610205), are distinguished by mutation in different genes (see below).

In about 94% of patients, AGS is caused by mutations in the gene, Jagged-1 (*JAG1*, OMIM 601920), encoding a cell surface membrane protein. It is a member of a family of ligands for transmembrane receptors that function in signal transduction in the Notch signaling pathway. This ligand is called Jagged-1 (*JAG1*) in humans. Its receptor, Notch, was originally identified in *Drosophila melanogaster* and the nematode *Caenorhabditis elegans* in functions relating to cell fate decisions in many different cell types. Four receptors, Notch 1, 2, 3, and 4, occur in mammals. Notch is unique by functioning both in cell–cell interactions and as a highly conserved signal transducing system downstream of ligand binding. Thus, mutations in the ligand *JAG1* have pleiotropic effects, involving many different cell types and tissues, which are consistent with a central role of the Notch signaling pathway in cell specification, tissue patterning, and morphogenesis throughout development in vertebrates and invertebrates.

JAG1 is a transmembrane protein with a single transmembrane domain, a small intracellular domain, a large extracellular domain consisting of 16 extracellular EGF-like (Epidermal Growth Factor) repeats (40–50 amino acids each), three highly conserved regions named DLS (for ligands Delta and Serrate in *Drosophila* and Lag-2 in *C. elegans*) of 40 amino acids, and a 21 amino acid signal peptide.

Functional disorders are a loosely defined, heterogeneous group of disorders. Hereditary pancreatitis, infantile hypertrophic pyloric stenosis, achalasia, intestinal pseudoobstruction, and intussusception are considered in the online version of this chapter.

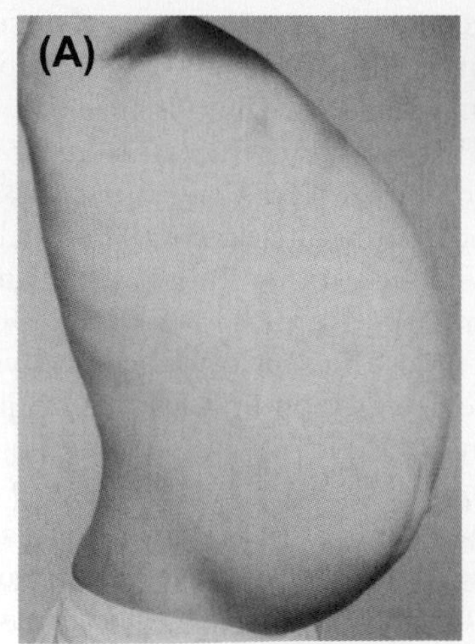

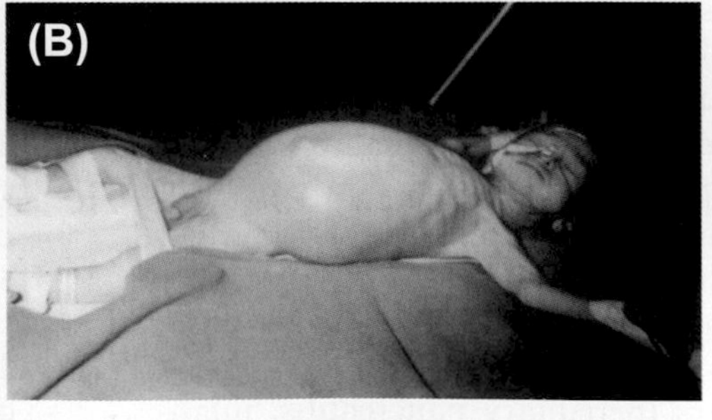

FIGURE 67-1 Hirschsprung disease. A. Distended abdomen due to megacolon. B. Infant with severe form of megacolon. *(Photo courtesy Dr. Lester Martin, Cincinnati.)*

TABLE 67-1 Syndromic Forms of Hirschsprung disease[a]

Disorder	OMIM
Multiple endocrine neoplasia type 2A	171400
Multiple endocrine neoplasia type 2B	162300
Familial medullary thyroid carcinoma	155240
Pheochromocytoma	171300
Waardenburg syndrome type 1 and type 2	193500/10
Shah–Waardenburg syndrome	277580
Cartilage hair dysplasia	250250
Smith–Lemli–Opitz syndrome	270400
Hypoventilation syndrome	209880
Goldberg–Shprintzen Syndrome	
Mowat–Wilson syndrome	235730
McKusick–Kaufman syndrome	236700
Polydactyly, renal dysgenesis, and deafness	235740
Brachydactyly type D	306980
Microcephaly and iris coloboma	235730
Dysmorphic facial features and nail hypoplasia	235760
Osteopetrosis, recessive type	259700
Bardet–Biedl syndrome	209900
Piebald trait	172800
Congenital deafness, isolated cases	277580

[a]Examples from OMIM (http://www.ncbi.nlm.nih.gov/Omim/). For additional data see Chakravarti & Lyonnet (Hirschsprung Disease. In: The Metabolic & Molecular Bases of Inherited Disease, 8th ed. Scriver, C., et al., Ed.; Mc Graw-Hill Med. Publishing Division, New York, pp. 6231–6255) and Amiel et al. (Hirschsprung Disease, Associated Syndromes and Genetics: A Review. *J. Med. Genet.* 45, 1–14).

TABLE 67-2 Risk of Pyloric Stenosis in Families

	Index Patient	
Affected Relatives (%)	Male	Female
First-degree		
Brothers	6.5	10.8
Sisters	2.8	3.8
Sons	5.5	18.9
Daughters	2.4	7.0
Second-degree		
Males	2.2	0.5
Females	4.3	1.7
General population	0.5	0.1

Data from Carter, C.O. The Inheritance of Congenital Pyloric Stenosis. *Brit. Med. Bull.* **1961**, *17*, 251–254 and Carter, C.O. Congenital Pyloric Stenosis. In *Principles and Practice of Medical Genetics*; Emery, A. E. H.; Rimoin, D. L., Eds.; 1st ed. Churchill Livingstone, Edinburgh, 1983; pp 879–883.

FURTHER READING

1. Amiel, J., et al. Hirschsprung Disease, Associated Syndromes and Genetics: A Review. *J. Med. Genet.* **2008**, *45*, 1–14.
2. Chakravarti, A.; Lyonnet, S., et al. Hirschsprung Disease. In *The Metabolic & Molecular Bases of Inherited Disease,* 8th ed.; Scriver, C. R., Ed.; Mc Graw-Hill Med. Publishing Division: New York, 2001; pp 6231–6255.
3. Emison, E. S.; McCallion, A. S.; Kashuk, C. S., et al. A Common Sex-Dependent Mutation in a RET Enhancer Underlies Hirschsprung Disease Risk. *Nature* **2005**, *434*, 857–863.
4. McCallion, A. S.; Chakravarti, A. RET and Hirschsprung Disease and Multiple Endocrine Neoplasia Type 2. In *Inborn Errors of Development. The Molecular Basis of Clinical Disorders of Morphogenesis,* 2nd ed.; Epstein, C. J.; Erickson, R. P.; Wynshaw-Boris, A., Eds.; Oxford University Press: Oxford, 2008; pp 512–520.
5. Passarge, E.; Stevenson, R. E. Small and Large Intestines. In *Human Malformations and Related Anomalies,* 2nd ed.; Stevenson, R. E.; Hall, J. G.; Goodman, R. M., Eds.; Oxford University Press, 2006, Vol. II; pp 1097–1114.
6. Passarge, E. Gastrointestinal Tract: Molecular Genetics of Hirschsprung Disease. In: *Encyclopedia of Life Sciences (ELS)*. John Wiley & Sons, Ltd: Chichester. DOI: http://dx.doi.org/10.1002/9780470015902.a0005518.pub2.
7. Spinner, N. B.; Krantz, I. D. JAG1 and NOTCH2 and the Alagille Syndrome. In *Inborn Errors of Development,* 2nd ed.; Epstein, C. J.; Erickson, R. P.; Wynshaw-Boris, A., Eds.; Oxford University Press: Oxford, 2008; pp 552–559.

CHAPTER

68

Bile Pigment Metabolism and Its Disorders

Namita Roy Chowdhury and Jayanta Roy Chowdhury

Departments of Medicine and Genetics, Albert Einstein College of Medicine, Bronx, NY, USA

Yesim Avsar

Department of Medicine, Division of Gastroenterology and Liver Diseases,
Albert Einstein College of Medicine, Bronx, NY, USA

Bilirubin is the end product of degradation of the heme moiety of hemoglobin, myoglobin, and various heme-containing enzymes. Bilirubin is water insoluble and potentially toxic. However, it is normally rendered harmless by albumin binding, rapid extraction by hepatocytes, enzyme-catalyzed conjugation with glucuronic acid, and active excretion into the bile. Hyperbilirubinemia is an important clinical marker of liver dysfunction. Patients with very high plasma concentrations of unconjugated hyperbilirubin are at risk for bilirubin-induced brain damage (kernicterus), which is seen in severe neonatal jaundice and in inherited disorders associated with marked unconjugated hyperbilirubinemia.

Breakdown of hemoglobin accounts for approximately 80% of the daily bilirubin production (*1*). The remainder is derived mainly from rapidly turning over hepatic hemoproteins. Bilirubin production is initiated by microsomal heme oxygenase-mediated opening of the α-methene bridge of the tetrapyrrole ring of heme, which requires oxygen and a reducing agent, such as NADPH. Inhibition of this rate limiting step by "dead-end" inhibitors, such as tin-protoporphyrin or tin-mesoporphyrin, reduces bilirubin production and lowers serum bilirubin levels in neonates (*2,3*). Opening of the heme tetrapyrrole generates a linear tetrapyrrole, biliverdin, and releases one mole each of CO and iron. The CO production, which can be measured by breath analysis, provides an estimate of bilirubin production and heme degradation. At a steady state, heme breakdown equals heme synthesis. Heme oxygenase and the antioxidant products of heme breakdown, biliverdin and bilirubin, may provide some protection against reactive oxygen species (*4,5*). The CO, the other product of heme oxygenase, may play a role in regulating the vascular tone in the liver and the heart, during stressful conditions. In most mammals, biliverdin is reduced to bilirubin by the action of biliverdin reductases.

Bilirubin IXα is water insoluble because of internal hydrogen bonding, which engages all its polar groups and protects the central methane bridge joining the two dipyrrolic halves. Disruption of the hydrogen bonds by enzyme-catalyzed glucuronidation converts bilirubin into a more polar and less bioactive conjugates and reacts with diazo reagents without any accelerator ("direct" reacting bilirubin), whereas unconjugated bilirubin reacts only when the hydrogen bonds are disrupted by a chemical accelerator ("total" bilirubin).

In the plasma, binding to albumin keeps unconjugated bilirubin in solution and prevents its toxicity by limiting the diffusion into tissues. In the hepatic sinusoids, bilirubin dissociates from albumin and is internalized by hepatocytes via facilitated diffusion. In the hepatocyte cytosol, bilirubin is bound to glutathione S-transferases (GSTs). Conjugation of bilirubin with glucuronic acid in the endoplasmic reticulum is catalyzed by uridinediphosphoglucuronate glucuronosyltransferase type 1 (UGT1A1), which transfers glucuronic acid from UDP-glucuronate to bilirubin. Bilirubin diglucuronide and monoglucuronide formed by this reaction are excreted into the bile canaliculi mainly by an energy-consuming process mediated by MRP2 (also termed ABCC2). In the intestine, bilirubin is degraded by intestinal bacteria into urobilinogen, most of which is reabsorbed and excreted in bile, while a small fraction is excreted in urine (*6*).

Increased bilirubin production caused by hemolytic disorders or ineffective erythropoiesis can lead to mild unconjugated hyperbilirubinemia (<5 mg/dl). On the other hand, inherited abnormalities of UGT1A1 can lead to complete or partial loss of bilirubin conjugation, leading to Crigler–Najjar syndrome type 1 (CN1) or Crigler–Najjar syndrome type 2 (CN2), respectively. In CN1, bilirubin concentrations of 20–40 mg/dl can lead to bilirubin encephalopathy and death. In CN2, the serum bilirubin level ranges from 5 to 20 mg/dl and is reduced by at least 25% after 7 days of phenobarbital therapy. CN2 is rarely associated with bilirubin encephalopathy. CN1 and CN2 are rare, autosomal recessive disorders, caused by genetic lesions within one of the five exons that constitute the coding region of UGT1A1 or their flanking intronic sequences (*7,8*). Gilbert syndrome is a common inherited disorder, in which a promoter polymorphism of UGT1A1 reduces the expression of the catalytically normal enzyme to ~30% of normal (*9*). The mild and usually intermittent hyperbilirubinemia in Gilbert syndrome, which is exacerbated on fasting, is innocuous by itself. However, subjects with Gilbert syndrome may have increased susceptibility to several drugs, such as irinotecan and acetaminophen.

Both conjugated and unconjugated bilirubin accumulate in plasma in acquired inflammatory or cholestatic diseases of the liver, as well as in several inherited disorders such as Dubin–Johnson syndrome (DJS), Rotor syndrome, progressive familial intrahepatic cholestasis syndromes, Alagille syndrome, and villin disease DJS is a benign condition, caused by inherited MRP2/ABCC2 deficiency. In DJS, conjugated bilirubin accumulates in plasma, but the total bile salt concentrations remain normal. The liver is black because of the accumulation of a dark brown pigment in hepatocytes. Total urinary coprophophyrin excretion is normal in DJS, but the ratio of isoform I to isoform III is reversed in favor of isoform I (*10*). Rotor syndrome is an autosomal recessive disorder, unrelated to DJS. In Rotor syndrome, hepatic bilirubin storage is abnormal and there is no pigment accumulation in hepatocytes. Urinary porphyrin excretion is increased and the ratio of isoforms I–III is approximately 1:1, which is similar to that found in many acquired liver diseases.

There are three major forms of progressive familial intrahepatic cholestasis (PFIC I, II and III) that are caused by genetic lesions of ATP8B1, ABCB11 (bile salt export pump, BSEP), and ABCB4 (multidrug resistance protein 3, MDR3), respectively (*11–13*). Bile salt concentration in canalicular bile is reduced in PFIC I and PFIC II, whereas in PFIC III, there is a marked reduction of biliary phospholipids excretion. In PFIC I and II, there is severe cholestasis, but serum gamma-glutamyl transpeptidase (GGT) levels are normal, whereas the GGT level is increased in PFIC III. Interestingly, some genetic lesions of ATP8B1 and ABCB11 are associated with two types of benign recurrent intrahepatic cholestasis (BRIC).

Two other genetic lesions, Alagille syndrome and abnormalities of Villin gene expression are associated with structural abnormalities of the biliary system and cholestasis. In Alagille syndrome, caused by genetic lesions of JAG-1, there is a paucity of interlobular bile ducts and associated cardiac anomalies, vertebral and facial abnormalities (*14*).

Inherited abnormality of the expression of villin, which is involved in organization of actin fibers, causes a biliary atresia-like condition (*15*) with serum conjugated and unconjugated hyperbilirubinemia and elevated serum GGT levels.

REFERENCES

1. London, I. M.; West, R.; Shemin, D.; Rittenberg, D. On the Origin of Bile Pigment in Normal Man. *J. Biol. Chem.* **1950**, *184*, 351.
2. Kappas, A.; Drummond, G. S. Direct Comparison of Tin-Mesoporphyrin, an Inhibitor of Bilirubin Production, and Phototherapy in Controlling Hyperbilirubinemia in Term and Near-Term Newborns. *Pediatrics* **1995**, *95*, 468.
3. Valaes, T.; Petmezaki, S.; Henschke, C.; Drummond, G. S.; Kappas, A. Control of Jaundice in Preterm Newborns by an Inhibitor of Bilirubin Production: Studies with Tin-Mesoporphyrin. *Pediatrics* **1994**, *93*, 1.
4. Elbirt, K. K.; Bonkovsky, H. L. Heme Oxygenase: Recent Advances in Understanding its Regulation and Role. *Proc. Assoc. Am. Physicians* **1999**, *111*, 438.
5. Hayashi, S.; Takamiya, R.; Yamaguchi, T.; Matsumoto, K.; Tojo, S. J.; Tamatani, T.; Kitajima, M.; Makino, N.; Ishimura, Y.; Suematsu, M. Induction of Heme Oxygenase-1 Suppresses Venular Leukocyte Adhesion Elicited by Oxidative Stress: Role of Bilirubin Generated by the Enzyme. *Circ. Res.* **1999**, *85*, 663.
6. Stoll, M. S.; Lim, C. D.; Gray, C. H. Chemical Variants of the Uroblins. In *Bile Pigments, Chemistry and Physiology*; Berk, P. D.; Berlin, N. I., Eds.; US Government Printing Office: Washington, DC, 1977; pp 483.
7. Kadakol, A.; Ghosh, S. S.; Sappal, B. S.; Sharma, G.; Roy-Chowdhury, J.; Roy-Chowdhury, N. Genetic Lesions of Bilirubin Uridinediphosphoglucuronate Glucuronosyl-Transferase Causing Crigler-Najjar and Gilbert's Syndromes: Correlation of Genotype to Phenotype. *Hum. Mutat.* **2000**, *16*, 297.
8. Gantla, S.; Bakker, C. T. M.; Deocharan, B.; Thummala, N. R.; Zweiner, J.; Sinaasappel, M.; Roy-Chowdhury, J.; Bosma, P. J.; Roy-Chowdhury, N. Splice Site Mutations: A Novel Genetic Mechanism of Crigler-Najjar Syndrome Type 1. *Am. J. Hum. Genet.* **1998**, *62*, 585.
9. Bosma, P. J.; Roy-Chowdhury, J.; Bakker, C.; Gantla, S.; DeBoer, A.; Oostra, B. A.; Lindhout, D.; Tytgat, G. N. J.; Jansen, P. L. M.; Oude Elferink, R. P. J.; Roy Chowdhury, N. A Sequence Abnormality in the Promoter Region Results in Reduced Expression of Bilirubin-UDP-Glucuronosyltransferase 1 in Gilbert Syndrome. *N. Engl. J. Med.* **1995**, *333*, 1171.
10. Kondo, T.; Kuchiba, K.; Shimizu, Y. Coproporphyrin Isomers in Dubin-Johnson Syndrome. *Gastroenterology* **1976**, *70*, 1117.
11. Eppens, E. F.; van Mil, S. W.; de Vree, J. M., et al. FIC1, the Protein Affected in Two Forms of Hereditary Cholestasis, is Localized in the Cholangiocyte and the Canalicular Membrane of the Hepatocyte. *J. Hepatol.* **2001**, *35*, 436.
12. Wang, L.; Soroka, C. J.; Boyer, J. L. The Role of Bile Salt Export Pump Mutations in Progressive Familial Intrahepatic Cholestasis Type II. *J. Clin. Invest.* **2002**, *110*, 965.
13. Jacquemin, E.; De Vree, J. M.; Cresteil, D., et al. The Wide Spectrum of Multidrug Resistance 3 Deficiency: From Neonatal Cholestasis to Cirrhosis of Adulthood. *Gastroenterology* **2001**, *120*, 1448.
14. Oda, T.; Elkahloun, A. G.; Meltzer, P. S.; Chandrasekharappa, S. C. Identification and Cloning of the Human Homolog (JAG1) of the Rat Jagged1 Gene from the Alagille Syndrome Critical Region at 20p12. *Genomics* **1997**, *43*, 376.
15. Phillips, M. J.; Azuma, T.; Meredith, S. L., et al. Abnormalities in Villin Gene Expression and Canalicular Microvillus Structure in Progressive Cholestatic Liver Disease of Childhood. *Lancet* **2003**, *362*, 1112.

CHAPTER

69

Cancer of the Colon and Gastrointestinal Tract

C Richard Boland

Department of Medicine, Division of Gastroenterology, GI Cancer Research Lab, Baylor University Medical Center, Dallas, TX, USA

Barbara Jung

Department of Medicine, Northwestern University, Feinberg School of Medicine, Division of Gastroenterology, Robert H. Lurie Comprehensive Cancer Center, Chicago, IL, USA

John M Carethers

Department of Internal Medicine, University of Michigan, Ann Arbor, MI, USA

69.1 FAMILIAL GASTROINTESTINAL CANCER SYNDROMES

Lynch syndrome is an autosomal dominant disease that predisposes to cancers of the colon and other organs, and is currently defined on the basis of documenting a germline mutation in a DNA mismatch repair (*MMR*) gene. The possible diagnosis of Lynch syndrome is supported when the family history meets the clinical Amsterdam criteria, which have been modified over the years to accommodate the fact that colorectal cancer is just one of the manifestations of the disease. Lynch syndrome is caused by a germline mutation in one of several DNA *MMR* genes, and its hallmark is the presence of microsatellite instability (MSI) in the tumor. In this disease, the manifestations are determined, to some degree, by which of the four key *MMR* genes is responsible: *hMSH2*, *hMLH1*, *hMSH6* and *hPMS2*.

The population prevalence of Lynch syndrome is approximately 1/300 in the population, making it possibly the most common inherited predisposition to cancer. Tumors develop as a result of local failure of the DNA MMR system, which monitors replication mismatches that occur from misincorporation of bases by DNA polymerase during DNA replication. The MMR system is complex, but loss of either of two "major" MMR proteins (hMSH2 or hMLH1) results in a total loss of DNA MMR activity. These two proteins are necessary for stabilization of the repair complex through heterodimerization with the "minor" DNA MMR proteins, hMSH6, hMSH3, and hPMS2.

Traditionally, the diagnosis of Lynch syndrome is made on the basis of family history, with an emphasis on early onset of proximal colon cancers clustering in a family. The Amsterdam criteria were developed to identify involved kindreds and required (1) three or more family members with histologically verified colorectal cancer, one of whom is a first-degree relative of the other two; (2) colorectal cancer involving at least two successive generations; (3) one or more colorectal cancer cases diagnosed before age 50; and (4) familial adenomatous polyposis (FAP) is excluded. The original Amsterdam criteria were too restrictive as they failed to acknowledge the contribution of endometrial and other extracolonic cancers, and made it relatively unlikely that a small family would fit the criteria. The modified Amsterdam criteria and the revised Bethesda criteria have been developed to improve our ability to diagnose this disease, and these acknowledge the presence of tumors in non-intestinal organs in this disease (most prominently endometrial cancer) and the use of molecular diagnostics on tumor tissue.

FAP is an autosomal dominant disease in which affected individuals develop a large number of adenomatous polyps, predominantly in the colon. Adenomatous polyps may also occur in the small intestine and, rarely, the stomach. Gastric polyps are common in FAP, but most of these are the result of fundic gland hyperplasia, and gastric cancer occurs in only ~0.5% of FAP patients. Colorectal cancer is nearly an inevitable consequence of FAP, as these benign neoplasms slowly grow and are subject to the same possible fates as sporadic adenomas. The polyps in FAP are typical adenomas, which individually have a relatively low likelihood of malignant progression. It is the very large number of them, and their early onset, that overwhelm the low risk for cancer in each individual polyp.

Some patients develop multiple colonic adenomas, but do not have a vertical family history of FAP; although there may be multiple polyps in more than one sibling, polyposis is not found in the parents. A systematic investigation of sibling clusters with multiple polyps in a British registry led to the discovery of an autosomal recessive genetic form of polyposis caused by germline mutations in the *MUTYH* gene, now called *MUTYH*-associated polyposis.

69.1.1 GI Hamartomatosis Syndromes

There are several recognizable familial syndromes that are associated with hamartomatous polyps of the GI tract and an increased risk for GI cancers, including Peutz–Jeghers syndrome (*STK11/LKB1* gene), Juvenile Polyposis syndrome (associated with *SMAD4*, *BMRPR1A*, and *ENG*), Cowden's syndrome and Bannayan–Riley–Ruvalcaba syndrome—these latter two jointly known as the PTEN Hamartoma Tumor Syndrome (associated with *PTEN*, and also *KILLIN*).

69.1.2 Hereditary Diffuse Gastric Cancer

There is a rare form of hereditary gastric cancer that can be distinguished by unique pathological features, termed hereditary diffuse gastric cancer (HDGC). This is caused by germline mutations in the *E-CADHERIN* (*CDH1*) gene, which predispose to early onset, diffuse type gastric cancers which may include a linitis plastica phenotype. Women with this syndrome are also at increased risk for lobular breast cancers.

69.1.3 Familial Pancreatic Cancer

There is an excess of pancreatic cancer in a variety of genetic diseases that predispose to other types of cancers; in many of these, pancreatic cancer may not be a highly penetrant feature. It has been projected that as many as 5% of pancreatic cancer may have a genetic basis and a large number of genes that might be involved. Thus, one must consider the possibilities of germline mutations in the familial breast cancer genes *BRCA1* and *BRCA2*, in familial atypical multiple mole/melanoma (FAMMM) syndrome with germline mutations in *p16* or *CDKN2a*, in Peutz–Jeghers syndrome families (*STK11* and *PRSS1*), in Lynch syndrome (both *hMSH2* and *hMLH1*), and in FAP. In the absence of a syndromic cancer syndrome, the presence of ≥3 first-degree relatives with pancreatic cancer is associated with a 40% lifetime risk for that disease, suggesting that there are other forms of this disease yet to be discovered.

69.1.4 Other GI Cancers

Gastrointestinal stromal tumors (GISTs) may occasionally be familial. In one large kindred, the risk for GISTs was linked to a mutation in the *KIT* gene. In another kindred with a germline mutation in the *PDGFRA* gene, three sisters had intestinal neurofibromatosis and developed five GIST tumors.

Familial gastrointestinal carcinoid tumors have been reported, are rare, and have yet to be linked convincingly to specific genetic loci, except for multiple endocrine neoplasia (MEN1) kindreds. MEN1 families have germline mutations in the *MENIN* gene, and some have increases in pancreatic islet cell tumors as a feature of the disease. Pancreatic islet cell tumors may also occur in the context of von Hippel-Lindau germline mutations.

TABLE 69-1 Clinical Features of Lynch Syndrome
Autosomal dominant inheritance
Increased risk for colorectal cancer (70% and 40% penetrance for men and women respectively)
Colon cancers have increased frequency of unique pathological features: 60–70% occur in the proximal colon, early age of onset (average age 40–45 years), mucinous or signet ring pathological appearance, Crohns-like lymphocytic infiltration of the tumor, poorly differentiated appearance
Multiple colonic malignancies (both simultaneous and metachronous)
Increased incidence of endometrial cancer (40–60%)
Increased risk for tumors of:
Stomach (observed/expected [o/e] = 4.1, $P < 0.001$)
Small intestine (o/e = 25.0, $P < 0.001$)
Ovary (o/e = 3.5, $P < 0.001$)
Kidney (o/e = 3.2, $P < 0.01$)
Ureter (o/e = 22.0, $P < 0.001$)
Brain tumors: glioblastoma multiforme
Increased risk for pancreatic and prostate cancers
No increase in the frequency of lung, breast, or prostate cancers

O/E = observed/expected frequency of the tumor.

TABLE 69-2 Clinical Features of Lynch Syndrome: By Germline Mutation

Germline Mutation	Unique Genetic and Clinical Features
hMSH2	Classic features of Lynch syndrome; common locus Nearly all colorectal cancers are MSI-H IHC shows absence of staining for hMSH2 and hMSH6 proteins Many of the germline mutations are large deletions (missed by DNA sequencing) May be associated with Muir–Torre syndrome (with skin lesions)
hMLH1	Classic features of Lynch syndrome; common locus Most colorectal cancers are MSI-H, but some mutations do not show this (i.e. D132H) IHC shows absence of staining for hMLH1 and hPMS2 proteins Some missense mutations in hMLH1 do not lead to loss of hMLH1 immunostaining Some mutations in hMLH1 will lead to isolated loss of hPMS2 immunostaining Must be distinguished from more common, acquired loss of hMLH1 (promoter methylation), which may be associated with mutations in BRAF
hMSH6	Less common locus for Lynch syndrome Produces an attenuated disease with delayed onset of colorectal cancers (average age 56 years) Penetrance same as with hMSH2 and hMLH1, but average age 10–20 years later Increased penetrance for endometrial cancer (71% by age 70) Tumors may be MSI-H (86%), but occasionally MSI-L (14%) MSI testing less sensitive in endometrial cancers IHC shows isolated loss of hMSH6 staining in 90%; hMSH2 staining intact Some mutations retain immunoreactivity, but lose function Families often fail to meet Amsterdam criteria because of later onset of cancers
hPMS2	Less common locus for Lynch syndrome Attenuated disease with later onset of cancer (50 years old) and reduced penetrance Most colorectal cancers are MSI-H IHC shows isolated loss of hPMS2 protein, with intact hMLH1, hMSH2, and hMSH6 Germline mutation detection problematic because of multiple (20) pseudogenes that mask mutations in the true gene

IHC, immunohistochemistry; MSI, microsatellite instability.

TABLE 69-3 Increased Suspicion for Lynch Syndrome: Clinical Criteria

A. Modified Amsterdam Criteria [for Lynch syndrome] (Vasen et al., 1999)
   (1) Three relatives with pathologically verified colorectal cancer, or other Lynch syndrome-associated cancer[a]
   (2) One patient is a first-degree relative of the other two
   (3) At least two successive generations are affected
   (4) FAP is excluded
   (5) One cancer has occurred <50 years

B. Revised Bethesda criteria [to prompt MSI testing in colorectal cancer] (Umar et al., 2004)
   (1) Colorectal cancer in a patient <50 years old
   (2) Multiple synchronous or metachronous Lynch syndrome-associated cancers[a]
   (3) Colorectal cancer with the "MSI-H" histology[b] in a patient <60 years old
   (4) Colorectal cancer in ≥1 first-degree relatives with a Lynch syndrome-related tumor,[a] in which one of the cancers is diagnosed in a person <50 years old
   (5) Colorectal cancer diagnosed in ≥2 first- or second-degree relatives with a Lynch syndrome-related tumor,[a] regardless of age

[a]Adenocarcinoma of the colon, rectum, endometrium, stomach, ovary, pancreas, small bowel, ureter or renal pelvis, and biliary tract; also, glioblastoma multiforme of the CNS or tumors of the Muir–Torre syndrome spectrum (sebaceous neoplasms and keratoacanthomas).

[b]The presence of tumor-infiltrating lymphocytes, Crohn-like lymphocytic reaction, mucinous/signet ring differentiation, or medullary growth pattern.

TABLE 69-4 Intestinal Polyposis Syndromes

Adenomatous Syndrome	Chromosomal Location(s)	Mutated Gene(s)	Inheritance Pattern
Familial adenomatous polyposis Gardner's variant Turcot's variant	5q21	*APC*	Autosomal dominant
MUTYH-associated polyposis	1p32-34	*MUTYH*	Autosomal recessive
Lynch syndrome Muir–Torre variant Turcot's variant	2p16, 3p21, 7p22, and 2p16	*hMSH2, hMLH1, hPMS2, hMSH6*	Autosomal dominant
Syndrome X	?	?	Autosomal dominant?

Hamartomatous Syndrome	Chromosomal Location	Mutated Gene	Frequency in Germline (%)
PTEN Hamartoma syndrome			
Cowden's disease	10q22-23	PTEN/MMAC1/TEP1	>80
Lhermitte–Duclos variant	10q22-23	KILLIN (methylation)	?
Bannayan–Riley–Ruvalcaba syndrome	10q22-23	PTEN/MMAC1/TEP1	~60
Juvenile polyposis syndrome	18q21.1	SMAD4	~20
(with HHT overlap)	10q22-23	BMPR1A/ALK3	~25
	9q34	ENG	?
Peutz–Jeghers syndrome	19p13.3	STK11/LKB1	70–90
Hereditary mixed polyposis syndrome	15q13-q14	? CRAC1	?
	15q21	? THBS1	?
	10q23	? BMPR1A/ALK3	?
Hyperplastic polyposis syndrome	?1p	?	?

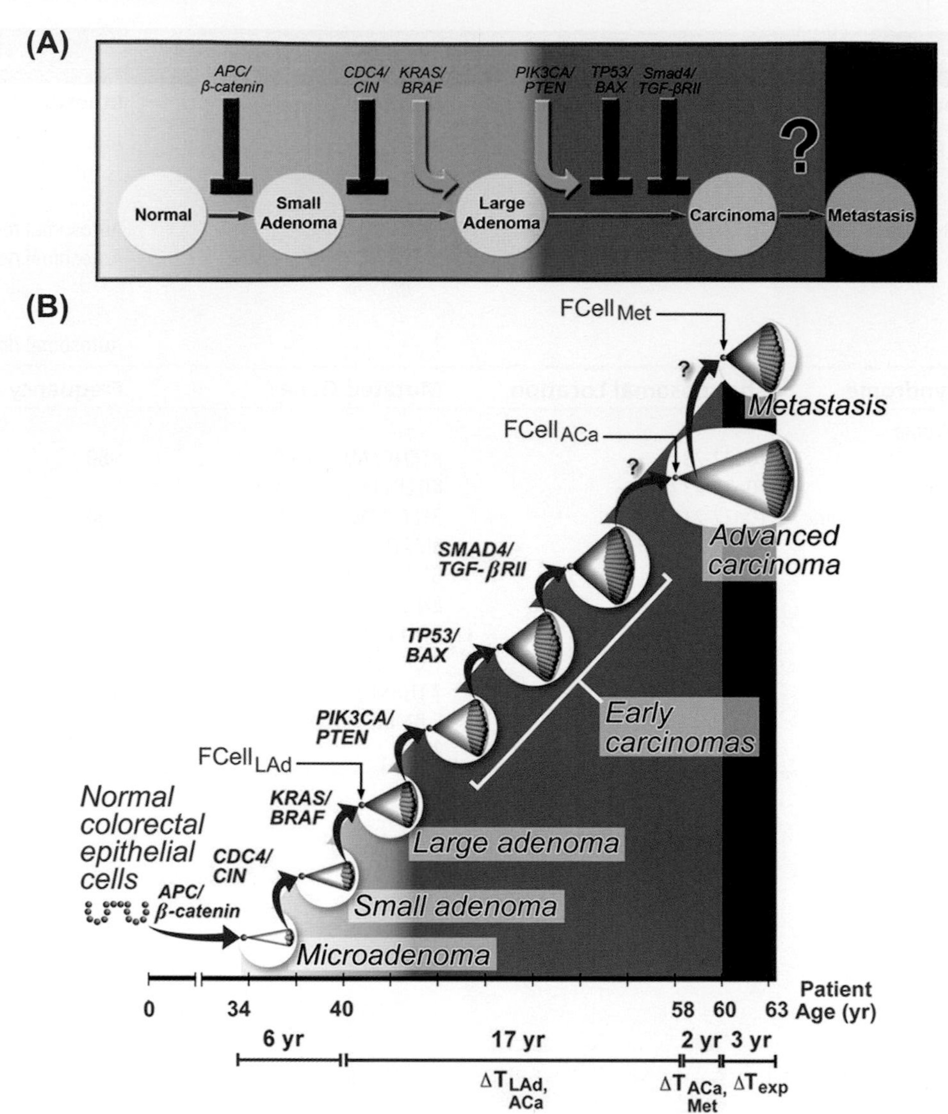

FIGURE 69-1 Sequential, multistep colorectal carcinogenesis. A. Fearon and Vogelstein proposed in 1990 that the sequential accumulation of genetic alterations was responsible for the morphological changes seen in tumor progression in the colon. Normal colonic tissue is represented on the left, and the sequential appearance of hyperproliferative epithelium, benign adenomatous polyps, and carcinoma are indicated with progression to the right. B. Colorectal cancers evolve by genetic alteration (mutation in this example), followed by clonal expansion, the generation of novel, diverse clones, and successive waves of expansion. However, the generation of a uniquely advantageous genetic alteration is a rare event, accounting for the long periods of latency that occur during the evolution of a lethal colorectal cancer. FCell indicates Founder Cells for advanced carcinoma or metastasis. *(Both A and B from Jones, S., et al. Comparative Lesion Sequencing Provides Insights into Tumor Evolution.* Proc. Natl. Acad. Sci. U.S.A. *2008*, 105, *4283–4288.*)

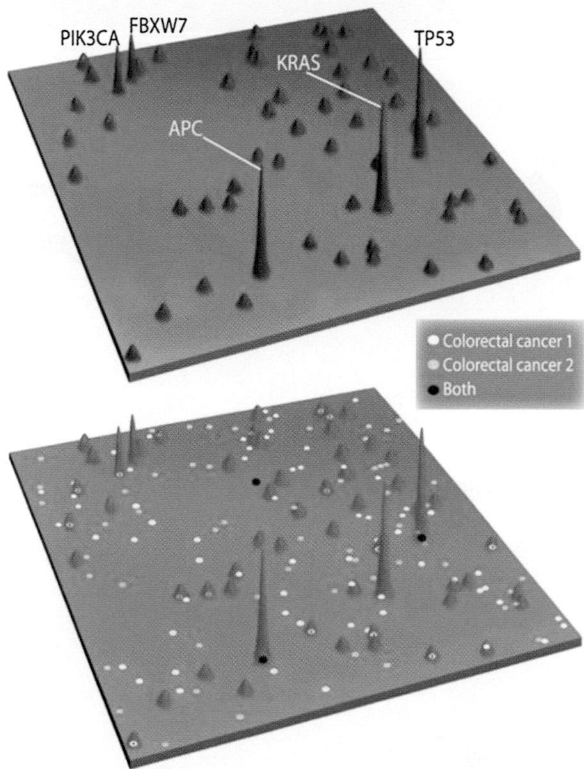

FIGURE 69-2 The genomic landscape of human colorectal cancer. Eleven colorectal cancers were subjected to total exomic sequencing, and the results plotted in two dimensions, representing all chromosomes, end to end, beginning with chromosome 1p in the left rear corner. Individual tumors had unique sets of mutations, as illustrated in the "landscape," and only a handful of mutations were common to many tumors (creating the "mountains" in the landscape) or fewer (creating "hills"). The heterogeneity of the mutational landscape helps to explain the variations in microscopic appearance, clinical behavior and therapeutic responses of individual colorectal cancers. *(From Wood, L. D., et al. The Genomic Landscapes of Human Breast and Colorectal Cancers.* Science *2007,* 318, *1108–1113.*)

(A) *Single mismatch*

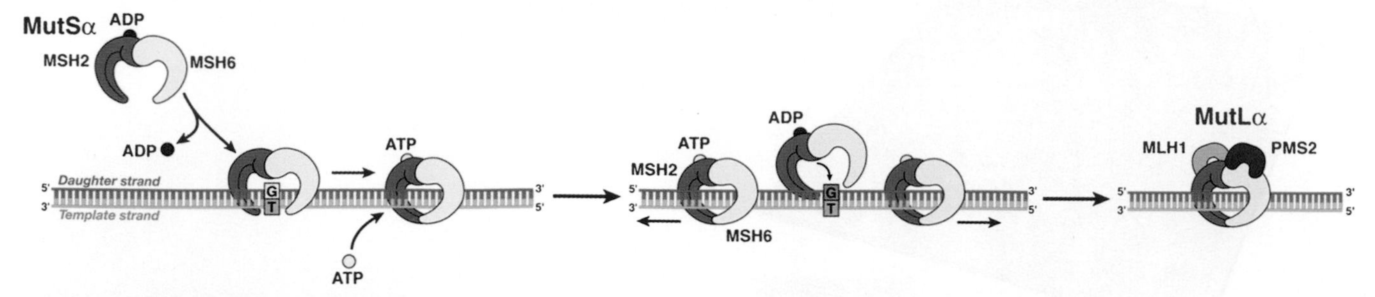

(B) *Exonuclease complex and resynthesis*

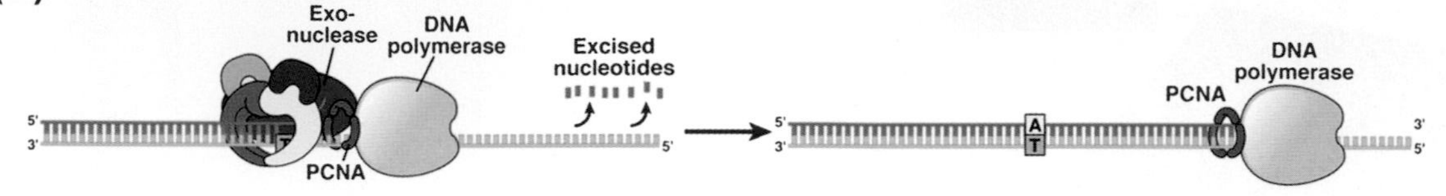

(C) *Insertion/deletion loop and variations in MutL complexes*

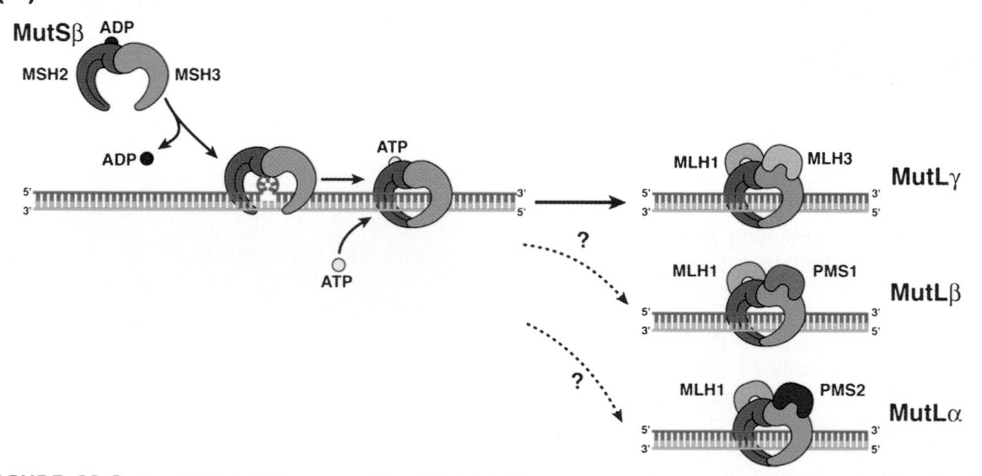

FIGURE 69-3 The DNA MMR system. The DNA MMR system requires one pair of heterodimerized proteins (either hMSH2 + hMSH6—or MutSα—as shown in the top panel or hMSH2 + hMSH3—called MutSβ, as shown in the bottom panel) to recognize a mismatch error in newly synthesized DNA. The mismatch triggers a conformational change in which the dimer forms a sliding clamp on the DNA, sliding away from the mispair toward the DNA polymerase complex. A second dimer [consisting of either hMLH1 + hPMS2 (MutLα, as shown on the top), hMLH1 + hMLH3 (MutLβ), or hMLH1 + hPMS1 (MutLγ, all three shown in the bottom panel)] which interact with PCNA and Exonuclease 1 to facilitate the excision of the mismatch, and ultimately, resynthesis to repair the error. *(From Boland, C. R.; Goel, A. Microsatellite Instability in Colorectal Cancer.* Gastroenterology *2010*, 138, *2073–2087.*)

FURTHER READING

1. Hitchins, M. P.; Wong, J. J.; Suthers, G.; Suter, C. M.; Martin, D. I.; Hawkins, N. J.; Ward, R. L. Inheritance of a Cancer-Associated MLH1 Germ-Line Epimutation. *N. Engl. J. Med.* 2007, *356*, 697–705.
2. Jones, S.; Chen, W. D.; Parmigiani, G.; Diehl, F.; Beerenwinkel, N.; Antal, T.; Traulsen, A.; Nowak, M. A.; Siegel, C.; Velculescu, V. E., et al. Comparative Lesion Sequencing Provides Insights into Tumor Evolution. *Proc. Natl. Acad. Sci. U.S.A.* 2008, *105*, 4283–4288.
3. Leggett, B.; Whitehall, V. Role of the Serrated Pathway in Colorectal Cancer Pathogenesis. *Gastroenterology* 2010, *138*, 2088–2100.
4. Boland, C. R.; Goel, A. Microsatellite Instability in Colorectal Cancer. *Gastroenterology* 2010, *138*, 2073–2087.
5. Greenman, C.; Stephens, P.; Smith, R.; Dalgliesh, G. L.; Hunter, C.; Bignell, G.; Davies, H.; Teague, J.; Butler, A.; Stevens, C., et al. Patterns of Somatic Mutation in Human Cancer Genomes. *Nature* 2007, *446*, 153–158.
6. Wood, L. D.; Parsons, D. W.; Jones, S.; Lin, J.; Sjoblom, T.; Leary, R. J.; Shen, D.; Boca, S. M.; Barber, T.; Ptak, J., et al. The Genomic Landscapes of Human Breast and Colorectal Cancers. *Science* 2007, *318*, 1108–1113.
7. Nielsen, M.; Morreau, H.; Vasen, H. F.; Hes, F. J. MUTYH-Associated Polyposis (MAP). *Crit. Rev. Oncol. Hematol.* 2010.
8. Boland, C. R.; Shike, M. Report from the Jerusalem Workshop on Lynch Syndrome-Hereditary Nonpolyposis Colorectal Cancer. *Gastroenterology* 2010, *138*, 2197.
9. Tomlinson, I. P.; Dunlop, M.; Campbell, H.; Zanke, B.; Gallinger, S.; Hudson, T.; Koessler, T.; Pharoah, P. D.; Niittymaki, I.; Tuupanen, S., et al. COGENT (COlorectal cancer GENeTics): An International Consortium to Study the Role of Polymorphic Variation on the Risk of Colorectal Cancer. *Br. J. Cancer* 2010, *102*, 447–454.
10. Wijnen, J. T.; Brohet, R. M.; van, E. R.; Jagmohan-Changur, S.; Middeldorp, A.; Tops, C. M.; van, P. M.; Ausems, M. G.; Gomez, G. E.; Hes, F. J., et al. Chromosome 8q23.3 and 11q23.1 Variants Modify Colorectal Cancer Risk in Lynch Syndrome. *Gastroenterology* 2009, *136*, 131–137.

11. Edelstein, D. L.; Axilbund, J.; Baxter, M.; Hylind, L. M.; Romans, K.; Griffin, C. A.; Cruz-Correa, M.; Giardiello, F. M. Rapid Development of Colorectal Neoplasia in Patients with Lynch Syndrome. *Clin. Gastroenterol. Hepatol.* **2011,** *9,* 340–343.
12. Ligtenberg, M. J.; Kuiper, R. P.; Chan, T. L.; Goossens, M.; Hebeda, K. M.; Voorendt, M.; Lee, T. Y.; Bodmer, D.; Hoenselaar, E.; Hendriks-Cornelissen, S. J., et al. Heritable Somatic Methylation and Inactivation of MSH2 in Families with Lynch Syndrome Due to Deletion of the 3' Exons of TACSTD1. *Nat. Genet.* **2009,** *41,* 112–117.
13. Lindor, N. M.; Petersen, G. M.; Hadley, D. W.; Kinney, A. Y.; Miesfeldt, S.; Lu, K. H.; Lynch, P.; Burke, W.; Press, N. Recommendations for the Care of Individuals with an Inherited Predisposition to Lynch syndrome: A Systematic Review. *JAMA* **2006,** *296,* 1507–1517.
14. Sinicrope, F. A.; Foster, N. R.; Thibodeau, S. N.; Marsoni, S.; Monges, G.; Labianca, R.; Yothers, G.; Allegra, C.; Moore, M. J.; Gallinger, S., et al. DNA Mismatch Repair Status and Colon Cancer Recurrence and Survival in Clinical Trials of 5-Fluorouracil-Based Adjuvant Therapy. *J. Natl. Cancer Inst.* **2011.**
15. Heald, B.; Mester, J.; Rybicki, L.; Orloff, M. S.; Burke, C. A.; Eng, C. Frequent Gastrointestinal Polyps and Colorectal Adenocarcinomas in a Prospective Series of PTEN Mutation Carriers. *Gastroenterology* **2010,** *139,* 1927–1933.
16. Kaurah, P.; MacMillan, A.; Boyd, N.; Senz, J.; De, L. A.; Chun, N.; Suriano, G.; Zaor, S.; Van, M. L.; Gilpin, C., et al. Founder and Recurrent CDH1 Mutations in Families with Hereditary Diffuse Gastric Cancer. *JAMA* **2007,** *297,* 2360–2372.

Hematologic Disorders

Hemoglobinopathies and Thalassemias, 283
Other Hereditary Red Blood Cell Disorders, 290
Hemophilias and Other Disorders of Hemostasis, 292
Rhesus and Other Fetomaternal Incompatibilities, 298
Leukemias, Lymphomas, and Other Related Disorders, 300
Immunologic Disorders Autoimmunity: Genetics and Immunologic Mechanisms, 302
Systemic Lupus Erythematosus, 304
Rheumatoid Disease and Other Inflammatory Arthropathies, 306
Amyloidosis and Other Protein Deposition Diseases, 309
Immunodeficiency Disorders, 312
Inherited Complement Deficiencies, 315

CHAPTER

70

Hemoglobinopathies and Thalassemias

John Old

National Haemoglobinopathy Reference Laboratory, Molecular Haematology, John Radcliffe Hospital, Oxford, UK

The hemoglobinopathies and the thalassemia syndromes are a diverse group of inherited disorders of hemoglobin synthesis that result from the qualitative defects (hemoglobinopathies) or quantitative defects (thalassemia syndromes) in globin synthesis. As a group, they are the most common and clinically important single-gene disorders in the world and pose a serious health problem in many countries. The only definitive cure for the hemoglobinopathies is bone marrow transplantation, and thus only supportive management is available for the treatment of the vast majority of affected patients. The methods of clinical management have improved considerably during the past few years and the life expectancy of affected individuals has been significantly increased with patients now surviving from the third to fifth decade. However, the treatment required is still very expensive, especially for many developing countries with a high incidence of β-thalassemia. Many countries now apply a program of antenatal screening of the population for carriers, identifying couples at risk, and providing prenatal diagnosis by chorionic villus DNA analysis. Research into using cell-free DNA from maternal serum for a noninvasive approach to prenatal diagnosis is gathering momentum with the application of the new technologies of digital polymerase chain reaction and next generation sequencing. Preimplantation diagnosis is now routinely performed in a growing number of centers worldwide for couples who do not wish to terminate a pregnancy, and has been used to allow the birth of a normal child so that an affected sibling can potentially be cured by stem cell transplantation.

More than 1000 abnormal hemoglobins have now been characterized by peptide sequencing, DNA sequencing, or mass spectrometry. The majority are due to point mutations resulting in a single amino acid substitution with nearly twice as many β-chain variants identified compared to α-chain variants. The amino acid substitution may affect the stability of the hemoglobin molecule causing premature red cell destruction; it may raise or lower oxygen binding affinity and cause polycythemia, or it may permit the oxidation of the heme molecule causing cyanosis. Sickle cell disease is one of the most common single gene disorders in the world, resulting from the polymerization of Hb S when deoxygenated and leading to erythrocyte rigidity and vaso-occlusion. Sickle cell disease is a genetically complex disorder with nine variant hemoglobins that sickle and with many different compound heterozygous genotypes, including Hb SC, SD-Punjab, SO-Arab, and Sβ-thalassemia. Hydroxyurea has proved to be the most effective antisickling agent and it now has an established role in ameliorating the disease and improving life expectancy by the induction of Hb F expression in adults. Newborn screening and close follow-up of homozygotes with prophylactic penicillin treatment has increased improved survival. Prevention by antenatal screening, genetic counseling of at-risk couples, and prenatal diagnosis by fetal DNA analysis has now been available in many countries for more than 30 years.

α-Thalassemia is inherited as an autosomal recessive disorder characterized by a microcytic hypochromic anemia. It is probably the most common monogenic gene disorder in the world and is especially frequent in Mediterranean countries, southeast Asia, Africa, Middle Eastern countries, and in the Indian subcontinent, and is now commonly found in North-European countries and Northern America because of demographic changes. The most common type is α^+-thalassemia, resulting from a deletion of one of the two α-genes on chromosome 16, or more rarely from a point mutation in one of the two α-genes. The more severe and clinically significant type, α^0-thalassemia, results from the deletion of both the α-genes. The clinical phenotype varies from almost asymptomatic to a lethal hemolytic anemia (Hb Bart's hydrops fetalis) in which no α-globin is synthesized. Carriers of α^+-thalassemia or α^0-thalassemia generally do not need treatment. Compound heterozygotes and some homozygotes have a moderate to severe form of α-thalassemia called Hb H disease. Hb H patients may require intermittent transfusion therapy especially during intercurrent illness. All affected individuals have a variable degree of anemia (low Hb), reduced mean corpuscular hemoglobin, and reduced mean corpuscular volume values, and a normal/slightly reduced level of Hb A_2. Molecular analysis is usually required to confirm the hematological observations (especially to distinguish homozygous α^+-thalassemia from α^0-thalassemia trait). The predominant features in Hb H disease are anemia with variable amounts of Hb H (0.8–40%). The type of mutation influences the clinical severity of Hb H disease, with a few nondeletion types interacting with a α^0-thalassemia deletion to cause Hb H hydrops fetalis syndrome. The distinguishing features of Hb Bart's hydrops fetalis syndrome are the presence of Hb Bart's and the total absence of Hb F. Genetic counseling is offered to couples at risk for severe Hb H disease or Hb Bart's hydrops fetalis syndrome. Most pregnancies in which the fetus is known to have the Hb Bart's hydrops fetalis syndrome are terminated due to the increased risk of both maternal and fetal morbidity, although a very small number survived following intrauterine transfusion therapy and subjected to the same treatment as for β-thalassemia major.

β-Thalassemias are a group of hereditary blood disorders characterized by anomalies in the synthesis of the β-chains of hemoglobin resulting in variable phenotypes ranging from severe anemia to clinically asymptomatic individuals. The group includes the δβ-thalassemias, the εγδβ-thalassemias and the hereditary persistence of fetal hemoglobin (HPFH) disorders. There are three main clinical phenotypes: thalassemia major, thalassemia intermedia and thalassemia trait (minor). Individuals with thalassemia major usually present within the first 2 years of life with severe anemia, requiring regular transfusions. Patients with thalassemia intermedia present later in life with moderate anemia and do not require regular transfusions. Thalassemia minor is clinically asymptomatic, but some subjects may have moderate anemia. β-Thalassemias are caused by more than 200 point mutations or, more rarely, deletions in the β-globin gene on chromosome 11, leading to reduced (β^+) or absent (β^0) synthesis of the β-chains of hemoglobin. The mutations may affect globin gene transcription, RNA processing or translation, RNA cleavage and polyadenylation, or result in a highly unstable globin chain. Frameshift and nonsense codon mutations have been observed in all three exons and RNA processing mutations have been found in both introns and the four splice junctions. β-Thalassemia is inherited as an autosomal recessive disorder, although a small number of dominant mutations have also been reported. Diagnosis of β-thalassemia is based on hematologic and molecular genetic testing. Differential diagnosis is usually straightforward, but may require molecular techniques for precise mutation identification. Prevention is by antenatal screening and the offer of prenatal diagnosis. Treatment of thalassemia major includes regular transfusions, iron chelation and management of secondary complications of iron overload. Two oral iron chelators, deferiprone and deferasirox, are now in use in many parts of the world in addition to the original chelator desferrioxamine, delivered by subcutaneous infusion. The induction of Hb F by pharmacological agents in thalassemia major patients has not proved very effective, although some thalassemia intermedia patients have shown a good response. Prognosis for individuals with β-thalassemia has improved substantially in the last 20 years following recent medical advances in transfusion, iron chelation and bone marrow transplantation therapy. However, cardiac disease remains the main cause of death in patients with iron overload. Bone marrow transplantation remains the only definitive cure currently available, and gene therapy using autologous bone marrow is a step closer following the correction of mouse β-thalassemia with a lentiviral vector and research is now focused on correcting of the molecular defect in hematopoietic stem cells or the use of homologous recombination instead of gene transfer.

TABLE 70-1 Human Hemoglobins

Hb	Stage of Development	Structure	Percent in Adults	Conditions in Which Increased
A	Adult	$\alpha_2\beta_2$	92	
A_{1c}		α_2(β-N-glucose)$_2$	5	Diabetes mellitus
A_2		$\alpha_2\delta_2$	2–3	β-thalassemia
H		β_4	0	Some α-thalassemias
F	Fetal	$\alpha_2\gamma_2$	<1	Newborn, δβ-, β-thalassemia, HPFH, and marrow stress
Bart's		γ_4	0	Some α-thalassemias
Gower I	Embryonic	$\zeta_2\varepsilon_2$	0	Early embryos (<8 weeks)
Gower II	Embryonic	$\alpha_2\varepsilon_2$	0	Early embryos (>8 weeks)
Portland	Embryonic	$\zeta_2\gamma_2$	0	(<8 weeks) and α°-thalassemia (hydrops fetalis)

TABLE 70-2 Molecular Basis of the Hb Variants

Mutation	Example Hb		Clinical Manifestation	Molecular Basis (Presumed)
Nucleotide Base Substitutions for				
One amino acid	Hb S	β6Glu→Val	Sickling	β:Cd 6 GAG→GTG
Two amino acids	Hb C-Harlem	β6Glu→Val + β73Asp→Asn	Sickling	β:Cd 6 GAG→GTG and β:Cd 73 GAT→AAT
Termination	Hb McKees Rocks	β145 Tyr→Termination	Increased oxygen affinity and polycythemia	β:Cd 145 TAT→TAA
Amino acid instead of termination	Hb Constant Spring	α2:142 Termination→Gln	Decreased synthesis (thalassemia-like)	α2:Cd 142 TAA→AAA
Nucleotide base deletions				
Single base deletion→frame shift	Hb Wayne	α2:139-146 Lys-Tyr-Arg→Asn + 7 residues	Normal	α2:Cd 139 (–A)
Triplet deletion →single amino acid	Hb Leiden	β6 or 7 Glu→0	Unstable	β:Cd 6 or 7 (–GAG)
Multiple codon	Hb Gun Hill	β91-95 Leu-His-Cys-Asp-Lys→0)	Unstable	β:Cd 91-95 (–15bp)
Crossover	Hb Lepore	δβ-fusion with segments of β and β lost	Decreased synthesis (thalassemia-like)	δβ:7.4kb deletion
Nucleotide base additions				
Two bases added→frame shift	Hb Cranston	β144 Tyr-His→Ser-Ile-Thr	Unstable	β:Cd 144/145 + CT
Multiple codon	Hb Grady	α118 (+Glu-Phe-Thr)	Normal	α2 or α1:Cd 118/119 (+9bp)

TABLE 70-3 The Number of Known Hemoglobin Variants

Type	Number
α-Chain variants (total)	378
β-Chain variants (total)	529
γ-Chain variants (total)	87
δ-Chain variants (total)	50
Variants with two amino acid replacements	19
Variants with hybrid chains	12
Variants with elongated chains	13
Variants with deletions, insertions, and deletions/insertions	27
α-Chain variants with same mutation on both the α1 and α2 gene	14
α-Chain variants with a different mutation on the α1 and α2 gene	2

TABLE 70-4 Clinical Manifestations of Hemoglobin Mutants

Type	Example	Clinical Manifestation
Sickling	Hb S	Sickling due to decreased solubility
Unstable	Hb Bristol	Anemia with Heinz body formation
Abnormal oxygen affinity		
Decreased	Hb Kansas	Mild anemia possible
Increased	Hb Chesapeake	Polycythemia due to decreased oxygen transport
M hemoglobin	Hb M-Boston	Cyanosis due to ferric hemoglobin
Decreased synthesis	Hb Lepore	Thalassemia

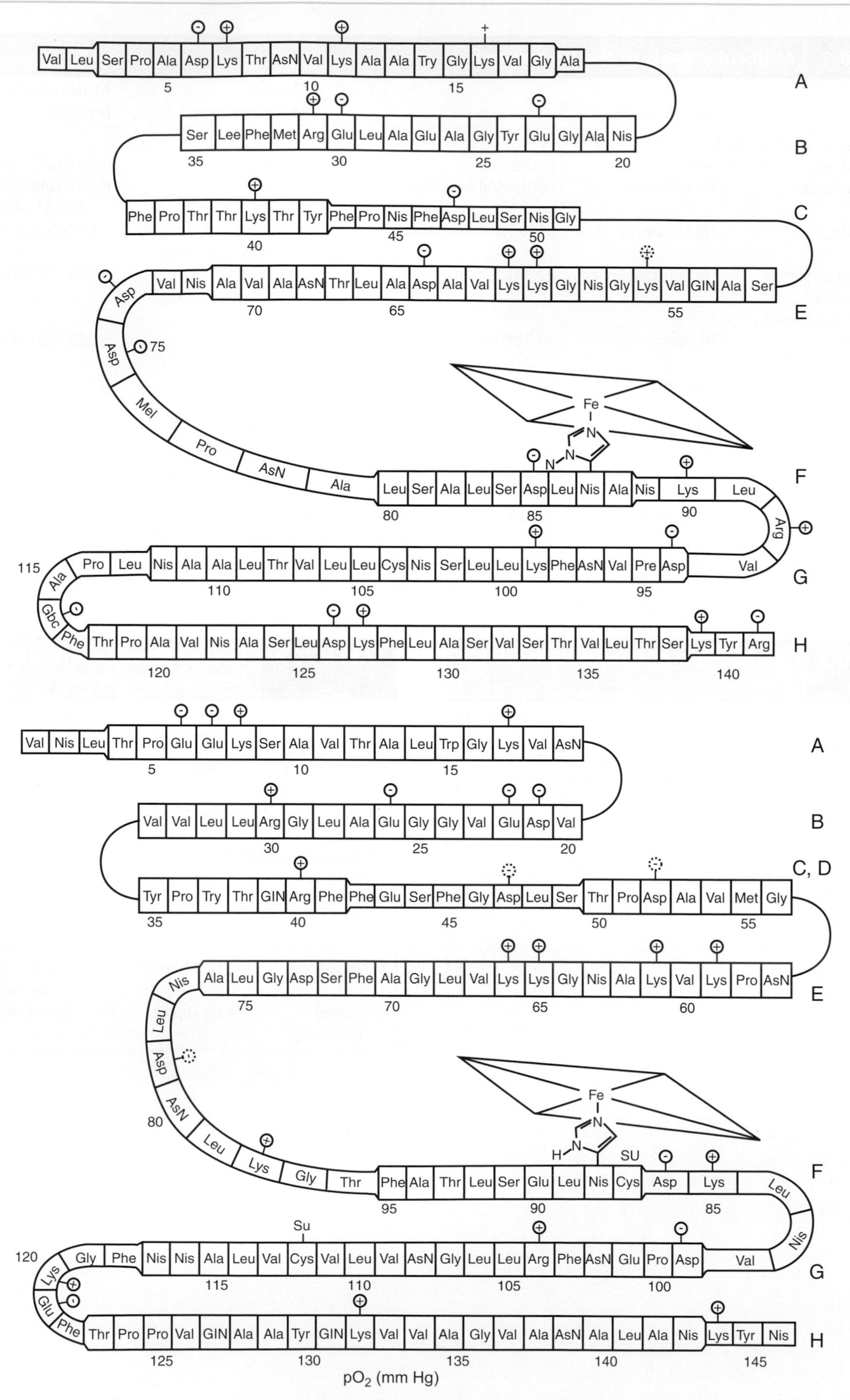

FIGURE 70-1 The primary and secondary structures of globin chains. (Top) α-chain. (Bottom) β-chain. Residues in squares are in α-helix configuration and nonhelical are in rectangles. *(From Murayama, M. Structure of Sickle Cell Hemoglobin and Molecular Mechanisms of the Sickling Phenomenom. In* Molecular Aspects of Sickle Cell Hemoglobin; *Nalbandian, R. M., Ed.; Charles C Thomas: Springfield, IL, 1971, with permission.)*

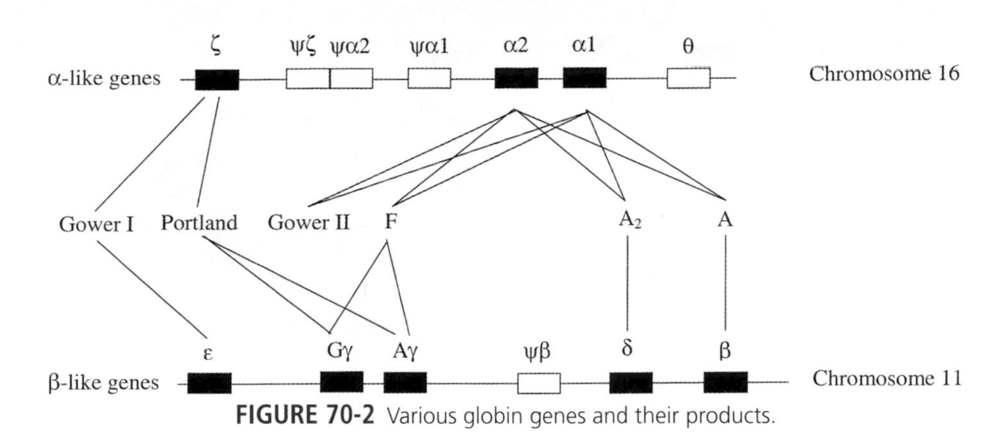

FIGURE 70-2 Various globin genes and their products.

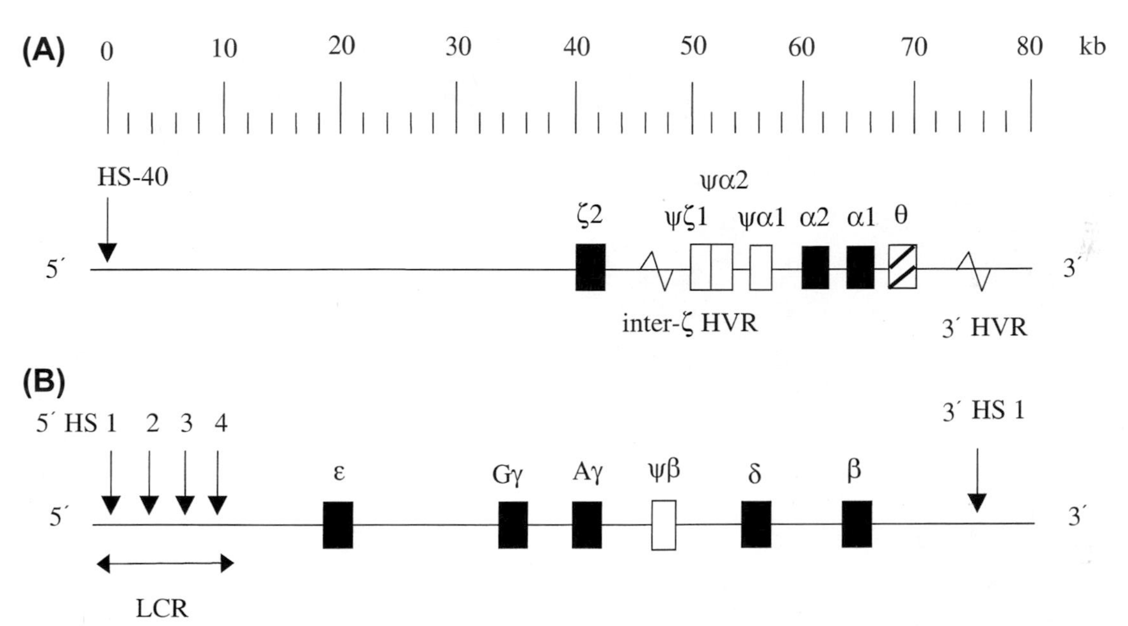

FIGURE 70-3 Globin genes complexes. A. β-gene complex on chromosome 11. B. α-gene complex on 16. Distances along the chromosome are measured in kilobases (kb) at top.

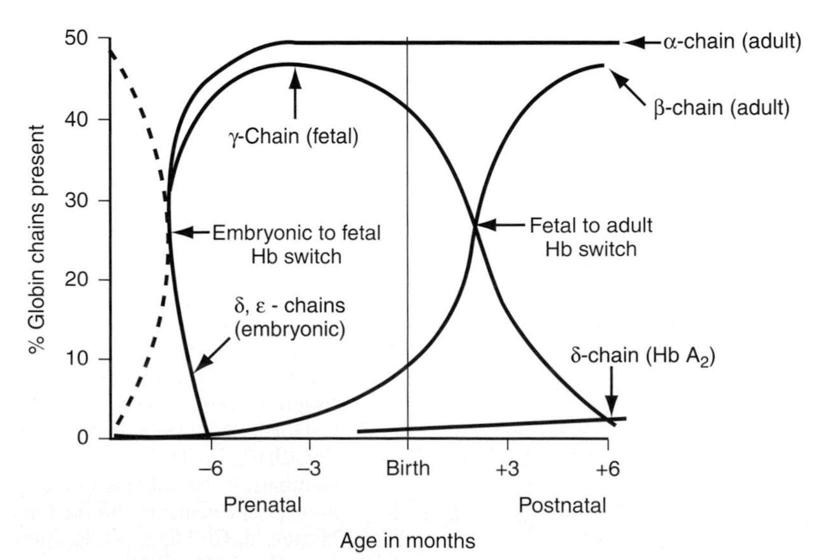

FIGURE 70-4 Qualitative and quantitative changes in globin chains during human development. Note that the percentage of β-chains accumulated in early fetal development is much less than the percentage of β-chains synthesized during fetal development. *(Modified from Bunn, H. F.; Forget, B. G.; Ranney, H. M. Hemoglobin Structure. In* Human Hemoglobins. *WB Saunders: Philadelphia, 1977; pp 4, with permission.)*

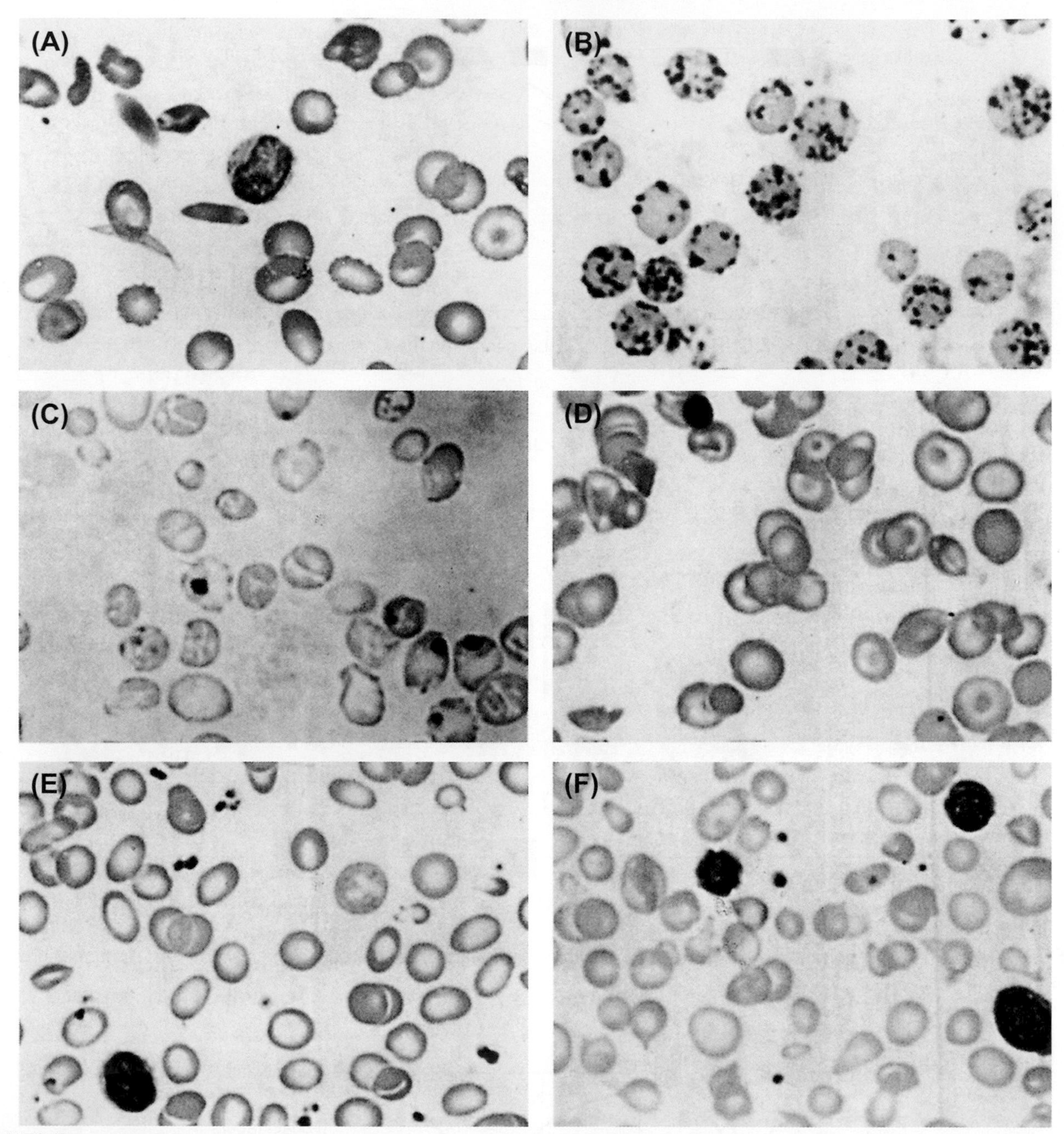

FIGURE 70-5 Peripheral blood smears from patients with various disorders of globin synthesis. A. Homozygous sickle cell anemia. B. Unstable Hb Zurich with Heinz bodies. C. Hb H disease. D. Sickle/β-thalassemia. E. β-thalassemia trait. F. Homozygous β-thalassemia. Figures B and C were prepared as follows: whole blood with EDTA was incubated at 41°C for 3–6 h, then a 1:1 mixture of blood and 0.5% rhodanile blue in 0.9% saline was made and immediately smeared. Hemoglobin precipitates formed secondary to heating are seen. *(Courtesy of Dr. William Zinkham.)*

FURTHER READING

1. Angelucci, E. Hematopoietic Stem Cell Transplantation in Thalassemia. *Hematology Am. Soc. Hematol. Educ. Program* **2010,** 456–462.
2. Arumugam, P.; Malik, P. Genetic Therapy for Beta-Thalassemia: From the Bench to the Bedside. *Hematology. Am. Soc. Hematol. Educ. Program* **2010,** 445–450.
3. Ataga, K. I. Novel Therapies in Sickle Cell Disease. *Hematol. Am. Soc. Hematol. Educ. Program* **2009,** 54–61.
4. Cappellini, M. D.; Musallam, K. M.; Taher, A. T. Insight onto the Pathophysiology and Clinical Complications of Thalassemia Intermedia. *Hemoglobin* **2009,** *33* (Suppl. 1), S145–S159.
5. Efremov, G. D. Dominantly Inherited Beta-Thalassemia. *Hemoglobin* **2007,** *31* (2), 193–207.
6. Fucharoen, S.; Viprakasit, V. Hb H Disease: Clinical Course and Disease Modifiers. *Hematology Am. Soc. Hematol. Educ. Program* **2009,** 26–34.
7. Galanello, G.; Origa, R. Beta-Thalassaemia. *Orphanet J. Rare Dis.* **2010,** *5* (11).
8. Gambari, R. Foetal Haemoglobin Inducers and Thalassaemia: Novel Achievements. *Blood Transfus.* **2010 Jan,** *8* (1), 5–7.
9. Harteveld, Cl; Higgs, D. R. Alpha-Thalassaemia. *Orphanet J. Rare Dis.* **2010,** *5* (13).
10. Inati, A.; Chabtini, L.; Mounayar, M.; Taher, A. Current Understanding in the Management of Sickle Cell Disease. *Hemoglobin* **2009,** *33* (Suppl 1), S107–S115.

11. Karimi, M. Hydroxyurea in the Management of Thalassemia Intermedia. *Hemoglobin* **2009**, *33* (Suppl 1), S177–S182.
12. Kutlar, F. Diagnostic Approach to Hemoglobinopathies. *Hemoglobin* **2007**, *31* (2), 243–250.
13. Petrou, M. Preimplantation Genetic Diagnosis. *Hemoglobin* **2009**, *33* (Suppl 1), S7–S13.
14. Porter, J. B. Optimizing Iron Chelation Strategies in Beta-Thalassaemia Major. *Blood Rev.* **2009**, *23* (Suppl 1), S3–S7.
15. Schwartz Cowan, R. *Heredity and Hope: The Case for Genetic Screening*; Harvard University Press: Cambridge, 2008.
16. Serjeant, G. R. One Hundred Years of Sickle Cell Disease. *Br. J. Haematol.* **2010**, *151* (5), 425–429.
17. Steinberg, M. H. Sickle Cell Anemia, the First Molecular Disease: Overview of Molecular Etiology, Pathophysiology, and Therapeutic Approaches. *Scientific World J.* **2008**, *25* (8), 1295–1324.
18. Thein, S. L. Genetic Modifiers of the Beta-Haemoglobinopathies. *Br. J. Haematol.* **2008**, *141* (3), 357–366.
19. Vichinsky, E. Hemoglobin E Syndromes. *Hematology Am. Soc. Hematol. Educ. Program* **2007**, 79–83.
20. Vichinsky, E. P. Alpha Thalassemia Major–New Mutations, Intrauterine Management, and Outcomes. *Hematology Am. Soc. Hematol. Educ. Program* **2009**, 35–41.
21. Viprakasit, V.; Lee-Lee, C.; Chong, Q. T.; Lin, K. H.; Khuhapinant, A. Iron Chelation Therapy in the Management of Thalassemia: The Asian Perspectives. *Int. J. Hematol.* **2009**, *90* (4), 435–445.
22. Ware, R. E.; Aygun, B. Advances in the Use of Hydroxyurea. *Hematology Am. Soc. Hematol. Educ. Program* **2009**, 62–69.

RELEVANT WEB PAGES

1. Hb Var Database of Globin Gene Mutations: http://globin.cse.psu.edu.
2. NHS Sickle Cell and Thalassemia Screening Programme: http://sct.screening.nhs.uk/.
3. Thalassemia Community Portal: http://www.ithanet.eu/.
4. Thalassemia International Federation: http://www.thalassaemia.org.cy/.
5. Genetic Information for Patients: http://www.chime.ucl.ac.uk/APoGI/.

CHAPTER

71

Other Hereditary Red Blood Cell Disorders

Bertil Glader

Department of Pediatrics and Pathology, Stanford University School of Medicine, Stanford, CA, USA

This chapter focuses on genetic red blood cell (RBC) disorders associated with hemolytic anemia (enzyme and membrane disorders) or impaired RBC production (bone marrow failure disorders, megaloblastic, dyserythropoietic, and sideroblastic anemias).

71.1 RBC ENZYME DISORDERS

The most common RBC enzyme abnormality associated with hemolysis is *glucose-6-phosphate dehydrogenase* (G6PD), affecting millions of people throughout the world, particularly in Mediterranean countries, Africa, and Asia (*1*). It is an X-linked disorder, primarily affecting males. G6PD is necessary to generate NADPH and thereby maintain reduced glutathione (GSH) levels in the face of oxidant stress. Acute hemolysis due to hemoglobin oxidation occurs when affected individuals are exposed to oxidants. The oxidant stresses encountered clinically are infections, certain drugs, and exposure to fava beans. Hemolytic anemias due to glycolytic enzymopathies are relatively rare, affecting only a few thousand individuals in the world. Abnormalities in virtually every glycolytic enzyme have been described, although over 90% of cases associated with hemolysis are due to *pyruvate kinase* (PK) *deficiency* (*1,2*). Most glycolytic enzymopathies, including PK deficiency, manifest an autosomal recessive pattern of inheritance.

71.2 RBC MEMBRANE DISORDERS

Hereditary RBC membrane disorders are due to alterations in the quantity and/or quality of RBC cytoskeleton proteins and their interactions with each other (*3*). *Hereditary spherocytosis* is due to an uncoupling of the cytoskeleton from the lipid bilayer, thereby leading to membrane instability with loss of lipids and some integral membrane proteins. It is the most common genetic RBC membrane disorder, occurring primarily in Caucasians of northern European extraction. It is an autosomal dominant disorder although autosomal recessive inheritance and new mutations occur. Jaundice is common in the newborn period sometimes requiring exchange transfusion. Beyond the neonatal period, jaundice is rarely intense and the degree of anemia often is minimal.

Hereditary elliptocytosis (HE) disorders are due to α-spectrin gene mutations, resulting in impaired spectrin dimer interactions and RBC fragmentation. These are autosomal dominant disorders that occur in all racial and ethnic groups. Three HE syndromes are recognized: *common HE*, *homozygous (doubly heterozygous) HE*, and *hereditary pyropoikilocytosis*.

71.3 CONGENITAL BONE MARROW FAILURE SYNDROMES

Diamond–Blackfan anemia (DBA) is due to an intrinsic defect of erythroid progenitors resulting in impaired erythroid production (*4*). Several hundred individuals have been identified in the medical literature and through registries in Europe and North America. DBA is characterized by a lifelong anemia with reticulocytopenia that usually presents in the first 2–3 months of life, occurring equally in both sexes and identified in all ethnic groups. Associated congenital abnormalities are common. Most DBA cases are sporadic although 10–15% of cases occur in kindred that have had more than one affected family member. DBA is a disorder of ribosome biogenesis and nine ribosomal protein gene mutations have been identified to date, accounting for 50% of recognized DBA cases (*5*).

Fanconi anemia (FA) is characterized by anemia, thrombocytopenia, and leukopenia with reduced hemopoietic precursor cells in the bone marrow. This is an autosomal recessive disorder occurring in all races and ethnic groups. In many FA patients, there are physical abnormalities such as short stature and skeletal abnormalities. Clinical manifestations begin in the first decade with thrombocytopenia, leukopenia, and/or macrocytic anemia. Pancytopenia becomes more severe over months to years, evolving into fully blown aplastic anemia, myelodysplasia, or acute leukemia. A characteristic feature of FA is increased chromosomal breaks in cultured skin fibroblasts and in peripheral blood lymphocytes stimulated in vitro with clastogenic drugs, and this is the basis of the current diagnostic test, which employs diepoxybutane. Thirteen FA genes have been identified (*5*).

71.4 MEGALOBLASTIC ANEMIAS

Megaloblasic anemias are due to impaired DNA synthesis in bone marrow erythroblasts, resulting in ineffective erythropoiesis and macrocytic anemia. Genetic causes of megaloblastic anemia are very rare and are mostly due to abnormalities of vitamin B12 metabolism (*6*). *Congenital intrinsic factor* (IF) *deficiency* is due to mutations in the gastric IF gene (*gif*). Vitamin B12 binds to IF secreted by gastric parietal cells, a reaction that is necessary before cobalamin is absorbed. *Imerslund–Grasbeck syndrome* is due to a receptor abnormality in the ileum where the IF-B12 complex is absorbed, a consequence of mutations in the cubulin (*cub*) or amnionless (*amn*) genes. *Transcobalamin II* deficiency is an autosomal recessive disorder due to deficiency of the major transport protein for cobalamin in the blood.

71.5 DYSERYTHROPOIETIC ANEMIAS

Congenital dyserythropoietic anemias (CDA) are hemolytic disorders with unique morphological abnormalities in marrow erythroblasts (multinuclearity, abnormal nuclear fragments, and intrachromatin bridges between cells) (*7*). Clinically, these disorders are characterized by variable degrees of anemia despite markedly increased marrow erythroid activity (i.e. ineffective erythropoiesis). *TYPE I CDA* is a very rare autosomal recessive disorder. The onset of anemia and/or jaundice may be noted at any age, and a characteristic feature is RBC macrocytosis with mean cell volumes ranging between 100 and 120 fl. *TYPE II CDA* is the most common variant encountered and several hundred cases of this autosomal recessive disorder are recognized (*8*). The gene for CDA II (*sec23b*) is associated with several different mutations. *TYPE III CDA* is the least common variant, characterized by autosomal dominant inheritance, mild-to-moderate anemia, and RBC macrocytosis. The bone marrow demonstrates erythroid hyperplasia, with many multinucleated erythroblasts containing up to 12 nuclei. The gene for CDA III has been mapped to *15q21*.

71.6 SIDEROBLASTIC ANEMIAS

The inherited sideroblastic anemias are rare genetic conditions presenting in childhood (*9*). In all sideroblastic anemias a common feature is impaired heme synthesis leading to iron accumulation within mitochondria. *X-linked sideroblastic anemia* usually occurs in males. It is due to abnormalities of the erythrocytic isozyme for 5-aminolevulinic acid synthetase (ALAS2). The gene for the erythrocyte-specific ALAS (*ALAS2*) is on the X chromosome and many different mutations have been identified. Severe anemia may be recognized in infancy or early childhood, whereas milder cases may not become apparent until adulthood. Clinical features are pallor, icterus, moderate splenomegaly, and hepatomegaly or both. *Pearson syndrome* is characterized by the early onset of transfusion-dependent anemia, neutropenia, and thrombocytopenia. In addition to the usual marrow abnormalities of sideroblastic anemia, these children also have vacuolization of RBC and myeloid precursors, and pancreatic fibrosis. It is now known that this syndrome is due to mitochondrial DNA deletions. The clinical features of Kearns–Sayre syndrome later develop in the few children who survive the hematologic consequences of this disorder.

REFERENCES

1. Glader, B. Hereditary Hemolytic Anemias Due to Red Blood Cell Enzyme Disorders. In *Wintrobe's Clinical Hematology*; Greer, J. P., et al. Eds.; Lippincott, Williams & Wilkins: Philadelphia, PA, 2009; pp 933–955.
2. Mentzer, W. Pyruvate Kinase Deficiency and Disorders of Glycolysis. In *Nathan and Oski's Hematology of Infancy and Childhood*, 7th ed.; Orkin, S.; Fisher, D. E.; Look, A. T.; Lux, S. E.; Ginsburg, D.; Nathan, D. G., Eds.; Elsevier Inc., Philadelphia, PA, 2009.
3. Gallagher, P.; Glader, B. Hereditary Sperocytosis, Hereditary Elliptocytosis, and Other Disorders Associated with Abnormalities of the Erythrocyte Membrane. In *Wintrobe's Clinical Hematology*; Greer, J. P., Ed.; Lippincott, Williams and Wilkins: Philadelphia, PA, 2009; pp 911–932.
4. Vlachos, A.; Ball, S.; Dahl, N., et al. Diagnosing and Treating Diamond Blackfan Anaemia: Results of an International Clinical Consensus Conference. *Br. J. Haematol.* **2008**, *142* (6), 859–876.
5. Dokal, I.; Vulliamy, T. Inherited Bone Marrow Failure Syndromes. *Haematologica* **2010**, *95* (8), 1236–1240.
6. Carmel, R. Megaloblastic Anemias: Disorders of Impaired DNA Synthesis. In *Wintrobe's Clinical Hematology*; Greer, J. P., Ed.; Lippincott, Williams and Wilkins: Philadelphia, PA, 2009; pp 1143–1172.
7. Wickramasinghe, S.; Glader, B. Congenital Dyserythropoietic Anemias. In *Wintrobe's Clinical Hematology*; Greer, J. P., Ed.; Lippincott, Williams & Wilkins: Philadelphia, PA, 2009; pp 1212–1220.
8. Heimpel, H.; Anselstetter, V.; Chrobak, L., et al. Congenital Dyserythropoietic Anemia Type II: Epidemiology, Clinical Appearance, and Prognosis Based on Long-Term Observation. *Blood* **2003**, *102* (13), 4576–4581.
9. Fleming, M. D. The Genetics of Inherited Sideroblastic Anemias. *Semin. Hematol.* **2002**, *39* (4), 270–281.

CHAPTER

72

Hemophilias and Other Disorders of Hemostasis

Jordan A Shavit and David Ginsburg

72.1 INHERITED DISORDERS OF THE COAGULATION CASCADE

The coagulation cascade was historically one of the first biologic systems associated with human disease to be studied in detail at the biochemical level. Investigations over many years identified an ordered cascade consisting of a plasma protease activating an inactive zymogen target to an active protease form, which subsequently acts on the next step in the cascade.

Inherited bleeding disorders due to deficiencies of factors within the coagulation cascade generally result in similar phenotypes. Hemorrhage in deep tissues, particularly the joints (hemarthroses), is characteristic, as well as increased bleeding following surgery or trauma. The pattern of bleeding can often be distinguished clinically from those that are associated with defective platelet function. The bleeding associated with coagulation cascade disorders is generally delayed compared to that of platelet defects. The latter is also more often from mucosal surfaces, particularly the nose, oral cavity, and gastrointestinal tract, in contrast to the deep tissue hemorrhage characteristic of abnormalities in the coagulation cascade.

72.2 HEMOPHILIAS

The term *hemophilia* is generally reserved for two specific inherited X-linked disorders, factor VIII (FVIII) deficiency (hemophilia A) and factor IX (FIX) deficiency (hemophilia B). The clinical manifestations of the hemophilias are indistinguishable from each other, and vary considerably from a severe bleeding disorder presenting at birth to a very mild condition that may be totally asymptomatic or only diagnosed late in life. Hemophilia should be suspected in any male with a severe congenital bleeding disorder and also in older males with mild bleeding. The diagnosis can usually be readily established by specific FVIII or FIX assays, with the level expected to correlate directly with the clinical course. The typical clinical presentation in hemophilia patients includes joint and muscle hemorrhages, easy bruising, and excessive, sometimes fatal, hemorrhage after trauma or surgery. Hemorrhage in joints is unusual until the child begins to walk, and the disease may go undiagnosed until that time. However, many severe hemophiliacs are diagnosed around the time of birth due to hemorrhage from the umbilicus or following circumcision.

A broad spectrum of genetic defects within the *FVIII* and *FIX* genes has been defined, including the unique recurrent intron 22 inversion that is responsible for approximately 45% of severe hemophilia A. As predicted by Haldane, about one-third of cases appear to be new mutations.

Many hemophiliacs are followed in comprehensive hemophilia treatment centers where they receive multidisciplinary care. Treatment for hemophilia rests on replacement of the deficient factor activity. Standard of care in the Western world now consists of prophylaxis with recombinant factor two to three times a week, to prevent long-term arthropathy. Minor bleeding in mild to moderate hemophilia A patients can also be treated with desmopressin (DDAVP), which will result in a two- to fivefold increase in FVIII and VWF (von Willebrand factor) levels. Approximately 20% of patients develop antibody inhibitors to FVIII or FIX that can dramatically complicate therapy. DNA sequencing of all exons and associated splice junctions can identify the molecular defect in the vast majority of cases, facilitating genetic counseling and prenatal diagnosis.

72.3 VON WILLEBRAND DISEASE

von Willebrand Disease (VWD) is due to either a quantitative or a qualitative defect in VWF, the major ligand facilitating binding of the platelet to the injured vessel wall. Clinically, VWD presents as a platelet-like bleeding disorder. In contrast to the hemophilias, where straightforward correlation between the level of factor activity and clinical phenotype can be made, the clinical manifestations of VWD are bewilderingly complex, with extensive phenotypic and genotypic heterogeneity. Although VWD is considerably more frequent than hemophilia, the most common forms are generally quite mild and frequently go undiagnosed. Multiple clinical subtypes of VWD have been described, with inheritance generally autosomal dominant with decreased penetrance. VWD patients suffer prolonged cutaneous or mucosal bleeding, such as spontaneous nosebleeds or gastrointestinal bleeding. More serious systemic bleeding is unusual but can complicate major surgery or trauma. Increased or prolonged menstrual bleeding is also a frequent complication in female patients who, as a result, are more likely to have the diagnosis established. A large number of genetic defects responsible for VWD have been reported, with a clear correlation between specific subtypes and the type or location of mutations.

The true prevalence of VWD is controversial. Initial large screening studies suggested that mild VWD may affect about 1% of individuals in several populations, although other analyses suggest that the prevalence may be 10- to 100-fold lower. However, there is clear overlap in VWF levels between normal and VWD patients, and the criteria for a VWD diagnosis have been questioned.

The treatment of choice for mild VWD is the vasopressin analog DDAVP with which adequate hemostasis can generally be obtained, even during major surgery. In patients who are unresponsive to DDAVP or in whom it is contraindicated, factor replacement is indicated.

72.4 OTHER FACTOR DEFICIENCIES

Other clinically relevant factor deficiencies include factor XI, fibrinogen, factor XIII, factor VII, factor X, factor V, and prothrombin (factor II), listed roughly in the order of decreasing prevalence. In contrast to hemophilia A or B, the plasma levels of these factors may not always closely correlate with the severity of bleeding.

Inherited deficiency of FXII (Hageman factor), prekallikrein (Fletcher factor), and high-molecular-weight kininogen (Fitzgerald factor) have all been reported. These disorders are primarily a laboratory curiosity, resulting in a markedly prolonged aPTT but generally not associated with clinically significant bleeding.

72.5 DEFECTS IN OTHER COAGULATION CASCADE PROTEINS

A number of rare cases of multiple clotting factor defects have been reported, including defects in the γ-carboxylase or vitamin K reductase pathways, as well as combined deficiency of factor V and factor VIII due to mutations in either of two components of a specific ER cargo receptor for these proteins.

72.6 PLATELET DISORDERS

72.6.1 Inherited Disorders of Platelet Function

Deficiencies in platelet number (thrombocytopenia) or platelet function (qualitative platelet disorders) are associated with abnormal bleeding, characterized by predominantly cutaneous mucosal bleeding and prolonged bleeding from minor injuries. Deeper tissue bleeding and hemarthroses are less common than in disorders of the coagulation cascade, although serious central nervous system (CNS) hemorrhage can occur. By far the most common genetic cause of platelet-type bleeding is VWD, with the remaining inherited platelet disorders quite uncommon.

Bernard–Soulier syndrome is a rare disorder associated with platelet-type bleeding, unusually large platelets, and thrombocytopenia due to abnormalities in the GPIb/IX/V complex on the platelet surface. Glanzmann thrombasthenia shows a profound defect in platelet aggregation due to absence or dysfunction of the GPIIb-IIIa ($\alpha_{IIb}\beta_{III}$) integrin receptor on the platelet surface. Both disorders display autosomal recessive inheritance.

The rare congenital platelet disorders known as storage pool deficiencies include the gray platelet syndrome, Hermansky–Pudlak syndrome, Chediak–Higashi syndrome, thrombocytopenia and absent radii (TAR) syndrome, and Wiskott–Aldrich syndrome, most of which also exhibit constitutional abnormalities in other organ systems.

Treatment of qualitative platelet disorders is primarily supportive, with platelet transfusion administered as necessary. Antiplatelet drugs such as nonsteroidal anti-inflammatory drugs should be avoided. DDAVP may be of benefit in some patients, although the mechanism is not well understood.

72.6.2 Defects of the Fibrinolytic System

Abnormal overactivity of the fibrinolytic system can also result in pathologic bleeding due to accelerated clot lysis. These disorders occur with rare deficiencies of the fibrinolytic inhibitors α_2-antiplasmin and plasminogen activator inhibitor-1.

72.7 INHERITED DISORDERS PREDISPOSING TO THROMBOSIS

Venous thrombosis, most commonly in the lower extremities, affects approximately 1 in 1000 individuals in the United States per year. The occurrence of venous thrombosis in patients under the age of 45, recurrent unexplained thromboses, or positive family history are all suggestive of an inherited predisposition to thrombosis or "thrombophilia."

72.7.1 Deficiency of the Natural Anticoagulants

The natural anticoagulants consist of antithrombin III (AT3), protein C and protein S. The hemostatic balance is very sensitive to variations in these factors, with levels in the range of 50% of normal or less associated with a significantly increased risk of thrombosis, accounting for the autosomal dominant nature of these disorders.

DNA diagnosis is not routinely performed for these disorders, with diagnosis generally based on clinical antigen and functional assays, though considerable overlap can be observed between genetically deficient patients and normal individuals. A number of cases of homozygous protein C or S deficiency have been reported, associated with the severe form of neonatal purpura fulminans, which is fatal if not treated promptly.

72.7.2 Factor V Leiden

An underlying mutation was identified in factor V that renders it resistant to inactivation. The substitution of glutamine for arginine 506 (R506Q), also known as factor V Leiden, results in FV resistance to inactivation by protein C. This variant is present in approximately 5% of individuals in European populations and results in a 10% lifetime risk of thrombosis.

Given the high prevalence of factor V Leiden, its relatively low penetrance for thrombosis, and the lack of a significant effect on life expectancy coupled with the significant morbidity of long-term anticoagulation with currently available drugs, no specific treatment can be recommended at this time for asymptomatic patients. In addition, treatment following an episode of venous thrombosis does not differ based on factor V Leiden status. Thus, routine screening is not currently recommended.

72.7.3 Thrombotic Thrombocytopenic Purpura

Thrombotic thrombocytopenic purpura (TTP) is a catastrophic, multisystem disorder characterized by the formation of platelet and VWF-rich microthrombi in vessels of multiple organs, leading to the classic pentad of microangiopathic hemolytic anemia, thrombocytopenia, fever, and renal and CNS dysfunction. A rare familial form typically presenting by early childhood is due to autosomal recessive deficiency for the VWF-cleaving metalloprotease, ADAMTS13. Most adult cases of TTP are associated with acquired autoantibodies against ADAMTS13.

TABLE 72-1 Components of the Classic Coagulation Cascade

Factor Number	Synonym	Chromosomal Localization
I	Fibrinogen	4q28
II	Prothrombin	11p11-q12
III	Tissue thromboplastin (TF and phospholipids)	1p21.3 (TF)
IV	Calcium	—
V	Proaccelerin	1q23
VI	Activated form of factor V (FVa; FVI no longer used)	—
VII	Proconvertin	13q34
VIII	Anti-hemophilic factor	Xq28
IX	Christmas factor	Xq27.1-q27.2
X	Stuart–Prower factor	13q34
XI	Plasma thromboplastin antecedent	4q35
XII	Hageman factor	5q33-qter
XIII	Fibrin stabilizing factor, Plasma transglutaminase	6p25-p24 (A subunit) 1q31-q32.1 (B subunit)

TABLE 72-2 Clinical Findings in Inherited Bleeding Disorders

	Platelet-Type Defects	Coagulation Cascade Defects
Timing of Bleeding	Early after trauma or spontaneous	Delayed
Sites	Skin and mucous membranes, petechiae, ecchymoses	Deep tissue hematomas, including joint, muscle and retroperitoneum
Inherited Disorders	von Willebrand disease, Glanzmann thrombasthenia, Bernard–Soulier syndrome, other inherited platelet defects	Hemophilia A and B, factor V deficiency, deficiency of Factors XI, VII, II, or X; afibrinogenemia

TABLE 72-3 Clinical Classification of Hemophilia A and B

Classification	FVIII or FIX Activity	Clinical Manifestations
Severe	<1%	Spontaneous hemorrhage beginning, in early infancy. Frequent hemarthroses; hemarthroses and other serious hemorrhages requiring factor replacement
Moderate	1–5%	Hemorrhaging following trauma; occasional spontaneous hemorrhage
Mild	5–25%	Bleeding generally only following significant trauma or surgery
	>25%	No significant bleeding

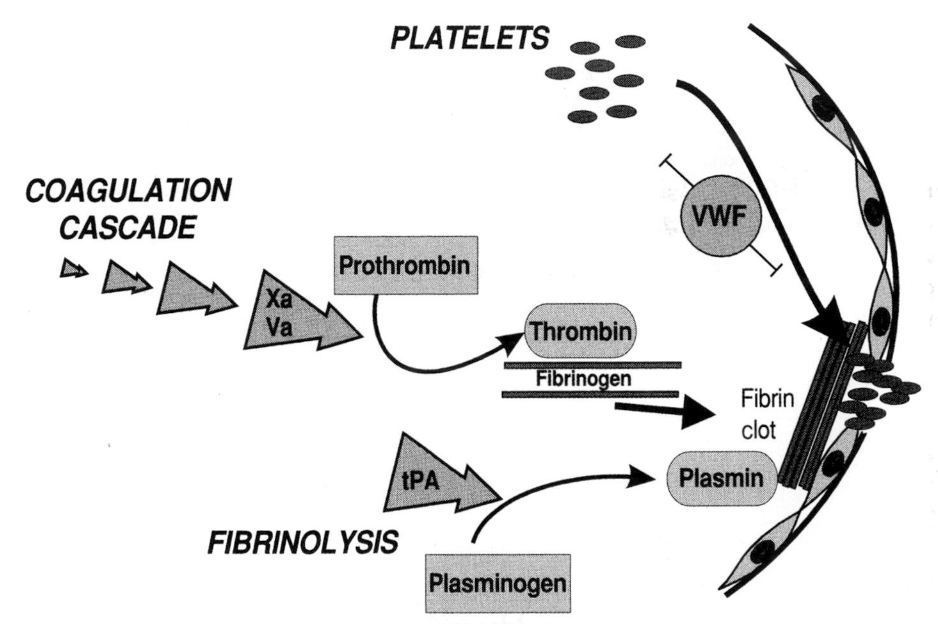

FIGURE 72-1 Overview of hemostasis. The platelet, coagulation cascade, and fibrinolysis limbs of hemostasis are illustrated schematically. Xa, factor Xa; Va, factor Va; VWF, von Willebrand factor; tPA, tissue plasminogen activator. See text for description.

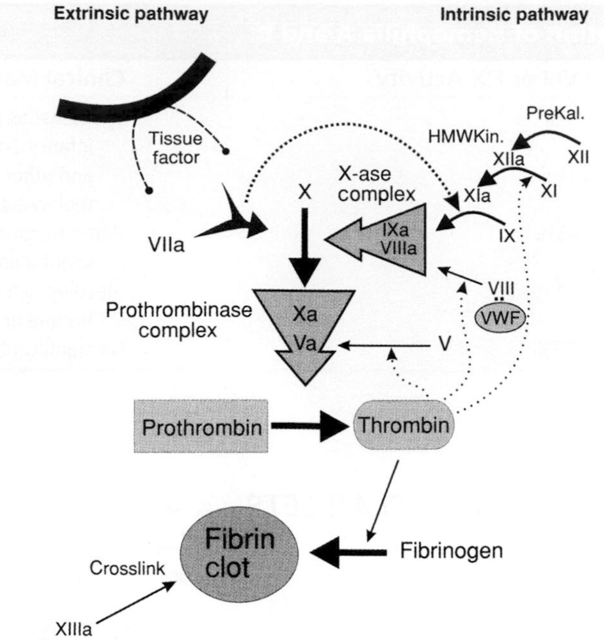

FIGURE 72-2 The coagulation cascade. See text for description. VWF, von Willebrand factor; HMWKin, high-molecular-weight kininogen; PreKal, prekallikrein. Other clotting factors are indicated by their Roman numeral designations.

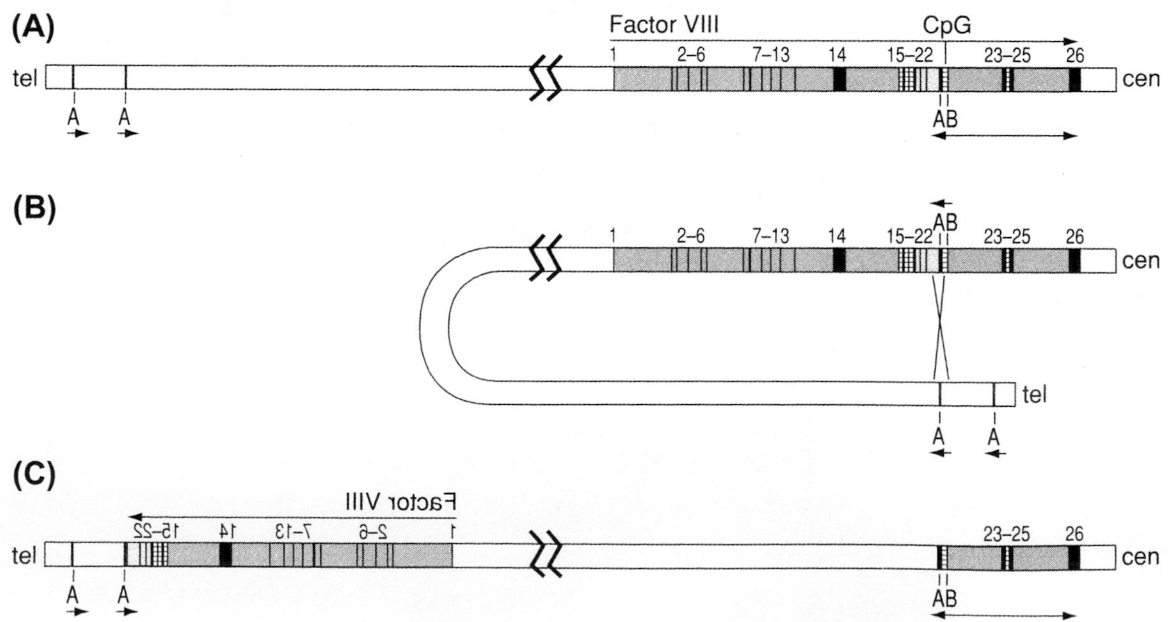

FIGURE 72-3 Model of factor VIII gene inversion. A. Structure of the normal factor VIII gene (tel, telomere; cen, centromere). A, the two copies of the *A* gene upstream of the factor VIII gene and the single copy within intron 22. B, another small locus that shares the *A* gene promoter. The arrows indicate the directions of transcription. B. Model for homologous recombination between the intron 22 copy of the *A* gene and one of the two upstream copies. The indicated crossover results in an inversion of the sequence between the two recombined *A* genes C. (*From Lakich, D.; et al. Inversions Disrupting the Factor VIII Gene Are a Common Cause of Severe Haemophilia A.* Nat. Genet. ***1993***, 5, *236–241,with permission.*)

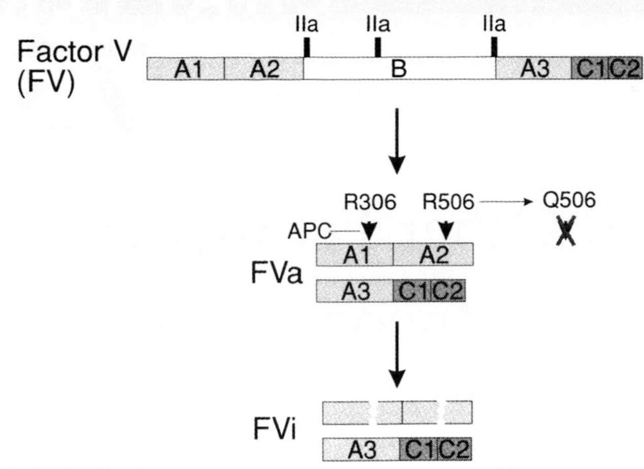

FIGURE 72-4 Domain structure and processing of factor V. Thrombin cleavage sites (IIa) are indicated. FVa, the active form of factor V; FVi, factor V which has been inactivated through cleavage by activated protein C (APC). The APC cleavage sites at Arg306 (R306) and Arg506 (R506) are indicated. The common factor V Leiden mutation, substitution of glutamine for Arg506 (Q506), results in APC resistance.

FURTHER READING

1. Asselta, R.; Peyvandi, F. Factor V Deficiency. *Semin. Thromb. Hemost.* **2009,** *35,* 382–389.
2. Dahlback, B. Advances in Understanding Pathogenic Mechanisms of Thrombophilic Disorders. *Blood* **2008,** *112,* 19–27.
3. Goodeve, A. C. The Genetic Basis of von Willebrand Disease. *Blood Rev.* **2010,** *24,* 123–134.
4. Hoffman, M.; Monroe, D. M., 3rd. A Cell-Based Model of Hemostasis. *Thromb. Haemost.* **2001,** *85,* 958–965.
5. Kujovich, J. L. Factor V Leiden Thrombophilia. *Genet. Med.* **2011,** *13,* 1–16.
6. Lancellotti, S.; De Cristofaro, R. Congenital Prothrombin Deficiency. *Semin. Thromb. Hemost.* **2009,** *35,* 367–381.
7. Levy, G. G.; Motto, D. G.; Ginsburg, D. ADAMTS13 Turns 3. *Blood* **2005,** *106,* 11–17.
8. Mariani, G.; Bernardi, F. Factor VII Deficiency. *Semin. Thromb. Hemost.* **2009,** *35,* 400–406.
9. Menegatti, M.; Peyvandi, F. Factor X Deficiency. *Semin. Thromb. Hemost.* **2009,** *35,* 407–415.
10. Montgomery, R. R.; Gill, J. C.; Di Paola, J. Hemophilia and von Willebrand Disease. In *Nathan and Oski's Hematology of Infancy and Childhood*; Orkin, S. H.; Nathan, D. G.; Ginsburg, D.; Look, A. T.; Fisher, D. E.; Lux, S. E., Eds.; 2009; pp 1488–1525.
11. Muszbek, L.; Bagoly, Z.; Cairo, A.; Peyvandi, F. Novel Aspects of Factor XIII Deficiency. *Curr. Opin. Hematol.* **2011,** *18,* 366–372.
12. Nichols, W. L.; Hultin, M. B.; James, A. H.; Manco-Johnson, M. J.; Montgomery, R. R.; Ortel, T. L.; Rick, M. E.; Sadler, J. E.; Weinstein, M.; Yawn, B. P. von Willebrand disease (VWD): Evidence-Based Diagnosis and Management Guidelines, the National Heart, Lung, and Blood Institute (NHLBI) Expert Panel Report (USA). *Haemophilia* **2008,** *14,* 171–232.
13. Patnaik, M. M.; Moll, S. Inherited Antithrombin Deficiency: A Review. *Haemophilia* **2008,** *14,* 1229–1239.
14. Wang, L.; McLeod, H. L.; Weinshilboum, R. M. Genomics and Drug Response. *N. Engl. J. Med.* **2011,** *364,* 1144–1153.

CHAPTER

73

Rhesus and Other Fetomaternal Incompatibilities

Gregory Lau

Cedars-Sinai Medical Center, Department of Obstetrics and Gynecology, Los Angeles, CA, USA

73.1 ETIOLOGY OF ALLOIMMUNIZATION

In 1953, Chown determined that the passage of Rh positive fetal erythrocytes into the maternal circulation was the cause of Rh immunization. The risk of Rh alloimmunization is 16% if Rh positive fetus is ABO-compatible with its mother. If not, the risk is approximately 2%. This risk increases if an invasive procedure, such as an amniocentesis, chorionic villus sampling, or any other procedure causes a fetomaternal blood exchange.

73.2 DETECTION OF FETOMATERNAL HEMORRHAGE

The Kleihauer–Betke acid elution test depends on the resistance of fetal hemoglobin to acid elution. Fetal erythrocytes appear as a series of dark cells set upon a background of maternal cells, which appear pale, "ghost" like. Given that this test depends upon the detection of fetal erythrocytes instead of Rh positive cells, there is the potential for a false positive result. Up to 2% of cells with fetal hemoglobin can be considered a false positive since it is not unusual for adults to circulate up to 2% fetal hemoglobin. It has been demonstrated that anywhere from 0.2% to 7% of adult erythrocytes contain fetal hemoglobin. This proportion is genetically determined. This is one of the drawbacks of the Kleihauer–Betke, which leads to the potential overtreatment of some adults. Given this concern with respect to the Kleihauer–Betke test, consideration for more rigorous means of evaluation should be considered.

73.3 Rh BLOOD GROUP SYSTEM

Individuals are considered as either Rh positive or negative depending upon the presence of the major D antigen on the surface of erythrocytes. RhD is considered the most immunogenic antigen followed by RhC. Other major antigens, C/c and E/e, are also considered clinically important. The upper and lower case letters indicate the presence of a serologically definable antigen. The genes responsible for these antigens are inherited together and are thought to be closely linked given that recombination rarely occurs between them.

73.4 MOLECULAR BASIS OF Rh ANTIGENS

There are approximately 105 distinct Rh antigens to each erythrocyte with C/c, D, and E contributing to approximately one-third of the total. Each of these polypeptides is distinct, but they do have similarities. Most of what is known about these proteins comes from cloning studies of Rh cDNA. As a result of these studies, the Rh gene locus was localized to chromosome 1. The first cDNA clone proved to encode both the C/c and E/e proteins. Each of these polypeptides contains 417 amino acids. RhD and RhCE differ in sequence by 30–35 amino acids. There is only an 8% difference in the sequence between these two proteins. Despite the homology, these proteins do not share antigens.

Genetic rearrangements have come about through gene deletion, microconversion or macroconversion, and antithetical missense mutations. Of all the variations, RhD negative (Rh negative) is the most clinically relevant. The incidence of the Rh-negative phenotype in whites is 15% (mostly because of total gene deletion). In other ethnic groups (Asian and some African races), other mechanisms are responsible for the Rh negative phenotype.

73.5 PRENATAL Rh GENOTYPING

Until recently, two approaches were used to identify the fetus, which is at risk for Rh disease. These approaches include fetal blood sampling or serial amniocentesis. Fetal blood sampling carries a 1–2% risk of fetal loss and a 40% risk of fetomaternal hemorrhage. Serial amniocentesis is less accurate because it is unable to distinguish between an RhD negative fetus from a mildly affected Rh-D positive fetus. It also exposes the mother to multiple invasive procedures as well as increases her risk for Rh sensitization. Now, the cloning and sequencing of the Rh C/c, E/e, and D cDNAs have allowed for Rh genotyping of the fetus through the use of a small amount of DNA from sources such as amniocytes, chorionic villi, and fetal cells/free fetal DNA in the maternal circulation.

73.6 NONINVASIVE PRENATAL DIAGNOSIS

Approximately 3% of cell-free DNA in the maternal plasma during the first trimester is of fetal origin. This proportion increases to 6% by the end of the third trimester. Fetal Rh genotyping using cell-free fetal DNA appears to be cost-effective and readily acceptable for clinical practice. Reliable methods to accomplish this feat include real-time quantitative PCR with Tagman chemistry. This technique involves the use of primers and probes to detect exons 4,5, and 10 of RhD, but not RhCE.

73.7 ALLOIMMUNE THROMBOCYTOPENIA

Platelet alloantigens are located on the glycoproteins found within platelet membranes. Five platelet antigens have been identified: HPA 1–5. Maternal alloantibodies are IgG. In whites, these antibodies are directed against HPA-1a.

Alloimmune thrombocytopenia results from maternal antibodies, which cross the placenta and affect fetal platelets, resulting in thrombocytopenia (AITP). AITP carries a risk of severe fetal hemorrhage with intracranial bleeding occurring in 20% of cases. Screening is not routinely performed; therefore, antenatal treatment is usually provided for those with a previously affected child. AITP generally presents in the first pregnancy. Various in utero treatments have been attempted: IVIG, steroids, or fetal platelet transfusions.

73.8 MANAGEMENT AND PREVENTION OF ALLOIMMUNIZATION

Prevention is the most important aspect of managing alloimmunization because sensitization (or active immunization) leads to lifelong alloimmunization.

Prevention can be achieved by the following:

1. Prevention of conception of a fetus with target antigens.
2. Preventing implantation of embryos positive for the antigen.
3. Preventing sensitization to the RhD antigen by immunization with anti-D immunoglobulin.

Despite these measures, sensitization still occurs most likely as a result of spontaneous fetomaternal hemorrhage.

73.9 TREATMENT

A variety of treatment strategies have been developed. Early delivery removes the fetus from the adverse intrauterine environment and from exposure to the antibodies. However, delivery increases the risks due to prematurity. Exchange transfusion is another option. The aims of exchange transfusion include lowering unconjugated bilirubin, preventing kernicterus, washing out maternal anti-D, reducing hemolysis, and treating fetal anemia (replacing RhD positive erythrocytes with those which are Rh negative). Phototherapy is also an acceptable means of lowering unconjugated bilirubin and the need for exchange transfusion. The most successful in utero treatment of severe Rh alloimmunization has been the transfusion of RhD-negative RBCs in utero. With this treatment, fetal/neonatal survival can be as high as 90–95%. Fetal transfusion is considered safe, but cord accidents, premature rupture of membranes, and preterm labor can occur.

The optimal treatment for a patient with a severely affected fetus is controversial. Noninvasive therapies such as IVIG or steroids are of uncertain efficacy. Invasive therapies such as fetal transfusion are effective, but carry the aforementioned risks and require serial transfusions given the short half-life of platelets.

FURTHER READING

1. Allen, F. H. Attempts at Prevention of Intrauterine Death in Erythroblastosis Fetalis. *N. Engl. J. Med.* **1963**, *269*, 1344–1249.
2. Bowman, J. M.; Pollock, J. M.; Penston, L. E. Feto-Maternal Transplacental Haemorrhage During Pregnancy and after Delivery. *Vox Sang.* **1986**, *51*, 117.
3. Chown, B. Anaemia from Bleeding of the Fetus into the Mothers Circulation. *Lancet* **1954**, *1*, 1213.
4. Clarke, C. A. Preventing Rhesus Babies: The Liverpool Research and Follow-Up. *Arch. Dis. Child* **1989**, *64*, 1734–1740.
5. Daniels, G.; Finning, K.; Martin, P. Fetal RhD Genotyping: A More Efficient Use of Anti-D Immunoglobulin. *Transfus. Clin. Biol.* **2008**, *14*, 568–571.
6. Daniels, G.; Finning, K.; Martin, P. Noninvasive Fetal Blood Grouping: Present and Future. *Clin. Lab. Med.* **2010**, *30*, 431–442.
7. DeBoer, I. P.; Zeestraten, E. C.M.; Lopriore, E., et al. Pediatric Outcome in Rhesus Hemolytic Disease Treated With and Without Intrauterine Transfusion. *Am. J. Ob. Gynecol.* **2008**, *198*, 54.31–54.34.
8. Kamphuis, M. M.; Paridaans, N.; Porcelijn, L.; De Haas, M., et al. Screening in Pregnancy for Fetal or Neonatal Alloimmune Thrombocytopenia: Systematic Review. *BJOG* **2010**, *117*, 1335–1343.
9. Margulies, M.; Voto, L. S.; Mathet, E., et al. High-Dose Intravenous IgG for the Treatment of Severe Rhesus Alloimmunisation. *Vox Sang.* **1991**, *61*, 18–189.

CHAPTER

74

Leukemias, Lymphomas, and Other Related Disorders

Yanming Zhang

Department of Pathology, Northwestern University, Chicago, IL, USA

Janet D Rowley

Section of Hematology/Oncology, University of Chicago, Chicago, IL, USA

About 25% of acute myeloid and lymphoid leukemia (AML and ALL) and non-Hodgkin lymphoma (NHL) are characterized by various balanced chromosome translocations and inversions, such as t(8;21) and inv(16) in core binding factor (CBF) AML, t(14;18) in follicular lymphoma, and t(2;5) in anaplastic large cell lymphoma, whereas myelodysplastic syndrome (MDS) and myeloproliferative neoplasm (MPN) often display deletion or gain of chromosome material and variable genomic imbalances, such as del(5q) and trisomy 8 in MDS and MPN. Additional chromosome abnormalities, including amplification and deletion, accompany disease progression and transformation. Conventional cytogenetics, fluorescence in situ hybridization, array comparative genomic hybridization, and expression and genomic microarrays are the specific techniques used to detect disease-related chromosome aberrations and genomic imbalances. In conjunction with molecular analyses, they can detect gene mutations and implicate abnormalities of gene regulation. Detection of these changes is highly relevant in clinical practice. Specific chromosome abnormalities are associated with subtypes of leukemia and lymphoma, and can establish a precise diagnosis. Certain chromosome translocations are disease- and lineage-specific, such as t(15;17) in APL and t(12;21) in precursor B-cell ALL, and some chromosome abnormalities occur in both myeloid and lymphoid leukemia, such as t(9;11) and t(11;19). Deletion or loss of chromosomes 5 and 7 is the most frequent abnormality in therapy-related AML and MDS with previous exposure to alkylating agents and/or radiation for primary cancer or immunological diseases, whereas balanced chromosome translocations, mainly involving *MLL* and *AML1/RUNX1*, are often associated with prior treatment with topoisomerase II inhibitors. Moreover, chromosome abnormalities and gene mutations are one of the most reliable predictors of disease prognosis, and may determine treatment choice; for instance, t(9;22) in pediatric ALL is strongly associated with a poor prognosis, in comparison with t(12;21) or a hyperdiploid karyotype. Similarly, all-trans retinoic acid treatment in combination with chemotherapy or arsenic trioxide is very efficient in patients with acute promyelocytic leukemia (APL) with t(15;17), leading to a complete remission for 93% and a 10-year overall survival of 81%. Gene expression profiling in diffuse large B-cell lymphoma distinguishes activated B-cell and germinal B-cell-like groups from others. Some chromosome abnormalities are either diagnostic or prognostic, such as t(8;14) in Burkitt lymphoma, and in high-grade B-cell NHL. Therefore, the detection of chromosome abnormalities in leukemia and lymphoma importantly impacts patient care.

Chromosome translocations in myeloid leukemia often result in a chimeric fusion protein with novel functions, such as t(9;11), and other *MLL* translocations in AML. Chromosome abnormalities in lymphoid disorders usually involve immunoglobulin heavy or light chain loci or T-cell receptor genes and lead to overexpression of an oncogene or anti-apoptosis gene, such as t(8;14) in Burkitt lymphoma, t(14;18) in follicular and other B-cell lymphoma, or the *HOX* gene translocations in T-cell leukemia and lymphoma. Various gene mutations identified with molecular analysis contribute to the initiation and progression of myeloid and lymphoid leukemia. Activating tyrosine kinase mutations, such as *JAK2* mutations in MPN, *TET2* and *IDH1/2* mutations in MDS, *c-KIT* and *FLT3* mutations in ITD, *NPM1* and *CEBPA* mutations in AML, and the *BCR/ABL* fusion in CML, confers cell proliferative and survival advantage, whereas mutations involving transcription factors or tumor suppressor genes, such as the *AML1/ETO* and *CEBF/MYH11* fusions in CBF leukemia, and mutations in *MLL* and *EVI1* impair hematopoietic differentiation and subsequent apoptosis. It is evident that these two types of mutations cooperate to lead to acute leukemia. In addition, point mutations, insertions or deletions of the *NOTCH* gene that is required for commitment to T-cell lineage, and proliferation are frequent in T-ALL. Dysregulation of the cell cyclin D (D1, D2, and D3) due to various IGH translocations is a hallmark of multiple myeloma, which facilitates activity of CDK4 or CDK6, and phosphorylation and inactivation of RB leading to activation of E2F, and to the G1 > S cell cycle progression. Furthermore, epigenetic dysregulation of target genes through microRNA and histone modification in both DNA and RNA levels contributes to leukemogenesis. Loss of miR-15a and miR16-1, which are located in the commonly deleted region of 13q14.3 in CLL, is an early genetic event in the pathogenesis of CLL.

The detection of t(9;22), characterization of the *BCR/ABL* fusion in CML, and the discovery of imatinib (Gleevec) have elegantly reflected the success of translational cancer research. CML was the first disease that was associated with a unique chromosome abnormality, a balanced chromosome translocation leading to the Philadelphia chromosome, namely t(9;22). Cloning the involved genes, *BCR* and *ABL1*, led to CML being the first cancer with a rationally designed drug that directly targeted the molecular consequence responsible for the pathogenesis of the disease. Thus, CML has served as a valid research model in cancer biology, normal and abnormal hematopoiesis, and targeted drug development and therapy. With many other powerful techniques including microarray and next-generation sequencing, we will soon discover more genetic abnormalities and understand how various gene mutations and epigenetic deregulation cooperate to induce leukemia and lymphoma; these insights will greatly expand personalized medicine.

FURTHER READING

1. Bergsagel, P. L.; Kuehl, W. M.; Zhan, F., et al. Cyclin D Dysregulation: An Early and Unifying Pathogenic Event in Multiple Myeloma. *Blood* **2005**, *106*, 296–303.
2. Calin, G. A.; Cimmino, A.; Fabbri, M., et al. MiR-15a and miR-16-1 Cluster Functions in Human Leukemia. *Proc. Natl Acad. Sci. U.S.A.* **2008**, *105*, 5166–5171.
3. Chen, J.; Odenike, O.; Rowley, J. D. Leukemogenesis: More than Mutant Genes. *Nat. Rev. Cancer* **2010**, *10*, 23–36.
4. Drucker, B. J.; Guilhot, F.; O'Brien, S. G., et al. Five-Year Follow-Up of Patients Receiving in Imatinib for Chronic Myeloid Leukemia. *N. Engl. J. Med.* **2006**, *355*, 2408–2417.
5. Greenberg, P.; Cox, C.; LeBeau, M. M., et al. International Scoring System for Evaluating Prognosis in Myelodysplastic Syndromes. *Blood* **1997**, *89*, 2079–2088.
6. Grimwade, D.; Hills, R. K.; Moorman, A. V., et al. Refinement of Cytogenetic Classification in Acute Myeloid Leukemia: Determination of Prognostic Significance of Rare Recurring Chromosomal Abnormalities among 5876 Younger Adult Patients Treated in the United Kingdom Medical Research Council Trials. *Blood* **2010**, *116*, 354–365.
7. Haase, D.; Germing, U.; Schanz, J., et al. New Insights into the Prognostic Impact of the Karyotype in MDS and Correlation with Subtypes: Evidence from a Core Dataset of 2124 Patients. *Blood* **2007**, *110*, 4385–4395.
8. Heim, S.; Mitelman, F. *Cancer Cytogenetics*, 3rd ed.; John Wiley & Sons, Inc.: New Jersey, 2009.
9. Powell, B. L.; Moser, B.; Stock, W., et al. Arsenic Trioxide Improves Event-Free and Overall Survival for Adults with Acute Promyelocytic Leukemia: North American Leukemia Intergroup Study C9710. *Blood* **2010**, *116*, 3751–3757.
10. Morin, R. D.; Mendez-Lago, M.; Mungall, A. J., et al. Frequent Mutation of Histone-Modifying Genes in Non-Hodgkin Lymphoma. *Nature* **2011**, *476*, 298–303.
11. Mullighan, C. G.; Phillips, L. A.; Su, X., et al. Genomic Analysis of the Clonal Origins of Relapsed Acute Lymphoblastic Leukemia. *Science* **2008**, *322*, 1377–1380.
12. Swerdlow, S. H.; Campo, E.; Harris, N., et al. *WHO Classification of Tumors of Hematopoietic and Lymphoid Tissue*, 4th ed.; IARC Press: Lyon, France, 2008.

CHAPTER

75

Immunologic Disorders

Autoimmunity: Genetics and Immunologic Mechanisms

Nancy L Reinsmoen

HLA Laboratory, Cedars-Sinai Medical Center; David Geffen School of Medicine, University of California, Los Angeles, CA, USA

Kai Cao

HLA Laboratory, Department of Laboratory Medicine, Division of Pathology, University of Texas MD Anderson Center, Houston, TX, USA

Chih-Hung Lai

HLA Laboratory, Comprehensive Transplant Center, Cedars-Sinai Health Systems, Los Angeles, CA, USA

The autoimmune response is a breakdown of the regulatory processes that maintain the normal immune response to specific antigens. Both genetic and environmental factors are involved in the development of autoimmune diseases. Many of these autoimmune disorders are associated with specific alleles of the human leukocyte antigen (HLA) loci, which are the major histocompatibility complex (MHC) in humans. The adaptive immune response consists of T and B lymphocytes responding to peptides presented by HLA molecules. The two general categories of HLA molecules are HLA class I molecules (HLA-A, B, C) and HLA class II molecules (HLA-DR, DQ, and DP). The NK cells are involved in the antigen nonspecific innate immune response. Autoimmunity is the result of activation of a self-peptide specific T cell or NK cells resulting in tissue damage mediated by autoantibodies, autoreactive T cells, or NK cells.

Several studies indicate many autoimmune disorders that show linkage to, or association with, specific alleles of the HLA loci. The HLA class I and II genes are the most polymorphic coding sequences in the human genome numbering over 6400 alleles. Linkage of *HLA* genes to several autoimmune diseases has been demonstrated by co-segregation studies in families, or with nonparametric approaches such as haplotype sharing among affected sibling pairs as well as disease association studies within ethnic groups. HLA alleles that are positively associated with a disease are referred to as susceptible, while negatively associated alleles are referred to as protective. A strong association with a given autoimmune disease suggests that the associated HLA locus is very close to the disease locus or that it confers susceptibility. The genome project has provided valuable new information about the genetics of autoimmune diseases and has allowed the identification of a number of new non-*HLA* genes involved in autoimmunity.

Some of the more common autoimmune diseases associated with specific HLA alleles discussed in this chapter include the following disorders: *Ankylosing spondylitis* has been found associated with HLA-B*27 and there appears to be a hierarchy of association with some of the B*27 alleles. Although HLA-B*27 may be a dominant genetic component, the presence of B*27 alone is not sufficient for the disease to develop. In certain ethnic groups in the Middle East and along the ancient silk routes, HLA-B*51 and A*26 have been found associated with *Behcet's disease*. Studies have also shown an independent MHC class I association telomeric to HLA-B in the region from HLA-A to HLA-E. In *multiple sclerosis*, both class II (HLA-DRB1*02) and class I (HLA-B*44:02) associations have been reported along with non-MHC components. Different associations have been reported in various ethnic groups. For *type 1 diabetes*, there is a hierarchy of risk associated with different alleles and genotypes with DRB1*03/DRB1*04-DQB1*03:02 resulting in the highest relative risk. Negative associations suggesting a protective role has been reported for DR2 haplotypes. *Pemphigus vulgaris* is an autoimmune disease caused by antibodies specific for desmogleins resulting in loss of keratinocyte cell adhesion and blisters. There are specific class II associations reported unique in different ethnic groups. *Narcolepsy* is a sleep disorder which has been associated with the HLA-DRB1*15:01-DQB1*06:02 haplotype in Japanese patients. More recently, other non-HLA genes have been found associated with narcolepsy. *Rheumatoid arthritis* is characterized by the presence of rheumatoid factor, an autoantibody to the Fc portion of IgG. Antibodies to citrullinated proteins, which bind with high affinity to DRB1*04:01, have also been detected in these patients. Another HLA association seen is with DQA1*03:01. *Celiac disease*, an inflammatory autoimmune disease of the intestinal mucosa after ingestion of wheat gluten, has been reported to be associated with DRB1*03:01 and DRB1*07:01 extended haplotypes. These HLA-associated autoimmune diseases as well as recent developments in the genetic analysis of non-*HLA* genes will likely have a significant impact on the ability to assess disease predisposition. Taken together, these associations may also provide insight into the autoimmune disease process and the progression of disease severity.

CHAPTER

76

Systemic Lupus Erythematosus

Yun Deng, Bevra H Hahn and Betty P Tsao

Department of Medicine, David Geffen School of Medicine, University of California, Los Angeles, CA, USA

Epidemiologic studies of systemic lupus erythematosus (SLE) have led to great interests in studying genetic basis of the disease. Before 2007, two main approaches used to explore genes predisposing to disease risk are genome-wide linkage studies and candidate gene case-control studies. Linkage studies, which localize chromosome regions co-transmitted with disease in families containing multiple affected members, have successfully established 9 loci (1q23, 1q31-32, 1q41-43, 2q37, 4p16, 6p11-21, 10q22-23, 12q24, and 16q12-13) linked to SLE. But the identified linkage intervals are usually large and contain many potential candidate genes, which limit the further localization of susceptibility genes. Candidate gene studies commonly assess whether a test genetic marker is present at a higher frequency among patients with SLE than in ethnically matched healthy control individuals. Due to the variations in sample sizes, ethnicities, and numbers of genetic markers studied among individual reports conflicting results often appear, which result in a limited number of confirmed SLE susceptibility genes, including MHC class II (*DR2* and *DR3*), class III (*C2* and *C4*), *C1Q*, and a cluster of *Fcγ* receptor genes at 1q23 (*FCGR2A*, *FCGR3A*, *FCGR3B*, and *FCGR2B*). Rapid advances in technology for high-throughput genotyping of genetic markers (known as single nucleotide polymorphisms, SNPs), along with collaborative efforts in establishment of large collections of individuals, have facilitated the large-scale candidate gene association and genome-wide association studies in mapping complex disease loci. Since 2007, six genome-wide association studies and a series of candidate replication studies in SLE have been published, which have expanded the number of established genetic associations with SLE to more than 30. Such associations are consistent with the common disease–common variant hypothesis and each of the identified variants has a modest risk with an odds ratio in the range of 1.1–2.3, accounting for a fraction of the overall genetic risk for SLE. Most of these susceptibility genes can be assigned into three pathogenic pathways: (1) clearance and processing of immune complex (*ITGAM*), (2) toll-like receptor (TLR)/type I interferon signaling in innate immune response (*IRF5*, *STAT4*, *PHRF1/IRF7*, *TNFAIP3*, *TNIP1*, and *TLR7*), and (3) lymphocyte activation and regulation in adaptive immune response (*PTPN22*, *TNFSF4*, *BLK*, *BANK1*, *LYN*, *ETS1*, *PRDM1*, *IKZF1*, and *IL10*). Some novel loci have not yet been fully characterized or their functions have no obvious connection to known SLE pathways (*PXK*, *XKR6*, *JAZF1*, *UHRF1BP1*, *RASGRP3*, and *WDFY4*). Some risk loci are shared between SLE and other autoimmune diseases (*HLA class II* with multiple autoimmune diseases, *PTPN22* with SLE, RA, GD, SSc and T1D, *TNFSF4* with SLE and SSc), providing the possibility of common immunological mechanisms underlying these disease processes. Devoted efforts in the coming years will certainly reveal additional new variants that contribute to SLE, including rare SNPs and structural variations. Future studies that focus on the integration of the genetic findings into a functional context will help obtain a more comprehensive picture of the biological pathways perturbed by each SLE-associated locus. It will be also informative to explore the potential roles of gene–gene and/or gene–environment interactions in predisposing to SLE. Results of these studies are likely to suggest innovative molecular targets for strategic development of more effective therapeutics in SLE.

FURTHER READING

1. Pearson, T. A.; Manolio, T. A. How to Interpret a Genome-Wide Association Study. *JAMA* **2008**, *299* (11), 1335–1344.
2. Manolio, T. A. Genomewide Association Studies and Assessment of the Risk of Disease. *N. Engl. J. Med.* **2010**, *363* (2), 166–176.
3. Spencer, C. C.; Su, Z.; Donnelly, P., et al. Designing Genome-Wide Association Studies: Sample Size, Power, Imputation, and the Choice of Genotyping Chip. *PLoS Genet.* **2009**, *5* (5), e1000477.

4. Wang, K.; Dickson, S. P.; Stolle, C. A., et al. Interpretation of Association Signals and Identification of Causal Variants from Genome-Wide Association Studies. *Am. J. Hum. Genet.* **2010,** *86* (5), 730–742.
5. Wren, J. D. A Global Meta-Analysis of Microarray Expression Data to Predict Unknown Gene Functions and Estimate the Literature-Data Divide. *Bioinformatics* **2009,** *25* (13), 1694–1701.
6. Alkan, C.; Coe, B. P.; Eichler, E. E. Genome Structural Variation Discovery and Genotyping. *Nat. Rev. Genet.* **2011,** *12* (5), 363–376.
7. Warden, M.; Pique-Regi, R.; Ortega, A., et al. Bioinformatics for Copy Number Variation Data. *Methods Mol. Biol.* **2011,** *719,* 235–249.
8. Xi, R.; Kim, T. M.; Park, P. J. Detecting Structural Variations in the Human Genome Using Next Generation Sequencing. *Brief. Funct. Genomics* **2010,** *9,* 405–415.
9. Kariuki, S. N.; Niewold, T. B. Genetic Regulation of Serum Cytokines in Systemic Lupus Erythematosus. *Transl. Res.* **2010,** *155* (3), 109–117.
10. Rubtsov, A. V.; Rubtsova, K.; Kappler, J. W., et al. Genetic and Hormonal Factors in Female-Biased Autoimmunity. *Autoimmun. Rev.* **2010,** *9* (7), 494–498.
11. Tower, C.; Crocker, I.; Chirico, D., et al. SLE and Pregnancy: The Potential Role for Regulatory T Cells. *Nat. Rev. Rheumatol.* **2011,** *7* (2), 124–128.
12. Molokhia, M.; McKeigue, P. Systemic Lupus Erythematosus: Genes Versus Environment in High Risk Populations. *Lupus* **2006,** *15* (11), 827–832.

CHAPTER

77

Rheumatoid Disease and Other Inflammatory Arthropathies

Sarah Keidel, Catherine Swales, and Paul Wordsworth

Nuffield Department of Orthopaedics, Rheumatology and Musculoskeletal Sciences, Nuffield Orthopaedic Centre, Headington, Oxford, UK

The inflammatory arthropathies described in this chapter are multifactorial polygenic disorders. Associations with human leukocyte antigens (HLA) were first described in 1973, although the precise mechanisms by which these cause disease are still unclear. In recent years, the application of genome-wide association studies (GWAS) has played an important part in identifying many of the genes involved.

Rheumatoid arthritis (RA) is an inflammatory polyarthropathy with a prevalence of approximately 1%. The recognition that anti-citrullinated protein antibodies (ACPA) occur with high frequency in RA has led to their development as diagnostic aids in early inflammatory arthritis. ACPA have a similar sensitivity to rheumatoid factor (RF) for RA (~80%) but are more specific (~98%). Historically, despite active treatment with standard disease-modifying drugs, fewer than 50% of those with RA have been able to work full-time after 10 years of disease. RA also carries a highly significant excess mortality, mainly as a result of increased cardiovascular disease and infections. New approaches to treatment with early, intensive use of combination disease-modifying therapy and the addition of biologic agents in refractory cases have had a dramatic effect on the joint disease in RA. The disease is approximately three times more common in women than men, but this varies with the age of onset. Based on twin studies, broad sense heritability has been estimated to be about 55% (*1*). Environmental factors include smoking, which is associated with seropositive RA. In most populations that have been studied, there is a strong association with certain *HLA-DRB1*04* alleles (*HLA-DRB1*0401*, **0404*, **0408*, and **0405*). These alleles have in common a highly conserved sequence between amino acids 67 and 74 along the α-helix derived from the DRβ chain (the shared epitope), which forms one side of the antigen-binding site of the DR molecule. Although the HLA association is by far the strongest, it has been estimated that the contribution to heritability is no greater than 40% of the whole genetic component. Initial studies using whole genome linkage approaches had limited success in elucidating the genetic risk, but more recent GWAS have proved very effective, identifying more than 30 genetic regions likely to be involved in the disease. It appears that there are no other major genetic effects of the size of the major histocompatibility complex (MHC); the majority of loci have small effect sizes, with odds ratios of ≤1.05. It is interesting that many genetic variants conferring RA risk are associated with multiple autoimmune or inflammatory diseases. Currently, there are 14 non-HLA risk loci common to both celiac disease and RA; 7 of these are also common to type 1 diabetes (*2,3*). The themes of these common risk alleles include innate immunity (e.g. *TRAF/C5*), T cell differentiation (*STAT4*), and T cell signaling (*PTPN22*) (*4*). It has been postulated that the unique and dominant HLA associations result in presentation of disease-specific auto-antigens to T cells; the common non-HLA risk determinants may in turn influence the response of the immune cells to these auto-antigens (*3*).

The seronegative spondyloarthropathies are a group of inflammatory conditions that include ankylosing spondylitis, reactive arthritis, enteropathic arthritis, and psoriatic arthritis. They are characterized by a number of common features, including inflammation of the entheses; prominent axial and asymmetric lower limb peripheral large joint arthritis; extra-articular features including uveitis; the formation of new bone at the site of inflammation; and association with the MHC class I gene HLA-B*27. Ankylosing spondylitis (AS) is the hallmark spondyloarthropathy, with a prevalence of approximately 0.5%. Overall, men are affected more commonly than women (ratio ~2.8:1). The prevalence of AS roughly parallels the prevalence of HLA-B*27 in different populations. In the United Kingdom, more than 90% of patients with AS carry HLA-B*27 (relative risk ~160). Many theories have been proposed to explain the association with HLA-B*27, but none is universally accepted (*5*). Most likely, the role of HLA-B*27 in antigen presentation plays an important part. Several relatively small twin studies have suggested a significant genetic component to AS; heritability has been estimated in excess of 92%, of which HLA-B*27 accounts for <50%. GWAS studies have therefore been instrumental in identifying loci conferring modest risk of AS; at least 14 have been identified to date. The association of AS with several genes in the IL-17 producing (Th17) lymphocyte subset has already marked this as a potential therapeutic target. The highly significant association of AS with *ERAP1* has provided an important example of gene–gene interaction in susceptibility to a complex disease, since the association is entirely restricted to those individuals with AS who also carry HLA-B*27 (*6*). ERAP1 variants associated with protection against AS have been demonstrated to have reduced ability to trim peptide antigens to optimal length for binding to HLA class 1 molecules (*7*). This raises the possibility that inhibitors of ERAP1 could be protective against AS.

It has been estimated that juvenile idiopathic arthritis (JIA) has a prevalence of between 20 and 120 per 100,000 and an annual incidence of 10–20 per 100,000. Under the current classification, JIA has been divided into multiple subtypes, including oligoarticular, seropositive and seronegative polyarticular, psoriatic, enthesitis-related, and undifferentiated arthritis. The available family studies provide support for a genetic component to JIA, but many of these studies predate the more accurate classification of JIA. HLA-DR alleles have been estimated to confer almost 20% of the total sibling recurrence risk in JIA, although the risks associated with HLA alleles differ between the JIA subtypes (*8*). *HLA-DRB1*08* (a rare allele in most white populations) is consistently found in between 25% and 50% of patients with oligoarticular JIA (relative risk ~12). *HLA-DRB1*04*, which is associated with RA in adults, is associated with seropositive polyarthritis (OR 3.2), but is protective against many other JIA subtypes. *DRB1*07* is associated with decreased risk of JIA (*9*). There is considerable evidence for a marked compound heterozygote effect with predisposing DRB1 alleles (*10*). Conducting studies of sufficient power to examine the risk contribution of non-HLA alleles is made difficult by the rarity of JIA and the phenotypic heterogeneity of the disease. While studies encompassing JIA as an umbrella disease may mask genetic associations with individual JIA subsets, subset sub-analysis may fail to detect genetic risk alleles with small effects (*11*). Therefore, approaches to identifying candidate genes have included investigating loci associated with other autoimmune diseases (*12*). More than 100 non-MHC risk alleles have been examined (*11*), although only a small number of these have been replicated to date. The *PTPN22* and *IL2RA* gene associations have reached genome-wide significance. Other significant associations include genes implicated in T cell signaling and activation (*STAT4*, *VTCN1*), innate immunity (*TNFA*, *TNFAIP3*, *TRAF1/C5*, *MIF*, *SCL11A1*), and cartilage homeostasis (*WISP3*) (*13*).

Although great strides have been made in the genetic basis of inflammatory arthropathies, in many cases additional work is required, including Fine mapping or deep resequencing of the genetic regions of interest, to precisely identify the causative allelic variants underlying the single-nucleotide polymorphism associations. In addition, despite these advances, more than 50% of the genetic contribution to RA remains unexplained. This may be due to as yet unidentified rare genetic variants with large effect or large numbers of risk alleles of small effect not yet reliably identifiable by current methods (*14*). Further large-scale association studies with saturation mapping of the relevant loci are nearing completion and should further advance the field in the near future.

REFERENCES

1. MacGregor, A. J.; Snieder, H.; Rigby, A. S.; Koskenvuo, M.; Kaprio, J.; Aho, K.; Silman, A. J. Characterizing the Quantitative Genetic Contribution to Rheumatoid Arthritis Using Data from Twins. *Arthritis Rheum.* **2000**, *43*, 30–37.
2. Eyre, S.; Hinks, A.; Bowes, J.; Flynn, E.; Martin, P.; Wilson, A. G.; Morgan, A. W.; Emery, P.; Steer, S.; Hocking, L. J., et al. Overlapping Genetic Susceptibility Variants between Three Autoimmune Disorders: Rheumatoid Arthritis, Type 1 Diabetes and Coeliac Disease. *Arthritis Res. Ther.* **2010**, *12*, R175.
3. Zhernakova, A.; Stahl, E. A.; Trynka, G.; Raychaudhuri, S.; Festen, E. A.; Franke, L.; Westra, H. J.; Fehrmann, R. S.; Kurreeman, F. A.; Thomson, B., et al. Meta-Analysis of Genome-Wide Association Studies in Celiac Disease and Rheumatoid Arthritis Identifies Fourteen Non-HLA Shared Loci. *PLoS Genet.* **2011**, *7*, e1002004.
4. Zhernakova, A.; van Diemen, C. C.; Wijmenga, C. Detecting Shared Pathogenesis from the Shared Genetics of Immune-Related Diseases. *Nat. Rev. Genet.* **2009**, *10*, 43–55.
5. Alvarez, I.; Lopez de Castro, J. A. HLA-B27 and Immunogenetics of Spondyloarthropathies. *Curr. Opin. Rheumatol.* **2000**, *12*, 248–253.
6. Evans, D. M.; Spencer, C. C.; Pointon, J. J.; Su, Z.; Harvey, D.; Kochan, G.; Oppermann, U.; Dilthey, A.; Pirinen, M.; Stone, M. A., et al. Interaction between ERAP1 and HLA-B27 in Ankylosing Spondylitis Implicates Peptide Handling in the Mechanism for HLA-B27 in Disease Susceptibility. *Nat. Genet.* **2011**, *43*, 761–767.
7. Kochan, G.; Krojer, T.; Harvey, D.; Fischer, R.; Chen, L.; Vollmar, M.; von Delft, F.; Kavanagh, K. L.; Brown, M. A.; Bowness, P., et al. Crystal Structures of the Endoplasmic Reticulum Aminopeptidase-1 (ERAP1) Reveal the Molecular Basis for N-Terminal Peptide Trimming. *Proc. Natl. Acad. Sci. U.S.A* **2011**, *108*, 7745–7750.
8. Prahalad, S.; Ryan, M. H.; Shear, E. S.; Thompson, S. D.; Giannini, E. H.; Glass, D. N. Juvenile Rheumatoid Arthritis: Linkage to HLA Demonstrated by Allele Sharing in Affected Sibpairs. *Arthritis Rheum.* **2000**, *43*, 2335–2338.

9. Thomson, W.; Barrett, J. H.; Donn, R.; Pepper, L.; Kennedy, L. J.; Ollier, W. E.; Silman, A. J.; Woo, P.; Southwood, T. Juvenile Idiopathic Arthritis Classified by the ILAR Criteria: HLA Associations in UK Patients. *Rheumatology (Oxford)* 2002, *41*, 1183–1189.
10. Hollenbach, J. A.; Thompson, S. D.; Bugawan, T. L.; Ryan, M.; Sudman, M.; Marion, M.; Langefeld, C. D.; Thomson, G.; Erlich, H. A.; Glass, D. N. Juvenile Idiopathic Arthritis and HLA Class I and Class II Interactions and Age-at-Onset Effects. *Arthritis Rheum.* 2010, *62*, 1781–1791.
11. Prahalad, S.; Glass, D. N. A Comprehensive Review of the Genetics of Juvenile Idiopathic Arthritis. *Pediatr. Rheumatol. Online J.* 2008, *6*, 11.
12. Hinks, A.; Cobb, J.; Sudman, M.; Eyre, S.; Martin, P.; Flynn, E.; Packham, J.; Barton, A.; Worthington, J.; Langefeld, C. D., et al. Investigation of Rheumatoid Arthritis Susceptibility Loci in Juvenile Idiopathic Arthritis Confirms High Degree of Overlap. *Ann. Rheum. Dis.* 2012.
13. Macaubas, C.; Nguyen, K.; Milojevic, D.; Park, J. L.; Mellins, E. D. Oligoarticular and Polyarticular JIA: Epidemiology and Pathogenesis. *Nat. Rev. Rheumatol.* 2009, *5*, 616–626.
14. Okada, Y.; Terao, C.; Ikari, K.; Kochi, Y.; Ohmura, K.; Suzuki, A.; Kawaguchi, T.; Stahl, E. A.; Kurreeman, F. A.; Nishida, N., et al. Meta-Analysis Identifies Nine New Loci Associated with Rheumatoid Arthritis in the Japanese Population. *Nat. Genet.* 2012, *44*, 511–516.

CHAPTER

78

Amyloidosis and Other Protein Deposition Diseases

Merrill D Benson

Professor of Pathology and Laboratory Medicine, Professor of Medical and Molecular Genetics, and Professor of Medicine, Indiana University School of Medicine, Indianapolis, IN, USA

Hereditary amyloidoses are a group of protein deposition diseases in which β-structured protein fibrils are deposited in the extracellular space of tissues. These deposits (amyloid) displace normal tissue structures causing cell death and organ dysfunction. Mutations in a number of proteins can cause amyloidosis and each specific protein denotes a separate disease. The general theme for hereditary amyloidosis pathogenesis is that mutation in a normal biologic protein causes misfolding, altered metabolic processing and resultant aggregation to form insoluble β-structured fibrils. By definition these fibril deposits are characterized by their extracellular location and exhibition of green birefringence when stained with Congo red and viewed by polarization microscopy.

At present mutations in seven different proteins are known to cause hereditary systemic amyloidosis (*1*,*2*). All are normally present in serum and can give deposits in all organ systems and, while several cause specific disease phenotypes, they often can be distinguished from the more common systemic immunoglobulin light chain (AL) type of amyloidosis only with difficulty. This is important since all too often patients with hereditary amyloidosis have mistakenly received a diagnosis of AL amyloidosis and subjected to inappropriate chemotherapy.

Transthyretin (TTR) amyloidosis is the most common form of hereditary amyloidosis. Greater than 100 mutations in TTR have been reported to cause amyloidosis (*3*,*4*). Most are single-nucleotide changes and a number of codons for this single-chain 127-amino acid residue protein have multiple mutations. TTR amyloidosis is most often characterized by peripheral neuropathy and cardiomyopathy; although gastrointestinal and, less common, renal involvement may occur (*5*). Approximately 15 TTR mutations are associated with leptomeningeal and/or vitreous amyloid deposition. In addition, senile cardiac amyloidosis or senile systemic amyloidosis may occur in individuals, mostly men, over the age of 60 years (*6*). In this condition the amyloid deposits contain only normal TTR. No mutation in TTR is required to initiate this disease.

Apolipoprotein AI (apo-AI) amyloidosis is not as common as TTR amyloidosis but can give disease phenotypes similar to other forms of amyloidosis. Fifteen different mutations in apo-AI have been reported to give systemic amyloidosis. Disease phenotype varies with the location of the mutation. Mutations in the amino terminal part (amino acid residues 1–75) cause renal or hepatic pathology whereas mutations from amino acid 90 to the carboxyl end of the molecule cause laryngeal, cutaneous and cardiac amyloid deposition.

Fibrinogen Aα-chain (AFibA) has been associated with nine different mutations all in the protease-sensitive carboxyl-terminal portion of the molecule. Pathology is typically renal with clinical phenotype of proteinuria followed by hypertension and azotemia over a 1- to 10-year period. All reported patients have been heterozygous for an amyloid mutation but unlike TTR the fibril deposits contain only the mutated protein. The renal pathology is typical with glomerular effacement and loss of tubule structures. The majority of mutations in Fibrinogen Aα-chain are single-nucleotide substitutions but single-nucleotide deletions with shift of reading frame and novel peptide sequences have been reported. This type of amyloidosis has been reported in the second decade of life, but most patients are 30–60 years old at disease onset. Penetrance is quite variable.

Lysozyme amyloidosis is associated with renal or hepatic pathology. Six different mutations have been identified with this disease. One has been associated with cardiac amyloidosis. While lysozyme is an enzyme, the disease is obviously not related to its enzymatic function. The amyloidosis is an autosomal dominant trait as are all the structural protein forms of amyloidosis.

Apolipoprotein AII (apo-AII) amyloidosis is the result of mutation at the stop codon of apo-AII. This results in a 21-amino acid peptide extension of mature apo-AII and subsequent fibril formation. The entire mutant apo-AII is found in the amyloid but no normal apo-AII. The amyloidosis is typically renal, although most organ systems may be involved in advanced stages of the disease. apo-AII amyloidosis is rare but so far five different mutations in the stop codon have been discovered and all lead to amyloid formation.

Gelsolin amyloidosis (AGel) is caused by mutation in the actin-modulating protein gelsolin. Two mutations at the same gene site can give this disease. The most common was discovered in Finnish families. Disease manifestations include lattice corneal dystrophy in early adult life followed by cranial nerve palsy and cutis laxa involving the face and upper torso. Renal and cardiac amyloid deposition may occur and is more pronounced in homozygous patients.

Cystatin C (ACys) amyloidosis was originally discovered in Iceland. Cerebrovascular amyloid deposition is the main pathological feature. This causes repeated intracranial hemorrhage and death in early to mid-life.

78.1 DNA DIAGNOSIS

DNA analysis has been developed for each type of hereditary amyloidosis but first it is necessary to consider the diagnosis and then use knowledge of the various disease phenotypes to select appropriate tests. Genomic DNA analysis is available for TTR mutations from several commercial laboratories. Testing for mutations in the other forms of hereditary amyloidosis is not readily available and consultation with one of the amyloidosis centers is recommended (http://www.iupui.edu/~amyloid/; http://www.mayoclinic.com/health/amyloidosis/DS00431; http://www.bu.edu/amyloid/).

78.2 TREATMENT OF HEREDITARY AMYLOIDOSIS

At the present time symptomatic treatment is the most we have to offer for most patients with hereditary amyloidosis. Treatment of neuropathy, heart failure, renal failure and gastrointestinal dysfunction is the same as with other forms of systemic amyloidosis (AL, AA).

Liver transplantation is a specific therapy for TTR amyloidosis since it stops production of mutant TTR. This has helped many patients but others have had progression of disease by continued amyloid made from normal TTR (*3*). Liver transplantation for AFib does stop the disease since fibrinogen is only synthesized by the liver and only the variant fibrinogen participates in fibril formation. Recently liver transplantation before advanced renal failure has been proposed and promises to help avert the need for concurrent renal transplantation (*7*). Since disease penetrance is variable, liver transplantation should not be done before documented amyloid deposition has started. Liver transplantation for apo-AI amyloid has been met with variable results probably because not all apo-AI is of hepatic origin.

Solid organ transplantation for ALys and apo-AII has given prolongation of life for a number of patients but this is not a form of specific therapy.

Recognition of systemic forms of hereditary amyloidosis even when there is lack of family history is very important to a timely and correct diagnosis. While we do not have much to offer for specific therapies for these diseases we can avoid incorrect diagnosis which can lead to inappropriate treatments.

FIGURE 78-1 Amyloid in the vitreous of the eye, which causes progressive loss of vision, can be seen as fluffy deposits on funduscopic examination.

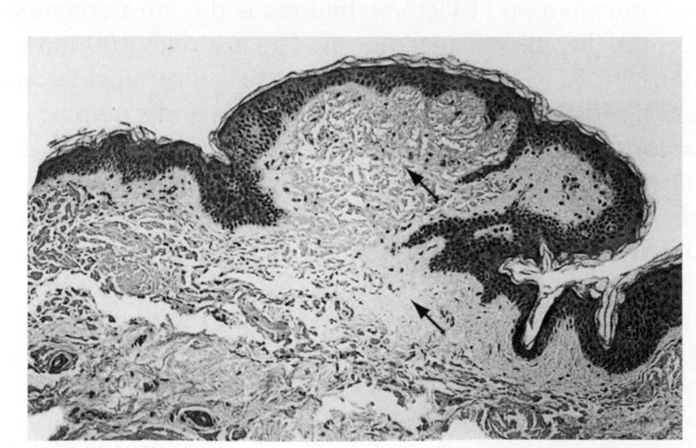

FIGURE 78-2 Apolipoprotein A-I amyloid deposits in the skin are characteristic of the Leu90Pro and Arg173Pro mutations.

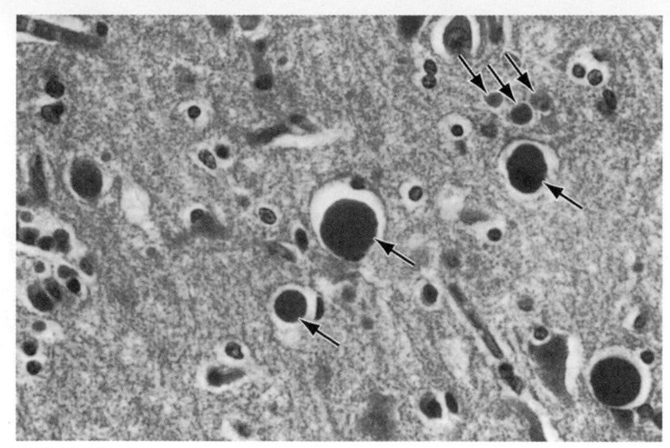

FIGURE 78-3 Intraneuronal nonamyloid deposits contain mutant neuroserpin in a patient with hereditary myoclonus epilepsy and dementia.

REFERENCES

1. Benson, M. D. Other Systemic Forms of Amyloidosis. In *Amyloidosis: Diagnosis and Treatment*; Gertz, M. A.; Rajkumar, S. V., Eds.; Humana Press—Springer Science + Business Media: LLC, 2010, Chapter 15; pp 205–225.
2. Westermark, P.; Benson, M. D.; Buxbaum, J. N., et al. A Primer of Amyloid Nomenclature. *Amyloid* **2007**, *14,* 179–183.
3. Zeldenrust, S. R. ATTR: Diagnosis, Prognosis, and Treatment. In *Amyloidosis: Diagnosis and Treatment*; Gertz, M. A.; Rajkumar, S. V., Eds.; Humana Press—Springer Science + Business Media: LLC, 2010, Chapter 14; pp 191–204.
4. Benson, M. D. Genetics: Clinical implications of TTR Amyloid. In *Recent Advances in Transthyretin Evolution, Structure and Biological Functions*; Richardson, S.; Cody, V., Eds.; Springer Publications, 2009, Chapter 11; pp 173–189.
5. Benson, M. D.; Kincaid, J. C. Invited Review—The Molecular Biology and Clinical Features of Amyloid Neuropathy. *Muscle Nerve* **2007**, *36,* 411–423.
6. Zeldenrust, S. R.; Benson, M. D. Familial and Senile Amyloidosis Caused by Transthyretin. In *Protein Misfolding Diseases: Current and Emerging Principles and Therapies, Part IV*; Ramirez-Alvarado, M.; Kelly, J.; Dobson, C., Eds.; Copyright © 2010 Wiley[Imprint], Inc., 2010, Chapter 36; pp 795–815.
7. Stangou, A. J.; Banner, N. R.; Hendry, B. M., et al. Hereditary Fibrinogen Aα-Chain Amyloidosis: Phenotypic Characterization of a Systemic Disease and the Role of Liver Transplantation. *Blood* **2010**, *115,* 2998–3007.

CHAPTER

79

Immunodeficiency Disorders

Rochelle Hirschhorn, Kurt Hirschhorn and Luigi D Notarangelo

Primary immunodeficienciency diseases (PIDs) comprise a heterogeneous group of disorders that affect development and/or function of the immune system. Most forms of PIDs are genetically determined disorders and follow a Mendelian pattern of inheritance.

PIDs are classified according to the nature of the defect. In particular, the following subgroups are recognized in the most recent classification of PIDs, elaborated by the International Union of Immunological Societies (*1*):

(1) Combined Immunodeficiencies: These include disorders with impaired development and/or function of both T and B lymphocytes;
(2) Well-defined syndromes with immunodeficiency: In these disorders, immunodeficiency is one aspect of the complex clinical phenotype;
(3) Predominantly antibody deficiencies: In this group of disorders, antibody production is affected as the result of impaired development and/or function of B lymphocytes;
(4) Diseases of immune dysregulation: Inability to extinguish immune responses and genetically determined autoimmunity is a prominent feature of this group of PIDs;
(5) Congenital defects of phagocyte number, function, or both: This group of PIDs include forms of severe congenital neutropenia, diseases with functional abnormalities of neutrophils and macrophages, disorders with defects of the interleukin-12 (IL-12)/interferon-γ (IFN-γ) axis leading to mycobacterial disease, and genetic defects of dendritic cells;
(6) Defects of innate immunity: These include abnormalities of Toll-like receptor (TLR) signaling pathways and of NF-κB signaling pathway, as well as chronic mucocutaneous candidiasis due to defects of IL-17-mediated signaling;
(7) Autoinflammatory disorders: This group of diseases is characterized by genetically determined susceptibility to hyper-inflammatory responses. Although they are clearly distinct from classical forms of PIDs, these diseases are part of the PID classification to highlight their nature of genetic defects of the immune system;
(8) Complement deficiencies: Disorders that affect the classical, alternative, and lectin pathways of complement are included in this subgroup.

The clinical phenotype of PIDs is strictly related to the nature and the severity of the underlying immune defect. Recognition of PIDs is based on careful analysis of the medical and family history; specific tests to investigate the immune system should be planned according to the clinical history.

The most severe forms of PID are represented by defects that affect development of T lymphocytes. In these patients, there is also an impairment of antibody responses (because of the lack of T–B cell interaction); hence, these conditions are also known as severe combined immune deficiency (SCID). SCID may be caused by various genetic defects that affect specific stages in T-cell development; some of these defects also affect differentiation of B and/or NK lymphocytes. Patients with SCID present early in life with life-threatening infections (often sustained by opportunistic pathogens), chronic diarrhea and failure to thrive. SCID is inevitably fatal within the first years of life, unless immune reconstitution is achieved, mostly with hematopoietic stem cell transplantation (HSCT). Survival rate after HSCT for SCID exceeds 90% if the transplant is performed from an HLA-matched related donor or if the transplant is performed early after birth, before infections develop. Recently, newborn screening for SCID has been introduced in the United States and is anticipated to facilitate recognition and prompt and effective treatment for this group of disorders. Adenosine deaminase (ADA) deficiency represents a form of SCID caused by a metabolic disorder; patients with ADA deficiency may also be treated with enzyme replacement therapy. Gene therapy has been successfully used in infants with X-linked SCID and with SCID due to ADA deficiency; however, leukemic proliferation due to insertional mutagenesis has been observed in five of these patients, prompting the development of novel and safer vectors.

X-linked agammaglobulinemia (XLA) is the prototype of disorders of humoral immunity. Patients with XLA lack circulating B cells and are unable to produce antibodies. They suffer from recurrent bacterial infections and are also at higher risk for enteroviral meningoencephalitis. Common variable immune deficiency (CVID) includes a heterogeneous group of conditions characterized by low immunoglobulin levels and impaired antibody production. Genetic defects have been identified only in a minority of patients with CVID. In addition to recurrent infections, CVID is characterized also by an increased occurrence of autoimmune manifestations and an increased risk of lymphoma. Management of antibody deficiency syndromes is based on regular intravenous or subcutaneous administration of immunoglobulins, and on prompt and aggressive diagnosis and treatment of infections. Use of immunosuppressive drugs may be necessary in patients with CVID and active autoimmune disease.

Defects in the mechanisms of central and peripheral tolerance are responsible for some rare forms of autoimmunity. Immune dysregulation–polyendocrinopathy–enteropathy–X-linked (IPEX) syndrome is an X-linked disorder caused by mutations of *FOXP3*, a gene that encodes for a transcription factor that plays a key role in development of regulatory T (Treg) cells. Typically, patients with IPEX present early in life with intractable diarrhea, skin rash, and insulin-dependent diabetes mellitus. This is a life-threatening disease that requires immune suppression but may often need HSCT. Autoimmune lymphoproliferative syndrome (ALPS) includes various conditions characterized by defective apoptosis and autoimmune cytopenias, associated with lymphadenopathy and splenomegaly.

Immune dysregulation is also observed in patients with defects of cell-mediated cytotoxicity, such as various forms of hemophagocytic lymphohistiocytosis (HLH) and X-linked lymphoproliferative disease. In these patients, inability to eliminate viruses may cause persistent activation of the immune system, with increased production of inflammatory cytokines (IFN-γ), cytopenias, severe liver disease and coagulopathy. Definitive treatment is based on HSCT.

Numerical and functional defects of neutrophils cause increased susceptibility to severe bacterial and fungal infections. Administration of granulocyte colony-stimulating factor (G-CSF) is the mainstay of treatment of severe congenital neutropenia, which may be due to a variety of genetic defects that affect development and survival of neutrophils. Patients with chronic granulomatous disease (CGD) have impairment of respiratory burst in neutrophils and macrophages, leading to defective killing of bacteria and fungi. Furthermore, these patients are often prone to inflammatory complications of hollow organs. Treatment of CGD is based on regular antibacterial and antifungal prophylaxis and aggressive management of active infections; however, HSCT may be needed in the most severe forms of the disease.

Patients with defects in early components of the classical pathway of complement are susceptible to invasive bacterial infections and to autoimmune disease. Late defects (C5–C9) predispose to recurrent meningitis due to Neisseria. Invasive bacterial infections may also be seen in patients with X-linked properdin deficiency. Immunization against *Streptococcus pneumoniae, Haemophilus* and *Neisseria meningitidis* represents the mainstay to prevent these infections in patients with defects of complement.

Finally, some forms of PIDs are characterized by extra-immune manifestations. Defective repair of DNA double-strand breaks and increased cellular radiosensitivity are the hallmark of ataxia telangiectasia. These patients suffer from progressive cerebellar involvement, recurrent infections and increased risk of leukemia and other tumors. Patients with the X-linked Wiskott–Aldrich syndrome suffer from eczema and have increased susceptibility to infections, autoimmunity, leukemia and lymphoma.

Although PIDs are considered rare disorders, their study has been essential to unravel molecular and cellular mechanisms that play a critical role in development and function of the immune system. Furthermore, it is being increasingly recognized that different mutations in the same PID-associated gene may cause a spectrum of clinical phenotypes, thus adding to the complexity of the diagnostic approach. Finally, PIDs have served an important role in the development of novel therapeutic approaches to human diseases, including the first successful experiences with HSCT and gene therapy in infants with SCID.

REFERENCE

1. Al-Herz, W.; Bousfiha, A.; Casanova, J. L.; Chapel, H.; Conley, M. E.; Cunningham-Rundles, C.; Etzioni, A.; Fischer, A.; Franco, J. L.,; Geha, R. S., et al. Primary Immunodeficiency Diseases: An Update on the Classification from the International Union of Immunological Societies Committee for Primary Immunodeficiency. *Front. Immunol.* **2011**, Epub Nov 8.

FURTHER READING

Casanova, J. L.; Abel, L.; Quintana-Murci, L. Human TLRs and IL-1Rs in Host Defense: Natural Insights from Evolutionary, Epidemiological, and Clinical Genetics. *Annu. Rev. Immunol.* **2011**, *29*, 447–491.

Chapel, H.; Cunningham-Rundles, C. Update in Understanding Common Variable Immunodeficiency Disorders (CVIDs) and the Management of Patients with These Conditions. *Br. J. Haematol.* **2009**, *145* (6), 709–727.

Conley, M. E.; Dobbs, A. K.; Farmer, D. M.; Kilic, S.; Paris, K.; Grigoriadou, S.; Coustan Smith, E.; Howard, V.; Campana, D. Primary B Cell Immunodeficiencies: Comparisons and Contrasts. *Annu. Rev. Immunol.* **2009**, *27*, 199–227.

Filipovich, A. H. The Expanding Spectrum of Hemophagocytic Lymphohistiocytosis. *Curr. Opin. Allergy Clin. Immunol.* **2011**, *11* (6), 512–516.

Fischer, A.; Hacein-Bey-Abina, S.; Cavazzana-Calvo, M. Gene Therapy for Primary Adaptive Immune Deficiencies. *J. Allergy Clin. Immunol.* **2011**, *127* (6), 1356–1359.

Gennery, A. R.; Slatter, M. A.; Grandin, L.; Taupin, P.; Cant, A. J.; Veys, P.; Amrolia, P. J.; Gaspar, H. B.; Davies, E. G.; Friedrich, W., et al. Inborn Errors Working Party of the European Group for Blood and Marrow Transplantation; European Society for Immunodeficiency. Transplantation of Hematopoietic Stem Cells and Long-Term Survival for Primary Immunodeficiencies in Europe: Entering a New Century, Do We Do Better? *J. Allergy Clin. Immunol.* **2010,** *126* (3), 602–610.

Kang, E. M.; Marciano, B. E.; DeRavin, S.; Zarember, K. A.; Holland, S. M.; Malech, H. L. Chronic Granulomatous Disease: Overview and Hematopoietic Stem Cell Transplantation. *J. Allergy Clin. Immunol.* **2011,** *127* (6), 1319–1326.

Klein, C. Genetic Defects in Severe Congenital Neutropenia: Emerging Insights into Life and Death of Human Neutrophil Granulocytes. *Annu. Rev. Immunol.* **2011,** *29,* 399–413.

Notarangelo, L. D. Primary Immunodeficiencies. *J. Allergy Clin. Immunol.* **2010,** *125* (2 Suppl. 2), S182–S194.

Puck, J. M. Laboratory Technology for Population-Based Screening for Severe Combined Immunodeficiency in Neonates: The Winner Is T-Cell Receptor Excision Circles. *J. Allergy Clin. Immunol.* **2012,** *129* (3), 607–616.

Skattum, L.; van Deuren, M.; van der Poll, T.; Truedsson, L. Complement Deficiency States and Associated Infections. *Mol. Immunol.* **2011,** *48* (14), 1643–1655.

Slatter, M. A.; Gennery, A. R. Primary Immunodeficiencies Associated with DNA-Repair Disorders. *Expert Rev. Mol. Med.* **2010,** *12,* e9.

Torgerson, T. R. Immune Dysregulation in Primary Immunodeficiency Disorders. *Immunol. Allergy Clin. North Am.* **2008,** *28* (2), 315–327.

CHAPTER

80

Inherited Complement Deficiencies

Kathleen E Sullivan

Division of Allergy Immunology, The Children's Hospital of Philadelphia, Philadelphia, PA, USA

The complement system is an ancient system of soluble proteins and cell receptors whose main functions in host defense are to opsonize bacteria, provide costimulation for B cells, and to directly kill gram-negative bacteria. The complement system also provides important host protective functions such as protection of vascular endothelial cells and clearance of apoptotic debris. Complement also plays an important role in establishing B cell tolerance.

The complement system is traditionally divided into the soluble members of the complement cascade, the regulators of the complement cascade, and the receptors for the complement components or their cleavage products. In addition, the soluble components are divided into three activation arms: the classical pathway, the lectin activation pathway, and the alternative pathway (Figure 80-1). Each of these activation arms converges on C3, leading to cleavage and activation of the membrane attack components. These latter components (C6, C7, C8, C9) perform one major function and that is to penetrate and destabilize cell membranes. On eukaryotic cell membranes, the damage to the cell membrane is typically repaired; however, red cells do not have sufficient protein synthetic processes to effectively repair cell membranes and they are uniquely susceptible to lysis by complement. Bacteria have varying susceptibilities to complement in vitro, but Neisseria seems to the pathogen that requires complement for host defense.

80.1 CLINICAL ASPECTS OF COMPLEMENT DEFICIENCIES

There are five main phenotypes associated with inherited complement deficiencies and the phenotypes are predictable based on the known functions of the deficient components. Deficiencies of the early components of the classical pathway (C1, C4, C2) are associated with systemic lupus erythematosus and/or infections with encapsulated organisms. Although the lupus phenotype is more prevalent and is most often the manifestation leading to diagnosis, infection is the major cause of death and life expectancy is thought to be consequently somewhat shortened. C3 deficiency represents a slight variation on the theme, with very profound infectious susceptibility and glomerulonephritis rather than lupus.

Deficiencies of many of the regulatory proteins (FH, FI, MCP) are associated with atypical hemolytic uremic syndrome. The term atypical refers to the fact that the patients present with hemolytic uremic syndrome in the absence of a diarrheal prodrome. The prognosis is highly variable as is the penetrance, but most patients have recurrent episodes in the absence of treatment. MCP deficiency is conceptually different in that it is a membrane-bound protein and renal transplantation is curative.

Heterozygous deficiency of the regulatory protein C1 esterase inhibitor is associated with hereditary angioedema. This disorder is characterized by recurrent episodes of angioedema. Unlike allergic angioedema, there is no associated urticaria and the triggering episodes are most often trauma. Treatment with recombinant replacement protein is available as are other interventions to ameliorate swelling.

Terminal component deficiencies (C5, C6, C7, C8, C9) are associated with a unique susceptibility to Neisseria. Sepsis, meningitis, and chronic meningococcemia can all be seen. The susceptibility to Neisseria can be ameliorated by repeated vaccination but that strategy only addresses the serotypes contained in the vaccine.

80.2 GENETICS

Deficiencies of most complement components are inherited in autosomal recessive fashion (Table 80-1). C1 esterase inhibitor is inherited in an autosomal dominant fashion and properdin deficiency is the sole X-linked deficiency. Deficiencies associated with hemolytic uremic syndrome can be either recessive or dominant and adding additional diagnostic confusion are the lack of full penetrance and the presence of a phenocopy due to autoantibodies directed against similar proteins. Although deficiencies of the regulatory proteins are most notable for incomplete penetrance, in general, most complement deficiencies have incomplete penetrance.

80.3 DIAGNOSIS

The two main screening tests are the CH50 and the AH50. The CH50 tests for the intactness of the classical pathway from C1 through C9. The AH50 tests for the intactness of the alternative pathway from FB through C9. Both assays are technically difficult and performed at only a few laboratories. In addition, sample handling represents a significant source of artifactually abnormal results. Most inherited deficiencies of the cascade are associated with CH50 or AH50 results of zero or near zero. Low but detectable levels are most often due to sample handling, immune complex activation, immaturity, or liver disease. When those entities have been eliminated and the CH50 or the AH50 remains low, defects in regulatory proteins or C9 should be considered. If both the CH50 and the AH50 are low, C9 deficiency is likely, as only C5, C6, C7, C8, and C9 are shared in these two assays, and C9 deficiency is commonly associated with residual lytic ability. Once an inherited deficiency is suspected, the phenotype and the race can lead to a presumptive diagnosis. In some cases, there is a direct antigen or functional assay; however, most complement deficiencies are identified through a laborious mixing procedure. The mixing assays are only performed in the complement reference laboratory.

TABLE 80-1 Complement Regulatory Proteins

Protein	Localization	Function
C1 inhibitor	Serum	Binds to C1r and C1s and dissociates the C1 complex
C4 binding protein	Serum	Cofactor for factor I cleavage of C4b
Factor I	Serum	Cleaves C3b and C4b
Factor H	Serum	Defines activator surface
S-protein	Serum	Inhibits the insertion of the membrane attack complex into the cell membrane
Decay accelerating factor (DAF)	Ubiquitous-cell membrane	Dissociates both C3 and C5 convertases
Membrane cofactor protein	Hematopoietic cells except erythrocytes	Cofactor for C3b cleavage by factor I
C8 binding protein	Most hematopoietic cells	Binds to C8 and prevents interaction with C9
CD59	Hematopoietic cells, endothelial cells, epithelial cells, glomerular cells	Inhibits the membrane attack complex

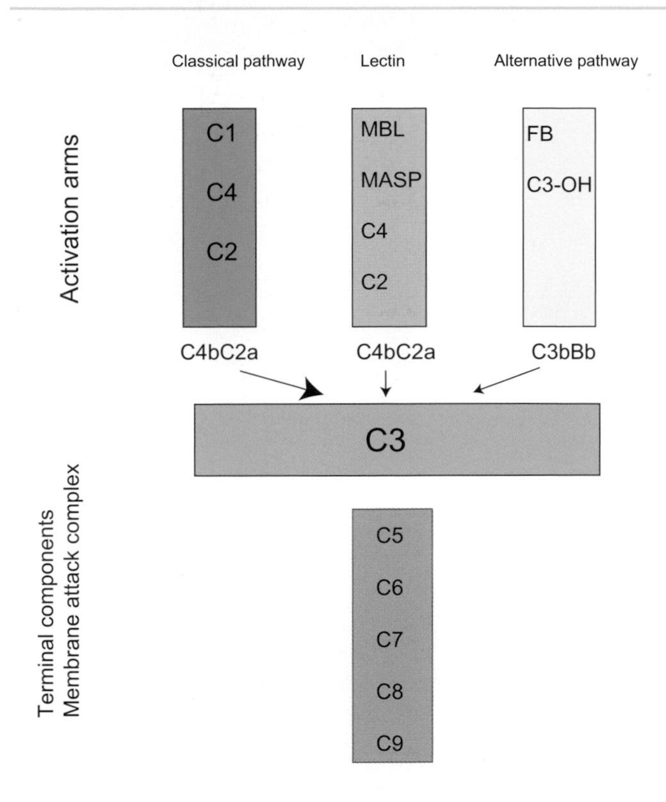

FIGURE 80-1 The three activation arms of the complement system. Activation of C3 is important for opsonization and B cell costimulation. These two major functions of host defense are dependent on C3 and its appropriate activation. The terminal components or membrane attack complex are required for lysis of bacteria.

FURTHER READING

1. Botto, M.; Kirschfink, M.; Macor, P.; Pickering, M. C.; Würzner, R.; Tedsco, F. Complement in Human Diseases: Lessons from Complement Deficiencies. *Mol. Immunol.* **2009,** *46* (14), 2774–2783.
2. Zimmerhackl, L. B.; Hofer, J.; Cortina, G.; Mark, W.; Würzner, R.; Jungraithmayr, T. C.; Khursigara, G.; Kliche, K. O.; Radauer, W. Prophylactic Eculizumab after Renal Transplantation in Atypical Hemolytic-Uremic Syndrome. *N. Engl. J. Med.* **2010,** *362* (18), 1746–1748.
3. Hirt-Minkowski, P.; Dickenmann, M.; Schifferli, J. A. Atypical Hemolytic Uremic Syndrome: Update on the Complement System and What Is New. *Nephron. Clin. Pract.* **2010,** *114* (4), c219–c235.
4. Garcia-Laorden, M. I.; Sole-Violan, J.; Rodriguez de Castro, F.; Aspa, J.; Briones, M. L.; Garcia-Saavedra, A.; Rajas, O.; Blanquer, J.; Caballero-Hidalgo, A.; Marcos-Ramos, J. A., et al. Mannose-Binding Lectin and Mannose-Binding Lectin-Associated Serine Protease 2 in Susceptibility, Severity, and Outcome of Pneumonia in Adults. *J. Allergy Clin. Immunol.* **2008,** *122* (2), 368–374.
5. Wen, L.; Atkinson, J. P.; Giclas, P. C. Clinical and Laboratory Evaluation of Complement Deficiency. *J. Allergy Clin. Immunol.* **2004,** *113* (4), 585–593.
6. Lipsker, D.; Hauptmann, G. Cutaneous Manifestations of Complement Deficiencies. *Lupus* **2010,** *19* (9), 1096–1106.
7. Reis, E. S.; Falcao, D. A.; Isaac, L. Clinical Aspects and Molecular Basis of Primary Deficiencies of Complement Component C3 and its Regulatory Proteins Factor I and Factor H. *Scand. J. Immunol.* **2006,** *63* (3), 155–168.
8. Würzner, R. Deficiencies of the Complement *MAC II* Gene Cluster (C6, C7, C9): Is Subtotal C6 Deficiency of Particular Evolutionary Benefit? *Clin. Exp. Immunol.* **2003,** *133* (2), 156–159.

Endocrinologic Disorders

Disorders of Leukocyte Function, 321
Genetic Disorders of the Pituitary Gland, 325
Thyroid Disorders, 330
Parathyroid Disorders, 334
Diabetes Mellitus, 338
Genetic Disorders of the Adrenal Gland, 350
Disorders of the Gonads, Genital Tract, and Genitalia, 352
Cancer of the Breast and Female Reproductive Tract, 361

CHAPTER

81

Disorders of Leukocyte Function

Harry R Hill, Attila Kumánovics and Kuender D Yang

Department of Pathology, University of Utah, Salt Lake City, UT, USA

Disorders of leukocyte function conventionally include the primary (or genetic) immunodeficiencies of phagocytes (granulocytes and monocytes). The first disorder of leukocyte function was described about 60 years ago by Kostmann as infantile genetic agranulocytosis (or severe congenital neutropnia, Kostmann syndrome), and since that an entire spectrum of phagocyte disorders, including adhesion, movement, ingestion, and bactericidal defects have been recognized and their genetic bases have been determined. With the exception of chronic granulomatous disease, these are individually rare diseases, but as a group they represent about 12% of all congenital immunodeficiency patients (*1*).

Phagocyte progenitor cells arise from hematopoietic stem cells in the bone marrow. These precursors can give rise to mature phagocytes, including polymorphonuclear granulocytes and mononuclear phagocytes, under the influence of various colony-stimulating factors and transcription factors (*2*). Neutrophil granulocytes (or polymorphonuclear leukocytes, PMNs) are the most abundant leukocytes in the bloodstream and are critical for host defense. The cytoplasmic granules contain cytotoxic molecules that are available for immediate release. Granulocytes are short-lived, terminally differentiated cells whose numbers are maintained in a narrow range. Empirically, a blood granulocyte concentration less than 1500/mm^3 is defined as granulocytopenia, and 500/mm^3 or less is usually considered severe, life-threatening granulocytopenia. It should be noted, however, that the circulating blood granulocyte level appears to be genetically determined, and varies among various ethnic groups (*3*). Circulating blood monocytes may develop into different forms of dendritic cells and macrophages, possessing diverse morphologies and functions depending on environmental and immunological factors. Mononuclear phagocytes, similar to polymorphonuclear granulocytes, can move toward foreign invaders and denatured tissues and destroy them by ingestion, degranulation, and respiratory burst activity as well as by the generation of nitric oxide from arginine. In addition, macrophages and monocyte-derived dendritic cells can also take up and present the antigens to lymphocytes for adaptive immune responses.

81.1 NEUTROPHIL NUMBER DISORDERS

Patients with congenital granulocytopenia usually have recurrent infections such as omphalitis, septicemia, and abscess formation early in life. Those who survive infancy frequently show progressive periodontitis. There are now at least 15 gene defects known to cause severe congenital neutropenia (SCN), but in about 30% of cases the genetic etiology is still unknown (*4,5*). Congenital granulocytopenias may occur as isolated deficiencies or as part of a syndrome. The majority of autosomal dominant nonsyndromic granulocytopenias are caused by mutations in ELA2 (or ELANE) that encodes neutrophil elastase. ELA2 mutations can lead to severe congenital neutropenia or cyclic neutropenia. Cyclic neutropenia is distinguished from SCN by the regular oscillations (approximately 21 day cycles) of blood neutrophil counts. Autosomal recessive SCN can be caused by mutations in HAX1, SLC7A4, G6PC3, or GFI1. X-linked neutropenias are caused by activating mutations in Wiskott–Aldrich syndrome (WAS) protein. The unifying feature of the molecular pathogenesis of SCN is the maturation arrest of precursor cells due to increased apoptosis. Increased apoptosis might be caused by the unfolded protein response (UPR) due to accumulation of misfolded proteins (e.g. ELA2 and G6PC3), mitochondrial membrane potential (HAX1), or defective mitosis and cytokinesis (WAS).

Chediak–Highasi syndrome, type 2 Hermansky Pudlak syndrome, Griscelli syndrome, and p14/ROBLD3 deficiency all lead to neutropenia associated with pigmentation defects caused by mutations in genes encoding proteins involved in intracellular trafficking of organelles. Cohen, Cartilage Hair Dysplasia, Reticular dysgenesis, Schwachmann–Diamond, Barth, and Pearson syndromes present with neutropenia. WHIM syndrome is a chemotaxis defect (see below) that also leads to neutropenia (*4,5*).

81.2 LEUKOCYTE ADHESION DEFICIENCIES

Leukocyte adhesion deficiency (LAD) disorders are characterized by the inability of leukocytes to emigrate from the circulation to the sites of injury. β2 integrins expressed on the leukocyte surface mediate tight binding to the endothelium (*6*). LAD-I is caused by mutations in the gene encoding the common β-chain of the β2 integrin family (CD18). LAD-II is caused by the deficiency in fucosylation required for the function of adhesion molecules and also known as Congenital Disorder of Glycosylation, Type IIc. It is caused by mutations in the *SLC35C1* gene encoding a Golgi ADP-fucose transporter (*7*). LAD-III is caused by deficient cytokine activation of the adherent leukocytes leading to diminished integrin expression. LAD-III is caused by homozygous mutation in the *FERMT3* (or *KINDLIN3*) gene (*8*). Patients with LAD-I frequently have delayed separation of the umbilical cord, perirectal abscesses, and recurrent staphylococcal and gram-negative bacterial infections. Patients with LAD-II also have mental retardation, neutrophilia, recurrent infections, and periodontitis, but they do not have delayed separation of the umbilical cord. Patients with LAD-III were first described as LAD-I variants who suffered from a bleeding tendency due to platelet aggregation deficiency.

81.3 DISORDERS OF GRANULOCYTE CHEMOTAXIS

WHIM syndrome is an autosomal dominant disorder of Warts, Hypogammaglobulinemia, Infections, and Myelokathexis caused by heterozygous gain-of-function mutations in the gene encoding the CXCR4 chemokine receptor (*9*). Chemotaxis defects also include the previously mentioned deficiency of the adhesion molecules and defects associated with impaired degranulation (e.g. Chédiak–Higashi syndrome). Hyper-IgE syndrome (or Job syndrome) caused by mutations in *STAT3* is usually associated with a granulocyte chemotactic defect and high levels of IgE, altered T-cell activity, and low IFN-γ and IL17 production (*10*).

81.4 DISORDERS OF MICROBICIDAL ACTIVITY

Disorders of oxygen-dependent microbicidal activity disorders are relatively common and include chronic granulomatous disease (CGD), glucose-6-phosphate dehydrogenase (G6PD) deficiency, myeloperoxidase (MPO) deficiency, and glutathione synthetase and reductase deficiencies. Patients with defects in oxygen-dependent bactericidal activity suffer from recurrent cutaneous, organ, and tissue abscesses and recurrent sinopulmonary infections. Granulomas may form at the site of tissue infections, resulting in obstructive disorders involving the gastrointestinal and urinary systems. Infections are most often due to catalase-positive microorganisms.

CGD is caused by the deficiency of the phagocyte-specific NADPH oxidase (or NOX2; (*11*)). The neutrophil NADPH oxidase is composed of two membrane and four cytosolic components. Two-thirds of CGD cases are transmitted in an X-linked recessive pattern owing to the absence of cytochrome b558 heavy chain (p91phox, encoded by CYBB). The rest are inherited by an autosomal recessive mechanism. Mutations in p47phox are found in about 30% of CGD patients. Mutations in p22phox and p67phox each are less than 5% of CGD cases. Only one CGD patient is described with mutation in NCF4 that encodes p40phox (*12*). Rac2 mutations have not been reported as a cause of CGD.

G6PD deficiency is one of the most common enzyme deficiency in humans, with over 140 different mutations described (*13*). The most common mutations in G6PD (seen in patients with African ancestry) lead a mild to moderate enzyme deficiency (20–60% activity) and to intermittent hemolysis usually secondary to infections or drugs. Few patients with severe forms of G6PD deficiency can present with a CGD-like clinical picture, as the profound deficiency of G6PD leads to a defect in NADPH production, causing impaired respiratory burst activity.

Granulocytes with glutathione reductase or glutathione synthetase deficiency have a normal early respiratory burst, whereas subsequent continuous production of toxic oxygen products, which are normally handled by glutathione, results in auto-oxidative damage and defective microbicidal activity (*14*). Both glutathione reductase and glutathione synthetase deficiencies are transmitted by autosomal recessive inheritance. Patients with these deficiencies can have hemolytic disorders in addition to the abnormal bactericidal activity.

Granulocytes in MPO deficiency (MPO; (*15*)) have normal production of superoxide but defective generation of hypochlorite ion. Patients with this disorder are not uncommon (1 in 2000 to 1 in 4000 in the general population), but are frequently asymptomatic or at most have delayed granulocyte killing activity and recurrent candidal infections.

Disorders of oxygen-independent microbicidal activity arise from the absence of granules or from defective degranulation. Patients with Specific Granule Deficiency have autosomal recessive inheritance caused by mutations in CCAAT/enhancer-binding protein epsilon (*C/EBPE*) gene (*16*), whereas Hermansky–Pudlak, Chediak–Higashi and Griscelli syndromes lead to granulocyte dysfunction due to the failure of the granule constituents to fuse with the phagosome (*17*).

81.5 MONOCYTE/MACROPHAGE FUNCTION DISORDERS

Monocytes/macrophages share the same ancestor cells as granulocytes, and most of the functional defects present in granulocyte function disorders also occur in monocytes/macrophages, including LAD, respiratory burst defects such as CGD, degranulation defects such as in Chediak–Higashi syndrome, and specific granule deficiency. Primary macrophage function defects are relatively rare; most of the macrophage function defects are secondary.

Monocytes, macrophages express toll-like receptors (TLRs) that are pattern-recognition receptors (PRRs). PPRs recognize microorganisms or microbial products, and mediate signals for innate immunity and adaptive T-cell differentiation. Mutations in the shared intracellular signaling molecules of TLRs (IRAK4 and MyD88) lead to recurrent pyogenic bacterial infections, including invasive pneumococcal disease (*18*).

Macrophage phagosomes are both the site of elimination of mycobacteria and the site of mycobacterial replication (*19,20*). Mendelian susceptibility to mycobacterial diseases (MSMD) and Salmonella infection are caused by deficiencies in the IL-12/IL-23/IFN-γ cytokines and their signaling pathways (IFN-γR, IFN-γR2, IL-12p40, IL-12Rβ1, STAT1, or NEMO deficiency). Interestingly, the autosomal recessive form of the STAT1 signaling molecule deficiency leads to susceptibility to both mycobacteria (IFN-γ-mediated immunity) and viruses (IFN-α/β-mediated immunity), whereas the autosomal dominant STAT1 deficiency leads to MSMD only. NEMO deficiency is an X-linked MSDS.

Tissue resident macrophages have specialized functions. The best example is in the bone. Osteopetrosis (*21*) is due to dysfunction of osteoclasts (specialized macrophages), resulting in impaired resorption of mineralized cartilage and bone remodeling. Osteopetrosis, transmitted as an autosomal recessive trait, is called osteopetrosis congenita, and it occurs in infants and children. Osteopetrosis tarda is an autosomal dominant trait, which occurs in adults. Six genes are known as causative for autosomal recessive osteopetrosis (*TCIRG1*, *ClC7*, *OSTM1*, *PLEKHM1*, *TNFSF11*, and *TNFRSF11A*). A rare X-linked recessive anhidrotic ectodermal dysplasia with osteopetrosis and immunodeficiency is caused by mutations in *IKBKG* gene (encoding NEMO). An autosomal dominant form of anhidrotic ectodermal dysplasia with T cell immunodeficiency is caused by gain-of-function mutation of IKBA.

Another group of hereditary macrophage function defects include the metabolic storage disorders, especially those with lysosomal enzyme deficiencies (discussed elsewhere).

REFERENCES

1. Gathmann, B.; Grimbacher, B.; Beaute, J.; Dudoit, Y.; Mahlaoui, N.; Fischer, A.; Knerr, V.; Kindle, G. The European Internet-Based Patient and Research Database for Primary Immunodeficiencies: Results 2006–2008. *Clin. Exp. Immunol.* **2009**, *157* (Suppl. 1), 3–11.
2. Friedman, A. D. Transcriptional Control of Granulocyte and Monocyte Development. *Oncogene* **2007**, *26* (47), 6816–6828.
3. Nalls, M. A.; Wilson, J. G.; Patterson, N. J.; Tandon, A.; Zmuda, J. M.; Huntsman, S.; Garcia, M.; Hu, D.; Li, R.; Beamer, B. A., et al. Admixture Mapping of White Cell Count: Genetic Locus Responsible for Lower White Blood Cell Count in the Health ABC and Jackson Heart studies. *Am. J. Hum. Genet.* **2008**, *82* (1), 81–87.
4. Bouma, G.; Ancliff, P. J.; Thrasher, A. J.; Burns, S. O. Recent Advances in the Understanding of Genetic Defects of Neutrophil Number and Function. *Br. J. Haematol.* **2010**, *151* (4), 312–326.
5. Boztug, K.; Klein, C. Novel Genetic Etiologies of Severe Congenital Neutropenia. *Curr. Opin. Immunol.* **2009**, *21* (5), 472–480.
6. Ley, K.; Morris, M. Signals for Lymphocyte Egress. *Nat. Immunol.* **2005**, *6* (12), 1215–1216.
7. Helmus, Y.; Denecke, J.; Yakubenia, S.; Robinson, P.; Luhn, K.; Watson, D. L.; McGrogan, P. J.; Vestweber, D.; Marquardt, T.; Wild, M. K. Leukocyte Adhesion Deficiency II Patients with a Dual Defect of the GDP-Fucose Transporter. *Blood* **2006**, *107* (10), 3959–3966.
8. Svensson, L.; Howarth, K.; McDowall, A.; Patzak, I.; Evans, R.; Ussar, S.; Moser, M.; Metin, A.; Fried, M.; Tomlinson, I., et al. Leukocyte Adhesion Deficiency-III Is Caused by Mutations in KINDLIN3 Affecting Integrin Activation. *Nat. Med.* **2009**, *15* (3), 306–312.
9. Kawai, T.; Malech, H. L. WHIM Syndrome: Congenital Immune Deficiency Disease. *Curr. Opin. Hematol.* **2009**, *16* (1), 20–26.
10. Woellner, C.; Gertz, E. M.; Schaffer, A. A.; Lagos, M.; Perro, M.; Glocker, E. O.; Pietrogrande, M. C.; Cossu, F.; Franco, J. L.; Matamoros, N., et al. Mutations in STAT3 and Diagnostic Guidelines for Hyper-IgE Syndrome. *J. Allergy Clin. Immunol.* **2010**, *125* (2), 424–432, e8.
11. Kuhns, D. B.; Alvord, W. G.; Heller, T.; Feld, J. J.; Pike, K. M.; Marciano, B. E.; Uzel, G.; DeRavin, S. S.; Priel, D. A.; Soule, B. P., et al. Residual NADPH Oxidase and Survival in Chronic Granulomatous Disease. *N. Engl. J. Med.* **2010**, *363* (27), 2600–2610.
12. Matute, J. D.; Arias, A. A.; Wright, N. A.; Wrobel, I.; Waterhouse, C. C.; Li, X. J.; Marchal, C. C.; Stull, N. D.; Lewis, D. B.; Steele, M., et al. A New Genetic Subgroup of Chronic Granulomatous Disease with Autosomal Recessive Mutations in p40 phox and Selective Defects in Neutrophil NADPH Oxidase Activity. *Blood* **2009**, *114* (15), 3309–3315.
13. Beutler, E.; Vulliamy, T. J. Hematologically Important Mutations: Glucose-6-Phosphate Dehydrogenase. *Blood Cells Mol. Dis.* **2002**, *28* (2), 93–103.
14. Njalsson, R.; Ristoff, E.; Carlsson, K.; Winkler, A.; Larsson, A.; Norgren, S. Genotype, Enzyme Activity, Glutathione Level, and Clinical Phenotype in Patients with Glutathione Synthetase Deficiency. *Hum. Genet.* **2005**, *116* (5), 384–389.
15. Marchetti, C.; Patriarca, P.; Solero, G. P.; Baralle, F. E.; Romano, M. Genetic Characterization of Myeloperoxidase Deficiency in Italy. *Hum. Mutat.* **2004**, *23* (5), 496–505.
16. Gombart, A. F.; Koeffler, H. P. Neutrophil Specific Granule Deficiency and Mutations in the Gene Encoding Transcription Factor C/EBP(epsilon). *Curr. Opin. Hematol.* **2002**, *9* (1), 36–42.

17. Huizing, M.; Helip-Wooley, A.; Westbroek, W.; Gunay-Aygun, M.; Gahl, W. A. Disorders of Lysosome-Related Organelle Biogenesis: Clinical and Molecular Genetics. *Annu. Rev. Genomics Hum. Genet.* **2008**, *9*, 359–386.
18. Picard, C.; von Bernuth, H.; Ghandil, P.; Chrabieh, M.; Levy, O.; Arkwright, P. D.; McDonald, D.; Geha, R. S.; Takada, H.; Krause, J. C., et al. Clinical Features and Outcome of Patients with IRAK-4 and MyD88 Deficiency. *Medicine* **2010**, *89* (6), 403–425, (Baltimore).
19. Bustamante, J.; Arias, A. A.; Vogt, G.; Picard, C.; Galicia, L. B.; Prando, C.; Grant, A. V.; Marchal, C. C.; Hubeau, M.; Chapgier, A., et al. Germline CYBB Mutations that Selectively Affect Macrophages in Kindreds with X-linked Predisposition to Tuberculous Mycobacterial Disease. *Nat. Immunol.* **2011**, *12* (3), 213–221.
20. Cottle, L. E. Mendelian Susceptibility to Mycobacterial Disease. *Clin. Genet.* **2011**, *79* (1), 17–22.
21. Stark, Z.; Savarirayan, R. Osteopetrosis. *Orphanet J. Rare Dis.* **2009**, *4*, 5.

CHAPTER

82

Genetic Disorders of the Pituitary Gland

Amy Potter

Ian M. Burr Division of Pediatric Endocrinology and Diabetes, Division of Endocrinology, Metabolism, and Diabetes, Vanderbilt University Medical Center, Nashville, TN, USA

John A Phillips

Division of Medical Genetics and Genomic Medicine, Vanderbilt University of Medicine, Medical Center North, Nashville, TN, USA

David L Rimoin

Medical Genetics Institute, Cedars-Sinai Medical Center, Los Angeles, CA, USA

82.1 INTRODUCTION

The pituitary gland is composed of two developmentally and functionally distinct units, the anterior pituitary and the posterior pituitary. Disease processes usually involve only one of these units unless both are affected because of their anatomic proximity or because of hypothalamic involvement. Thus disorders of anterior and posterior pituitary function will be discussed separately.

82.2 ANTERIOR PITUITARY

82.2.1 Normal Pituitary Development and Function

The anterior pituitary is derived from an epithelial invagination of the roof of the posterior pharynx, Rathke's pouch, while the future posterior pituitary arises from the third ventricle. The pituitary stalk, composed primarily of neurohypophyseal tissue surrounded by nerves and blood vessels, connects the pituitary with the hypothalamus, allowing sensitive hypothalamic control of pituitary function.

The anterior pituitary contains several distinct cell types responsible for secretion of the pituitary hormones: growth hormone (GH), thyroid stimulating hormone (TSH), adrenocorticotropic hormone (ACTH), luteinizing hormone (LH), follicle-stimulating hormone (FSH), prolactin (PRL), and melanocyte-stimulating hormone (MSH). Control of hormone secretion is mediated by a variety of hypothalamic-releasing hormones (e.g. thyrotropin-releasing hormone (TRH) stimulates TSH secretion) and inhibitory hormones (e.g. somatostatin (SST) inhibits GH secretion). The interplay between specific releasing and inhibitory hormones controls secretion of each of the pituitary hormones. In turn, the secretion of the hypothalamic inhibitory and releasing hormones is modulated by a variety of humoral and central nervous system factors. Once released into the plasma, the pituitary hormones exert their effects on their targets, either specific endocrine glands (e.g. thyroid) or a variety of end organs. Because of the complexity of this hypothalamo-pituitary axis, a variety of pathogenetic mechanisms can operate at each level of the system, resulting in a heterogeneous group of disorders with similar symptoms of pituitary insufficiency.

A complex cascade of transcription factors guides development of the pituitary. Important factors in early pituitary development include PITX1/2, HESX1, SHH, and LHX3/4, while genes such as *PROP1* and *POU1F1* (also called PIT1) play an important role in the differentiation of anterior pituitary hormone–secreting cells. In general, disruptions of genes involved in early pituitary development result in more severe defects and often brain anomalies, while defects in genes involved in later aspects of pituitary development lead to varying combinations of pituitary hormone deficiencies (*1*).

GH is a multifunctional hormone which promotes postnatal growth of skeletal and soft tissues through a variety of effects. A major portion of the effects of GH are mediated through the actions of GH-dependent insulin-like growth factor 1 (IGF1, also called somatomedin C). IGF1 is produced in many tissues, primarily the liver, and acts through its own receptor, IGF1R, to enhance the proliferation and maturation of many tissues, including bone, cartilage, and skeletal muscle. Each GH molecule binds two GH receptor (GHR) molecules, causing them to dimerize. Dimerization of the two GH-bound GHR molecules leads to signal transduction through the JAK-STAT pathway (2).

82.2.2 Developmental Anomalies and Syndromes

The holoprosencephalies are developmental anomalies associated with impaired midline cleavage of the embryonic forebrain, aplasia of the olfactory bulbs and tracts, and midline dysplasia of the face, ranging from cyclopia to cleft lip and palate with hypotelorism. Anomalies of the pituitary gland have been described in all forms of holoprosencephaly, and varying degrees of pituitary insufficiency may occur. Mutations have been described in a number of genes, including *SHH*, *ZIC2*, *SIX3*, and *TGIF* (see Chapter 114) (7).

Cases of transsphenoidal encephalocele associated with variable hypopituitarism have been reported. Associated features include an epipharyngeal or nasopharyngeal mass, hypertelorism, midfacial or midline craniocerebral anomalies, and optic nerve abnormalities. Diagnosis of this syndrome is important, as iatrogenic hypopituitarism had been described following extirpation of a nasopharyngeal mass containing pituitary tissue.

Septo-optic dysplasia (SOD) is a rare malformation which classically includes optic nerve hypoplasia, midline brain defects (most commonly absent or cavum septum pellucidum or agenesis of the corpus callosum), and hypopituitarism, sometimes with hypoplasia of the anterior pituitary and/or ectopic posterior pituitary. Affected neonates may present with signs of hypopituitarism including hypoglycemia and prolonged jaundice; patients may also have defective vision, developmental delay, hypotonia, and seizures. In mildly affected cases, the child may present with proportionate short stature and pendular nystagmus, with or without amblyopia. Intelligence may be normal or mildly to moderately affected. Ophthalmologic examination reveals bilateral hypoplasia of the optic nerves, with small optic discs and irregular field defects. This syndrome should be considered in any child with hypopituitarism and nystagmus or abnormalities of the optic disc. A small percentage of patients with SOD and pituitary defects have been found to have HESX1 mutations (3).

Midline cleft lip and palate can be associated with GH deficiency, diabetes insipidus (DI), holoprosencephaly, and optic nerve hypoplasia with absence of the septum pellucidum. Pituitary insufficiency has been described in a number of individuals with cleft lip and palate without other facial or neurologic abnormalities. Pituitary insufficiency varies from complete combined pituitary hormone deficiency (CPHD), associated with congenital aplasia of the pituitary, to isolated growth hormone deficiency (IGHD).

82.2.3 Combined Pituitary Hormone Deficiency

CPHD is defined as deficiency of any two or more of the pituitary hormones (most commonly GH plus at least one other hormone). Hereditary forms of CPHD represent a genetically heterogeneous group of disorders, mostly resulting from disruption of normal pituitary development. The clinical features of hereditary CPHD are dependent on which hormones are deficient.

POU1F1. In Snell (dw) and Jackson (dw) dwarf mice, mutations in Pit1, a tissue-specific POU-domain transcription factor, lead to the absence of somatotroph, lactotroph and thyrotroph cells. Multiple mutations in the human gene *POU1F1* have been found in a subtype of CPHD associated with GH, PRL, and TSH deficiency. Some patients present with all hormone deficiencies and growth failure at a young age, while others present with GH deficiency initially and develop TSH deficiency later in childhood. The appearance of the pituitary on MRI can be either normal or hypoplastic. Although most cases of CPHD due to *POU1F1* mutations have autosomal recessive inheritance, some mutations lead to dominant inheritance via a dominant negative mechanism.

PROP1. Ames dwarf (df/df) mice have CPHD and hypocellular anterior pituitaries lacking somatotropes, lactotropes, and thyrotropes. The Ames dwarf phenotype is caused by defects in the *Prop1* gene, which encodes a pituitary-specific homeodomain factor. Human *PROP1* mutations causing CPHD have also been identified. Development of hormone deficiencies is sequential, following a predictable pattern: GH, TSH, FSH, LH, and ACTH. PRL deficiency is variable. ACTH deficiency may not develop until the fourth or fifth decade of life.

HESX1 functions as a transcriptional repressor. Its role in pituitary development was first discovered in mice; mice null for the mouse homolog *Hesx1* were found to have a phenotype similar to SOD in humans, leading to analysis of the human *HESX1* gene. Pituitary deficiencies range from IGHD to CPHD.

82.2.4 Disorders of the Growth Hormone Pathway

The various disorders of growth hormone secretion and function can be classified based on (1) the level of the defect, (2) mode of inheritance, (3) number of hormone deficiencies, (4) complete or partial absence of GH versus defective GH, and (5) receptor function or resistance (Table 82-1).

82.2.4.1 Familial Isolated Growth Hormone Deficiency (IGHD). IGHD IA, the most severe form, is autosomal recessive. Short stature develops in infancy but responds readily to GH. The initial good response is frequently followed by development of anti-GH antibodies which block response to treatment. Initially, all individuals with IGHD IA were found to be homozygous for

GH1 gene deletions and developed anti-GH antibodies with treatment. Additional cases with *GH1* gene deletions have been described who also have complete GH deficiency but respond well to GH replacement. Frameshift and nonsense mutations have also been found in subjects with the IGHD IA phenotype.

IGHD IB is the second autosomal recessive form. GH is deficient but detectable. Growth responds consistently to GH therapy and patients do not develop high titers of blocking antibodies. Both splicing mutations and deletion/frameshift mutations have been shown to cause IGHD IB.

IGHD II is an autosomal dominant disorder usually caused by heterozygosity for splicing mutations that increase GH1 exon 3 skipping (*4*). IGHD II cases vary in severity but respond well to exogenous GH and do not develop neutralizing antibodies. The first IGHD II mutations identified were point mutations at the 5 prime or donor splice site of IVS3 of GH1. When Δexon3hGH was targeted to somatotrophs in transgenic mice, it induced autosomal dominant GHD, with mild to severe pituitary hypoplasia and dwarfism. Pituitary somatotrophs in Δexon3hGH mice showed disruption of cellular morphology, due to accumulation of protein complexes in the Golgi apparatus and endoplasmic reticulum. This triggers the misfolded protein response, resulting in cell death (*5*). In humans, severity of IGHD II severity has been correlated with ratios of mutant to normal transcripts, with higher levels of mutant proteins leading to shorter stature (*6,7*).

IGHD III is an X-linked disorder, with some kindreds having hypogammaglobulinemia in association with growth hormone deficiency. IGHD III has been associated with contiguous gene defects of Xq21.3-q22, interstitial deletions of Xp22.3, or duplication of Xq13.3-q21.2.

82.2.4.2 Growth Hormone Resistance. Laron dwarfism type 1 is an autosomal recessive disorder caused by target resistance to the action of GH. Patients have clinical features of pituitary dwarfism associated with high plasma concentrations of immunoreactive GH. These individuals have the clinical appearance of patients with IGHD to an exaggerated extent, with severe growth retardation, severely pinched faces, high-pitched voices, and small male genitalia. Fasting plasma GH concentrations are usually elevated, but may fluctuate from normal levels to over 100 ng/mL in the same patient. There is further elevation of plasma immunoreactive GH concentration following insulin-induced hypoglycemia and arginine infusion. Plasma IGF1 levels are low and, unlike those in GH-deficient patients, do not respond to GH administration. Multiple point mutations have been detected within the *GHR* gene. These may cause GH resistance by affecting intracellular GHR transport, binding affinity, receptor expression, dimerization, or post-receptor signaling. Other mutations may affect splice sites, resulting in partial deletions.

Patients with Laron dwarfism II have elevated serum GH, normal GHBP levels, and respond well to treatment with IGF1, indicating that their growth deficiency is due to a post-GHR defect. This phenotype has been shown to be caused by defects in IGF1, STAT5B, or IGF1R.

82.2.5 Disorders of the Gonadotropin Pathway

82.2.5.1 Hypogonadotropic Hypogonadism. Kallmann Syndrome (KS) is a syndrome of hypogonadotropic hypogonadism coupled with anosmia, caused by failure of migration of GNRH neurons from the developing olfactory lobe to the hypothalamus. X-linked, autosomal dominant, and autosomal recessive forms are recognized. The X-linked form is the most common and is caused by mutations in *KAL1* which encodes anosmin-1, an extracellular matrix protein which acts as a chemoattractant for migrating neurons. Anosmin-1 also interacts with fibroblast growth factor signaling, and FGFR1 has been shown to the causative gene in autosomal dominant KS. Another gene, *CHD7*, has also been implicated in idiopathic hypogonadotropic hypogonadism as well as in CHARGE association.

82.2.5.2 Central Precocious Puberty. Onset of puberty is a complex process, in which onset of pulsative hypothalamic GNRH secretion is a key step. One regulator of this process is signaling of kisspeptin through its receptor (KISS1R, also called GPR54). A case of central precocious puberty caused by an activating mutation of KISS1R has been reported.

82.2.6 Disorders of Pituitary Hypersecretion and/or Neoplasia

82.2.6.1 Familial Isolated Pituitary Adenoma (FIPA). Although the majority of pituitary tumors are sporadic, many families have been reported in which multiple members are affected. Initially, it was thought that familial cases involved only one type of tumor (usually GH-secreting); however, it is now recognized that affected families may manifest many different types of tumors (*8*). Prolactinomas are most common, followed by somatotropinomas, and then nonfunctioning tumors. Kindreds have been described in which all affected family members have the same type of tumor, while other kindreds have multiple different types. Approximately, 15% of kindreds with FIPA have mutations in the aryl hydrocarbon receptor interacting protein gene (*AIP*). *AIP* appears to function as a tumor suppressor gene in pituitary tissue.

82.2.6.2 Multiple Endocrine Neoplasia Type 1 (MEN1). MEN1 is an autosomal dominant familial disorder characterized by multiple tumors or hyperplasia of the endocrine glands, most commonly parathyroid, pancreas, and pituitary. The clinical manifestations of pituitary disease are dependent on the type of tumor. The *MEN1* gene encodes a nuclear protein, menin, and a large number of inactivating mutations have been recognized.

82.2.6.3 Multiple Endocrine Neoplasia Type 4 (MEN4). MEN4 is an autosomal dominant neoplastic syndrome similar to MEN1, consisting primarily of pituitary tumors and hyperparathyroidism. Mutations in the cyclin-dependent kinase inhibitor p27 (CDKN1B), another tumor suppressor gene, have been identified.

82.2.6.4 Carney Complex. Carney complex is a syndrome consisting of mucocutaneous and cardiac myxomas, spotty skin pigmentation, schwannomas, and endocrine overactivity. Pituitary adenomas occur in >20% of patients and are always GH secreting. Carney complex 1 is related to mutations in *PRKAR1A*, inactivation of which leads to enhanced activity of the GH-releasing, hormone-induced signal transduction pathway. Carney complex type II is linked to chromosome 2p, but the gene is still unknown.

82.3 POSTERIOR PITUITARY: GENETIC DISORDERS OF AVP DEFICIENCY

82.3.1 Autosomal Dominant Neurohypophyseal Diabetes Indipidus (ADNDI)

The signs and symptoms of ADNDI are similar to those of other forms of DI, but with significant intrafamilial variability in the clinical severity and age of onset of the disease. Typical MRI studies of the brain show absence of the pituitary bright spot; however, the bright spot may still be present in young children. AVP secretion may be normal for the first few years of life, but then decreases rapidly to very low or undetectable levels. Mutations causing ADNDI generally result in abnormalities of protein trafficking or folding. Accumulation of cytotoxic precursors leads to cell death, which at least partially explains the delayed onset of ADNDI in most kindreds (*9*).

82.3.2 Wolfram Syndrome

Wolfram syndrome (WFS) is an autosomal recessive syndrome consisting of *DI*, *D*iabetes *M*ellitus, *O*ptic *A*trophy, and neurosensory *D*eafness (DIDMOAD). Diabetes mellitus in WFS is caused by degeneration of the insulin-producing cells of the pancreas and is a constant feature. Optic atrophy is typically the next sign to appear (occurs in 100% of patients), and may be present at the time of diabetes diagnosis. Bilateral neurosensory occurs in about 60% of patients. Neurohypophyseal DI occurs in more than one-third of patients with this syndrome (*10*). The gene, *WFS1*, has been identified and a variety of mutations have been described. *WFS1* codes for a novel transmembrane protein called wolframin, an endoglycosidase H-sensitive glycoprotein, which helps regulate calcium homeostasis and the unfolded protein response. A second autosomal recessive form of Wolfram syndrome (WFS2) is caused by mutations in the zinc finger protein CISD2.

TABLE 82-1 Genetic Disorders of the Growth Hormone (GH) Pathway

Phenotype	Disorder	OMIM	Mode	Locus	Gene/OMIM	Endogenous GH	Response to GH Therapy
Isolated GH deficiency	IGHD IA	262400	AR	17q22-q24	GH1/139250	Absent	Often temporary due to antibody development
	IGHD IB	612781	AR	17q2-q24	GH1/139250	Decreased	Present
			AR	7p15-P14	GHRHR/139191	Decreased	Present
	IGHD II	173100	AD	17q22-q24	GH1/139250	Decreased	Present
	IGHD III	307200	XL	Xq21.3-q22	BTK/300300	Decreased	Present
	Mental retardation X-linked with isolated GH deficiency	300123	XL	Xq26.3	SOX3/313430	Decreased	Present
	Kowarski syndrome ("bioinactive" GH)	262650	AR	17q22-q24	GH1/139250	Present (but non functional)	Present
Short stature with or without GH deficiency	Short stature, idiopathic, autosomal	604271	AR	3q26.3	GHSR/601898	Present or Absent	Present in at least one patient
GH resistance	Laron dwarfism 1	262500	AR	5p13-P12	GHR/600946	Normal or increased	Absent or decreased (may respond to exogenous IGF1)
	GH insensitivity with immunodeficiency	245590	?AR	17q11.2	STAT5B/604260	Normal or increased	Absent or decreased (may respond to exogenous IGF1)
	IGF1 deficiency	608747	AR	12q22-Q24.1	IGF1/147440	Normal or increased	Absent or decreased (may respond to exogenous IGF1)
	IGF1 resistance	270450	AR	15q25-Q26	IGF1R/147370	Normal or increased	Absent or decreased (may respond to exogenous IGF1)

IGHD, isolated growth hormone deficiency; AR, autosomal recessive; AD, autosomal dominant; XL, X-linked.
Online Mendelian Inheritance in Man, OMIM (TM). McKusick–Nathans Institute of Genetic Medicine, Johns Hopkins University (Baltimore, MD) and National Center for Biotechnology Information, National Library of Medicine (Bethesda, MD), {Accessed October 2010–March 2011}. World Wide Web URL: http://www.ncbi.nlm.nih.gov/omim/.

REFERENCES

1. Kelberman, D.; Rizzoti, K.; Lovell-Badge, R., et al. Genetic Regulation of Pituitary Gland Development in Human and Mouse. *Endocr. Rev.* **2009,** *30,* 790–829.
2. Dattani, M.; Preece, M. Growth Hormone Deficiency and Related Disorders: Insights in Causation, Diagnosis, and Treatment. *Lancet* **2004,** *363,* 1977–1987.
3. Kelberman, D.; Dattani, M. T. Genetics of Septo-Optic Dysplasia. *Pituitary* **2007,** *10,* 393–407.
4. Binder, G.; Keller, E.; Mix, M., et al. Isolated GH Deficiency with Dominant Inheritance: New Mutations, New Insights. *J. Clin. Endocrinol. Metab.* **2001,** *86,* 3877–3881.
5. McGuinness, L.; Magoulas, C.; Sesay, A. K., et al. Autosomal Dominant Growth Hormone Deficiency Disrupts Secretory Vesicles In Vitro and In Vivo in Transgenic Mice. *Endocrinology* **2003,** *144,* 720–731.
6. Ryther, R. C.; McGuinness, L. M.; Phillips, J. A., III, et al. Disruption of Exon Definition Produces a Dominant-Negative Growth Hormone Isoform that Causes Somatotroph Death and IGHD II. *Hum. Genet.* **2003,** *113,* 140–148.
7. Hamid, R.; Phillips, J. A., III; Holladay, C., et al. A Molecular Basis for Variation in Clinical Severity of Isolated Growth Hormone Deficiency Type II. *J. Clin. Endocrinol. Metab.* **2009,** *94,* 4729–4734.
8. Daly, A. F.; Tichomirow, M. A.; Beckers, A. Update on Familial Pituitary Tumors: From Multiple Endocrine Neoplasia Type 1 to Familial Isolated Pituitary Adenoma. *Horm. Res.* **2009,** *71,* 105–111.
9. Christensen, J. H.; Rittig, S. Familial Neurohypophyseal Diabetes Insipidus-an Update. *Semin. Nephrol.* **2006,** *26,* 209–223.
10. Kumar, S. Wolfram Syndrome: Important Implications for Pediatricians and Pediatric Endocrinologists. *Pediatr. Diabetes* **2010,** *11,* 28–37.

CHAPTER

83

Thyroid Disorders

Michel Polak

Department of Pediatric Endocrinology, Gynecology, and Diabetology, Centres de référence des maladies endocriniennes rares de la croissance, Université Paris Descartes, INSERM U 485, Hôpital Necker Enfants Malades, Paris, France

Gabor Szinnai

Department of Pediatric Endocrinology, and Diabetology, University Children's Hospital Basel (UKBB), University Basel, Basel, Switzerland

The thyroid gland consists of two endocrine cell types. (1) Thyroid follicular cells or thyrocytes (TFC) represent the large majority of endocrine cells of the thyroid. They form the thyroid follicles and are responsible for thyroid hormone synthesis, storage and secretion, upon TSH stimulation. (2) Parafollicular C-cells represent less than 1% of cells in the human thyroid and produce calcitonin, a hormone involved in calcium homeostasis. The mammalian TFC precursor cells derive from the foregut endoderm and form the median anlage of the thyroid, while C-cells are of neuroectodermal origin and are localized within the paired ultimobranchial bodies of the caudal pharyngeal pouches. After a migratory process, the median anlage fusions with the paired ultimobranchial bodies at the definitive pretracheal position. Thyroid development is accomplished with the onset of thyroid hormone synthesis in a significant amount at the end of the first trimester in the human embryo.

Abnormal thyroid gland development or thyroid dysgenesis is the most common cause of permanent peripheral or thyroidal congenital hypothyroidism in the human in iodine sufficient areas of the world.

Inborn errors of thyroid hormone synthesis, called thyroid dyshormonogenesis, occur at any step of thyroid hormone synthesis and usually cause familial goitrous congenital hypothyroidism. Central or hypothalamic-pituitary congenital hypothyroidism is rare. Finally, genetic disorders may explain the absence of transport of the thyroid hormone within the cells, or the absence of thyroid action with impaired binding to their receptors.

83.1 HYPOTHALAMIC-PITUITARY CONGENITAL HYPOTHYROIDISM

Defects of pituitary development result in various forms of impaired secretion of pituitary hormones and are reviewed in Chapter 82. To date, no mutations have been described in the human *TRH* gene. Resistance to thyrotropin-releasing hormone has been reported in two families linked to mutation in the human *TRH* receptor gene. Isolated TSH deficiency due to TSH-beta subunit defects is a rare autosomal recessive cause of congenital central hypothyroidism (*1,2*).

83.2 THYROIDAL CONGENITAL HYPOTHYROIDISM

Permanent congenital hypothyroidism affects about 1:3000–1:4000 newborns. Primary or thyroidal congenital hypothyroidism outweighs by far the rare central or hypothalamic-pituitary congenital hypothyroidism cases. Girls are more commonly affected than boys (female to male ratio 2:1–4:1) and is more common in Hispanic infants (1:2000) than in white infants, and is thought to be less common in black infants (1:11,000).

In iodine sufficient areas, the most common etiology of primary congenital hypothyroidism is thyroid dysgenesis, representing 80–85% of all cases. Thyroid dysgenesis includes a spectrum of developmental abnormalities of the thyroid gland ranging from (1) "agenesis" or athyreosis (20–30%) due to a defect in survival of the thyrocyte precursors, (2) ectopic thyroid gland (50–60%) mostly located in a sublingual position as a result of premature arrest of its migratory process, or (3) hypoplasia of an orthotopic gland (5%). Based on the phenotype of transgenic mouse models, mutational screening of cohorts of thyroid dysgenesis patients led to the identification of mutations in genes involved in thyroid development: *PAX8*, *FOXE1*, *NKX2-1*, and *NKX2-5*. However, mutations in these four genes are only found in a small percentage of patients with thyroid dysgenesis, even when screening large cohorts for each of the genes. So far, TSH receptor mutations causing TSH resistance are the most frequent cause of thyroid dysgenesis with over 60 patients identified so far followed by *NKX2-1* mutations with about 50 published cases. *PAX8* mutations have been reported in about 30 patients, and mutations in *FOXE1* and *NKX2-5* in less than 10 families so far (*3–5*).

In 10–15% of patients, primary congenital hypothyroidism is the consequence of defects of thyroid hormone synthesis (thyroid dyshormonogenesis) characterized by a normal or enlarged thyroid gland in normal position.

These include defects in (1) iodine uptake into thyrocytes due to mutations in the sodium/iodide symporter (*NIS*, encoded by the *SLC5A5* gene), (2) organification defects due to mutations in the thyroid peroxidase (*TPO*), dual oxidase 2 (*DUOX2*) and dual oxidase maturation factor 2 (*DUOXA2*) genes, or mutations in the Pendrin gene (*SLC26A4*), (3) defects in thyroglobulin (TG), a glycoprotein responsible for synthesis, storage, or release of thyroid hormones, and (4) defective iodotyrosine deiodinase (IYD/DEHAL1) activity leading to failure of iodide recycling in the thyrocyte (*6,7*).

83.2.1 Inheritance

While thyroid dyshormonogenesis is inherited in an autosomal recessive way, recent work has revealed some arguments for genetic basis of thyroid dysgenesis, which is assumed to be sporadic. First, although representing only a minority of cases of thyroid dysgenesis, familial cases were observed in a significantly higher proportion (>15 fold) than would be expected by chance alone. Second, an increased frequency of minor thyroid abnormalities in first-degree relatives of patients with thyroid dysgenesis has been described. Third, thyroid dysgenesis is associated with an increased incidence of extrathyroidal malformations. A fourfold increase in congenital malformations was found in a population based study of 1420 infants with congenital hypothyroidism (CH) (8.4%) compared to the control infant population (1–2%). Nevertheless, non-Mendelian mechanisms need to be considered to explain thyroid dysgenesis in most cases. A multigenic origin of thyroid hypoplasia and hemiagenesis has been shown in transgenic mouse models, but discordance in monozygotic twins excludes multigenic origin of thyroid dysgenesis as the exclusive mechanism in the human. Early somatic mutations or epigenetic modifications could play a role in the development of thyroid dysgenesis, in analogy with other human diseases. A two-hit model has been proposed: a germline mutation (first hit) and a somatic mutation or epigenetic modification (second hit) of genes involved in critical steps of thyroid development (*8–12*).

83.3 DISORDERS OF THYROID HORMONE TRANSPORT PROTEINS, MEMBRANE TRANSPORTERS AND THYROID HORMONE ACTION

Disorders of serum thyroid binding proteins result in alteration of measured thyroid hormone concentrations but do not alter thyroid hormone–dependent metabolic state of the body and do not cause thyroid disease. Thus, patients are clinically euthyroid but present with altered total T3 and T4 levels (but normal free T3 and free T4) in the context of normal TSH.

Many thyroid hormone transporters have been identified at the molecular level. So far, mutations have only been described in the *MCT8* gene which is located on the chromosome Xq13.2. The hallmark of *MCT8* gene mutations is the combination of thyroid and neurological dysfunction. Patients present during infancy with central hypotonia, poor head control, evolving to spastic quadriplegia, nystagmus, and severe mental retardation with absence of speech. There are no other signs of hypothyroidism. The key laboratory finding is elevated T3 concentration in the context of low T4 and free T4. TSH levels are in the upper normal range (*13*).

Intracellular metabolism of T4, which serves as prohormone, and availability of the active T3 is regulated by the three selenoprotein iodothyronine deiodinases. The presence of high serum T4 and TSH concentrations, low T3 concentrations, and markedly elevated rT3 concentrations led to the identification of an inherited selenocysteine incorporation defect, caused by a homozygous missense mutation in the selenocysteine insertion sequence binding protein 2 (*SECISBP2*) gene. *SECISBP2* serves as a cofactor for deiodinase synthesis (*14*).

In the nucleus, thyroid hormones are effective by binding to the thyroid hormone receptor (TR). So far, mutations were only identified in the *TR beta* gene, which is localized on chromosome 3. The clinical presentation is heterogeneous. Some patients have no or minor symptoms, others can present with symptoms of hypo- or hyperthyroidism. The characteristic diagnostic features of resistance to thyroid hormone are (1) elevated serum levels of fT3 and fT4, (2) normal or slightly increased TSH levels that respond to TRH, (3) absence of signs and symptoms of hyperthyroidism, and (4) goiter (*14*).

83.4 GENETIC HYPERTHYROIDISM

Non-autoimmune hyperthyroidism can be caused by activating TSH receptor mutations as well as activating

G-protein mutations in the context of McCune Albright syndrome (*15*).

83.5 THYROID DISEASE ASSOCIATED WITH CHROMOSOMAL DISORDERS AND CONTIGOUS GENE DELETION SYNDROMES

Thyroid disorders have been reported to have a prevalence rate of 3–54% in individuals with Down syndrome. Individuals with Turner syndrome have an increased risk for autoimmune thyroiditis. Patients with Williams syndrome and DiGeorge syndrome have a high prevalence of thyroid hypoplasia (50–70%) and compensated hypothyroidism (25–30%).

83.6 GENETIC BASIS OF AUTOIMMUNE THYROID DISEASE

Autoimmune thyroid disease (AITD) susceptibility genes have been identified. They can be divided into immune-modulating genes and thyroid-specific genes.

The autoimmune polyglandular syndromes (APS) are a group of diseases characterized by the presence of a combination of multiple autoimmune disorders. AITD occurs frequently in these patients. APS-1 also called autoimmune polyendocrinopathy candidiasis ectodermal dystrophy is a rare monogenic disease caused by mutations in the autoimmune regulator gene. APS-2 is the most common APS, a combination of Addison's disease with either AITD (Hashimoto's thyroiditis or Grave's disease) or type 1 diabetes mellitus and its inheritance is complex with multiple loci involved (*16*).

83.7 GENETIC BASIS OF THYROID CARCINOMA

Medullary thyroid cancer (MTC) derives from calcitonin-producing parafollicular C-cells. MTC can be sporadic but also familial, inherited in an autosomal dominant trait, in the context of Multiple Endocrine Neoplasia Type 2 (MEN 2). The common genetic basis of the MEN 2 subtypes is linked to gain-of-function germline point mutations of the rearranged during transfection (RET) proto-oncogene in patients with MEN 2A, MEN 2B and familial MTC (*17*).

Thyroid follicular cell tumors mostly occur sporadically, but familial forms do exist. The distinct forms of thyroid carcinomas are characterized by different genetic alterations in signaling pathways. Papillary thyroid carcinomas are essentially characterized by mutations of the *BRAF* and *RAS* genes and RET/PTC rearrangement—all effectors of the MAPK signaling pathway—leading to constitutive activation of the pathway. Follicular thyroid carcinoma is mainly associated with mutations of the *RAS* gene and the PAX8-PPARgamma rearrangement, while in poorly dedifferentiated and anaplastic thyroid carcinomas mutations in *RAS*, *BRAF*, *TP53* and *CTNNB1* are found. Familial non-medullary thyroid cancer is rare, either non-syndromic or syndromic in the context of familial tumor syndromes.

REFERENCES

1. Bonomi, M.; Busnelli, M.; Beck-Peccoz, P.; Costanzo, D.; Antonica, F.; Dolci, C.; Pilotta, A.; Buzi, F.; Persani, L. A Family with Complete Resistance to Thyrotropin-Releasing Hormone. *N. Engl. J. Med.* **2009**, *360*, 731.
2. Ramos, H.; Labedan, I.; Carré, A.; Castanet, M.; Guemas, I.; Tron, E.; Madhi, F.; Delacourt, C.; Maciel, R. M.; Polak, M. New Cases of Isolated Congenital Central Hypothyroidism Due to Homozygous Thyrotropin Beta Gene Mutations: A Pitfall to Neonatal Screening. *Thyroid* **2010**, *20*, 639.
3. Castanet, M.; Mallya, U.; Agostini, M.; Schoenmakers, E.; Mitchell, C.; Demuth, S.; Raymond, F. L.; Schwabe, J.; Gurnell, M.; Chatterjee, V. K. Maternal Isodisomy for Chromosome 9 Causing Homozygosity for a Novel FOXE1 Mutation in Syndromic Congenital Hypothyroidism. *J. Clin. Endocrinol. Metab.* **2010**, *95*, 4031.
4. De Felice, M.; DiLauro, R. Thyroid Development and its Disorders: Genetics and Molecular Mechanisms. *Endocr. Rev.* **2004**, *25*, 722.
5. Guillot, L.; Carré, A.; Szinnai, G.; Castanet, M.; Tron, E.; Jaubert, F.; Broutin, I.; Counil, F.; Feldmann, D.; Clement, A., et al. NKX2-1 Mutations Leading to Surfactant Protein Promoter Dysregulation Cause Interstitial Lung Disease in "Brain-Lung-Thyroid Syndrome". *Hum. Mutat.* **2010**, *31*, E1146.
6. Hoste, C.; Rigutto, S.; Van Vliet, G.; Miot, F.; De Deken, X. Compound Heterozygosity for a Novel Hemizygous Missense Mutation and a Partial Deletion Affecting the Catalytic Core of the H_2O_2-Generating Enzyme DUOX2 Associated with Transient Congenital Hypothyroidism. *Hum. Mut.* **2010**, *31*, E1304.
7. Moreno, J. C.; Klootwijk, W.; van Toor, H.; Pinto, G.; D'Allessandro, M.; Leger, A.; Goudie, D.; Polak, M.; Grueters, A.; Visser, T. J. Mutations in the Iodotyrosine Deiodinase Gene and Hypothyroidism. *N. Engl. J. Med.* **2008**, *358*, 1811.
8. Castanet, M.; Lyonnet, S.; Bonaïti-Pellié, C.; Polak, M.; Czernichow, P.; Léger, J. Familial Forms of Thyroid Dysgenesis Among Infants with Congenital Hypothyroidism. *N. Engl. J. Med.* **2000**, *343*, 441.
9. Deladoëy, J.; Vassart, G.; Van Vliet, G. Possible Non-Mendelian Mechanisms of Thyroid Dysgenesis. *Endocr. Dev.* **2007**, *10*, 29.
10. Léger, J.; Marinovic, D.; Garel, C.; Bonaïti-Pellié, C.; Polak, M.; Czernichow, P. Thyroid Developmental Anomalies in First Degree Relatives of Children with Congenital Hypothyroidism. *J. Clin. Endocrinol. Metab.* **2002**, *87*, 575.
11. Olivieri, A.; Stazi, M. A.; Mastoiacovo, P.; Fazzini, C.; Medda, E.; Spagnolo, A.; De Angelis, S.; Grandolo, M. E.; Taruscio, D.; Cordeddu, V., et al. A Population Based Study on the Frequency of Additional Congenital Malformations in Infants with Congenital Hypothyroidism: Data from the Italian Registry for Congenital Hypothyroidism (1991–1998). *J. Clin. Endocrinol. Metab.* **2002**, *87*, 557.
12. Perry, R.; Heinrichs, C.; Bourdoux, P.; Khoury, K.; Szöts, F.; Dussault, J. H.; Vassart, G.; Van Vliet, G. Discordance of Monozygotic Twins for Thyroid Dysgenesis: Implications for Screening and for Molecular Pathophysiology. *J. Clin. Endocrinol. Metab.* **2002**, *87*, 4027.

13. Visser, W. E.; Friesema, E. C.; Visser, T. J. Minireview: Thyroid Hormone Transporters: The Knowns and the Unknowns. *Mol. Endocrinol.* **2011,** *25,* 1.
14. Cheng, S. Y.; Leonard, J. L.; Davis, P. J. Molecular Aspects of Thyroid Hormone Actions. *Endocr. Rev.* **2010,** *31,* 139.
15. Hébrant, A.; van Staveren, W. C. G.; Meanhaupt, C.; Dumont, J. E.; Leclère, J. Genetic Hyperthyroidism: Hyperthyroidism Due to Activating TSHR Mutations. *Eur. J. Endocrinol.* **2011,** *164,* 1.
16. Michels, A. W.; Gottlieb, P. A. Autoimmune Polyglandular Syndromes. *Nat. Rev. Endocrinol.* **2010,** *6,* 270.
17. Kloos, R. T.; Eng, C.; Evans, D. B.; Francis, G. L.; Gagel, R. F.; Gharib, H.; Moley, J. F.; Pacini, F.; Ringel, M. D.; Schlumberger, M., et al. Medullary Thyroid Cancer: Management Guidelines of the American Thyroid Association. *Thyroid* **2009,** *19,* 565.

CHAPTER

84

Parathyroid Disorders

Geoffrey N Hendy

Calcium Research Laboratory, McGill University Hospital Center, Montreal, QC, Canada

Murat Bastepe

Harvard Medical School, Boston, MA, USA

David E C Cole

Department of Laboratory Medicine, University of Toronto, Toronto, ON, Canada

This chapter describes the genetic basis for disordered parathyroid gland function and parathyroid hormone (PTH) target organ responsiveness. These disorders include the PTH overproduction of primary hyperparathyroidism (PHPT), the hormone deficiency of primary hypoparathyroidism, the end-organ hyperresponsiveness of Jansen's chondrodysplasia, and the end-organ unresponsiveness of pseudohypoparathyroidism. Knowledge of the genetic factors involved is important for diagnosis and genetic counseling, for understanding the underlying pathophysiology, and for optimum management.

PHPT is a common cause of hypercalcemia in adults: 85% of cases involve a single benign parathyroid adenoma, 15% hyperplasia, and <1% carcinoma. In one PHPT case, an R58X mutation in the *PTH* gene leads to a truncated mutant PTH that is secreted from the gland and is active at target tissues, bones and kidney. Alterations in two genes—the proto-oncogene, cyclin D1/*CCND1*, and the tumor suppressor gene, *MEN1*—have been implicated in the development of some parathyroid adenomas. Ten percent of PHPT cases are hereditary, occurring as an isolated form or with other abnormalities. The conditions (and genes) discussed are familial hypocalciuric hypercalcemia/neonatal severe hyperparathyroidism (calcium-sensing receptor, *CASR*), multiple endocrine neoplasia type 1 (*MEN1*), type 2 (MEN2, *RET*), and type 4 (MEN4, *CNKN1B*), the hyperparathyroidism-jaw tumor syndrome (HPT-JT, *HRPT2/CDC73*) and familial isolated hyperparathyroidism (*FIHP*, gene(s) unknown). The incidence of parathyroid carcinoma is much higher in HPT-JT than in sporadic PHPT and MEN syndromes. Sporadic carcinomas show somatic mutations of the tumor suppressor *HRPT2/CDC73* gene and loss of nuclear staining of the protein product, parafibromin.

Autosomal dominant Jansen metaphyseal chondrodysplasia—a form of pseudohyperparathyroidism—manifests with severe growth plate abnormalities, hypercalcemia and hypophosphatemia. It is caused by heterozygous gain-of-function mutations in the parathyroid hormone receptor (*PTHR1*).

Most genetic forms of hypoparathyroidism are sporadic, but familial isolated hypoparathyroidism (FIH) may show dominant, recessive, or X-linked transmission. In the few cases with mutations in the *PTH* gene, both dominant and recessive inheritance has been observed. In multigeneration families with the X-linked form, a deletion–insertion mutation may affect *SOX3* expression and embryonic development of the parathyroid glands. *CASR* gain-of-function mutations inhibit PTH secretion and cause autosomal dominant hypocalcemia, although de novo mutations are also quite common. Less frequently, inherited FIH can occur with mutations of the glial cells missing-2 gene (*GCM2*) that encodes a transcription factor essential for the development of the parathyroid glands in terrestrial vertebrates. *GCM2* transactivates the *CASR* gene promoter. Homozygous and heterozygous inactivating mutations in *GCM2* cause autosomal recessive and dominant hypoparathyroidism, respectively.

Idiopathic hypoparathyroidism is also found associated with diverse developmental abnormalities such as lymphedema, nephropathy, nerve deafness, or cardiac malformation. Hypoparathyroidism due to parathyroid hypoplasia is a frequent feature of 22q11 microdeletions, the common cause of DiGeorge syndrome. Failure of the third and fourth pharyngeal pouches to develop leads to hypoplasia of the parathyroid glands, thymus, and conotruncal heart defects. Hypoparathyroidism is also part of the Barakat or HDR (Hypoparathyroidism, nerve Deafness, Renal dysplasia) syndrome. Deletions of two nonoverlapping regions of chromosome 10p contribute to a DiGeorge-like phenotype and the HDR syndrome. The latter is due to haploinsufficiency of GATA3, a transcription factor essential for normal embryonic development of the parathyroids, auditory system and kidney.

The recessive Kenny–Caffey and Sanjad–Sakati syndromes characterized by congenital hypoparathyroidism, growth and developmental retardation and characteristic dysmorphic features map to 1q42-43. Causative mutations in the tubulin chaperone E gene (*TBCE*) were identified in this condition.

Hypoparathyroidism due to metabolic disease is a variable component of the neuromyopathies such as Kearns–Sayre and Pearson marrow pancreas syndromes caused by mitochondrial gene defects. Long-chain hydroxyacyl-CoA dehydrogenase deficiency is an inborn error of oxidative fatty acid metabolism that has been associated with hypoparathyroidism. Parathyroid insufficiency may be seen in inherited disorders causing excess metal storage of iron (thalassemia, Diamond–Blackfan anemia, and hemochromatosis) or copper (Wilson disease).

Hypoparathyroidism also occurs as an autoimmune disorder either alone or with other endocrine deficiencies in a pluriglandular autoimmune syndrome. One parathyroid protein selectively associated with the autoimmune process is the NACHT leucine-rich-repeat protein 5 (NALP5). Elevated antibody titres occur in half the patients with autoimmune hypoparathyroidism. Autoantibodies against the extracellular domain of the parathyroid CASR occur in the recessive type 1 autoimmune polyglandular syndrome (APS-1), also known as autoimmune polyendocrinopathy-candidiasis-ectodermal dystrophy (APECED), as well as in acquired hypoparathyroidism associated with autoimmune hypothyroidism. Patients with APS-1 have mutations in the autoimmune regulator (*AIRE*) gene that encodes a transcriptional regulator. When *AIRE* is absent tissue-specific self-antigens are not expressed in the thymus and multiorgan autoimmunity develops.

Syndromes of end-organ resistance to PTH are well known, but uncommon and often complex in their pathophysiology. Mutations in the imprinted *GNAS* gene encoding the Gsα protein that transduces signals between G-protein coupled receptors (like the *PTHR1*) and adenylate cyclase are responsible for a series of autosomal dominant disorders: pseudohypoparathyroidism-1a (PHP-1a), pseudopseudohypoparathyroidism (PPHP) and PHP-1b. PHP-1a is characterized by end-organ resistance to PTH and other Gsα-coupled hormone receptor systems, as well as the distinctive physical appearance of Albright hereditary osteodystrophy (AHO). This is a specific pattern of physical stigmata, typically, short stature, round facies, brachydactyly, obesity, and osteoma cutis. In the same PHP-1a family, some individuals only express AHO and not the hormone resistance and are said to have PPHP. Maternal or paternal inheritance of mutations in the Gsα exons causes PHP-1a or PPHP, respectively, due to differential imprinting effects. Like PHP-1a, PHP-1b is inherited maternally, but Gsα exons are normal and the epigenetic defect lies in *GNAS* imprinting control regions that regulate the expression of Gsα in renal proximal tubule cells. Paternally inactivating *GNAS* mutations occur in the connective tissue conditions congenital osteoma cutis and progressive osseous heteroplasia (POH). Rare cases of multiple hormone resistance and acrodysostosis, a form of skeletal dysplasia that resembles AHO, have a truncation mutation in the gene encoding the regulatory subunit of protein kinase A (*PRKAR1A*) that blocks binding of cAMP to PKA.

Inactivating *PTHR1* mutations are found in Blomstrand's lethal chondrodysplasia and other skeletal dysplasias and dental abnormalities. Dominantly acting heterozygous *PTHR1* mutations occur in enchondromatosis (Ollier's Disease), characterized by multiple benign cartilage tumors and a predisposition to malignant chondrosarcomas. Dominantly inherited symmetrical enchondromatosis is associated with duplication of 12p11 that includes the *PTHLH* gene encoding PTHrP. In a few patients with nephrolithiasis or bone demineralization, mutations occur in the sodium–hydrogen exchanger regulatory factor 1 (NHERF1) that bridges the renal proximal tubule *PTHR1* and sodium phosphate cotransporter, NPT2a leading to reduced phosphate transport in response to PTH.

TABLE 84-1 Familial Hypercalcemic Syndromes

	FHH	NSHPT	MEN1	MEN2A	HPT-JT	FHPT
OMIM #	145980	239200	131100	171400	145001	145000
Mode of inheritance	AD	AR	AD	AD	AD	AD
Genetic locus	3q13.3-21	3q13.3-21	11q13	10q11.2	1q21-31	2p13.3-14
Mutated gene	*CASR*	*CASR*	*MEN1*	*RET*	*HRPT2*	Not known
Gene product	*CASR*	*CASR*	*MENIN*	*RET*	PARAFIBRO-MIN/CDC73	—
Associated conditions	Chondro-calcinosis Pancreatitis	—	Pancreatic islet and pituitary tumors; other neuroendo-crine tumors	MTC and pheochromo-cytoma	Jaw fibromas; renal and uterine tumors	—

FHH, familial hypocalciuric hypercalcemia; NSHPT, neonatal severe primary hyperparathyroidism; MEN, multiple endocrine neoplasia; HPT-JT, hyperparathyroidism-jaw tumor; FHPT, familial hyperparathyroidism; AD, autosomal dominant; AR, autosomal recessive.

TABLE 84-3 Biochemical Characteristics of Hypoparathyroidism and Pseudohypoparathyroidism

		Response to PTH Infusion					
	Serum PO_4	PTH	25(OH)D	$1,25(OH)_2D$	U_{cAMP}	U_{PO4}	Multiple Endocrine Defects
Hypoparathyroidism	↑	↓	→	↓	→	→	Yes/no[a]
Pseudohypoparathyroidism							
Type 1a	↑	↑	→	↓	↓	↓	Yes
Type 1b	↑	↑	→	↓	↓	↓	No/yes[b]
Type 1c	↑	↑	→	↓	↓	↓	Yes
Type 2	↑	↑	→	↓	→	↓	No

↑, increased; ↓, decreased; →, normal.
[a]Depending upon the etiology.
[b]Variable defects of the thyroid and somatotropin axes are seen.

TABLE 84-2 Forms of Hypoparathyroidism Having a Genetic Basis

1. Isolated
   1.1. Autosomal dominant
      1.1.1. PreproPTH signal peptide mutation
      1.1.2. CASR activating mutation
      1.1.3. GCM2 mutation (dominant negative)
   1.2. Autosomal recessive
      1.2.1. PreproPTH RNA splice-site mutation
      1.2.2. GCM2 mutation
   1.3. X-linked
2. Congenital multisystem syndromes
   2.1. DiGeorge & Velocardiofacial (22q11)
   2.2. Barakat/HDR
   2.3. Kenny–Caffey
3. Metabolic disease
   3.1. Mitochondrial neuromyopathies
   3.2. Long-chain hydroxyacyl-CoA dehydrogenase deficiency
   3.3. Heavy-metal storage disorders
4. Autoimmune disease
   4.1. Autoimmune polyendocrine syndrome type 1 (APS-1 or APECED)
5. PTH resistance syndromes
   5.1. Pseudohypoparathyroidism
   5.2. Blomstrand chondrodysplasia and related PTH receptor defects
   5.3. Hypomagnesemia

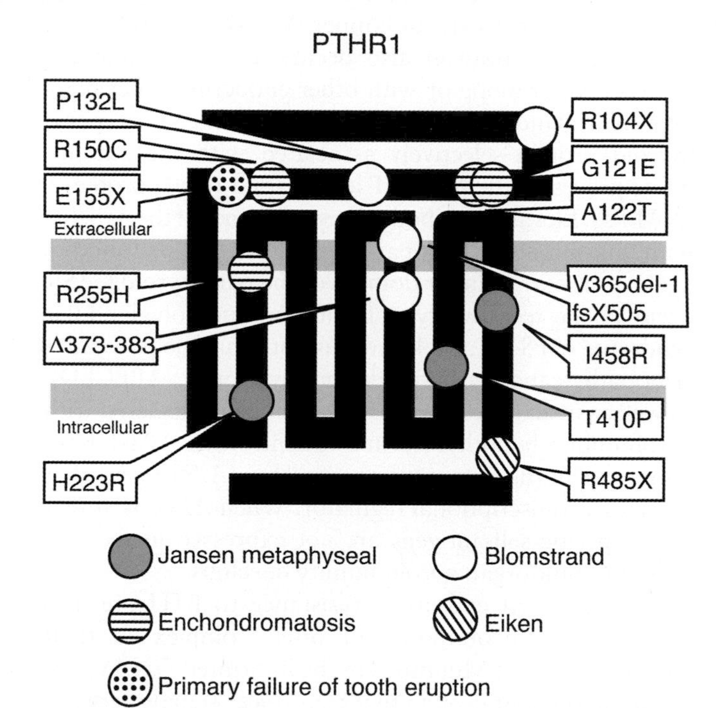

FIGURE 84-1 Schematic representation of the human *PTHR1*. The locations of the *H223R*, *T410P*, and *I458R*-activating mutations identified in patients with Jansen metaphyseal chondrodysplasia, the R104X, P132L, V365del-1fsX505, and Δ373–383 inactivating mutations found in patients with Blomstrand chondrodysplasia, the R485X Eiken syndrome mutation, the G121E, A122T, R150C, and R255H enchondromatosis mutations, and the E155X primary failure of tooth eruption (PFE) mutation are indicated. Splice-site mutations that would result in predicted mutant C351fsX485 and E182fsX203 proteins have been identified in additional PFE cases.

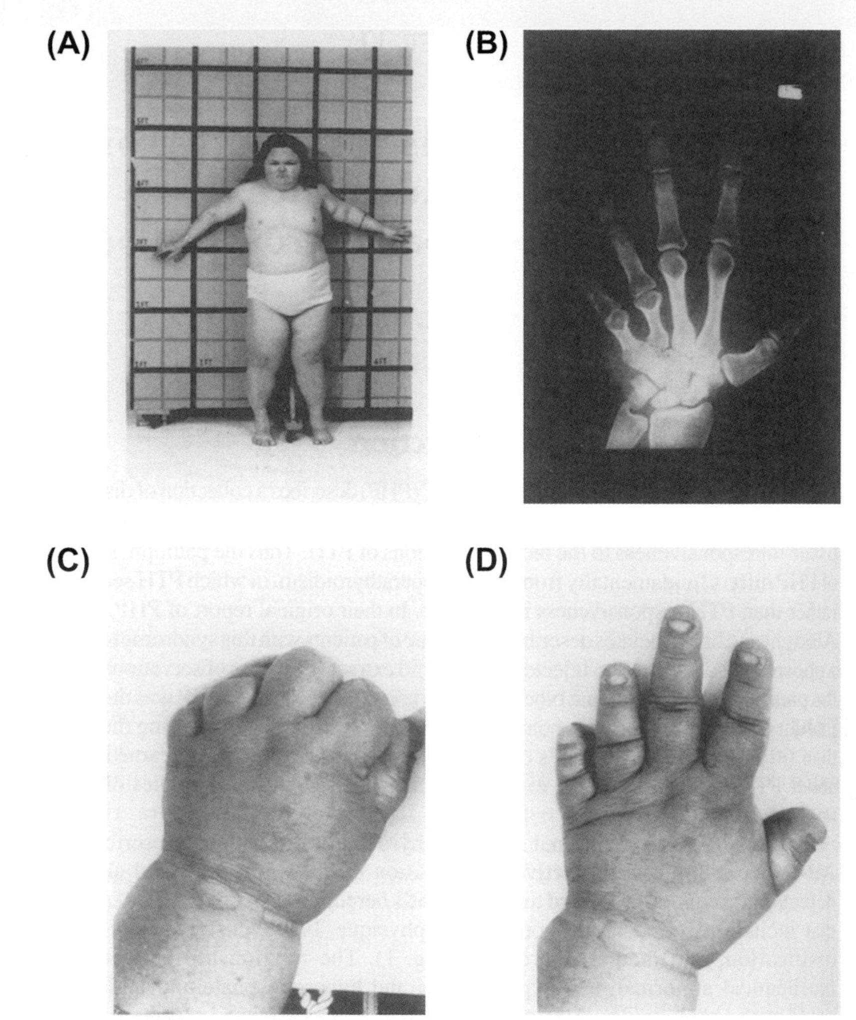

FIGURE 84-2 Features of Albright hereditary osteodystrophy (AHO). A. Young woman with short stature (~3rd centile), disproportionate shortening of the limbs, generalized obesity, and round, flattened face. B. Radiograph of the hand showing the shortened fourth and fifth metacarpals. C. Fist with the characteristic "dimples" over the third, fourth, and fifth digits replacing the knuckles formed by the distal head of normally sized metacarpal bones (Archibald sign). D. Brachydactyly of the hand, with the short fourth and fifth digits, the greatly foreshortened terminal first digit, and very short, wide thumbnail (potter's thumb). *(From Levine, M. A. In* The Genetics of Osteoporosis and Metabolic Bone Disease*; Econs, M. J., Ed.; Humana Press: New York, 2000; pp 179–209.)*

FURTHER READING

1. Arnold, A.; Lauter, K. Genetics of Hyperparathyroidism Including Parathyroid Cancer. In *Genetic Diagnosis of Endocrine Disorders*; Weiss, R. E.; Refetoff, S., Eds.; Elsevier B.V., 2010; pp 141–148.
2. Bastepe, M.; Juppner, H. GNAS Locus and Pseudohypoparathyroidism. *Horm. Res.* **2005,** *63,* 65–74.
3. Brown, E. M. Clinical Lessons from the Calcium-Sensing Receptor. *Nat. Clin. Pract. Endocrinol. Metab.* **2007,** *3,* 122–133.
4. Brown, E. M. Anti-Parathyroid and Anti-Calcium Sensing Receptor Antibodies in Autoimmune Hypoparathyroidism. *Endocrinol. Metab. Clin. North Am.* **2009,** *38,* 437–445.
5. Georgitsi, M. MEN-4 and Other Multiple Endocrine Neoplasia Due to Cyclin-Dependent Kinase Inhibitors (p27Kip1 and p18INK4C) Mutations. *Best Pract. Res. Clin. Endocrinol. Metab.* **2010,** *24,* 425–437.
6. Hendy, G. N.; Guarnieri, V.; Canaff, L. Calcium-Sensing Receptor and Associated Diseases. *Prog. Mol. Biol. Transl. Sci.* **2009,** *89C,* 31–95.
7. Jüppner, H.; Schipani, E.; Silve, C. Genetic Disorders Caused by PTH/PTHrP Receptor Mutations. In *Principles of Bone Biology* 3rd ed.; Bilezikian, J. P.; Martin, T. J.; Raisz, L. G., Eds.; Academic Press: San Diego, 2008; pp 1431–1452.
8. Marx, S. J. Hyperparathyroid Genes: Sequences Reveal Answers and Questions. *Endocr. Pract.* **2011,** *17* (Suppl 3), 18–27.
9. Newey, P. J.; Bowl, M. R.; Thakker, R. V. Parafibromin–Functional Insights. *J. Intern. Med.* **2009,** *266,* 84–98.
10. Sharretts, J. M.; Simonds, W. F. Clinical and Molecular Genetics of Parathyroid Neoplasms. *Best Pract. Res. Clin. Endocrinol. Metab.* **2010,** *24,* 491–502.
11. Shoback, D. Hypoparathyroidism. *N. Engl. J. Med.* **2008,** *359,* 391–403.
12. Shore, E. M.; Kaplan, F. S. Inherited Human Diseases of Heterotopic Bone Formation. *Nat. Rev. Rheumatol.* **2010,** *6,* 518–527.
13. Thakker, R. V. Multiple Endocrine Neoplasia Type 1 (MEN1). *Best Pract. Res. Clin. Endocrinol. Metab.* **2010,** *24,* 355–370.
14. Weinstein, L. S.; Collins, M. T. Diseases Resulting from Defects in the G Protein Gsα. In *Principles of Bone Biology* 3rd ed.; Bilezikian, J. P.; Martin, T. J.; Raisz, L. G., Eds.; Academic Press: San Diego, 2008; pp 1453–1477.

CHAPTER

85

Diabetes Mellitus

Leslie J Raffel

Medical Genetics Institute, Cedars-Sinai Medical Center, Pacific Theatres, Los Angeles, CA, USA

Mark O Goodarzi

Division of Endocrinology, Diabetes and Metabolism, Cedars-Sinai Medical Center, Los Angeles, CA, USA

85.1 INTRODUCTION

Diabetes mellitus is a diagnostic term for a group of disorders characterized by abnormal glucose homeostasis resulting in elevated blood sugar. It is among the most common of chronic disorders, affecting up to 5–10% of the adult population of the Western world. The prevalence of diabetes is increasing dramatically; it has been estimated that the worldwide prevalence will increase by more than 50% between the years 2000 and 2030 (*1*). It is clearly established that diabetes mellitus is not a single disease, but a genetically heterogeneous group of disorders that share glucose intolerance in common. The concept of genetic heterogeneity (i.e. that different genetic and/or environmental etiologic factors can result in similar phenotypes) has significantly altered the genetic analysis of this common disorder.

85.2 TYPE 1 DIABETES MELLITUS

85.2.1 Introduction

Type 1 diabetes mellitus is characterized by low or absent levels of endogenous insulin production. This is secondary to destruction of the insulin-producing beta cells of the pancreas, and is the single characteristic that most decisively separates type 1 and type 2 diabetes. It is estimated that 5–10% of all US patients with diabetes have type 1 diabetes, and that the US incidence in children 0–14 years is in the range of 12 per 20/100,000 (*2*).

85.2.2 Evidence of a Genetic Contribution to Type 1 Diabetes

Prospective studies of monozygotic twins have estimated the concordance rate for type 1 diabetes to be about 35%, suggesting that there is substantial genetic susceptibility to this form of diabetes. Given that the concordance is much less than 100%, however, it appears likely that environmental triggers are also involved in the development of overt type 1 diabetes. Both family studies and animal models of type 1 diabetes suggest that it is a multifactorial disease, involving a number of susceptibility loci in addition to environmental factors.

85.2.3 Genetic Approaches in Type 1 Diabetes

With the discovery of HLA antigen associations with type 1 diabetes, the genetic region that provides the major (but by no means only) genetic susceptibility to type 1 diabetes was identified. Given the autoimmune nature of this disease, it is not surprising that genes in the HLA region contribute to susceptibility. Because of the extensive linkage disequilibrium across the HLA region, it has proven difficult to identify exactly which genes are directly involved, but it appears likely that multiple genes within this region contribute to the development of type 1 diabetes.

85.2.4 Candidate Genes and Type 1 Diabetes

While HLA is the major locus contributing to type 1 diabetes, other genes are also involved. Using candidate gene studies, several other loci were identified. The insulin gene was the first non-HLA gene to be implicated. A variable number tandem repeat region near the 5′ regulatory region of INS affects expression of insulin mRNA in the fetal thymus and thus influences the development of tolerance to antigenic determinants of insulin. Other genes identified from candidate gene studies include CTLA4, a member of the immunoglobulin superfamily that appears to be important in normal Treg function; PTPN22 which encodes the lymphoid-specific protein tyrosine phoshatase (LYP), another inhibitor of T-cell activation; and IL2RA, the interleukin 2 receptor A gene (also known as CD25) which may influence CD4+ memory cell responsiveness to IL-2 and to TCR-mediated activation.

85.2.5 Type 1 Diabetes Genes Identified by Genome-Wide Association Studies

Genome-wide association studies (GWAS) have resulted in the identification of a substantial number of genes for type 1 diabetes (Table 85-1). This has been facilitated by cooperation between multiple centers, both in the acquisition of large numbers of cases and controls and in data sharing to confirm association signals. The first GWAS for type 1 diabetes was published in 2006, reporting an association with the innate immunity viral RNA receptor gene region, IFIH1 (*3*). This was followed rapidly by a number of other GWAS that increased the number of loci demonstrating association. Additional loci have been identified by GWAS meta-analysis using very large combined study populations. Many of the GWAS and meta-analysis findings have also been confirmed in independent samples and as of early 2011, there are more than 50 loci with quite convincing evidence for association with type 1 diabetes (Table 85-1); see also http://www.t1dbase.org (*4*).

85.3 TYPE 2 DIABETES MELLITUS

85.3.1 Introduction

Type 2 diabetes is characterized by a relative disparity between endogenous insulin production and insulin requirements, leading to an elevated blood glucose. The primary pathogenetic lesion in type 2 diabetes has yet to be discovered. Primary insulin resistance of the peripheral tissues has been suggested by many as the initial event. Similarly, insulin secretion abnormalities have been argued as the primary defect in type 2 diabetes. It is likely that both phenomena are important in the development of type 2 diabetes, and genetic defects predisposing to both are likely to be important contributors to the disease process. GWAS in recent years appear to suggest that failure of insulin secretion is the key event that initiates type 2 diabetes.

85.3.1.1 Evidence of a Genetic Contribution to Type 2 Diabetes. Several lines of evidence suggest the importance of genetic susceptibilities underlying the development of type 2 diabetes. Genetic epidemiologic studies provide convincing descriptive data including population and ethnic differences, studies of familial aggregation, familial transmission patterns, and comparisons of twin concordance rates. Animal models of type 2 diabetes and studies of specific genetic syndromes that feature glucose intolerance provide further data supporting the etiologic role of genetic factors in the pathogenesis of type 2 diabetes. Finally, the genetic etiologies for type 2 diabetes have also been supported by association and linkage studies of genetic markers in populations and families.

85.3.2 Genetic Approaches in Type 2 Diabetes

In dealing with heterogeneity in type 2 diabetes, there are several possible research strategies that can potentially be employed. In general, there are three options: (1) start with observable physiological differences and then work backward to determine if these differences can be explained by different genetic defects (working from the phenotype down); (2) start with a candidate gene or allele proposed to be related to diabetes, establish a genetic relationship, and work forward (working from the genotype up) to determine if specific physiological traits are associated with this gene or gene defect; and (3) use genome-wide linkage or GWAS approaches to identify chromosomal regions likely to contain diabetes-related traits or clinical diabetes itself and then assess if there are detectable physiological differences between those individuals (or families) displaying linkage and/or linkage disequilibrium and those individuals or families that do not appear to be linked to the particular genomic locus. In actuality, research often involves the sequential application of all these approaches, as is well demonstrated by the investigations of MODY.

85.3.3 Candidate Genes and Type 2 Diabetes

A large number of candidate genes have been tested for possible roles in the etiology of type 2 diabetes with mostly negative results. Those genes with the strongest evidence of association are peroxisome proliferator activated receptor gamma (PPARG), the pancreatic beta cell inwardly rectifying potassium channel Kir 6.2 (KCNJ11), transcription factor 2, hepatic (TCF2, also known as HNF1B), and Wolfram syndrome 1 (WFS1). Of note, rare, severe mutations in all four of these genes lead to syndromic diabetes, and two of these genes (PPARG, KCNJ11) code for proteins that are targets of oral antidiabetic medications.

Of note, the odds of diabetes in carriers of particular variants in these genes range from 1.1 to 1.2; such moderate effects are detectable only by sufficiently powered studies. This likely contributed to the limited success of the candidate gene approach.

85.3.4 Identifying Type 2 Diabetes Genes by Genome-Wide Linkage Scans

In 2006, TCF7L2 (transcription factor 7 like 2), the locus with the strongest effect on the risk of type 2 diabetes, was discovered by investigators following up a linkage signal found on chromosome 10 in Icelandic individuals (*5*). Each risk allele at this locus confers 1.4 odds of type 2 diabetes in Europeans. A large number of subsequent studies reproduced association of TCF7L2 variation with type 2 diabetes, most commonly in Europeans but also in other racial/ethnic groups.

85.3.5 Type 2 Diabetes Genes Identified by Genome-Wide Association Studies

GWAS have resulted in the discovery of a dramatic number of genes for type 2 diabetes (Table 85-2). The first wave of GWAS in type 2 diabetes consisted of five studies published in 2007, all of which were conducted in European-origin subjects (*6–10*). Several additional GWAS and GWAS meta-analysis, conducted mainly in European and Asian case/control cohorts, identified additional loci. For most of these loci, the causal gene and variant have not been established; the associated variants may be in linkage disequilibrium with functional variants in the nearest gene (typically used to label the locus) or elsewhere. As displayed in Table 85-2, in most cases the associated variants are in introns or intergenic regions; only a handful of functional missense variants have been discovered. Despite the fact that over 40 loci for type 2 diabetes have been described, they appear to explain only 10% of the inherited basis of diabetes.

Another avenue to discovery of type 2 diabetes loci was GWAS for fasting glucose, which revealed that some loci influence both type 2 diabetes and fasting glucose (e.g. TCF7L2, ADCY5), while other loci influence fasting glucose with no effect on diabetes risk. These point to the complexity of glucose regulation and how genes that affect glucose may or may not influence the risk of diabetes.

Several studies, typically in nondiabetic subjects, have examined the type 2 diabetes loci for association with indexes of insulin resistance and insulin secretion. The overwhelming finding of these studies is that the majority of type 2 diabetes loci are associated with impaired beta cell function, either in the development of beta cells or the insulin response to glucose or to the incretin hormones (Figure 85-1). Only a few genes appear to act via insulin resistance. This may reflect a unique genetic architecture of insulin resistance, or that insulin resistance may not be well reflected by the quantitative traits commonly examined. Alternatively, insulin resistance may be more influenced by environmental and lifestyle factors, such as diet and weight gain. The predominance of diabetes genes associating with beta cell function suggests a model wherein genetically determined robustness of the beta cell's ability to maintain compensatory insulin secretion in the face of insulin resistance is the main gateway by which subjects develop type 2 diabetes. Those individuals who maintain sufficient insulin secretion do not develop diabetes, even if substantial insulin resistance is present.

REFERENCES

1. Wild, S.; Roglic, G.; Green, A.; Sicree, R.; King, H. Global Prevalence of Diabetes: Estimates for the Year 2000 and Projections for 2030. *Diabetes Care* **2004,** *27* (5), 1047–1053.
2. Incidence and Trends of Childhood Type 1 Diabetes Worldwide 1990–1999. *Diabet. Med.* **2006,** *23* (8), 857–866.
3. Smyth, D. J.; Cooper, J. D.; Bailey, R.; Field, S.; Burren, O.; Smink, L. J.; Guja, C.; Ionescu-Tirgoviste, C.; Widmer, B.; Dunger, D. B., et al. A Genome-Wide Association Study of Nonsynonymous SNPs Identifies a Type 1 Diabetes Locus in the Interferon-Induced Helicase (IFIH1) Region. *Nat. Genet.* **2006,** *38* (6), 617–619.
4. Burren, O. S.; Adlem, E. C.; Achuthan, P.; Christensen, M.; Coulson, R. M.; Todd, J. A. T1DBase: Update 2011, Organization and Presentation of Large-Scale Data Sets for Type 1 Diabetes Research. *Nucleic Acids Res.* **2011,** *39,* D997–D1001, (Database issue).
5. Grant, S. F.; Thorleifsson, G.; Reynisdottir, I.; Benediktsson, R.; Manolescu, A.; Sainz, J.; Helgason, A.; Stefansson, H.; Emilsson, V.; Helgadottir, A., et al. Variant of Transcription Factor 7-Like 2 (TCF7L2) Gene Confers Risk of Type 2 Diabetes. *Nat. Genet.* **2006,** *38* (3), 320–323.
6. Saxena, R.; Voight, B. F.; Lyssenko, V.; Burtt, N. P.; de Bakker, P. I.; Chen, H.; Roix, J. J.; Kathiresan, S.; Hirschhorn, J. N.; Daly, M. J., et al. Genome-Wide Association Analysis Identifies Loci for Type 2 Diabetes and Triglyceride Levels. *Science* **2007,** *316* (5829), 1331–1336.
7. Scott, L. J.; Mohlke, K. L.; Bonnycastle, L. L.; Willer, C. J.; Li, Y.; Duren, W. L.; Erdos, M. R.; Stringham, H. M.; Chines, P. S.; Jackson, A. U., et al. A Genome-Wide Association Study of Type 2 Diabetes in Finns Detects Multiple Susceptibility Variants. *Science* **2007,** *316* (5829), 1341–1345.
8. Sladek, R.; Rocheleau, G.; Rung, J.; Dina, C.; Shen, L.; Serre, D.; Boutin, P.; Vincent, D.; Belisle, A.; Hadjadj, S., et al. A Genome-Wide Association Study Identifies Novel Risk Loci for Type 2 Diabetes. *Nature* **2007,** *445* (7130), 881–885.
9. Steinthorsdottir, V.; Thorleifsson, G.; Reynisdottir, I.; Benediktsson, R.; Jonsdottir, T.; Walters, G. B.; Styrkarsdottir, U.; Gretarsdottir, S.; Emilsson, V.; Ghosh, S., et al. A Variant in CDKAL1 Influences Insulin Response and Risk of Type 2 Diabetes. *Nat. Genet.* **2007,** *39* (6), 770–775.
10. Zeggini, E.; Weedon, M. N.; Lindgren, C. M.; Frayling, T. M.; Elliott, K. S.; Lango, H.; Timpson, N. J.; Perry, J. R.; Rayner, N. W.; Freathy, R. M., etal. Replication of Genome-Wide Association Signals in UK Samples Reveals Risk Loci for Type 2 Diabetes. *Science* **2007,** *316* (5829), 1336–1341.

TABLE 85-1 Susceptibility Genes for Type 1 Diabetes

Locus: Nearest Gene(s)	Chromosomal Location	Marker(s)	Variant Type(s)	Source of Initial Discovery	Diabetes Effect: or (95% CI)	Other Genome-Wide Associated Disorder(s)	Protein
PTPN22	1p13.2	rs2476601	Arg620Trp	Candidate gene study in North American Caucasians	2.05	Crohn's, Graves', RA, SLE, vitiligo	Protein tyrosine phosphatase, non-receptor type 22 (lymphoid)
RGS1	1q31.2	rs2816316	8 kb upstream	Candidate gene study following up a Coeliac GWAS hit European Caucasian	0.89	Coeliac, MS	Regulator of G-protein signaling 1
CD55, IL10	1q32.1	rs3024505	1 kb downstream of IL10	GWAS meta-analysis in European and North American Caucasians	0.84	Crohn's, SLE, UC	CD55 antigen Interleukin 10
AFF3	2q11.2	rs9653442 rs1160542	66 kb upstream, 73 kb upstream	GWAS meta-analysis in European and North American Caucasians	1.11	JIA, RA	AF4/FMR2 family, member 3
IFIH1	2q24.2	rs1990760	Ala946Thr	GWAS analysis in European Caucasians	0.81–0.86	Graves', SLE	Interferon induced with helicase C domain 1; receptor for dsRNA from viral infections
STAT4	2q32.2	rs7574865 rs6752770	Intronic	Candidate gene study following up a RA and SLE GWAS/ linkage hits European Caucasians	1.1–1.11	RA, SLE, SSc, pSS	Signal transducer and activator of transcription 4
CTLA4	2q33.2	rs3087243	Immediately downstream	Candidate gene study	0.82–0.88	Coeliac, RA	Cytotoxic T-lymphocyte-associated protein 4
CCR5	3p21.31	rs11711054 rs333	66 kb upstream, 32 bp insertion–deletion variant	Candidate gene study in European Caucasians; multiple prior small studies with equivocal results	0.85	Celiac	Human C–C chemokine receptor gene
	4p15.2	rs10517086	Intergenic	GWAS meta-analysis in European and North American Caucasians	1.09		
IL2	4q27	rs2069762, rs2069763, rs4505848	Immediately upstream, Leu38Leu, 240 kb downstream	GWAS meta-analysis in European and North American Caucasians	0.89 1.13	UC	Interleukin 2
HLA-DQB1, HLA-B, HLA-DRB1, HLA-C	MHC			Candidate gene studies in multiple populations	3.05		

Continued

TABLE 85-1 Susceptibility Genes for Type 1 Diabetes—*Cont'd*

Locus: Nearest Gene(s)	Chromosomal Location	Marker(s)	Variant Type(s)	Source of Initial Discovery	Diabetes Effect: or (95% CI)	Other Genome-Wide Associated Disorder(s)	Protein
BACH2	6q15	rs11755527	Intronic	Candidate SNP follow-up of suggestive GWAS associations in European and North American Caucasians	1.13		
	6q22.32	rs9388489	Intergenic	GWAS meta-analysis in European and North American Caucasians	1.17		
TNFAIP3	6q23.3	rs10499194, rs2327832, rs6920220	188 kb upstream, 216 kb upstream, 182 kb upstream	Candidate gene study following up a RA GWAS in North American Caucasians	0.9 1.09	Coeliac, RA, SLE, UC	
TAGAP	6q25.3	rs1738074	5′ untranslated region	Candidate gene study following up a Coeliac GWAS in European Caucasians	0.92	Coeliac	T-cell activation Rho GTPase-activating protein
SKAP2	7p15.2	rs7804356	Intronic	GWAS in North American Caucasians	0.88		src family associated phosphoprotein 2
IKZF1	7p12.2	rs10272724	4 kb downstream	Candidate gene study following up an acute leukemia GWAS and borderline T1D GWAS in European Caucasians	0.87		IKAROS family zinc finger 1
	7p12.1	rs4948088	Intergenic	GWAS meta-analysis in European and North American Caucasians	0.77		
GLIS3	9p24.2	rs7020673	Intronic	GWAS meta-analysis in European and North American Caucasians	0.88		GLIS family zinc finger 3
IL2RA	10p15.1	rs11594656, rs12722495	18 kb upstream, intronic	Candidate gene study in European and North American Caucasians	0.84–0.87 0.62–0.63	MS, RA, vitiligo	Interleukin 2 receptor, alpha chain
PRKCQ	10p15.1	rs11258747, rs947474	Arg616Arg, 79 kb downstream	GWAS meta-analysis in European and North American Caucasians	0.69 0.88–0.91		Protein kinase C, theta
ZMIZ1	10q22.3	rs1250550, rs1250558	Intronic	GWAS meta-analysis in European and North American Caucasians		IBD	Zinc finger, MIZ-type containing 1
RNLS	10q23.31	rs10509540	11kb downstream	GWAS meta-analysis in European and North American Caucasians	0.75		Renalase, FAD-dependent amine oxidase

INS	11p15.5	rs689	5′ untranslated region	Candidate gene studies in multiple populations	0.42		Insulin
CD69	12p13.31	rs4763879	Intronic	GWAS meta-analysis in European and and North American Caucasians	1.09		CD69 antigen
CYP27B1	12q13.3	rs10877012	Immediately upstream of gene	Candidate gene study in European Caucasian families	1.22	MS	Cytochrome P450, family 27, subfamily B, polypeptide 1
ERBB3	12q13.2	rs2292239	Intronic	GWAS meta-analysis in European and North American Caucasians	1.31		v-erb-b2 erythroblastic leukemia viral oncogene homolog 3
SH2B3	12q24.12	rs3184504	Trp262Arg	GWAS in European Caucasians	1.28	Coeliac, MS	SH2B adapter protein 3
GPR183	13q32.3	rs9585056	122kb upstream	Candidate gene study in European Caucasians following up rat expression data and borderline T1D GWAS in European Caucasians	1.15		G protein-coupled receptor 183
	14q24.1	rs1465788	Intergenic	GWAS meta-analysis in European & North American Caucasians	0.86		
	14q32.2	rs4900384	Intergenic	GWAS meta-analysis in European and North American Caucasians	1.09		
DLK1	14q32.2	rs941576	105 kb downstream	GWAS meta-analysis in European and North American Caucasians	0.90		Delta-like 1 homolog
RASGRP1	15q14	rs17574546, rs7171171	45 kb upstream, 50 kb upstream	Family based GWAS meta-analysis in North American Caucasians	1.21		RAS guanyl releasing protein 1
CTSH	15q25.1	rs3825932	Intronic	GWAS meta-analysis in European and North American Caucasians	0.86		Cathepsin H
CLEC16A	16p13.13	rs12708716	Intergenic	GWAS meta-analysis in European and North American Caucasians	0.81	MS, SLE	C-type lectin domain family 16, member A
	16p13.13	rs12927773	Intergenic	GWAS in European Caucasians	0.81	Coeliac	
IL27	16p11.2	rs4788084	22kb upstream	GWAS meta-analysis in European and North American Caucasians	0.86	Crohn's, IBD	Interleukin 27
	16q23.1	rs7202877	Intergenic	GWAS meta-analysis in European and North American Caucasians	1.28		
GSDMB, ORMDL3	17q12	rs2290400	Intron of GSDMB	GWAS meta-analysis in European and North American Caucasians	0.87	Crohn's, UC	Gasdermin B ORM1-like 3

Continued

TABLE 85-1 Susceptibility Genes for Type 1 Diabetes—*Cont'd*

Locus: Nearest Gene(s)	Chromosomal Location	Marker(s)	Variant Type(s)	Source of Initial Discovery	Diabetes Effect: or (95% CI)	Other Genome-Wide Associated Disorder(s)	Protein
	17q21.2	rs7221109	Intergenic	GWAS meta-analysis in European and North American Caucasians	0.95		
PTPN2	18p11.21	rs45450798, rs478582	Intronic	GWAS in European Caucasians	1.28 0.83	Coeliac, Crohn's	Protein tyrosine phosphatase, non-receptor type 2
CD226	18q22.2	rs763361	Ser307Gly	GWAS meta-analysis in European and North American Caucasians	1.16	MS	CD226 antigen
TYK2	19p13.2	rs2304256	Val362Phe	GWAS meta-analysis in European and North American Caucasians	0.86		Tyrosine kinase 2
PRKD2	19q13.32	rs425105	Intronic	GWAS meta-analysis in European and North American Caucasians	0.86		Protein kinase D2
FUT2	19q13.4	rs602662	Gly258Ser	GWAS meta-analysis in European and North American Caucasians		Crohn's	Fucosyltransferase 2
SIRPG	20p13	rs2281808	Intronic	GWAS meta-analysis in European and North American Caucasians	0.90		Signal-regulatory protein gamma
UBASH3A	21q22.3	rs3788013	Intronic	Family-based GWAS in North American Caucasians	1.13	Vitiligo	Ubiquitin associated and SH3 domain containing, A
AIRE	21q22.3	rs760426	Intronic	Candidate gene causing APECED	1.12		Autoimmune regulator
	22q12.2	rs5753037	Intergenic	GWAS meta-analysis in European and North American Caucasians	1.10	Crohn's, IBD	
IL2RB	22q12.3	rs3218253	Intronic	GWAS meta-analysis in European and North American Caucasians		RA	Interleukin 2 receptor, beta chain
C1QTNF6	22q13.1	rs229527, rs229541	Gly21Val, 7 kb upstream	GWAS meta-analysis in European and North American Caucasians	1.11–1.12	Vitiligo	C1q and tumor necrosis factor related protein 6
TLR7, TLR8	Xp22.2	rs5979785	Intergenic	GWAS meta-analysis in European and North American Caucasians	0.86	Coeliac	Toll-like receptors 7, 8
GAB3	Xq28	rs2664170	Intronic	GWAS meta-analysis in European and North American Caucasians	1.16		GRB2-associated binding protein 3

Adapted from http://t1dbase.org/ accessed 3/20/2011.

TABLE 85-2 Susceptibility Genes for Type 2 Diabetes

Locus: Nearest Gene(s)	Chromosomal Location	Marker(s)	Variant Type(s)	Source of Initial Discovery	Genome-Wide Associated trait(s)	Diabetes Effect: OR (95% CI)	Protein	Apparent Effect of Risk Allele(s)
ADAMSTS9	3p14.3-p14.2	rs4607103	38 kb upstream	European GWAS meta-analysis	T2DM	1.09 (1.06–1.12)	ADAM metallopeptidase with thrombospondin type 1 motif, 9	Insulin resistance
ADCY5	3q13.2-q21	rs2877716, rs11708067	Intronic	European fasting measures GWAS meta-analysis, 2-hour measures GWAS meta-analysis	FG, HOMA-B, 2-hour G, T2DM, birth weight	1.12 (1.09–1.15)	Adenylate cyclase 5	Unknown
BCL11A	2p16.1	rs243021	99 kb downstream	European GWAS meta-analysis	T2DM	1.08 (1.06–1.10)	B-cell CLL/lymphoma 11A (zinc finger protein)	β-cell dysfunction
C2CD4A/4B	15q22.2	rs7172432	Intergenic	Japanese GWAS	T2DM	1.13 (1.09–1.18)	C2 calcium-dependent domain containing 4B	Unknown
CDC123/CAMK1D	10p13	rs12779790	Intergenic	European GWAS meta-analysis	T2DM	1.11 (1.07–1.14)	Cell division cycle 123 homolog (Saccharomyces cerevisiae)/calcium/calmodulin-dependent protein kinase type 1D	β-cell dysfunction
CDKAL1	6p22.3	rs7754840, rs10946398	Intronic	Multiple European GWAS	T2DM	1.12 (1.08–1.16)	CDK5 regulatory subunit associated protein 1-like 1	β-cell dysfunction
CDKN2A/2B	9p21	rs10811661	125 kb upstream	Multiple European GWAS	T2DM	1.20 (1.14–1.25)	Cyclin-dependent kinase inhibitor 2A/2B	β-cell dysfunction
CENTD2	11q13.4	rs1552224	5′ UTR	European GWAS meta-analysis	T2DM	1.14 (1.11–1.17)	Centaurin, delta 2	β-cell dysfunction
DGKB/TMEM195	7p21.2	rs2191349	Intergenic	European fasting measures GWAS meta-analysis	FG, T2DM	1.06 (1.04–1.08)	Diacylglycerol kinase, beta 90 kDa/transmembrane protein 195	β-cell dysfunction
DUSP9	Xq28	rs5945326	8 kb upstream	European GWAS meta-analysis	T2DM	1.27 (1.18–1.37)	Dual specificity phosphatase 9	Unknown
FTO	16q12.2	rs8050136, rs9939609	Intronic	Multiple European GWAS	BMI, T2DM	1.15 (1.09–1.22)	Fat mass and obesity-associated protein	BMI-dependent insulin resistance
GCK	7p15.3-p15.1	rs4607517, rs1799884 (30G>A)	36 kb upstream, promoter	European fasting measures GWAS meta-analysis	FG, HbA1c, T2DM	1.07 (1.05–1.10)	Glucokinase (hexokinase 4)	β-cell dysfunction

Continued

TABLE 85-2 Susceptibility Genes for Type 2 Diabetes—*Cont'd*

Locus: Nearest Gene(s)	Chromosomal Location	Marker(s)	Variant Type(s)	Source of Initial Discovery	Genome-Wide Associated trait(s)	Diabetes Effect: OR (95% CI)	Protein	Apparent Effect of Risk Allele(s)
GCKR	2p23	rs780094, rs1260326	Leu446Pro, intronic	European fasting measures GWAS meta-analysis, 2-hour measures GWAS meta-analysis	FG, FI, 2-hour G, HOMA-IR, T2DM	1.06 (1.04–1.08)	Glucokinase (hexokinase 4) regulator	Hepatic insulin resistance
HHEX/IDE	10q23/10q23-q25	rs1111875	7.7 kb downstream	French GWAS	T2DM	1.13 (1.08–1.17)	Hematopoietically expressed homeobox/insulin-degrading enzyme	β-cell dysfunction
HMGA2	12q15	rs1531343	43 kb upstream	European GWAS meta-analysis	T2DM	1.10 (1.07–1.14)	High mobility group AT-hook 2	Unknown
HNF1A	12q24.2	rs7957197	20 kb downstream	European GWAS meta-analysis	T2DM	1.07 (1.05–1.10)	HNF1 homeobox A	β-cell dysfunction
TCF2 (HNF1B)	17cen-q21.3	rs757210, rs4430796	Intronic	Candidate gene	T2DM	1.12 (1.07–1.18)	Hepatocyte nuclear factor 1-beta	β-cell dysfunction
IGF2BP2	3q27.2	rs4402960	Intronic	Multiple European GWAS	T2DM	1.17 (1.10–1.25)	Insulin-like growth factor 2 mRNA binding protein 2	β-cell dysfunction
IRS1	2q36	rs2943641	502 kb upstream	French, Danish GWAS	T2DM	1.19 (1.13–1.25)	Insulin receptor substrate 1	Insulin resistance
JAZF1	7p15.2-p15.1	rs864745	Intronic	European GWAS meta-analysis	T2DM	1.10 (1.07–1.13)	Juxtaposed with another zinc finger protein 1	β-cell dysfunction
KCNJ11/ABCC8	11p15.1	rs5219	Glu23Lys in KCNJ11	Candidate gene	T2DM	1.15 (1.09–1.21)	K inwardly-rectifying channel, subfamily J, member 11/ATP-binding cassette, sub-family C (CFTR/MRP), member 8	β-cell dysfunction
KCNQ1	11p15.5	rs2237892	Intronic	Japanese, Korean, Chinese GWAS	T2DM	1.40 (1.34–1.47)	K voltage-gated channel, KQT-like subfamily, member 1	β-cell dysfunction, decreased incretin secretion
KCNQ1	11p15.5	rs231362	Intronic	European GWAS meta-analysis	T2DM	1.08 (1.06–1.10)	K voltage-gated channel, KQT-like subfamily, member 1	Unknown

KLF14	7q32.3	rs972283	47 kb upstream	European GWAS meta-analysis	T2DM	1.07 (1.05–1.10)	Kruppel-like factor 14	Insulin resistance
MTNR1B	11q21-q22	rs10830963	Intronic	European GWAS meta-analysis	FG, HOMA-B, HbA1c, T2DM	1.09 (1.06–1.12)	Melatonin receptor 1B	Increased melatonin inhibition of insulin secretion
NOTCH2	1p13-p11	rs10923931	Intronic	European GWAS meta-analysis	T2DM	1.13 (1.08–1.17)	Neurogenic locus notch homolog protein 2 (Drosophila)	Unknown
PPARG	3p25	rs1801282	Pro12Ala	Candidate gene	T2DM	1.14 (1.08–1.20)	Peroxisome proliferator-activated receptor-γ	Insulin resistance
PRC1	15q26.1	rs8042680	Intronic	European GWAS meta-analysis	T2DM	1.07 (1.05–1.09)	Protein regulator of cytokinesis 1	Unknown
PROX1	1q32.2-q32.3	rs340874	2 kb upstream	European fasting measures GWAS meta-analysis	FG, T2DM	1.07 (1.05–1.09)	Prospero homeobox 1	β-cell dysfunction
SLC30A8	8q24.11	rs13266634	Arg325Trp	French GWAS	T2DM, FG, HbA1c	1.12 (1.07–1.16)	Solute carrier family 30 (zinc transporter), member 8	β-cell dysfunction
TCF7L2	10q25.3	rs7903146, rs7901695	Intronic	Icelandic linkage region	T2DM, FG, HbA1c	1.37 (1.28–1.47)	Transcription factor 7-like 2 (T-cell-specific, HMG-box)	β-cell dysfunction, decreased incretin-stimulated insulin secretion
THADA	2p21	rs7578597	Thr1187Ala	European GWAS meta-analysis	T2DM	1.15 (1.10–1.20)	Thyroid adenoma associated	β-cell dysfunction
TLE4 (CHCHD9)	9q21.31	rs13292136	234 kb upstream	European GWAS meta-analysis	T2DM	1.11 (1.07–1.15)	Transducin-like enhancer of split 4 (E(sp1) homolog, Drosophila)	Unknown
TP53INP1	8q22	rs896854	Intronic	European GWAS meta-analysis	T2DM	1.06 (1.04–1.09)	Tumor protein p53 inducible nuclear protein 1	Unknown
TSPAN8/LGR5	12q14.1-q21.1/12q22-q23	rs7961581	Intergenic	European GWAS meta-analysis	T2DM	1.09 (1.06–1.12)	Tetraspanin 8/leucine-rich repeat-containing G protein-coupled receptor 5	β-cell dysfunction
WFS1	4p16	rs1801214, rs10010131	Intronic	Candidate gene	T2DM	1.13 (1.07–1.18)	Wolfram syndrome 1 (wolframin)	β-cell dysfunction

Continued

TABLE 85-2 Susceptibility Genes for Type 2 Diabetes—*Cont'd*

Locus: Nearest Gene(s)	Chromosomal Location	Marker(s)	Variant Type(s)	Source of Initial Discovery	Genome-Wide Associated trait(s)	Diabetes Effect: OR (95% CI)	Protein	Apparent Effect of Risk Allele(s)
ZBED3	5q13.3	rs4457053	41 kb upstream	European GWAS meta-analysis	T2DM	1.08 (1.06–1.11)	Zinc finger, BED-type containing 3	Unknown
ZFAND6	15q25.1	rs11634397	1.5 kb down-stream	European GWAS meta-analysis	T2DM	1.06 (1.04–1.08)	Zinc finger, AN1-type domain 6	Unknown
SPRY2	13q13.1	rs1359790	193 kb down-stream	Asian GWAS	T2DM	1.15 (1.10–1.20)	Sprouty homolog 2 (Drosophila)	Unknown
SRR	17p13	rs391300	Intronic	Taiwanese GWAS	T2DM	1.28 (1.18–1.39)	Serine racemase	Unknown
PTPRD	9p23-p24.3	rs17584499	Intronic	Taiwanese GWAS	T2DM	1.57 (1.36–1.82)	Protein tyrosine phos-phatase, receptor type, D	Unknown
UBE2E2	3p24.2	rs7612463	Intronic	Japanese GWAS	T2DM	1.19 (1.12–1.26)	Ubiquitin-conjugating enzyme E2E 2 (UBC4/5 homolog, yeast)	β-cell dysfunction
RBMS1/ITGB6	2q24.2	rs7593730	Intron 3 of RBMS1	European GWAS meta-analysis	T2DM	0.90 (0.86–0.93)	RNA binding motif, single stranded interacting protein 1/integrin, beta 6	Possibly improved insulin sensitivity

FG, fasting glucose; T2DM, type 2 diabetes mellitus; HbA1c, hemoglobin A1c; HOMA-IR, homeostasis model assessment of insulin resistance; HOMA-B, homeostasis model assessment of beta cell function; 2-hour G, two-hour glucose level on oral glucose tolerance test.

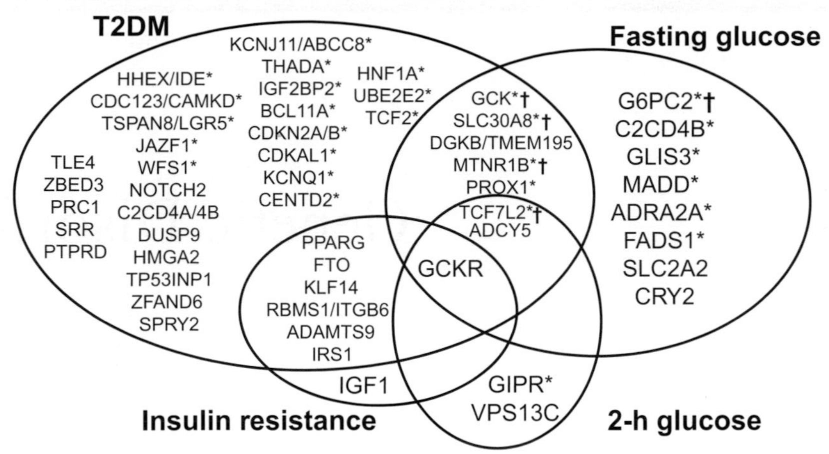

FIGURE 85-1 Established loci for type 2 diabetes, fasting glucose, 2-h glucose, and insulin resistance. Those loci also associated with insulin secretion (apparent effects on beta cell development/mass or glucose/incretin-stimulated insulin release) and hemoglobin A1c are indicated with * and †, respectively.

CHAPTER

86

Genetic Disorders of the Adrenal Gland

Karen Lin-Su

Department of Pediatrics, Division of Pediatric Endocrinology, Weill Medical College of Cornell University, New York, NY, USA

Oksana Lekarev

Department of Pediatrics, Division of Adrenal Steroid Disorders, Mount Sinai School of Medicine, New York, NY, USA

Maria I New

Department of Pediatrics, Department of Genetics and Genomic Sciences, Division of Adrenal Steroid Disorders, Mount Sinai School of Medicine, New York, NY, USA

The adrenal cortex produces numerous steroids, secreted widely in different amounts, each of which has a different potency of hormonal action. The adrenal cortex is divided into three regions: the outer zona glomerulosa, the wide middle zona fasciculata, and the more compact inner zona reticularis. Synthesis of the mineralocorticoid aldosterone is dependent on enzymatic activity limited to the zona glomerulosa, while production of cortisol and androgens requires the enzymes found in the zonas fasciculata and reticularis (*1*). Corticotropin-releasing hormone, produced in the hypothalamus, and vasopressin, produced in the posterior pituitary, synergistically stimulate adrenocorticotropic hormone (ACTH) production (*2*). In turn, ACTH exercises acute and chronic effects on adrenocortical cell processes, driving steroidogenesis (*3*). Cortisol, the principal glucocorticoid in humans, is the main secretory product of the adrenal cortex and is indispensable for proper carbohydrate metabolism and capacity to withstand physiologic stress. The primary regulation of aldosterone synthesis is via the renin–angiotensin system, and the major factors that affect aldosterone synthesis are renin (via angiotensin II), K^+, Na^+ and ACTH (*4, 5*).

Congenital adrenal hyperplasia (CAH) refers to a group of autosomal recessive disorders which, due to various enzymatic defects, result from defective steroidogenesis at any of the enzyme-mediated steps. Each deficient enzymatic step produces characteristically abnormal adrenal hormone and precursor levels. 21- and 11β-hydroxylase deficiencies, occurring distal to the common precursor stages, cause channeling of these precursor steroids into the androgen pathway, resulting in virilization of females and hyperandrogenic effects of both sexes (*5*). In 3β-hydroxysteroid dehydrogenase deficiency, there is glandular production of Δ^5 steroids, which are relatively inactive. While lack of testicular production of Δ^4 androgens produces pseudovaginal hypospadias in the male, enormously high secreted levels of DHEA may undergo peripheral conversion to more potent androgens, which are then able to cause external virilization in females (*6*). In 17α-hydroxylase deficiency, there is extremely high production of 17-deoxysteroids: corticosterone (B), which has some glucocorticoid function, and (11-) deoxycorticosterone (DOC), which causes hypertension. Blocked formation of C_{19}/C_{18} steroids in both the 17α-hydroxylase/17,20-lyase deficiency and the isolated 17,20-lyase defect (in which cortisol production is intact) causes undervirilization in males and sexual infantilism in females (*7*). Congenital lipoid adrenal hyperplasia results from a complete inability to synthesize steroids causing lipid accumulation in adrenal cells, and along with the very serious effects of adrenal insufficiency, undervirilization occurs in genetic males (*8*).

21-hydroxylase deficiency accounts for 90–95% of all cases of CAH. The gene encoding the enzyme is CYP21A2, which is situated within HLA, on the short arm of chromosome 6. The overall worldwide frequency is 1 per 13,000–15,000 live births (*9*). Of the classical form, 75% of cases are of the salt-wasting type (SW-CAH) and the rest are of the simple virilizing type (SV-CAH). Non-classical 21-hydroxylase deficiency (NC-CAH) is due to a less severe enzymatic defect and is much more common, particularly in the Ashkenazi Jewish population, in which up to 1 in 27 are reported to be affected (*10*).

The cardinal features of classic 21-hydroxylase deficiency are genital masculinization in affected females (it is the most common cause of ambiguous genitalia), progressive virilization, advanced somatic development that leads to bone age advancement, and short final adult height. The disorder can lead to polycystic ovarian syndrome (PCOS), infertility, hirsutism, acne and male-pattern baldness in adults. Patients with 11β-hydroxylase deficiency develop similar clinical features of hyperandrogenemia. Patients with SW-CAH have the additional feature of aldosterone deficiency, which leads to hyponatremia with hyperkalemia, and concomitantly elevated plasma renin activity. NC-CAH is an attenuated, late-onset form of 21-hydroxylase deficiency that has a varied clinical presentation, which may include premature adrenarche, precocious puberty, bone age advancement and short final adult height. It is an important and frequently unrecognized cause of female infertility. PCOS, hirsutism, acne and male-pattern baldness may be presenting features as well. By definition, genital development is normal. Some individuals with NC-CAH, confirmed by genetic testing, have no signs or symptoms of the disease. Treatment for CAH consists of glucocorticoid therapy in all patients and additional mineralocorticoid and salt replacement therapy in those with the salt-wasting form.

Genetic testing of all forms of adrenal disorders is important not only for proper diagnosis of the disease, but also for genetic counseling for future pregnancies. The development of genital masculinization of females affected with 21- and 11β-hydroxylase deficiencies can be prevented prenatally by adminstering dexamethasone orally to the mother during pregnancy. For this reason it is important to establish the genetic diagnosis and sex of the fetus prenatally (affected males do not require treatment) via chronionic villous sampling or amniocentesis. Treatment with dexamethasone should be started before the 9th week of gestation, that is before the genitalia begins to form, yet the results of the prenatal testing are not available until at least the 14th week. Therefore, unnecessary dexamethasone treatment is administered to 7 out of 8 fetuses for some time. In order to avoid unnecessary prenatal treatment and invasive diagnostic procedures, an innovative method of noninvasive prenatal diagnosis, where fetal DNA is extracted from the mother's blood early in first trimester, is currently under investigation.

FURTHER READING

1. Hornsby, P. Physiological and Pathological Effects of Steroids on the Function of the Adrenal Cortex. *J. Steroid Biochem.* **1987,** *27,* 1161.
2. Ganong, W. Neurotransmitters and Pituitary Function: Regulation of ACTH Secretion. *Fed. Proc.* **1980,** *39,* 2923.
3. Waterman, M.; Simpson, E. Cellular Mechanisms Involved in the Acute and Chronic Actions of ACTH. In *Adrenal Cortex*; Anderson, D.; Winter, J., Eds.; Butterworths: London, 1985; pp 50.
4. Lifton, R. P.; Gharavi, A. G.; Geller, D. S. Molecular Mechanisms of Human Hypertension. *Cell* **2001,** *104,* 545.
5. Speiser, P. W.; Dupont, J.; Zhu, D., et al. Disease Expression and Molecular Genotype in Congenital Adrenal Hyperplasia due to 21-Hydroxylase Deficiency. *J. Clin. Invest.* **1992,** *90,* 584.
6. Bongiovanni, A. M. Congenital Adrenal Hyperplasia due to 3b-Hydroxysteroid Dehydrogenase. In *Adrenal Diseases in Childhood*; New, M. I.; Levine, L. S., Eds.; Karger: Basel, 1984; pp 72.
7. Biason-Lauber, A.; Kempken, B.; Werder, E., et al. 17Alpha-Hydroxylase/17,20-Lyase Deficiency as a Model to Study Enzymatic Activity Regulation: Role of Phosphorylation. *J. Clin. Endocrinol. Metab.* **2000,** *85,* 1226.
8. Miller, W. L. Steroidogenic Acute Regulatory Protein (StAR), a Novel Mitochondrial Cholesterol Transporter. *Biochim. Biophys. Acta* **2007,** *1771,* 663–676.
9. NNSIS 2009 National Newborn Screening Information System. http://www2.uthscsa.edu/nnsis.
10. Speiser, P. W.; Dupont, B.; Rubinstein, P., et al. High Frequency of Nonclassical Steroid 21-Hydroxylase Deficiency. *Am. J. Hum. Genet.* **1985,** *37,* 650.

CHAPTER

87

Disorders of the Gonads, Genital Tract, and Genitalia

Joe Leigh Simpson

Herbert Wertheim College of Medicine, Florida International University, Miami, FL, USA

Disorders of sex development (DSD) result from chromosomal abnormalities or from perturbations of genes controlling sexual differentiation (Mendelian inheritance). The latter are the focus of this chapter. The spectrum of clinical disorders is broad and typically stratified into several general categories depending on whether external genitalia are ambiguous or opposite than expected on the basis of karyotype (sex reversal). Pathogenesis of these disorders can usually be deduced on the basis of embryonic deviation from the pathways expected for normal male (46,XY) or female (46,XX) sex differentiation.

Initially, the embryo has indifferent gonads and is morphologically indistinguishable whether 46,XY or 46,XX. Testes become morphologically identifiable 7–8 weeks after conception (9–10 weeks gestational or menstrual weeks). Fetal Leydig cells produce testosterone, which stabilizes Wolffian ducts and permits differentiation of the vasa deferentia, epididymides, and seminal vesicles. Following conversion of testosterone by 5α-reductase to dihydrotestosterone (DHT), external genitalia are virilized. Fetal Sertoli cells produce anti-Müllerian hormone (AMH), a glycoprotein that diffuses locally to cause regression of Müllerian derivatives (uterus and fallopian tubes). Absent embryonic testes, differentiation occurs along female lines (ovary, external genitalia, uterus and other Müllerian derivatives).

Genetic control of male sex differentiation involves SRY (Y chromosome), which apparently de-represses an autosomal region to allow SOX9 (17q24.1) to be expressed. In the absence of SRY and SOX9, the ovary develops under the influence of FOXL2 (3q21). Haploinsufficiency of SOX9 in 46,XY embryos results in XY sex reversal (male genotype to female phenotype), whereas duplication of the SOX9 region results in the converse (XX sex reversal) even if SRY is lacking. Thus, SRY alone neither suffices for male sex differentiation nor is obligatory. Two intact X chromosomes are required for normal ovarian development (maintenance), but not for initial differentiation as shown by 45,X fetuses having oocytes in utero. Other X-linked and autosomal genes must remain intact.

87.1 46,XX SEX REVERSAL (XX MALES)

Sex-reversed males are 46,XX individuals with bilateral testes. Their penis and scrotum may be small but usually well differentiated. Androgen deficiency is evident, resembling 47,XXY Klinefelter syndrome. Among XX males *not* having genital ambiguity, 80% show SRY. The cytologic basis is crossing over between Xp and Yp during paternal meiosis that extends beyond the pseudoautosomal region, encompassing SRY and thus becoming translocated from the paternal Y to the maternal X.

87.2 XY SEX REVERSAL (46,XY FEMALES)

In XY sex reversal disorders, genetic males (46,XY) are characterized by normal female external genitalia, vagina, uterus, fallopian tubes, and streak gonads in lieu of functioning gonads. Secondary sexual development fails to occur at puberty. Height is normal, and in most cases somatic anomalies are present. Because functioning gonads are almost always absent, FSH and LH are elevated; estrogens are decreased. XY gonadal dysgenesis contrasts with XX gonadal dysgenesis in that approximately 20–30% of XY gonadal dysgenesis patients develop a dysgerminoma or gonadoblastoma as a result of the GBY locus on Yq.

One explanation for XY sex reversal is, predictably, deletions or perturbations of SRY. XY gonadal dysgenesis may also segregate in the fashion expected of an X-linked recessive disorder, or may be due to duplication of an X-linked region encoding a dose sensitive sex reversal gene, *DAX1*. Perturbations of various autosomal loci cause XY sex reversal: *MAP3K1*, *DHH*, *SF1(NR5A1)*, and *LHR*. Other genes that if perturbed may be causative are *WT1*, *SOX9* and *ATX*; autosomal chromosomal deletions (2q,9p,10q) have similar effect. In rare cases,

XY females have shown oocytes, in particular compound heterozygosity for *CBX2*.

87.3 46,XY DSD (MALE PSUDOHERMAPHRODITISM)

In these disorders, individuals with a Y chromosome fail to virilize external genitalia. One obvious explanation is 45,X/46,XY mosaicism, in which phenotypes range from almost normal male (cryptorchidism or penile hypospadias) through genital ambiguity to phenotypically normal female save streak gonads (sex reversal). In individuals who have ambiguous external genitalia and a 45,X/46,XY complement, a uterus is usually (90%) present. 46,XY disorders of development (46DSD) may also result from deficiencies of various adrenal or gonadal enzymes: 3β-ol-dehydrogenase (HSD3β2), 17α-hydroxylase/17,20 desmolase (CYP17), P450 oxidoreductase, 17-ketosteroid reductase (HSD17β2), as well as enzymes necessary to convert cholesterol to pregnenolone in particular (P450scc or StAR). These disorders are autosomal recessive and are generally caused by missense or nonsense mutations throughout any of their several exons. No predominant mutation exists. The common pathogenesis involves low testosterone, specifically levels inadequate to virilize external genitalia. Congenital adrenal lipoid hyperplasia is the result of perturbation of the gene encoding steroidogenic acute regulatory protein (StAR); StAR facilitates entry of cholesterol into mitochondria. In 3β-ol-dehydrogenase type II, (HSD3β2), deficiency synthesis of both androgens and estrogens is decreased, the major androgen produced being dihydroepiandrosterone (DHEA). A relatively weaker androgen than testosterone, DHEA alone is not capable of adequately virilizing the male fetus; thus, genital ambiguity occurs. CYP17 governs both 17α-hydroxylase and 17,20 desmolase (lyase) activities. 46,XY individuals deficient in this enzyme usually show ambiguous external genitalia. However, in some cases, only 17α-hydroxylase activity seems deficient with sex reversal (female phenotype) occurring. In deficiency of 17β-ol-dehydrogenase deficiency type 3 (17HSD3), pubertal virilization may be greater than in other enzyme deficiencies. Finally, in 5α-reductase deficiency type II (SRD5A2), genetic men show ambiguous external genitalia at birth yet at puberty undergo virilization like normal men. This is predictable given 5α-reductase required to convert testosterone (T) to DHT; at puberty testosterone alone suffices to virilize. Different ethnic groups typically show different mutations for SRDA2. Virilization is pronounced at puberty and gender change at puberty is not uncommon.

87.3.1 46,XY DSD Due to Defects of Androgen Action

In these forms of 46,XY, DSD testosterone is produced but cellular response and, hence, functional androgen action is impeded. Affected 46,XY individuals develop phenotypically as female. In complete androgen insensitivity syndrome (CAIS), affected individuals show bilateral testes, female external genitalia, a blindly ending vagina, and no Müllerian derivatives (uterus). Affected individuals manifest breast development. Because AMH is synthesized normally as predicted given presence of Sertoli cells, the body responds appropriately to AMH as result of which there is no uterus.

Some 46,XY individuals with androgen insensitivity feminize (i.e. breast development), but their external genitalia are nonetheless characterized by phallic enlargement and partial labioscrotal fusion. This condition—partial or incomplete androgen insensitivity (PAIS)—must be excluded before a male sex of rearing can be assigned because ability to respond to exogenous androgens is obviously necessary. Both CAIS and PAIS result from perturbation of the same gene (*Xq11*) encoding the intranuclear androgen receptor. Molecular heterogeneity is extensive, although mutations can be identified in most cases of CAIS and in about 75% of PAIS.

87.4 OVOTESTICULAR DSD (TRUE HERMAPHRODITISM)

In ovotesticular DSD both ovarian and testicular tissue exist. Gonads may consist of a separate ovary and testis or one or both ovotestes. A uterus is usually (90%) present. Genital ambiguity is typical. This condition accounts for less than 5% of all DSD cases, the exception occurring in Southern Africa where approximately half of all DSD cases have this condition. Most cases have a 46,XX chromosomal complement; however, 46,XX/46,XY, 46/XY, 46,XX/ 47,XXY and other complements may exist. In 46,XX/46,XY cases, chimerism is responsible. In 46,XX true hermaphrodites, depression of autosomal testis-determining gene(s) seems to be the most likely explanation. Consistent with this are kindreds characterized by either multiple siblings with XX true hermaphroditism or kindreds in which some members were 46,XX males and others 46,XX true hermaphrodites.

87.5 46,XX OVARIAN DYSGENESIS/ PREMATURE OVARIAN FAILURE (POF)

In XX ovarian dysgenesis (XX gonadal dysgenesis), phenotypic female have ovaries devoid of oocytes (streak gonads). Unlike Turner syndrome (45,X) there are no somatic anomalies and height is normal. "XX ovarian dysgenesis" is genetically heterogeneous, and more specific diagnosis will increasingly become available. Inheritance is usually autosomal recessive. Variable expression is not uncommon for many disorders, and relatively few cases studied mutations molecularly in FSH-β, FSHR and LHR are known but rare. An exception exists in Finland, where Ala566Val is a common explanation for primary amenorrhea. Other genes controlling reproductive hormones that may, if perturbed, cause XX gonadal dysgenesis are CYP17, CYP19 (aromatase), and inhibin A (INHA). The pivotal ovarian-determining gene *FOXL2* is perturbed (nonsense or frameshift mutations) in the autosomal dominant syndrome Blepharophimosis-Ptosis-Epicanthus (BPE) type II. Absent somatic features (BPES), however, *FOXL2* mutations are uncommon explanations for POF. More common is premutation of FMR1 (fragile X syndrome). Approximately 15% of women with the FRAXA premutation (CGG repeats >55) show POF, specifically when the number of CGG repeats is 80–99 repeats. No further increased risk occurs after >100 repeats. Ovarian failure also occurs in galactosemia (galactose 1-phosphate uridyl transferase (GALT) deficiency) and in type 1 carbohydrate-deficient-glycoprotein (CDG) deficiency.

Many pleiotropic genes have ovarian failure as one component. Noteworthy are AIRE, NOG (symphangism), Perrault syndrome (neurosensory deafness) and the disorder cerebellar ataxia and ovarian dysgenesis. Based on either known roles in folliculogenesis or phenotypes of murine knockout models, many candidate genes have been interrogated to determine their role in human nonsyndromic POF. In addition to FSHβ, LHR, CYP17, and premutation of FMR1, functionally proven causative mutations have been found for NOBOX, FIGLA, POU51, ADAMTS, PTHB1, MSH5 and SMC1. No perturbations have been found in a dozen other plausible candidates. However, sample sizes are typically too small to exclude any lacking a role, and usually restricted to a single ethnic group.

87.6 INTERNAL GENITAL DUCT ANOMALIES (MÜLLERIAN OR WOLFFIAN)

Both females (46,XX) and males (46,XY) may have abnormalities restricted to internal genital ducts. Disorders in females include absence or failure of the fusion of the embryonic heri-uteri (incomplete Müllerian fusion), imperforate hymen, transverse vaginal septum (McKusick–Kaufman syndrome or MKS), and vaginal atresia. MKS and vaginal atresia (VA) may present with primary amenorrhea due to lack of egress of menstrual blood. External genitalia and gonads are normal. Transverse vaginal septa (TVS) are about 2 cm thick and usually located near the junction of the upper third and lower two-thirds of the vagina. In the Amish, an autosomal recessive gene cosing for a chaperonin (MKS) is responsible MKS. In VA, the lower portion of the vagina (typically one-fifth to one-third of the total length) is replaced by 2–3 cm of fibrous tissue. The embryonic basis presumably involves failure of the urogenital sinus to contribute the caudal portion of the vagina. Ovaries are normal and the uterus is likewise, albeit with obstruction to menstrual blood flow.

Aplasia of the Müllerian ducts (MA) leads to absence of the uterine corpus, uterine cervix, and upper (superior) vagina. This is the converse embryologically of VA. In MA, ovaries are normal and, hence, secondary sexual development normal. Because most of the vaginal length is contributed by Müllerian derivatives, the vagina may be shortened to 1–2 cm. MA can thus be presumed polygenic/multifactorial in etiology. No molecular abnormalities have been found in interrogated candidate genes, save WNT4 in atypical cases characterized by both MA and virilization.

Incomplete Müllerian fusion (IMF) is a relatively common gynecological condition in which the paired Müllerian ducts fail to fuse and canalize. Recurrence risk for first-degree relatives is consistent with predictions based on polygenic/multifactorial etiology. Perturbation of *HOXA13* causes hand–foot–genital (HFG) syndrome, an autosomal dominant disorder characterized by IMF, skeletal anomalies, and urologic anomalies. Molecular perturbations have not been detected in isolated IMF.

Internal genital abnormalities also occur in males (46,XY). The uterus and fallopian tubes (Müllerian derivatives) may persist (persistent Müllerian derivatives or PMD). External genitalia, Wolffian derivatives, and testes develop as expected for males, and pubertal virilization occurs. Etiology involves perturbation of one of the two genes integral for precluding Müllerian development in men: AMH or AMH receptor (AMHR). If AMH is not detected, a mutation in that structural gene can usually be demonstrated. Such cases are most often homozygous mutations in individuals of North Africa (Arab) or Mediterranean descent. If AMH is elevated, an AMHR mutation is likely. The most common perturbation is a 27bp deletion (del6331-6357).

Almost all men with cystic fibrosis (CF) are infertile, as a result of congenital absence of the vas deferens (CAVD). The great majority of men with CVAD have one or more CFTR mutations. The most common CFTR mutations causing CAVD are ΔF508 and W128X. However, in addition to mutations, that cause CF with pancreatic and pulmonary pathology, less deleterious mutations exist that play roles in CVAD. A polymorphism of specific relevance to CAVD is 5T (5-thymidines) in intron 8. CFTR function is impeded but not to the extent that pulmonary and pancreatic function is affected. However, if 5T "polymorphism" is homozygous then bilateral CAVD (CBAVD) occurs. The same phenotype occurs in many compound heterozygous genotypes if 5T is trans to a CF causing CFTR mutation (e.g. AF508/5T).

TABLE 87-1 Malformations Syndromes in Which Male Pseudohermaphroditism (46,XY DSD) Is One Component

Syndrome	OMIM Number	Prominent Features	Etiology
Ablepharon-macrostomia	200110	Absent eyelids, eyebrows, eyelashes, external ears; fusion defects of the mouth; ambiguous genitalia; absent or rudimentary nipples; parchment-like skin; delayed development of expressive language	Autosomal recessive
Aniridia-Wilms' tumor association	194070	Moderate to severe mental deficiency, growth deficiency, microcephaly, aniridia, nystagmus, ptosis, blindness, Wilms' tumor, ambiguous genitalia, gonadoblastoma	Chromosomal or autosomal dominant
AntleyBixter	210750	Craniosynostoses, synostoses, radius and humerus, Bowing of ulna and femur, arachnodactyly, joint contractures	Autosomal recessive? (P450 Oxidoreductase deficiency)
Asplenia, cardiovascular	208530	Hypoplasia or aplasia of the spleen complex cardiac position and development of the abdominal organs, agenesis of corpus callosum, imperforate anus, ambiguous genitalia, contractures of the lower limb	Autosomal recessive
Beemer	209970	Hydrocephalus, dense bones, cardiac malformation, bulbous nose, broad nasal bridge, ambiguous genitalia	Autosomal recessive
Jacobsen	147791	Trigonocephaly, flat and broad nasal bridge, micrognathia, carp mouth, hypertelorism, low-set ears, severe congenital heart disease, anomalies of limbs, external genitalia	Autosomal dominant or Chromosomal
Denys–Drash	194080	Wilms' tumor, nephropathy, ambiguous genitalia with 46,XY karyotype	Unknown
Fraser	219000	Cryptophthalmia, defect of auricle, hair growth on lateral forehead to lateral eyebrow, hypoplastic nares, mental deficiency, partial cutaneous syndactyly, urogenital malformation	Autosomal recessive
Lethal acrodysgenital dysplasia	270400	Failure to thrive, facial dysmorphism, ambiguous genitalia, syndactyly, postaxial polydactyly, Hirschprung disease, cardiac and renal malformations	Autosomal recessive
Rutledge	270400	Joint contractures, cerebellar hypoplasia, renal hypoplasia, ambiguous genitalia, urologic anomalies, tongue cysts, shortness of limbs, eye abnormalities, heart defects, gallbladder agenesis, ear malformations	Autosomal recessive
SCARF	312830	Skeletal abnormalities, cutis laxa, craniosynostosis, ambiguous genitalia, psychomotor retardation, facial abnormalities	Uncertain
Short rib–polydactyly, Majewski-type	263520	Short stature; short limbs; cleft lip and palate; ear anomalies; limb anomalies, including preaxial and postaxial polysyndactyly; narrow thorax; short horizontal ribs; high clavicles; ambiguous genitalia	Autosomal recessive
Smith-Lemli-Opitz	270400	Microcephaly, mental retardation, hypotonia, ambiguous genitalia, abnormal facies	Autosomal recessive (deficiency 7-OH cholesterol dehydrogenase)
Trimethadione, teratogenicity	N/A	Mental deficiency, speech disorders, prenatal onset growth deficiency, brachycephaly, midfacial hypoplasia, broad and upturned nose, prominent forehead, eye anomalies, cleft lip and palate, cardiac defects, ambiguous genitalia	Teratogeneicity
VATER association	192350	Vertebral, anal, tracheoesophageal, and renal; anomalies, subjects with ambiguous genitalia as part of the cloacal anomalies	Unknown (if valid entity); alleged progestational teratogenicity unproved

Updated from Simpson, J. L.; Elias, S. *Genetics in Obstetrics and Gynecology*, 3rd ed.; WB Saunders: Philadelphia, 2003.

TABLE 87-2 Mendelian Disorders Associated with Ovarian Failure

Disorder	OMIM Number	Somatic Features	Ovarian Anomalies	Etiology
Cockayne syndrome	216400	Dwarfism, microcephaly, mental retardation, pigmentary retinopathy senility. Sensitivity to ultraviolet light	Ovarian atrophy and fibrosis	Autosomal recessive and photosensitivity, premature
Martsolf syndrome		Short stature, microbrachycephaly, cataracts, abnormal facies with relative prognathism due to maxillary hypoplasia	"Primary hypogonadism"	Autosomal recessive
Nijmegen syndrome	251260	Chromosomal instability, immunodeficiency, hypersensitivity to ionizing radiation, malignancy	Ovarian failure (primary)	Autosomal recessive (7;14 rearrangement)
Werner syndrome	277700	Short stature, premature senility, skin changes (scleroderma)	Ovarian failure	Autosomal recessive
Rothmund–Thompson Syndrome	268400	Skin abnormalities (telangiectasia, erythrema, irregular pigmentation), short stature, cataracts, sparse hair, small hands and feet, mental retardation, osteosarcoma	Ovarian failure (primary hypogonadism or delayed puberty)	Autosomal recessive
Carbohydrate-deficient type 1 (phosphomannomutase deficiency)		Neurologic abnormalities (e.g. unscheduled eye movements), ataxia, hypotonial/hyporeflexia strokes, joint contractures	Ovarian failure (hypogonadism) (Kristiansson et al., 1995)	Autosomal recessive glycoprotein syndrome
Ataxia–telangiectasia	208900	Cerebellar ataxia, multiple telangiectasias (eyes, ears, flexar surface of extremities), immunodeficiency, chromosomal breakage, malignancy, X-ray hypersensitivity	"Complete absence of ovaries," "absence of primary follicles"	Autosomal recessive
Bloom syndrome	210900	Dolichocephaly, growth deficiency, sun-sensitive facial erythema, chromosomal instability (increased sister chromatical exchange), increased malignancy	Ovarian failure	Autosomal recessive

Updated from Simpson, J. L.; Elias, S. *Genetics in Obstetrics and Gynecology*, 3rd ed.; WB Saunders: Philadelphia, 2003.

TABLE 87-3 Malformation Syndromes with Müllerian Aplasia[a]

Syndrome	OMIM Number	Somatic Anomalies	Etiology
Fraser	219000	Cryptophthalmia, nose and external ear anomalies, stenotic larynx, skeletal defects, syndactyly, renal agenesis, large clitoris and labia majora, mental retardation	Autosomal recessive
Meckel–Gruber	249000	Microcephaly, posterior encephalocele, eye anomalies, cleft palate, polydactyly, polycystic kidneys	Autosomal recessive
MURCS Association	601076	Renal aplasia, cervicothoracic somite dysplasia, Klippel–Feil anomaly, deafness, short stature	Unknown; probably heterogeneous
Thalidomide teratogenicity		Nasal hemangioma, neurosensory hearing loss, ear anomalies, limb reduction defects, visceral anomalies	Teratogen
Urogenital adysplasia, hereditary (hereditary renal adysplasia; bilateral renal agenesis)		Oligohydramnios, flattened (Potter) facies, pulmonary hypoplasia, unilateral or bilateral absent kidneys, limb deformities	Autosomal dominant
Winter	247990	Lacrimal duct stenosis, external and middle ear anomalies, renal agenesis	Autosomal recessive
Kumar		Skeletal anomalies	Unknown

[a]Several of these syndromes could be heterogenous, that is, more than one entity. In particular, Klippel–Feil and Müllerian aplasia above could be a distinct entity, as could deafness and Müllerian aplasia.

From Simpson, J. L. Genetics of the Female Reproductive Ducts. *Am. J. Med. Genet*. **1999**, *89*, 224–239.

TABLE 87-4 Selected Multiple Malformation Syndromes in which Incomplete Müllerian Fusion

Syndrome	OMIM Number	Somatic Anomalies	Etiology
Bardet–Biedl	209900	Retinal pigmentary degeneration (retinitis pigmentosa), polydactyly, obesity, mental deficiency	Autosomal recessive
Beckwith-Wiedemann	130650	Macroglossia, omphalocele, macrosomia	Autosomal dominant, after uniparental disomy
Donohue (leprechaunism)	24200	Elfin facies with thick lips; large, low-set ears; prominent breasts and external genitalia; hirsutism; abnormal carbohydrate metabolism; failure to thrive; motor and mental retardation	Autosomal recessive
Fraser	219000	Cryptophthalmia, external ear and nose anomalies, laryngeal stenosis, syndactyly, skeletal defects, renal agenesis, large clitoris and labia majora, mental retardation	Autosomal recessive
Hand-foot-genital (HFG)	140000	Metacarpal and metatarsal anomalies, malformed thumbs, displaced urethral meatus, urinary incontinence	Autosomal dominant
Johanson-Blizzard	243800	Deafness, hypoplastic alae nasi, primary hypothyroidism, mental retardation	Autosomal recessive
Laryngeal atresia	607132	Hydrocephaly, complete or partial laryngeal obstruction, tracheoesophageal fistula or atresia, renal hypoplasia, varus deformity of feet	Unknown
Meckel-Gruber	24900, 603194 607301	Microcephaly, posterior encephalocele, eye anomalies, cleft palate, polycystic kidneys, polydactyly	Autosomal recessive
Roberts	268300	Sparse, silvery blond hair; midfacial hemangioma; cleft lip with or without cleft palate; limb reduction defect; intrauterine growth retardation	Autosomal recessive
Cavalcanti		Tibial aplasia, triphalangeal thumb, micratia, scaliosis, club foot (Müllerian aplasia also reported)	Unknown
Rudiger	268650	Bifid uvula, coarse facies, absent ear cartilage, hydronephrosis secondary to ureterovesical stenosis, short digits	Autosomal recessive
Thalidomide teratogenicity		Nasal hemangioma, neurosensory hearing loss, ear anomalies, limb reduction defects, visceral anomalies	Teratogen
Trisomy 18		Prominent occiput, malformed ears, micrognathia, short sternum, cardiac defects, horseshoe kidney, overlapping fingers, intrauterine growth retardation, severe development retardation	Chromosomal aneuploidy
Trisomy 13		Microcephaly, microphthalmia, malformed ears, cleft lip and palate, cardiac anomalies, polydactyly, intrauterine growth retardation, severe developmental retardation	Chromosomal aneuploidy
Urogenital adysplasia, hereditary (hereditary renal agenesis)		Oligohydramnios, flattened (Potter) facies, pulmonary hypoplasia, unilateral or bilateral absent kidneys, limb deformities	Autosomal dominant

From Simpson, J. L. Genetics of the Female Reproductive Ducts. *Am. J. Med. Genet*. **1999**, 89, 224–239.

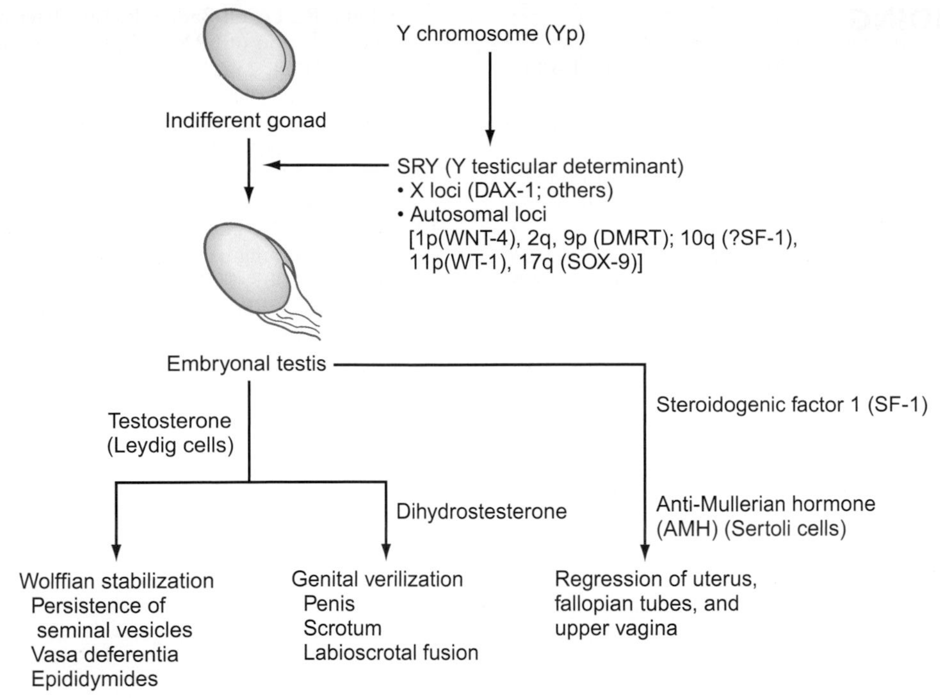

FIGURE 87-1 Schematic diagram illustrating embryonic differentiation in the normal male. *(Modified from Simpson, J. L.; Elias, S. Disorders of Sex Differentiation: Gonadal Abnormalities and Hypogonadotropic Hypogonadism. In* Genetics in Obstetrics and Gynecology, *3rd ed.; WB Saunders: Philadelphia, 2003.)*

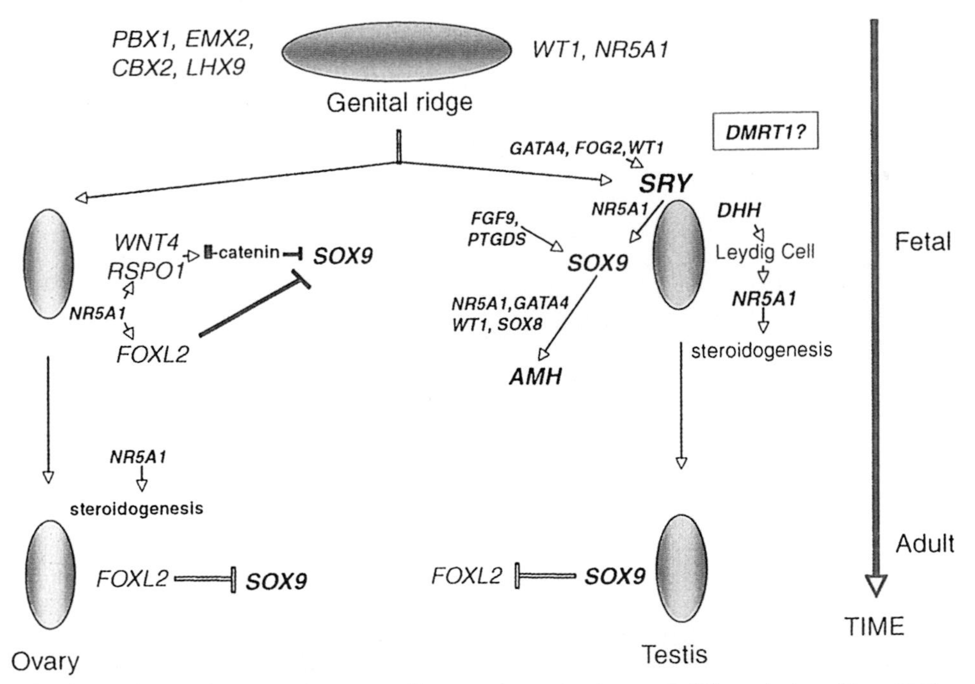

FIGURE 87-2 The molecular and genetic events in mammalian sex determination and differentiation. *(From McElreavey, K.; Bashamboo, A. Genetic Disorders of Sex Differentiation.* Adv. Exp. Med. Biol. ***2011***, 11, *707, 91–99.)*

FURTHER READING

1. Hughes, I. A.; Houk, C.; Ahmed, S. F.; Lee, P. A.; LWPES Consensus Group; ESPE Consensus Group Consensus Statement on Management on Intersex Disorders. *Arch. Dis. Child* **2006,** *91,* 554–563.
2. Josso, N.; Belville, C.; di Clemente, N., et al. AMH and AMH Receptor Defects in Persistent Müllerian Duct Syndrome. *Hum. Reprod. Update* **2005,** *11,* 351–356.
3. Mendonca, B. B.; Domenice, S.; Arnhold, I. J. P., et al. 46, XY Disorders of Sex Development (DSD). *Clin. Endocrinol.* **2009,** *70,* 173–187.
4. Persani, L.; Rossetti, R.; Cacciatore, C. Genes Involved in Human Premature Ovarian Failure. *J. Mol. Endocrinol.* **2010,** *45,* 257–279.
5. Qin, Y.; Choi, Y.; Zhao, H., et al. NOBOX Homeobox Mutation Causes Premature Ovarian Failure. *Am. J. Hum. Genet.* **2007,** *81,* 576–581.
6. New, M.; Simpson, J. L. *Hormonal and Genetic Basis of Sexual Differentiation Disorders and Hot Topics in Endocrinology: Proceedings of the 2nd World Conference*; Springer, 2011.
7. Sekido, R.; Lovell-Badge, R. Sex Determination Involves Synergistic Action of SRY and SF1 on a Specific Sox9 Enhancer. *Nature* **2008,** *453,* 930–934.
8. Simpson, J. L. Genetic and Phenotypic Heterogeneity in Ovarian Failure: Overview of Selected Candidate Genes. *Ann. N. Y. Acad. Sci.* **2008,** *1135,* 146–154.
9. Simpson, J. L.; Elias, S. *Genetics in Obstetrics and Gynecology*, 3rd ed.; WB Saunders: Philadelphia, 2003.
10. Uhlenhaut, N. H.; Jakob, S.; Anlag, K., et al. Somatic Sex Reprogramming of Adult Ovaries to Testes by FOXL2 Ablation. *Cell* **2009,** *139,* 1130–1142.
11. Wiersma, R. True Hermaphroditism: The Clinical Picture. *Pediatr. Surg. Int.* **2004,** *20* (5), 363–368.
12. Zhao, H.; Chen, Qin Y, et al. Transcription Factor FIGLA Is Mutated in Patients with Premature Ovarian Failure. *Am. J. Hum. Genet.* **2008,** *82,* 1342–1348.

CHAPTER

88

Cancer of the Breast and Female Reproductive Tract

Ora Karp Gordon

GenRisk Adult Genetics Program, Cedars-Sinai Medical Center, Los Angeles, CA, USA

Over the last two decades, the field of clinical cancer genetics has moved from recognition of familial susceptibility in almost all common female cancers to identification of multiple single genes, which confer many orders of magnitude higher risk for cancer than seen in the general population. Most notably is the single genes for breast, ovarian, and endometrial cancer risk. This has led to intensive investigations into the frequency of mutations in various populations, guidelines for genetic testing as well as specific screening and prevention recommendations that now have associated outcomes data. In addition, there is greater understanding of the molecular pathways involved in all cancer development, both sporadic and inherited. The ability to specifically identify an at risk, unaffected woman and provide enhanced surveillance, medical and surgical prevention allows for genetics to dictate clinical care for the first time.

The monumental discovery of BRCA1 and BRCA2 in rapid succession changed the face of risk assessment for cancer. Though there are many similarities between BRCA1 and BRCA2 phenotypes, there are some substantial differences. The cancer spectrum in BRCA1 is largely confined to women, with lifetime risks for breast (56–81%) and ovarian cancer (27–54%)—breast and ovary being the main organs at risk—compared to BRCA2 which holds a similar risk for breast cancer (45–84%), albeit later in onset generally, a lower overall risk for ovarian cancer (11–27%), and a risk for gastrointestinal cancers including pancreatic, esophageal and gastric. Male BRCA2 carriers have a substantial risk for both breast (6%) and prostate cancer (20%) in addition to the gastrointestinal cancers. The reasons for the large variability in lifetime risk estimations are related mostly to variation in study design and carrier ascertainment but are also related to genetic modifiers such as RAD51. BRCA1 and BRCA2 breast tumors mirror sporadic cancers in terms of the distribution of ductal and carcinomas, but BRCA1 cancers less often have an in situ component. BRCA1 tumors frequently have a specific pathologic profile, characterized by highly aggressive, proliferative, hormone non-responsive tumors, referred to as a basal phenotype and research is ongoing to determine if a particular histologic and molecular profile could be used to predict germ line BRCA1 carriers in pathology specimens.

The frequency of BRCA mutations varies greatly by ethnic group with Ashkenazi Jewish individuals having up 1 in 40 risk of being a carrier, whereas the estimated carrier frequency ranges from 1 in 200 to 1 in 500 in other ethnic groups. A number of other significant founder mutations (Icelandic, West African, French Canadian and Dutch) have been described and influence testing practices worldwide. Sequencing is the mainstay of mutation detection in the BRCA genes, but there are challenges due to the size of these genes. Newer technologies including multiplex ligation-dependent probe amplification have increased the detection rate of mutations considerably, but clinically testing remains complex and convoluted because of the BRCA1/2 patent issues in North America and the United Kingdom.

Rare familial syndromes and moderate risk single genes including hereditary diffuse gastric and lobular cancer (CDH1 mutations), Li Fraumeni syndrome (P53), Cowden syndrome (PTEN), Peutz–Jeghers syndrome (STK11), CHEK2, and PALB2 are all rare moderate penetrant genes that have specific clinical features or associated unusual cancers in addition to breast cancer. The lifetime risk for breast cancer for most of these disorders is in the 40% range. All are autosomal dominant with reduced penetrance associated with a distinctive array of cancer risks or clinical stigmata.

Despite all of these advances, a large component of familial risk for breast cancer is unaccounted for and thought to be polygenic. Genome-wide association studies have revealed eight single nucleotide polymorphisms (SNPs) that are highly significant in terms of their association with increased or decreased risk of breast cancer, but the contribution of each risk allele is very small and not clinically relevant (RR ranging from 0.89 to 1.6).

To date the addition of genomic information to classic risk models such as the Gail have provided very modest improvements in the ability to predict individual risk. Many more polymorphisms will need to be discovered before a sizable portion of the population attributable risk is accounted for. Many of these SNPs are likely in significant linkage disequilibrium with a disease-causing locus, but they are not causal for breast cancer.

The genetics of hereditary endometrial and ovarian cancer focus primarily on mutations in the mismatch repair family of genes as the most important genetic susceptibility to endometrial cancer. In hereditary nonpolyposis colorectal cancer (HNPCC or Lynch) syndrome, the lifetime risks for uterine cancer are up to 60% and often the presenting cancer in women because of the earlier age of onset, even though the colon cancer risk is higher (up to 80%). The ovarian cancer risk is about 10% lifetime. Gynecologic cancers related to mismatch repair defects are unique in that immunohistochemistry stains for the *MLH1*, *MSH2*, *MSH6* and *PMS2* mutations can be used as a pathology screen to predict risk for germ line mutations. Recommendations have been made toward universal screening of all colon cancers for mismatch repair gene deficits and for endometrial cancer as well.

There is now clear evidence that the preventive interventions, particularly for hereditary breast–ovarian cancer syndrome, decrease mortality. The most powerful protective intervention is prophylactic salpingo-oophorectomy. National guidelines for enhanced surveillance for breast cancer include annual MRI alternating with annual mammogram beginning at age 25, consideration of anti-hormonal medication and prophylactic risk reducing mastectomy and oophorectomy, each with benefits and limitations. Finally, advances in targeted therapy including mounting evidence for the selective sensitivity of BRCA-related tumors to poly-ADP ribose polymerase (PARP1) inhibitors and potentially platinum-based chemotherapies herald great promise for personalized medicine. If efficacy is sustained, it will usher in the new era where germ line genetic status, and not just histology or tumor genomics, dictates therapy.

Metabolic Disorders

Disorders of the Body Mass, 365
Genetic Lipodystrophies, 367
Amino Acid Metabolism, 372
Disorders of Carbohydrate Metabolism, 374
Congenital Disorders of Protein Glycosylation, 376
Purine and Pyrimidine Metabolism, 379
Lipoprotein and Lipid Metabolism, 382
Organic Acidemias and Disorders of Fatty Acid Oxidation, 384
Vitamin D Metabolism or Action, 387
Inherited Porphyrias, 391
Inherited Disorders of Human Copper Metabolism, 393
Iron Metabolism and Related Disorders, 394
Mucopolysaccharidoses, 397
Oligosaccharidoses: Disorders Allied to the Oligosaccharidoses, 399
Sphingolipid Disorders and the Neuronal Ceroid Lipofuscinoses or Batten Disease (Wolman Disease, Cholesterylester Storage Disease, and Cerebrotendinous Xanthomatosis), 403
Peroxisomal Disorders, 409

CHAPTER

89

Disorders of the Body Mass

Patricia A Donohoue and Omar Ali

Medical College of Wisconsin, Milwaukee, WI, USA

Overview: There are numerous human conditions that produce both elevated and reduced body mass, and extremes at both ends of the body size spectrum are associated with increased relative risk of mortality. This chapter is devoted to increased body mass associated with excess fat stores (obesity). Obesity-associated comorbidities include significantly increased risks for diabetes, cardiovascular disease, the "metabolic syndrome", cancer, respiratory disease (asthma, sleep apnea), infertility, degenerative joint disease, depression, anxiety, and discrimination both in social life and in the workplace. The metabolic syndrome, defined as the combined presence of obesity, hyperinsulinemia, hypertension, and hyperlipidemia, is increasing in prevalence in parallel with obesity in adults and youth. For the first time in known history the life expectancy in the United States may potentially decline, due in large part to the obesity epidemic. The etiology of human obesity is undoubtedly multifactorial, reflecting complex interactions between genetic background, environmental conditions, and developmental processes.

Definition: The CDC defines individuals as obese whose body mass index (BMI) (weight in kilograms divided by the square of the height in meters; kg/m^2) exceeds the age- and sex-specific 95th percentile.

The regulation of body fat stores: This complex process is controlled by the interaction of environment and genetic background. Environmental changes have increased obesity prevalence over the past 35 years, as this is a time period too brief for a significant change in the gene pool. The prevalence of obesity is increasing at a dramatic rate as illustrated in a dramatic animation of US obesity prevalence by states compiled by the Centers for Disease Control and Prevention (http://www.cdc.gov/obesity/data/trends.html).

Body fat may be preferentially located in the abdomen (visceral fat of android or central obesity pattern) or surrounding the hips and thighs (gynoid obesity pattern). The android pattern is more strongly associated with dyslipidemia, hypertension, and glucose intolerance.

Genetic architecture of obesity: The detailed genetic architecture of obesity risk has not yet been precisely defined. Even the true magnitude of the heritability of obesity is not yet settled and estimates range from as low as 20% to as much as 80%.

Animal models and human correlates (spontaneous mutations): In rodents, multiple examples of spontaneous single-gene mutations producing obesity form the basis for a candidate gene approach to identify the genes responsible for human obesity.

Leptin and its receptor: The prototypic obese mice with single gene defects are the obese (*ob/ob*, Lep^{ob}) and diabetes (*db/db*, $Lepr^{db}$) autosomal recessive mutations. Other rodent obesity syndrome models include the tubby (*tub*) mutation in mice, the fatty (*fat/fat*) mouse, and the yellow mutation of Agouti mice. The human correlate of the Agouti mutation involves mutations in the *MC4R* gene.

Energy expenditure: Hormonal and genetic control: Defects specific to this system are also important as independent contributors to obesity. These defects involve thermogenesis through UCPs, adrenergic action through beta adrenergic receptor variants, and calorie storage through multiple mechanisms.

Control of feeding: Energy intake and expenditure is under the control of complex interactions between effector systems and neuroendocrine systems. Acute responses to feeding represent peripheral signaling from the GI tract and pancreas, and long-term energy balance is impacted by the central actions of leptin and neurotransmitters. The hypothalamic-pituitary-adrenal axis is intimately involved in this process.

Adipose tissue development and function: Adipose tissue is a biologically active organ, involved not only with energy storage and release, but with autocrine and paracrine effects that are widespread. These effects are manifested in endocrine/hormonal function, immune function including its role in cardiovascular risk, and the central nervous system communication through leptin, tumor necrosis factor alpha (TNFα), interleukin 6 (IL6), adiponectin, resistin, and other adipokines. Adipocytes play a pivotal role in the maintenance of energy balance, and their differentiation is influenced by peroxisome proliferator-activated receptors. Adipocytes also secrete hormones (adipokines) that impact their own fate, such as leptin and adiponectin.

TABLE 89-1 Rodent Obesity Mutations, Human Regions of Synteny, and Human Homologs

	Mutation	Gene	Mode of Inheritance (Autosomal)	Rodent Chromosome	Human Syntenic Region	Human Mutation Described	Mutant Protein
MOUSE	agouti	*Ay*	Dominant	2	20q11	No[a]	ASP
	diabetes	*db*	Recessive	4	1p31	Yes	Lepr
	fat	*fat*	Recessive	8	4q21	No[b]	Carboxypeptidase E
	obese	*ob*	Recessive	6	7q31	Yes	Leptin
	tubby	*tub*	Recessive	7	11p15	No[c]	Phosphodiesterase
RAT	fatty	*fa*	Recessive	5	1p31	Yes	Lepr
	corpulent	*faK*	Recessive	5	1p31	Yes	Lepr

ASP, Agouti signaling protein; Lepr, leptin receptor; CPE, carboxypeptidase E.
[a]ASP prevents binding of αMSH to its receptor MC4R. Several dominant MC4R mutations are associated with human obesity.
[b]CPE is required for normal prohormone processing by prohormone convertase 1 (PC1). PC1 mutations are associated with human obesity.
[c]Phenotype is similar to human Bardet-Biedl and Usher Syndromes.

Polygenic models of obesity in animals: Multiple QTLs within individual rodent strains have been identified that impact obesity, plasma cholesterol levels, specific deposition of body fat depots and propensity, as well as a tendency toward development of high leptin levels and obesity on a high fat diet. Polygenic impact of background strains determines the effect of even profound single gene defects including leptin deficiency and leptin receptor defects. Polygenic models may more closely resemble the human obesity phenotypes; however, the single gene defects producing recessive traits, dominant traits, promoter alterations, and those subject to parental imprinting must continue to be considered candidates for genetic effects in human obesity.

Gene–environment interactions: There are a number of examples of environmental factors influencing genotype expression. These include impact of diet and physical activity on certain gene effects such as those influenced by adrenergic receptor and UCP genotypes.

Non-mammalian models: Insights into energy storage and metabolism are being gained from studies of Drosophila, *C. elegans*, Zebrafish and other models.

Mendelian disorders associated with increased BMI in humans: These disorders are rare, and for the most part the diagnoses do not lead to specific treatments that prevent or alleviate the obesity. Examples of single gene defects include mutations in the leptin and leptin receptor genes, MC4R, POMC, and PC1. Syndromic etiologies include Prader–Willi, Bardet–Biedl, and WAGR syndromes.

Linkage and candidate gene studies of the genetics of human obesity: In general, linkage analyses have not led to remarkable insights in obesity studies, but genome-wide association studies are shedding more light. Specific loci implicated by GWAS include FTO, PPARG, UCP and adrenergic receptor genes.

Addressing the human obesity problem: Predictive factors, prevention strategies, and treatments: Factors that predict adult obesity include childhood obesity, parental obesity (particularly maternal obesity), early adiposity rebound, and lack of physical activity. Prevention strategies that are most effective are instituted in early childhood, and are community based rather than physician based. Medical therapies for obesity have been disappointing and often risky. Significant and sustained lifestyle changes impacting caloric intake and physical activity are required for most patients with obesity. Bariatric surgery is effective but has risks.

SUGGESTED READING

1. Strawbridge, W. J.; Wallhagen, M. I.; Shema, S. J. New NHLBI Clinical Guidelines for Obesity and Overweight: Will They Promote Health? *Am. J. Public Health* **2000,** *90* (3), 340–343.
2. Ogden, C. L.; Carroll, M. D. *Prevalence of Overweight, Obesity, and Extreme Obesity among Adults: United States, Trends 1960–1962 through 2007–2008*; National Center for Health Statistics, Centers for Disease Control and Prevention, 2010.
3. Kissebah, A. H.; Krakower, G. R. Regional Adiposity and Morbidity. *Physiol. Rev.* **1994,** *74,* 761–811.
4. Bouchard, C. Defining the Genetic Architecture of the Predisposition to Obesity: A Challenging but Not Insurmountable Task. *Am. J. Clin. Nutr.* **2010,** *91* (1), 5–6.
5. Clement, K.; Ruiz, J.; Cassard-Doulcier, A. M.; Bouillaud, F.; Ricquier, D.; Basdevant, A.; Guy-Grand, B.; Froguel, P. Additive Effect of A to G (-3826) Variant of the Uncoupling Protein Gene and the Trp64Arg Mutation of the b3-Adrenergic Receptor Gene on Weight Gain in Morbid Obesity. *Int. J. Obes.* **1996,** *20,* 1062–1066.
6. Seeley, R. J.; Schwartz, M. W. Neuroendocrine Regulation of Food Intake. *Acta Paediatr. Suppl.* **1999,** *88* (Suppl 428), 58–61.
7. Myers, M. G.; Leibel, R. L.; Seeley, R. J.; Schwartz, M. W. Obesity and Leptin Resistance: Distinguishing Cause from Effect. *Trends Endocrinol. Metab.* **2010,** *21* (11), 643–651.
8. Bell, C. G.; Walley, A. J.; Froguel, P. The Genetics of Human Obesity. *Nat. Rev. Genet.* **2005,** *6,* 221–234.
9. Korner, J.; Leibel, R. L. To Eat or Not to Eat—How the Gut Talks to the Brain. *N. Engl. J. Med.* **2003,** *349,* 926–928.
10. Diamond, F. The Endocrine Function of Adipose Tissue. *Growth Genet. Horm.* **2002,** *18,* 17–22.
11. Chung, W. K.; Zheng, M.; Chua, M.; Kershaw, E.; Power-Kehoe, E.; Tsuji, M.; Wupeng, X. S.; Williams, J.; Chua, S. C.; Leibel, R. L. Genetic Modifiers of Lepr (fa) Associated with Variability in Insulin Production and Susceptibility to NIDDM. *Genomics* **1997,** *41,* 332–344.
12. *Energy Metabolism and Obesity: Research and Clinical Applications. Contemporary Endocrinology*; Donohoue, P. A., Ed.; Humana Press: Totowa, NJ, 2008.

CHAPTER

90

Genetic Lipodystrophies

Abhimanyu Garg

Division of Nutrition and Metabolic Diseases, Department of Internal Medicine, Center for Human Nutrition, UT Southwestern Medical Center at Dallas, Dallas, TX, USA

Lipodystrophies are disorders of adipose tissue characterized by selective loss of body fat and a predisposition to develop insulin resistance and its complications. The prevalence of metabolic and other manifestations of insulin resistance, such as impaired glucose tolerance, diabetes, hyperinsulinemia, dyslipidemia, hepatic steatosis, acanthosis nigricans, polycystic ovarian disease and hypertension, varies among the different subtypes of lipodystrophies and is generally determined by the extent of adipose tissue loss. The loss of adipose tissue is readily apparent in females and thus the diagnosis of lipodystrophies is relatively easy in them but it may be missed in affected males as many normal men look muscular and have much less body fat than women. Lipodystrophies can be classified into genetic syndromes and acquired disorders due to various causes. The inherited lipodystrophy syndromes can be subclassified into autosomal recessive or autosomal dominant disorders (Table 90-1). There is considerable phenotypic heterogeneity among various types of lipodystrophies.

A great deal of progress has been made recently in elucidating the genetic basis of many types of inherited lipodystrophies. The two most common types of genetic lipodystrophies are autosomal recessive congenital generalized lipodystrophy (CGL) and autosomal dominant familial partial lipodystrophy (FPL) (Figure 90-1). In the last few years, several genes for CGL (*AGPAT2, BSCL2, CAV1* and *PTRF*), FPL (*LMNA, PPARG, AKT2, CIDEC* and *PLIN1*), mandibuloacral dysplasia associated lipodystrophies (*LMNA* and *ZMPSTE24*) and an autoinflammatory lipodystrophy (*PSMB8*) have been identified.

The 1-acylglycerol-3-phosphate-*O*-acyltransferase 2 (AGPAT2) is a critical enzyme involved in the biosynthesis of triglyceride and phospholipids in the adipose tissue (Figure 90-2). The Berardinelli-Seip Congenital Lipodystrophy 2 (*BSCL2*)-encoded protein, seipin, appears to play a role in lipid droplet formation and in adipocyte differentiation. Caveolin 1 is an integral component of caveolae, which are specialized microdomains seen in abundance on adipocyte membranes. PTRF (also known as cavin) is involved in biogenesis of caveolae and regulates expression of caveolins 1 and 3.

Lamin A/C (*LMNA*) is an integral component of nuclear lamina and zinc metalloproteinase (*ZMPSTE24*) is involved in posttranslational proteolytic processing of prelamin A to mature lamin A (Figure 90-3). Peroxisome proliferator-activated receptor γ (*PPARG*) is a key transcription factor involved in adipocyte differentiation; v-AKT murine thymoma oncogene homolog 2 (*AKT2*) is involved in downstream insulin signaling; and cell death-inducing DNA fragmentation factor a-like effector c (*CIDEC*) and perilipin 1 (*PLIN1*) are involved in lipid droplet formation. Proteasome subunit, beta-type, 8 (*PSMB8*) is part of immunoproteasome assembly involved in generating immunogenic epitopes presented by major histocompatibility complex class I molecules.

Many lipodystrophy patients lack mutations in the known lipodystrophy genes suggesting additional loci. The genetic basis of many extremely rare varieties of lipodystrophies associated with SHORT (*s*hort stature, *h*yperextensibility/*h*ernias, *o*cular depression, *R*ieger anomaly and *t*eething delay), neonatal progeroid and MDP (*m*andibular hypoplasia, *d*eafness and *p*rogeroid features) syndromes remains unknown. The discoveries of the molecular mechanisms underlying various types of genetic lipodystrophies has added to our understanding of adipocyte biology and has revealed potential pathways which may be implicated in development of insulin resistance in these disorders.

The clinical management of these patients poses many challenges and includes cosmetic surgery and early identification and treatment of metabolic and other complications with diet, exercise, hypoglycemic drugs and lipid-lowering agents.

TABLE 90-1 Classification of Genetic Lipodystrophies

A. Autosomal recessive syndromes:
1. Congenital generalized lipodystrophy (CGL; Berardinelli-Seip Syndrome)
   a. CGL Type 1: *AGPAT2* (1-acylglycerol-3-phosphate-*O*-acyltransferase 2) mutations
   b. CGL Type 2: *BSCL2* (Berardinelli-Seip congenital lipodystrophy 2) mutations
   c. CGL Type 3: *CAV1* (Caveolin 1) mutation
   d. CGL Type 4: *PTRF* (polymerase I and transcript release factor) mutations
   e. Other varieties
2. Mandibuloacral dysplasia (MAD) associated lipodystrophy
   a. Partial lipodystrophy (Type A pattern): *LMNA* (lamin A/C) mutations
   b. Generalized lipodystrophy (Type B pattern) *ZMPSTE24*(zinc metalloproteinase) mutations
   c. Other varieties: Unknown
3. Autoinflammatory syndromes
   a. *J*oint contractures, *M*uscle atrophy, *M*icrocytic anemia, and *P*anniculitis-induced lipodystrophy (JMP) syndrome: *PSMB8* (proteasome subunit, beta-type, 8) mutations
   b. *C*hronic *A*typical *N*eutrophilic *D*ermatosis with *L*ipodystrophy and *E*levated Temperature (CANDLE) syndrome: *PSMB8* mutations
4. Familial partial lipodystrophy (FPL) : *CIDEC* (cell death–inducing DNA fragmentation factor a-like effector c) mutation
5. SHORT (*S*hort stature, *H*yperextensibility or inguinal hernia, *O*cular depression, *R*ieger anomaly and *T*eething delay) syndrome: Unknown
6. *M*andibular hypoplasia, *D*eafness, *P*rogeroid features (MDP)-associated lipodystrophy: Unknown
7. Neonatal progeroid (Wiedemann–Rautenstrauch) syndrome: Unknown

B. Autosomal dominant syndromes:
1. Familial partial lipodystrophy (FPL)
   a. FPL, type 1, Kobberling variety: Unknown
   b. FPL, type 2, Dunnigan variety (FPLD): *LMNA* (lamin A/C) mutations
   c. FPL type 3: *PPARG* (peroxisome proliferator-activated receptor-γ) mutations
   d. FPL type 4: *AKT2* (v-AKT murine thymoma oncogene homolog 2) mutation
   e. FPL, type 5: *PLIN1* (perilipin 1) mutations
   f. Other varieties: Unknown
2. Atypical progeroid Syndrome: *LMNA* mutations
3. Hutchinson–Gilford progeria Syndrome: *LMNA* mutations
4. SHORT syndrome: Unknown

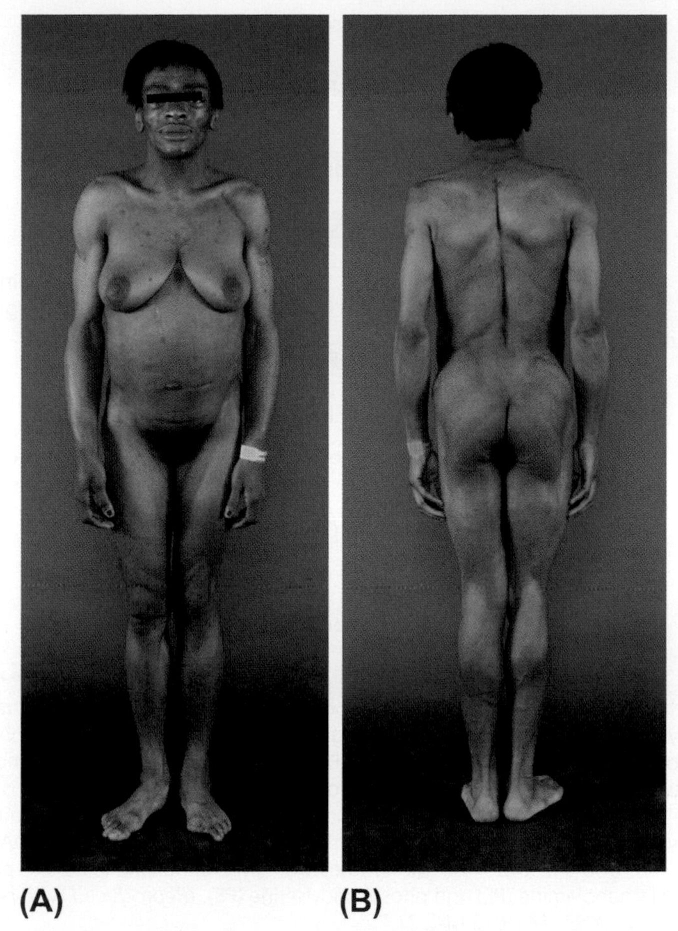

FIGURE 90-1 Congenital generalized lipodystrophy. Panels A and B show the anterior and posterior views of a 37-year-old female of African-American origin with congenital generalized lipodystrophy type 1, showing generalized lack of fat, extreme muscularity and acromegaloid features. She developed diabetes mellitus at the age of 17 years. Acanthosis nigricans was present in the neck, axillae and groin. She had a homozygous mutation (IVS4-2A>G resulting in prematurely truncated protein Gln196fsX228) in the *AGPAT2* gene.

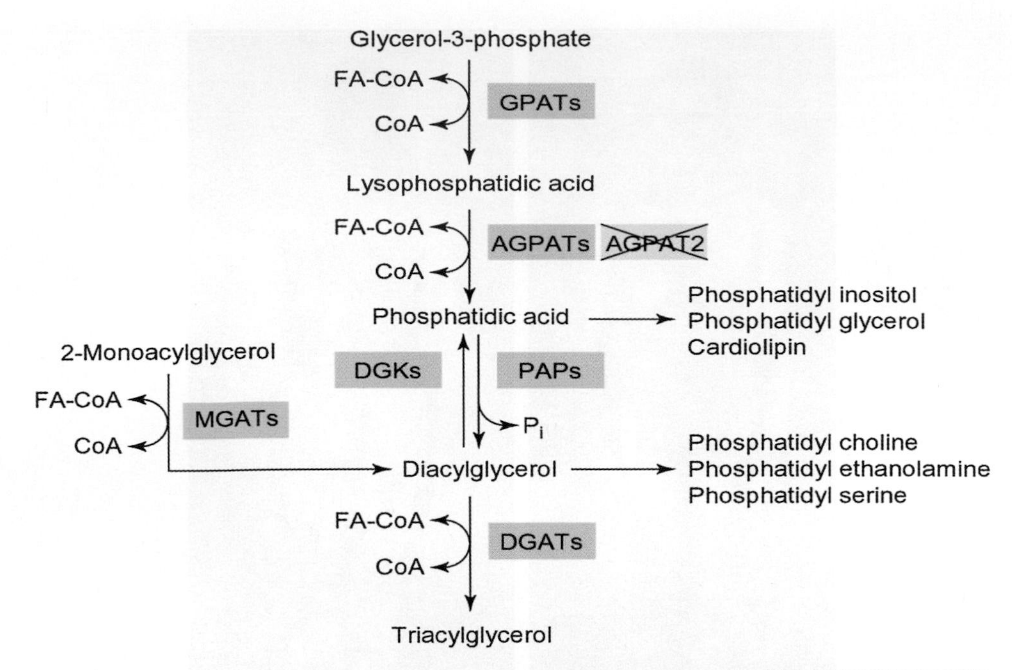

FIGURE 90-2 Pathways for biosynthesis of triacylglycerol: Glycerol-3-phosphate (G3P) is the initial substrate for acylation at sn-1 position by the enzyme glycerol-3-phosphate acyltransferase (GPAT), to form 1-acylglycerol-3-phosphate or lysophosphatidic acid (LPA). LPA is further acylated at sn-2 position by 1-acylglycerol-3-phosphate acyltransferase (AGPAT, aka LPAAT) to form phosphatidic acid (PA). In the next step, phosphate group is removed by phosphatidate phosphohydrolase (PAP) to produce diacylglycerol (DAG). DAG is further acylated at sn-3 position by diacylglycerol acyltransferase (DGAT) to produce triacylglycerol (TG). In addition, TG can be synthesized via the acylation of 2-monoacylglycerol by the enzyme monoacylglycerol-acyltransferase (MGAT), which is highly expressed in small intestine. DAG kinase (DGK) can phosphorylate DAG to synthesize PA. PA and DAG are also substrates for the synthesis of glycerophospholipids like phosphatidylinositol (PI), cardiolipin, phosphatidylcholine (PC), phosphatidylethanolamine (PE) and phosphatidylserine (PS). *(Reproduced with permission from Elsevier Agarwal, A. K. and Garg, A.* Trends Endocrinol. Metab. *2003, 14 (5), 214–221.)*

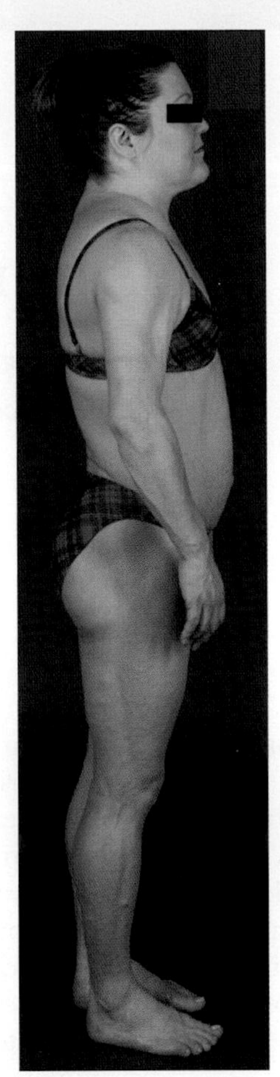

FIGURE 90-3 Familial partial lipodystrophy. A 38-year-old white woman with familial partial lipodystrophy, Dunnigan variety due to heterozygous missense mutation in the *LMNA* gene. She had loss of fat from the extremities and trunk beginning at puberty and had excess fat accumulation in the face and neck region. She has had surgical removal of fat from the chin, neck, axillae and mons pubis previously. She had acanthosis nigricans in the axillae and groins. *(Reproduced with permission from Elsevier Inc. Garg, A.; Misra, A. Lipodystrophies: Rare Disorders Causing Metabolic Syndrome.* Endocrinol. Metab. Clin. N. Am. *2004, 33, 305–331.)*

CHAPTER

91

Amino Acid Metabolism

Raymond Y Wang, William R Wilcox, and Stephen D Cederbaum

Division of Metabolic Disorders, CHOC Children's, Orange, CA, USA

Most disorders of amino acid metabolism are conceptually identical to organic acidemias, as both are caused by inherited deficiencies of processing enzymes that lead to accumulation of toxic substrates. In general, amino acid processing disorders are more proximal in the catabolic pathways and therefore result in accumulation of the amino acid itself. In contrast, organic acidemias tend to be more distal, accumulate the organic acid intermediates of amino acid degradation, and rarely have elevated levels of the parent amino acid.

General categories of amino acid processing defects considered in this chapter include disorders of aromatic amino acid metabolism, glycine metabolism, metabolic defects of sulfur-containing amino acids, disorders of branched-chain amino acid metabolism, disorders of proline processing, urea cycle disorders and related conditions, and finally serine biosynthetic defects. For each disorder, the natural history, typical biochemical findings, and information about mutations in the causative gene are discussed. When available, parameters for treatment and long-term outcomes are also detailed.

A deficiency in phenylalanine hydroxylase (PAH), the first enzyme of phenylalanine processing, leads to accumulation of the amino acid in body fluids. Depending on the severity of the PAH enzymatic defect, phenylalanine levels may exceed a critical threshold that leads to permanent cognitive defects. Consequently, patients whose pretreatment levels exceed that threshold require life-long adherence to a low-phenylalanine diet; such treatment preserves normal cognition. Tyrosine metabolic defects can manifest as an oculocutaneous form characterized by corneal and palmar/plantar tyrosine crystallization, or a hepatorenal form with early-onset liver synthetic failure and kidney dysfunction. A low phenylalanine/tyrosine diet, together with nitisinone administration for the hepatorenal form to block synthesis of toxic intermediates, can mitigate the symptoms. Finally, defects in generation of tetrahydrobiopterin, a necessary cofactor in the generation of neurotransmitters derived from aromatic amino acids, leads to developmental delay, seizures, and movement disorders. Supplementation with synthetic tetrahydrobiopterin and neurotransmitter precursors can help some of the symptoms, but these patients often continue to have psychomotor delay and neurologic complications.

Genetic deficiencies in the enzymes of the glycine cleavage system result in a similar constellation of symptoms, known collectively as glycine encephalopathy. Clinical presentation is fairly uniform, with highly elevated levels of glycine in plasma and cerebrospinal fluid (CSF); the ratio of CSF to plasma glycine is also abnormally elevated. These disorders manifest primarily with neonatal seizures, severe hypotonia, apnea, and often to neonatal death. Overwhelmingly, those who survive infancy have profound mental retardation and frequently suffer from refractory seizures. Efforts to reduce plasma glycine levels and antagonize its neurotransmitter effects in the central nervous system rarely result in reduction of seizure severity and frequency; "treated" children still remain profoundly retarded.

Disorders affecting the metabolism of sulfur-containing amino acids can be grouped into deficiencies of the catabolic enzymes themselves, and deficiencies in the processing of cobalamin, a cofactor required for the conversion of homocysteine to methionine. Most of these disorders share a common biochemical abnormality, the buildup of reactive homocysteine within the body. In addition to increased risk of thromboembolic events, many of these disorders can be associated with psychomotor retardation, seizure disorders, and megaloblastic anemia. Treatment aims to reduce thromboembolic risk, homocysteine levels, and normalize methionine levels. Supplemental hydroxocobalamin is used as well for the cobalamin processing disorders; supplemental pyridoxine is effective to reduce homocysteine levels in certain patients with cystathionine-β-synthase deficiency. For patients with certain types of molybdenum cofactor deficiency, which results in severe neonatal seizures, progressive cystic brain lesions, and early death, a promising new treatment with a cofactor precursor called cPMP is currently under clinical investigation.

Maple syrup urine disease (MSUD) is the primary aminoacidopathy affecting branched-chain amino acid metabolism. There are four catalytic subunits that comprise the branched-chain amino acid dehydrogenase enzyme complex, which decarboxylates the transaminated 2-oxoacid derivatives of leucine, isoleucine, and valine into branched-chain organic acids. A deficiency of the $E_1\alpha$, $E_1\beta$, or E_2 subunits of the dehydrogenase complex results in accumulation of the three branched-chain amino acids and their 2-oxoacids. E_3 subunit deficiency has similar biochemical findings with additional lactic acidosis and α-ketoglutaric aciduria because the subunit is also a component of the pyruvate dehydrogenase and α-ketoglutaryl CoA dehydrogenase complexes. MSUD patients can present in infancy with ketoacidosis, hyperammonemia, altered level of consciousness, and basal ganglia injury. Alternatively, "intermittent" forms exist where patients with residual enzymatic activity are biochemically normal until an intercurrent illness results in catabolism and accumulation of toxic intermediates. Treatment of MSUD with a low-branched chain amino acid diet and prompt attention during illnesses to prevent ketoacidosis can preserve cognition, but cognitive deficits and movement disorders are permanent in those who have had episodes of severe metabolic decompensation.

The imino acid proline is catabolized via a dehydrogenation reaction, followed by a reversible ring-opening reaction. Proline can then be converted to ornithine, serving an anaplerotic role for the urea cycle, or to glutamic acid. The pathogenicity of disorders in this pathway was uncertain, but evidence indicates that proline dehydrogenase deficiency may be a risk factor in the development of schizophrenia and Δ^1-pyrroline-5-carboxylic reductase deficiency may predispose to seizures. Prolidase deficiency is a disorder of proline dipeptide hydrolysis, leading to mental retardation, ulcerative skin lesions, and urinary excretion of iminodipeptides. Δ^1-pyrroline-5-carboxylate synthetase deficiency is a very rare condition of proline synthesis causing joint laxity, microcephaly, failure to thrive, and extrapyramidal symptoms due to systemic proline deficiency.

Urea cycle disorders are aminoacidopathies that prevent the conversion of toxic ammonia to urea. Five catalytic enzymes (carbamyl phosphate synthetase I, ornithine transcarbamylase, argininosuccinate synthase, argininosuccinate lyase, and arginase) and one enzyme that synthesizes an allosteric activator (*N*-acetylglutamate synthase) are involved in this pathway. Deficiencies in these enzymes result in hyperammonemia and subsequent alteration in level of consciousness, vomiting, lethargy, and eventually death. Treatment for urea cycle disorders requires restriction of dietary protein, conjugation of ammonia (in the form of the amino acids glycine or glutamine) to excretable forms, and prevention of catabolism. Neurocognitive complications can still occur even in well-controlled patients. This chapter also discusses deficiencies of two transporters (the mitochondrial ornithine/citrulline antiporter and the mitochondrial aspartate/glutamate antiporter) that result in the hyperornithinemia, hyperammonemia, homocitrullinuria syndrome and citrin deficiency, respectively.

While most of the other disorders discussed in this chapter result in impaired degradation of amino acids, or in synthesis of cofactors that assist in amino acid catabolism, defects of serine metabolism are a novel group of inborn errors of metabolism that lead to impaired production of serine. As serine is converted into phospholipids, one-carbon unit for methylation reactions, and into neurotransmitters, its deficiency usually manifests as early-onset seizures, microcephaly, and developmental delay. Prenatal and postnatal supplementation with serine has been attempted, with positive results upon cognition and prevention of seizures when used presymptomatically.

FURTHER READING

1. The Online Metabolic and Molecular Bases of Inherited Disease. www.ommbid.com.
2. Scriver, C. R.; Beaudet, A. L.; Sly, W. S.; Valle, D., et al. *Metabolic and Molecular Bases of Inherited Disease*; 8th ed..; McGraw-Hill: New York, 2004www.ommbid.com2004.
3. Online Mendelian Inheritance in Man database OMIM. http://www.ncbi.nlm.nih.gov/omim.
4. GeneTests Database. http://www.ncbi.nlm.nih.gov/sites/GeneTests.
5. Applegarth, D. A.; Toone, J. R. Glycine Encephalopathy (Nonketotic Hyperglycinaemia): Review and Update. *J. Inherit. Metab. Dis.* **2004**, *27*, 417–422.
6. Knerr, I.; Weinhold, N.; Vockley, J.; Gibson, K. M. Advances and Challenges in the Treatment of Branched-Chain Amino/Keto Acid Metabolic Defects. *J. Inherit. Metab. Dis.* **2011**, E-pub: Feb 3.
7. Mitsubuchi, H.; Nakamura, K.; Matsumoto, S.; Endo, F. Inborn Errors of Proline Metabolism. *J. Nutr.* **2008**, *138*, 2016S–2020S.
8. Mudd, S. H. Hypermethioninemias of Genetic and Non-Genetic Origin: A Review. *Am. J. Med. Genet. C Semin. Med. Genet.* **2011**, *157*, 3–32.
9. Watkins, D.; Rosenblatt, D. S. Inborn Errors of Cobalamin Absorption and Metabolism. *Am. J. Med. Genet. C Semin. Med. Genet.* **2011**, *157*, 33–44.
10. Scott, C. R. The Genetic Tyrosinemias. *Am. J. Med. Genet. C Semin. Med. Genet.* **2006**, *15*, 121–126.
11. Tabatabaie, L.; Klomp, L. W.; Berger, R.; de Koning, T. J. L-Serine Synthesis in the Central Nervous System: A Review on Serine Deficiency Disorders. *Mol. Genet. Metab.* **2010**, *99*, 256–262.

CHAPTER

92

Disorders of Carbohydrate Metabolism

Priya S Kishnani and Yuan-Tsong Chen

Division of Medical Genetics, Duke University Medical Center, Durham, NC, USA

Inborn errors of carbohydrate metabolism covered in this chapter include disaccharidase deficiencies, disorders of monosaccharide metabolism, glycogen storage diseases (GSDs), and gluconeogenic disorders. A number of strides have been made in understanding the clinical course, variability, molecular dissection, and treatment interventions for these disorders. This chapter focuses mainly on clinical aspects, genetics and current treatments pertaining to inborn errors of carbohydrate metabolism.

Disaccharidase deficiencies are characterized by the defective digestion of dietary sugars, starch, lactose and sucrose. Congenital lactase deficiency is a rare autosomal recessive disorder where infants, as a result of being unable to break down lactose in milk or other foods due to absence of the enzyme lactase, present with severe gastrointestinal symptoms. Late-onset or adult-type lactase deficiency (hypolactasia) is quite common in comparison to lactase persistence. Sucrase-isomaltase deficiency usually causes severe symptoms in younger children who ingest sucrose-containing foods. Dietary modification results in successful treatment outcome.

Several disorders arise due to transport defects, including Fanconi–Bickel syndrome and arterial tortuosity syndrome. Glucose-galactose malabsorption marks a defect in the *SGLT 1* gene, which codes for concentrative glucose transporter proteins. Defects in *GLUT* genes, however, mark abnormalities in facilitative glucose transporters. *GLUT1*, *GLUT2* (Fanconi-Bickel syndrome), and *GLUT10* (arterial tortuosity syndrome) are clinically significant.

Disorders of galactose metabolism stem from defects in the three main galactose enzymes involved in the Leloir pathway. Galactokinase deficiency usually results in cataracts and pseudotumor cerebri. Galactose-1-phosphate uridyltransferase deficiency (classic galactosemia) is the most severe disorder. If not diagnosed early, galactosemia can result in hypotonia, hepatomegaly, jaundice, cataracts, reduced immune function, hypergonadotropic hypogonadism in women, physical retardation and mental delay. Elimination of dietary galactose with calcium supplementation can significantly alter outcome. Finally, uridine diphosphate galactose-4-epimerase deficiency (epimerase deficiency) presents in both a benign and a rare but severe form that elicit symptoms similar to those in classic galactosemia.

Fructose metabolism disorders include essential fructosuria, a generally benign condition, and hereditary fructose intolerance, a deficiency of liver fructose-1-phosphate aldolase. Hereditary fructose intolerance results in generalized aminoaciduria, hypoglycemia, and a range of clinical manifestations dependent upon age and amount of fructose or sucrose ingested.

Dehydrogenases fructose-6-phosphate and glucose-6-phosphate are products of pentose metabolism, a process in the minor hexose monophosphate pathway of glucose metabolism. Glucose-6-phosphate dehydrogenase deficiency, the most prominent genetic disorder in humans, is a well-known cause of hemolytic anemia (*1*). All 10 individuals from six families identified as having transaldolase deficiency (TALDO) presented with liver disease (*2*). TALDO has broad phenotypic heterogeneity, ranging from fetal hydrops to slow-progressing liver cirrhosis. Only one case of ribose-5-phosphate isomerase deficiency has been identified, possibly because of its complex molecular etiology (*3*). Essential pentosuria is a benign disorder not requiring treatment.

To date, there are over 12 glycogenoses, or glycogen metabolism disorders, that have been cataloged. GSDs, a major category of glycogenoses, are categorized by type of tissue involved: primarily liver, muscle, and/or cardiac. In some, there is accumulation of normally structured glycogen while in others (GSD III, IV), the structure of glycogen is abnormal. Hepatic GSDs include types I, III, IV, VI, IX, IX, and XI. Muscle GSDs include types II, III, V, VII and IX. GSD type Ia (Von Gierke disease) results from deficiency of a catalytic subunit of the glucose-6-phosphatase enzyme and causes severe, often life-threatening hypoglycemia when not managed. GSD type Ib, marked by neutropenia and neutrophil dysfunction, is due to the absence of glucose-6-phosphate translocase, T1. GSD III (amylo-1,6-glucosidase (debrancher) deficiency) has two forms: GSD IIIa involving both liver and muscle, and GSD IIIb involving only liver. The defective debranching enzyme, which functions to break down glycogen, causes glycogen buildup in both types and muscle weakness in type IIIa. GSD type VI (hepatic phosphorylase deficiency, Hers disease) results from mutations in PYGL, or the liver isoform of phosphorylase. Type IV GSD (Andersen's disease, brancher deficiency) has several presentations. The hepatic form is accompanied by hypotonia and fatal liver dysfunction appearing between ages 2 and 4 years. The neuromuscular form of GSD-IV has four forms: the first two causing neonatal death, the second causing late childhood symptoms of myopathy or cardiomyopathy, and the third an adult form characterized by central and peripheral nervous system dysfunction and polyglucosan body disease. GSD XI (Fanconi–Bickel syndrome) involves a mutation in the facilitative glucose transporter GLUT2.

GSD II, or Pompe disease, derives from acid alfa-glucosidase deficiency (GAA). The only GSD where glycogen accumulates in the lysosomes, Pompe disease is characterized by a range of phenotypes each including myopathy but differing in age at onset, organ involvement, and clinical severity. The advent of enzyme replacement therapy with alglucosidase alfa has significantly advanced GSD II: infantile patients experience respiratory and cardiac improvement and adult-onset patients experience stabilization of skeletal muscle function and pulmonary disease (4,5). GSD V (McArdle disease) appears in late adolescence or in the second decade of life with pain and muscle stiffness due to muscle phosphorylase deficiency. GSD type VII (Tarui disease) has several features in common with McArdle disease but includes an infantile form with neurologic involvement. Muscle phosphorylase kinase deficiency, characterized by muscle weakness and atrophy, is due to mutations in the *PHKA1* gene. In addition, two GSDs mimic hypertrophic cardiomyopathy: Glycogen can amalgamate in heart and skeletal muscle due to malfunctioning lysosomal associated membrane 2 proteins (LAMP2, classified as Danon's disease) and AMP-activated kinase gamma 2 proteins (PRKAG2).

Two rare disorders—hepatic and muscle glycogen synthase deficiencies—are caused by a lack of glycogen synthase. Although both appear in the glycogen storage disorders section, neither involves excessive glycogen storage but rather the lack of due to synthase deficiency. Hepatic glycogen synthase deficiency is also called GSD type 0. Muscle glycogen synthase deficiency (GYS1) has only been reported in four children. If diagnosed early on, it can be managed.

The final section covers gluconeogenic disorders associated with lactic acidosis, for, in fasting conditions, blood glucose is derived mainly from glycogen breakdown (glycogenolysis) and from the conversion of lactic acid and certain amino acids to glucose (gluconeogenesis). These gluconeogenic disorders include fructose-1,6-diphosphatase deficiency, pyruvate carboxylase deficiency, phosphoenolpyruvate carboxykinase deficiency, and pyruvate dehydrogenase complex deficiencies. Fructose-1,6-diphosphatase deficiency and pyruvate dehydrogenase complex deficiencies often cause cognitive difficulties. Hypoglycemia can present in all four gluconeogenic disorders, whereas hypotonia is seen primarily in pyruvate carboxylase deficiency, phosphoenolpyruvate carboxykinase deficiency, and pyruvate dehydrogenase complex deficiencies.

REFERENCES

1. Perl, A.; Hanczko, R.; Telarico, T.; Oaks, Z.; Landas, S. Oxidative Stress, Inflammation and Carcinogenesis Are Controlled Through the Pentose Phosphate Pathway by Transaldolase. *Trends Mol. Med.* **2011**, Epub ahead of print.
2. Balasubramaniam, S.; Wamelink, M.; Ngu, L.; Talib, A.; Salomons, G., et al. Novel Heterozygous Mutations in *TALDO1* Gene Causing Transaldolase Deficiency and Early Infantile Liver Failure. *J. Pediatr. Gastroenterol. Nutr.* **2011**, *52*, 113–116.
3. Wamelink, M.; Grüning, N.-M.; Jansen, E.; Bluemlein, K.; Lehrach, H., et al. The Difference Between Rare and Exceptionally Rare: Molecular Characterization of Ribose 5-Phosphate Isomerase Deficiency. *J. Mol. Med.* **2010**, *88*, 931–939.
4. Kishnani, P.; Corzo, D.; Leslie, N. D.; Gruskin, D.; Van der Ploeg, A.; Clancy, J. P.; Parini, R.; Morin, G.; Beck, M.; Bauer, M. S., et al. Early Treatment with Alglucosidase Alpha Prolongs Long-Term Survival of Infants with Pompe Disease. *Pediatr. Res.* **2009**, *66*, 329–335.
5. Hobson-Webb, L. D.; DeArmey, S.; Kishnani, P. S. The Clinical and Electrodiagnostic Characteristics of Pompe Disease with Post-Enzyme Replacement Therapy Findings. *Clin. Neurophysiol.* **2011**, Epub ahead of print.

CHAPTER

93

Congenital Disorders of Protein Glycosylation

Jaak Jaeken

Center for Metabolic Diseases, University Hospital Gasthuisberg, Leuven, Belgium

Congenital disorders of protein glycosylation (CDG) are genetic diseases caused by defective glycosylation of glycoproteins. Recently, a novel CDG nomenclature was introduced. It consists of the official gene symbol (unitalicized) followed by "-CDG." Forty seven disorders of glycoprotein glycosylation are actually known. This CDG group comprises disorders of *N*-glycosylation, disorders of *O*-glycosylation, and disorders with a combined *N*- and *O*-glycosylation defect. The *N*-glycosylation pathway encompasses three cellular compartments: the cytosol, the endoplasmic reticulum (ER) and the Golgi. It starts in the cytosol with the formation of the mannose donor GDP-mannose from fructose 6-phosphate, an intermediate of the glycolytic pathway. In the ER, the dolichol-linked oligosaccharide $GlcNAc_2Man_9Glc_3$ is assembled and subsequently transferred from dolichol to selected asparagines of nascent proteins. Still in the ER this glycan starts to be processed by trimming off the three glucoses and one of the mannoses. This processing is continued in the Golgi by trimming off five mannoses and replacing them by two residues each of *N*-acetylglucosamine, galactose and finally sialic acid. *O*-glycosylation is usually limited to the Golgi and has no processing pathway. The *O*-glycans are linked to the hydroxyl group of selected threonines and serines.

The first report on CDG appeared in 1980 and was on patients who were subsequently shown to have an *N*-glycosylation disorder, phosphomannomutase 2 deficiency (PMM2-CDG). It is a cytosolic assembly defect and turned out to be, by far, the most frequent *N*-glycosylation disorder with around 600 patients known worldwide. They show mild to severe neurological disease and variable involvement of many other organs. Dysmorphy ranges from mild and aspecific to a characteristic abnormal subcutaneous adipose tissue distribution with fat pads and nipple retraction. Mortality is about 20% in the first years of life. The second most frequent *N*-glycosylation disorder is an ER assembly defect, glucosyltransferase 1 deficiency (ALG6-CDG). It is mainly a neurological disorder but milder than that of PMM2-CDG. The third most reported *N*-glycosylation disorder is again a cytosolic assembly defect, phosphomannose isomerase deficiency. It is a very remarkable disorder not only because of its pure hepatic-intestinal presentation but particularly because it is still the only efficiently treatable CDG (with oral mannose). The 17 remaining *N*-glycosylation disorders have been reported in less than 10 patients each and show variable combinations and degrees of neurological disease, other organ involvement and dysmorphy: mannosyltransferase 6 deficiency (ALG3-CDG), mannosyltransferase 8 deficiency (ALG12-CDG), glucosyltransferase 2 deficiency (ALG8-CDG), mannosyltransferase 2 deficiency (ALG2-CDG), UDP-GlcNAc:Dol-P-GlcNAc-P transferase deficiency (DPAGT1-CDG), mannosyltransferase 1 deficiency (ALG1-CDG), mannosyltransferase 7–9 deficiency (ALG9-CDG), mannosyltransferase 4–5 deficiency (ALG11-CDG), flippase of Man5GlcNAc2-PP-Dol deficiency (RFT1-CDG), *N*-acetylglucosaminyltransferase 2 deficiency (MGAT2-CDG), glucosidase 1 deficiency (GCS1-CDG), oligosaccharyltransferase subunit TUSC3 deficiency (TUSC3-CDG), and magnesium transporter 1 deficiency (MAGT1-CDG).

The known protein *O*-glycosylation disorders can be divided on the basis of the monosaccharide which links the glycan to the protein. Five groups have been described:

1. Defects in *O*-xylosylglycan synthesis (EXT1/EXT2-CDG, CHSY1-CDG, B4GALT7-CDG)
2. Defects in *O*-*N*-acetylgalactosaminylglycan synthesis (GALNT3-CDG)
3. Defects in *O*-xylosyl-/*N*-acetylgalactosaminylglycan synthesis (SLC35D1-CDG)
4. Defects in *O*-mannosylglycan synthesis (POMT1/POMT2-CDG, POMGNT1-CDG)
5. Defects in O-fucosylglycan synthesis (LFNG-CDG, B3GALTL-CDG)

Nearly all these diseases have also a descriptive name, based on their typical clinical presentation: multiple cartilaginous exostoses (EXT1/EXT2-CDG), familial tumoral calcinosis (GALNT3-CDG), Schneckenbecken dysplasia

(SLC35D1-CDG), muscle–eye–brain disease (POMGNT1-CDG), and spondylocostal dysostosis type 3 (LFNG-CDG). Waardenburg syndrome (POMT1/POMT2-CDG) is also a muscle–eye–brain disease but with a more severe expression than POMGNT1-CDG while Peters plus syndrome (B3GALTL-CDG) shows peculiar anterior eye chamber abnormalities besides many other symptoms.

Eighteen disorders have a combined *N*- and *O*-glycosylation defect. These comprise defects in dolichol phosphomannose synthesis (DPM1-CDG, DPM3-CDG), dolichol phosphomannose utilization (MPDU1-CDG), galactosylation (B4GALT1-CDG), sialylation (SLC35A1-CDG), fucosylation (SLC35C1-CDG), dolicholphosphate synthesis (DK1-CDG, SRD5A3-CDG), ER–Golgi intermediate compartment proteins (SEC23B-CDG), and in other Golgi-associated proteins (COG7-CDG, COG1-CDG, COG8-CDG, COG4-CDG, COG5-CDG, COG6-CDG, GMAP210-CDG, and ATP6V0A2-CDG). The latter two groups comprise defects in proteins that are not only involved in glycosylation but also in other functions (such as endosome pH regulation). The symptomatology of patients with a combined protein glycosylation defect is very heterogeneous: neurological, muscular, ophthalmological, cardiological, hepatic, skeletal, hematological, dermatological, etc.

All CDG are inherited in an autosomal recessive way except MAGT1-CDG (X-linked), and EXT1/EXT2-CDG (autosomal dominant).

Essentially two techniques are available for CDG screening: serum transferrin isofocusing (IEF) detects protein *N*-glycosylation disorders associated with sialic acid deficiency, and serum apolipoprotein C-III (IEF) detects core 1 mucin type *O*-glycosylation defects. Serum transferrin IEF can be replaced by other techniques such as capillary zone electrophoresis. In case of an abnormal transferrin IEF profile, first an artifact, a protein polymorphism and a secondary CDG (fructosemia, galactosemia, alcohol abuse, bacterial sialidase, etc.) should be excluded. Two types of abnormal transferrin IEF profiles can be distinguished: a type 1 pattern, characterized by an increase of di- and/or asialoprotein (CDG-I), and a type 2 pattern, characterized by an increase of tri-, di-, mono- and/or asialoprotein (CDG-II). A type 1 pattern points to an assembly or transfer defect of the dolichol-linked glycan (in the cytosol or the ER). Measurement of the phosphomannomutase activity in fibroblasts or leukocytes is the next step because PMM2-CDG is by far the most frequent *N*-glycosylation defect. In case of a purely hepatointestinal presentation, the activity should be measured of phosphomannose isomerase, deficient in MPI-CDG, a treatable disease. Finding a normal activity of these enzymes necessitates analysis of the lipid-linked oligosaccharides (LLO) in fibroblasts. A type 2 pattern indicates a glycan processing defect (in the ER or the Golgi). Processing defects that are not associated with sialic acid deficiency (GCS1-CDG and SLC35C1-CDG) show, of course, a normal transferrin IEF profile. CDG-II patients are further investigated by means of mass spectrometry of isolated serum *N*-glycans. In a small minority of patients, this will lead to the identification of specific defects such as MGAT2-CDG and B4GALT1-CDG, but in the majority of patients a specific glycan profile will be found. In the latter situation, the possibility of an associated mucin type *O*-glycosylation defect should be investigated by IEF of serum apolipoprotein C-III. In patients with an abnormal IEF of transferrin as well as of apolipoprotein C-III, it is recommended to look for a defect in one of the COG (conserved oligomeric Golgi complex) subunits by mutation analysis of the eight COG subunit genes. However, if the patient presents a cutis laxa syndrome, mutation analysis of the *ATP6V0A2* gene is indicated.

TABLE 93-1 Genetic Protein *N*-Glycosylation Disorders

Disorder	Affected Protein	Defective Gene	OMIM Number
PMM2-CDG	Phosphomannomutase 2	*PMM*	212065
MPI-CDG	Phosphomannose isomerase	*MPI*	602579
ALG6-CDG	Glucosyltransferase 1	*ALG6*	603147
ALG3-CDG	Mannosyltransferase 6	*ALG3*	601110
ALG12-CDG	Mannosyltransferase 8	*ALG12*	607143
ALG8-CDG	Glucosyltransferase 2	*ALG8*	608104
ALG2-CDG	Mannosyltransferase 2	*ALG2*	607906
DPAGT1-CDG	UDP-GlcNAc:Dol-P-GlcNAc-P-transferase	*DPAGT1*	608093
ALG1-CDG	Mannosyltransferase 1	*ALG1*	608540
ALG9-CDG	Mannosyltransferase 7-9	*ALG9*	608776
ALG11-CDG	Mannosyltransferase 4-5	*ALG11*	613611
RFT1-CDG	Flippase of $Man_5GlcNAc_2$.PP-Dol	*RFT1*	612015
TUSC3-CDG	Oligosaccharyl transferase tusc 3	*TUSC3*	611093
MAGT1-CDG	Magnesium transporter 1	*MAGT1*	300716, 300853
GCS1-CDG	Glucosidase 1	*GCS1*	606056
MAN1B1-CDG	Mannosidase 1B1	*MAN1B1*	614202
MGAT2-CDG	N-acetylglucosaminyl-transferase 2	*MGAT2*	212066
ST3GAL3-CDG	Beta-galactoside-alpha-2,3-sialyltransferase 3	*ST3GAL3*	611090

TABLE 93-2 Genetic Protein O-Glycosylation Disorders

Disease	Affected Protein	Defective Gene	OMIM Number
Defect in *O-N*-Acetylgalactosaminylglycan Synthesis			
GALNT3-CDG	N-acetylgalactosaminyl-transferase 3	*GALNT3*	211900
Defects in *O*-Xylosylglycan Synthesis			
B4GALT7-CDG	Beta-1,4-galactosyltransferase 7	*B4GALT7*	130070
EXT1-CDG	Exostosin 1	*EXT1*	130700
EXT2-CDG	Exostosin 2	*EXT2*	130701
CHSY1-CDG	Chondroitin sulfate synthase	*CHSY1*	605282
Defect in *O-N*-Acetylgalactosaminylglycan and *O*-Xylosylglycan Synthesis			
SLC35D1-CDG	UDP-glucuronic acid/UDP-N-acetylgalactosamine dual transporter	*SLC35D1*	269250
Defects in *O*-Mannosylglycan Synthesis			
POMT1-CDG	O-mannosyltransferase 1	*POMT1*	236670, 613155, 609308
POMT2-CDG	O-mannosyltransferase 2	*POMT2*	613150, 613156, 613158
POMGNT1-CDG	O-mannose beta-1,2-*N*-acetyl-glucosaminyltransferase	*POMGNT1*	253280, 613151, 613157
Defects in *O*-Fucosylglycan Synthesis			
LFNG-CDG	O-fucose-specific beta-1,3-*N*-acetylglucosaminyltransferase	*LFNG*	609813
B3GALTL-CDG	O-fucose-specific beta-1,3-*N*-glucosyltransferase	*B3GALTL*	261540

TABLE 93-3 Genetic Protein Combined *N*- and *O*-Glycosylation Disorders

Disease	Affected Protein	Defective Gene	OMIM Number
DPM1-CDG	GDP-Man:Dol-P-mannosyl-transferase 1	*DPM1*	608799
DPM3-CDG	GDP-Man:Dol-P-mannosyl-transferase 3	*DPM3*	612937
MPDU1-CDG	Dol-P-Man utilization 1	*MPDU1*	609180
B4GALT1-CDG	Beta-1,4-galactosyltransferase 1	*B4GALT1*	607091
GNE-CDG	UDP-GlcNAc epimerase/kinase	*GNE*	600737, 605820
SLC35A1-CDG	CMP-sialic acid transporter	*SLC35A1*	603585
SLC35C1-CDG	GDP-fucose transporter	*SLC35C1*	266265
DK1-CDG	Dolichol kinase	*DK1*	610768
SRD5A3-CDG	Steroid 5-alpha-reductase	*SRD5A3*	612379
DHDDS-CDG	Dehydrodolichyl-PP synthase	*DHDDS*	613861
COG7-CDG	COG complex 7	*COG7*	608779
COG1-CDG	COG complex 1	*COG1*	611209
COG8-CDG	COG complex 8	*COG8*	611182
COG4-CDG	COG complex 4	*COG4*	613489
COG5-CDG	COG complex 5	*COG5*	613612
COG6-CDG	COG complex 6	*COG6*	606977
ATP6V0A2-CDG	Vesicular H^+-ATPase a2	*ATP6V0A2*	219200, 278250
SEC23B-CDG	COPII component SEC23B	*SEC23B*	224100

CHAPTER

94

Purine and Pyrimidine Metabolism

Naoyuki Kamatani

Center for Genomic Medicine, RIKEN Yokohama Institute, Kanagawa, Japan

H A Jinnah

Departments of Neurology, Human Genetics and Pediatrics, Emory University, Atlanta, GA, USA

Raoul C M Hennekam

Department of Pediatrics and Translational Genetics, Academic Medical Center, University of Amsterdam, Amsterdam, The Netherlands

André B P van Kuilenburg

Laboratory of Genetic Metabolic Disease, Academic Medical Center, University of Amsterdam, Amsterdam, The Netherlands

Purine and pyrimidine nucleotides are essential for a vast number of biological processes, such as RNA and DNA synthesis, signal transduction, phospholipids, glycogen, sialylation and glycosylation of proteins, cofactor synthesis and the supply of high-energy phosphate esters (e.g. ATP) in phosphate transfer reactions. In addition, purines and pyrimidines play an important role in the regulation of the central nervous system.

Before markers for the human genome were obtained, responsible genes for genetic disorders in purine and pyrimidine metabolism were identified by the candidate gene approach, that is, on the basis of functional, metabolic, or molecular abnormalities. After such markers became available, genome-wide approaches were used to clarify the associations between genomic mutations or variations and diseases. First, linkage analysis was used to identify the uromodulin gene responsible for familial juvenile hyperuricemic nephropathy and the thymidine phosphorylase gene responsible for mitochondrial neurogastrointestinal encephalomyopathy. Second, genome-wide association studies (GWASs) have successfully identified genes associated with serum urate concentrations many of which code for renal transporters and related proteins. Lastly, dihydroorotate dehydrogenase mutations were found to cause Miller syndrome by whole exome sequencing.

Two types of diseases are associated with purine and pyrimidine metabolism. One type is the polygenic or multifactorial disease gout and the other type is a group of monogenic (Mendelian) diseases. Gout is a very common disease that is characterized by the elevation of serum urate levels and the formation of urate crystals, and the main symptoms include acute arthritis, tophi and renal stones. Genetic factors influencing serum urate levels include transporters in the renal tubules and related proteins. Most of the monogenic diseases in purine and pyrimidine metabolism are genetic enzyme abnormalities in either the purine or pyrimidine pathway. Although most of the enzyme abnormalities are deficiencies, some of them exhibit superactivities. The latter enzyme abnormalities include the superactivity of phosphoribosyl pyrophosphate (PRPP) synthetase, adenosine deaminase (ADA) and cytosolic 5′-nucleotidase.

Since purine and pyrimidine metabolism is essential in all types of cells with a variety of functions, the symptoms caused by the enzyme abnormalities are diverse. At least 30 monogenic abnormalities in purine and pyrimidine metabolism have been reported. Among the enzyme abnormalities in purine pathways, hypoxanthine guanine phosphoribosyltransferase (HGPRT) deficiency and PRPP synthetase superactivity cause hyperuricemia with or without psychological, motor and neurological abnormalities. Adenine phosphoribosyltransferase deficiency and type I and type II xanthine oxidase deficiencies cause urolithiasis, while ADA deficiency and purine nucleoside phosphorylase deficiency cause immunodeficiency. In adenylosuccinate lyase deficiency, ATIC deficiency and molybdenum cofactor deficiency, psychological, motor, and neurological symptoms are the main disorders, while deoxyguanosine kinase deficiency causes progressive liver damage. Thiopurine methyltransferase deficiency is of particular interest because it is associated with drug-related adverse events caused by 5-mercaptopurine, thioguanine and azathioprine. Finally, myoadenylate deaminase deficiency may cause muscle abnormalities.

Among the monogenic abnormalities in the pyrimidine pathways, UMP synthase deficiency causes developmental delay and anemia, while β-ureidopropionase deficiency and cytosolic 5′-nucleotidase superactivity cause psychological, motor and neurological symptoms. Pyrimidine 5′-nucleotidase deficiency causes anemia, while thymidine phosphorylase deficiency causes peripheral neuropathy and gastrointestinal dysmotility. Dihydroorotate dehydrogenase deficiency causes Miller syndrome, which is characterized by postaxial acrofacial dysostosis. Thymidine kinase 2 deficiency causes neonatal hypotonia, lactic acidosis, subsequent encephalopathy, liver failure, and renal tubulopathy. Dihydropyrimidine dehydrogenase deficiency and dihydropyrimidinase deficiency are associated with neurological dysfunction and can cause drug-induced adverse events to 5-fluorouracil and related compounds, while cytidine deaminase deficiency causes toxicity to ara-C and gemcitabine.

Diagnosis of monogenic disorders in purine and pyrimidine metabolism is performed either by the measurement of metabolites, enzyme assays, protein detection, cell-based methods, or by DNA analysis. The identification of the carrier state and prenatal diagnoses are possible in some diseases. The biological mechanisms of hyperuricemia or hypouricemia, anemia and the formation of urinary stones in purine and pyrimidine metabolic abnormalities are not difficult to explain, while the mechanisms of other symptoms remain to be clarified (Figure 94-2). Abnormalities in immune function in ADA deficiency and purine nucleoside phosphorylase deficiency may be explained by the accumulation of deoxynucleotides in lymphocytes, resulting in either the dysfunction or destruction of these cells, although other mechanisms have been proposed. The mechanisms underlying the psychological, neurological, and motor symptoms in patients with abnormalities in the purine and pyrimidine pathways are still unclear, but they may be explained by abnormalities in neurotransmitters or developmental abnormalities in the central nervous system. Studies using knockout mice deficient in HGPRT have suggested that a decrease in dopamine and an increase in serotonin in the basal ganglia may be responsible for the symptoms. Treatments for the acute arthritis of gout include colchicine, corticosteroid, and nonsteroidal anti-inflammatory drugs. Treatments for hyperuricemia include xanthine oxidase inhibitors, uricosuric drugs, and uricases, in addition to the improvement of habits. Recurrent urolithiasis and renal insufficiency in adenine phosphoribosyltransferase deficiency patients can be prevented by the administration of allopurinol, a xanthine oxidase inhibitor. Big progress has been made for the treatment of ADA deficiency. When HLA-matched donors are available, bone marrow transplantation is the recommended treatment; otherwise, enzyme replacement therapy and somatic gene therapy targeted at hematopoietic stem cells have drastically changed the prognosis of ADA deficiency. Although various therapeutic attempts have been made to address the psychological, neurological and motor disorders, no evident beneficial effects have been reported.

TABLE 94-1 PRPP Synthetase Disorders

	PRS1 Superactivity	Arts Syndrome	CMTX5	DFN2
Hearing loss	Variable	Yes	Yes	Yes
Neuropathy	Variable	Yes	Yes	No
Optic atrophy	No	Yes	Yes	No
Psychomotor retardation	Variable	Yes	No	No
Recurrent infections	No	Yes	No	No
Uric acid overproduction	Yes	No	No	No
PRS1 enzyme activity	High	Absent	Low	Low
Reported mutations	D52H, N114S, L129I, D182H, A189V, H192L, H192Q	Q133P, L152P	E43D, M115T	D65N, A87T, I290T, G306R

The reported mutations follow standard nomenclature with one-letter abbreviations to designate the amino acid change.

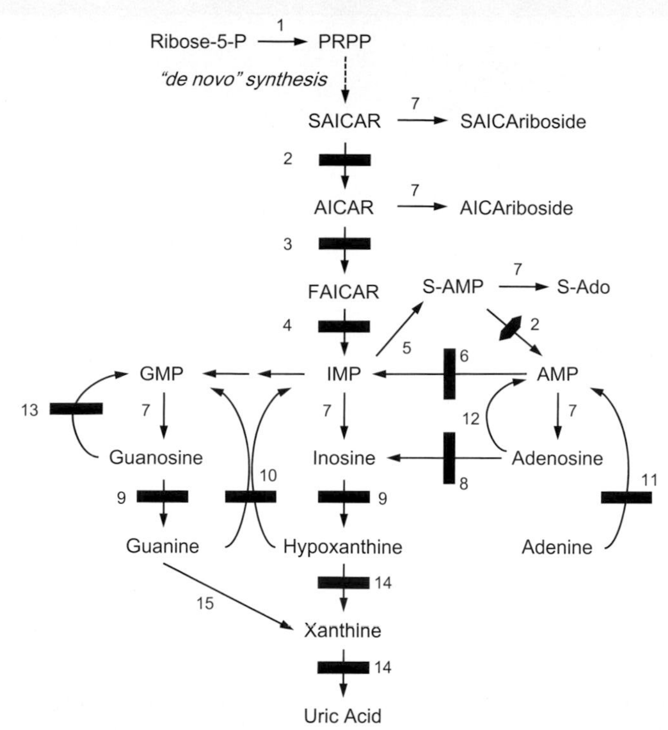

FIGURE 94-1 Pathways of purine metabolism. AICAR, aminoimidazole carboxamide ribotide; AMP, adenosine monophosphate; FAICAR, formylaminoimidazole carboxamide ribotide; GMP, guanosine monophosphate; IMP, inosine monophosphate; P, phosphate; PRPP, phosphoribosyl pyrophosphate, S-Ado, succinyladenosine; SAICAR, succinyl aminoimidazole carboxamide ribotide; S-AMP, adenylosuccinate, XMP, xanthosine monophosphate. 1, PRPP synthetase; 2, adenylosuccinate lyase (adenylosuccinase); 3, AICAR transformylase; 4, IMP cyclohydrolase (3 and 4 form ATIC); 5, adenylosuccinate synthetase; 6, AMP deaminase; 7, 5′-nucleotidase(s), 8, adenosine deaminase; 9, purine nucleoside phosphorylase; 10, hypoxanthine-guanine phosphoribosyltransferase; 11, adenine phosphoribosyltransferase; 12, adenosine kinase; 13, guanosine kinase; 14, xanthine oxidase (dehydrogenase); 15, guanine deaminase. Enzyme defects are indicated by solid bars.

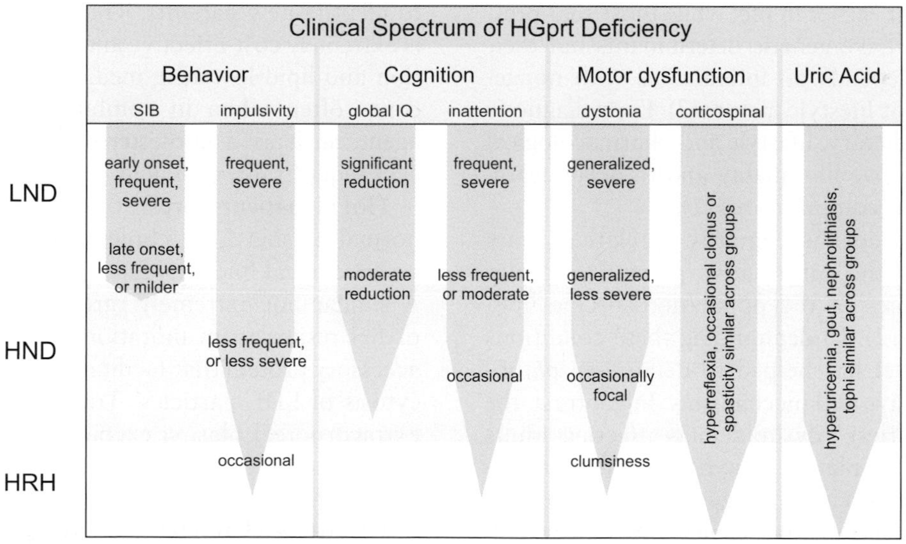

Clinical Spectrum of HGprt Deficiency

	Behavior		Cognition		Motor dysfunction		Uric Acid
	SIB	impulsivity	global IQ	inattention	dystonia	corticospinal	
LND	early onset, frequent, severe late onset, less frequent, or milder	frequent, severe	significant reduction moderate reduction	frequent, severe less frequent, or moderate	generalized, severe generalized, less severe	hyperreflexia, occasional clonus or spasticity similar across groups	hyperuricemia, gout, nephrolithiasis, tophi similar across groups
HND		less frequent, or less severe		occasional	occasionally focal		
HRH		occasional			clumsiness		

FIGURE 94-2 Schematic representation of the continuous spectrum of clinical features across LND and its variants. The main clinical problems are listed across the top. Patients are subdivided into three typical groups along the left side. The most severely affected group is LND. The least severely affected group is designated HPRT-related hyperuricemia (HRH). An intermediate phenotype is designated HPRT-related neurological dysfunction (HND). The frequency and severity of each of the clinical problems is depicted by the thickness and taper of the gray bar.

CHAPTER

95

Lipoprotein and Lipid Metabolism

Robert A Hegele

Robarts Research Institute and Schulich School of Medicine and Dentistry, University of Western Ontario, London, ON, Canada

Lipoproteins transport hydrophobic lipids and fat-soluble vitamins through plasma from their site of origin (intestine or liver) to their site of uptake and disposition. Abnormal levels of certain plasma lipids and lipoproteins increase the risk of cardiovascular disease (CVD) end points, such as myocardial infarction and stroke (*1*). A wide range of genetic and environmental factors contribute to interindividual variation in plasma concentrations of lipids and lipoproteins (*2*). Dyslipidemias that have a monogenic basis typically present earlier in life, while those that present later in life also have genetic determinants but their expression further depends on interactions with nongenetic environmental or lifestyle factors (*2*). Early diagnosis is central to specific dietary, lifestyle and pharmacological interventions to improve life quality and to delay death, disability and medical complications (*1*).

Monogenic dyslipidemias comprise a relatively rare but very important and informative patient subgroup found at the extremes of the population-specific lipoprotein distribution (*2*). Understanding these conditions at the molecular level has helped to define key pathophysiological pathways and mechanisms. In contrast, the common "garden variety" dyslipidemias affecting adults have a much more complex genetic basis with contributions from both common and rare genetic variants, compounded by a large overlay of nongenetic factors. Several causative genes with rare mutations have resurfaced as loci with common small-effect genetic variants that underlie lipoprotein variation in population studies (*3*).

Disorders characterized primarily by elevated LDL cholesterol include heterozygous and homozygous familial hypercholesterolemia (HeFH and HoFH, respectively) (*4*). In HeFH, patients have twice-normal plasma LDL cholesterol levels and present with clinical stigmata such as corneal arcus, xanthelasmas, extensor tendon xanthomas and early vascular disease. Classical HeFH is caused by >1000 documented mutations in the *LDLR* gene encoding the LDL receptor (LDLR). Two additional very rare forms of autosomal dominant hypercholesterolemia are clinically similar to HeFH but result from either heterozygous mutations in *APOB* affecting the LDLR-binding domain of apo B-100 and from gain-of-function mutations in *PCSK9* encoding a protease that normally short circuits LDLR recycling. Screening for HeFH and related conditions in families is a cost-effective approach to identify new patients. Treatment for these conditions is extremely cost-effective and involves lifestyle modification and lipid-lowering medications, in particular statin drugs often taken in combination with a second-line agent, such as a cholesterol absorption inhibitor, bile acid sequestrant, or niacin.

HoFH patients present with more than four times normal plasma LDL cholesterol levels and much more severe clinical findings, especially cardiovascular disease. A similar but extremely rare autosomal recessive form of FH results from mutations in *LDLRAP1*, an LDLR accessory protein that facilitates receptor-mediated endocytosis of LDL particles. Treatment for HoFH involves extracorporeal plasma exchange or selective removal of LDL, since these patients are minimally responsive to currently available medications (*4*).

Disorders characterized primarily by low-to-absent plasma levels of LDL include abetalipoproteinemia (ABL) and homozygous familial hypobetalipoproteinemia (FHBL), which result from mutations in the *MTP* and *APOB* genes, respectively (*5*). These conditions are characterized clinically by the consequences of fat-soluble vitamin deficiencies including retinopathy, bone abnormalities, neuromyopathy and coagulopathy, in tandem with such specific manifestations as acanthocytosis and hepatosteatosis. Treatment includes supplementation with very high oral doses of fat-soluble vitamins.

Obligate heterozygotes for *MTP* mutations have no biochemical or clinical phenotype while obligate heterozygotes for *APOB* mutations in FHBL have half-normal plasma LDL cholesterol levels and may be protected from cardiovascular disease. Loss-of-function mutations in *PCSK9* (*6*) are also associated with low LDL cholesterol.

Elevated plasma concentrations of HDL cholesterol result from rare mutations in *CETP* and *SRB1* genes encoding cholesteryl ester transfer protein and scavenger receptor type B1, respectively (*7*). These disturbances are not typically associated with any symptoms and treatment is generally not required. In contrast, several genetic disorders lead to reduced plasma HDL cholesterol, including Tangier disease due to homozygous mutations in *ABCA1* encoding ATP binding cassette protein type A1; LCAT deficiency, due to homozygous mutations in *LCAT* encoding lecithin:cholesterol ester and apo A-I deficiency due to mutations in the *APOA1* gene (*8*). These low-to-absent HDL states are associated with various clinical manifestations and association with increased risk of CVD is inconsistent. Currently available medications, such as niacin or fibrates, are minimally effective at raising plasma HDL cholesterol. Treatment is thus directed toward lowering LDL cholesterol to reduce CVD risk.

Severe hypertriglyceridemia (HTG) due to hyperchylomicronemia is most often due to one of two monogenic disorders affecting the peripheral metabolism of TG-rich, intestinally derived chylomicron particles, namely lipoprotein lipase (LPL) deficiency due to homozygous mutations in the *LPL* gene and apo C-II deficiency due to mutations in the *APOC2* gene encoding a key cofactor for LPL (*9*). In addition to cutaneous xanthomas, lipemia retinalis, hepatosplenomegaly and a range of other clinical manifestations, the main risk to health is recurrent pancreatitis. Mutations in three other genes that are involved in the lipolytic pathway, including *APOA5* encoding apo A-V, which is thought to be essential for lipolysis, *LMF1*, which encodes lipase maturation factor type 1 (*10*), and *GPIHBP1* which encodes a chaperone that delivers LPL to the capillary lumen (*11*), are also associated with severe HTG. Restriction of dietary fat is the cornerstone of treatment, while currently available TG-lowering medications, such as fibrates, niacin and fish oil, are variably effective in correcting the biochemical disturbance.

Other rare recessive dyslipidemias are associated with multiple lipoprotein disturbances, and include dysbetalipoproteinemia due to both common and rare variations in the *APOE* gene, hepatic lipase deficiency, due to mutations in the *LIPC* gene, and familial combined hypolipidemia, due to mutations in the *ANGPTL3* gene. Sitosterolemia is a recessive trait that results from mutations in the *ABCG5* and *ABCG8* genes encoding intestinal half-transporters for sterols (*12*). Lipoprotein (a) is a polymorphic macromolecular complex whose plasma levels are determined by size variation at the *LPA* locus. These disorders are associated with increased risk of atherosclerosis end points and treatments selected from among currently available medications are directed at risk reduction.

Much of the dyslipidemia that is encountered in adult lipid clinics is polygenic and often secondary to a wide range of medical conditions and medications (*2*). It is felt that the susceptibility toward developing secondary dyslipidemia is genetically determined, since not all individuals develop dyslipidemia when exposed to a particular stressor. There are well-established evidence-based guidelines for the treatment of dyslipidemia, beginning with lifestyle modification, maintenance of a healthy weight, appropriate physical activity, and then progressing to medications. A range of emerging treatments for dyslipidemia are under evaluation and one or more of these may soon become available to provide additional treatment options for monogenic and polygenic dyslipidemia (*8*).

REFERENCES

1. Baigent, C.; Blackwell, L.; Emberson, J.; Holland, L. E.; Reith, C.; Bhala, N.; Peto, R.; Barnes, E. H.; Keech, A.; Simes, J., et al. Efficacy and Safety of More Intensive Lowering of LDL Cholesterol: A Meta-Analysis of Data from 170,000 Participants in 26 Randomised Trials. *Lancet* **2010**, *376* (9753), 1670–1681.
2. Hegele, R. A. Plasma Lipoproteins: Genetic Influences and Clinical Implications. *Nat. Rev. Genet.* **2009**, *10* (2), 109–121.
3. Teslovich, T. M.; Musunuru, K.; Smith, A. V.; Edmondson, A. C.; Stylianou, I. M.; Koseki, M.; Pirruccello, J. P.; Ripatti, S.; Chasman, D. I.; Willer, C. J., et al. Biological, Clinical and Population Relevance of 95 Loci for Blood Lipids. *Nature* **2010**, *466* (7307), 707–713.
4. Yuan, G.; Wang, J.; Hegele, R. A. Heterozygous Familial Hypercholesterolemia: An Underrecognized Cause of Early Cardiovascular Disease. *CMAJ* **2006**, *174* (8), 1124–1129.
5. Burnett, J. R.; Hooper, A. J. Common and Rare Gene Variants Affecting Plasma LDL Cholesterol. *Clin. Biochem. Rev.* **2008**, *29* (1), 11–26.
6. Cohen, J.; Pertsemlidis, A.; Kotowski, I. K.; Graham, R.; Garcia, C. K.; Hobbs, H. H. Low LDL Cholesterol in Individuals of African Descent Resulting from Frequent Nonsense Mutations in PCSK9. *Nat. Genet.* **2005**, *37* (2), 161–165.
7. Vergeer, M.; Korporaal, S. J.; Franssen, R.; Meurs, I.; Out, R.; Hovingh, G. K.; Hoekstra, M.; Sierts, J. A.; Dallinga-Thie, G. M.; Motazacker, M. M., et al. Genetic Variant of the Scavenger Receptor BI in Humans. *N. Engl. J. Med.* **2011**, *364* (2), 136–145.
8. Rader, D. J.; Daugherty, A. Translating Molecular Discoveries into New Therapies for Atherosclerosis. *Nature* **2008**, *451* (7181), 904–913.
9. Johansen, C. T.; Kathiresan, S.; Hegele, R. A. Genetic Determinants of Plasma Triglycerides. *J. Lipid Res.* **2011**, *52* (2), 189–206.
10. Peterfy, M.; Ben-Zeev, O.; Mao, H. Z.; Weissglas-Volkov, D.; Aouizerat, B. E.; Pullinger, C. R.; Frost, P. H.; Kane, J. P.; Malloy, M. J.; Reue, K., et al. Mutations in LMF1 Cause Combined Lipase Deficiency and Severe Hypertriglyceridemia. *Nat. Genet.* **2007**, *39* (12), 1483–1487.
11. Beigneux, A. P.; Weinstein, M. M.; Davies, B. S.; Gin, P.; Bensadoun, A.; Fong, L. G.; Young, S. G. GPIHBP1 and Lipolysis: An Update. *Curr. Opin. Lipidol.* **2009**, *20* (3), 211–216.
12. Lee, M. H.; Lu, K.; Hazard, S.; Yu, H.; Shulenin, S.; Hidaka, H.; Kojima, H.; Allikmets, R.; Sakuma, N.; Pegoraro, R., et al. Identification of a Gene, ABCG5, Important in the Regulation of Dietary Cholesterol Absorption. *Nat. Genet.* **2001**, *27* (1), 79–83.

CHAPTER

96

Organic Acidemias and Disorders of Fatty Acid Oxidation

Jerry Vockley

University of Pittsburgh, Children's Hospital of Pittsburgh of UPMC, Pittsburgh, PA, USA

Most organic acidemias are disorders of amino acid metabolism and can present in many ways. They include (i) features of ketotic hyperglycinemia; (ii) an unusual body odor; (iii) acute disease in infancy, especially when associated with metabolic acidosis, hypoglycemia, or hyperammonemia; (iv) chronic or recurrent metabolic acidosis, with or without an anion gap; (v) static or progressive extrapyramidal movement disorder in childhood; (vi) Recurrent Reye syndrome–like episodes; (vii) the combination of ataxia, alopecia, and rash; and (viii) seizure disorder, movement disorder, and/or developmental delay of unknown cause. Because most organic acids are effectively cleared from the bloodstream by the kidney, urine is the preferred fluid for analysis.

The largest group of organic acidemias derives from defects in the mitochondrial-based degradation of the branched chain amino acids leucine, isoleucine and valine (*1–3*). The branched chain amino acids enter mitochondria via a combined, reversible deamination and transport step is performed by the vitamin B6-dependent branched chain amino acid aminotransferases. Subsequently, the branched chain keto acids undergo oxidative decarboxylation by a single-branched chain keto acid dehydrogenase active with all three substrates, another reversible reaction. Deficiency of this dehydrogenase leads to maple syrup urine disease. The remaining steps of the pathway are unique to the metabolites of each of the amino acids. The convergence of many metabolic pathways especially onto the distal end of branched chain amino acid catabolism has led to much confusion in diagnosis over the years. Diseases named for metabolites ultimately proved to be imprecise because of the accumulation of the same metabolites in more than one condition. Thus, it is best to discuss this group of disorders based on the enzymatic and/or molecular defect rather than metabolite accumulation. Of this group of disorders, isovaleric acidemia (isovaleryl-CoA dehydrogenase deficiency), propionic acidemia (propionyl-CoA carboxylase deficiency), and methylmalonic acidemia (methylmalonyl-CoA mutase deficiency or defects in cobalamin metabolism) are particularly common (though the latter defects also encompass methionine, threonine, and cholesterol metabolism) (*4*). Multiple carboxylase deficiency due to defects of biotin is also relatively frequently identified. 3-Methylglutaconic acidemia has at least five different etiologies and diagnostic care in dealing with this metabolite finding is necessary (*5*). Several defects in these pathways appear to lead to a biochemical phenotype with minimal or no clinical symptoms including 3-methylglutaconyl-CoA carboxylase, isobutyl-CoA dehydrogenase, and short/branched chain acyl-CoA dehydrogenase deficiencies (*6*).

Of the other organic acidemias, glutaric acidemia type 1 (glutaryl-CoA dehydrogenase deficiency) and ethylmalonic aciduria (a biochemical finding of mixed etiology) can be particularly challenging to identify or treat. Glutaric acidemia type 1 requires vigilance especially in the first year of life to prevent episodes of metabolic acidosis and its subsequent basal ganglia damage. Ethylmalonic aciduria is problematic as patients can have a clinical picture ranging from asymptomatic (due to short chain acyl-CoA dehydrogenase) to the devastating ethylmalonic encephalopathy caused by a deficiency of a mitochondrial sulfur dioxygenase (*7*). Treatment for the latter condition has only recently emerged. Succinic semialdehyde dehydrogenase deficiency leading to 4-hydroxybutyric aciduria is a defect in the metabolism of the neurotransmitter GABA and presents primarily with neurologic manifestations including a movement disorder. Abnormal movements are also seen in L- and D-2-hydroxybutyrate dehydrogenase deficiencies.

There are three goals for therapy of almost any organic acidemia (*1,4*). The first is prevention of metabolic decompensation by careful clinical observation of the patient. During times of metabolic stress (including illness and fasting) endogenous substrate (leucine in IVA) from protein catabolism adds significantly to the production of metabolic intermediates. Achieving or maintaining anabolism is the main therapeutic approach to counter this problem. Reducing, but not eliminating, natural protein in the diet for 12–24 h may help in this regard, but only if additional other calories to promote anabolism can be given. The second goal is long-term reduction of the production of toxic metabolites from general catabolism through dietary manipulation. Total protein and caloric intake must be adequate to support normal growth in children and maintain an anabolic state, but this may require the use of an artificial protein source restricted in the appropriate amino acid for a portion of the protein requirement. The third goal of therapy is to prevent the accumulation of toxic metabolites by enhancing alternative metabolic pathways that produce alternative nontoxic compounds that are readily excreted. Carnitine is commonly used for this purpose in many of the organic acidemias. Cofactor supplementation may be curative (pyridoxine responsive seizures and biotin recycling defects) or augment deficient function (cobalamin metabolism defects, FAD-dependent dehydrogenases).

Mitochondrial fatty acid oxidation is a complex process involving transport of activated acyl-CoA moieties into the mitochondria, and sequential removal of two carbon acetyl-CoA units. It is the main source of energy for many tissues including heart and skeletal muscle and is critically important during times of fasting or physiologic stress (*8*). While short- and medium-chain fatty acids (C_4–C_{12}) diffuse freely across plasma and mitochondrial membranes, the transport of longer chain species (C_{14}–C_{20}) depends at least in part on active transport, a high-affinity mechanism of major physiological importance in skeletal muscle, liver, and adipocytes. Each cycle of the pathway produces a molecule of acetyl-CoA and a fatty acid with two fewer carbons. Under physiological conditions, the latter reenters the cycle until it is completely consumed. In peripheral tissues, the acetyl-CoA is terminally oxidized in the Krebs cycle for ATP production. In the liver, the acetyl-CoA from fatty acid oxidation can instead be utilized for the synthesis of ketones, 3-hydroxybutyrate, and acetoacetate, which are then exported for final oxidation by brain and other tissues. At least 25 enzymes and specific transport proteins are responsible for carrying out the steps of mitochondrial fatty acid metabolism, some of which have only recently been recognized. Of these, defects in at least 22 have been shown to cause disease in humans. In general, the disorders of long chain fat metabolism, including the fatty acid transport defects, show a more severe phenotype that the medium and short chain defects. Multiple acyl-CoA dehydrogenase deficiency, caused by mutations in the genes for electron transfer flavoprotein and its dehydrogenase, represents a mixed disorder of fatty acid oxidation and branched chain amino acid metabolism.

Patients can present throughout life. In the first week of life, cardiac arrhythmias, hypoglycemia, sudden death, and occasionally with facial dysmorphism and malformations including renal cystic dysplasia are seen. Symptoms later in infancy and early childhood may relate to the liver or cardiac or skeletal muscle dysfunction, and include fasting or stress-related hypoketotic hypoglycemia or Reye-like syndrome, conduction abnormalities, arrhythmias or dilated or hypertrophic cardiomyopathy, and muscle weakness or fasting- and exercise-induced rhabdomyolysis. Prior to the incorporation of tandem mass spectroscopy into newborn screening, presentation with sudden unexpected death or life-threatening events was common. Adolescent or adult onset muscular symptoms in milder cases include rhabdomyolysis, muscle pain, myopathy, and cardiomyopathy. Medium chain acyl-CoA dehydrogenase deficiency is now recognized as one of the most common inborn errors of metabolism in Caucasians. Chronic muscle symptoms are rare and risk for hypoglycemia decreases (though never disappears) with increasing age. Short chain acyl-CoA dehydrogenase deficiency may be identified by a secondary accumulation of ethylmalonic acid in blood and urine, but is most often asymptomatic in patients identified through newborn screening. Thus, its clinical relevance remains in question. Fatty acid oxidation defects historically have been an important cause of sudden infant death, but identification of these disorders through newborn screening has dramatically reduced this clinical presentation.

Treatment of the acute encephalopathy of hypoketotic hypoglycemia is by intravenous glucose and L-carnitine (*9*). Long-term therapy involves replenishing carnitine stores with L-carnitine, and preventing hypoglycemia. In some cases, this can be done by providing a snack, glucose polymers, or uncooked cornstarch before bedtime, but in others requires continuous intragastric feeding. Supplementation with medium chain triglyceride oil provides a fat source that can be utilized by patients with long chain defects but only after ruling out a diagnosis of medium chain acyl-CoA dehydrogenase deficiency. Experimental therapies to induce increased expression of defective enzymes and replete intermediates of the tricarboxylic acid cycle have been proposed but are unproven.

Diagnosis of organic acidemias and fatty acid oxidation defects depends on the methods to separate and identify organic acids (*10*). One of the most widely used of these is gas chromatography-mass spectrometry, in which organic acid derivatives are separated by gas chromatography and then passed into a mass spectrometer for identification. Tandem mass spectrometry is a related technology in which compounds are separated by molecular weight by one mass spectrometer, fragmented as they exit, and identified on the basis of their fragments by a second mass spectrometer. This technique is used to measure the acylcarnitine esters that accumulate in various fatty acid oxidation and organic acid disorders and serves the basis for effective newborn screening for many of these disorders. Diagnosis can usually be established even when the patient is asymptomatic, though analysis of samples during acute illness can uncover some mild or intermittent cases.

Many disorders of organic acid and fatty acid oxidation metabolism are now identified through newborn screening by tandem mass spectrometry of carnitine esters in blood spots. Early identification and prospective screening are changing our perspective of this group of diseases. Most of them were originally identified in symptomatic patients and thus perceived as serious, often life-threatening conditions. While this undoubtedly remains true in some diseases (notably propionic and methylmalonic acidemias), others identified through newborn screening have shown a much broader spectrum of symptoms than in patients identified through symptoms (e.g. isovaleric acidemia). In the latter situation, asymptomatic mothers may actually be diagnosed for the first time due to an abnormal screen from their newborn baby (e.g. carnitine transporter and 3-methylcrotonyl-CoA carboxylase deficiencies). Finally, the clinical relevancy of some biochemical defects has even been called into question based on newborn screening data. For example, short/branched chain acyl-CoA dehydrogenase deficiency was initially described in a patient with a severe neonatal episode of acidosis, but since has been found in numerous babies through newborn screening (as well as their family members) who have remained asymptomatic. While the possibility of later onset symptoms cannot be dismissed, this condition clearly carries little or no risk in childhood and thus is more appropriately viewed as biochemical phenotype rather than a disease.

Most organic acidemias and fatty acid oxidation defects are inherited as autosomal recessive traits. Two are X-linked recessive. Prenatal diagnosis is typically accomplished through measurement of enzyme activity in cultured amniotic cells or a chorionic villus sample or through measurement of diagnostic metabolites excreted by the affected fetus into the amniotic fluid. Molecular diagnosis is also possible when a proband's genetic defect has been identified.

REFERENCES

1. Vockley, J.; Ensenauer, R. Isovaleric Acidemia: New Aspects of Genetic and Phenotypic Heterogeneity. *Am. J. Med. Genet. C Semin. Med. Genet.* **2006**, *142*, 95–103.
2. Korman, S. H. Inborn Errors of Isoleucine Degradation: A Review. *Mol. Genet. Metab.* **2006**, *89*, 289–299.
3. Alfardan, J.; Mohsen, A. W.; Copeland, S.; Ellison, J.; Keppen-Davis, L.; Rohrbach, M.; Powell, B. R.; Gillis, J.; Matern, D.; Kant, J.; Vockley, J. Characterization of New ACADSB Gene Sequence Mutations and Clinical Implications in Patients with 2-Methylbutyrylglycinuria Identified by Newborn Screening. *Mol. Genet. Metab.* **2010**.
4. Deodato, F.; Boenzi, S.; Santorelli, F. M.; Dionisi-Vici, C. Methylmalonic and Propionic Aciduria. *Am. J. Med. Genet. C Semin. Med. Genet.* **2006**, *142C*, 104–112.
5. Arn, P.; Funanage, V. L. 3-Methylglutaconic Aciduria Disorders: The Clinical Spectrum Increases. *J. Pediatr. Hematol. Oncol.* **2006**, *28*, 62–63.
6. Frazier, D. M.; Millington, D. S.; McCandless, S. E.; Koeberl, D. D.; Weavil, S. D.; Chaing, S. H.; Muenzer, J. The Tandem Mass Spectrometry Newborn Screening Experience in North Carolina: 1997–2005. *J. Inherit. Metab. Dis.* **2006**, *29*, 76–85.
7. Tiranti, V.; Viscomi, C.; Hildebrandt, T.; Di Meo, I.; Mineri, R.; Tiveron, C.; Levitt, M. D.; Prelle, A.; Fagiolari, G.; Rimoldi, M.; Zeviani, M. Loss of ETHE1, a Mitochondrial Dioxygenase, Causes Fatal Sulfide Toxicity in Ethylmalonic Encephalopathy. *Nat. Med.* **2009**, *15*, 200–205.
8. Rinaldo, P.; Matern, D.; Bennett, M. J. Fatty Acid Oxidation Disorders. *Annu. Rev. Physiol.* **2002**, *64*, 477–502.
9. Vockley, J.; Singh, R. H.; Whiteman, D. A. Diagnosis and Management of Defects of Mitochondrial Beta-Oxidation. *Curr. Opin. Clin. Nutr. Metab. Care* **2002**, *5*, 601–609.
10. Rinaldo, P.; Hahn, S.; Matern, D. Clinical Biochemical Genetics in the Twenty-First Century. *Acta Paediatr. Suppl.* **2004**, *93*, 22–26, discussion 7.

CHAPTER

97

Vitamin D Metabolism or Action

Elizabeth A Streeten

Associate Professor of Medicine, Division of Endocrinology, Diabetes, and Nutrition, University of Maryland School of Medicine, Baltimore, MD, USA

Michael A Levine

Medical Director, Center for Bone Health at the Children's Hospital of Philadelphia, PA, USA

97.1 HISTORY

Rickets and osteomalacia were widespread problems until the discovery of the calciferols by Mellanby in 1919, after which they were used for prevention and treatment. However, some cases did not respond to the usual doses of calciferols, and multiple genetic and other causes were subsequently recognized. In 1937, Albright reported a child with calciferol-resistant rickets and suggested hereditary calciferol resistance. In 1961, Prader and associates reported hereditary rickets associated with hypocalcemia, which they called pseudodeficiency rickets. In 1971, $1\alpha,25(OH)_2D_3$ was shown to be the active metabolite of vitamin D_3 that accumulated in the nuclei of target tissues (*1–3*). This discovery quickly led to the development of assays for serum $1\alpha,25(OH)_2D$, characterization of defects in $1\alpha,25(OH)_2D$ synthesis and action, and an understanding of the roles of 1α-hydroxylated and other analogs for therapy.

97.2 NORMAL PHYSIOLOGY OF CALCIFEROLS

Vitamin D_3 is a secosteroid produced via opening of the B-ring of 7-dehydrocholesterol precursors present in skin keratinocytes in response to UVB sunlight. To become active, both endogenous D_3 and exogenous D_3 (supplements) and D_2 (ergocalciferol, from some vegetables) need to be sequentially hydroxylated by two mitochondrial cytochrome P450 enzymes. The first, 25-hydroxylation (*CYP2R1*) occurs in the liver and the second, 1α-hydroxylation (*CYP27B1*) in the kidneys, forming active $1\alpha,25(OH)_2D$. The primary source of circulating α $1,25(OH)_2D$ is the kidney and its formation is largely controlled by parathyroid hormone (PTH). However, most nonrenal tissues also express *CYP27B1* and the locally produced $1\alpha,25(OH)_2D$, not controlled by PTH, serves as a differentiating factor for tissues. Both $1\alpha,25(OH)_2D$ and 25(OH)D are catabolized to water soluble metabolites for excretion by 24-hydroxylase, *CYP24A1*, one of the most highly inducible enzymes in the body. *CYP24A1*, present in all cells that express *CYP27B1*, can be induced 20,000-fold over basal levels (*4*), the likely reason that the therapeutic window for vitamin D is very broad.

97.3 TRANSCRIPTIONAL AND NONGENOMIC EFFECTS OF 1,25(OH)2D

Active $1\alpha,25(OH)_2D$ circulates in the blood bound to vitamin D binding protein (*DBP*) and its biological actions are mediated through two different mechanisms. The first is through binding to the high-affinity intranuclear vitamin D receptor (*VDR*); the second, "nongenomic effects," are through binding to putative plasma membrane VDR. After binding of $1\alpha,25(OH)_2D$ to the nuclear VDR, the $1\alpha,25(OH)_2D$-VDR complex dimerizes with the retinoid X receptor (RXR) and this $1\alpha,25(OH)_2D$-VDR-RXR heterodimer then binds to vitamin D response elements (VDRE) present in target genes, thereby regulating gene transcription. The "nongenomic" effects of $1\alpha,25(OH)_2D$ are very rapid and do not involve gene transcription. Activation of this signaling pathway leads to stimulation of MAP-kinase and involves crosstalk with the nuclear VDR, interaction with ion channels and other signal transduction pathways (e.g. protein kinase C) (*5*).

The most important physiological action of $1\alpha,25(OH)_2D$ is stimulation of active, saturable calcium transport across the duodenal lumen into the bloodstream, which requires the vitamin D–dependent protein calbindin. Passive, nonsaturable calcium absorption also occurs in the jejunum and ileum. 1α,25$(OH)_2$D also has direct effects on bone, skin, hair, parathyroid gland and other tissues. Although the most important effect of 1α,25$(OH)_2$D is to maintain extracellular fluid levels of calcium (Ca) and phosphate for bone mineralization, it also directly affects osteoblast differentiation and indirectly activates osteoclasts to resorb bone, a process that requires the presence of osteoblasts and the RANKL system. In parathyroid, 1α,25$(OH)_2$D inhibits PTH production and secretion and induces expression and transcription of the calcium sensing receptor (*CASR*) present on parathyroid cells, making the parathyroids more sensitive to Ca-mediated reduction in PTH secretion.

97.4 HEREDITARY VITAMIN D–DEPENDENT RICKETS TYPES 1 AND 2

Hereditary defects in calciferol metabolism or action lead to the same features as environmental deficiency (due to lack of sunlight and inadequate dietary intake) including bone pain (osteomalacia-unmineralized osteoid), fracture and muscle weakness in adults and rickets (bowing of long bones, splaying of epiphyses), delay in tooth eruption and respiratory failure if severe, in growing children. Hereditary vitamin D–dependent rickets type 1 (VDDR-1) is an autosomal recessive disorder due to mutations in *CYP27B1*, leading to reduced or absent conversion of 25(OH)D into active 1α,25$(OH)_2$D (*6*). VDDR-1 has been reported in approximately 100 individuals from a variety of ethnic groups but is unusually common in French-Canadian families in Quebec (*7*). Clinical features typically become apparent between ages 2 and 24 months, including seizures due to hypocalcemia and serum levels of 1α,25$(OH)_2$D are low. Treatment with 1α,25$(OH)_2$D (calcitriol) or its analogs corrects the hypocalcemia and rickets, leading to normal growth (*8*). Hereditary vitamin D–dependent rickets type 2 (VDDR-2) is a rare autosomal recessive disorder due to mutations in *VDR*, leading to true resistance to 1α,25$(OH)_2$D, reported in around 100 individuals (*9*). Mutations reported include absent and reduced binding and reduced affinity for 1α,25$(OH)_2$D. Most patients respond to extraordinarily high doses of calciferols but some, including those with alopecia, require high oral and intravenous calcium. Symptoms are similar to those in VDDR-1 but may also include alopecia.

97.5 CALCIFEROL EXCESS STATE

One calciferol excess state has been described, autosomal recessive infantile hypercalcemia due to inactivating mutations in *CYP24A1*, usually presenting as failure to thrive within the first year of life, nephrocalcinosis and normal to mildly elevated 25(OH)D and 1α,25$(OH)_2$D but can be asymptomatic (*10*). One reported patient with homozygous CYP24A1 mutation presented at age 19 with nephrolithiasis and frankly elevated 25(OH)D and 1α,25$(OH)_2$D (*11*).

TABLE 97-1 Vitamin D-Dependent Rickets

OMIM	25(OH)D	1,25$(OH)_2$D	i PTH	Inheritance	Gene Defect
VDDR-1	N/I	D	I	A.R	1-α-hydroxylase, CYP27B1 (264700)
VDDR-2	N/I	N/I	I	A.R	Vitamin D receptor, *VDR*

VDDR, Vitamin D-dependent rickets.

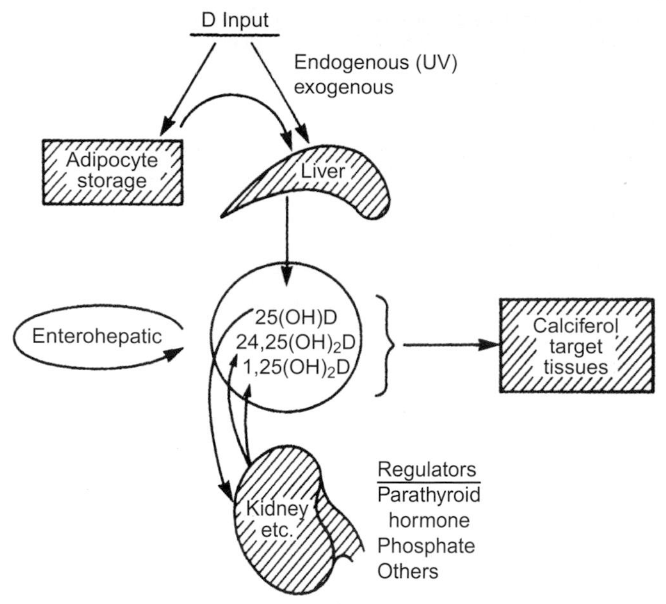

FIGURE 97-1 Metabolic pathways for activation of cholecalciferol or ergocalciferol. *(From Aurbach, G. D.; Marx, S. J; Spiegel, A. M. Parathyroid Hormone, Calcitonin, and the Calciferols. In Williams Textbook of Endocrinology; Wilson, J. D., Foster, D. W., Eds., 7th ed. WB Saunders, Philadelphia, 1985 pp 1137–1217.)*

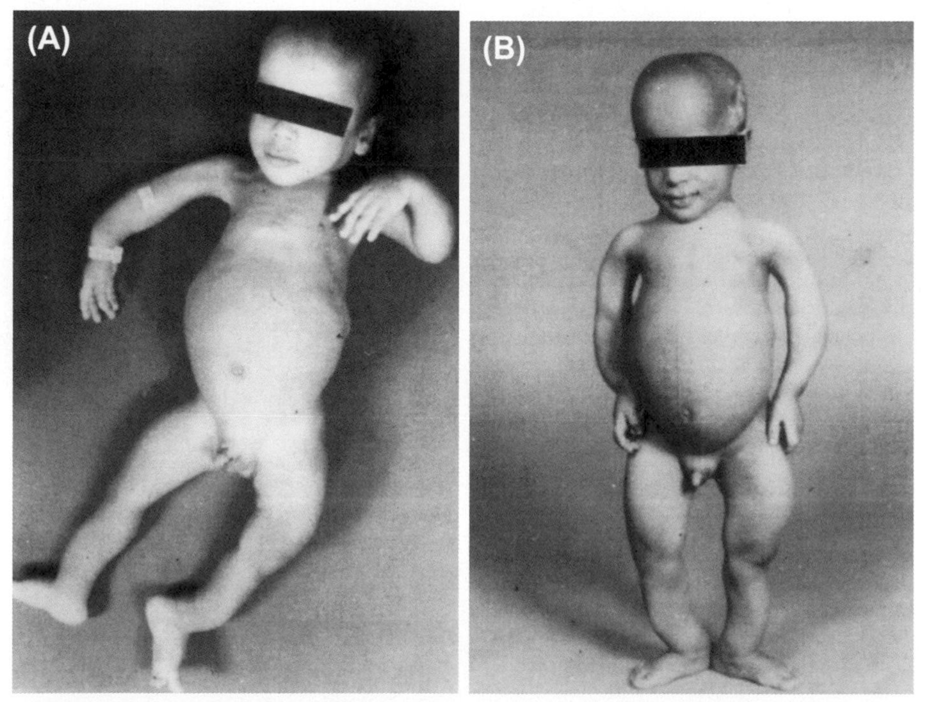

FIGURE 97-2 A 5-year-old boy with hereditary resistance to 1,25$(OH)_2$D. He was unable to sit without support at age 5. A. Despite being unresponsive to high doses of calciferols, he showed improvement 2 years later following treatment with high doses of calcium orally. B. Although the weakness and deformity improved, alopecia worsened during treatment. *(From Sakati, N.; Woodhouse, N.J.Y.; Niles, N., et al. Hereditary Resistance to 1,25-Dihydroxyvitamin D: Clinical and Radiological Improvement During High-Dose Oral Calcium Therapy. Horm. Res. 1986, 24, 280–287.)*

REFERENCES

1. Holick, M. F.; Schnoes, H. K.; DeLuca, H. F. Identification of 1,25-Dihydroxycholecalciferol, a Form of Vitamin D_3 Metabolically Active in the Intestine. *Proc. Natl. Acad. Sci. U.S.A.* **1971,** *68,* 803–804.
2. Lawson, D. E. M.; Fraser, D. R.; Kodicek, E., et al. Identification of 1,25-Dihydroxycholecalciferol, a New Kidney Hormone-Controlling Calcium Metabolism. *Nature* **1971,** *230,* 228–230.
3. Norman, A. W.; Midgett, R. J.; Myrtle, J. F., et al. 1, 25-Dihydrocholecalciferol: Identification of the Proposed Active Form of Vitamin D_3 in the Intestine. *Science* **1971,** *173,* 51.
4. Tashiro, K.; Abe, T.; Oue, N.; Yasui, W.; Ryoji, M. Characterization of Vitamin D-Mediated Induction of the CYP 24 Transcription. *Mol. Cell Endocrinol.* **2004,** *226,* 27–32.
5. Bikle, D. Nonclassic Actions of Vitamin D. *J. Clin. Endocrinol. Metab.* **2009,** *94* (1), 26–34.
6. Fraser, D.; Kooh, S. W.; Kind, H. P., et al. Pathogenesis of Hereditary Vitamin-D-Dependent Rickets: An Inborn Error of Vitamin D Metabolism Involving Defective Conversion of 25-Hydroxyvitamin D to 1-a, 25-Dihydroxyvitamin D. *N. Engl. J. Med.* **1973,** *289,* 817–822.
7. Arnaud, C.; Maijer, R.; Reade, T., et al. Vitamin D Dependency: An Inherited Postnatal Syndrome with Secondary Hyperparathyroidism. *Pediatrics* **1970,** *46,* 871.
8. Edouard, T.; Alos, N.; Chabot, G.; Roughley, P.; Glorieux, F. H.; Rauch, F. Short- and Long-Term Outcome of Patients with Pseudo-Vitamin D Deficiency Rickets Treated with Calcitriol. *J. Clin. Endocrinol. Metab.* **2011,** *96,* 82–89.
9. Brooks, M. H.; Bell, N. H.; Love, L., et al. Vitamin-D-Dependent Rickets Type II: Resistance of Target Organs to 1,25-Dihydroxyvitamin D. *N. Engl. J. Med.* **1978,** *298,* 996–999.
10. Schlingmann, K. P.; Kaufmann, M.; Weber, S.; Irwin, A.; Goos, C.; John, U.; Misselwitz, J.; Klaus, G.; Kuwertz-Broking, E.; Fehrenbach, H., et al. Mutations in CYP24A1 and Idiopathic Infantile Hypercalcemia. *N. Engl. J. Med.* **2011 Aug 4,** *365* (5), 410–421, Epub 2011 Jun 15.
11. Streeten, E. A.; Zarbalian, K.; Damcott, C. M. CYP24A1 Mutations in Idiopathic Infantile Hypercalcemia. *N. Engl. J. Med.* **2011 Nov 3,** *365* (18), 1741–1742, author reply 1742–1743.

CHAPTER

98

Inherited Porphyrias

R J Desnick and Manisha Balwani

Department of Genetics and Genomic Sciences, Mount Sinai School of Medicine of New York University, New York, NY, USA

Karl E Anderson

Departments of Preventive Medicine and Community Health (Division of Human Nutrition), Internal Medicine (Division of Gastroenterology) and Pharmacology and Toxicology, University of Texas Medical Branch/UTMB HealthGalveston, TX, USA

The inherited porphyrias are inborn errors of metabolism resulting from the deficient activity of specific enzymes in the heme biosynthetic pathway. These are inherited as autosomal dominant, autosomal recessive, or X-linked traits, with the exception of porphyria cutanea tarda (PCT) which is usually sporadic. The porphyrias are classified as either hepatic or erythropoetic depending on the primary site of overproduction and accumulation of their respective porphyrin precursors or porphyrins. They are also classified clinically as acute or cutaneous. The major manifestations of the acute porphyrias are neurologic, including neuropathic abdominal pain, peripheral motor neuropathy, and mental disturbances, with attacks often precipitated by steroid hormones, certain drugs and nutritional changes such as dieting. While hepatic porphyrias rarely present before puberty, rare homozygous variants of the autosomal dominant porphyrias may manifest in childhood. In contrast, the erythropoetic porphyrias, in which overproduction of heme pathway intermediates occur primarily in bone marrow erythroid cells, usually present with cutaneous photosensitivity at birth or in early childhood, or in the case of congenital erythropoetic porphyria even in utero as hydrops fetalis. Cutaneous sensitivity to sunlight results from excitation of excess porphyrins in the skin by long wave ultraviolet light, leading to cell damage, scarring, and disfigurement. Thus, the porphyrias are metabolic disorders in which genetic, physiologic, and environmental factors interact to cause disease.

Because many symptoms of the porphyrias are nonspecific, the diagnosis is often delayed. First line diagnostic testing involves the determination of the porphyrin precursors and/or porphyins in urine, plasma, or erythrocytes. A definitive diagnosis is based on the demonstration of the specific enzyme deficiency and/or gene mutation(s). The isolation and characterization of the genes encoding the heme biosynthetic enzymes have permitted identification of the mutations causing each porphyria. Molecular genetic analyses now make it possible to provide precise heterozygote or homozygote identification and prenatal diagnosis in families with known mutations.

Porphyrias can cause high morbidity and mortality in both children and adults and it is important to initiate treatment as soon as the diagnosis is established. The management of acute porphyria attacks includes narcotic analgesics for pain, phenothiazines for nausea and vomiting, and benzodiazapines for insomnia and restlessness. Carbohydrate loading usually with intravenous glucose may be effective for milder attacks. Hemin should be used initially for more severe attacks and for milder attacks which do not respond to glucose. Recovery depends on the degree of neuronal damage and is usually rapid if therapy is started early. All the acute porphyrias (acute intermittent porphyria, variegate porphyria and hereditary coproporphyria, ALA-dehydratase deficient porphyria) are managed similarly and it is not necessary to establish the type of acute porphyria before initiating treatment.

The mainstay of management of cutaneous porphyrias is avoiding sunlight exposure and use of protective clothing. In erythropoetic protoporphyria (EPP), oral beta-carotene may improve tolerance to sunlight in some patients. The late hepatic complications of EPP may be significant and may require liver transplant. PCT can be sporadic in a majority of cases. Management includes discontinuation of the precipitating factor if known. Phlebotomy to reduce hepatic iron is the standard treatment and low dose chloroquine can be used as an alternative. In congenital EPP, severe cases often require blood transfusions. Protection from sunlight is essential and minor skin trauma should be avoided. Early bone marrow and cord blood transplantation have proven curative used in severe transfusion-dependent cases.

In summary, the porphyrias are a diverse group of inherited disorders which require genetic as well as environmental factors for disease expression. The manifestations range from mild photosensitivity to severe neurologic symptoms which may be life threatening. Diagnosis is difficult due to the clinical heterogeneity and complexity in interpreting results. Early initiation of treatment is important to prevent morbidity and mortality.

FURTHER READING

COMPREHENSIVE REVIEWS

1. Anderson, K. E.; Sassa, S.; Bishop, D. F.; Desnick, R. J. Disorders of Heme Biosynthesis: X-Linked Sideroblastic Anemias and the Porphyrias. In *The Metabolic and Molecular Basis of Inherited Disease*, 8th ed.; Scriver, C. R.; Beaudet, A. L.; Sly, W. S.; Valle, D., Eds.; McGraw-Hill: New York, 2001; pp 2991–3062.
2. Kauppinen, R. Porphyrias. *Lancet* **2005**, *365*, 241–252.
3. Layer, G.; Reichelt, J.; Jahn, D.; Heinz, D. W. Structure and Function of Enzymes in Heme Biosynthesis. *Protein Sci.* **2010**, *19*, 1137–1161, Overview—Heme Biosynthetic Enzymes.
4. Puy, H.; Gouya, L.; Deybach, J. C. Poprhyrias. *Lancet* **2010**, *375*, 924–937, Overview.

ACUTE PORPHYRIAS

5. Anderson, K. E.; Bloomer, J. E.; Bonkovsky, H. L.; Kushner, J.; Pierach, C.; Pimstone, N.; Desnick, R. J. Recommendations for the Diagnosis and Treatment of the Acute Porphyrias. *Ann. Intern. Med.* **2005**, *142*, 439–450, Consensus Recommendations.
6. Solis, C.; Martinez-Bermejo, A.; Naidich, T. P.; Kaufmann, W. E.; Astrin, K. H.; Bishop, D. F.; Desnick, R. J. Acute Intermittent Porphyria: Studies of the Severe Homozygous Dominant Disease Provides Insights into the Neurologic Attacks in Acute Porphyrias. *Arch. Neurol.* **2004**, *61*, 1764–1770, Homozygous AIP.
7. Meyer, U. A.; Schuurmans, M. M.; Lindberg, R. L. P. Acute Porphyrias: Pathogenesis of Neurological Manifestations. *Semin. Liver Dis.* **1998**, *18*, 43–52, Neurologic Aspects of Acute Porphyrias.
8. Wahlin, S.; Harper, P.; Sardh, E.; Andersson, C.; Andersson, D. E.; Ericzon, B. G. Combined Liver and Kidney Transplantation in Acute Intermittent Porphyria. *Transpl. Int.* **2010**, *23*, e18–e21, Transplantation in AIP.
9. Yasuda, M.; Bishop, D. F.; Fowkes, M.; Cheng, S. H.; Gan, L.; Desnick, R. J. AAV8-Mediated Gene Therapy Prevents Induced Biochemical Attacks of Acute Intermittent Porphyria and Improves Neuromotor Function. *Mol. Ther.* **2010**, *18*, 17–22, Experimental Gene Therapy for AIP.
10. Innala, E.; Andersson, C. Screening for Hepatocellular Carcinoma in Acute Intermittent Porphyria: A 15-year Follow-Up in Northern Sweden. *J. Intern. Med.* **2011**, *269*, 538–545, Management of AIP.
11. Hift, R. J.; Thunell, S.; Brun, A. Drugs in Porphyria: From Observation to a Modern Algorithm-Based System for the Prediction of Porphyrogenicity. *Pharmacol. Ther.* **2011**, *132*, 158–169, Drugs Trigger Acute Attacks.

ERYTHROPOIETIC PORPHYRIAS

12. Desnick, R. J.; Astrin, K. H. Congenital Erythropoietic Porphyria: Advances in Pathogenesis and Treatment. *Br. J. Haematol.* **2002**, *117*, 779–795, Overview of CEP.
13. Whatley, S. D.; Ducamp, S.; Gouya, L.; Grandchamp, B.; Beaumont, C.; Badminton, M. N.; Elder, G. H.; Holme, S. A.; Anstey, A. V.; Parker, M., et al. C-Terminal Deletions in the ALAS2 Gene Lead to Gain of Function and Cause X-Linked Dominant Protoporphyria Without Anemia or Iron Overload. *Am. J. Hum. Genet.* **2008**, *83*, 408–414, X-Linked Protoporphyria.

CHAPTER
99

Inherited Disorders of Human Copper Metabolism

Stephen G Kaler

Unit on Human Copper Metabolism, Molecular Medicine Program, *Eunice Kennedy Shriver*, National Institute of Child Health, and Human Development, Bethesda, MD, USA

Seymour Packman

Departments of Pediatrics and Genetics, University of California, School of Medicine, San Francisco, CA, USA

Copper is an essential trace element that is necessary for numerous critical biological processes, including iron transport, connective tissue and blood vessel formation, pigmentation of hair, retina, and skin, detoxification of reactive oxygen species, mitochondrial electron transport chain function, amidation of neuroendocrine peptides, and catecholamine biosynthesis. The critical involvement of copper-transporting ATPase 1 (ATP7A) in axonal outgrowth, synapse integrity and neuronal activation underlines the fundamental role of copper metabolism in neurological function. Mutations in ATP7A yield three distinct clinical syndromes: Menkes disease, occipital horn syndrome, and isolated distal motor neuropathy, each of which has distinct neurological effects. Menkes disease is characterized by infantile onset cerebral and cerebellar neurodegeneration, failure to thrive, coarse hair, and connective tissue abnormalities, coupled with a characteristic biochemical phenotype (low copper levels in the blood and brain, and abnormal plasma and cerebrospinal fluid neurochemical levels). Menkes disease is lethal in infancy if left untreated, but normal neurodevelopmental outcomes are sometimes possible if therapy can be administered on the basis of neonatal diagnosis. In occipital horn syndrome, leaky splice junction or hypomorphic missense mutations in ATP7A allow considerable ATP7A-mediated copper transport, thereby sparing the CNS, but cuproenzyme deficiencies can cause dysautonomia and connective tissue problems. A recently discovered ATP7A phenotype, adult-onset distal motor neuropathy, shares no clinical or biochemical abnormalities with Menkes disease or OHS, and results from unique missense mutations in ATP7A. The basic defect underlying this phenotype is under investigation.

Wilson disease is an autosomal recessive disorder of copper transport. Affected individuals accumulate abnormal levels of copper in the liver and later in the brain as a consequence of mutations in both alleles of the Wilson disease gene (ATP7B). This gene was also identified in 1993 and encodes a copper-transporting ATPase expressed primarily in the liver, where its major function is excretion of hepatic copper into the biliary tract. The clinical condition was first described in 1912 by S. A. K. Wilson, an American neurologist working in England. Thirty-six years later, pathologist J. N. Cummings proposed an etiologic connection with copper overload. Therapy with copper chelation by penicillamine was introduced in 1956 by J. M. Walshe, a British physician working at Boston City Hospital in Massachusetts. More recent medical treatments also rely upon copper chelation.

Idiopathic hepatic copper toxicosis refers to a condition observed mainly in clusters in infants and children in isolated rural districts of India and Northern Europe. Family studies imply autosomal recessive inheritance. Fulminant, rapidly progressive liver disease with distinctive pathology, including ballooning hepatocyte degeneration with Mallory hyaline bodies, is typical. The molecular bases for this condition have not been identified and a multifactorial etiology is possible, including environmental contributions that favor increased copper exposure.

A panoply of useful animal models provides opportunities for further exploration of human copper metabolism and evaluation of potential disease remedies for inherited copper transport disorders, including gene therapy.

CHAPTER

100

Iron Metabolism and Related Disorders

Kaveh Hoda and Christopher L Bowlus

Division of Gastroenterology and Hepatology, University of California, Davis School of Medicine, Sacramento, CA, USA

Thomas W Chu

Clinical Research and Exploratory Development, Palo Alto, CA, USA

Jeffrey R Gruen

Departments of Pediatrics, Genetics, and Investigative Medicine, Yale Child Health Research Center, Yale University School of Medicine, New Haven, CT, USA

100.1 INTRODUCTION

Iron is vital to life due to its ability to transfer electrons in redox reactions. However, this makes iron toxic through the generation of hydroxyl radicals. A complex system has evolved to maintain iron homeostasis. Central to iron regulation is hepcidin, a peptide that is produced by the liver and inhibits the iron exporter ferroportin in intestinal cells and macrophages (*1*). In addition to iron, erythropoiesis, inflammation, and hypoxia affect hepcidin production through a variety of pathways. The cause of iron overload is misregulation of dietary iron absorption in the duodenum, the site of highest iron absorption in humans. Normally, homeostasis of body iron levels is maintained by regulating intestinal iron absorption. When iron stores are adequate, liver hepcidin expression is induced and hepcidin circulates to the duodenum where it binds ferroportin, which is then degraded preventing the export of iron into the circulation. Several syndromes leading to iron overload have been described and the genes involved identified (*2*). In most cases, the defect results in a relative deficiency of hepcidin or hepcidin signaling (Table 100-1).

100.2 CLINICAL CONDITIONS

The most common syndrome of iron overload, hereditary hemochromatosis (HH), is an autosomal recessive condition resulting from a single mutation (C282Y) in *HFE* and affecting 1 in 250 individuals of northern European descent (*3*). A second mutation (H63D) is common in the most populations, is in complete linkage disequilibrium with C282Y, and results in mild iron loading only in the compound C282Y/H63D heterozygous state. About 75% of men and 50% of women homozygous for C282Y show evidence of iron loading by serum iron indices. Although previous estimates ranged as high as 50%, newer data suggest the frequency of clinical sequelae of iron overload is particularly low in women and approximately 25% in men (*4*). Factors that contribute to the varying expression of disease manifestations include genetic factors that influence iron metabolism, inflammation, or fibrosis as well as age, diet, insulin resistance, and alcohol intake. The initial phase of treatment involves removal of 500 ml of blood once or twice each week until iron-limited erythropoiesis occurs. Iron depletion can be confirmed by a markedly decreased transferrin saturation and serum ferritin concentration (<15% and <20 mg/l, respectively) followed by lifelong maintenance phlebotomy therapy at intervals of 2–6 months. The serum ferritin concentration are measured each year to estimate body iron stores and to adjust the frequency of phlebotomies.

In addition to HH, other, less common syndromes due to mutations in hemojuvelin (*HJV*), hepcidin (*HAMP*), and transferrin receptor 2 (*TFR2*) result in hepcidin deficiency and accumulation of iron within the liver and other organs (*5,6*). Mutations in ferroportin

TABLE 100-1 Hereditary Disorders of Iron Metabolism

	HFE-Related Adult-Onset Hemochromatosis	Juvenile Hemochromatosis		TFR2-Related Hemochromatosis	Ferroportin-Related Iron Overload	Rare Disorders of Local or Cellular Iron Overload					
OMIM classification	Type 1	Type 2A	Type 2B	Type 3	Type 4	Atransferrinemia	PKAN, HARP	Aceruloplasminemia	Neuroferritinopathy	Anemia, Sideroblastic, Spinocerebellar ataxia	Friedreich's Ataxia
Gene	*HFE*	*HJV*	*HAMP*	*TFR2*	*SCL40A1*	*TF*	*PANK2*	*CP*	*FTL*	*ABCB7*	*FRDA*
Aliases	HLA-H	JH, HFE2A, RGMC	HEPC, LEAP-1		MTP1, IREG1, FPN1, HFE4						
Location	6p21.3	1q21	19q13.1	7q22	2q32	3q21	20p13-p12.3	3q23-q24	19q13.3-q13.4	Xq13.1-q13.3	9q13
Gene product	HFE	Hemojuvelin	Hepcidin	Transferrin receptor 2	Ferroportin	Transferrin	Pantothenate kinase 2	Ceruloplasmin	Ferritin light chain	ATP-binding cassette, subfamily B, member 7	Frataxin
Function	Regulation of hepcidin expression through interaction with TFR1 and TFR2	Modulation of hepcidin expression acting as co-receptor for BMP receptor	Inhibition of iron export from intestine and macrophages by FPN	Uptake of iron by hepatocytes and regulation of hepcidin expression	Export of iron from enterocyte macrophages, placental cells, and hepatocytes	Circulating iron transport molecule	Brain specific enzyme; defect results in cysteine accumulation which may chelate iron	Oxidation of ferrous to ferric iron required for release of cellular iron to transferrin	Intracellular iron storage	Transport of FeS clusters from mitochondria to cytosol	Mitochondrial iron metabolism; FeS cluster assembly
Inheritance	Autosomal recessive	Autosomal recessive	Autosomal recessive	Autosomal recessive	Autosomal dominant	Autosomal recessive	Autosomal recessive; some dominant alleles	Autosomal recessive	Autosomal dominant	X-linked recessive	Autosomal recessive
Tissue iron accumulation	Liver, endocrine glands, heart	Liver, endocrine glands, heart	Liver, endocrine glands, heart	Liver, endocrine glands, heart	Liver, spleen	Liver, heart	Brain: basal ganglia	Liver, pancreas, brain	Brain: basal ganglia	Brain, erythroid cells	Brain, heart
Cellular iron accumulation	Parenchymal	Parenchymal	Parenchymal	Parenchymal	Reticulo-endothelial	Reticulo-endothelial	Parenchymal	Parenchymal	Parenchymal	Mitochondrial	Mitochondrial
Organ damage	Variable	High	High	Variable	Low	Variable	High	High	High	High	High
Anemia	No	No	No	No	Possibly menstruating women, post phlebotomy	Yes	No	No	No	Yes	No
Treatment	Phlebotomy-excellent response	Phlebotomy-excellent response	Phlebotomy-excellent response	Phlebotomy-excellent response	Phlebotomy-fair response	Human transferrin; transfusion-phlebotomy	Pantothenate?	Desferoxamine			Antioxidant therapy?
Decade onset	4th or 5th	2nd or 3rd	2nd or 3rd	4th or 5th	4th or 5th	1st–2nd	1st–2nd	5th–6th	5th–6th	1st	1st–2nd

Source: Modified from Pietrangelo (*2*).

(*SCL40A1*) that abrogate hepcidin binding have been reported to result in rare cases of iron overload (*7,8*). Other syndromes of abnormal iron metabolism include atransferrinemia, aceruloplasminemia, neuroferritinopathy, X-linked sideroblastic anemia with ataxia, and Friedreich ataxia.

REFERENCES

1. Knutson, M. D. Iron-Sensing Proteins That Regulate Hepcidin and Enteric Iron Absorption. *Annu. Rev. Nutr.* **2010,** *30,* 149–171.
2. Pietrangelo, A. Hereditary Hemochromatosis: Pathogenesis, Diagnosis, and Treatment. *Gastroenterology* **2010,** *139* (2), 393–408, 408 e1–e2.
3. Feder, J. N.; Gnirke, A.; Thomas, W.; Tsuchihashi, Z.; Ruddy, D. A.; Basava, A.; Dormishian, F.; Domingo, R., Jr.; Ellis, M. C.; Fullan, A., et al. A Novel MHC Class I-Like Gene Is Mutated in Patients with Hereditary Haemochromatosis. *Nat. Genet.* **1996,** *13* (4), 399–408.
4. Beutler, E.; Felitti, V. J.; Koziol, J. A.; Ho, N. J.; Gelbart, T. Penetrance of 845G→A (C282Y) HFE Hereditary Haemochromatosis Mutation in the USA. *Lancet* **2002,** *359* (9302), 211–218.
5. Babitt, J. L.; Huang, F. W.; Wrighting, D. M.; Xia, Y.; Sidis, Y.; Samad, T. A.; Campagna, J. A.; Chung, R. T.; Schneyer, A. L.; Woolf, C. J., et al. Bone Morphogenetic Protein Signaling by Hemojuvelin Regulates Hepcidin Expression. *Nat. Genet.* **2006,** *38* (5), 531–539.
6. Le Gac, G.; Mons, F.; Jacolot, S.; Scotet, V.; Ferec, C.; Frebourg, T. Early Onset Hereditary Hemochromatosis Resulting from a Novel TFR2 Gene Nonsense Mutation (R105X) in Two Siblings of North French Descent. *Br. J. Haematol.* **2004,** *125* (5), 674–678.
7. Nemeth, E.; Tuttle, M. S.; Powelson, J.; Vaughn, M. B.; Donovan, A.; Ward, D. M.; Ganz, T.; Kaplan, J. Hepcidin Regulates Iron Efflux by Binding to Ferroportin and Inducing Its Internalization. *Science* **2004.**
8. Montosi, G.; Donovan, A.; Totaro, A.; Garuti, C.; Pignatti, E.; Cassanelli, S.; Trenor, C. C.; Gasparini, P.; Andrews, N. C.; Pietrangelo, A. Autosomal-Dominant Hemochromatosis Is Associated with a Mutation in the Ferroportin (SLC11A3) Gene. *J. Clin. Invest.* **2001,** *108* (4), 619–623.

CHAPTER

101

Mucopolysaccharidoses

J Ed Wraith

Pediatric Inherited Metabolic Medicine, Genetic Medicine, St. Mary's Hospital, Manchester, UK

The mucopolysaccharidoses (MPS) are a family of disorders caused by inherited defects in the catabolism of sulfated components of connective tissue known as glycosaminoglycans (GAGs). The crude cumulative rate for all types of MPS is around 3.5 in 100,000 live births and generally the patients present in one of three ways:

1. As a dysmorphic syndrome (MPS IH, MPS II, MPS VI) often with early onset middle ear disease, deafness, or upper airways obstruction
2. With learning difficulties, behavioral disturbance and dementia and mild somatic abnormalities (MPS III)
3. As a severe bone dysplasia (MPS IV)

The diagnosis is based on clinical suspicion, supported by appropriate clinical and radiological examinations followed by urinary examination for GAG excretion and then specific enzyme assay usually on white blood cells.

Patients with MPS often have a facial appearance, which is characteristically labeled "coarse," although most parents find the term objectionable. A combination of subcutaneous storage and involvement of the facial bones in the dysostosis produces the typical appearance, seen in its most developed form in MPS IH (Figure 101-1).

Underdevelopment of the mid-facial skeleton and the firm puffiness associated with subcutaneous storage results in a flat nasal bridge and a "blurring" of the facial features. The lips and tongue are thickened and the hair is often abundant and dull. The persistent nasal discharge detracts further from the child's general appearance. A dark synophyrs is a characteristic finding and affected children are often hirsute. The facial phenotype is much less obvious in patients with MPS III and absent in patients with MPS IV.

All of the MPS disorders (with the exception of MPS IX where only one patient has been described) are heterogeneous with severe and more attenuated variants. These are often differentiated on the basis of survival or the presence or absence of CNS involvement. The reality is more complex with most disorders representing a complex spectrum with severely affected patients and attenuated patients at opposite ends.

Patients with the severe form of MPS I (Hurler syndrome, MPSIH), MPS II (Hunter syndrome) and MPS VI (Maroteaux–Lamy syndrome) present with facial dysmorphism and persistent respiratory disease in the early years of life. Many patients will have undergone surgical procedures for middle ear disease and a hernia before the diagnosis is established. Infants with MPS III (Sanfilippo A, B, C and D) present with learning difficulties and then develop a profound behavioral disturbance which is characteristic and often leads to the diagnosis. Somatic features are mild in these patients. Children with MPS IVA (Morquio A syndrome) have normal cognitive functions but are affected by a severe bone dysplasia which in most patients leads to extreme short stature and vertebral instability. MPS IVB (Morquio B syndrome) is much more variable in its effects it has some features of the bone dysplasia of MPS IVA but in addition most patients have learning difficulties. MPS VII (Sly syndrome) often presents with hydrops fetalis and in those patients who survive or who present later the clinical phenotype, and management issues are similar to MPS IH and H/S. So far only one patient with MPS IX (Natowicz syndrome) has been described.

The phenotype of patients with more attenuated forms of MPS, e.g. MPS IH/S or S (Hurler–Scheie or Scheie syndromes) is much more difficult to predict and treatment needs in this group of patients can be very variable. Because of the multisystem involvement in these patients, treatment is multidisciplinary and encompasses both "curative" and palliative elements. Those patients with severe central nervous system involvement (MPS IIIA-D, Sanfilippo syndrome) or severe bone dysplasia (MPS IVA, Morquio syndrome) present a particular challenge as current therapies are very poor in correcting the effects of the genetic lesion in brain and bone.

Attempts at "curative therapy" have previously centered on the use of hematopoetic stem cell transplant (HSCT), using either bone marrow or umbilical cord blood cells.

Although all MPS disorders have been treated by HSCT evidence for efficacy is strong in only MPS IH (Hurler syndrome) or MPS VI (Maroteaux–Lamy syndrome). The procedure is ineffectual in MPS IIIA-D (Sanfilippo syndrome) and MPS IV (Morquio syndrome) and too few patients with MPS II (Hunter syndrome) and MPS VII (Sly syndrome) have been transplanted to make a reasonable assessment. HSCT is not curative and most treated patients have multiple orthopedic complications and in addition corneal clouding and cardiac valve disease progresses.

The introduction of recombinant human enzyme replacement therapy is likely to make a major impact in the area of treatment in the years to come. ERT is available for the treatment of MPS I, II and VI and other enzyme strategies are in advanced stages of development with phase III clinical trials underway for MPS IV. In addition, alternative routes of enzyme delivery are being explored to try and circumvent the blood–brain barrier to allow prevention or treatment of CNS involvement.

Despite these advances for many patients with various MPS disorders, all that can be offered is palliative care. This should encompass a holistic approach with symptom control the main goal. Many different specialties both within and allied to medicine have a role to play as well as involvement of the voluntary sector. Adequate respite care is important for those families who have children with profound behavioral disturbance. Prenatal diagnosis is possible in MPS I to VII.

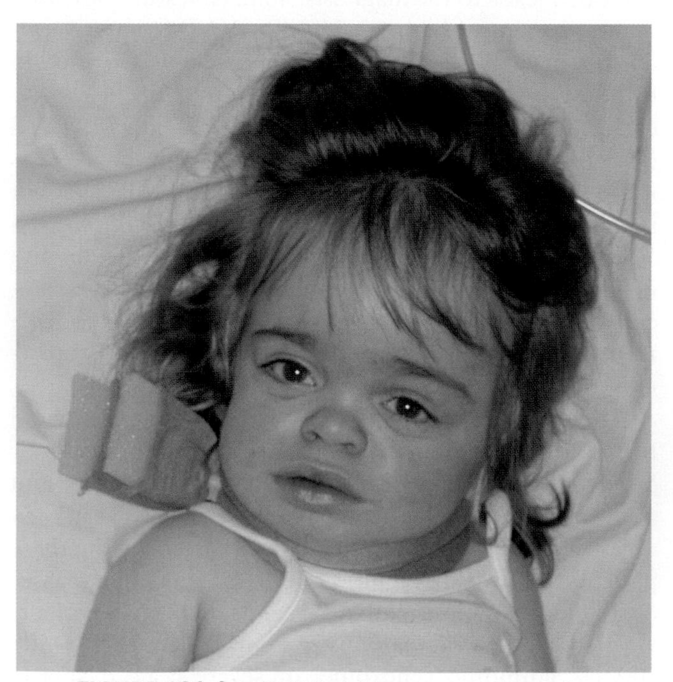

FIGURE 101-1 MPS IH facial features at 27 months.

FURTHER READING

1. Neufeld, E. F.; Meunzer, J. The Mucopolysaccharidses. In *The Metabolic and Molecular Bases of Inherited Disease,* 8th ed.; McGraw-Hill: New York, 2001.
2. Lachman, R.; Martin, K. W.; Castro, S.; Basto, M. A.; Adams, A.; Teles, E. L. Radiologic and Neuroradiologic Findings in the Mucopolysaccharidoses. *J. Pediatr. Rehabil. Med.* **2010,** *3,* 109–118.
3. Orchard, P. J.; Milia, C.; Braunlin, E.; DeFor, T.; Bjoraker, K.; Blazar, B. R.; Peters, C.; Wagner, J.; Tolar, J. Pre-Transplant Risk Factors Affecting Outcome in Hurler Syndrome. *Bone Marrow Transplant.* **2010,** *45,* 1239–1246.
4. Aldenhoven, M.; Boelens, J. J.; de Konig, T. J. The Clinical Outcome of Hurler Syndrome after Stem Cell Transplantation. *Biol. Blood Marrow Transplant.* **2008,** *14,* 485–498.
5. Muenzer, J.; Wraith, J. E.; Clarke, L. A. The International Consensus Panel on the Management and Treatment of Mucopolysaccharidosis I. *Pediatrics* **2009,** *123,* 19–29.
6. Muenzer, J.; Beck, M.; Eng, C. M.; Escolar, M. L.; Giugliani, R.; Guffon, N. H.; Harmatz, P.; Kamin, W.; Kampmann, C.; Koseoglu, S. T.; Link, B.; Martin, R. A.; Molter, D. W.; Munoz Rojas, M. V.; Ogilvie, J. W.; Parinin, R.; Ramaswami, U.; Scarpa, M.; Schwartz, I. V.; Wood, R. E.; Wraith, E. Multidisciplinary Management of Hunter Syndrome. *Pediatrics* **2009,** *124,* e1228–e1239.
7. Giugliani, R.; Harmatz, P.; Wraith, J. E.; Management Guidelines for Mucopolysaccharidosis VI. *Pediatrics* **2010,** *120,* 405–418.

CHAPTER

102

Oligosaccharidoses: Disorders Allied to the Oligosaccharidoses

Jules G Leroy

Greenwood Genetic Center, Greenwood, SC, USA

The oligosaccharidoses with excessive urinary excretion of sialyloligosaccharides as the common hallmark are a group of rare metabolic disorders with onset most often in infancy or early childhood. Various steps in the catabolism of N-linked glycans in glycoproteins and glycolipids are adversely affected with progressive clinical course, multisystemic morbidity and grave or moderately severe effect on life expectancy. All oligosaccharidoses are monogenic disorders with autosomal recessive mode of inheritance. In each one, the gene affected by various mutations has been identified and characterized. Molecular mutation screening plays an increasing role in formal confirmation of the clinical and/or biochemical diagnosis and in prenatal diagnosis. This group of disorders is usually listed among the lysosomal storage disorders, because in several of them the mutant genotype causes complete or partial deficiency of one among the many acid lysosomal hydrolases and failure of intralysosomal degradation of one or more macrocompounds. The resulting storage, not always clinically apparent, is readily demonstrable electronmicroscopically. All disorders point at the physiology importance of N-linked glycans and glycosylation, in general, in human tissues as in all of biology. Moreover in all disorders, the central nervous system and the connective tissues, including the skeleton, are prominent targets.

In section "Disorders Allied to the Oligosaccharidoses," a number of disorders, historically allied to the oligosaccharidoses on clinical grounds, are discussed. More recent investigations have shown that the mutant genotypes cause deleterious effects either upon the normal maturation and routing of nascent hydrolases or affect integral lysosomal membrane proteins.

102.1 OLIGOSACCHARIDOSES

- The sialidoses are due to the failing acid glycoprotein sialidase (sialidase 1). Even the most common type, childhood dysmorphic sialidosis, previously known as mucolipidosis (ML) I, is a rare disorder. Other clinically delineated entities include juvenile normosomatic sialidosis (JNS), also called cherry-red spot myoclonus syndrome with onset in adolescence or adulthood, and the early-onset sialidoses. Whereas the former lacks dysmorphic features and has an extremely slow clinical course, the latter represents an important differential diagnosis in the event of neonatal nonimmune hydrops often with fatal perinatal outcome. The gene *NEU1*, mapped at chromosome 6p21.3, encodes sialidase 1 (neuraminidase) which in homozygous or compound heterozygous genotypes comprising only "amorph" mutations is completely inactive. In the event of at least one missense (MS) mutation with "hypomorph" effect, there is some residual activity resulting in the late onset, more slowly evolving type of sialidosis, such as JNS.
- Three types of galactosialidosis have been delineated decades ago. Congenital galactosialidosis presents as nonimmune hydrops fetalis, with often perinatal fatal outcome, despite excellent neonatal intensive care. Both late infantile galactosialidosis and juvenile galactosialidosis or the Goldberg syndrome is slowly progressive, but severe disorders with multisystemic morbidity. In addition to cherry-red spot with decreasing visual acuity and myoclonus, a host of other morbid features such as facial coarsening, moderate dysostosis multiplex, mild to moderate intellectual disability, cardiac valvular dysfunction and cardiomyopathy become overt. The diagnosis of galactosialidosis is probable by demonstration of the combined deficiency in fresh leukocytes or cultured fibroblasts, of the lysosomal enzymes, β-D-galactosidase and glycoprotein sialidase. It is formally confirmed by finding a significant deficiency of cathepsin A (CTSA). The *CTSA* gene is located at 20q13.12 and encodes the lysosomal 54 kDa "protective" glycoprotein CTSA precursor.
- The *N*-acetylglucosamine phosphotransferase deficiency disorders include ML II, initially called

"Inclusion cells" (I-cells), ML III α/β (pseudo-Hurler polydystrophy) and ML III γ. These disorders are due to either complete or much reduced activity of GlcNAc-1-phosphotransferase (GNPT), the enzyme that normally catalyzes the first step in the synthesis of the mannose-6-phosphate (M6P) recognition marker onto the oligomannose type N-glycans in lysosomal enzymes. Mutant genotype associated with GNPT activity reduced to near zero result is ML II. In these circumstances, the acid lysosomal hydrolases lack the M6P marker and do not enter lysosomes because they cannot bind to the M6P receptors. Hence, the intracellular (intralysosomal) activity of all enzymes affected is considerably reduced in cultured fibroblasts that have the special phase-contrast microscopy phenotype of I-cells. In culture media, these enzymes are highly hyperactive as is the case in the patient's plasma and other body fluids. The specific activity of lysosomal enzymes is normal, except for a reduced activity of β-D-galactosidase, in postmortem parenchymatous organs. These in vitro and in vivo findings are most manifest in ML II, but are also present in ML III. The lysosomal enzymes have a normal activity in the patient's leukocytes that therefore cannot be used for diagnostic purposes. Clinically, the ML II patient is reminiscent of the patient with Hurler disease (MPS I), but has an earlier neonatal onset and more severe morbidity. Head circumference is proportional to stature. The facies remains flat, and the eyes and the periorbital veins are rather prominent. Cumulative growth slows soon after birth and ceases altogether in the second year of life. There is severe dysostosis multiplex in addition to delay of ossification and gradual osteopenia. The range of motion in all joints is severely limited due to soft connective tissue and bony dysplasia. Only an occasional patient reaches unaided walking. Speech development is deficient and intellectual disability is obvious; upper respiratory infections and otitis are frequent. ML II patients represent an increased risk to anesthesiologic in- and detubation. Few ML II patients survive beyond mid-childhood.

In vitro tests performed decades ago have shown that ML II and ML III α/β fibroblast strains failed to complement one another, whereas mutual complementation was observed between ML II. ML II and ML III α/β are allelic conditions and ML III α/β and ML III γ, clinically hardly distinguishable, are nonallelic disease entities. ML III has its clinical onset from the third year of life, when physical slowness, limited motion in major joints, especially the shoulders, is noticed. Neuromotor development is delayed, but all patients with ML III can walk unaided. Speech is normal in ML III, where intellectual disability is a rare feature. The dysostosis multiplex in ML III is quantitatively rather similar to that in ML II. The bones in hands and fingers are anatomically normal contrary to the strong diaphyseal widening during later infancy in ML II. Painful hip dysplasia progresses from late childhood. ML III patients have severe and painful osteopenia that may be treated by cyclic IV bisphosphonate infusions. The mutant *GNPTAB* genotype composed of two mutations (nonsense or reading frameshift type) with "amorph" phenotypic effect consistently result in ML II. At least one mutation with hypomorph effect (missense and some splice-type mutations) in the mutant *GNPTAB* genotype is associated with some residual GNPT activity and causes ML III α/β. Apparently, all mutant *GNPTG* genotypes cause ML III γ, a disorder so far most frequently observed in Middle-East populations.

- α-D-mannosidosis is due to the severe deficiency of lysosomal acid α-D-mannosidase (LAMAN) and usually causes a slowly progressive disorder. The variable age of clinical onset has been a reason for the previous subdivision in Type I and Type II disease. Facial coarsening is milder than in Hurler disease. Developmental delay, slow speech development and frequent URIs with otitis are the main initial parental complaints. The dysostosis multiplex in α-mannosidosis is usually mild. Neurosensory hearing deficit is a rather consistent feature and may adversely affect the learning disability. Late in the course the patients' mobility decreases due to muscular weakness, polyarthropathy and encephalopathic neurologic signs. The diagnosis is based on the finding of the specifically deficient LAMAN activity in leukocytes or cultured fibroblasts. The gene *MAN2B1* encoding LAMAN is on chromosome 19pcen-p13.1.
- The diagnosis of β-mannosidosis is challenging because of its apparent rarity, the lack of physical abnormality in the young patient and his rather unspecific neurologic and behavioral symptoms and signs including epilepsy. It has a slow progressive course and CNS atrophy can be documented by. Excessive oligosacchariduria may be absent, but specific beta-linked mannose containing disaccharides need being looked for. Lysosomal β-mannosidase is absent in leukocytes or cultured fibroblasts. The *MAMBA* gene, mapped to 4q22-q25 may be screened for mutations to obtain the most objective confirmation of the diagnosis.
- The clinical onset of fucosidosis follows after a symptom-free interval of about 1 year, with frequently recurring URIs, neuro- and psychomotor developmental delay and slightly coarsened facial features. Some patients never achieve unaided ambulation. Acquired neuromotor skills soon decline increasing intellectual deficiency, and progressive hypertonia and spasticity become the characteristics of fucosidosis. The second type of clinical course is less well defined, with early age of onset but slower progress. Angiokeratoma

and/or telangiectasia are sometimes prominent. Rather mild dysostosis multiplex type is detected. MRI imaging of the CNS confirms the encephalopathy. T2-weighted hypointensity in the globus pallidus has been found to be specifically associated with fucosidosis. Assay of α-L-fucosidase in leukocytes or fibroblasts establishes the diagnosis. The assay of the enzyme in plasma is without value because 12% of normal people have only 5% of the α-L-fucosidase activity found in the majority of the human population. In FUCA1, the gene encoding the lysosomal acid fucosidase and mapped to 1p34, many different mutations have been documented.

- Aspartylglucosaminuria (AGU), with high incidence in Finland, is a chronic metabolic disease that is hardly diagnosable before the age of 5 years. Psychomotor developmental delay and mild intellectual disability represent the initial parental worries. Gradual facial coarsening becomes apparent in later childhood. Inguinal herniae point at connective tissue involvement. Loss of acquired skills becomes more apparent from adolescence. AGU is the clinical consequence of deficient aspartyl-β-glucosaminidase (AGA) that causes accumulation of aspartylglucosamins and its excessive urinary excretion *AGA*, the gene encoding aspartylglucosaminidase has been mapped to 4q32-q33.
- Deficiency of lysosomal *N*-acetyl-α-galactosaminidase (NAGA) is the primary metabolic defect in two very rare, clinically different disorders: Schindler disease, the severe neuroaxonal dystrophy type, and Kanzaki disease, the milder variant with disseminated angiokeratoma. The trisaccharide GalNAcGalFuc is the most specific chemical that accumulates. The responsible mutant genotypes have been detected following the molecular characterization of *GALB*, located at 22q13, the gene encoding NAGA.

102.2 DISORDERS ALLIED TO THE OLIGOSACCHARIDOSES

- The clinical phenotypes associated with free sialic acid storage, the sialic acid storage disorders (SASDs), include infantile sialic acid storage disease and Salla disease with late infantile onset and chronic clinical course with high incidence in the Salla region of Finland. The former has a prenatal onset and manifests either as nonimmune hydrops fetalis or with significantly edema in the prematurely, small-for-dates neonate. This patient is weak, hypotonic, has a pale complexion and sometimes already a coarse facies. Refractory feeding difficulty, failure to thrive and neuromotor developmental delay characterize the seriously ill infant. Hepatomegaly is equivocal, but cardiomegaly, pericardial effusion and anemia are rather consistent. Signs of a worsening encephalopathy with failing mental development, spasticity and epileptic crises and respiratory infections lead to fatal outcome most often around the first birthday. Skeletal abnormalities remain mild, although hip dysplasia, kyphosis and extraskeletal calcified stippling may be present. Salla disease has its clinical onset before the patient's first birthday with concerns about slow neuromotor development, generalized hypotonia, but hyperreflexia in the legs. Speech development is quite deficient. Only about one-third of patients achieve independent ambulation. Facial coarsening is mild and statural growth is slightly impaired. At various ages, tonic/clonic seizures or "absence"-like episodes, refractory to treatment, compound the picture. The disorder becomes more stationary from adolescence. Life expectancy is reduced, but some patients have reached over 60 years of age. MRI neuroimaging records hypomyelination of white matter and abnormal density changes in the basal ganglia and later, generalized brain atrophy. Other organ systems, including the kidneys, are also affected in this slowly evolving SASD.

Urinary excretion of free sialic acid is 5–20 times normal. Egress of free sialic acid from lysosomes, where it is a product of normal degradation of glycans, is significantly impaired, because of the inactivity of the lysosomal membrane sialic acid transporter, called sialin. The gene *SLC17A5* that belongs to the SLC17 anion transporter family encodes sialin. *Sialuria* is a mild clinical condition with autosomal dominant inheritance that may even be present in "normal" individuals unaware of the metabolic disorder. It is due to overproduction of sialic acid because of failing retroinhibition of the rate-limiting bifunctional enzyme, UDP-GLcNAc 2-epimerase (GNE)/ManNAckinase (MNK) by the pathways' endproduct, CMP-sialic acid or CMP-Neu5Ac. There is no lysosomal storage of free sialic acid and no evidence of lysosomal damage. The gene *GNE* encoding GNE/MNK has been mapped to 9p13.3. Mutant genotypes composed of MS mutations anywhere else in the *GNE* gene cause the autosomal recessive hereditary inclusion body myopathy (HIBM) with adult onset, apparently due to hyposialylation of muscular proteins. These patients still have between 30% and 60% of normal GNE/MNK activity and no sialuria.

- Multiple sulfatase deficiency also called mucosulfatidosis, a rare metabolic disorder, is due to dysfunction of "all" sulfatases with several phenotypic variants. The completed clinical picture of mucosulfatidosis is a composite of features in some of the mucopolysaccharidoses, metachromatic leucodystrophy and X-linked ichtyosis. Clinical onset is in infancy, sometimes at birth. The course is inexorably progressive. Facial coarsening, increasing intellectual disability, slow growth rate and from early childhood dermal scaling or frank ichtyosis are consistent features. The more advanced neurologic syndrome includes gradual quadriplegia, ataxia, convulsions, nystagmus and decreased vision and hearing. The average fatal outcome occurs in mid-childhood. All lysosomal but also microsomal sulfatases have a deficient activity due to a mutation that severely interferes with the specific functional activation step, the chemical modification of a specific cysteine into formylglycine, during ER-based posttranslational maturation of the nascent enzymes. The enzyme normally catalyzing this modification is called the formylglycine generating enzyme (FGE) or sulfatase modifying factor 1 (SUMF1). The gene encoding this soluble FGE has been mapped to 3p26, called *SUMF1*.

- Mucolipidosis (ML) IV: Berman disease, clinically a homogeneous disease, has its high incidence in Ashkenazi-Jews, but has been observed worldwide. Onset is variable and occurs from birth until early childhood. Developmental delay and axial hypotonia are the presenting symptoms, but soon substituted by spasticity and significant mental disability. Corneal opacities, strabismus and visual deficit are also consistent components. The ophthalmologic signs are progressive. The disorder becomes nearly stationary from late childhood. Constitutive achlorhydria has been found in ML IV. Oromotor function and swallowing are much impaired. Life expectancy is considerably reduced and MRI shows hypoplasia of the corpus callosum and generalized demyelination of white matter. There is a milder form of ML IV, where the clinical course is slower and the neurologic components milder, but the ophthalmologic problems evolve rather similarly to the ones in the "classic" form of ML IV. The causal gene mutations have been mapped to chromosome 19p13.2-p13.3 called MCOLN1. Over 95% of Ashkenazi-Jewish patients are homozygous or compound heterozygous for one or both founder mutations in this ethnic group. The protein encoded by MCOLN1 is called mucolipin 1 and is structurally similar to proteins in the transient receptor potential (TRP) superfamily of cation channels, most of them Ca^{++} channels. Mucolipin 1 is called TRPML1 as it has been identified as an integral membrane protein in the vesicular membranes of the lysosomal cell compartment.
- Pyknodysostosis with typical facial characteristics, bone dysplasia with zones of osteopetrosis next to regions of osteolysis, increased bone fragility, and deficiency of statural growth are due to mutations in the *Cathepsin K* gene (*CTSK*), a lysosomal cysteine proteinase most intensely expressed in osteoclasts which regulates site-specific bone homeostasis because it governs normal apoptosis and senescence of osteoclasts.

Nodulosis–arthropathy–osteolysis syndrome, with multicentric osteolysis, cutaneous nodules and arthropathy, is the collective name for three very rare skeletal dysplasias that have been recognized as allelic conditions due to mutations in the matrix metalloproteinase 2, *MMP2* gene mapped to chromosome 16q13.

CHAPTER

103

Sphingolipid Disorders and the Neuronal Ceroid Lipofuscinoses or Batten Disease (Wolman Disease, Cholesterylester Storage Disease, and Cerebrotendinous Xanthomatosis)

Rose-Mary Boustany

Neurogenetics Program and Division of Pediatric Neurology, Departments of Pediatrics and Adolescent Medicine and Biochemistry, American University of Beirut Medical Center, Beirut, Lebanon, Departments of Pediatrics and Neurobiology, Duke University Medical Center, Durham, NC, USA

Ibraheem Al-Shareef

American University of Beirut Medical Center, Beirut, Lebanon

Sariah El-Haddad

American University of Beirut Medical Center, Beirut, Lebanon

The GM1-gangliosidoses: They are characterized by the accumulation of GM1-ganglioside, oligosaccharides, and keratan sulfate. The three types recognized are the infantile type I, the juvenile type II and the adult type III. Type I combines features of a neurolipidosis and a mucopolysaccharidosis with death at 2 years. Type II presents with ataxia, dystonia, and spasticity with death by 20 years. Type III presents with dementia, parkinsonism, and dystonia. Measuring β-galactosidase enzyme activity in leukocytes is diagnostic. Quantitative polymer chain reaction assays allow allelic discrimination.

GM2-gangliosidosis: They are lipid storage diseases caused by defects in HEXA, HEXB, or GM2-activator protein (GM2A). Their gene products are HEXA (α-subunit of β-hexosaminidase A), HEXB (β-subunit of β-hexosaminidase A), and the GM2-activator protein, all necessary for degradation of GM2-ganglioside. **Infantile acute GM2-gangliosidosis**: Defects in HEXA, HEXB, and GM2A genes cause infantile Tay–Sachs disease, Sandhoff disease, and ABO variant, respectively. Symptoms are hyperacusis, loss of psychomotor skills before 10 months, seizures, macular cherry-red spots and macrocephaly **late-onset forms**: Disease onset is from the late infantile period to adult age. Late-onset forms result from at least one allele (G269S mutation) associated with residual enzyme activity. **Subacute GM2-gangliosidosis**: Ataxia develops between 2 and 10 years of age, followed by dementia, spasticity and seizures. The subacute phenotype is described in patients with HEXA or HEXB defects. **Chronic GM2-gangliosidosis**: Tay–Sachs disease variants are most common.

The defect in GM2-gangliosidosis is caused by failure of α-hexosaminidase to hydrolyze GM2-ganglioside. Disease severity correlates inversely with level of residual enzyme activity. The acute forms of GM2-gangliosidosis are associated with complete HEXA deficiency, with activity levels ~5% in the juvenile/adult forms. Most mutations cause severe infantile onset disease. Screening programs targeting Ashkenazi-Jewish groups have resulted in reduction of the incidence by >90%. In the adult form in Ashkenazi Jews, patients carry the common Ashkenazi mutation on one allele, and a milder mutation on the other, defining the phenotype.

103.1 LOSS OF FUNCTION MUTATION OF GM3-SYNTHASE

Irritability and failure are noted before 3 months, followed by seizures, startle myoclonus and developmental regression. Patients end up mute, cortically blind, and nonambulatory, with choreo-athetoid movements. The protein catalyzes formation of GM3-ganglioside from lactosylceramide. Patients accumulate lactosylceramide. All cases have a nonsense mutation in the *SIAT9* gene (R232X). Seizures are difficult to control and most children require feeding tubes.

103.2 NIEMANN–PICK DISEASE

Niemann–Pick disease constitutes four clinical forms grouped because of sea-blue histiocytes in the reticuloendothelial system. Types a and b result from low acid sphingomyelinase activity, accumulation of sphingomyelin, and defects in the same gene. Type C1 accounts for 95% of Niemann–Pick C cases. A second gene, NP-C2, has been identified. Niemann–Pick C results from a lipid trafficking with LDL-derived cholesterol trapped in the lysosomal compartment.

Niemann–Pick disease, type A (infantile/acute form): Hepatosplenomegaly and moderate lymphadenopathy develop in the first months. Most cases of NP-A are Ashkenazi-Jewish. Hypotonia, feeding difficulties, vomiting, constipation and macular cherry-red spots are described. Death ensues by 4 years of age. **Niemann–Pick disease, type B (chronic, non-neuronopathic forms)**: Type B is the chronic, non-neuropathic form with hepatosplenomegaly, decreased pulmonary diffusion with dyspnea, broncho-pneumonias and cor pulmonale. The patients have cherry-red spots or gray pigmentation around the fovea that is halo-like. Cirrhosis/hepatic failure occurs. Adolescent patients with NP-B have delayed skeletal maturation/growth restriction and osteopenia.

Niemann–Pick, types C and D: Age of onset can be from infancy to adulthood. Manifestations can be neurologic/hepatic/psychiatric. Liver involvement is present in the first month of life. Hepatosplenomegaly, ataxia, vertical supranuclear palsy, dysarthria, dystonia, cataplexy, tremors, parkinsonism, seizures, and jaundice can be present. Antenatal NP-C disease is rapidly fatal with fetal ascites. NP-C may present with visceral disease without neurological symptoms. Pulmonary alveolar proteinosis due to NP-C2 disease can result in bronchiolitis or pneumonia unresponsive to treatments.

Cholesterol esterification assays in skin fibroblasts, and NP-C gene analysis lead to the diagnosis. NP-D types C1 and D are allelic variants classified as cellular lipid trafficking disorders. Type D refers to a genetic isolate from Nova Scotia.

NP-C is pan-ethnic with 95% with mutations in the *NP-C1* gene. The rest have mutations in the *NP-C2* gene. Diagnosis of NP-C is achieved by filipin staining of fibroblasts grown with LDL. Treatment is supportive and includes anticonvulsants for seizures and protriptyline/SSRIs for cataplexy. Substrate reduction therapy (SRT) using miglustat has been approved in Europe for treating NP-C. Miglustat (Zavesca) is a reversible inhibitor of glucosylceramide synthase, and crosses the BBB, targeting neurological symptoms.

Farber's disease: It is caused by defects in acid ceramidase, the enzyme that breaks down ceramide, stored in lysosomes. It is characterized by subcutaneous nodules over extensor joints, painful arthritic joints, and hoarseness. The nodules involve the eyelids, lips, and gums. A fatal neonatal form can present as hydrops fetalis. The classic variant presents between 4 months and 4 years with the triad. Most die within 1–2 years of onset. Joint nodular swellings lead to a misdiagnosis as juvenile idiopathic arthritis. Diagnosis is confirmed by deficient acid ceramidase activity. Bone marrow transplantation (BMT) results in regression of nodules/organomegaly without impact on neurologic deterioration. Patients die in the second decade because of respiratory insufficiency.

103.3 ACID LIPASE DEFICIENCY (WOLMAN DISEASE AND CHOLESTERYL ESTER STORAGE DISEASE)

Both diseases result from acid lipase defects which causes cholesteryl esters and triglycerides to accumulate in the reticuloendothelial system. Wolman disease (WD) manifests between 2 weeks and 1 year of age with hepatosplenomegaly, steatorrhea, failure to thrive, and adrenal calcifications. Patients die by 6 month. It is of high frequency in the Iranian Jewish population with 1 in 4200 newborns affected. Cholesteryl ester storage disease (CESD) has a prevalence of <1/2000. It is benign and not detected before adulthood. Hyperbetalipoproteinemia and a large liver are the presenting symptoms at diagnosis. CESD presents during the first–second decade of life, with abdominal pain, growth failure, chronic diarrhea, hepatosplenomegaly and malabsorption and/

or gallbladder dysfunction. Long-term complications include premature atherosclerosis/stroke, bone marrow suppression, and testicular storage. Human acid lipase plays a crucial role in hydrolysis of triglyceride/cholesterol esters and its absence results in accumulation of nonhydrolyzed cholesterol esters/triglycerides and upregulation of endogenous cholesterol/LDL synthesis. Diagnosis is made because of low acid lipase activity in leukocytes/cultured fibroblasts. WD or CESD is allelic, autosomal recessive variants. The former is due to total absence of enzyme activity. CESD patients have 5% acid lipase activity. Therapy of CESD consists of a low-fat diet and lovastatin and cholestyramine. There is increased efficacy by adding ezetimibe to statins. Patients may require liver transplantation.

27-Hydroxylase deficiency or cerebrotendinous xanthomatosis (CTX): Defective 27-hydroxylase leads to blocked bile synthesis, elevated cholesterol and conversion of cholesterol to cholestanol. Deposition of cholestanol/cholesterol causes tendinous xanthomas, juvenile cataracts, neurological defects and death from arteriosclerosis. Liver disease presents as neonatal or prolonged jaundice in adults. Chenodeoxycholic acid/HMG-CoA reductase inhibitors instituted early will prevent neurologic deterioration.

103.4 GAUCHER DISEASE

Gaucher disease (GD) arises because of defective acid β-glucosidase glucosylceramide accumulates in mononuclear phagocytes causing hepatosplenomegaly. The three types of GD are non-neuronopathic/type I, acute neuronopathic or GD/type II and subacute neuronopathic or GD/type III.

GD type I: This non-neuronopathic form of GD is common with high prevalence in Ashkenazi Jews (1 per 850) compared to other populations (1 per 40,000). Age of presentation is from birth to eighth decade. Patients present with hepatosplenomegaly, anemia and thrombocytopenia and bony abnormalities. Splenomegaly appears in 90% and splenic infarction is a complication. Liver failure/cirrhosis/portal hypertension occurs in 10%. The GD heterozygote state is a risk factor for Parkinson disease or PD (4–5% of "sporadic PD"). GDI affects menarche, pregnancy, delivery/lactation and menopause. Enzyme replacement therapy (ERT) reduces menorrhagia/spontaneous abortions and complications of delivery. Neuronopathic variants, types II and III, manifest brain symptoms. Glucosylceramide is elevated and there is neuronal cell death. GD II has early onset and rapid progression. Bilateral fixed strabismus/bulbar signs, spasticity and choreoathetosis within 3–6 months of life, followed by neurological deterioration with convulsions and death during the first 2 years, are the norm. GD III has later onset and a protracted course. Visceral involvement is moderate in GD III, but course is severe with dementia/ataxia. Type 3A is the Norbotten form. Type 3B develops in early childhood with visceral disease ending in death due to pulmonary/portal hypertension. Acid β-glucosidase activity in leucocytes/fibroblasts is diagnostic but does not distinguish types. Plasma chitotriosidase is elevated 1000-fold in symptomatic patients. Monitoring plasma chitotriosidase levels helps in decisions regarding ERT/SRT.

ERT with Cerezyme® is central to GD management. The aim of ERT is to reverse clinical manifestations and prevent fibrosis of viscera. Imiglucerase (Cerezyme), recombinant human glucocerebrosidase, ameliorates hepatosplenomegaly, anemia, thrombocytopenia and skeletal abnormalities in patients with GD I, II and III. ERT does not repair bone in splenectomized GD. ERT is the standard of care for patients with GD I and is given to most GD III patients to alleviate systemic problems. ERT does not eliminate neurologic problems in GD III. SRT is limited to adult type I patients who cannot tolerate ERT. ERT is effective in ameliorating bone marrow involvement. Miglustat reduces bone pain and improves bone mineral density.

103.5 GALACTOSYLCERAMIDE LIPIDOSIS, GLOBOID CELL LEUKODYSTROPHY, OR KRABBE DISEASE: INTRODUCTION

Krabbe disease (KD) arises because of defects in the galactosylceramidase (galactocerebroside β-galactosidase/GALC) that degrades galactosylceramide (GalCer). Late infantile, juvenile, and adult forms exist and are characterized by spasticity, peripheral neuropathy, dementia, and blindness. Initial symptoms between ages 0 and 12 months are crying and irritability, stiffness, and seizures. Older children present with gait disturbances or loss of milestones. **Infantile form:** After normal development, infants become irritable and develop opisthotonic posturing with hands in a claw-like position. Psychomotor decline, microcephaly and optic atrophy develop. Peripheral neuropathy is present. Reflexes disappear. Seizures occur late. Patients never live past their second birthday. **Late infantile form:** Onset is from 6 months to 3 years with psychomotor decline, irritability, ataxia, spasticity, blindness, neuropathy and seizures with death 2 years later. Acquired obstructive hydrocephalus occurs due to high CSF protein and viscosity, or aseptic arachnoiditis. **Juvenile form:** Onset is between 3 and 8 years. Spasticity, ataxia, hemiparesis, and peripheral neuropathy are rare. Cognitive/motor decline ensues. They survive to the second decade. **Adult form:** Patients are intact during childhood, or have club feet, ataxia, tremor, or rigidity. Late-onset cognitive decline and peripheral neuropathy set in. Diagnosis is confirmed by low galactocerebrosidase activity in leukocytes/fibroblasts, and DNA molecular analysis in some families. When clinical course/neuroimaging is supportive of KD, but enzyme activity and gene sequence are normal, a defect in saposin A (SAP A) is suspected. GalCer and sulfatide are major constituents of myelin. Beneficial effects of BMT/cord blood transplants were in presymptomatic infants in the first 3 weeks of life or later onset cases and most eventually develop progressive neurologic/somatic growth impairment. BMT/cord blood transplants are not helpful in symptomatic cases, because these individuals cannot remyelinate.

The only treatment for early infantile KD is hematopoietic cell transplantation using umbilical cord blood. Hematopoietic stem cell transplantation (HSCT) significantly increases life span and ameliorates neurological outcome when performed before disease onset. Despite improvement in myelination after HSCT, children with KD still developed motor difficulty and cognitive/language deficits. Transplantation in late-onset KD may help if undertaken early.

Metachromatic leukodystrophy (MLD): MLD arises because of accumulation of sulfatide and other sulfated lipids in the white matter of the nervous system because of a deficiency in arylsulfatase A (ASA) enzyme. The pathologic hallmark is extensive demyelination in the CNS/PNS. Also, SAP B increases ASA activity severalfold. Defects in SAP B account for MLD with normal ASA activity. Multiple sulfatase deficiency has features of a leukodystrophy, ichthyosis, and a mucopolysaccaridosis. Late infantile MLD presents between 18 and 24 months. They have optic atrophy, cognitive decline and speech disturbances, become quadriparetic, develop seizures and do not survive the first decade. Mean age at death is 4.2 years. The juvenile variant can present between 4 and 6 years, or 6 and 16 years. Symptoms include difficulty in walking, ataxia, progressive spasticity, a peripheral neuropathy and slow cognitive decline. Presenting symptoms in the **adult variant** are psychiatric with slow mental decline. The course can be short or last decades. Patients become quadriparetic and may develop seizures. SAP B deficiency resembles MLD with normal ASA activities and is common in Arabs.

Multiple sulfatase deficiency (MSD): MSD is similar to late infantile MLD. There are white matter changes and a mucopolysaccaridosis (MPS) appearance. The common form is an infantile variant with mild MPS features, clear corneas, ichthyosis, optic atrophy, retinal degeneration and a cherry-red macula. The diagnosis is based on ASA enzyme activity, urine sulfatide measurement, sulfatide turnover in fibroblasts and molecular DNA analysis.

MLD, SAP B and multiple sulfatase deficiency variants are inherited as autosomal recessive traits. MLD is due to mutations in ASA. The defect in multiple sulfatase deficiency is unknown, but affects serine 69 in ASA, which normally becomes formylglycine and is necessary for catalytic function of sulfatases.

Understanding the genetics of ASA pseudodeficiency is crucial for interpretation of ASA test results. The pseudoallele codes for low enzyme activity (5–10% of normal) without clinical consequences. Patients with prosaposin precursor defects with SAP B deficiency present with an MLD-like disorder but with normal ASA.

No available treatment reverses the outcome and therapy is supportive. BMT can stop the progress of juvenile/adult MLD with benefit. Umbilical cord blood transplantation has stopped progression of the disease neurophysiologically and pathologically.

Fabry disease (FD): FD is an X-linked recessive lysosomal storage disorder with deposition of neutral glycosphingolipids (GSLs) due to defective α-galactosidase A. Clinical onset is in childhood/adolescence, but can be delayed. FD complications are left ventricular hypertrophy/conduction abnormalities/vascular spasms/proteinuria and renal insufficiency. Female carriers can display all manifestations. Excruciating burning pain starting at the palms and soles is termed acroparesthesia. Fever, exercise, fatigue, and stress trigger painful crises. Patients experience hypohidrosis/anhydrosis and fevers. A hallmark is development of angiokeratoma corporis diffusum in a bathing trunk distribution. Mitral valve insufficiency, left ventricular enlargement and faulty conduction system are described. Patients may present with angina/dyspnea/palpitations/syncope, and ultimately may die of heart failure. There is damage to distal interphalangeal joints with limitations, avascular necrosis of the head of femur/talus and involvement of metacarpal/metatarsal and temporomandibular joints. Low α-galactosidase A causes accumulation of neutral GSLs consisting of globotriaosylceramide/ceramide trihexoside in endothelial/smooth muscle cells, and epithelial and perithelial cells. Lysosomal α-galactosidase A deficiency causes FD. Affected males have normal plasma but deficient leukocyte α-galactosidase A. Activity in leukocytes/cultured fibroblasts is diagnostic without distinguishing affected males from carrier females. ERT should start early, some suggesting it should be given to all male patients to prevent organ damage. Kidney transplantation and dialysis are sometimes necessary. Two different forms of α-galactosidase A ERT are available, one made from human cell lines (agalsidase α, Replagal, Shire) and the other produced in a Chinese hamster ovary cell lines (agalsidase β, Fabrazyme, Genzyme). Both reduce plasma, urinary sediment and tissue GB3. Reports support a decrease in the frequency of pain crises, reduction in left ventricular mass, and stabilization of renal function, when started before tissue damage has occurred. Therapy is for a patient's entire lifetime. Statins are effective in ameliorating vasculopathic changes in FD.

Neuronal ceroid lipofuscinosis (NCL), Batten disease: NCLs are a group of heterogenous neurodegenerative diseases with an incidence of 1 in 12,500. Clinical features are visual loss, cognitive/motor deterioration, seizures and death. The NCLs comprise 10 variants termed CLN1–10, until CLN9 turned out to be a juvenile CLN5 variant. The clinical types are (1) congenital, (2) infantile, (3) classical late infantile, (4) variant late infantile/early juvenile, (5) classical juvenile (JNCL/CLN3), (6) epilepsy with mental retardation (EPMR)/CLN8, (7). Diagnosis is based on clinical course supplemented by neuroradiologic and electrophysiologic studies and established by measuring PPT1, TTP1, or cathepsin D (CTSD) activity in CLN1/CLN2 and CLN10 disease, respectively, or by DNA tests. Skin biopsy EM is a mainstay of diagnosis, especially for delineation of new variants. Prenatal diagnosis is possible by CVS enzyme testing, mutation analysis, or EM examination of cells/tissues. The term neuronal ceroid lipofuscinosis describes the autofluorescent, waxy, dusky lipid deposits seen in neuronal endosomes.

103.6 MAJOR BATTEN/NEURONAL CEROID LIPOFUSCINOSIS SYNDROMES

Infantile NCL (CLN1 disease, infantile variant, Haltia–Santavuori variant, palmitoyl-protein thioesterase, or PPT1-deficient): CLN1 disease and infantile variant is caused by defects in PPT1. Its function is to remove fatty acids attached in thioester linkages to cysteine residues in proteins. Most cases of CLN1 disease in the Finnish population have an infantile onset. Only 50% of CLN1 cases have an infantile onset in the United States. Development is normal until 10–18 months of age, with slowed head growth beginning at 5 months. Developmental stagnation occurs, fine motor skills are impaired, hypotonia, ataxia, microcephaly and blindness by age 2 years follows. Myoclonic jerks/generalized seizures and Rett-like knitting movements are observed but vanish by 2 years. At 3 years, children are nonambulatory, spastic, hypotonic, and irritable. Severe flexion contractures, acne, hirsutism, and precocious puberty develop, with death between 7 and 13 years. Onset can be delayed to 4 years or to adolescence. Measurement of PPT1 activity in leukocytes, from a dried blood spot or cultured fibroblasts less than 5% of normal, is diagnostic.

Late infantile neuronal ceroid lipofuscinosis (CLN2 disease, late infantile variant, Jansky–Bielchowski, tripeptidyl peptidase, or TPP1-deficient): CLN2 disease is caused by defects in TPP1. In the late infantile phenotype, the disease presents with seizures and ataxia between 2.5 and 3.5 years. Vision, motor and cognitive skills deteriorate. Patients are blind by age 4 years, and bedridden and mute by 5 years requiring tube feeding. Myoclonic jerks are prominent in the face, trunk and extremities. Spasticity leads to contractures. In CLN2 disease, juvenile phenotype, onset is delayed to 6–8 years. Cognitive decline is followed by seizures, ataxia and motor dysfunction. Blindness is variable with survival up to the fourth decade. Contractures are minimized with physical therapy. BMT has failed.

Variant late infantile forms (CLN5 disease; CLN6 disease; CLN8 disease including epilepsy with mental retardation-EPMR, and Turkish variant late infantile; CLN7 disease/MFSD8): Variant late infantile types with onset between 5 and 8 years similar to late infantile type, but with a protracted course are described. Four different genes are identified. *Finnish variant CLN5 disease* has a late infantile phenotype with initial symptoms clumsiness and difficulty concentrating at age 4.5 years. Cognitive decline at age 6 years and generalized/myoclonic epilepsy at age 8 years follow with blindness and loss of walking by age 10. Patients die between 14 and 34 years. **Early juvenile phenotype (CLN5 disease)**: There is decreased vision at age 4 years, cognitive decline at 6 and ataxia and rigidity by age 9 years. Dysarthria, scanning speech and mutism set in by age 12. **Northern epilepsy or epilepsy with mental retardation (CLN8 disease)**: The disease manifests with frequent, generalized tonic–clonic convulsions/complex partial seizures and cognitive decline. After puberty, slow movements, diminishing seizures, clumsiness, ataxia, and impaired vision set in. **Turkish CLN8 disease**: The clinical phenotype is more severe than in EPMR. Patients present between 2 and 5 years with seizures, intellectual decline and blindness. Behavioral problems are prominent by age 8–9 years. Most patients are wheelchair-bound by 10 years. **Costa-Rican/Portuguese/Lake-Cavanaugh variant (CLN6 disease)**: Onset is between 1.5 and 8 years with ataxia and speech difficulties. Retinitis pigmentosa, myoclonic jerks/other seizures and intellectual decline follow. Loss of motor skills occurs between 4 and 10 years. Patients die in early-mid teens. *CLN6* mutations are reported in Kufs type A. **Turkish late infantile NCL (CLN7 disease/MFSD8-deficient)**: Disease onset and progression are similar to CLN6 disease. The occurrence of seizure and motor difficulties before visual failure is peculiar to both CLN6 and CLN7 disease.

Juvenile neuronal ceroid lipofuscinosis/Batten Disease/CLN3 disease, Spielmeier–Vogt–Batten–Mayou: Most cases of JNCL have Northern European ancestry. It is the most common NCL in the United States. Early development is normal and the first symptom is decreased vision due to retinitis pigmentosa at 4–6 years of age. Patients are blind starting at age 10 years. Many cases manifest insomnia and difficult behavior between 7 and 9 years. By age 10, cognitive decline is apparent followed by seizures at 12–14 years. Cogwheel rigidity and a stooped, shuffling gait are reminiscent of Parkinson disease. Growth and physical maturity are normal and sexual development becomes problematic. Seizures increase to 150–200 a day and become impossible to control. Patients succumb in the early-mid twenties to uncontrollable seizures or cardiopulmonary arrest, with few surviving into the fourth decade. Vacuolated lymphocytes are a hallmark of juvenile CLN3 disease. Initial seizure control is achieved, but later there are over 100 seizures per day. Antipsychotics and mood stabilizers lower seizure threshold and aggravate parkinsonian symptoms. Death in the early-mid twenties is common with some dying early at 13 years and few surviving to age 40 years.

Kufs disease or adult NCL: Kufs disease represents 1.3% of the NCLs. Symptoms occur by age 30 years or earlier. Autosomal recessive and dominant inheritance patterns are described. Type A, the predominant form, is characterized by myoclonic epilepsy followed years later by dementia, ataxia, and pyramidal/extrapyramidal symptoms. Mutations in the *CLN6* gene are responsible for a number of cases of Kufs type A disease. In type B, onset is delayed beyond 50 years. Behavioral abnormalities and dementia are followed by ataxia/extrapyramidal/suprabulbar symptoms. Enzyme testing of leukocytes/fibroblasts may reveal deficient PPT1 activity and gene analysis reveals CLN1 mutations for few cases. The genetic defects underlying most cases are unknown.

Congenital NCL/CNCL or CLN10 disease: CNCL is the earliest and most aggressive form of NCL. Clinical features include congenital microcephalus with overriding sutures, receding forehead and low-set ears. Deceleration of head growth begins during the third trimester and jerky fetal movements are noted. Spasticity, status epilepticus and respiratory insufficiency occur soon after birth. Death follows in a few hours/days/weeks after birth. Another, more protracted form of the disease with a course resembling variant late infantile is described. CLN10 disease should be suspected when an infant presents with intractable seizures and microcephaly. The later protracted form presents with visual disturbances and ataxia. Mature CTSD is a ubiquitously expressed, only symptomatic treatment is available for patients with later onset CTSD deficiency.

CHAPTER

104

Peroxisomal Disorders

Ronald J A Wanders

Departments of Clinical Chemistry and Pediatrics, Laboratory Genetic Metabolic Diseases, Academic Medical Center, University of Amsterdam, Emma Children's Hospital, Amsterdam, The Netherlands

The group of peroxisomal disorders constitutes some 15 different inborn errors of metabolism in which there is an impairment in one or more peroxisomal functions. Zellweger syndrome (ZS) is the prototypic peroxisomal disorder, since ZS patients essentially lack peroxisomes due to a defect in peroxisome biogenesis disorder (PBD). Thorough investigations in body fluids and tissues of ZS patients have led to the identification of multiple metabolic aberrations including the accumulation of very long-chain fatty acids, pristanic acid, and di- and trihydroxycholestanoic acid, the deficiency of plasmalogens—a special class of phospholipids—and docosahexaenoic acid (DHA) plus the accumulation of phytanic acid, pipecolic acid, as well as glycolic and oxalic acid. All these metabolic abnormalities have been traced back to the actual metabolic functions of peroxisomes which include: (1) fatty acid beta-oxidation; (2) etherphospholipid biosynthesis; (3) fatty acid alpha-oxidation; and (4) glyoxylate detoxification. Importantly, these detailed studies in ZS patients have led to the development of a peroxisomal biomarker panel, which has subsequently been used to identify new peroxisomal disorders which now amount to some 15 different diseases.

The group of PDs is usually subdivided into two groups including the disorders of peroxisome biogenesis with ZS as prototype and the disorders of peroxisome function with X-linked adrenoleukodystrophy (X-ALD) as best known and most frequent representative.

Another major breakthrough in recent years has been the elucidation of the molecular basis of the different PDs notably the disorders of peroxisome biogenesis. Genetic complementation studies had already shown that the genetic basis of the PBDs was markedly heterogeneous with the involvement of at least 12 different genes. The identity of all these 12 genes has been worked out in recent years so that the true molecular defect can now be determined in any PBD patient with obvious consequences for prenatal diagnosis and genetic counseling.

The availability of highly sophisticated laboratory methods, including analysis of the peroxisome biomarker panel, has also led to the identification of new unexpected phenotypes in patients lacking most of the clinical features originally thought to be characteristic. In this respect, the discovery of isolated cerebellar ataxia in patients who turned out to have a defect in peroxisome biogenesis is worth mentioning. The recent introduction of whole exome sequencing and whole genome sequencing has also brought new peroxisomal phenotypes to light with Perrault syndrome as typical example. In these patients, the defect turned out to be at the level of D-bifunctional protein which in its classical presentation is associated with multiple severe abnormalities resembling ZS.

One aspect of peroxisome research which has not held up to its promises is the development of treatment strategies. This is true for all peroxisomal disorders with limited success only for Refsum disease (dietary treatment) and hyperoxaluria Type 1 (pyridoxine supplementation). Another positive exception is X-ALD in which allogeneic hematopoietic stem cell transplantation (HCT) has been shown to arrest or even reverse cerebral demyelination provided the procedure is performed at an early age. Human stem cell (HSC) therapy may very well become a true alternative for HCT, especially since two X-ALD patients have recently been successfully treated by *HSC* gene therapy.

TABLE 104-1 The Peroxisomal Disorders

Disorder	Abbreviation	MIM	Defective Mutant		
			Defective Protein	Mutant Gene	Locus
Disorders of peroxisome biogenesis	PBD				
• PBD-group A:					
Zellweger spectrum disorders	ZSD				
(1) Zellweger syndrome	ZS	214100	PEX1	*PEX1*	7q21.2
(2) Neonatal adrenoleukodystrophy	NALD	214110	PEX2	*PEX2*	8q21.1
(3) Infantile Refsum disease	IRD	202370	PEX3	*PEX3*	6q24.2
			PEX5	*PEX5*	12p13.3
			PEX6	*PEX6*	6p21.1
			PEX10	*PEX10*	1p36.32
			PEX12	*PEX12*	17q12
			PEX13	*PEX13*	2p14–p16
			PEX14	*PEX14*	1p36.22
			PEX16	*PEX16*	11p11.2
			PEX19	*PEX19*	1q22
			PEX26	*PEX26*	22q11.21
• PBD-group B:					
(4) Rhizomelic chondrodysplasia type 1	RCDP-1	215100	PEX7p	*PEX7*	6q21–q22.2
Disorders of peroxisome function	PFD				
• Fatty acid beta-oxidation					
(5) X-linked adrenoleukodystrophy	X-ALD	300100	ALDP	*ABCD1*	Xq28
(6) Acyl-CoA oxidase deficiency	ACOX-deficiency	264470	ACOX1	*ACOX1*	17q25.1
(7) D-Bifunctional protein deficiency	DBP deficiency	261515	DBP/MFP2/MFEII	*HSD17B4*	5q2
(8) Sterol-carrier-protein X deficiency	SCPx-deficiency	–	SCPx	*SCP2*	1p32
(9) 2-Methylacyl-CoA racemase deficiency	AMACR-deficiency	604489	AMACR	*AMACR*	5p13.2–q11.1
• Etherphospholipid biosynthesis					
(10) Rhizomelic chondrodysplasia puncatata type 2	RCDP-2	222765	DHAPAT	*GNPAT*	1q42.1–42.3
(11) Rhizomelic chondrodysplasia puncatata type 3	RCDP-3	600121	ADHAPS	*AGPS*	2q33
• Fatty acid alpha-oxidation					
(12) Refsum disease	ARD/CRD	266500	PHYH/PAHX	*PHYH/PAHX*	10p15–p14
• Glyoxylate metabolism					
(13) Hyperoxaluria type 1	PH-1	259900	AGT	*AGTX*	2q37.3
• Bile acid synthesis (conjugation)					
(14) Bile acid-CoA: amino acid N-acyltransferase deficiency	BAAT-deficiency		BAAT	*BAAT*	
• H_2O_2-metabolism					
(15) Acatalasemia		115500	Catalase	*CAT*	11p13

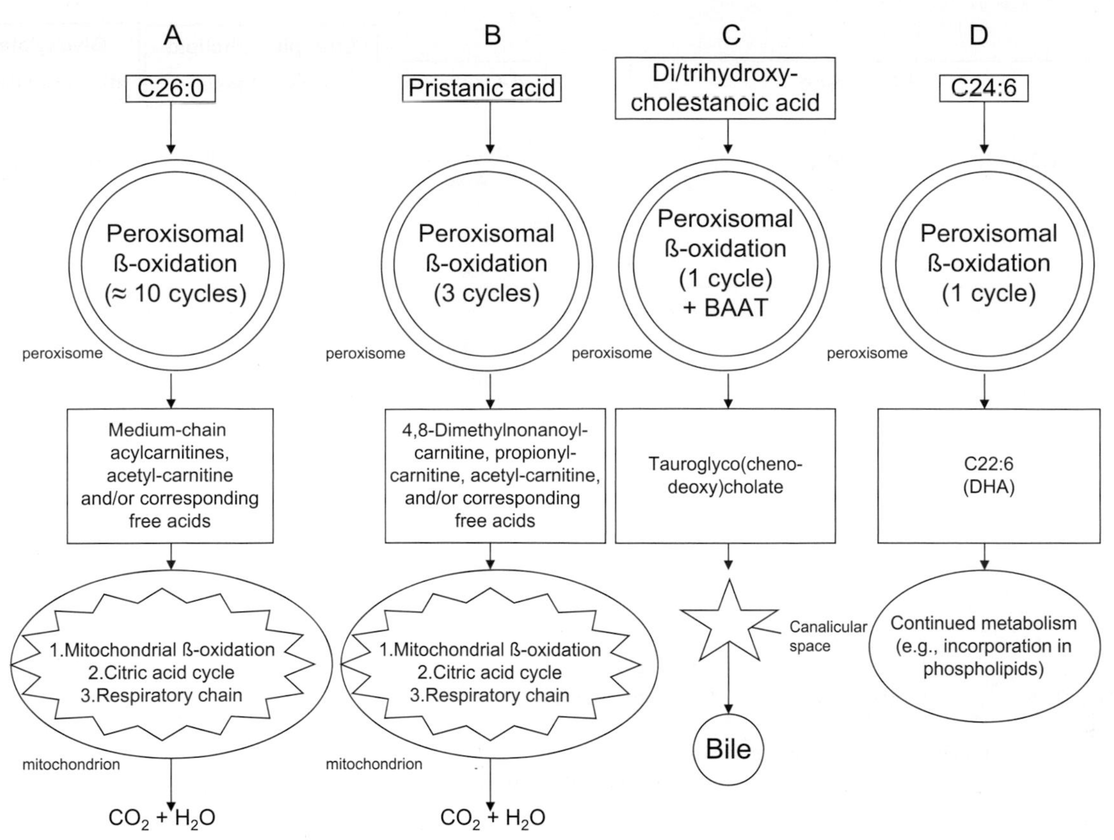

FIGURE 104-1 Schematic diagram depicting the fate of the main fatty acid substrates oxidized in peroxisomes which include (A) the very long-chain fatty acid C26:0 (hexacosanoic acid), (B) the branched-chain fatty acid pristanic acid (2,6,10,14-tetramethylpentadecanoic acid), (C) the bile acid synthesis intermediates di- and trihydroxycholestanoic acid (DHCA and THCA), and (D) the polyunsaturated fatty acid tetracosahexaenoic acid (C24:6n-3). See text for further details.

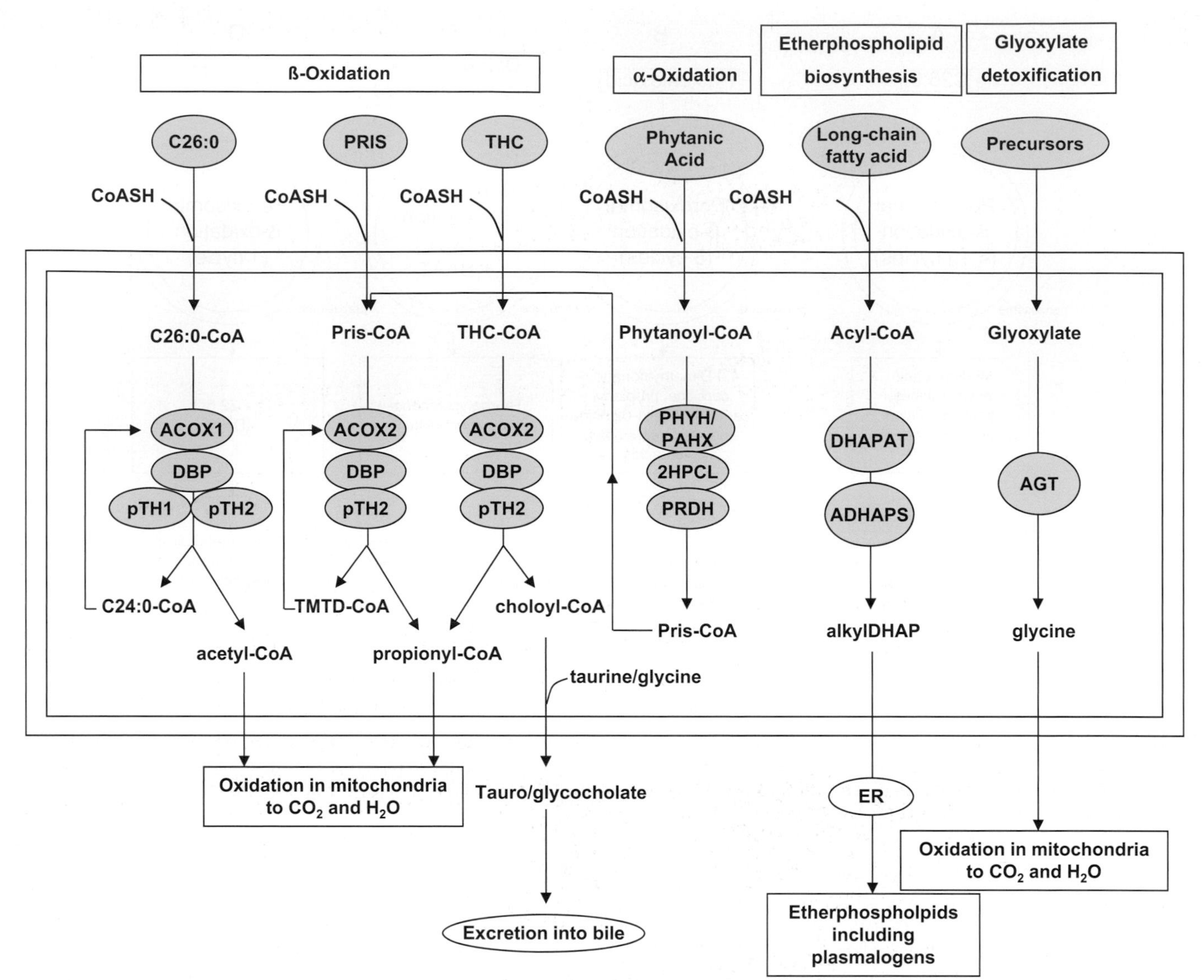

FIGURE 104-2 Schematic diagram showing the four major metabolic functions of peroxisomes which include: (1) fatty acid beta-oxidation; (2) fatty acid alpha-oxidation; (3) etherphospholipid biosynthesis; and (4) glyoxylate detoxification plus the individual peroxisomal enzymes involved and the end products of each of the four pathways in peroxisomes. See text for further details.

Mental and Behavioral Disorders

The Genetics of Personality, 415
Fragile X Syndrome and X-linked Intellectual Disability, 417
Dyslexia and Related Communication Disorders, 420
Attention-Deficit/Hyperactivity Disorder, 423
Autism Spectrum Disorders, 425
Genetics of Alzheimer Disease, 427
Schizophrenia and Affective Disorders, 429
Addictive Disorders, 431

CHAPTER 105

The Genetics of Personality

Matt McGue

Department of Psychology, University of Minnesota, Minneapolis, MN, USA
Department of Epidemiology, University of Southern Denmark, Odense, Denmark

Lindsay K Matteson

Department of Psychology, University of Minnesota, Minneapolis, MN, USA

Personality is one of psychology's core constructs. It has long been recognized that individuals stably differ in the characteristic ways they behave alone as well as interact with others. Personality is hypothesized to be the basis for these enduring aspects of our behavior. While personality is too broad to be comprehensively captured by a small number of traits, most personality researchers today recognize the core importance of five major dimensions of personality (i.e. the Five Factor Model): Neuroticism, Extraversion, Agreeableness, Conscientiousness, and Openness to Experience. Because differences in personality are predictive of differences in a large number of important life outcomes, including social achievements, physical health and mortality, and mental health, there is a broad array of researchers, including psychologists, psychiatrists, and social scientists, interested in understanding the nature and origins of individual differences in personality.

It was once thought that personality was largely shaped by socialization experiences, primarily those that take place within the rearing home. Twin and adoption studies have, however, convincingly documented the existence of genetic influences on personality. For example, twin studies of personality have consistently reported substantially greater similarity among monozygotic (MZ) than dizygotic (DZ) twins and heritability estimates for the Big Five as well as other dimensions of personality based on twin resemblance are generally about 50%. Nonetheless, parent–offspring correlations for personality are typically between 0.10 and 0.20, noticeably less than the expected half the heritability estimate, suggesting the existence of nonadditive genetic effects. Indeed biometric analysis of combined twin and family data suggest that as much as 20–30% of personality scale score variance may be attributed to nonadditive genetic effects.

Twin and adoption studies have also radically changed our understanding of how environments shape personality. It has been repeatedly shown that adopted relatives who grow up in the same home but who are not genetically identical to one another bear very little similarity in their personalities, challenging psychology's long-held belief that our personalities are largely shaped by the homes we grow up in. Alternatively, personality similarity between MZ twins, who are effectively genetically identical and reared together, is typically only about 0.50. Apparently, it is the experiences that are unique to each member of a twin pair (what behavioral geneticists call the non-shared environment), rather than those that they share by virtue of their common rearing, that constitute the major source of environmental influence on personality. Many candidates have been proposed as possible sources for the non-shared environmental influence on personality, including intrauterine factors, peer group characteristics, epigenetics, and even stochastic processes. Nonetheless, the specific environmental factors that shape personality remain largely unknown.

As was the case with many complex inherited human phenotypes, human genetic research on the origins of personality from the end of the twentieth century through the beginning of the twenty first century was dominated by candidate-gene studies. These studies were loosely based on neurobiological models of personality and focused primarily on polymorphisms (usually functional) in a small set of genes involved in neurotransmission. Although there were several initial promising reports of genetic association, few of these associations have held up over time. We now know that initial reports likely overstate the evidence supporting the existence of an association, a phenomenon that has come to be known as the "winner's curse." To correct for the bias inherent to the "winner's curse," geneticists have increasingly turned to meta-analysis to effectively pool findings from multiple studies. As applied to studies of personality, meta-analysis has failed to confirm some initially prominent associations (e.g. a variable number of tandem repeat polymorphism in DRD4 and approach-related behavior) while supporting the existence of others (e.g. the association between a polymorphism in the promoter region of the serotonin transporter gene and neuroticism). Significantly, even when supported by meta-analysis, the estimated strength of the association is modest.

There have been a small number of genome-wide association studies (GWAS) of personality. To date, no association between a single nucleotide polymorphism (SNP) and a personality measure has met the GWAS significance threshold of $p < 5 \times 10^{-8}$. The failure of GWAS to identify significant associations is likely due to the small effect any single SNP will have on individual differences in personality. In principle, "success" may only be a matter of achieving larger GWAS sample sizes. Alternatively, others have investigated the aggregate effect of multiple variants by combining individual SNPs into polygenic scores. These scores have been found to be significantly associated with personality, although large samples will also be needed here to properly weight the contribution of individual SNPs to the aggregate and maximize predictability.

Personality does not conform to the standard biometric assumption of independence and additivity of genetic and environmental effects. There is both evidence of gene–environment correlation and gene–environment interaction. Gene–environment correlation arises because individuals who inherit a specific dispositional tendency (e.g. towards extraversion) are likely to create environments that reinforce and amplify that tendency (e.g. by participating in social gatherings). Gene–environment interaction arises because genetic effects on personality can either be dampened or amplified by the environment (e.g. genetic effects on aggression may be minimized in a psychologically protective environment). The complex interplay of genetic and environmental effects in its development makes personality a challenging phenotype for genetic analysis, but it also makes it a prototype for understanding how genetics contribute to the diversity of our behavior.

CHAPTER

106

Fragile X Syndrome and X-Linked Intellectual Disability

Kathryn B Garber, Stephen T Warren, and Jeannie Visootsak

Department of Human Genetics, Emory University School of Medicine, Atlanta, GA, USA

Intellectual disability (ID), formerly known as mental retardation, is a broad diagnostic category with a variety of underlying etiologies that include both environmental and genetic factors. Disruption of many different genes across the genome can impede normal intellectual development, but the X chromosome has an overabundance of known *ID* genes, such that X-linked ID (XLID) is estimated to account for 10–12% of ID cases in males. Because males have a single X chromosome, they are more sensitive to mutations that reduce the expression or activity of genes encoded therein. This means that the abundance of *ID* genes on the X chromosome is at least partly responsible for the ~1.4-fold excess of males affected with ID compared to females. Mutations in many different genes have been identified in individuals with both syndromic and nonsyndromic forms of XLID. The most common form of XLID is Fragile X Syndrome (FXS). Besides this, each *XLID* gene individually accounts for a small fraction of ID overall. In this chapter, we will summarize what is broadly known about the genes that contribute to XLID with particular emphasis on the FXS.

106.1 FRAGILE X SYNDROME

With an estimated 1 in 4000 males affected, FXS is the most common inherited form of ID. Approximately 85% males and 25–30% females with FXS will have an intelligence quotient (IQ) below 70, which falls in the ID range. Although originally considered to be a nonsyndromic form of ID because its associated physical characteristics are relatively subtle, affected males tend to have a long narrow face, prominent ears, and joint hypermobility; macro-orchidism is present from puberty. Behavioral and psychological issues are prominent and wide-ranging in individuals with FXS. These features are typically grouped into the following symptom clusters: (1) Attention deficit/hyperactivity disorder (ADHD)-like symptoms of hyperactivity, distractibility, impulsivity, and overarousal; (2) anxiety-related symptoms, including obsessive compulsive disorder (OCD)-like and perseverative behaviors; (3) emotional lability; and (4) aggressive and self-aggressive behavior.

FXS is the most common single gene cause of autism, resulting in 2–6% of all cases of autism. Based on the Autism Diagnostic Observation Scale and the Autism Diagnostic Interview, ~30% of males with FXS have autism spectrum disorder (ASD) and another 20–30% have pervasive developmental disorder, not otherwise specified (PDD-NOS). Even those children with FXS who do not meet the criteria for ASD diagnosis usually have one or more autistic characteristics, such as hand flapping, poor eye contact, repetitive behaviors and language, and tactile defensiveness.

Females with FXS typically have milder behavioral and cognitive problems than males with FXS because they have one normal copy of the relevant gene. Approximately 25% of females with FXS have an IQ < 70; the remaining have a high propensity for learning and/or behavioral issues including depression, withdrawal, social anxiety, and executive dysfunction.

In the vast majority of cases, FXS is caused by a CGG trinucleotide expansion in the 5′ untranslated region of *FMR1*. This leads to hypermethylation and transcriptional silencing of the gene, which normally encodes the Fragile X Mental Retardation Protein, FMRP. Expanded *FMR1* alleles arise from meiotic instability of so-called premutation alleles, which have a higher number of repeated units than those found in the general population. There are four allelic classes for the *FMR1* trinucleotide repeat, and these categories are based on observations of allele instability in affected families.

They are as follows: normal alleles, which range in size up to 40 copies; intermediate alleles, which range in size from 45 to 54 repeats and may exhibit some instability; premutation alleles, which can expand to a full mutation and range from 55 to 200 repeats; and full mutations, which are >200 repeats in length and are hypermethylated. Premutation alleles of *FMR1* are unmethylated alleles that are transcribed and translated. When passed from females, they can also undergo massive expansion into the full mutation range in a single generation, and the likelihood with which this occurs is governed by the size of the premutation allele. This gives rise to a characteristic inheritance pattern in families in which the likelihood that FXS will be penetrant in a family increases down through the generations.

Individuals who carry a premutation allele at *FMR1* do not have FXS, but they are at risk of other *FMR1*-related disorders. The best documented of these are Fragile X Tremor Ataxia syndrome and Fragile X-related Primary Ovarian Insufficiency/Premature Ovarian Failure. Whereas FXS is due to the lack of FMRP, the premutation-associated disorders result from the accumulation of CGG-repeat containing *FMR1* transcripts.

Genetic testing for FXS and for *FMR1* premutations involves an assessment of the size and methylation status of the CGG-repeat tract in the 5′ UTR of *FMR1*. Due to the fairly high prevalence of FXS and the subtlety of the syndromic features, it is recommended that genetic testing for FXS be included as a first line genetic test for all individuals with developmental delays, learning issues, ID and/or ASD. This is estimated to have a diagnostic yield of ~1–2%. Females with a history of infertility should also be screened for premutations in *FMR1*.

The current psychopharmacology treatment in FXS is symptom-based combined with supportive strategies, which include speech therapy, occupational therapy, behavioral interventions, and a specialized educational program. This includes treatment aimed at managing ADHD-like symptoms, anxiety, aggression, and mood instability. Treatments aimed specifically at the pathways disrupted in FXS are currently in clinical trials.

106.2 OTHER FORMS OF XLID

Nearly 100 of more than 800 annotated protein-coding genes on the X chromosome have been implicated in XLID, and they are estimated to be as many as-yet-unidentified putative *XLID* genes. Other than *FMR1*, each accounts for only a small fraction of cases of ID. A thorough evaluation must be undertaken before XLID is considered as a likely diagnosis, and this diagnosis is typically made in males once other causes of ID have been excluded. A recommended first tier of laboratory testing for developmental delay includes fragile X testing and chromosomal microarray analysis (whole genome array-comparative genomic hybridization—array CGH), which has supplanted karyotyping in the cytogenetic analysis of individuals with developmental delay due to its increased diagnostic yield. Provided no diagnosis is achieved following this evaluation, the family history must be evaluated with particular attention to individuals with learning problems, autism, psychiatric problems, or ID itself because they could share an underlying genetic etiology and could suggest a diagnosis. Although in classic XLID pedigrees, one would expect multiple-affected males who are related through unaffected females, it may not be possible to document an extensive family history in every family, nor are families always large enough to observe this pattern. Thus, XLID should be considered in males with ID of unknown etiology, and it should not be ruled out in females with ID.

Due to the large number of *XLID* genes and the lack of a defined phenotype associated with many of them, individual gene testing for XLID has not been used widely in the genetic evaluation of ID, but this is becoming increasingly common as we gain a more complete understanding of the *XLID* genes and as DNA sequencing costs decrease. Because many of the *XLID* genes have been discovered and evaluated through large consortia that required more than one affected male in a family for inclusion in the study, the likelihood of identifying a causative mutation in sporadic cases of ID is going to be lower than reported in the literature. Although our ability to identify the causative mutations for ID is increasing rapidly, they will remain undetected in a significant fraction of patients, even with extensive sequence and deletion/duplication analysis. The high prevalence of nonrecurring, private genetic variation on the X chromosome in families with ID complicates interpretation of sequence variation, even when the variation is truncating or de novo. However, we should not downplay the fact that we can uncover the relevant genetic change in an increasing number of cases of ID, both inherited and sporadic, and that sequence and deletion/duplication analysis of a large number of *XLID* genes is available on clinical testing panels. The more we use these XLID-testing panels, the better our interpretion of this genetic variation will be.

FURTHER READING

1. Bassell, G. J.; Warren, S. T. Fragile X Syndrome: Loss of Local mRNA Regulation Alters Synaptic Development and Function. *Neuron* 2008, *60,* 201–214.
2. Bauters, M.; Weuts, A.; Vandewalle, J.; Nevelsteen, J.; Marynen, P.; Van Esch, H.; Froyen, G. Detection and Validation of Copy Number Variation in X-Linked Mental Retardation. *Cytogenet. Genome Res.* 2008, *123,* 44–53.
3. de Brouwer, A. P. M.; Yntema, H. G.; Kleefstra, T.; Lugtenberg, D.; Oudakker, A. R.; de Vries, B. B. A., et al. Mutation Frequencies of X-Linked Mental Retardation Genes in Families from the EuroMRX Consortium. *Hum. Mutat.* 200710.1002/humu.9482, [online] Mutation in Brief #953.

4. Garber, K. B.; Visootsak, J.; Warren, S. T. Fragile X Syndrome. *Eur. J. Hum. Genet.* **2008**, *16*, 666–672.
5. Kaufman, L.; Ayub, M.; Vincent, J. B. The Genetic Basis of Non-syndromic Intellectual Disability: A Review. *J. Neurodev. Disord.* **2010**, *2*, 182–209.
6. Penagarikano, O.; Mulle, J. G.; Warren, S. T. The Pathophysiology of Fragile X Syndrome. *Annu. Rev. Genomics Hum. Genet.* **2008**, *8*, 109–129.
7. Ropers, H. H. Genetics of Intellectual Disability. *Curr. Opin. Genet. Dev.* **2008**, *18*, 241–250.
8. Shen, Y.; Wu, B.-L.; Gusella, J. F. Large-Scale Medical Resequencing for X-Linked Mental Retardation. *Clin. Chem.* **2010**, *56*, 339–341.
9. Stevenson, R. E.; Schwartz, C. E. X-Linked Intellectual Disability: Unique Vulnerability of the Male Genome. *Dev. Disabil. Res. Rev.* **2009**, *15*, 361–368.
10. Tarpey, P. S.; Smith, R.; Pleasance, E.; Whibley, A.; Edkins, S.; Hardy, C., et al. A Systematic, Large-Scale Resequencing Screen of X-Chromosome Coding Exons in Mental Retardation. *Nat. Genet.* **2009**, *41*, 535–543.

CHAPTER

107

Dyslexia and Related Communication Disorders

Angela Friend and Bruce F Pennington

University of Denver, Denver, CO, USA

Shelley D Smith

University of Nebraska Medical Center, Nebraska Medical Center, Omaha, NE, USA

Jeffrey W Gilger

Purdue University, West Lafayette, IN, USA

Specific reading disability is an unexpected difficulty learning to read (and/or spell) despite adequate intelligence and opportunity and without demonstrable sensory, psychiatric, or neurologic factors that could explain the disorder. This definition, like that for many learning disorders, is exclusionary in that other causes of reading problems have been ruled out. Overall, typical estimates of prevalence of dyslexia in school children in third grade or beyond have been 5–10%. If a broader definition of poor reading is used that does not require a significant discrepancy with nonreading abilities, the prevalence may be as high as 20–30%; in contrast, populations in which the written language is more orthographically transparent, such as Italian, the frequency of diagnosis of dyslexia is lower. Specific reading disability (SRD) is persistent, and while those with milder cases may compensate as adults, it is rare for even compensated adults to fully compensate in terms of reading fluency or underlying phonological problems.

The familial nature of reading disability was noted around the turn of the century, and numerous reports of multiplex families followed (e.g. Reference; see reviews by Fisher and DeFries, 2002; DeFries and Gillis, 1993; DeFries and Alarcón, 1996; and Schulte-Korne, 2001). The majority of genetic studies have focused on the first loci that were reported to show linkage, 15q and 6p. These localizations have been refined to the regions 15q21 and 6p22.2, and linkage has been confirmed by separate, independent studies. Additional studies, including several full genome screens, have suggested linkage for at least a dozen loci (Table 107-1), and five of these loci have also been replicated in independent samples (e.g. 1q36, 2p16-15, 3p12-q13, 18p11, and possibly Xq27.3). Gene designations have been given to the seven loci for which linkage has been replicated: DYX1, 15q21; DYX2, 6p22.2; DYX3, 2p16-p15; DYX5, 3p12-q13; DYX6, 18p11.2; DYX8, 1p36; and DYX9, Xp27.3. DYX4 and DYX7 had been assigned to separate loci on chromosome 15, but currently these are subsumed under DYX1. DYX4 has also been applied unofficially to the 6q13-q16.2 locus. The evidence so far supports the idea that both normal reading and SRD can be influenced by genetic factors. Segregation analyses support the hypothesis that a limited number of loci are involved, with at least some having a major gene effect, and linkage analyses have identified at least seven candidate loci. Some of the ambiguity in results or the lack of findings may reflect the fact that both segregation and linkage analysis are dependent on the definition of the phenotype. As the phenotypes are more closely defined and new techniques for analysis of quantitative trait loci are developed, the mode of inheritance and the role of individual loci should become clearer.

SRD is frequently comorbid with other language disabilities, including specific language impairment (SLI), and speech sound disorder (phonologic disorder). For example, familial segregation has been demonstrated in 20–40% of first-degree relatives of children with speech sound disorder (SSD). Moreover, segregation

TABLE 107-1 Gene Localization Studies with Reading Disability

Region	Supportive Results	Negative Results
	Population Reference	Reference Population
1p36-p34 (DYX8)	9 extended families (US)	Rabin et al., 1993
	8 extended families (US)	Grigorenko et al., 2001
	100 families (Canada)	Tzenova et al., 2004
	291 families (Canada)	Couto et al., 2008
2p16-p15 (DYX3)	1 extended family (Norway)	Fagerheim et al., 1999
	111 nuclear families (US)	Chapman et al.,2004
	89 nuclear families (UK),	Fisher et al., 2002
	119 nuclear families (US)	Francks et al., 2002
	96 families (Canada)	Petryshen et al., 2002
2p12-q13 (DYX3)	11 families (Finland)	Anthoni et al., 2007
	251 families (Germany)	Anthoni et al., 2007
2p11	11 families (Finland)	Kaminen et al., 2003
	11 families (Finland)	Peyrard-Janvid et al., 2004
3p12-q13 (DYX5)	1 extended family (Finland)	Nopola-Hemmi et al., 2001
	89 nuclear families (UK)	Fisher et al., 2002
	119 nuclear families (US)	
6p22.2 (DYX2)	18 extended families (US)	Smith et al., 1991
	79 families (Canada)	Field and Kaplan, 1998
	18 families; 41 sib-pairs (US)	Cardon et al., 1994, 1995
	7 families (Germany)	Nöthen et al., 1999
	6 families (US)	Grigorenko et al., 1997
	79 families (Canada)	Petryshen et al., 2000
	82 families (UK)	Fisher et al., 1999
	111 nuclear families (US)	Chapman et al., 2004
	79 nuclear families (US)	Gayán et al., 1999
	89 nuclear families (UK),	Fisher et al., 2002
	8 families (US)	Grigorenko et al., 2000
	119 nuclear families (US)	
	104 nuclear families (US)	Kaplan et al., 2002
	178 parent/proband trios (Wales)	Turic et al., 2003
	89 families (UK)	Marlow et al., 2003
	349 nuclear families (US)	Deffenbacher et al., 2004
	264 nuclear families (UK),	Francks et al., 2004
	159 nuclear families (US)	
6q13-q16.2 (DYX4)	96 families (Canada)	Petryshen et al., 2001
7q32	11 families (Finland)	Kaminen et al., 2003
11p15.5 (DYX7)	100 families (Canada)	Hsiung et al., 2004
12p13.3	200 children with SRD (US)	Roeske et al., 2009
	186 children with SRD (US)	
13q21	5 families. Specific language impairment (Canada)	Bartlett et al., 2002
5q21 (DYX1C1)	9 extended families (US)	Smith et al., 1983
	9 extended families (US)	Rabin et al., 1993 (includes Smith et al., 1983)
	9 extended families	Smith et al., 1991
		Smith et al., 1990,
	19 extended families	Fulker et al., 1991
		Smith et al., 1990,
	6 extended families (US)	Grigorenko et al., 1997
	7 extended families (Germany)	Nöthen et al., 1999
	*100 parent/proband trios (UK)	Morris et al., 2000
	2 families (Finland)	Nopola-Hemmi et al., 2000
	1 family with translocation;	Taipale et al., 2003
	23 families and 33 singletons	
	111 nuclear families (US)	Chapman et al., 2004
18p11.2 (DYX6)	89 nuclear families (UK)	Fisher et al., 2002
	111 nuclear families (US)	Chapman et al., 2004
	119 nuclear families (US),	
	84 nuclear families (UK)	
Xq27.3 (DYX9)	89 nuclear families (UK),	Fisher et al., 2002
	119 nuclear families (US)	
	3 extended families, 67 male sib pairs (Holland)	de Kovel et al., 2004

analyses have also confirmed familial aggregation of SSD and supported both a major gene and polygenic modes of transmission of the disorder. Results suggest that shared developmental deficits in phonological representation or retrieval affect both disorders. Like SRD, SSD also contributes to adverse outcomes in terms of personal achievement as measured by education, social, and occupational status. There are several major hypotheses regarding the causes of SLI, and some show overlap with the hypothesized causes of SRD. However, linkage studies suggest that the etiologic overlap that exists may be small. For example, a genome screen of SLI phenotypes did not show localization to any known SRD loci, but further studies of a candidate locus on 16q that was identified in this screen showed that SRD phenotypes also linked to this region. In another study of families selected for SLI, Bartlett and colleagues found linkage of SRD phenotypes (but not SLI phenotypes) to chromosome 13q21. Neither of these regions was identified in genome screens of SRD.

Literacy research has identified the basic cognitive component areas or processes needed for the development of reading skill. These processes or abilities include phonemic awareness, phonics, fluency, vocabulary, and comprehension strategies, each of which is essential for the accurate and fluent identification of words and the ultimate construction of meaning (comprehension). Effective identification and remediation strategies based on literacy research focus on these basic cognitive component areas.

WEBSITES

1. www.ASHA.org.
2. www.whatworks.ed.gov.
3. www.interdys.org.

CHAPTER

108

Attention-Deficit/Hyperactivity Disorder

Stephen V Faraone

Medical Genetics Research Center and Department of Psychiatry and Behavioral Sciences, SUNY Upstate Medical University, Syracuse, NY, USA

Alysa E Doyle

The Center for Human Genetics Research and Department of Psychiatry, Massachusetts General Hospital and Harvard Medical School, Boston, MA, USA

Attention-deficit/hyperactivity disorder (ADHD) is a common childhood-onset behavioral disorder characterized by excessive inattention and/or hyperactivity and impulsivity (*1*). The condition has a considerable negative public health impact; impairing symptoms are associated with conduct problems, substance abuse and academic and career underachievement that can last into adulthood (*2*).

108.1 BEHAVIORAL GENETICS

Family, twin and adoption studies based on early and current versions of the diagnostic criteria are remarkably consistent in suggesting that ADHD runs in families and that familial resemblance is due to the effects of genes. Summaries of twin studies place heritability at .76, suggesting that ADHD is one of the most heritable psychiatric conditions (*3*). Although there is considerable genetic overlap between its symptom dimensions of inattention and hyperactivity/impulsivity, several, large studies indicate nonoverlapping genetic influences on these two dimensions as well (*4*). Twin studies also implicate environmental factors, particularly factors unique to individuals rather than those shared among family members. To date, the most compelling of these include low birth weight, maternal smoking and alcohol use during pregnancy. Behavioral genetic studies also provide information about the relationship between ADHD and other conditions. For example, ADHD that co-occurs with bipolar disorders may be a distinct familial subtype (*5*). Additionally, there appears to be a heritable liability for externalizing psychopathology that cuts across diagnoses of ADHD, conduct disorder (CD), and oppositional defiant disorder (*6*). Also, there appear to be mutual genetic influences on ADHD inattention symptoms and measures of reading (*7*).

108.2 MOLECULAR GENETICS

Biological candidate genes for ADHD predominantly relate to the catecholamine and serotonin systems. Meta-analyses have shown small but statistically significant pooled odds ratios, ranging from 1.1 to 1.5, for the dopamine (DA) transporter (*SLC6A3*), the DA 4 receptor gene (*DRD4*), synaptosomal-associated protein 25 (*SNAP-25*), the DA 5 receptor gene (*DRD5*), the gene coding for dopamine beta-hydroxylase (*DBH*), the serotonin transporter (*5HTT*) and the serotonin receptor 1B (*HTR1B*) (*3*).

Results of genome-wide linkage and genome-wide association studies (GWAS) indicate genes of moderately large effect are unlikely to exist in ADHD. Genome-wide linkage scans have not yielded replicated genome-wide significant findings using strict criteria; however, regions highlighted by multiple studies, including 5p13, 17p11 and 16q23, are of considerable interest for further investigation given that these studies individually had low power to detect linkage to genes of small effects. A genome scan meta-analysis of published linkage data reported genome-wide significant linkage at 16q23.1 that appeared to result from a contribution, though in some cases modest, from the majority of the seven studies examined (*8*).

Individual GWAS have not achieved genome-wide significant results ($P < 5 \times 10^{-8}$). Exploratory analyses within GWAS data sets are also noteworthy because they highlight biological processes with a potential role in ADHD. Among the most interesting are those relevant to cellular adhesion and synaptic plasticity. For example, cadherin 13 (CDH13), which resides in the 16p23 region highlighted in linkage studies, was implicated in a pooled GWAS analysis of ADHD adults and controls (*9*) and trait-specific analyses in the International Multi Center ADHD Genetics project (*10*). A meta-analysis combining

GWAS in some promising findings on chromosomes 7 and 8, though no results surpassed genome-wide criteria for significance (*11*). As the authors note, this meta-analysis is the largest ADHD sample analyzed to date; yet, given sample sizes needed to achieve genome-wide significance for other complex psychiatric disorders, including schizophrenia and bipolar disorder, larger numbers are needed to achieve power to detect variants of the full range of expected effect sizes.

Given evidence that large (>500 kb) rare structural deletions or duplications of DNA substantially increase risk for other psychiatric and neurodevelopmental disorders, two recent studies have investigated the role of these copy number variants (CNVs) in ADHD. Although the first study was largely negative, a team from Cardiff (*12*) identified an excess of rare deletions and duplications >500 kb in length in their ADHD sample (14%) versus controls (7%). Although ADHD youth with intellectual disability (ID) had a particularly high rate of these rare variants, an excess of large CNVs was still found in ADHD youth without ID compared to controls. Within this restricted group, an excess of a particular schizophrenia-related duplication (16p13.11) was found, and this finding replicated in an Icelandic sample. There are multiple genes of interest within this 16p13.11 region, with nuclear distribution gene E homologue 1 (*NDE1*) being of particular prior interest for schizophrenia because of its role in neurodevelopment and binding with the disrupted in schizophrenia 1 (*DISC1*) gene.

Given the complexity of the ADHD phenotype, future studies will benefit from the use of large samples as well as efforts to target more homogenous and/or familial forms of the phenotype, including ADHD that is comorbid with CD or BPD. Additionally, "endophenotypes" that lie in the pathway from genes to behavior provide a way of targeting a phenotype that may be closer to gene products and less complex than the disorder as a whole. Although considerably more work is needed to clarify their utility in molecular genetic analyses, endophenotypes based on neurocognitive and neuroimaging data appear promising (*13*).

108.3 CONCLUSIONS

As with other neuropsychiatric disorders, the majority of genetic variance in the disorder remains unexplained. Yet field of ADHD genetics is poised for considerable advances in the coming years due to recent leads combined with large collaborative efforts that are currently underway and new opportunities for sequencing to further understand the role of rare variants.

REFERENCES

1. *Diagnostic and Statistical Manual of Mental Disorders*, 4th ed.; American Psychiatric Association: Washington D.C, 2000.
2. Biederman, J.; Monuteaux, M. C.; Mick, E.; Spencer, T.; Wilens, T. E.; Silva, J. M., et al. Young Adult Outcome of Attention Deficit Hyperactivity Disorder: A Controlled 10-year Follow-Up Study. *Psychol. Med.* **2006 Feb,** *36* (2), 167–179.
3. Faraone, S. V.; Perlis, R. H.; Doyle, A. E.; Smoller, J. W.; Goralnick, J. J.; Holmgren, M. A., et al. Molecular Genetics of Attention-Deficit/Hyperactivity Disorder. *Biol. Psychiatry* **2005 Jun 1,** *57* (11), 1313–1323.
4. Greven, C. U.; Rijsdijk, F. V.; Plomin, R. A Twin Study of ADHD Symptoms in Early Adolescence: Hyperactivity-Impulsivity and Inattentiveness Show Substantial Genetic Overlap But Also Genetic Specificity. *J. Abnorm. Child Psychol.* **2011 Feb,** *39* (2), 265–275.
5. Doyle, A. E.; Faraone, S. V. Familial Links Between Attention Deficit Hyperactivity Disorder, Conduct Disorder, and Bipolar Disorder. *Curr. Psychiatry Rep.* **2002 Apr,** *4* (2), 146–152.
6. Tuvblad, C.; Zheng, M.; Raine, A.; Baker, L. A. A Common Genetic Factor Explains the Covariation Among ADHD ODD and CD Symptoms in 9–10 Year Old Boys and Girls. *J. Abnorm. Child Psychol.* **2009 Feb,** *37* (2), 153–167.
7. Paloyelis, Y.; Rijsdijk, F.; Wood, A. C.; Asherson, P.; Kuntsi, J. The Genetic Association between ADHD Symptoms and Reading Difficulties: The Role of Inattentiveness and IQ. *J. Abnorm. Child Psychol.* **2010 Nov,** *38* (8), 1083–1095.
8. Zhou, K.; Dempfle, A.; Arcos-Burgos, M.; Bakker, S. C.; Banaschewski, T.; Biederman, J., et al. Meta-Analysis of Genome-Wide Linkage Scans of Attention Deficit Hyperactivity Disorder. *Am. J. Med. Genet. B Neuropsychiatr. Genet.* **2008 Dec 5,** *147B* (8), 1392–1398.
9. Lesch, K. P.; Timmesfeld, N.; Renner, T. J.; Halperin, R.; Roser, C.; Nguyen, T. T., et al. Molecular Genetics of Adult ADHD: Converging Evidence from Genome-Wide Association and Extended Pedigree Linkage Studies. *J. Neural. Transm.* **2008 Nov,** *115* (11), 1573–1585.
10. Lasky-Su, J.; Neale, B. M.; Franke, B.; Anney, R. J.; Zhou, K.; Maller, J. B., et al. Genome-Wide Association Scan of Quantitative Traits for Attention Deficit Hyperactivity Disorder Identifies Novel Associations and Confirms Candidate Gene Associations. *Am. J. Med. Genet. B Neuropsychiatr. Genet.* **2008 Dec 5,** *147B* (8), 1345–1354.
11. Neale, B. M.; Medland, S. E.; Ripke, S.; Asherson, P.; Franke, B.; Lesch, K. P., et al. Meta-Analysis of Genome-Wide Association Studies of Attention-Deficit/Hyperactivity Disorder. *J. Am. Acad. Child Adolesc. Psychiatry* **2010 Sep,** *49* (9), 884–897.
12. Williams, N. M.; Zaharieva, I.; Martin, A.; Langley, K.; Mantripragada, K.; Fossdal, R., et al. Rare Chromosomal Deletions and Duplications in Attention-Deficit Hyperactivity Disorder: A Genome-Wide Analysis. *Lancet* **2010 Oct 23,** *376* (9750), 1401–1408.
13. Doyle, A. E.; Willcutt, E. G.; Seidman, L. J.; Biederman, J.; Chouinard, V. A.; Silva, J., et al. Attention-Deficit/Hyperactivity Disorder Endophenotypes. *Biol. Psychiatry* **2005 Jun 1,** *57* (11), 1324–1335.

CHAPTER

109

Autism Spectrum Disorders

Sunil Q Mehta

Department of Psychiatry and Biobehavioral Sciences, UCLA Semel Institute for Neuroscience and Human Behavior, David Geffen School of Medicine, University of California at Los Angeles, CA, USA

Daniel H Geschwind

Department of Neurology; Department of Psychiatry and Center for Autism Research, UCLA Semel Institute for Neuroscience and Human Behavior; Department of Human Genetics, David Geffen School of Medicine, University of California at Los Angeles, CA, USA

Autism spectrum disorders (ASDs) encompass a range of illnesses that share deficits in three core domains: social interaction, restricted interests/repetitive behaviors, and language (*1*). In the current clinical conception, there are three major categories of ASD, autistic disorder, Asperger syndrome, and pervasive developmental disorder not otherwise specified. Current estimates about the prevalence of ASDs vary from 0.6% to 0.9%, but there is growing consensus that at least part of the increase in prevalence during the last four decades is due to a standardization of criteria and increased referral to specialized evaluators on the part of families, schools, and primary care physicians (*2*). In addition to behavioral symptoms, individuals with ASDs can have variable impairments in sensory, motor, and medical domains, including epilepsy (see (*3*) for review). Classically, it was thought that 60–70% of individuals with autistic disorder have intellectual disability, but more recent estimates range between 34% and 55%.

Some of the most important evidences supporting a major role for genes in ASDs come from twin studies (reviewed in (*4*)). These studies suggest that the contribution of genetic factors to risk for ASDs is significant and consistent with a heritability of 0.7. The most recent large scale, multi-site prospective study of recurrence risk in siblings found it to be about 20% for autism (*5*), similar to the risk observed in DZ twins. There is also evidence that environment can play a role in the occurrence of ASDs. In the 1960s, an association was found between maternal viral infection during pregnancy (particularly rubella) and autism (*6*). Prenatal exposure to the anti-epileptic drug sodium valproate also significantly increased the risk of developing an ASD (*7*). If we assume the general population prevalence for ASDs to be about 0.8%, we can calculate the rough odds ratios for prenatal rubella infection or valproate exposure leading to ASDs, as at least 9.9 and 6.0, respectively, which is higher than common genetic risk factors.

Microarray studies have showed that de novo nonrecurrent CNV caused about 5–10% of ASDs. Although advances in array technology have been steadily increasing the resolution of rearrangements that can be identified (reviewed in (*8*)), the proportion of ASD that can be explained by large (>250 kb) de novo CNV remains between 5% and 10%. The identification of these syndromes is important from a clinical standpoint as many of them have specific medical comorbidities that can be screened for. The success of identification of de novo structural variation coupled with the lack of strong replication of common variants associated with ASD indicates that rare variation is likely to play a larger role than previously suspected. This indicates that whole exome and whole genome sequencing approaches are likely to be very informative in identifying new causes of ASD in the near future.

Although multiple highly powered studies looking for linkage peaks in ASD have been performed in the last decade, only two regions of linkage have been replicated (7q and 17q11-q21). The lack of replication of large linkage studies is thought to be due to the heterogeneity present in ASDs. One way to tackle heterogeneity is to identify intermediate or endophenotypes that may relate to more homogeneous subsets of patients. For example, those with epilepsy, large heads, specific psychiatric comorbidities, intellectual disability, language delay, or males and females may have distinct genetic risk factors. Another approach uses the power of continuous quantitative traits of ASD features to find chromosomal regions of interest (reviewed in (*9*)). Several groups have adopted the idea that the overlapping genetic risk across ASDs, other psychiatric disorders, and the general population represent a continuum of neurodevelopmental vulnerabilities and function. The existence of a "broad autism phenotype," or subclinical features of ASDs found in family members of patients also argues that ASDs are an amalgam of traits (*10*) and the clinical heterogeneity of ASDs comes from the particular mix of traits observed in a given individual. Considerable progress has been made in defining those traits, determining their heritability, and developing psychometric instruments to quantitatively measure them in the past decade (reviewed in (*11*)).

Several gemome-wide associated studies in ASD have been performed, but they have not revealed any loci with major effect sizes. Efforts at replicating findings across studies have been unsuccessful, in contrast to the CNV studies mentioned earlier that have largely been consistent in their findings. To date, the list of genes that have been preliminarily associated with ASD (affected by CNVs or mutations) has over 100 members, representing a wide range in the level of supporting evidence (see (*8*) for review). As multiple groups complete whole genome and exome sequencing projects on patients with ASDs, this list is certain to increase significantly.

As our genetic knowledge advances, the classic boundaries that separate psychiatric diseases are being called into question; multiple genes and chromosomal regions implicated in ASDs have also been implicated in other psychiatric disorders such as schizophrenia, ADHD, and OCD (*12*). It is important to understand mechanisms of variable expressivity in order to understand how a given genetic defect can lead to different illnesses in different people. The clinical and genetic heterogeneity of ASDs have led to the suggestion that ASDs encompass hundreds of unique "autisms" (*3*), but this may be true of many neuropsychiatric disorders. As the many genes implicated in ASDs are organized and validated in pathways, biomarkers may be found for their dysfunction and specific therapeutic targets can be identified. A robust understanding of the genetics will finally give us improved diagnostic clarity, improved family and genetic counseling, as well as help guide treatment and prognostication.

REFERENCES

1. Association, A. P. *Diagnostic and Statistical Manual of Mental Disorders*, Revised 4th ed.; American Psychiatric Association: Washington, DC, 2000.
2. Prevalence of Autism Spectrum Disorders—Autism and Developmental Disabilities Monitoring Network, United States, 2006. *MMWR Surveill. Summ.* **2009**, *58* (10), 1–20.
3. Geschwind, D. H. Advances in Autism. *Annu. Rev. Med.* **2009**, *60*, 367–380.
4. Ronald, A.; Hoekstra, R. A. Autism Spectrum Disorders and Autistic Traits: A Decade of New Twin Studies. *Am. J. Med. Genet. B Neuropsychiatr. Genet.* **2011**, *156B* (3), 255–274.
5. Ozonoff, S., et al. Recurrence Risk for Autism Spectrum Disorders: A Baby Siblings Research Consortium Study. *Pediatrics* **2011**.
6. Chess, S. Autism in Children with Congenital Rubella. *J. Autism. Child Schizophr.* **1971**, *1* (1), 33–47.
7. Rasalam, A. D., et al. Characteristics of Fetal Anticonvulsant Syndrome Associated Autistic Disorder. *Dev. Med. Child Neurol.* **2005**, *47* (8), 551–555.
8. Betancur, C. Etiological Heterogeneity in Autism Spectrum Disorders: More Than 100 Genetic and Genomic Disorders and Still Counting. *Brain. Res.* **2011**, *1380*, 42–77.
9. Abrahams, B. S.; Geschwind, D. H. Advances in Autism Genetics: On the Threshold of a New Neurobiology. *Nat. Rev. Genet.* **2008**, *9* (5), 341–355.
10. Losh, M., et al. Neuropsychological Profile of Autism and the Broad Autism Phenotype. *Arch. Gen. Psychiatry* **2009**, *66* (5), 518–526.
11. Constantino, J. N. The Quantitative Nature of Autistic Social Impairment. *Pediatr. Res.* **2011**, *69* (5 Pt 2), 55R–62R.
12. Geschwind, D. H. Genetics of Autism Spectrum Disorders. *Trends Cogn. Sci.* **2011**.

CHAPTER

110

Genetics of Alzheimer Disease

Adam C Naj, Regina M Carney, and Susan E Hahn

John P Hussman Institute for Human Genomics, Miller School of Medicine, University of Miami, Miami, FL, USA

Margaret A Pericak-Vance

John P Hussman Institute for Human Genomics; Department of Human Genetics, Miller School of Medicine, University of Miami, Miami, FL, USA

Michael A Slifer

Department of Human Genetics, Miller School of Medicine, University of Miami, Miami, FL, USA

Jonathan L Haines

Center for Human Genetics Research, Vanderbilt University, Nashville, TN, USA

Alzheimer Disease (AD) is the most common cause of dementia, a progressive and irreversible impairment of cognitive function. It is the leading cause of dementia in the elderly, affecting more than 5.4 million people in the United States, and prevalence increases with age, from 0.3–0.5% at age 60 to 11–15% at age 80.

In 1906, Bavarian Psychiatrist Alois Alzheimer presented the first case of dementia characterized by histopathological signs of senile (neuritic) plaques and neurofibrillary tangles (NFTs), thought at that time to be a rare cause of senility; an autopsy series by Blessed et al. (1968) revealed that, among hundreds of brains affected with "normal senility," a majority were affected by plaques and NFT lesions characteristic of AD. Subsequent work has since shown that most dementia results from specific pathological processes, not simply aging.

Several lines of evidence suggest that AD has a heritable component, including familial aggregation studies and twin studies. These studies have shown that familial forms tend to have an average age-at-onset below 60 years of age, classing these as early-onset AD (EOAD), whereas late-onset AD (LOAD) is more sporadic, though risk is still strongly affected by family history of AD. Twin studies have also demonstrated a genetic component to AD in multiple studies, the concordance rate among monozygotic twins (22–83%), who share all their genes, was higher than the concordance rate among dizygotic twins (0–50%), who share only half of their genes on average. The lack of perfect concordance in AD status among identical twins implies that not all AD risk is genetic effects. A variety of environmental factors have been implicated in AD risk including highest level of education completed, several risk factors for cardiovascular disease, lead exposure, history of head injury, and female gender.

The complexity of LOAD genetics has required a multitude of advanced statistical and laboratory approaches to identify candidate loci. Laboratory approaches such as molecular cloning have been instrumental in identifying some monogenic causes of AD, like the chromosome 21 *APP* gene. Statistical approaches such as linkage analysis have also been used to identify genetic risk factors. When examined in families with high burden of AD using linkage analysis, linkage to the chromosome 21 region containing APP was found in some but not all families, suggesting other causes of familial ADD. Subsequent investigations using linkage analysis have identified missense mutations in the *APP*, *PSEN1*, and *PSEN2* genes which cause EOAD. However, individuals in families with these autosomal-dominant mutations constitute less than 2% of AD cases.

More expanded use of statistical approaches, including linkage and association studies, have proven highly effective in identifying genetic risk factors, including the *APOE* ε4 polymorphism which carries the strongest risk of any known variation in LOAD, and modulates both risk and age of onset of LOAD. Statistical approaches include regional linkage studies, genome-wide linkage studies, candidate genes association studies, and genome-wide association studies (GWAS), and meta-analytic approaches, which combine results from multiple studies for greater power to identify genes of modest effect. Multiple regions have been identified through linkage studies of AD on chromosomes 1 (p31.1-q31.1), 3 (q12.3-q25.31), 6 (p21.1-q15), 7 (pter-q21.11), 8 (p22-p21.1), 9 (p22.3-p13.3; q21.31-q32), 10 (p14-q24), 12 (p12.3-q13.13), 17 (q24.3-qter), and 19 (p13.3-qter). Follow-up fine-mapping in these regions have not identified candidate genes in most of these regions, particularly the heavily examined regions on chromosomes 8, 9, and 12. This has led to the use of genetic association studies to identify potentially causal genes and variants.

Most recently, with the development in the mid-2000s of high-density genotyping chips capturing genotypes at 500,000 single nucleotide polymorphisms or more throughout the genome, GWAS on thousands of cases and controls became feasible and have been used to identify a multitude of risk genes for AD. Outside of the strong association with AD risk of the *APOE* ε4 allele, nine additional genes or genomic regions have been identified as contributing to AD risk: *CR1* on chromosome 1, *BIN1* on chromosome 2, *CD2AP* on chromosome 6, *EPHA1* on chromosome 7, *CLU* on chromosome 8, *PICALM* and *MS4A* gene cluster on chromosome 11, and *ABCA7* and *CD33* on chromosome 19.

Novel laboratory techniques include deep resequencing of the exome (coding portions of the genome) and large chromosomal segments on large numbers of subjects to identify rare variants which may, through modified linkage and association, be found to cosegregate with or be associated with AD.

These genetic findings raise questions in clinical scenarios, such as whether individuals should be tested for AD risk genes. The benefits of foreknowledge of disease risk must be weighed against the risks of possible negative outcomes including discriminatory practices. Important in this balance is the fact that effective treatments and preventive measures for AD are still lacking. Pharmacogenomic studies aim to provide information that will allow for personalized clinical treatments based on an individual's genetic makeup. Targeted treatments are a promising direction for the future of AD patient care.

FURTHER READING

1. Goldman, J. S.; Hahn, S. E.; Catania, J. W., et al. Genetic Counseling and Testing for Alzheimer Disease: Joint Practice Guidelines of the American College of Medical Genetics and the National Society of Genetic Counselors. *Genet. Med.* **2011**, *13*, 597–605.
2. Holtzman, D. M.; Morris, J. C.; Goate, A. M. Alzheimer's Disease: The Challenge of the Second Century. *Sci. Transl. Med.* **2011**, *3* (77), 77, sr1.
3. O'Brien, R. J.; Wong, P. C. Amyloid Precursor Protein Processing and Alzheimer's Disease. *Annu. Rev. Neurosci.* **2011**, *34*, 185–204.
4. Lambert, J. C.; Amouyel, P. Genetics of Alzheimer's Disease: New Evidences for an Old Hypothesis? *Curr. Opin. Genet. Dev.* **2011**, *21* (3), 295–301.
5. Ertekin-Taner, N. Genetics of Alzheimer Disease in the Pre- and Post-GWAS Era. *Alzheimers Res. Ther.* **2010**, *2* (1), 3.
6. Berkis, L. M.; Yu, C. -E.; Bird, T. D.; Tsuang, D. W. Review Article: Genetics of Alzheimer Disease. *J. Geriatr. Psychiatry Neurol.* **2010**, *23* (4), 213–227.
7. Citron, M. Alzheimer's Disease: Strategies for Disease Modification. *Nat. Rev. Drug. Discov.* **2010**, *9*, 387–398.
8. Bettens, K.; Sleegers, K.; Van Broeckhoven, C. Current Status on Alzheimer Disease Molecular Genetics: From Past, to Present, to Future. *Hum. Mol. Genet.* **2010**, *19* (R1), R4–R11.
9. Hardy, J.; Guerreiro, R.; Wray, S.; Ferrari, R.; Momeni, P. The Genetics of Alzheimer's Disease and Other Tauopathies. *J. Alzheimers Dis.* **2011**, *23*, S33–S39.
10. Avramopoulos, D. Genetics of Alzheimer's Disease: Recent Advances. *Genome Med.* **2009**, *1*, 34.

CHAPTER

111

Schizophrenia and Affective Disorders

Jonathan D Picker

Division of Genetics, Departments of Medicine and Child and Adolescent Psychiatry, Childrens Hospital, Boston, MA, USA

Psychiatric disorders have an enormous impact on health at all levels, from the individual and their family to other contacts and through society. Amongst psychiatric disorders, perhaps the most profound are schizophrenia and the affective disorders of bipolar disease (BPD) and major depression (MDD). These disorders cost the society billions of dollars on top of the morbidity and mortality associated with them. The human genome project has raised hopes of identifying gene mutations to better understand and treat these disorders (*1*). The process has been significantly more challenging than expected, however. This has been surprising given that both schizophrenia and BPD have over 80% heritability rates (*2,3*) and even MDD has a significant level of heritability of between 37% and 50% (*4*). In part, the difficulty in identifying genes for these disorders may reflect our uncertainty as to what these diseases actually are.

Through history, people have sought explanations for psychopathology. Unlike most physical illnesses, however, the determination of what constitutes a psychiatric illness is mutable and depends on the beliefs of the society then extant. The modern definitions of schizophrenia and the affective disorders (*5*) have evolved through the twentieth century and continue to do so. At this time, schizophrenia is considered to be a disorder involving combinations of delusions, hallucinations, disorganized speech and behavior, as well as negative symptoms, including flattened affect, poverty of speech or avolition, coupled with a failure to maintain daily functions of life. The classification of MDD includes having a period of depression (characterized by anhedonia or dysphoria, and may include changes in sleep, appetite, loss of sense of self-worth, guilt and ideations of suicidality) that is not secondary to BPD, substances, or medical illnesses. BPD includes a period of mania (expansive, persistently elevated or irritable mood, increased self-esteem, grandiosity, talkativeness, distractibility, disinhibited potentially reckless or goal directed activities, and decreased need for sleep). BPD may include periods of depression. These definitions are, however, observational, and it is unclear if the described disorders represent or approximate to discrete phenotypes; end symptomatology of multiple different disorders or arbitrary definitions within interrelated pathology that may be on a psychopathology continuum.

Within the constraints of this limitation, however, an intense search for genes of interest has been undertaken for all these disorders, particularly schizophrenia. Studies have included linkage analysis, genome-wide association studies, and candidate gene approaches. Results have been conflicting, even amongst different meta-analyses of the studies. The sum of these studies appears to be that no single gene or small group of genes is responsible for either schizophrenia or the affective disorders. The working hypothesis is that these disorders are multifactorial in origin. In the majority of cases, it is likely that many genes of small effect interacting with environmental factors (GXE) result in the disorder. In a small proportion of cases, it appears likely that there are rare genes of larger effect. Furthermore, many of the candidate genes, including *COMT* and *DISC1*, have evidence for a role in both schizophrenia and BPD (*6,7*). Similarly, there appears to be some overlap in candidate genes and chromosomal loci between BPD and MDD. These findings, while not definitive, support some hypothesized pathological overlaps between the disorders. This is not inconsistent with the phenotypic overlap seen between them as well as in intermediate conditions such as schizoaffective disorder. While there has been a lack of definitive finding, the investigations have nonetheless provided directions for developing more targeted drug treatments and thus remain of utility. In addition, pharmacogenomic investigations into the enzyme systems responsible for metabolizing the drugs currently used has clinically applicability now, particularly the cytochrome P450 oxidizing enzymes (*8*). These are responsible for a significant proportion of antidepressant metabolism, as well as having a role with antipsychotics.

Given the lack of hard quantifiable data, counseling families for risk and recurrence remains largely empiric. It should be done by trained genetic counselors, as the raw data do not fully take into account variables such as age of onset, severity of symptoms, number of affected family members, and other factors, that mean an individual's risk may be quite different from the published risk tables (*9,10*).

While much remains unclear, the field may be advancing and it is to be hoped that even if a comprehensive understanding remains elusive, the data will still provide tangible benefits.

REFERENCES

1. Insel, T. R.; Collins, F. S. Psychiatry in the Genomics Era. *Am. J. Psychiatry* **2003,** *160,* 616–620.
2. Smoller, J. W.; Finn, C. T. Family, Twin, and Adoption Studies of Bipolar Disorder. *Am. J. Med. Genet. C Semin. Med. Genet.* **2003,** *123C,* 48–58.
3. Sullivan, P. F.; Kendler, K. S.; Neale, M. C. Schizophrenia as a Complex Trait: Evidence from a Meta-Analysis of Twin Studies. *Arch. Gen. Psychiatry* **2003,** *60,* 1187–1192.
4. Lohoff, F. W. Overview of the Genetics of Major Depressive Disorder. *Curr. Psychiatry Rep.* **2010,** *12,* 539–546.
5. American Psychiatric Association. *Diagnostic Criteria from DSM-IV-TR*; American Psychiatric Association: Washington, D.C, 2000.
6. Abdolmaleky, H. M.; Cheng, K. H.; Faraone, S. V., et al. Hypomethylation of MB-COMT Promoter is a Major Risk Factor for Schizophrenia and Bipolar Disorder. *Hum. Mol. Genet.* **2006,** *15,* 3132–3145.
7. Brandon, N. J.; Millar, J. K.; Korth, C., et al. Understanding the Role of DISC1 in Psychiatric Disease and During Normal Development. *J. Neurosci.* **2009,** *29,* 12768–12775.
8. Seeringer, A.; Kirchheiner, J. Pharmacogenetics-Guided Dose Modifications of Antidepressants. *Clin. Lab. Med.* **2008,** *28,* 619–626.
9. Finn, C. T.; Smoller, J. W. Genetic Counseling in Psychiatry. *Harv. Rev. Psychiatry* **2006,** *14,* 109–121.
10. Tsuang, D. W.; Faraone, S. V.; Tsuang, M. T. Genetic Counseling for Psychiatric Disorders. *Curr. Psychiatry Rep.* **2001,** *3,* 138–143.

CHAPTER

112

Addictive Disorders

David Goldman

National Institute of Alcohol Abuse and Alcoholism, Chief of the Laboratory of Neurogenetics, National Institutes of Health, Rockville, MD, USA

Paola Landi

Dipartimento di Psichiatria, Neurobiologia, Farmacologia e Biotecnologi, University of Pisa, Pisa, Italy

Francesca Ducci

Institute of Psychiatry, Kings College, London, UK

Addictions are common multistep pathologies that progress from exposure to the addictive agent, often occurring in adolescence, to maladaptive patterns of use and dependence. Addictions are frequently chronic, lifelong diseases with a relapsing/remitting course. The probability of initiation and the probability of developing a substance use disorder are both influenced by individual characteristics (e.g. genotype, sex, age, and psychopathology), environmental factors (e.g. drug availability, social support, stress, socioeconomic status), and the nature of the addictive agent (e.g. addictive liability, mode of use or administration, physiological response, secondary pathology). Twin studies have shown that genetic factors account for 40–70% of the overall variance in addiction liability. There is little evidence for large influences on overall population vulnerability from any single gene. Instead, multiple genetic loci are likely to be involved, each with a small attributable risk. Genetic and environmental influences modulating risk of addiction change during the lifespan. Longitudinal twin studies have shown that the relative importance of genetic factors tends to increase from adolescence to adulthood. In addition, genetic risk factors for initiation only partially overlap with genetic determinants for developing substance use disorders.

Gene discovery related to addiction is in its infancy. Susceptibility loci for addictions include both drug-specific genes (e.g. alcohol-metabolizing genes) and loci moderating neuronal pathways, such as reward, behavioral control, and stress resiliency. Some of these genes (e.g. *MAOA* and *COMT*) also moderate risk of other psychiatric disorders. In recent years, progress has been made in identification of genes influencing addiction liability by the use of intermediate phenotypes. Phenotypes such as task-related brain activation confer the opportunity of exploring the neuronal mechanisms through which genetic variation is translated into behavior. Fundamental to the detection of gene effects has also been the understanding that genetic and environmental factors jointly shape vulnerability to addiction. Environmental factors such as severe childhood stress and neglect increase vulnerability to addiction and to addiction-related psychiatric diseases including antisocial personality disorder, conduct disorder, anxiety disorders, and depression. The risk of addictions and the risks of these other psychiatric diseases are elevated severalfold in the stress exposed. However, not all people exposed to early life stress develop addiction or other psychiatric diseases, indicating wide variation in stress resiliency. Several loci that have been shown to partially account for interindividual differences in stress resiliency include functional loci at *MAOA*, *SLC6A*, *COMT*, *NPY*, and *FKBP5*.

Genome-wide association studies (GWAS) and family-based linkage studies have both enabled the whole genome to be interrogated in hypothesis-free fashion for loci influencing vulnerability. GWAS has been particularly useful to identity genetic determinants of nicotine addiction. It was thereby discovered that a relatively common missense variant within the *CHRNA5* gene influences the risk of smoking and the development of tobacco-related diseases such as lung cancer.

Genetic studies of addiction, as well as other common disorders, have focused on common genetic variation and the contribution of rare variants (minor allele frequency <1%) is still largely unknown. However, recent advances in sequencing technologies have allowed the sequencing of genome, exome (the expressed portion of the genome), or targeted gene panels at low cost, and opened the way for extensive searches for rare variants.

Rare variants moderating impulsivity, an important risk factor for addiction, have been found within both *MAOA* and *HTR2B*.

Although the genetic bases of addiction remain largely unknown, progress is likely to be more rapid in the future. Understanding genetic influences is an important step toward personalized medicine. Genotypes may be used to predict risk, develop new treatments, and target treatments. This potential is already exemplified by the use of an OPRM1 variant to target treatment in alcoholism, and by studies revealing that alcohol metabolic gene variants common in East Asians are partially protective against alcoholism, but increase the risk of upper GI cancer in those carriers of the alleles who do drink moderately.

FURTHER READING

1. Kendler, K. S.; Prescott, C. A.; Myers, J.; Neale, M. C. The Structure of Genetic and Environmental Risk Factors for Common Psychiatric and Substance Use Disorders in Men and Women. *Arch. Gen. Psychiatry* **2003**, *60*, 929–937.
2. Goldman, D.; Oroszi, G.; Ducci, F. The Genetics of Addictions: Uncovering the Genes. *Nat. Rev. Genet.* **2005**, *6*, 521–532.
3. Ducci, F.; Enoch, M. A.; Hodgkinson, C.; Xu, K.; Catena, M.; Robin, R. W., et al. Interaction between a Functional MAOA Locus and Childhood Sexual Abuse Predicts Alcoholism and Antisocial Personality Disorder in Adult Women. *Mol. Psychiatry* **2008**, *13*, 334–347.
4. Thomasson, H. R.; Edenberg, H. J.; Crabb, D. W.; Mai, X. L.; Jerome, R. E.; Li, T. K., et al. Alcohol and Aldehyde Dehydrogenase Genotypes and Alcoholism in Chinese Men. *Am. J. Hum. Genet.* **1991**, *48*, 677–681.
5. Zhou, Z.; Zhu, G.; Hariri, A. R.; Enoch, M. A.; Scott, D.; Sinha, R., et al. Genetic Variation in Human NPY Expression Affects Stress Response and Emotion. *Nature* **2008**, *452*, 997–1001.
6. Hong, L. E.; Hodgkinson, C. A.; Yang, Y.; Sampath, H.; Ross, T. J.; Buchholz, B.; Salmeron, B. J.; Srivastava, V.; Thaker, G. K.; Goldman, D., et al. A Genetically Modulated, Intrinsic Cingulate Circuit Supports Human Nicotine Addiction. *Proc. Natl. Acad. Sci. U.S.A.* **2010**, *107* (30), 13509–13514.
7. Egan, M. F.; Goldberg, T. E.; Kolachana, B. S.; Callicott, J. H.; Mazzanti, C. M.; Straub, R. E., et al. Effect of COMT Val108/158 Met Genotype on Frontal Lobe Function and Risk for Schizophrenia. *Proc. Natl. Acad. Sci. U.S.A.* **2001**, *98*, 6917–6922.
8. Thorgeirsson, T. E.; Geller, F.; Sulem, P.; Rafnar, T.; Wiste, A.; Magnusson, K. P., et al. A Variant Associated with Nicotine Dependence, Lung Cancer and Peripheral Arterial Disease. *Nature* **2008**, *452*, 638–642.
9. Bevilacqua, L; Doly, S; Kaprio, J; Yuan, Q; Tikkanen, R; Paunio, T; et al. A Population-Specific HTR2B Stop Codon Predisposes to Severe Impulsivity. *Nature 468*, 1061–1066.
10. Caspi, A.; McClay, J.; Moffitt, T. E.; Mill, J.; Martin, J.; Craig, I. W., et al. Role of Genotype in the Cycle of Violence in Maltreated Children. *Science* **2002**, *297*, 851–854.

Neurologic Disorders

Neural Tube Defects, 435
Genetic Disorders of Cerebral Cortical Development, 438
Genetic Aspects of Human Epilepsy, 440
Basal Ganglia Disorders, 443
The Hereditary Ataxias, 445
Hereditary Spastic Paraplegia, 447
Autonomic and Sensory Disorders, 454
The Phakomatoses, 458
Multiple Sclerosis and Other Demyelinating Disorders, 464
Genetics of Stroke, 467
Primary Tumors of the Nervous System, 469

CHAPTER

113

Neural Tube Defects

Richard H Finnell

Departments of Nutritional Sciences, Chemistry and Biochemistry, Dell Pediatric Research Institute, The University of Texas at Austin, Dell Children's Medical Center, Austin, TX, USA

Timothy M George

Pediatric Neurosurgery Center, Dell Children's Medical Center, Austin, TX, USA

Laura E Mitchell

The University of Texas School of Public Health, Houston, TX, USA

113.1 EMBRYOLOGY

Neurulation is the critical morphogenetic event occurring during the fourth week of human gestation, converting the previously developed neural plate into the ectoderm covered neural tube that will eventually differentiate into the brain and spinal cord. Genetic regulation of the events involved in mammalian neural tube morphogenesis is a complicated process that involves a multitude of genes. These genes have vital functions in a range of biological activities that are thought to include signaling molecules, transcription proteins and factors, cytoskeletal and gap junction proteins, growth factors, and tumor suppressor genes (*1–4*). The neural tube, formed by the cell shape changes and movements, ultimately fuses at several discrete points.

113.2 DEFINITION

Abnormalities of neural tube closure may result from alterations in any of the processes that are involved in the formation of the primary neural tube (e.g. elevation and bending, apical constriction, convergent extension). As these processes are under the control of a multitude of genes, the molecular basis of neural tube closure defects is likely to be quite heterogeneous. Indeed, in the mouse, defects of neural tube closure have been shown to result from genetic mutations affecting cell proliferation and apoptosis, convergent extension, elevation/apposition of the neural folds, and neural tube fusion (*5–8*). Despite the complexity of the processes that are required for normal neural tube closure, the gross phenotypic consequences of abnormalities in these processes are relatively homogenous. In general, failure of neural tube closure is associated with defects in the overlying bony structures (i.e. cranial vault and neural arches) such that the underlying neural tissue is exposed to the body surface. Consequently, defects of primary neural tube closure are often referred to as "open" neural tube defects (NTDs). Further classifications of the open NTDs are based on the location and extent of the defect. In addition to the open NTDs, there are also a number of "closed" or skin covered conditions that involve the neural tube, including encephalocele, iniencephaly and lesions often referred to as occult spinal dysraphisms such as meningocele (that may be partially skin covered), spinal lipomas (lipomyelomeningocele or lipomeningocele), myelocystocele, spilt cord malformations, and various forms of sacral agenesis. In general, these conditions are not thought to result from defects in primary neural tube closure, but may result from defects in secondary neural tube development. Nonetheless, there is evidence that open and closed NTDs may have some common etiological underpinnings.

113.3 RISK FACTORS

The open NTDs are recognized as being etiologically heterogeneous. A small proportion of affected individuals have an associated chromosomal or Mendelian malformation syndrome (*9*) and there are rare families in which NTDs segregate in patterns consistent with X-linked or autosomal recessive inheritance. In addition, a small proportion of cases can be linked to an established risk factor. The most notable of these factors are maternal pregestational, insulin-dependent diabetes and maternal use of medications, in particular, specific anticonvulsant drugs.

While many environmental agents are suspected of possessing a teratogenic potential based on data from studies in animals, relatively few have been confirmed as human teratogens. This may be due to several factors, including differences in dose; laboratory studies performed on mice generally utilize doses that are much higher than encountered by humans; exposure assessment, environmental exposures are more difficult to quantify in studies involving humans than in mice kept under laboratory conditions; and degree of experimental control, genetic and environmental differences that may influence disease susceptibility are more easily controlled in studies of laboratory animals than in studies of human populations. However, several environmental exposures have emerged as potential risk factors for NTDs.

Studies of the recurrence patterns in families ascertained through an individual affected with anencephaly or spina bifida indicate that these two conditions tend to co-segregate within families and are, therefore, likely to share a common etiology. In addition, family studies have demonstrated that the relatives of an affected individual are at increased risk of having an NTD compared with the general population. The relative risk ratio (i.e. risk to relative vs. risk in general population) for the sib of an affected individual is 30 to 50. This is much lower than the relative risks exhibited by Mendelian conditions, and higher than expected for conditions that are determined solely by environmental factors (*10*). Hence, it is commonly believed that the risk of NTDs is determined by multiple risk factors, including both genes and environmental agents.

Epidemiological studies of potential environmental risk factors for NTDs have a relatively long history, and have identified several factors (reviewed previously) that appear to be related to NTD risk. In contrast, the study of genetic risk factors for NTDs has a relative short history. Although traditional genetic linkage approaches (e.g. LOD score and affected relative pair analyses) have been applied to the study of other conditions since the 1980s, the data required by such approaches (i.e. DNA from multiple affected individuals within families) are largely unavailable for NTDs. Consequently, studies designed to identify genes that influence the risk of NTDs did not become feasible until the 1990s, when the Human Genome Project began to provide new information regarding the variability within the human genome. Such advances in our understanding of the human genome have provided new opportunities for studying the genetic contribution to conditions such as NTDs (e.g. case–control and family-based genetic association studies of non-Mendelian conditions).

The relative lack of success in identifying genetic variants that are related to the risk of NTDs in humans is somewhat surprising given the large number of mouse models that provide strong evidence of the genetic contribution of NTD. To date, over 240 gene mutations have been associated with NTD in mouse lines (*11*), yet none of the 155 genes implicated by these models has been identified as a determinant of NTD risk in humans. In general, the lack of success in past studies in humans is likely to reflect a number of factors including issues of sample size and power, etiologic heterogeneity, and failure to adequately capture the etiological complexities that are likely to underlie human NTDs. These complexities are likely to include gene–gene and gene–environment interactions; maternal and embryonic genotypic effects; the involvement of both rare and common variants; and consideration of pathways (as opposed to individual genes).

113.4 DIAGNOSIS, TREATMENT, AND OUTCOME

Screening to identify pregnant women who are at an increased risk of carrying an NTD affected fetus can be achieved by the evaluation of maternal serum alpha-fetoprotein (MSAFP) levels and/or ultrasound imaging (*12–14*). Follow-up studies for women with a positive result on either screening test include amniocentesis and/or detailed ultrasound evaluation. When amniocentesis is performed, evaluation of amniotic fluid alpha-fetoprotein and acetyl cholinesterase levels can be used to confirm the presence of an open fetal malformation and differentiate between open ventral wall and open NTDs. In addition, the fetal karyotype can be evaluated to rule out chromosomal anomalies. Detailed ultrasound can also be used to differentiate between open defects of the ventral wall and the neural tube, and to identify the presence of other associated anomalies (*15*). When a diagnosis of spina bifida is confirmed, ultrasound and prenatal magnetic resonance imaging (using ultrafast T2-weighted sequences) can be used to identify spontaneous leg and foot motion, leg and spine deformities, and the presence of a Chiari II malformation (*16*).

The majority of individuals with spina bifida have their spinal lesion closed postnatally, usually within 72 h of birth (*17,18*). However, a small proportion of fetuses have had their lesion closed in utero. A randomized clinical trial, designed to compare the outcome of infants treated postnatally with those treated in utero was initiated in 2003 and stopped in December 2010 based on the evidence for efficacy of prenatal surgery (*19*). Based on data from 158 patients who underwent a 12-month evaluation, the primary outcome (death or need for cerebrospinal fluid shunt by 12 months) was significantly less common in infants undergoing prenatal as compared to postnatal surgery (relative risk 0.70; 97.7% CI 0.58, 0.84). The prenatal surgery group also had lower rates of hindbrain herniation, brain-stem kinking and syringomyelia and, at 30 months, the difference between the functional and anatomical level of the lesion was significantly better as compared to the postnatal surgery group.

In the United States and many other areas of the world, it is recommended that all women of childbearing age consume a daily, multivitamin supplement containing at least 0.4mg of folic acid in order to reduce their risk of having a child with an NTD. Women who have had a child affected with an NTD may be counseled to take higher doses of folic acid as per the initial recommendations from the Centers for Disease Control (*20*). Specific recommendations for other categories of high-risk women (e.g. sisters of women who have had an affected pregnancy, women take antiepileptic drugs such as valproic acid, and diabetics) are not available.

As folic acid supplementation does not prevent 100% of all NTDs, women who are at increased risk for having an affected child (whether or not they are taking folic acid supplements) should be informed of the availability of MSAFP screening/ultrasound/amniocentesis for the prenatal identification of affected fetuses. At this time, there are no recommended genetic screening or testing procedures for nonsyndromic NTDs. Evaluation of genetic variants, such as the MTHFR C677T polymorphism, is not currently recommended due to the lack of consistent associations across studies, the relatively low proportion of cases that are likely to be attributable to any single genetic factor, and the fact that it would not change preconceptional advice regarding the utility of vitamin supplements or pregnancy management (*21*).

REFERENCES

1. Copp, A. J.; Green, N. D.E.; Murdoch, J. N. The Genetic Basis of Mammalian Neurulation. *Nat. Rev. Genet.* **2003,** *4,* 784–793.
2. Ewart, J. L.; Cohen, M. F.; Meyer, R. A. Heart and Neural Tube Defects in Transgenic Mice Overexpressing the Cx43 Gap Junction Gene. *Development* **1997,** *124,* 1281–1292.
3. Sah, V. P.; Attardi, L. D.; Mulligan, G. J.; Williams, B. O.; Bronson, R. T.; Jacks, T. A Subset of p53-Deficient Embryos Exhibit Exencephaly. *Nat. Genet.* **1995,** *10,* 175–180.
4. Zhang, J.; Hagopian-Donaldson, S.; Serbedzija, G. Neural Tube, Skeletal and Body Wall Defects in Mice Lacking Transcription Factor AP-2. *Nature* **1996,** *381,* 238–241.
5. Blom, H. J.; Shaw, G. M.; den Heijer, M.; Finnell, R. H. Neural Tube Defects and Folate: Case Far from Closed. *Nat. Rev. Neurosci.* **2006,** *7* (9), 724–731, PMID:16924261.
6. Copp, A. J.; Greene, N. D. Genetics and Development of Neural Tube Defects. *J. Pathol.* **2010,** *220,* 217–230.
7. Lee, S.; Jurata, L. W.; Nowak, R.; Lettieri, K.; Kenny, D. A.; Pfaff, S. L.; Gill, G. N. The LIM Domain-Only Protein LM04 Is Required for Neural Tube Closure. *Mol. Cell. Neurosci.* **2005,** *28,* 205–214.
8. Massa, V.; Savery, D.; Ybot-Gonzalez, P.; Ferraro, E.; Rongvaux, A.; Cecconi, F.; Flavell, R.; Greene, N. D.; Copp, A. J. Apoptosis Is Not Required for Mammalian Neural Tube Closure. *Proc. Natl. Acad. Sci. U.S.A.* **2009,** *106* (20), 8233–8238, Epub 2009 May 6. PMID:19420217.
9. Lynch, S. A. Non-multifactorial Neural Tube Defects. *Am. J. Med. Genet. C Semin. Med. Genet.* **2005,** *135,* 69–76.
10. Khoury, M. J.; Beaty, T. H.; Liang, K. Y. Can Familial Aggregation of Disease Be Explained by Familial Aggregation of Environmental Factors? *Am. J. Epidemiol.* **1988,** *127,* 674–683.
11. Harris, M. J.; Juriloff, D. M. An Update to the List of Mouse Mutants with Neural Tube Closure Defects and Advances Toward a Complete Genetic Perspective of Neural Tube Closure. *Birth Defects Res. A Clin. Mol. Teratol.* **2010,** *88,* 653–669.
12. Drugan, A.; Weissman, A.; Evans, M. I. Screening for Neural Tube Defects. *Clin. Perinatol.* **2001,** *28,* 279–287.
13. Krantz, D. A.; Hallahan, T. W.; Sherwin, J. E. Screening for Open Neural Tube Defects. *Clin. Lab. Med.* **2010,** *30* (3), 721–725, Epub 2010 Jun 15. PMID: 20638584.
14. Wald, N. J. Prenatal Screening for Open Neural Tube Defects and Down Syndrome: Three Decades of Progress. *Prenat. Diagn.* **2010,** *30,* 619–621.
15. Sepulveda, W.; Corral, E.; Ayala, C.; Be, C.; Gutierrez, J.; Vasquez, P. Chromosomal Abnormalities in Fetuses with Open Neural Tube Defects: Prenatal Identification with Ultrasound. *Ultrasound Obstet. Gynecol.* **2004,** *23,* 352–356.
16. Mangels, K.; Tulipan, N.; Tsao, L.; Alarcon, J.; Bruner, J. P. Fetal MRI in the Evaluation of Intrauterine Myelomeningocele. *Pediatr. Neurosurg.* **2000,** *32,* 124–131.
17. Bowman, R. M.; McLone, D. G.; Grant, J. A.; Tomita, T.; Ito, J. A. Spina Bifida Outcome: A 25-Year Prospective. *Pediatr. Neurosurg.* **2001,** *34,* 114–120.
18. Carmichael, S. L.; Shaw, G. M.; Neri, E.; Schaffer, D. M.; Selvin, S. Physical Activity and Risk of Neural Tube Defects. *Matern. Child Health J.* **2002,** *6,* 151–157.
19. Adzick, N. S.; Thom, E. A.; Spong, C. Y.; Brock, J. W., III; Burrows, P. K.; Johnson, M. P.; Howell, L. J.; Farrell, J. A.; Dabrowiak, M. E.; Sutton, L. N., et al. A Randomized Trial of Prenatal Versus Postnatal Repair of Myelomeningocele. *N. Engl. J. Med.* **2011,** *364,* 993–1004.
20. Centers for Disease Control Use of Folic Acid for Prevention of Spina Bifida and Other Neural Tube Defects—1983–1991. *MMWR* **1991,** *40,* 513–516.
21. Finnell, R. H.; Shaw, G. M.; Lammer, E. J.; Volcik, K. A. Does Prenatal Screening for 5,10-Methylenetetrahydrofolate Reductase (MTHFR) Mutations in High-Risk Neural Tube Defect Pregnancies Make Sense? *Genet. Test.* **2002,** *6,* 47–52.

CHAPTER

114

Genetic Disorders of Cerebral Cortical Development

Ganeshwaran H Mochida

Harvard Medical School, Boston, MA, USA; Division of Genetics, Department of Medicine, Boston Children's Hospital, Boston, MA, USA; Massachusetts General Hospital, Boston, MA, USA

Annapurna Poduri

Harvard Medical School, Boston, MA, USA; Division of Epilepsy and Clinical Neurophysiology Boston Children's Hospital, Boston, MA, USA

Christopher A Walsh

Harvard Medical School, Boston, MA, USA; Division of Genetics, Department of Medicine, Boston Children's Hospital, Boston, MA, USA; Howard Hughes Medical Institute, MD, USA

Patients with malformations of cerebral cortical development typically present with epilepsy and cognitive difficulty. Modern magnetic resonance imaging has allowed for the identification of a wide range of brain malformations, from subtle disorders involving a single gyrus to abnormalities of the entire cortex. Malformations of cortical development occur when there are disruptions of the normal processes of brain development, specifically, progenitor cell proliferation, neuronal migration, and cortical organization (*1*). Each of these steps is under the control of genes expressed during early development, many of which have been discovered through the study of human brain malformations.

We first consider disorders attributed to abnormal neuronal and glial proliferation, the prototype of which is microcephaly. Microcephaly, defined by small head size for age, can result from a variety of genetic causes all of which lead to a net reduction in the normal number of neurons produced. Microcephaly is a feature of many syndromes, but when it occurs as an isolated clinical feature it may be called microcephaly vera (*2*). The genes associated with autosomal recessive microcephaly vera include *MCPH1* (also known as *microcephalin*), *WDR62*, *CEP152*, *ASPM*, *CENPJ*, and *STIL* (*3*). Other autosomal recessive microcephaly genes include *PNKP*, *NDE1*, *TRAPPC9*, and *CASK*. Other disorders of neuronal and glial proliferation include tuberous sclerosis complex, caused by mutations in the genes *TSC1* and *TSC2* (*4*), as well as focal cortical dysplasia and hemimegalencephaly, most cases of which have not yet been attributed to specific genetic causes.

Disorders of neuronal migration include lissencephaly (or "smooth brain"), subcortical band heterotopia, and periventricular heterotopia. Lissencephaly, sometimes also called agyria-pachygyria, is caused by defects in the genes *LIS1*, *DCX* in males, *ARX* chiefly in males, and *TUBA1A* (*5,6*). The gene *DCX* is also associated with subcortical band heterotopia (or "double cortex") in females typically, though some cases of males with this pattern have been reported. A condition called "cobblestone dysplasia" is a form of lissencephaly associated with muscle and eye involvement and falling into the spectrum of dystroglycanopathy, the most severe form of which is Walker–Warburg syndrome. The genes responsible for this form of lissencephaly include *POMT1*, *POMT2*, *POMGNT1*, *FKTN*, *FKRP*, and *LARGE* (*7,8*). The classic malformation of periventricular heterotopia occurs in females and may be due to mutations in *FLNA* (*9*). There are also several copy number abnormalities that have been associated with periventricular heterotopia.

The brain malformations that are thought to arise from a disruption of post-migration cortical organization include polymicrogyria and schizencephaly. Polymicrogyria can occur in any distribution but most commonly affects the perisylvian regions. Genetic causes of polymicrogyria include *GPR56* when the bilateral frontoparietal regions are involved (*10*), deletion 22q11 when the perisylvian regions are involved (*11*), and *LAMC3* when the occipital regions are involved (*12*). In addition, polymicrogyria has been attributed to defects in genes encoding tubulins, including *TUBA1A*, *TUBB2B*, *TUBB3*, and *TUBA8* (*13*). Schizencephaly, while it has reported to occur in a family, has not been conclusively associated with a gene.

Genetic disorders of cortical development include both inherited and de novo mutations in genes involved in early brain development. There are some brain malformations for which there are both genetic and nongenetic etiologies described, and there are some for which a cause is not yet established. A genetic diagnosis presents not only a diagnostic conclusion and an opportunity for genetic counseling but also the ability to discuss the clinical spectrum associated with the gene involved. As more cases of brain malformations are identified and the number of genes associated with brain malformations expands, we can look forward to understanding better human brain development and providing more precise diagnoses to patients and families.

REFERENCES

1. Barkovich, A. J.; Kuzniecky, R. I.; Jackson, G. D.; Guerrini, R.; Dobyns, W. B. A Developmental and Genetic Classification for Malformations of Cortical Development. *Neurology* **2005,** *65* (12), 1873–1887.
2. Mochida, G. H.; Walsh, C. A. Molecular Genetics of Human Microcephaly. *Curr. Opin. Neurol.* **2001,** *14* (2), 151–156.
3. Thornton, G. K.; Woods, C. G. Primary Microcephaly: Do All Roads Lead to Rome? *Trends Genet.* **2009,** *25* (11), 501–510.
4. Orlova, K. A.; Crino, P. B. The Tuberous Sclerosis Complex. *Ann. N. Y. Acad. Sci.* **2010,** *1184,* 87–105.
5. Manzini, M. C.; Walsh, C. A. What Disorders of Cortical Development Tell us about the Cortex: One Plus One does Not Always Make Two. *Curr. Opin. Genet. Dev.* **2011,** *21* (3), 333–339.
6. Mochida, G. H. Genetics and Biology of Microcephaly and Lissencephaly. *Semin. Pediatr. Neurol.* **2009,** *16* (3), 120–126.
7. Clement, E.; Mercuri, E.; Godfrey, C.; Smith, J.; Robb, S.; Kinali, M.; Straub, V.; Bushby, K.; Manzur, A.; Talim, B., et al. Brain Involvement in Muscular Dystrophies with Defective Dystroglycan Glycosylation. *Ann. Neurol.* **2008,** *64* (5), 573–582.
8. Manzini, M. C.; Gleason, D.; Chang, B. S.; Hill, R. S.; Barry, B. J.; Partlow, J. N.; Poduri, A.; Currier, S.; Galvin-Parton, P.; Shapiro, L. R., et al. Ethnically Diverse Causes of Walker-Warburg Syndrome (WWS): FCMD Mutations Are a More Common Cause of WWS Outside of the Middle East. *Hum. Mutat.* **2008,** *29* (11), E231–E241.
9. Lu, J.; Sheen, V. Periventricular Heterotopia. *Epilepsy Behav.* **2005,** *7* (2), 143–149.
10. Piao, X.; Hill, R. S.; Bodell, A.; Chang, B. S.; Basel-Vanagaite, L.; Straussberg, R.; Dobyns, W. B.; Qasrawi, B.; Winter, R. M.; Innes, A. M., et al. G protein-Coupled Receptor-Dependent Development of Human Frontal Cortex. *Science* **2004,** *303* (5666), 2033–2036.
11. Bingham, P. M.; Lynch, D.; McDonald-McGinn, D.; Zackai, E. Polymicrogyria in Chromosome 22 Delection Syndrome. *Neurology* **1998,** *51* (5), 1500–1502.
12. Barak, T.; Kwan, K. Y.; Louvi, A.; Demirbilek, V.; Saygi, S.; Tuysuz, B.; Choi, M.; Boyaci, H.; Doerschner, K.; Zhu, Y., et al. Recessive LAMC3 Mutations Cause Malformations of Occipital Cortical Development. *Nat. Genet.* **2011,** *43* (6), 590–594.
13. Jaglin, X. H.; Chelly, J. Tubulin-Related Cortical Dysgeneses: Microtubule Dysfunction Underlying Neuronal Migration Defects. *Trends Genet.* **2009,** *25* (12), 555–566.

CHAPTER

115

Genetic Aspects of Human Epilepsy

Asuri N Prasad

Departments of Pediatrics and Clinical Neurosciences, Western University, Pediatric Neurologist, Children's Hospital of Western Ontario, London Health Sciences Centre, London, Ontario, Canada

Chitra Prasad

Department of Pediatrics, Western University, Genetics Program of South Western Ontario, Children's Hospital of Western Ontario, London Health Sciences Centre, London, Ontario, Canada

An epileptic seizure represents an endpoint of the interaction between differing anatomical, genetic, molecular and electrophysiological changes operative in the generation of abnormal cortical hyperexcitability and spread of excitation along different pathways (*1*). "Epilepsy" is recognized as a disorder in which recurrent seizures occur, accompanied by a variety of clinical phenomena involving motor, sensory, cognitive, psychic and autonomic manifestations, with or without loss of consciousness. The occurrence of at least two unprovoked seizures is necessary to fulfill generally accepted criterion to establish the diagnosis of epilepsy. Currently, human epilepsies are organized into several categories: distinct electroclinical syndromes, constellations, epilepsy with structural-metabolic causes, and epilepsy with unknown cause. These concepts are elaborated in detail in a special report of the ILAE Commission on Classification and Terminology 2005–2009 (*2*). Advances in our understanding of genetics, and molecular genetics are shaping new insights into our understanding of inheritance patterns and phenotypic variability in human epilepsy. Support for a heritable basis to epilepsy is currently based on findings from (i) aggregation studies in families and twins, (ii) linkage analysis and positional cloning of susceptibility genes, (iii) association studies, and study of (iv) single gene defects associated with human epilepsy and single gene defects in animal models of epilepsy (*3*). A number of single gene defects underlying epilepsy syndromes are now recognized, since the identification of the first human epilepsy gene (mutation involving a nicotinic cholinergic receptor subunit) associated with an idiopathic partial epilepsy syndrome (autosomal dominant nocturnal frontal lobe epilepsy) was identified in 1995 (*4*). Traditional Mendelian inheritance patterns leading to human epilepsy syndromes are rare in comparison to more common genetic (idiopathic) epilepsies. Genetic aspects of epileptogenesis seem to be characterized by involvement of multitude of genes, and other epigenetic processes involved in the regulation of gene expression. Ultimately, as a consequence of a mutational change, there results a change in the balance between excitability and inhibition in the nervous system. There is firm evidence for mechanistic diversity in the pathogenesis of idiopathic epilepsies. The pathways involve "channelopathy genes" as well as other non–ion channel genes. Mutations in different genes may encode for subunits of the same channel, resulting in an epilepsy phenotype that is essentially related to the channel dysfunction in a nonspecific manner; secondly, mutations affecting different channels may result in the same phenotype, by net effects converging on a common pathway (convergent effect); and finally, a situation where mutations affecting subunits of the same channel that may produce two entirely different epilepsy phenotypes (divergent effect).

The majority of the causative mutations identified seem to affect ion (sodium or K channels) or ligand gated channels (cholinergic receptor, GABA receptors) with few exceptions. Even within monogenic epilepsy syndromes, there is considerable variability. Mutations involving different gene loci may be expressed through a single phenotype (locus heterogeneity), while mutations involving the same gene may be expressed through completely different phenotypes at an individual level (variable expressivity). These are illustrated with examples in the online chapter. Nontraditional inheritance models also account for disorders with an inherited basis, where epilepsy is a prominent feature of the overall phenotype. These include imprinting disorders (differential expression of genes based on a parent of origin effect) (e.g. Angelman syndrome), maternal (mitochondrial) inheritance (e.g. myoclonic epilepsy with ragged red fibers, MERRF) and triplet repeat expansion (Dentato-rubro-pallidoluysian atrophy, DRPLA) and dodecamer repeats (progressive myoclonic epilepsy). "Complex epilepsy", that is, epilepsy with complex genetics, multiple susceptibility genes, and a triggering effect from the environment, may be involved. There are no clear segregation patterns and the phenomena of epistasis (intragenic interactions) influencing gene expression are possible. Pleiotropic effects and epigenetic modification of expression also influence the final phenotype. Inheritance patterns in the latter can be explained by a polygenic heterogeneity model, involving a dosage effect resulting from the combined effect of both common and rare variants of susceptibility genes (*5*). Copy number variants (CNVs) are associated with many disorders with complex inheritance patterns (intellectual disability, schizophrenia, and autism), and now have come under the scanner for complex genetically determined epilepsies (*6*). At least three microdeletions (15q11.2, 16p13.11 and 15q13.3) are now known to be associated with epilepsy; their combined frequency in populations with genetically generalized epilepsy is about 3%. The 15q13.3 microdeletion appears to be more specific for generalized epilepsy syndromes, while the other two have been described with both generalized and focal epilepsy syndromes (*7*). The likelihood that rare non-recurrent CNVs are also associated with epilepsy is a finding confirmed in recent studies (*8*). It has been suggested that the recurrent CNVs described above are providing an outline that encompasses a group of different electroclinical phenotypes within a "constellation" of genetic generalized epilepsies (childhood absence epilepsy, juvenile absence epilepsy, juvenile myoclonic epilepsy) syndromes. In addition to recurrent CNVs, there are other genetic mechanisms (somatic mutations, modifier genes, epigenetic effects) that add to the diversity as well as phenotypic variability in human epilepsy. An age-dependent approach to description of clinical features and seizure semiology, EEG findings for selected genetic epilepsy syndromes has been employed.

The clinical approach involved in the genetic evaluation of patients with epilepsy becomes very important when it comes to the question of which genetic investigations should be considered. With the commercial availability of several gene tests with a direct to consumer approach, several issues gain critical importance. The clinical history should focus on elucidating information on age of onset, seizure type, triggers or precipitating factors, prior history of central nervous system insult, and existence of comorbid neurological problems (mental retardation, developmental delay, risk factors of stroke, tumor, trauma, CNS infection). In the evaluation of the epilepsy phenotype, the assistance of a pediatric neurologist/epileptologist is essential. Seizure semiology and the electroencephalogram are critical in helping narrow an electroclinical syndrome. The age of onset of seizures, seizure type at onset, evolution and change in semiology must be documented. Thus, collecting clinical data, the nature of the epilepsy, progression, and presence of associative features are all essential steps in identifying an epilepsy phenotype. In the family history, information on all affected individuals should be collected along the same lines as for the probands. The ethnic background could alert physicians to a founder effect, consanguinity will suggest autosomal recessive inheritance. While a matrilineal inheritance pattern points toward mitochondrial cytopathies, these disorders can show autosomal recessive or dominant inheritance patterns. Other valuable features include a positive family history for seizures/any history of developmental delay/developmental regression/other associated symptoms such as organomegaly/vision and hearing involvement.

The physical examination of the affected individual is directed toward assessing cognitive and developmental abilities, and the presence or absence of focal neurological deficits, and lateralizing signs (preferably conducted by a neurologist in consultation). Anthropometry with special reference to head circumference can provide clues to certain genetic/metabolic syndromes (e.g. macrocephaly in Fragile X syndrome and microcephaly in Rett syndrome, Angelman syndrome and other inborn errors of metabolism). A careful search for dysmorphic features is warranted, as they may suggest association with specific chromosomal rearrangements. Examination and documentation of any findings in the eyes, skin, and other systems are equally important. The neurological history and examination must be supplemented by EEG data to characterize the seizure type and electroclinical epilepsy syndrome. The investigative work up should be individualized for each patient. The seizure types can be classified by age of onset, type, and with/without fever, frequency, time of occurrence, and association with sleep. In the idiopathic/genetic generalized epilepsies, an imaging study is not usually necessary. Video EEG studies may be considered to characterize the seizures and provide electroclinical correlation. The need for other investigations arises in the case of a focal- or localization-related epilepsy to rule out a structural lesion or when a symptomatic etiology is suspected. If it is deemed that the clinical profile is not consistent with the well-established clinical phenotypes of genetic generalized epilepsies, other investigations may be selected on a case by case basis. The role of special neuroimaging (MRI, MRS), and specific biochemical investigations when considering inborn errors of metabolism associated with epilepsy are also addressed.

The ILAE commission on genetic testing has brought out a detailed special report looking at the clinical contexts of genetic testing (diagnostic, predictive, prenatal and carrier), the molecular methods (sequencing, mutation scanning, targeted FISH and mutation analysis, arrayCGH, SNP arrays, and MPLA), and evaluating their potential benefits and harms. The special report specifically looks at the clinical validity and utility of diagnostic testing in individuals affected with epilepsy, and in unaffected relatives, as well as the ethical, social and legal implications of such testing. For the present, the key message is that despite availability of gene testing technologies, and the many genes identified, few have current clinical utility in the context of the common genetic epilepsies. The report summarizes the current status of such testing and it is likely that with time and advances in the field, recommendations may change (*9*). Genetic testing may be particularly important when epilepsy is associated with syncopal events and sudden death (*10*). Genetic counseling has become a central component of the comprehensive care of the family with a child who has epilepsy. By providing the family with accurate genetic risks for specific epilepsy syndromes, and empirical risks for complex polygenic epilepsies, the physician enables the family to make informed life decisions. The central issues relating to genetic counseling involve the heritability of epilepsy syndrome, the risk of epilepsy in family members and offspring, and the value of genetic tests in epilepsy (*3,9*).

The genetic makeup of an individual influences the response to a medication. In the era of personalized medicine, pharmacogenomics is emerging as a major field of interest. The medical challenges in treating women with epilepsy extend to the fetus and the risk of teratogenic defects particularly when antiepileptic medications are used in pregnancy is also described. Finally, new gene sequencing technologies, proteomics, and epigenomics hold the promise of opening many windows of understanding, and hopefully will lead to methods of early detection, counseling and designing effective therapies for affected individuals and their family stricken by this complex disorder.

REFERENCES

1. (a) Dichter, M. A.; Ayala, G. F. Cellular Mechanisms of Epilepsy: A Status Report. *Science* **1987**, *237* (4811), 157–164.
   (b) Blume, W. T. Invited Review: Clinical and Basic Neurophysiology of Generalised Epilepsies. *Can. J. Neurol. Sci.* **2002**, *29* (1), 6–18.
   (c) Dalby, N. O.; Mody, I. The Process of Epileptogenesis: A Pathophysiological Approach. *Curr. Opin. Neurol.* **2001**, *14* (2), 187–192.
   (d) Blumenfeld, H. Cellular and Network Mechanisms of Spike-Wave Seizures. *Epilepsia* **2005**, *46* (Suppl 9), 21–33.
2. Berg, A. T.; Berkovic, S. F.; Brodie, M. J.; Buchhalter, J.; Cross, J. H.; van Emde Boas, W.; Engel, J.; French, J.; Glauser, T. A.; Mathern, G. W., et al. Revised Terminology and Concepts for Organization of Seizures and Epilepsies: Report of the ILAE Commission on Classification and Terminology, 2005–2009. *Epilepsia* **2010**, *51* (4), 676–685.
3. Winawer, M. R. Epilepsy Genetics. *Neurologist* **2002**, *8* (3), 133–151.
4. Steinlein, O. K.; Mulley, J. C.; Propping, P.; Wallace, R. H.; Phillips, H. A.; Sutherland, G. R.; Scheffer, I. E.; Berkovic, S. F. A Missense Mutation in the Neuronal Nicotinic Acetylcholine Receptor Alpha 4 Subunit Is Associated with Autosomal Dominant Nocturnal Frontal Lobe Epilepsy. *Nat. Genet.* **1995**, *11* (2), 201–203.
5. (a) Dibbens, L. M.; Heron, S. E.; Mulley, J. C. A Polygenic Heterogeneity Model for Common Epilepsies with Complex Genetics. *Genes Brain Behav.* **2007**, *6* (7), 593–597.
   (b) Di Rienzo, A. Population Genetics Models of Common Diseases. *Curr. Opin. Genet. Dev.* **2006**, *16* (6), 630–636.
6. Miller, D. T.; Adam, M. P.; Aradhya, S.; Biesecker, L. G.; Brothman, A. R.; Carter, N. P.; Church, D. M.; Crolla, J. A.; Eichler, E. E.; Epstein, C. J., et al. Consensus Statement: Chromosomal Microarray Is a First-Tier Clinical Diagnostic Test for Individuals with Developmental Disabilities or Congenital Anomalies. *Am. J. Hum. Genet.* **2010**, *86* (5), 749–764.
7. de Kovel, C. G.; Trucks, H.; Helbig, I.; Mefford, H. C.; Baker, C.; Leu, C.; Kluck, C.; Muhle, H.; von Spiczak, S.; Ostertag, P., et al. Recurrent Microdeletions at 15q11.2 and 16p13.11 Predispose to Idiopathic Generalized Epilepsies. *Brain* **2010**, *133* (Pt 1), 23–32.
8. (a) Heinzen, E. L.; Radtke, R. A.; Urban, T. J.; Cavalleri, G. L.; Depondt, C.; Need, A. C.; Walley, N. M.; Nicoletti, P.; Ge, D.; Catarino, C. B., et al. Rare Deletions at 16p13.11 Predispose to a Diverse Spectrum of Sporadic Epilepsy Syndromes. *Am. J. Hum. Genet.* **2010**, *86* (5), 707–718.
   (b) Mefford, H. C.; Muhle, H.; Ostertag, P.; von Spiczak, S.; Buysse, K.; Baker, C.; Franke, A.; Malafosse, A.; Genton, P.; Thomas, P., et al. Genome-Wide Copy Number Variation in Epilepsy: Novel Susceptibility Loci in Idiopathic Generalized and Focal Epilepsies. *PLoS Genet.* **2010**, *6* (5), e1000962.
9. Ottman, R.; Hirose, S.; Jain, S.; Lerche, H.; Lopes-Cendes, I.; Noebels, J. L.; Serratosa, J.; Zara, F.; Scheffer, I. E. Genetic Testing in the Epilepsies–Report of the ILAE Genetics Commission. *Epilepsia* **2010**, *51* (4), 655–670.
10. (a) Goldman, A. M.; Glasscock, E.; Yoo, J.; Chen, T. T.; Klassen, T. L.; Noebels, J. L. Arrhythmia in Heart and Brain: KCNQ1 Mutations Link Epilepsy and Sudden Unexplained Death. *Sci. Transl. Med.* **2009**, *1* (2), 2ra6.
    (b) Taggart, N. W.; Haglund, C. M.; Tester, D. J.; Ackerman, M. J. Diagnostic Miscues in Congenital Long-QT Syndrome. *Circulation* **2007**, *115* (20), 2613–2620.

CHAPTER

116

Basal Ganglia Disorders

Andrew B West, Michelle Gray and David G Standaert

Department of Neurology, Center for Neurodegeneration and Experimental Therapeutics, University of Alabama at Birmingham, AL, USA

The term *Basal ganglia disorder* infers dysfunction of subcortical brain structures that include the neostriatum (caudate nucleus and putamen), globus pallidus (external and internal parts), subthalamic nucleus, and substantia nigra. Clinically, extrapyramidal disorders can be divided into hypokinetic (associated with a paucity of movement) and hyperkinetic (associated with excessive movement) syndromes. The former for the most part is represented by forms of parkinsonism, and is characterized by varying components of hypokinesia, bradykinesia, and rigidity, while the latter includes dystonia, athetosis, chorea, tremor, and tics. Although the motor aspects are the signature feature of these disorders, and the one that usually brings them to clinical attention, almost all patients with extrapyramidal diseases manifest cognitive or behavioral abnormalities, although sometimes of a subtle nature, that reflect the important role of the basal ganglia in cognitive function.

Parkinson disease (PD), a neurodegenerative condition, is by far the most common cause of parkinsonism in most clinical populations. In the case of a patient with parkinsonism, the major diagnostic considerations are PD or the associated condition dementia with Lewy bodies, drug-induced parkinsonism, progressive supranuclear palsy, multiple systems atrophy, Wilson disease, and corticobasal ganglionic degeneration. The most commonly used clinical criteria for the diagnosis of PD require the presence of bradykinesia along with at least one additional feature: muscular rigidity, 4–6 Hz rest tremor, or postural instability. An interesting development in recent years has been the recognition that in addition to these late, nonmotor complications of PD, there is also a prodromal syndrome preceding the classic motor features. PD was long considered to have little or no significant genetic contribution to disease susceptibility. The recent identification of specific genetic causes of PD has revolutionized research in neurodegeneration. The first gene identified in familial PD, a-synuclein, is likely the principal modifier and most important protein for disease. Additional genetic factors, including dominant genes, recessive genes, and risk factors, have been identified and are contributing to understanding the fundamental mechanisms of the disease.

Dystonia is a clinical syndrome, identified by its characteristic features: sustained muscle contractions, twisting, and abnormal postures. Dystonia may be the main, or only, clinical symptom, or it may be seen in association with a variety of other disorders, such as stroke, PD, or other degenerative disorders of the brain. Most of the cases of primary dystonia encountered by clinicians are sporadic in nature, and of uncertain etiology. Dystonia, in contrast to the other disorders described, lacks distinctive neuropathology. As in PD, however, in recent years genetic etiologies for primary dystonia have been identified, and these have begun to shed some light on the fundamental mechanisms responsible for these disorders.

Huntington disease (HD) is caused by a mutation in the gene for the protein *huntingtin*, and is inherited in an autosomal dominant pattern. The mutation is an expansion of a CAG triplet repeat within the protein-coding domain. The penetrance approaches 100% but the age of onset is variable, depending on the size of the expansion as well as other factors (some of which are likely heritable as well). In HD, there is atrophy of the caudate and putamen, accounting for the characteristic movement disorder, as well as more general atrophy of the cerebral hemispheres.

Tourette syndrome (TS) is a tic disorder that appears most often in childhood or adolescence. A tic is a sudden, rapid, recurrent, nonrhythmic, stereotyped motor movement or vocalization. In addition to the movements, TS is also associated with neuropsychiatric features, particularly obsessive compulsive disorder, which compounds the burden of the condition. Although there is no evidence for neurodegeneration in TS, recent morphometric studies have suggested developmental abnormalities of neuronal migration and differentiation. Familial clustering of TS is common and suggests an important contribution of heritability. Some candidate genes have been identified in specific pedigrees but in most cases the genes involved remain elusive.

Together, basal ganglia disorders represent a range of different pathologies that result in symptoms related to disordered movement. The discovery of genetic causes for these conditions has provided insight into fundamental mechanisms, but bridging the gap between the genetic defects and the systems-level abnormalities in brain function remains a major challenge for the field.

FURTHER READING

Jankovic, J.; Kurlan, R. Tourette Syndrome: Evolving Concepts. *Mov. Disord.* **2011,** *26* (6), 1149–1156.

Breakefield, X. O.; Blood, A. J.; Li, Y.; Hallett, M.; Hanson, P. I.; Standaert, D. G. The Pathophysiological Basis of Dystonias. *Nat. Rev. Neurosci.* **2008,** *9* (3), 222–234.

Langston, J. W. The Parkinson's Complex: Parkinsonism Is Just the Tip of the Iceberg. *Ann. Neurol.* **2006,** *59* (4), 591–596.

Ross, C. A.; Tabrizi, S. J. Huntington's Disease: From Molecular Pathogenesis to Clinical Treatment. *Lancet Neurol.* **2011 Jan** *10* (1), 83–98.

CHAPTER

117

The Hereditary Ataxias

Puneet Opal

Davee Department of Neurology, and Department of Cell and Molecular Biology, Northwestern University Medical School, Chicago, IL, USA

Huda Zoghbi

Departments of Pediatrics, Neurology, Neuroscience, and Molecular and Human Genetics, Baylor College of Medicine, Houston, TX, USA

The hereditary ataxias are a group of genetic neurological disorders characterized by motor in-coordination resulting from dysfunction of the cerebellum and its connections. These syndromes have traditionally been divided into childhood ataxias characterized by ataxia as a result of underlying enzymatic defects and late onset progressive degenerative ataxias. Both groups constitute a broad class of diseases that can be further divided based on the mode of inheritance.

Childhood metabolic ataxias are typically recessive disorders and usually present at least initially as intermittent ataxias that correlate with the severity of the metabolic dysfunction. They are usually diagnosed by screening biochemical tests when the neurological abnormalities are first noticed. Apart from the supportive therapy of hydration and dietary restriction for many of the metabolic ataxias, there is yet no treatment for the underlying enzyme deficiency. Gene therapy for these diseases is still in its infancy.

The progressive ataxias are typically later in onset. A subset of recessive ataxias can be caused by deficiency in factors that cause biochemical abnormalities and are thus also disorders of metabolism. However, unlike the intermittent childhood ataxias, the cumulative damage must reach a threshold before the clinical signs appear. The persistence of the biochemical abnormalities explains the progressive nature of the symptoms. Storage disorders, Wilson disease, and ataxia with isolated vitamin E deficiency predominate in this group of ataxias. These disorders are important to diagnose because measures can be taken that have significant impact on the natural history of these disorders. In addition, to the metabolic progressive ataxias, there are a group of recessive ataxias that are caused by other genetic defects. The most common recessive ataxia is Friedreich's ataxia. This ataxia is caused by a nucleotide repeat expansion, a mutational mechanism that also underlies many of the autosomal dominant ataxic syndromes. Other progressive ataxias are caused by mutations involved in proteins that regulate DNA repair mechanisms (of which ataxia telangiectasia is the most common). Finally, there are several less common progressive recessive spinocerebellar ataxias that are being classified as spinocerebellar ataxias with recessive inheritance. The progressive recessive ataxias typically have multisystem involvement or associated non-ataxic neurological features and they are all difficult to treat.

The autosomal dominant spinocerebellar ataxias are referred to as SCAs. All progressive ataxias are typically associated with cerebellar atrophy. The genetics of the progressive recessive and dominant ataxias is slowly being elucidated and there are now commercial genetic tests for several. Many of these disorders as alluded to earlier are tri-, tetra-, or pentanucleotide repeat disorders. In addition to these ataxias, there are several dominantly inherited episodic ataxias. Not all the mutations have been identified, but those that have appeared to be caused by mutations that affect the function of ion channels. Finally, there are rare genetic causes of adult-onset ataxia that are X linked or caused by mutations in the mitochondrial genome. Although the current status of treatment for all these progressive ataxias is still symptomatic, we have stressed the importance of basic science research that promises to bring advances to the bedside.

WEBSITES

http://www.ncbi.nlm.nih.gov/omim: Online Mendelian Inheritance in Man. This is the premiere website for genetic information and references to original scientific publications across the spectrum of human genetic diseases.

http://www.ncbi.nlm.nih.gov/sites/GeneTests/: The GeneTests website is a publicly funded medical genetics information resource developed for physicians, other healthcare providers, and researchers.

http://www.uptodate.com/index: A subscription-based website that serves as an online textbook. This is written by experts and is constantly updated.

http://clinicaltrials.gov/: This website provides a registry of federally and privately supported clinical trials conducted in the United States and around the world.

http://www.ataxia.org/: This is the official website of the National Ataxia Foundation. This is an excellent site to refer patients and families suffering from rare ataxias to learn more about the disease.

CHAPTER

118

Hereditary Spastic Paraplegia

John K Fink

Department of Neurology, University of Michigan and Geriatric Research Education and Care Center,
Ann Arbor Veterans Affairs Medical Center, Ann Arbor, MI, USA

118.1 CLINICAL OVERVIEW

The hereditary spastic paraplegias (HSPs) are clinically and genetically diverse disorders that share the primary feature of progressive, lower extremity spastic weakness (for HSP reviews, see Refs. (*1,2*)). Each of the genetic types of HSP represents a separate disorder in which spastic gait is a major syndromic feature and disability factor. HSP syndromes are classified clinically according to the mode of inheritance and whether progressive spasticity occurs in isolation ("uncomplicated HSP") or with other neurologic abnormalities ("complicated HSP") such as muscle wasting (*3*), peripheral neuropathy, cataracts, ataxia, mental retardation, or dementia. Although progressive worsening of spastic gait is common for most types of HSP, it is important to note that some forms of HSP begin in very early childhood (e.g. SPG3A HSP due to atlastin mutation), may have little clinical progression, and resemble spastic diplegic cerebral palsy (*4*).

There may be marked clinical variation between genetic types of HSP as well as between individuals in the same family who have the same *HSP* gene mutation. There is increasing awareness, for example that a given type of HSP may manifest as either an "uncomplicated" or a "complicated" HSP syndrome.

HSP diagnosis is based on clinical signs and symptoms and exclusion of alternate disorders. Family history is important but may be absent (e.g. recessive HSP, new mutation, nonpaternity). Genetic testing is increasingly available and presently can confirm the diagnosis in ~75% of subjects with dominantly inherited HSP; and ~25–50% of subjects with autosomal recessive HSP. In addition, depending on the extent of genetic testing, ~10–20% of subjects with no family history of HSP ("apparently sporadic spastic paraplegia," discussed below) can be shown to have a potentially pathogenic mutation in an *HSP* gene for which testing is clinically available (*5,6*).

The differential diagnosis (discussed in References (*2,7*)) should be considered carefully in every subject (particularly in subjects with no obvious family history and those with variant presentations) including structural disorders affecting brain or spinal cord; spinal cord arteriovenous malformation, leukodystrophies (including B12, adrenomyeloneuropathy, Krabbe, metachromatic leukodystrophy, multiple sclerosis), dopa-responsive dystonia, amyotrophic lateral sclerosis, primary lateral sclerosis, and tropical spastic paraparesis.

118.2 GENETIC HETEROGENEITY

There are more than 50 genetic types of HSPs (designated spastic paraplegia [SPG] loci in order of their discovery) including autosomal dominant, autosomal recessive, X-linked, and maternally inherited forms.

118.3 NEUROPATHOLOGY

The major neuropathologic feature of uncomplicated HSP is axon degeneration that is maximal in the terminal portions of the longest descending and ascending tracts (crossed and uncrossed corticospinal tracts to the legs and *fasciculus gracilis*, respectively). Spinocerebellar fibers are involved to a lesser extent. In general, neuronal cell bodies of degenerating fibers are preserved although minor decrease in anterior horn cell number is observed in some cases. Axon degeneration in uncomplicated HSP involves both motor and sensory fibers (corticospinal tract fibers and *fasciculus gracilus* fibers). One obvious feature shared by these degenerating axons is their length. These are among the longest fibers in the central nervous system (CNS). Degeneration is maximal in the distal axons of these fibers. Complicated forms of HSP have additional, syndrome-specific neuropathology.

118.4 MOLECULAR BASIS

HSP gene discoveries are yielding important insights into molecular processes underlying HSP. The 25 *HSP* genes discovered to date have diverse functions, indicating that long axon degeneration in HSP can arise from disturbance in microtubule dynamics (e.g. SPG4/spastin), axonal transport (e.g. SPG10/KIF5A), mitochondrial

TABLE 118-1 Genetic Types of HSP

Spastic Gait (SPG) Locus	HSP Syndrome	Protein Name and Function	Gene Testing
Autosomal dominant HSP			
SPG3A (14q11-q21)	Uncomplicated HSP: symptoms usually begin in childhood (and may be nonprogressive); symptoms may also begin in adolescence or adulthood and worsen insidiously. Genetic nonpenetrance reported. De novo mutation reported presenting as spastic diplegic cerebral palsy	Atlastin: dynamin family GTPase, interacts with spastin and REEP1; contributes to endoplasmic reticulum morphology	ADL
SPG4 (chr.2p22)	Uncomplicated HSP, symptom onset in infancy through senescence, single most common cause of autosomal dominant HSP (~40%); some subjects have late onset cognitive impairment	Spastin: cytosolic (and possibly nuclear) protein with AAA domain; (AAA domain is also present in paraplegin); interacts with microtubules and has microtubule severing properties; interacts with atlastin and REEP1 and contributes to endoplasmic reticulum morphology	ADL[a]
SPG6 (15q11.1)	Uncomplicated HSP: prototypical late-adolescent, early adult onset, slowly progressive uncomplicated HSP	"Not imprinted in Prader Willi/Angelman 1" (NIPA1): nine alternating hydrophobic–hydrophilic domains predicts integral membrane localization; NIPA1 binds to BMP-II receptor to inhibit BMP signaling; NIPA1 transcription is induced by low extracellular Mg^{++}; NIPA1 expression causes inwardly directly Mg^{++} conductance	ADL
SPG8 (8q23-q24)	Uncomplicated HSP	KIAA0196/Strumpellin, mutations may be pathogenic through protein aggregation: Strumpellin binds to valosin-containing protein (VCP; also known as p97, TER ATPase and Cdc48p); VCP-positive inclusions occur in a wide variety of neuro-degenerative disorders including Parkinson's disease, Lewy body disease, Huntington's disease, amyotrophic lateral sclerosis, and spinocerebellar ataxia type III (Machado-Joseph disease). Strumpellin may also regulate actin dynamics through its interaction with WASH	ADL
SPG9 (10q23.3-q24.2)	Complicated: spastic paraplegia associated with cataracts, gastroesophageal reflux, and motor neuronopathy	Unknown	No
SPG10 (12q13)	Uncomplicated HSP or complicated by distal muscle atrophy	Kinesin heavy chain (KIF5A), molecular motor involved in axonal transport	Research
SPG12 (19q13)	Uncomplicated HSP	Unknown	
SPG13 (2q24-34)	Uncomplicated HSP: adolescent and adult onset	*HSPD1* gene encodes Chaperonin 60 (also known as heat shock protein 60, HSP60), the large subunit of the mitochondrial Hsp60/Hsp10 chaperonin complex; functions in binding and sequestration, and refolding of unfolded proteins	Research
SPG17 (11q12-q14)	Complicated: spastic paraplegia associated with amyotrophy of hand muscles (Silver syndrome)	BSCL2/seipin: Endoplasmic reticulum transmembrane protein; mutations appear pathogenic through induction of endoplasmic reticulum stress-mediated apoptosis	Research
SPG19 (9q33-q34)	Uncomplicated HSP	Unknown	No
Autosomal dominant HSP			
SPG29 (1p31.1-21.1)	Complicated: spastic paraplegia associated with hearing impairment; persistent vomiting due to hiatal hernia inherited	Unknown	No

TABLE 118-1 Genetic Types of HSP—cont'd

Spastic Gait (SPG) Locus	HSP Syndrome	Protein Name and Function	Gene Testing
SPG31 (2p12)	Uncomplicated HSP or occasionally associated with peripheral neuropathy	Receptor expression enhancing protein 1 (REEP1), structurally related to the DP1/Yop1p family of endoplasmic reticulum-shaping proteins; required for ER network formation in vitro; forms complexes with atlastin and spastin within endoplasmic reticulum; also binds to microtubules and promotes ER alignment along the microtubule cytoskeleton in vitro	ADL
SPG33 (10q24.2)	Uncomplicated HSP	There is controversy whether ZFYVE27/protrudin: mutations identified by Mannan et al. are pathogenic. Martignoni et al. felt they may be benign polymorphisms that did not alter biochemical function. ZFYVE27/protruding has a role in Rab-11-mediated membrane trafficking and promotes neurite outgrowth	Research
SPG36 (12q23-24)	Onset age 14–28 years, associated with motor sensory neuropathy	Unknown	No
SPG37 (8p21.1-q13.3)	Uncomplicated HSP	Unknown	No
SPG38 (4p16-p15)	One family, five affected subjects, onset age 16–21 years. Subjects had atrophy of intrinsic hand muscles (severe in one subject at age 58)	Unknown	No
SPG40 (locus unknown)	Uncomplicated spastic paraplegia, onset after age 35, known autosomal dominant HSP loci excluded	Unknown	No
SPG41 (11p14.1-p11.2	Single Chinese family with adolescent onset, spastic paraplegia associated with mild weakness of intrinsic hand muscles	Unknown	No
SPG42 (3q24-q26)	Uncomplicated spastic paraplegia reported in single kindred, onset age 4–40 years, possibly one instance of incomplete penetrance	Acetyl CoA transporter (SLC33A1)	No
Autosomal recessive HSP			
SPG5 (8p)	Uncomplicated or complicated by axonal neuropathy, distal or generalized muscle atrophy, and white matter abnormalities on MRI	CYP7B1	Research
SPG7 (16q)	Uncomplicated or complicated: variably associated with mitochondrial abnormalities on skeletal muscle biopsy and dysarthria, dysphagia, optic disc pallor, axonal neuropathy, and evidence of "vascular lesions," cerebellar atrophy, or cerebral atrophy on cranial MRI	Paraplegin: mitochondrial metalloprotease	ADL et al
SPG11 (15q)	Uncomplicated or complicated: spastic paraplegia variably associated with thin corpus callosum, mental retardation, upper extremity weakness, dysarthria, and nystagmus; may have "Kjellin syndrome": childhood onset, progressive spastic paraplegia accompanied by pigmentary retinopathy, mental retardation, dysarthria, dementia, and distal muscle atrophy; juvenile, slowly progressive ALS reported in subjects with SPG11 HSP; 50% of autosomal recessive HSP is considered to be SPG11	Spatacsin (KIAA1840): function is unknown	Research

Continued

TABLE 118-1 Genetic Types of HSP—cont'd

Spastic Gait (SPG) Locus	HSP Syndrome	Protein Name and Function	Gene Testing
SPG14 (3q27-28)	Single consanguineous Italian family, three affected subjects, onset age ~30 years; Complicated spastic paraplegia with mental retardation and distal motor neuropathy (sural nerve biopsy was normal)	Unknown	No
SPG15 (14q)	Complicated: spastic paraplegia variably associated with associated with pigmented maculopathy, distal amyotrophy, dysarthria, mental retardation, and further intellectual deterioration (Kjellin syndrome)	Spastizin/ZFYVE26	Research
SPG18 (8p12-p11.21)	Two families described with spastic paraplegia complicated by mental retardation and thin corpus callosum	Unknown	No
SPG20 (13q)	Complicated: spastic paraplegia associated with distal muscle wasting (Troyer syndrome)	Spartin: N-terminal region similar to spastin; homologous to proteins involved in the morphology and trafficking of endosomes; also localizes to mitochondria	Research
SPG21 (15q21-q22)	Complicated: spastic paraplegia associated with dementia, cerebellar and extrapyramidal signs, thin corpus callosum, and white matter abnormalities (Mast syndrome)	Maspardin: protein localizes to endosome/trans-golgi vesicles, may function as protein transport and sorting	Research
SPG23 (1q24-q32)	Complicated: childhood onset HSP associated with skin pigment abnormality (vitiligo), premature graying, characteristic facies; Lison syndrome	Unknown	No
SPG24 (13q14)	Complicated: childhood onset HSP variably complicated by spastic dysarthira and pseudobulbar signs	Unknown	No
SPG25 (6q23-q24.1)	Consanguineous Italian family, four subjects with adult (30–46 years) onset back and neck pain related to disk herniation and spastic paraplegia; surgical correction of disk herniation ameliorated pain and reduced spastic paraplegia. Peripheral neuropathy also present	Unknown	No
SPG26 (12p11.1–12q14)	Single consanguineous Bedouin family with five affected subjects. Complicated: childhood onset (between 7 and 8 years), progressive spastic paraparesis with dysarthria and distal amyotrophy in both upper and lower limbs, nerve conduction studies were normal; mild intellectual impairment, normal brain MRI	Unknown	No
SPG27 (10q22.1-q24.1)	Complicated or uncomplicated HSP. Two families described. In one family (seven affected subjects), uncomplicated spastic paraplegia began between ages 25 and 45 years. In the second family (three subjects described), the disorder began in childhood and included spastic paraplegia, ataxia, dysarthria, mental retardation, sensorimotor polyneuropathy, facial dysmorphism and short stature	Unknown	No

TABLE 118-1 Genetic Types of HSP—cont'd

Spastic Gait (SPG) Locus	HSP Syndrome	Protein Name and Function	Gene Testing
SPG28 (14q21.3-q22.3)	Uncomplicated: childhood onset progressive spastic gait	Unknown	No
SPG29 (14q)	Uncomplicated HSP, childhood onset	Unknown	No
SPG30 (2q37.3)	Complicated: spastic paraplegia, distal wasting, saccadic ocular pursuit, peripheral neuropathy, mild cerebellar signs	Unknown	No
SPG32 (14q12-q21)	Mild mental retardation, brainstem dysraphia, clinically asymptomatic cerebellar atrophy	Unknown	No
SPG35 (16q21-q23)	Childhood onset (6–11 years), spastic paraplegia with extrapyramidal features, progressive dysarthria, dementia, seizures. Brain white matter abnormalities and brain iron accumulation; an Omani and a Pakistani kindred reported	Fatty acid 2-hydroxylase (FA2H)	Research
SPG39 (19p13)	Complicated: spastic paraplegia associated with wasting of distal upper and lower extremity muscles	Neuropathy target esterase: phospholipase localized to ER membranes; implicated in toxic organophosphorus compound induced chronic neurodegeneration; may function to regulate cyclic AMP-dependent protein kinase	Research
SPG43 (19p13.11-q12)	Two sisters from Mali, symptom onset 7 and 12 years, progressive spastic paraplegia with atrophy of intrinsic hand muscles and dysarthria (one sister)	Unknown	No
SPG44 (1q41)	Allelic with "Pelizeaus-Merzbacher-like disease" (PMLD, early onset dysmyelinating disorder with nystagmus, psychomotor delay, progressive spasticity, ataxia). GJA/GJC2 mutation I33M causes a milder phenotype: late onset (first and second decades), cognitive impairment, slowly progressive, spastic paraplegia, dysarthria, and upper extremity involvement. MRI and MR spectroscopy imaging consistent with a hypomyelinating leukoencephalopathy	Gap junction protein connexin47 (Cx47) encoded by GJA12/GJC2 gene	
SPG45 (10q24.3-q25.1)	Single consanguineous kindred from Turkey, five subjects described: affected subjects had mental retardation, infantile onset lower extremity spasticity and contractures, one subject with optic atrophy, two subjects with pendular nystagmus; MRI in one subject was normal	Unknown	No
SPG46 (9p21.2-q21.12)	Dementia, congenital cataract, ataxia, thin corpus callosum	Unknown	No
SPG47 (1p13.2-1p12)	Two affected siblings from consanguineous Arabic family with early childhood onset slowly progressive spastic paraparesis, mental retardation, and seizures; one subject had ventriculomegaly; the other subject had thin corpus callosum and periventricular white matter abnormalities	Unknown	No

Continued

TABLE 118-1 Genetic Types of HSP—cont'd

Spastic Gait (SPG) Locus	HSP Syndrome	Protein Name and Function	Gene Testing
SPG48 (7p22.1)	Analysis of *KIAA0415* gene in 166 unrelated spastic paraplegia subjects (38 recessive, 64 dominant, 64 "apparently sporadic") and control subjects revealed homozygous mutation in two siblings with late-onset (6th decade) uncomplicated spastic paraplegia; and heterozygous mutation in one subject with apparently sporadic spastic paraplegia	KIAA0415 encodes a putative helicase involved in DNA repair; interacts with SPG11/Spatacsin and SPG15/ZFYVE26	Research
"SPOAN" syndrome (11q23)	Complicated: Spastic paraplegia, optic atrophy, neuropathy (SPOAN)	Unknown	No
5p15.31-14.1 No SPG designation	Complicated spastic paraplegia associated with mutilating sensory neuropathy	Epsilon subunit of the cytosolic chaperonin-containing t-complex peptide-1 (*Cct5*) gene	Research
X-linked HSP			
SPG1 (Xq28)	Complicated: associated with mental retardation, and variably, hydrocephalus, aphasia, and adducted thumbs	L1CAM	Research
SPG2 (Xq28)	Complicated: variably associated with MRI evidence of CNS white matter abnormality; may have peripheral neuropathy	Proteolipid protein	Several labs[b]
SPG16 (Xq11.2-q23)	Uncomplicated; or complicated: associated with motor aphasia, reduced vision, nystagmus, mild mental retardation, and dysfunction of the bowel and bladder	Unknown	No
SPG22 (Xq21)	Complicated (Allan–Herndon–Dudley syndrome): congenital onset, neck muscle hypotonia in infancy, mental retardation, dysarthria, ataxia, spastic paraplegia, abnormal facies	*MCT8* gene encodes a thyroid hormone transporter, results in elevated serum triiodothyronine (T3) levels	Research
SPG34	Uncomplicated, onset 12–25 years		
Maternal inheritance (Mitochondrial genome)			
	Adult onset, progressive spastic paraplegia, mild to severe symptoms, variably associated with axonal neuropathy, late-onset dementia, cardiomyopathy	Mitochondrial *ATP6* gene (mutations also cause Leigh's disease, familial striate necrosis)	Research

[a]ADL: Athena Diagnostics Laboratory, Boston.
[b]Several laboratories including Dupont Nemours Clinic and Baylor University.
Modified from [6].

function (e.g. SPG7/paraplegin, SPG13/chaperonin 60, SPG31/REEP1, and mATP6 which are mitochondrial proteins), corticospinal tract development (SPG1/L1CAM), myelination abnormality (SPG2/proteolipid protein and SPG42/GJA12), endoplasmic reticulum morphology (e.g. SPG3A/atlastin, SPG4/spastin, and SPG31/REEP1), membrane trafficking disturbance (*8*), and protein accumulation/endoplasmic reticulum stress response (e.g. SPG8/Strumpellin, SPG17/BSCL2, Seipin).

118.5 TREATMENT

Presently, there are no treatments to stop, prevent, or reverse disability in HSP. Rehabilitative approaches are recommended including exercise, gait and balance training, spasticity reducing medications (e.g. oral or intrathecal Lioresal), ankle–foot orthotic devices, and medication to reduce urinary urgency (e.g. Oxybutyin).

REFERENCES

1. Fink, J. K. Hereditary Spastic Paraplegia. *Curr. Neurol. Neurosci. Rep.* **2006**, *6*, 65–76.
2. Fink, J. K. Hereditary Spastic Paraplegia. In *Emery and Rimoin's Principles and Practice of Medical Genetics*, 5th ed.; 2007; pp 2771–2801.
3. Rainier, S.; Albers, J. W.; Richardson, R. J. et al. Autosomal Recessive Progressive Spastic Paraplegia with Distal Muscle Wasting due to Neurotoxic Esterase Gene Mutation: Clinical Features of the Index Family. Manuscript submitted. 2006.

4. Rainier, S.; Sher, C.; Reish, O., et al. De novo Occurrence of Novel *SPG3A*/Atlastin Mutation Presenting as Cerebral Palsy. *Arch. Neurol.* **2006,** *63,* 445–447.
5. Brugman, F.; Scheffer, H.; Wokke, J. H. J., et al. Paraplegin Mutations in Apparently Sporadic Adult-Onset Upper Motor Neuron Syndromes. *Neurology* **2008,** *71,* 1500–1505.
6. Brugman, F.; Wokke, J. H.; Scheffer, H., et al. Spastin Mutations in Sporadic Adult-Onset Upper Motor Neuron Syndromes. *Ann. Neurol.* **2005,** *58,* 865–869.
7. Fink, J. K. Hereditary Myelopathies. *Continuum (N. Y.)* **2008,** *14,* 58–74.
8. Blackstone, C.; O'Kane, C. J.; Reid, E. Hereditary Spastic Paraplegias: Membrane Traffic and the Motor Pathway. *Nat. Rev. Neurosci.* **2011,** *12,* 31–42.

CHAPTER

119

Autonomic and Sensory Disorders

Felicia B Axelrod

NYU Dysautonomia Center, NYU Langone Medical Center, New York, NY, USA

The hereditary sensory and autonomic neuropathies (HSANs) are a rare group of disorders that illustrate the intimate relationship of the development and maintenance of sensory and autonomic neuronal populations. A variable degree of analgesia is common to all the disorders. The perturbations in autonomic dysfunction are even more variable, as the autonomic nervous system is pervasive and integrates multiple secondary functions, resulting in widespread and confounding symptoms (*1*). Diagnosis depends on clinical and biochemical evaluations, with pathologic examinations serving to further confirm differences (*2*).

In recent years, specific genetic mutations have been identified for some HSANs. With the exception of hereditary sensory radicular neuropathy (HSAN type I), which is a dominant disorder presenting in the second decade, the other HSANs are autosomal recessive disorders whose mutations impede normal neural crest migration and differentiation and result in decreased neuronal populations (*2,3*). Familial dysautonomia (Riley–Day syndrome or HSAN type III) and congenital sensory neuropathy (HSAN type IV) are two such disorders which have been particularly well described and categorized and for which genetic mutations have been identified. Table 119-1 summarizes which sensory and autonomic functions are affected in each of the HSAN disorders, as well as the type of inheritance (*4,5*).Table 119-2 lists the specific genetic mutations.

119.1 FAMILIAL DYSAUTONOMIA (RILEY–DAY SYNDROME OR HSAN TYPE III)

The most intensively studied of the HSAN disorders is familial dysautonomia. Signs of the disorder are present from birth and neurological function slowly deteriorates with age (*4,6*). Abnormalities of autonomic function include lack of overflow tears, supersensitivity of the pupil to autonomic drugs, postural hypotension, episodic hypertension, skin blotching, inappropriate hyperhidrosis, and episodic hyperadrenergic vomiting crisis. Sensory dysfunction results in decreased pain and temperature perception, as well as afferent baroreflex failure (*7*). The latter further destabilizes autonomic dysfunction.

Earliest signs include feeding difficulties with a poor suck, uncoordinated swallowing, and frequent misdirection causing aspiration pneumonias. Gastroesophageal reflux further increases the risk of aspiration. Hyperadrenergic vomiting crises associated with irritability, negativistic behavior, hypertension, tachycardia, blotchy erythema, diaphoresis and elevated circulating serum catecholamines are observed in 40% of patients (*7*). Insensitivity to hypoxia and hypercarbia can result in sleep disorders, difficulty in tolerating respiratory insults, and even hypoxemia at high altitudes or during plane travel (*8*).

The consistent lack of overflow emotional tearing is pathognomonic of the disorder and leads to its inclusion as one of the clinical cardinal diagnostic criteria along with absent lingual fungiform papillae, decreased deep tendon reflex and absent axon flare following intradermal histamine phosphate (*4*). Definitive diagnosis is made by finding mutations in *IKBKAP* (*9*). All cases have at least one splicing mutation (a T to C change located at base pair 6 of intron 20 (IVS20+6T>C); over 99% of cases are homozygous for this mutation (*9*).

Supportive treatments include fundoplication and gastrostomy to provide nutrition and avoid aspiration and medications to control labile blood pressure and hyperadrenergic crises. Although respiratory infections are decreasing and increasing numbers of patients are reaching adulthood, renal disease and sudden death remain risks (*10*).

119.2 CONGENITAL SENSORY NEUROPATHY WITH ANHIDROSIS (HSAN TYPE IV)

Congenital sensory neuropathy with anhidrosis (HSAN type IV) is characterized by decreased pain perception, anhydrosis and learning disabilities (*2,5*). The earliest and most common sign of the disorder is usually hyperpyrexia due to anhydrosis. With time, the skin may become thick and calloused with lichenification of palms, dystrophic nails, and areas of hypotrichosis on the scalp. As further evidence of autonomic dysfunction, patients exhibit miosis with dilute intraocular methacholine and have mild postural hypotension, but in contrast to patients with familial dysautonomia, there is compensatory tachycardia.

Decrease in pain and temperature sensation is extensive and profound, and touch also may be affected. Self-inflicted injury is more prominent feature than in familial dysautonomia. Injuries to soft tissues, bones, and joints may go unrecognized with subsequent severe secondary changes. Repeated fractures of the lower extremities and poor healing result in aseptic necrosis, Charcot joints, and even frequent osteomyelitis. Some neurological features are progressive, as reflexes may decrease and vibration and joint position sense may become impaired.

Definitive diagnosis is made by finding mutations in the neurotrophic tyrosine receptor kinase 1 gene (NTRK1) located on chromosome 1 (1q21–q22) (2,5). In contrast to familial dysautonomia that only has three mutations described, numerous NTRK1 mutations have been described for HSAN type IV.

Treatment is supportive. Extreme care must be taken to guard against inadvertent trauma and to treat fevers vigorously. Treatment of the hyperactivity remains difficult with poor response to stimulants.

As illustrated by the two HSAN disorders described above, the HSAN disorders affect both autonomic and sensory function, but there are definite phenotypic differences. Common to all the HSAN disorders is the lack of an axon flare after intradermal histamine (2). By refining clinical studies of the sensory and autonomic function, and with further information regarding neuropathology, distinctions between the other HSAN disorders should become clearer and permit accurate classification. The schematic outline in Figure 119-1 provides a helpful preliminary approach to differential diagnosis, but it is anticipated that this will be modified as new entities are described. As more of the genetic mutations are identified, diagnosis can be more exact and facilitate genetic counseling.

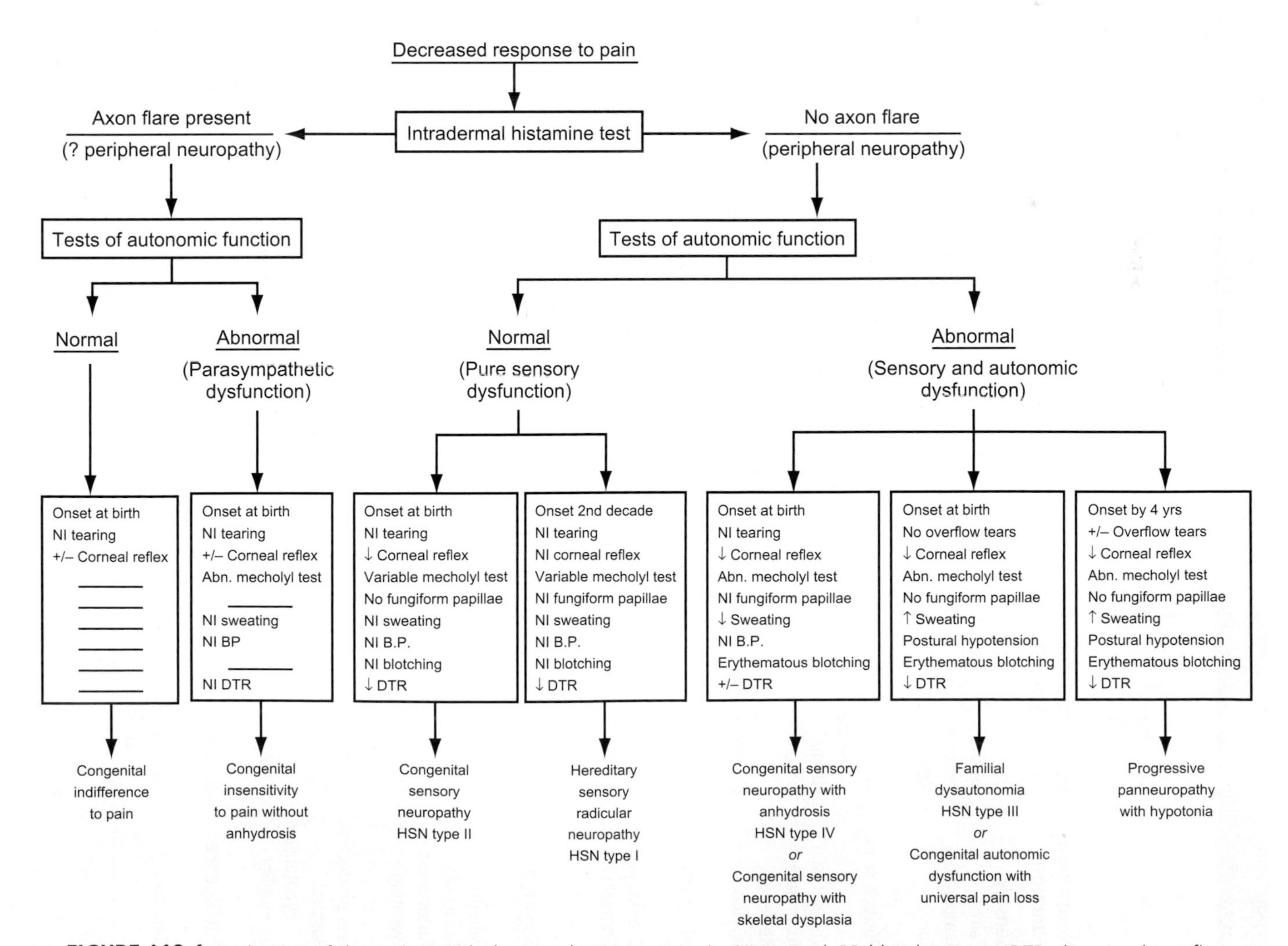

FIGURE 119-1 Evaluation of the patient with decreased response to pain. Nl, normal; BP, blood pressure; DTR, deep tendon reflex.

TABLE 119-1 Affected Neurologic Functions and Genetics of Nine Congenital Sensory Disorders

Neurologic Function	Congential Sensory Familial Dysautonomia (HSAN Type III)	Congenital Insensitivity Neuropathy with Anhidrosis (HSAN Type IV)	Congenital to Pain without Anhidrosis (HSAN Type V)	Congenital Autonomic Sensory Neuropathy (HSAN Type II)	Sensory Neuropathy Dysfunction with Universal Pain Loss	Progressive with Skeletal Dysplasia	Hereditary Panneuropathy with Hypotonia	Radicular Neuropathy (HSAN Type I)	Congenital Indifference to Pain
Sensory									
Pain	Reduced	Absent	Absent	Absent	Absent	Absent	Reduced	Distal absent	Normal
Temperature	Reduced	Reduced	Normal	Absent	Absent	Absent	Reduced	Distal absent	Normal
Touch	Normal	Reduced	Normal	Absent	Reduced	Reduced	Normal	Reduced or normal	Normal
Position sense	Reduced to normal	Normal	Normal	Reduced	?	?	Reduced	Reduced	Normal
Visceral pain	Normal	Reduced	?	?	Absent	Reduced	Normal	?	Normal
Axon flare	Absent	Absent	Normal	Absent	Absent	Absent	Absent	Locally absent	Normal
Reflexes									
Tendon	Hypoactive or absent	Hypoactive or normal	Normal	Hypoactive	Absent	Absent	Absent	Hypoactive or normal	Normal
Superficial	Hypoactive or normal	Hypoactive or normal	?	Hypoactive	Absent	Hypoactive	?	Normal	Normal
Cranial nerves									
Corneal (V)	Normal, diminished or absent	Absent	Variable	Diminished	Absent	Absent	Diminished	?	Normal or diminished
Gag (IX, X)	Diminished or normal	Normal	?	Diminished	Diminished	Diminished	Diminished	?	Normal
Pain (V)	Reduced	Reduced	?	Absent	Absent	Absent	Reduced	?	Normal
Taste (XII)	Reduced	Normal	Normal	Reduced	?	?	Reduced	?	Normal
Autonomic									
Sweating	Increased (with stress)	Absent	Normal	Reduced or normal	Increased (with stress)	Reduced	Increased	Reduced or normal	Normal
Tear production	Decreased	Normal	Decreased	Normal	Decreased	Normal	Normal, then reduced	Normal	Normal
Postural hypotension	Present	May be present	Not present	Not present	Present	?	Present	Not present	Not present
G1 motility	Abnormal	Normal	Abnormal	Variably abnormal	Variably abnormal	Abnormal	Abnormal	?	Normal
Intelligence	Normal	Decreased	Decreased	Decreased	Decreased	Decreased	Normal	Mild decrease or normal	Normal
Genetics	Autosomal recessive	Autosomal recessive	?Autosomal recessive	?Autosomal recessive	?Autosomal recessive	?	Autosomal recessive	Autosomal recessive	Autosomal recessive

TABLE 119-2 Genetic Mutations Identified for the Hereditary Sensory and Autonomic Neuropathies

Gene	HSAN	Type	Chromosome
Familial dysautonomia	III9q31	9q31	*IKBKAP*
Congenital sensory neuropathy with anhidrosis	IV	1q21–22	TrkA/NGF-receptor
Congenital insensitivity to pain without anhidrosis	V	1p11.2–p13.2	NGF-beta chain –
Congenital sensory neuropathy	II	?	?
Congenital autonomic dysfunction with universal pain loss	—	?	?
Congenital sensory neuropathy with skeletal dysplasia	—	?	?
Progressive panneuropathy with hypotonia	—	?	?
Hereditary radicular neuropathy	I	9q22.1–22.3	*SPTLC1*
		3q13–22	

REFERENCES

1. Axelrod, F. B.; Chelimsky, G.; Weese-Mayer, D. Pediatric Autonomic Disorders. State of the Art. *Pediatrics* **2006**, *118*, 309–321.
2. Axelrod, F. B. Hereditary Sensory and Autonomic Neuropathies: Familial Dysautonomia and Other HSANs. *Clin. Auton. Res.* **2002**, *12* (Suppl. 1), 2–14.
3. Bejaoui, K.; Uchida, Y.; Yasuda, S., et al. Hereditary Sensory Neuropathy Type 1 Mutations Confer Dominant Negative Effects on Serine Palmitoyltransferase, Critical for Sphingolipid Synthesis. *J. Clin. Invest.* **2002**, *110*, 1301–1308.
4. Axelrod, F. B. Familial Dysautonomia (Invited Review). *Muscle Nerve* **2004**, *29*, 352–363.
5. Indo, Y. Genetics of Congenital Insensitivity to Pain with Anhidrosis (CIPA) or Hereditary Sensory and Autonomic Neuropathy Type IV. *Clin. Auton. Res.* **2002**, *12* (Suppl), 20–32.
6. Axelrod, F. B.; Iyer, K.; Fish, I., et al. Progressive Sensory Loss in Familial Dysautonomia. *Pediatrics* **1981**, *67*, 517–522.
7. Norcliffe-Kaufmann, L.; Axelrod, F.; Kaufmann, H. Familial Dysautonomia: A Genetic Disorder with Complete Afferent Baroreflex Failure. *J. Neurol.* **2010**, *75*, 1904–1191.
8. Bernardi, L.; Hilz, M.; Stemper, B., et al. Respiratory and Cerebrovascular Responses to Hypoxia and Hypercapnia in Familial Dysautonomia. *Am. J. Respir. Crit. Care. Med.* **2002**, *167*, 141–149.
9. Slaugenhaupt, S. A.; Blumenfeld, A.; Gill, S. P., et al. Tissue-Specific Expression of a Splicing Mutation in the IKBKAP Gene Causes Familial Dysautonomia. *Am. J. Hum. Genet.* **2001**, *68*, 598–605.
10. Axelrod, F. B.; Goldberg, J. D.; Ye, X. Y.; Maayan, C. Survival in Familial Dysautonomia: Impact of Early Intervention. *J. Pediatr.* **2002**, *141*, 518–523.

CHAPTER

120

The Phakomatoses

Susan M Huson

Manchester Academic Health Science Centre, Central Manchester University Hospitals, NHS Foundation Trust, Saint Mary's Hospital, Manchester, UK

Bruce R Korf

Department of Genetics, Heflin Center for Genomic Sciences, University of Alabama at Birmingham, Birmingham, AL, USA

The phakomatoses are a set of distinct disorders that have in common the occurrence of "spotty" manifestations, including multiple hamartomas, dysplastic lesions, and in some cases benign and/or malignant tumors. For the most part, these are autosomal dominant disorders due to mutations in distinct tumor suppressor genes (Table 120-1), with high penetrance (though age-dependent for specific manifestations), variable expression, and a high rate of new mutation.

The neurofibromatoses include NF1, NF2, and schwannomatosis. Diagnostic criteria are provided in Table 120-2. NF1 is the most common, usually presenting with multiple café-au-lait macules in early childhood. The defining feature is the occurrence of neurofibromas, benign tumors of Schwann cell origin, that arise from peripheral and cranial nerves. Plexiform neurofibromas involve multiple branches of larger nerves and may cause overgrowth and disfigurement. Neurofibromas of nerve roots can invade the spinal canal and compress the spinal cord. Tumors of the optic nerve are common, arise in young children, but are often asymptomatic. There is an approximately 8–13% risk of malignant peripheral nerve sheath tumor, usually arising from a preexisting plexiform or nodular neurofibroma and presenting with pain or sudden growth. Other features of NF1 include neurocognitive disorders (learning disabilities, attention deficit disorder, behavioral problems) and skeletal dysplasia (long bone or sphenoid dysplasia) (Table 120-3). The defining feature of NF2 is bilateral vestibular schwannomas (Table 120-4). In addition, schwannomas occur on other cranial nerves, spinal nerves, and peripheral nerves (Table 120-5). Meningiomas and ependymomas also occur commonly. The one non-tumor manifestation of NF2 is cataract. Schwannomatosis involves the occurrence of schwannomas of peripheral nerves, usually associated with pain, though meningiomas can occur as well. NF1 is associated with mutation in the *NF1* gene, which encodes a GTPase-activating protein; NF2 is associated with mutation of *NF2*, which encodes a cytoskeletal protein; schwannomatosis is associated with mutation of *SMARCB1*, which encodes a chromatin remodeling protein. Genetic testing is available for all three disorders, and clinical trials are underway to treat many of the complications.

Tuberous sclerosis complex (TSC) most often presents in childhood with hypomelanotic macules and often presents with seizures, including infantile spasms. Diagnostic criteria are presented in (Table 120-6). Intellectual disability tends to occur in those with early-onset poorly controlled seizures. Central nervous system lesions include subependymal nodules, dysplastic cortical lesions ("tubers"), white matter abnormalities, and subependymal giant cell astrocytomas (SEGAs). Other organs that may be involved include kidneys (angiomyolipomas [AML] and cysts), heart (rhabdomyomas), lungs (lymphangioleiomyomatosis), skin (angiofibroma, collagen plaque, shagreen patch, ungual fibromas) (Table 120-7). Mutation in either *TSC1* or *TSC2* can cause tuberous sclerosis complex. The gene products interact and negatively regulate mTOR signaling. Genetic testing is available, and treatment of SEGA and symptomatic AML is now possible using an mTOR inhibitor.

von Hippel–Lindau syndrome (VHL) is characterized by hemangioblastomas of the cerebellum, brainstem, spinal cord, and retina. Other features include renal cysts and renal cell carcinoma, pheochromocytoma, and endolymphatic sac tumors (Table 120-8). The gene associated with VHL is involved in the control of angiogenesis. Mutation testing is available on a clinical basis and can help to distinguish individuals at risk for pheochromocytoma and renal cell carcinoma.

Other rare phakomatoses include the PTEN hamartoma syndrome, which has variable manifestations depending on the specific mutation; Proteus syndrome, associated with mosaic mutation of the *AKT1* gene, Klippel–Trenaunay–Weber syndrome, and Sturge–Weber syndrome.

TABLE 120-1 The Phakomatoses

Disorder	Frequency	Gene	Major Features
NF1	1:3000	*NF1; GAP protein*	Café-au-lait macules; neurofibromas; optic gliomas; malignant peripheral nerve sheath tumors; learning disabilities; skeletal dysplasias
NF2	1:33,000	*NF2*; cytoskeletal protein	Vestibular schwannomas; meningiomas; ependymomas; cataracts
Schwannomatosis	Unknown	*SMARCB1*; chromatin remodeling protein	Schwannomas of peripheral nerves; pain
Tuberous sclerosis complex	1:6000	*TSC1/TSC2*; mTOR regulator	Hypomelanotic macules; angiofibroma; cortical tubers; seizures; intellectual disability; renal angiomyolipomas; cardaic rhabdomyomas
von Hippel Lindau syndrome	1:40,000	*VHL*; angiogenesis regulator	Hemangioblastomas of cerebellum, brainstem, spinal cord, retina; renal cell Carcinoma; pheochromocytoma
PTEN hamartoma syndrome	Rare	*PTEN*; cell signaling protein	
Proteus syndrome	Rare	*AKT1*; cell signaling protein	Bony and cutaneous overgrowth; disfigurement

TABLE 120-2 Diagnostic Criteria for NF1

1. Six or more café-au-lait macules of over 5mm is the greatest diameter in prepubertal individuals and over 15mm is the greatest diameter in postpubertal individuals
2. Two or more neurofibromas of any type or one plexiform neurofibroma
3. Freckling in the axillary or inguinal regions
4. Optic glioma
5. Two or more Lisch nodules (iris hamartomas)
6. A distinctive osseous lesion such as sphenoid dysplasia or long bone dysplasia
7. A first-degree relative (parent, sibling, or offspring) with NF1 by the above criteria

Individuals with Legius syndrome or homozygous mutation of mismatch repair genes may also fulfill NF1 diagnostic criteria.

TABLE 120-3 NF1 Clinical Features and Age of Onset

Disease Feature	Frequency (%)[a]	Age at Presentation
Café-au-lait spots	>99	0–2 y
Skinfold freckling	67	3–5 y
Peripheral neurofibromas	>99	≥7 y
Plexiform neurofibromas		
All lesions	30	0–18 y
Head and neck	1.2	0–3 y
Trunk/extremities with hypertrophy	5.8	0–5 y
Lisch nodules	90–95	≥3 y
Macrocephaly	45	Birth
Short stature	31.5	Birth
Intellectual disability		
Severe	0.8	0–5 y
Moderate	2.4	0–5 y
Minimal/learning difficulties	29.8	0–5 y
Attention deficit disorder ± hyperactivity	48[b]	Childhood
Epilepsy		
No known cause	4.4	Lifelong
Secondary to disease complications	2.2	Lifelong
Hypsarrhythmia	1.5	0–5 y
CNS tumors		
Optic glioma	1.5[c]	0–20 y
Other CNS tumors	1.5	Lifelong
Spinal neurofibromas	1.5	Lifelong
Aqueductal stenosis	1.5	Lifelong
Malignancy		
Malignant peripheral nerve sheath tumor	1.5[d]	Lifelong
Rhabdomyosarcoma	1.5	0–5 y
Juvenile myelomonocytic leukemia	<1	0–18 y
Orthopedic		
Scoliosis, requiring surgery	4.4	0–18 y
Scoliosis, not requiring surgery	5.2	0–18 y
Long bone dysplasia	3.7	0–2 y
Vertebral scalloping	10[e]	Lifelong
GI tumors (neurofibromas and GISTs)[f]	2.2	Lifelong
Renal artery stenosis	1.5	0–20 y
Pheochromocytoma	0.7	≥10 y
Duodenal carcinoid	1.5	≥10 y
Juvenile xanthogranuloma	0.7	0–1 y
Congenital glaucoma	0.7	0–1 y
Pulmonic stenosis	<1	Childhood
Sphenoid wing dysplasia	<1	Congenital
Lateral meningocele	<1	Lifelong
Cerebrovascular disease (moyamoya)	<1	Childhood
Glomus tumors of nailbeds	<1	Adulthood

[a]Data from Huson and colleagues (Huson, S. M.; Harper, P. S.; Compston, D. A. *Brain*. **1988**, 111, Pt 6, 1355–1381).

[b]Data from Mautner and colleagues (Mautner, V. F.; Kluwe, L.; Thakker, S. D.; Leark, R. A. *Dev. Med. Child Neurol.* **2002**, 44, 3, 164–170).

[c]Optic glioma is seen in 15% of children if imaging study is done (Listernick, R.; Louis, D. N.; Packer, R. J.; Gutmann, D. H. *Ann. Neurol.* **1997**, 41, 2, 143–149).

[d]Lifetime risk may be 5–10% (Evans, D. G. R.; Baser, M. E.; McGaughran, J.; Sharif, S.; Howard, E.; Moran, A. *J. Med. Genet.* **2002**, 39, 5, 311–314).

[e]Data from Riccardi and Eichner (Riccardi, V. M.; Eichner, J. E. Neurofibromatosis: Phenotype, Natural History, and Pathogenesis. Johns Hopkins University Press: Baltimore, 1992).

[f]See Takazawa and colleagues (Takazawa, Y.; Sakurai, S.; Sakuma, Y.; Ikeda, T.; Yamaguchi, J.; Hashizume, Y., et al. *Am. J. Surg. Pathol.* **2005**, 29, 6, 755–763).

TABLE 120-4 Diagnostic Criteria for NF2

Bilateral Vestibular Schwannomas
First degree relative with NF2 and – Unilateral VS or – Any two of meningioma, schwannoma, glioma, neurofibroma, and posterior subcapsular lenticular opacities
Unilateral vestibular schwannoma and any two of meningioma schwannoma, glioma, neurofibroma, and posterior subcapsular lenticular opacities
Multiple meningioma (two or more) plus unilateral VS or any two of glioma, schwannoma, neurofibroma, and posterior subcapsular lenticular opacities

TABLE 120-5 NF2 Clinical Features

Feature	Frequency (%)
Tumors	
Vestibular schwannoma (bilateral)	85
Vestibular schwannoma (unilateral)	6
Cranial meningioma[a]	45–58
Spinal tumors (extramedullary)[b]	63–90
Spinal tumors (intramedullary)[c]	18–53
Peripheral neuropathies	10[d]
Peripheral schwannomas	
Overall	68
>10 tumors	10
NF2 plaques	48
Nodular schwannomas	43
NF1-like dermal neurofibromas	27
Plexiform schwannomas	<1
Café-au-lait spots	
1–2 spots	35
3–4 spots	7
6 spots	1
Ophthalmologic features	
Cataracts	60–81
Epiretinal membranes	12–40
Retinal hamartomas	6–22

[a]A higher frequency of tumors is found on cranial and spinal MRI; a significant proportion of which will never require treatment.
[b]Dorsal root schwannomas and, less frequently, meningiomas.
[c]Ependymomas, infrequently astrocytomas and schwannomas.
[d]Frequency of symptomatic neuropathies in one series (Evans, D. G.; Huson, S. M.; Donnai, D.; Neary, W.; Blair, V.; Newton, V.; Harris, R. *Q. J. Med.* **1992**, *84*, 304, 603–618), up to 66% have electrophysiological evidence of a neuropathy (Sperfeld, A. D.; Hein, C.; Schroder, J. M.; Ludolph, A. C.; Hanemann, C. O. *Brain*. **2002**, *125*, Pt 5, 996–1004).
Data from Evans et al. (*Q. J Med.* **1992**, *84*, 304, 603–618) and Asthagiri et al. (Lancet. **2009**, *373*, 9679, 1974–1986).

TABLE 120-6 Diagnostic Criteria for TSC

1. Major features: facial angiofibromas or forehead plaque, nontraumatic ungual or periungual fibroma, hypomelanotic macules (at least 3), shagreen patch, multiple retinal nodular hamartomas, cortical tuber, subependymal nodule, subependymal giant cell astrocytoma, cardiac rhabdomyoma (single or multiple), lymphangiomyomatosis, renal angiomyolipoma
2. Minor features: multiple pits in dental enamel, hamartomatous rectal polyps, bone cysts, cerebral white matter radial migration lines, gingival fibromas, nonrenal hamartoma, retinal achromic patch, "confetti" skin lesions, multiple renal cysts

Definite diagnosis requires either two major features or one major feature with two minor features. Probable TSC results when one major feature and one minor feature are present. The occurrence of either one major feature or two or more minor features raises suspicion of possible TSC, but does not establish the diagnosis. If cortical dysplasia and white matter migration tracts occur together, they should be counted as only a single feature; if lymphangiomyomatosis and renal angiomyolipomas are both present, one other feature of TSC should be present before a diagnosis is established.

TABLE 120-7 Clinical Features of TSC

Feature	Frequency (%)	Age of Onset	Comments
Seizures	78		
Infantile	54	0–1 y	Half infantile spasms
Childhood	21	1–18 y	
Adulthood	3	>18 y	
Neurocognitive			
Learning disability	53	0–5	Related to seizures autistic sleep disorders, hyperactivity, aggressive behavior
Behavior disorder	>50	Childhood	
CNS lesions			
Subependymal nodules	95	Birth	
Cortical tubers	93	Birth	
White matter abnormality	93	Birth	
Subependymal giant cell astrocytoma	6–7	1–31 y	
Cutaneous lesions			
Hypomelanotic macules	78–87	Birth	
Facial angiofibromas	80–90	5 y–puberty	
Shagreen patch	21–41	>10 y	
Forehead plaque	26	0–10 y	
Dental enamel pits	48	>6 y	
Ungual fibromas	17–47	>15 y	
Renal lesions			
Angiomyolipomas	37–67	Childhood	May become symptomatic in adulthood rarely symptomatic with PKD1 deletion
Cysts	10	Childhood	
Polycystic kidney disease	1.5	Childhood	
Renal cell carcinoma	Rare	Adulthood	
Retinal hamartoma	50	Childhood	
Cardiac rhabdomyoma	50	Prenatal	Regress early in childhood
Pulmonary lymphangiomyomatosis	0.8 (symptomatic)	>18 y	Mostly in females
Liver hamartomas	45	Childhood	Usually asymptomatic
Microhamartomatous rectal polyps	78	Childhood	Usually asymptomatic

See text for sources of data.

TABLE 120-8 von Hippel–Lindau Disease: Clinical Features and Age at Presentation

Feature	Frequency (%)	Mean Age at Presentation
Hemangioblastoma		
Retinal	59	24 (4–46)
Cerebellar	59	29 (1–36)
Spinal cord	13	33.9 (11–60)
Brain stem	4	NA
Elsewhere	<1	NA
Renal lesions		
Carcinoma	28	46.2 (20–69)
Cysts	37	34.6 (25–50)
Pheochromocytoma	7–19	20.2 (12–36)
Pancreatic lesions		
Cysts	40	NA*
Microcystic adenomas	4	NA*
Islet cell tumors	2	NA*
Carcinoma	0.7	NA
Epididymal lesions		
Cystadenomas	10–26	NA*
Cysts	NA	NA*

NA, data not available; NA*, usually asymptomatic (detected through screening as opposed to presenting with symptoms).

Data from Maher and colleagues (*Q. J. Med.* **1990**, *77*, 283, 1151–1163) and Choyke and coworkers (*Radiology*. **1995**, *194*, 3, 629–642).

FURTHER READING

1. Boyd, K.; Korf, B.; Theos, A. Neurofibromatosis Type 1. *J. Am. Acad. Dermatol.* **2009 Jun 30,** *61* (1), 1–14, quiz 15–16.
2. Ferner, R. E. Neurofibromatosis 1 and Neurofibromatosis 2: A Twenty First Century Perspective. *Lancet Neurol.* **2007, Mar 31,** *6* (4), 340–351.
3. Gutmann, D. H.; Aylsworth, A.; Carey, J. C.; Korf, B.; Marks, J.; Pyeritz, R. E., et al. The Diagnostic Evaluation and Multidisciplinary Management of Neurofibromatosis 1 and Neurofibromatosis 2. *JAMA* **1997,** *278* (9207339), 51–57.
4. Evans, G. R.; Lloyd, S. K.; Ramsden, R. T. Neurofibromatosis Type 2. *Adv. Otorhinolaryngol.* **2011,** 7091–7098.
5. MacCollin, M.; Chiocca, E. A.; Evans, D. G.; Friedman, J. M.; Horvitz, R.; Jaramillo, D., et al. Diagnostic Criteria for Schwannomatosis. *Neurology* **2005 Jun 14,** *64* (11), 1838–1845.
6. Crino, P. B. The Pathophysiology of Tuberous Sclerosis Complex. *Epilepsia* **2010, Feb,** *51* (Suppl.), 127–129.
7. Maher, E. R.; Neumann, H. P.; Richard, S. von Hippel–Lindau Disease: A Clinical and Scientific Review. *Eur. J. Hum. Genet.* **2011 Mar 9,** 19617–19623.
8. Hobert, J. A.; Eng, C. PTEN Hamartoma Tumor Syndrome: An Overview. *Genet. Med.* **2009 Oct,** *11* (10), 687–694.
9. Lindhurst, M. J.; Sapp, J. C.; Teer, J. K.; Johnston, J. J.; Finn, E. M.; Peters, K., et al. A Mosaic Activating Mutation in AKT1 Associated with the Proteus Syndrome. *N. Engl. J. Med.* **2011, Aug 18,** *365* (7), 611–619.
10. Pascual-Castroviejo, I.; Pascual-Pascual, S. I.; Velazquez-Fragua, R.; Viano, J. Sturge–Weber Syndrome: Study of 55 Patients. *Can. J. Neurol. Sci.* **2008 Jul,** *35* (3), 301–307.

CHAPTER

121

Multiple Sclerosis and Other Demyelinating Disorders

A Dessa Sadovnick

Professor of Medical Genetics and Neurology, Director, endMS Western Pacific Research and Training Centre, University of British Columbia, Vancouver, BC, Canada

The most common inherited demyelinating disease is multiple sclerosis (MS), which affects approximately 0.1% of Caucasians of northern and central European ancestry. The female:male ratio now approaches 3:1. MS is an inflammatory disease of the central nervous system characterized by myelin loss, varying degrees of axonal pathology, and progressive neurological dysfunction. The most recent diagnostic criteria (*1*) take into account findings on magnetic resonance imaging in the definition of dissemination in time and space, pediatric onset MS and MS in non-Caucasians.

While several genes have been implicated in MS susceptibility, the one repeatedly replicated is HLA class II extended *HLA-DRB1*15* haplotype (*HLA-DRB5*0101–HLA-DRB1*1501–HLA-DQA1*0102–HLA-DQB1*0602*) has remained the most consistent association (*2*).

Even with respect to HLA, the genetics of MS remain complex and topics continuing to be studied include epistatic effects and parent origin effects to name a few [see (*3–6*)]. This is clearly shown in Figure 121-1 (*7*) as it is the HLA haplotype and not the specific allele that influences risk/susceptibility to MS (e.g. 15/15 versus 15/04; see Figure 121-1).

A number of strategies have been employed to dissect the environmental from the genetic components (i.e. nature vs. nurture) underlying MS susceptibility, for example, half-sibling, adoptee, step-sib and conjugal MS studies. Taken together, these have clearly shown no evidence for environmental factors operative within the familial microenvironment, either in childhood or in adulthood. Thus, DNA sharing is responsible for most, if not all, of the familial aggregation of MS.

A maternal effect is now recognized in MS but the mechanism of the increased risk conferred maternally remains to be elucidated but epigenetic (*8*). At present, there are three major environmental factors associated with the risk of developing MS (*9*): (i) latitude and sunshine exposure and the resulting vitamin D levels which are closely related, (ii) Epstein–Barr virus seropositivity, and (iii) smoking.

121.1 PRACTICAL APPLICATIONS OF GENETIC STUDIES

121.1.1 Genetic Counseling

Genetic counseling for MS must take into consideration the known facts, the complexity of the disease susceptibility and the as-yet unknown factors and stochastic events. Up to 30% of MS patients have at least one biological relative with MS based on data from the longitudinal, population-based Canadian Collaborative Project on Genetic Susceptibility to MS (CCPGSMS) (e.g. see (*2–9*)).

Genetic epidemiological data suggest that genetic counseling must take into account individual family structure, that is, family-specific factors such as gender, age of MS onset for affected relatives, twin status and whether any (or both) parent have MS.

Lifetime risks for biological relatives of persons with MS, controlling for the amount of genetic sharing, are given in Table 121-1.

Depending on the amount of genetic (DNA) sharing, the recurrence risks for MS can range from 2/1000, that is the general population lifetime risk for Caucasians (e.g. adopted sibs, step-sibs) to 340/1000 (female monozygotic co-twins of MS probands with whom they share 100% of their genetic material).

When counseling full sibs of affected individuals, risks can be refined from those given in Figure 121-2 when information is available on gender of the MS proband and the sib (see Table 121-2).

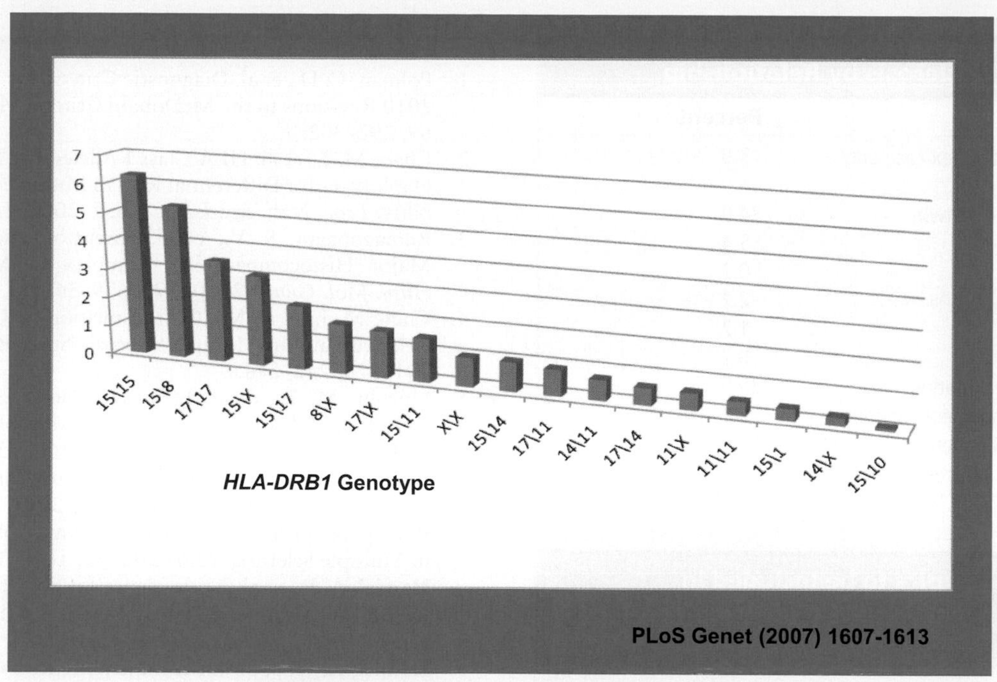

FIGURE 121-1 Genotype relative risk for DRB1.

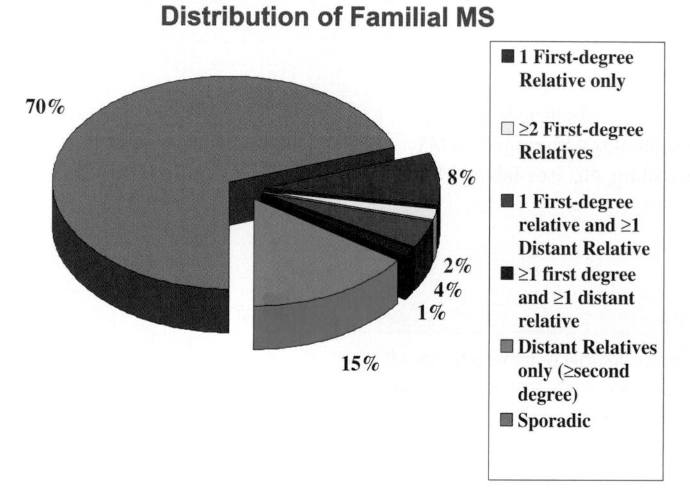

FIGURE 121-2 Demographics of CCS participants distribution of familial MS.

TABLE 121-1 Recurrence Risk of Multiple Sclerosis by Relationship

Relationship	Percent
First-degree relative; index case only MS	3.0
Monozygotic female co-twin	34.0
Dizygotic co-twin	5.4
Adopted sib	0.2
Maternal $^1/_2$ sib	2.2
Paternal $^1/_2$ sib	1.2
Step-sib	0.2
Offspring of conjugal mating	30.5
Offspring of consanguineous mating	9.0
Sister of female MS patient with onset under 30 plus 1 affected parent	12.7

TABLE 121-2 Topics to be Included in Reproductive Counseling for Individuals with Multiple Sclerosis

Effects of MS on Pregnancy
Contraception[a]
Conception[a]
Pregnancy management
Pregnancy outcome[a]
Effects of Pregnancy on MS
Short-term
Long-term
Recurrence Risks[a]
Factors to include in determining recurrence risks include gender, ages, family structure, simplex or multiplex with respect to MS, ethnicity, consanguinity, conjugal mating, etc. (see Table 121-1).
Teratogenicity of MS Treatments[a]
Symptom specific
Relapses
Disease modifying
Psychosocial Issues[a]

[a]Important to discuss regardless of whether it is mother or father who has MS.
Data from Dwosh, E.; Guimond, C. G.; Sadovnick, A. D. Reproduction Counselling in MS: A Guide for Healthcare Professionals. Int. MS J. **2003**, *10*, 67.

REFERENCES

1. Polman, C. H., et al. Diagnostic Criteria for Multiple Sclerosis: 2010 Revisions to the McDonald Criteria. *Ann. Neurol.* **2011**, *69*, 292–302.
2. Chao, M. J., et al. HLA Class I Alleles Tag HLA-DRB1*1501 Haplotypes for Differential Risk in Multiple Sclerosis Susceptibility. *Proc. Natl. Acad. Sci. U.S.A.* **2008**, *105*, 13069–13074.
3. Ramagopalan, S. V., et al. Parent-of-Origin Effects at the Major Histocompatibility Complex in Multiple Sclerosis. *Hum. Mol. Genet.* **2010**, *19*, 3679–3689.
4. Chao, M. J., et al. MHC Transmission by Gender of Affected Relative Pairs in Multiple Sclerosis Susceptibility. *Neurology* **2011**, *76*, 242–246.
5. Lincoln, M. R., et al. Epistasis Among HLA-DRB1, HLA-DQA1 and HLA-DQB1 Determines Multiple Sclerosis Susceptibility. *Proc. Natl. Acad. Sci.* **2009**, *106*, 7542–7547.
6. Chao, M. J., et al. MHC Transmission: Insights into Gender Bias in MS Susceptibility. *Neurology* **2011**, *76*, 242–246.
7. Ramagopalan, S. V., et al. The Inheritance of Resistance Alleles in Multiple Sclerosis. *PLoS* **2007**, *3*, 1607–1613.
8. Handel, A. E., et al. No Evidence for an Effect of DNA Methylation on Multiple Sclerosis Severity at HLA-DRB1*15 or HLA-DRB5. *J. Neuroimmunol.* **2010**, *223*, 120–123.
9. Handel, A. E., et al. Environmental Factors and Their Timing in Adult-Onset Multiple Sclerosis. *Nat. Rev. Neurol.* **2010**, *6*, 156–166.

CHAPTER

122

Genetics of Stroke

Mateusz G Adamski and Alison E Baird

SUNY Downstate Medical Center, Department of Neurology, Brooklyn, NY, USA

122.1 INTRODUCTION

Strokes result from a focal reduction of blood flow to the brain. Around 83% of strokes are due to arterial vascular occlusion (termed *ischemic stroke*) and around 17% are due to vascular rupture (termed *hemorrhagic stroke*) (*1*). Arterial vascular rupture may occur into the brain parenchyma resulting in intracerebral hemorrhage (ICH) or into the subarachnoid space results in subarachnoid hemorrhage. In about 1% of cases, strokes occur in the venous system (termed cerebral venous thrombosis). Strokes are a leading cause of morbidity and mortality in the United States. There is a substantial additional burden of asymptomatic cerebrovascular disease.

Evidence for a genetic basis for stroke comes from twin and family studies. Furthermore, a number of monogenic disorders cause stroke, either as a primary or a secondary manifestation. The contribution of these factors has been increasingly recognized over the past 10–20 years with the description of monogenic disorders such as cerebral autosomal dominant arteriopathy with subcortical infarcts and leukoencephalopathy (CADASIL). Advances in genomics methods, including genome-wide association studies are permitting possible new genetic risk markers to be explored.

122.2 GENETICS AND STROKE RISK

122.2.1 Family History of Stroke

In twin studies, the concordance rate for stroke in monozygotic twins is 17.7% and for dizygotic twins 3.6% (*2*). The heritability of stroke has varied in family studies, but in general it appears that a family history of stroke increases an individual's stroke risk by two- to threefold. ICH in a first-degree relative increases an individual's odds by as much as two- to sixfold. A family history of aneurysmal SAH is associated with an increased risk of intracranial aneurysms of around fourfold, with the risk highest in siblings.

122.2.2 Genetic Polymorphisms

Single-nucleotide polymorphisms in two genes increase the risk of cerebral venous thrombosis. A polymorphism in the gene coding for coagulation factor V results in a form of factor V variant—factor V Leiden—that is resistant to degradation by activated protein C resulting in a hypercoagulable state. The risk for cerebral venous thrombosis is increased almost eightfold in individuals carrying one copy of the variant. The prothrombin G20210A polymorphism confers an almost tenfold increased risk of cerebral venous thrombosis (*3*). The risk is further magnified with the use of the birth control pill and in the postpartum period. Both of these factors may be associated with a mildly increased risk of arterial ischemic stroke.

122.3 MONOGENIC DISORDERS CAUSING STROKE

Monogenic disorders are uncommon and account for less than 1% of all strokes. They are typically associated with stroke in childhood or young adulthood, with certain stroke types and subtypes, with the absence of other stroke risk factors and with specific phenotypes of the associated diseases including ocular and skin manifestations.

122.3.1 Stroke as a Primary Manifestation

CADASIL was initially recognized in the 1990s as an adult-onset hereditary disorder of dementia and stroke and its genetic and molecular basis was subsequently unraveled (*4*). CADASIL is characterized by recurrent subcortical ischemic strokes starting at ages 30–40, progressive or stepwise cognitive decline, white matter changes on magnetic resonance imaging and in some patients, migraine with aura and mood disturbances. In addition to genetic analysis, skin biopsy is used to aid in diagnosis. Cerebral autosomal recessive arteriopathy with subcortical infarcts and leukoencephalopathy is an autosomal recessive disorder associated with stroke, alopecia, psychiatric disturbances, and progressive mental and motor retardation (*5*). Retinal vasculopathy and cerebral leukodystrophy (includes hereditary endotheliopathy, retinopathy, nephropathy and stroke) is an autosomal dominant disorder caused by mutations in the *TREX1* gene (*6*). Monogenic disorders associated with hemorrhagic stroke include cerebral cavernous malformations (*7*) and the hereditary cerebral amyloid angiopathies. Moyamoya disease is a disorder with autosomal dominant inheritance and variable penetrance and is associated with large artery intracranial occlusive arteriopathy and with both ischemic and hemorrhagic strokes (*8*).

122.3.2 In Other Monogenic Disorders Stroke is a Secondary Manifestation

Fabry disease (X-linked recessive inheritance) is caused by a defect in the α-galactosidase A gene and is associated with large and small vessel ischemic strokes, angiokeratoma, neuropathic pain, renal and cardiac failure, hypohidrosis, cataracts, corneal opacities, gastrointestinal dysmotility and acroparesthesia (*9*). Fabry disease may be treated with enzyme replacement therapy. Ehlers–Danlos syndrome type IV (*10*) and pseudoxanthoma elasticum (*11*) are associated with large artery vasculopathy and arterial dissections. Marfan syndrome (*12*) and familial cardiomyopathies may be associated with cardioembolic stroke. Homocystinuria (autosomal recessive) is associated with premature atherosclerosis and small and large artery disease (*13*). Mitochondrial myopathy, encephalopathy, lactic acidosis and stroke-like episodes (MELAS, a mitochondrial disorder) is associated with ischemic strokes, migraine, seizures, weakness, developmental delay, cognitive decline and hearing loss (14). Sickle-cell disease is associated with both ischemic and hemorrhagic stroke (*15*). This disorder has autosomal recessive inheritance and causes large and small vessel disease in addition to pain crises, pulmonary and abdominal crises, and bacterial infection. Strokes may be prevented by transfusions. Hereditary hemorrhagic telangiectasia is also associated with ischemic and hemorrhagic strokes (*16*).

122.4 GENOME-WIDE ASSOCIATION STUDIES AND GENOMICS

Advances in genetics and genomics may permit new insights. In recent genome-wide association studies, a number of single-nucleotide polymorphisms have been associated with specific stroke subtypes and major stroke risk factors such as diabetes and atrial fibrillation but these have yet to be replicated. Studies of messenger ribonucleic acid expression have also shown promise for the development of genomic signatures for stroke classification.

122.5 SUMMARY

At present, the contribution of genetic factors to stroke etiology and risk is small, involving familial predisposition, a small number of monogenic disorders such as CADASIL—the prototype genetic disorder associated with stroke—and polymorphisms associated with cerebral venous thrombosis. Possible new associations are being explored in genome-wide association studies but no markers have yet emerged.

REFERENCES

1. Roger, V. L.; Go, A. S.; Lloyd-Jones, D. M.; Benjamin, E. J.; Berry, J. D.; Borden, W. B.; Bravata, D. M.; Dai, S.; Ford, E. S.; Fox, C. S., et al. Heart Disease and Stroke Statistics—2012 Update: A Report from the American Heart Association. *Circulation* **2012**, *125*, e2–e220.
2. Brass, L.; Isaacsohn, J.; Merikangas, K. A Study of Twins and Stroke. *Stroke* **1992**, *23*, 221–223.
3. Saposnik, G.; Barinagarrementeria, F.; Brown, R. D.; Bushnell, C. D.; Cucchiara, B.; Cushman, M.; DeVeber, G.; Ferro, J. M.; Tsai, F. Y. Diagnosis and Management of Cerebral Venous Thrombosis: A Statement for Healthcare Professionals from the American Heart Association/American Stroke Association. *Stroke* **2011**, *42*, 1158–1192.
4. Tournier-Lasserve, E.; Joutel, A.; Melki, J.; Weissenbach, J.; Lathrop, G. M.; Chabriat, H.; Mas, J. L.; Cabanis, E. A.; Baudrimont, M.; Maciazek, J., et al. Cerebral Autosomal Dominant Arteriopathy with Subcortical Infarcts and Leukoencephalopathy Maps to Chromosome 19q12. *Nat. Genet.* **1993**, *3*, 256–259.
5. Fukutake, T. Cerebral Autosomal Recessive Arteriopathy with Subcortical Infarcts and Leukoencephalopathy (CARASIL): From Discovery to Gene Identification. *J. Stroke Cerebrovasc. Dis.* **2011**, *20*, 85–93.
6. Richards, A.; van den Maagdenberg, A. M. J. M.; Jen, J. C.; Kavanagh, D.; Bertram, P.; Spitzer, D.; Liszewski, M. K.; Barilla-Labarca, M.-L.; Terwindt, G. M.; Kasai, Y., et al. C-terminal Truncations in Human 3′–5′ DNA Exonuclease TREX1 Cause Autosomal Dominant Retinal Vasculopathy with Cerebral Leukodystrophy. *Nat. Genet.* **2007**, *39*, 1068–1070.
7. McDonald, J.; Bayrak-Toydemir, P.; Pyeritz, R. E. Hereditary Hemorrhagic Telangiectasia: An Overview of Diagnosis, Management, and Pathogenesis. *Genet. Med.* **2011**, *13*, 607–616.
8. Kuroda, S.; Houkin, K. Moyamoya Disease: Current Concepts and Future Perspectives. *Lancet Neurol.* **2008**, *7*, 1056–1066.
9. Clarke, J. T. R. Narrative Review: Fabry Disease. *Ann. Intern. Med.* **2007**, *146*, 425–433.
10. Germain, D.; Herrera-Guzman, Y. Vascular Ehlers–Danlos syndrome. *Ann. Génét.* **2004**, *47*, 1–9.
11. Chassaing, N.; Martin, L.; Calvas, P.; Le Bert, M. Hovnanian, a Pseudoxanthoma Elasticum: A Clinical, Pathophysiological and Genetic Update Including 11 Novel ABCC6 Mutations. *J. Med. Genet.* **2005**, *42*, 881–892.
12. Dean, J. C. S. Marfan Syndrome: Clinical Diagnosis and Management. *Eur. J. Hum. Genet.* **2007**, *15*, 724–733.
13. Skovby, F.; Gaustadnes, M.; Mudd, S. H. A Revisit to the Natural History of Homocystinuria Due to Cystathionine Beta-Synthase Deficiency. *Mol. Genet. Metab.* **2010**, *99*, 1–3.
14. Testai, F.; Gorelick, P. Inherited Metabolic Disorders and Stroke Part 1: Fabry Disease and Mitochondrial Myopathy, Encephalopathy, Lactic Acidosis, and Strokelike Episodes. *Arch. Neurol.* **2010**, *67*, 19–24.
15. Rees, D. C.; Williams, T. N.; Gladwin, M. T. Sickle-Cell Disease. *Lancet* **2010**, *376*, 2018–2031.
16. Dupuis-Girod, S.; Bailly, S.; Plauchu, H. Hereditary Hemorrhagic Telangiectasia: From Molecular Biology to Patient Care. *J. Thromb. Haemost.* **2010**, *8*, 1447–1456.

CHAPTER

123

Primary Tumors of the Nervous System

Angel A Alvarez and Markus Bredel

Primary tumors of the central nervous system (CNS) are relatively rare, with an annual incidence of 24.55 out of 100,000 for adult cases and 48.8 out 1,000,000 for pediatric cases. Nonetheless, these tumors tend to be malignant in 33.7% of adult cases and 65.2% of pediatric cases, with 5-year survival for glioblastoma tumors ranging from 21% down to 1%, depending on age. The relative incidence and malignancy depends on several factors, including tumor type, location, size, as well as patient age. Overall, gliomas account for the majority of brain tumors, followed by meningiomas and schwannomas. The current standard for classification of CNS tumors is the World Health Organization (WHO) system, which classifies tumors based on histopathologic tumor typing and grading, from WHO grade I (benign tumors) to WHO grade IV (malignant tumors). New classification systems based on genetic alterations offer more accurate reflection of the intrinsic tumor characteristics and insight into prognoses and treatment options.

The majority of brain tumors appear to be sporadic in origin, and epidemiologic studies suggest the risk of developing a brain tumor is only slightly increased in relatives of brain cancer patients. A very small percentage of CNS tumors have been shown to result from inherited cancer syndromes including neurofibromatosis (NF) 1 and 2, von Hippel–Lindau disease, tuberous sclerosis, Li–Fraumeni syndrome, Gorlin syndrome, ataxia telangiectasia, Cowden syndrome, Werner syndrome, and Turcot syndrome. These cancer syndromes typically arise from mutations in critical tumor suppressor genes, such as *NF1*, *TP53*, and *PTEN*, which drive cellular transformation. Although the vast majority of CNS tumors are not the result of genetic inheritance, an increased familial risk may be associated with primitive neuroectodermal tumors (PNETs). In addition, a maternal family history of birth defects may increase the risk of childhood brain tumors slightly.

Gliomas comprise the most common of the CNS tumors. A genetic predisposition for the development of gliomas is seen in NF1 and NF2, Li–Fraumeni syndrome, tuberous sclerosis, Gorlin syndrome, Turcot syndrome, and ataxia telangiectasia. Familial inheritance of gliomas not associated with a specific cancer syndrome is rare, but has been reported. Gliomas comprise a heterogeneous collection of glia-resembling tumors that are subdivided into astrocytomas, oligodendrogliomas, oligoastrocytomas, and ependymomas. Of particular concern for treatment options, recurrent gliomas can be associated with a more malignant histologic type, and thus require more aggressive treatment. The most commonly diagnosed subtype of gliomas is the astrocytomas, of which glioblastomas are the most malignant and associated with the worst prognosis. Glioblastomas show complex genetic landscapes. Recent integrated genomic analyses have identified clinically relevant subtypes of glioblastoma characterized by abnormalities in the *PDGFRA*, *IDH1*, *EGFR*, *NF1*, and *NFKBIA* genes. These studies have shown that patient survival time may be related to specific genetic alterations, as evidenced, for example, by the shorter survival times reported for patients with *EGFR* gene amplifications or *NFKBIA* deletions, and the longer survival times reported for patients with Turcot syndrome and gliomas with *IDH1* or *IDH2* mutations. Aberrations of *EGFR* and loss of *CDKN2A* have been detected in the majority of primary (or de novo) glioblastomas, but other genetic alterations are commonly observed, including overexpression of *MDM2*, constitutive activation of NF-κB, and mutational loss of *PTEN* expression. *TP53* mutations and overexpression of PDGF ligands and receptors, as well as mutations in several other genes such as *IDH1*, have been identified in secondary glioblastomas, which develop through progression and malignant transformation from lower grade astrocytomas.

In contrast to glioblastomas, oligodendrogliomas are typically slow growing and less malignant. These tumors, most frequently diagnosed in adults, are usually sporadic, but familial clustering has been reported. Multiple chromosomal alterations have been linked to oligodendrogliomas, including loss of heterozygosity (LOH) and translocation of chromosome regions 1p and 19q, which are associated with chemosensitivity and longer periods of recurrence-free survival after treatment. Mutations in *IDH1* and overexpression of *EGFR* are commonly detected in oligodendrogliomas.

Tumors in the rarest subgroup of gliomas, the ependymomas, originate from ependymal cells and are primarily diagnosed in children. Familial ependymoma appears to be atypical, but has been reported. In addition, ependymomas are often diagnosed in patients with *NF2*, but loss or mutation of the *NF2* gene locus may not be involved in the development of all ependymomas. In contrast to other gliomas, *TP53* gene mutations appear to be uncommon in ependymomas.

PNETs include a broad range of undifferentiated tumors, predominately located in the cerebellum. These tumors have been shown to differentiate into neuronal, astrocytic, ependymal, muscular and melanotic tumors. The majority of PNETs are malignant medulloblastoma tumors found in children. Medullablastomas have been associated with genetic disorders including Gorlin syndrome, familial APC, and Turcot syndrome, but familial medulloblastoma is exceedingly rare. Other chromosomal alterations associated with PNETs include *Axin1* gene deletions and *PTCH* mutations. *TP53* mutations occur infrequently in medulloblastomas, although LOH has been detected on chromosome 17.

Intracranial and intraspinal schwannomas, benign, homogeneous tumors originating from the schwann cells comprising the nerve sheath, are often found in patients with NF2. Loss of the *NF2* tumor suppressor has been detected in most schwannomas and appears to be the principal mutagenic event leading to schwannoma formation. The condition of multiple schwannomas, or schwannomatosis, represents a unique class of NF.

Meningiomas include a variety of tumors arising from the meninges and are the most common of the benign CNS tumors, accounting for approximately 15% of all intracranial tumors and 25% of intraspinal tumors. The occurrence of meningiomas increases with age, and these tumors occur predominately in women. Meningiomas are frequently identified in patients with NF2, and occasionally in patients with Werner or Gorlin syndrome. Although undiagnosed NF2 is often suspected, meningiomas may also be dominantly inherited without NF2. Along with inactivating mutations in the *NF2* gene, alterations in *MN1*, *DAL1*, *ALPL* and protein 4.1R have been detected in meningiomas.

In aggregate, primary tumors of the CNS include a broad range of tumor types with diverse locations, genetic and biological characteristics, and prognoses. These tumors arise from a variety of cell types, often as the result of unique genetic alterations. These alterations typically regulate consistent cellular functions including proliferation, survival, angiogenesis, invasion, and energetics. Efforts to elucidate the molecular and genetic events underlying the dysregulation of signaling pathways in brain tumors continue to facilitate more accurate assessment of tumor origins and classification of subtypes. This research is paving the way for tailored treatment options with improved outcomes in patients with CNS tumors.

Neuromuscular Disorders

Muscular Dystrophies, 473
Hereditary Motor and Sensory Neuropathies, 476
Congenital (Structural) Myopathies, 478
Spinal Muscular Atrophies, 480
Hereditary Muscle Channelopathies, 482
The Myotonic Dystrophies, 484
Hereditary and Autoimmune Myasthenias, 486
Motor Neuron Disease, 489

CHAPTER

124

Muscular Dystrophies

Anna Sarkozy and Kate Bushby

Institute of Genetic Medicine, Newcastle University, International Centre for Life, Newcastle Upon Tyne, UK

Eugenio Mercuri

Department of Paediatric Neurology, Catholic University, Rome, Italy

Muscular dystrophies (MDs) are primary diseases of muscle characterized by muscle weakness and abnormalities of muscle fibers on histopathological examination.

MDs are clinically and genetically heterogeneous and can be classified based on their inheritance, genetic defect, protein involved, or on the prevalent patterns of muscle involvement and associated clinical features (*1*). MDs show variable onset, ranging from prenatal/birth (congenital muscular dystrophies, CMDs), childhood (Duchenne muscular dystrophy, DMD), early adulthood (most common age of presentation of the limb girdle muscular dystrophies, LGMDs), or late adult life (myofibrillar myopathies, MFMs). As DMD, the most common childhood onset MD, is X-linked, boys are overall more frequently affected by MDs.

Although all MDs share as common feature skeletal muscle involvement leading to variably progressive muscle weakness, single MDs may show distinguishing features such as the pattern of muscle involvement and associated clinical features that represent useful diagnostic handles in the differential diagnosis of these heterogeneous but overlapping conditions. Common signs of muscle weakness involving the pelvic girdle are waddling gait, difficulties getting up stairs and from the floor (the so-called "Gower's maneuver"), while scapular winging and difficulties raising arms and lifting objects reflect shoulder girdle and upper limb weakness. Facial muscle weakness causes difficulties closing the eyelids, puffing cheeks out, or whistling. Distal weakness can cause grip weakness, foot drop, and inability to walk on toes and on heels. Other frequently observed features are increased creatine kinase (CK) values (ranging from mildly elevated to values increased up to 100–200 times), muscle atrophy or pseudohypertrophy, contractures, rigid spine, and cardiac and respiratory involvement that represent common cause of death in these patients. Additional possible associated features that could help in the differential diagnosis of these overlapping conditions are central nervous system (CNS) involvement, predominant cardiac or respiratory problems, ocular anomalies, deafness, swallowing difficulties, skin abnormalities, learning difficulties and dementia.

CMDs usually present at birth or within the first few months of life, with hypotonia, muscle weakness, contractures, motor developmental delay, and possible CNS involvement (*2,3*). Dystrophinopathies, in particular DMD, are characterized by childhood onset symmetrical progressive skeletal muscle weakness starting from the pelvic girdle causing waddling gait and lumbar lordosis and invariably leading to loss of ambulation, in the absence of treatment, before the age of 13 years (*4*). DMD is also typically characterized at onset by pseudohypertrophy of calves. Becker muscular dystrophy is a milder form of dystrophinopathy often characterized at onset by calf pain and cramps. Ambulation is longer preserved and wheelchair dependency develops per definition after the age of 16 years and often very much later. Facioscapulohumeral muscular dystrophy (FSHD) is characterized by facial muscles weakness later expanding onto the scapular, humeral, truncal and lower extremity muscles, often with a prominent asymmetric pattern (*5*). LGMDs show predominant involvement of the proximal musculatures, i.e. the shoulder and pelvic girdle (*6*). Emery Dreifuss muscular dystrophies (EDMDs) are characterized by early contractures, slowly progressive muscle wasting and weakness with scapulo-humero-peroneal distribution and cardiac involvement. Finally, MFMs (which are named for the pathological findings on muscle biopsy) are characterized by slowly progressive, usually adult-onset proximal and distal weakness, often associated with cardiac and respiratory involvement (*7*).

Cardiac and respiratory involvement is commonly observed in MDs and is often the leading death cause, in particular, in dystrophinopathies, EDMD, MFM and specific subtypes of LGMD such as LGMD2I or sarcoglycanopathies. In laminopathies, use of an implantable defibrillator is currently the first line treatment in order to prevent sudden death (*8*). Respiratory insufficiency is usually observed after loss of ambulation in most MDs such as DMD, while in other conditions such as LGMD2I onset of respiratory insufficiency is typically reported in the ambulatory phase of the disease. Contractures are typical of EDMD, collagenopathies due to defects of the *COL6* genes (Ullrich CMD and Bethlem myopathy), and subtypes of LGMD such as LMGD2A. Contractures involving the spine (rigid spine, scoliosis) could severely affect ambulation and posture and orthopedic intervention are often indicated in these patients.

CNS involvement is not common in MDs and intelligence is usually preserved. However, some subtypes of MD can be associated with CNS involvement (especially some forms of CMD) and also learning difficulties. Patients with dystrophinopathies typically present a specific behavioral and cognitive pattern. The intelligence coefficient is typically 1 SD below the general population, but cognitive impairment is not progressive and not correlated with weakness. Verbal intelligence and verbal skills appear more affected than performance. Autistic spectrum disorders are also commonly reported.

Most of the congenital and childhood onset forms of MDs show autosomal recessive (AR) inheritance, while DMD, the most common form of childhood onset muscular dystrophy, is X-linked. LGMDs and EDMD show genetic heterogeneity and could either show dominant, recessive, or X-linked inheritance. Interestingly, late-onset LGMDs are often autosomal recessive. Conversely, great majority of later onset, slowly progressive myopathies such as the MFM are caused by dominant mutations.

Genetic splitting and lumping is frequently observed in MD. Indeed, mutations in the *LMNA* gene give rise to either a severe MD with congenital onset, or EDMD with autosomal dominant AR inheritance, or a slowly progressive late-onset LGMD with minor contractural phenotype (LGMD1B) (*9*). Conversely, hypoglycosylation of α-dystroglycan as observed in dystroglycanopathies could be caused by mutations in several genes encoding known or putative glycosyltransferase enzymes (*2,3*).

Diagnosis of MD is still based on clinical suspicion following presentation with a constellation of these clinical symptoms. Raised CK values, magnetic resonance imaging (MRI) and typical histological finding help with the differential diagnosis of MDs. Immunohistochemistry and immunoblot analysis of proteins commonly involved in MDs represents one of the most common diagnostic approach, although new high-throughput sequencing techniques are currently allowing fast and cheap(er) molecular screening of genes involved in MDs and definition of the causative mutation has become the gold standard for diagnosis. Dystrophin gene analysis, either deletion duplication analysis by multiplex ligation-dependent probe amplification (MLPA) or direct sequencing is now more widely available and has changed the diagnostic approach to DMD. In case of suspicion of DMD, in a young boy with clinical suspicion of DMD and highly raised CK values, *DMD* gene deletion/duplication analysis should be the first diagnostic step followed by whole gene sequencing or muscle biopsy analysis in case of negative MLPA analysis. Normal dystrophin labeling on immunohistochemistry and immunoblot analysis formally excludes a diagnosis of DMD (*10*). Conversely, improvement in the knowledge of underlying genetic defect and pathogenic mechanisms of conditions such as FSHD has further raised the complexity of molecular diagnosis for this condition. Development of chips designed to efficiently screen all known genes for all or specific subcategories of MDs would allow a faster and more cost-effective molecular genotyping, speeding up genetic confirmation of MDs and genetic counseling for patients and their families. Also, a precise diagnosis can be used to help predict prognosis and plan supportive therapies, which are proven to influence longevity and quality of life.

The clinical course and natural history of MDs have dramatically changed over the recent years since advances in medical management and care of patients have been made, with treatment of complications, in particular cardiac, respiratory and orthopedic, as well as enhancement of quality of life, timely supply of aids, adaptations and access to independent living (*10,11*). The same clinical care should be offered to patients independently by their geographic location and nihilistic approach should not be accepted. Use of steroids have been introduced in the treatment of DMD about a decade ago and currently represent the gold standard in DMD care, being the only currently available medication able to slow down the disease progression in terms of muscle strength and function. This also reduces risk of scoliosis, stabilizes respiratory function and improves cardiac function.

Different molecular strategies, such as cell and gene therapy, mutation-specific and cellular pathway approaches, have been tested, but at the current time stem and progenitor cell therapies appear more distant from therapeutic applications. Conversely, gene replacement therapy, mutation suppression (or stop codon read through), exon skipping and antisense oligonucleotides are currently the main fields of interest and research for different MDs and in particular for DMD (*12,13*).

In summary, MDs represent a widely variable and heterogeneous group of diseases, clinically with onset from infancy to late adult life, genetically, and at the level of the underlying protein defect. Precise diagnosis can be sought in the majority of cases. Current diagnostic strategies rely on a combination of review of clinical symptomatology, MRI studies, muscle biopsy analysis and direct mutation testing. Current novel technologies will most likely elucidate new causes for MDs, new diagnostic algorithms, and the hope is that they will also provide routes for specific therapies.

REFERENCES

1. Kaplan, J. C. The 2011 Version of the Gene Table of Neuromuscular Disorders. *Neuromuscul. Disord.* **2010,** *20* (12), 852–873.
2. Godfrey, C.; Foley, A. R.; Clement, E.; Muntoni, F. Dystroglycanopathies: Coming into Focus. *Curr. Opin. Genet. Dev.* **2011,** *21* (3), 278–285.
3. Muntoni, F. Journey into Muscular Dystrophies Caused by Abnormal Glycosylation. *Acta Myol.* **2004,** *23* (2), 79–84.
4. Muntoni, F.; Voit, T. 133rd ENMC International Workshop on Congenital Muscular Dystrophy (IXth International CMD Workshop) 21–23 January 2005, Naarden, The Netherlands. *Neuromuscul. Disord.* **2005,** *15* (11), 794–801.
5. Emery, A. E. H.; Muntoni, F. *Duchenne Muscular Dystrophy*, 3rd ed.; OUP: Oxford, 2003.
6. Tawil, R.; Van Der Maarel, S. M. Facioscapulohumeral Muscular Dystrophy. *Muscle Nerve* **2006,** *34,* 1–15.
7. Bushby, K. Diagnosis and Management of the Limb Girdle Muscular Dystrophies. *Pract. Neurol.* **2009,** *9* (6), 314–323.
8. Selcen, D. Myofibrillar Myopathies. *Curr. Opin. Neurol.* **2010,** *23* (5), 477–481.
9. Meune, C.; Van Berlo, J. H.; Anselme, F.; Bonne, G.; Pinto, Y. M.; Duboc, D. Primary Prevention of Sudden Death in Patients with Lamin A/C Gene Mutations. *N. Engl. J. Med.* **2006,** *354* (2), 209–210.
10. Worman, H. J.; Bonne, G. "Laminopathies": A Wide Spectrum of Human Diseases. *Exp. Cell Res.* **2007,** *313* (10), 2121–2133.
11. Bushby, K.; Finkel, R.; Birnkrant, D. J.; Case, L. E.; Clemens, P. R.; Cripe, L.; Kaul, A.; Kinnett, K.; McDonald, C.; Pandya, S., et al. DMD Care Considerations Working Group Diagnosis and Management of Duchenne Muscular Dystrophy, Part 1: Diagnosis, and Pharmacological and Psychosocial Management. *Lancet Neurol.* **2010,** *9* (1), 77–93.
12. Bushby, K.; Finkel, R.; Birnkrant, D. J.; Case, L. E.; Clemens, P. R.; Cripe, L.; Kaul, A.; Kinnett, K.; McDonald, C.; Pandya, S., et al. DMD Care Considerations Working Group Diagnosis and Management of Duchenne Muscular Dystrophy, Part 2: Implementation of Multidisciplinary Care. *Lancet Neurol.* **2010,** *9* (2), 177–189.
13. Guglieri, M.; Bushby, K. Molecular treatments in Duchenne muscular dystrophy. *Curr. Opin. Pharmacol.* **2010,** *10* (3), 331–337.

CHAPTER

125

Hereditary Motor and Sensory Neuropathies

Wojciech Wiszniewski

Department of Molecular and Human Genetics, Baylor College of Medicine, Houston, TX, USA

Kinga Szigeti

Department of Neurology, University at Buffalo, SUNY Buffalo, New York, NY, USA

James R Lupski

Department of Molecular and Human Genetics, Baylor College of Medicine, Houston, TX, USA

Hereditary motor and sensory neuropathies, or Charcot–Marie–Tooth (CMT) disease and related neuropathies, are a group of clinically and genetically heterogeneous disorders primarily affecting the peripheral nervous system. CMT is characterized by slowly progressive, length-dependent neuropathy manifesting as distal weakness of the legs progressing proximally, followed, in some cases, by hand involvement. The diagnosis is based on the presence of lower motor neuron signs and evidence of sensory involvement. Motor nerve conduction velocities (NCVs) distinguish two major CMT types: type 1 or the demyelinating (CMT1) form, which is characterized by symmetrically slowed NCV and type 2 or the axonal (CMT2) form, associated with normal or subnormal NCVs and reduced compound muscle action potential. The disease spectrum of CMT is a continuum, and can extend in the same family from severe infantile-onset disease to mild adult-onset disease. Traditional clinical classifications also include Dejerine–Sottas neuropathy and congenital hypomyelinating neuropathy. During the previous two decades, an enormous amount of information regarding peripheral nerve function and dysfunction have been obtained by the identification of genes responsible for disease in patients manifesting inherited peripheral neuropathies.

CMT and related neuropathies exhibit all forms of Mendelian inheritance—AD, AR, and XL; AD-CMT1 is the most frequent pattern observed. Thus far, over 40 different genetic loci have been linked to CMT; for 32 of these loci, specific genes have been identified. Mutations in genes encoding proteins required for proper development, maintenance, or function of the peripheral nerve may result in neuropathy. Several of the disease genes initially identified encode (i) structural proteins that are important in myelination (e.g. *PMP22, MPZ*), (ii) proteins involved with radial transport through the myelin membrane sheath (e.g. *Cx32*), (iii) proteins involved with axonal transport (e.g. *NEFL, GAN1*), (iv) transcription factors associated with onset of myelination (*EGR2, SOX10*), (v) endosome-related proteins/endocytic recycling (*RAB7, SH3TC2, FIG4, SIMPLE*) (vi) members of signal transduction pathways (e.g. *PRX, MTMR2, SBF2, NDGR1, GDAP1*), (vii) mitochondrial function–related protein (e.g. *MFN2*), (viii) protein synthesis (*GARS, KARS, AARS, YARS*), (ix) a gene involved in DNA single-strand break repair (*TDP1*), (x) chaperones (*HSP22, HSP27*), and (xi) other genes. The first molecular event discovered, responsible for the majority of CMT, was the duplication of the chromosomal segment harboring *PMP22*. This discovery introduced a novel molecular mechanism in human mutagenesis, nonallelic homologous recombination and defined a new group of disorders, the genomic disorders. The reciprocal molecular event, deletion of the same genomic interval, was predicted and found in patients manifesting hereditary neuropathy with pressure palsy. This molecular mechanism and the diseases provided substantial evidence for the presence of dosage-sensitive genes in the human genome.

Duplication of a chromosomal segment harboring *PMP22* (i.e. the CMT1A duplication) represents 43% of the total CMT cases, whereas the yield of duplication detection rises to 70% in CMT1. The population-based studies suggest that in patients with the CMT1 phenotype *MPZ* and *PMP22* mutations are the next most common after *PMP22* duplication and *Cx32* mutations. In the CMT2 group, recent data, although not population based, suggest that *MFN2* mutations are one of the most common causes of CMT2 present in up to 20% of patients, followed by *Cx32* and *MPZ* mutations in frequency. Mutations in other genes are responsible for the CMT phenotype in only a small minority of patients. The high frequency of de novo mutations in duplication/deletion (37–90%) illustrates that genetic disease is commonly sporadic in presentation, lacking a family history. Molecular tools have increased the possibility of establishing a specific diagnosis. The expense associated with evaluating multiple genes for disease-causing mutations has also escalated. The introduction of genomic approaches including whole genome or exome sequencing may dramatically change molecular diagnostic evaluations. Interestingly, one of the first patients ever diagnosed using whole-genome sequencing had CMT type I due to *SH3TC2* mutation.

CHAPTER

126

Congenital (Structural) Myopathies

Heinz Jungbluth

Evelina Children's Hospital, St Thomas Hospital, London and the Clinical Neuroscience Division, IOP, King's College, London, UK

Carina Wallgren-Pettersson

Department of Medical Genetics, Haartman Institute, University of Helsinki, the Folkhälsan Institute of Genetics, Helsinki and the Folkhälsan Department of Medical Genetics, Helsinki, Finland

The congenital myopathies are a clinically and genetically heterogeneous group of inherited neuromuscular conditions associated with certain structural abnormalities on muscle biopsy. The concept of the congenital myopathies is closely linked to the introduction of new histochemical techniques into the study of diseased muscle, leading to the description of the major entities, namely Central Core Disease (CCD), Nemaline Myopathy (NM), Centronuclear (Myotubular) Myopathy (CNM/MTM) and Multi-minicore Disease (MmD). Congenital Fiber Type Disrproportion (CFTD), a histopathological diagnosis describing marked size difference between hypotrophic type 1 and type 2 fibers, is the only abnormality in some patients at presentation. Specific congenital myopathies cannot be distinguished from each other or other congenital muscle disorders on clinical grounds alone and diagnosis is therefore dependent on muscle biopsy findings. Muscle imaging has recently emerged as a helpful diagnostic tool, as it may reveal distinct patterns of selective involvement associated with specific genetic backgrounds.

For many of the congenital myopathies, the causative genes have now been identified, making molecular confirmation of the diagnosis possible in many cases. Most genes implicated in the congenital myopathies to date encode sarcomeric proteins or proteins involved in calcium homeostasis and excitation-contraction coupling, making the congenital myopathies a suitable model to investigate the physiology and pathophysiology of these processes. Mutations in the skeletal muscle ryanodine receptor (*RYR1*) gene, encoding the principal sarcoplasmic reticulum calcium release channel, are most common and have been implicated in CCD as well as subgroups of MmD and CNM. NM, where most causative genes (i.e. nebulin, *NEB*; skeletal muscle α-actin, *ACTA1*; α-tropomyosin, *TPM3*; β-tropomyosin; troponin T1, *TNNT1*, cofilin-2, *CFL2*) encode proteins involved in sarcomeric assembly, is an example of mutations in different genes giving rise to similar phenotypes, due to functional association of the respective gene products. The major proteins implicated in CNM/MTM (i.e. myotubularin, *MTM1*; dynamin 2, *DNM2*; amphiphysin 2, *BIN1*) are all involved in membrane trafficking and may, if mutated, cause secondary ryanodine receptor (*RyR1*) abnormalities, providing an explanation for the observed clinico-pathological overlap with recessive *RYR1*-related myopathies.

The clinical presentation of most congenital myopathies is nonspecific but with certain genetic backgrounds, certain features are more common than others. Many patients present at birth with severe hypotonia and weakness, and have difficulties with respiration and feeding. Arthrogryposis is not a characteristic feature, with the exception of rare, very severe cases. The face is often elongated and expressionless, the mouth tent-shaped, and the palate high-arched. Extraocular muscle involvement is common in all forms of CNM and recessive *RYR1*-related myopathies. Orthopedic complications comprising foot deformities and hip dislocations are frequent in CCD. Most other patients present later in infancy or childhood with delayed attainment of motor milestones and a waddling gait, and some with a disturbance of speech articulation. Clinical presentation in adulthood is infrequent, and these cases may not be genetic in origin, but it is to be observed that mild congenital muscle weakness may go unnoticed and patients may thus present in adulthood with respiratory insufficiency.

Some infants die from respiratory complications, but even patients with severe floppiness and lack of spontaneous respiration at birth have been known to survive, some of them with little residual disability. Others may experience deterioration during the prepubertal period of rapid growth, and some will require a wheelchair from this time. Otherwise, the course of the disease is often only very slowly progressive, and most patients will be able to lead an active life. The main factors influencing prognosis may be respiratory capacity and the development of scoliosis, particularly common in MmD due to recessive mutations in the selenoprotein N (*SEPN1*) gene. The diaphragm is affected as well as the intercostal muscles and the accessory muscles of respiration; this is of special importance during sleep and when the patient is in the recumbent position. The monitoring of respiratory function and of the spine are essential elements in the ongoing care of these patients.

Cardiac involvement is uncommon but an initial evaluation including ultrasound imaging to exclude structural abnormalities or a cardiomyopathy is indicated. Cardiac follow-up is necessary when hypoxia carries the risk of cor pulmonale, and also in genetically unresolved patients and those with *ACTA1* mutations. Intelligence is usually normal.

On examination, the gait is usually waddling because of proximal weakness; some patients with distal involvement also have foot drop. Facial weakness may be overt or subtle, with inability to bury the eyelashes often the only sign. Extraocular muscle involvement is common in CNM and subgroups of MmD. The build is often slender but muscle bulk is not necessarily small, especially not in young children. The spine is hyperlordotic or sometimes rigid, particularly in NM and *SEPN1*-related MmD where scoliosis is also common. Tendon reflexes are weak or absent. Many patients have joint hypermobility, but contractures, mainly of the Achilles tendon, often develop with time.

Electromyography may be normal in young patients and mild cases but usually shows polyphasic motor unit potentials with small amplitude, a full interference pattern during weak effort, and normal fiber density. In addition to these "myopathic" features, electromyographic signs often interpreted as neurogenic may develop with time, especially in distal muscles. Severely affected neonates may even show spontaneous activity to the extent of mimicking spinal muscular atrophy. Ultrasonography often shows abnormally high echogenicity in affected muscles. Muscle MRI may reveal a typical pattern of selective involvement, prominent in congenital myopathies due to mutations in the *RYR1*, *DNM2* and *NEB* genes. Serum concentrations of creatine kinase are usually normal or only slightly elevated, particularly in some patients harboring *RYR1* mutations.

Histochemical staining of fresh-frozen muscle biopsy sections is necessary to reveal the specific characteristics of each of the congenital myopathies, but on routine staining these myopathies have many features in common. Predominance of often hypotrophic type 1 fibers is common, whilst type 2 fibers, if present, may be of more normal size. In some patients, marked size variability between type 1 and type 2 fibers consistent with the definition of CFTD is the only feature, whereas in others fiber typing is indistinct. There may be replacement of muscle fibers by fat and fibrous tissue, but necrotic and regenerating fibers are rarely seen. Marked inflammation is not typical. Internal nuclei may be numerous, and there may be occasional fiber splitting. Specific histochemical stains reveal the typical structural changes, that is, central cores, multi-minicores, central nuclei and nemaline rods, but recent genetic findings suggest that none of those are entirely specific and that there may be considerable histopathological overlap between genetically distinct entities. Electron microscopic studies of muscle ultrastructure may be necessary in confirming the histological findings characteristic of each congenital myopathy.

More than one mode of inheritance is known for most of the congenital myopathies. The proportion of new mutations and the incidence of germline mosaicism are uncertain. Where the causative gene has been identified and molecular genetic testing is available, this is recommended for confirmation of the diagnosis and for determining the mode of inheritance in each individual family. All families should be provided access to genetic counseling.

Overall, the process from the initial suspicion of a congenital myopathy to the diagnosis of underlying genetic defect is complex and ought to be informed by a combined appraisal of histopathological, clinical and, increasingly, muscle MRI data. In view of the rarity of the congenital myopathies, it is worth concentrating diagnostic investigations at centers that have sufficient experience with the clinical, histopathological and genetic investigation of these disorders.

FURTHER READING

1. Arbogast, S.; Ferreiro, A. Selenoproteins and Protection Against Oxidative Stress: Selenoprotein N as a Novel Player at the Crossroads of Redox Signaling and Calcium Homeostasis. *Antioxid. Redox Signal.* **2010**, *12* (7), 893–904.
2. Clarke, N. F. Congenital Fibre Type Disproportion—A Syndrome at the Crossroads of the Congenital Myopathies. *Neuromuscul. Disord.* **2011**, *21* (4), 252–253.
3. Clarke, N. F.; North, K. N. Congenital Fiber Type Disproportion—30 Years On. *J. Neuropathol. Exp. Neurol.* **2003**, *62* (10), 977–989.
4. Dowling, J. J.; Gibbs, E. M., et al. Membrane Traffic and Muscle: Lessons from Human Disease. *Traffic.* **2008**, *9* (7), 1035–1043.
5. Goebel, H. H.; Laing, N. G. Actinopathies and Myosinopathies. *Brain Pathol.* **2009**, *19*, 516–522.
6. Jungbluth, H. Central Core Disease. *Orphanet J. Rare Dis.* **2007**, *2*, 25.
7. Jungbluth, H. Multi-Minicore Disease. *Orphanet J. Rare Dis.* **2007**, *2*, 31.
8. Klingler, W.; Rueffert, H., et al. Core Myopathies and Risk of Malignant Hyperthermia. *Anesth. Analg.* **2009**, *109* (4), 1167–1173.
9. North, K. N.; Wang, C. H.; Clarke, N. et al. Approach to the Diagnosis of Congenital Myopathies. *Semin. Pediatr. Neurol.* **2011**, *18*, 216–220.
10. Rosenberg, H.; Davis, M., et al. Malignant Hyperthermia. *Orphanet J. Rare Dis.* **2007**, *2*, 21.
11. Selcen, D. Myofibrillar Myopathies. *Neuromuscul. Disord.* **2011**, *21* (3), 161–171.
12. Treves, S.; Anderson, A. A., et al. Ryanodine Receptor 1 Mutations, Dysregulation of Calcium Homeostasis and Neuromuscular Disorders. *Neuromuscul. Disord.* **2005**, *15* (9–10), 577–587.
13. Wang, C.; Dowling, J. J.; North, K. N., et al. Consensus Statement on Standard of Care for Congenital Myopathies. *J. Child Neurol.* **2012**, *27*, 363–382.

CHAPTER

127

Spinal Muscular Atrophies

Sabine Rudnik-Schöneborn and Klaus Zerres

Institute for Human Genetics, Medical Faculty, RWTH Aachen, Aachen, Germany

The term spinal muscular atrophy (SMA) comprises a clinically and genetically heterogeneous group of diseases characterized by degeneration and loss of the anterior horn cells in the spinal cord, and—depending on type and severity—sometimes also in the brainstem nuclei, resulting in muscle weakness and atrophy.

The criteria used for the subdivision of the SMAs into separate entities are age of onset, severity, distribution of weakness, the inclusion of additional features, and different modes of inheritance (Table 127-1). The classification has been and will be modified with increasing knowledge of the underlying defects.

About 90% of cases belong to autosomal recessive proximal SMA caused by deletions/mutations of the *SMN1* gene localized on chromosome 5q13 (*1*). The combined incidence is at least 1 in 10,000 in the Caucasian population (*2*). Distal SMA accounts for about 10% of all SMAs, while the definition of scapuloperoneal syndromes is still under debate. Progressive bulbar palsy is extremely rare, while spinobulbar neuronopathy type Kennedy has a prevalence of about 1 in 40,000.

127.1 DIAGNOSIS

Neurological investigations disclose progressive muscle weakness and atrophy and reduced tendon reflexes in most of the SMAs. Creatine kinase (CK) activity in the serum is normal or only mildly elevated. EMG and muscle biopsy studies reveal a neurogenic lesion in the muscle. Peripheral nerve conduction velocities are generally normal which helps to differentiate SMA from clinically similar motor neuropathies. Postmortem studies show a variable degeneration of motor neurons in the spinal cord and in the brain stem.

Diagnostic criteria for proximal SMA have been repeatedly defined by the International SMA Consortium (*3*). Genetic screening is now replacing invasive neurologic tests in the diagnosis of proximal SMA types I–III, the rare diaphragmatic type of SMA, and bulbospinal neuronopathy (Kennedy syndrome). However, the majority of genes responsible for the non-proximal SMAs and autosomal dominant forms of SMAs remain unidentified at the time of writing.

127.2 PROXIMAL SMA AND VARIANTS

The infantile- and juvenile-onset proximal SMAs (SMA types I–III) play the predominant role and are caused by defects of the *SMN1* gene on chromosome 5q13 (for current review see Lunn and Wang (*4*)) The clinical picture indicates a continuous spectrum with ages of onset ranging from before birth to adulthood (*5*). Although there is a spectrum of manifestations in SMA types I–III, with overlapping features between the subgroups, SMA type IV or adult SMA (onset >30 years) is a more distinct and heterogeneous entity. Lower motor neuron disease is also seen in series of patients with amyotrophic lateral sclerosis, that is a distinction is not always possible.

Variants of infantile SMA include diaphragmatic SMA or SMA with primary respiratory distress, SMA with pontocerebellar hypoplasia or atrophy, SMA with arthrogryposis multiplex and SMA with myoclonic epilepsy (*6*).

127.2.1 Genetics

SMA types I–III follow an autosomal recessive mode of inheritance, and more than 90% of patients show homozygous deletions of the *SMN1* gene, enabling a fast and reliable molecular diagnosis. There is an inverse correlation between disease severity and *SMN2* gene copies (*7*). In 3–4% of patients, a compound heterozygous mutation can be detected with a *SMN1* gene deletion on one chromosome and with a subtle mutation on the other chromosome, while in consanguineous families, a homozygous subtle mutation has to be taken into account. Meanwhile, improved methods for heterozygosity testing have become available allowing a reliable risk stratification of relatives and their spouses. The *SMN1* copy number can be measured by quantitative methods and can reliably identify heterozygous carriers of *SMN1* gene deletions. The test sensitivity does not exceed 95%,

TABLE 127-1 Classification of Spinal Muscular Atrophies

1. Proximal SMA (80–90%)
1.1 Autosomal recessive SMA
– Infantile SMA (SMA I–III)
– Adult SMA (SMA IV)
1.2 Autosomal dominant SMA (juvenile and adult forms)
2. Variants of infantile SMA
2.1 Diaphragmatic SMA (SMARD) (ar)
2.2 SMA plus cerebellar hypoplasia (ar)
2.3 SMA plus arthrogryposis and bone fractures (ar, XI)
2.4 SMA plus myoclonic epilepsy (ar)
3. Nonproximal SMA
3.1 Distal SMA (ad, ar, xi)
3.2 Scapuloperoneal SMA (ad, ar)
4. Bulbar palsy
4.1 Adult-onset bulbar palsy (ad)
4.2 Progressive bulbar palsy of childhood type Fazio–Londe (ar)
4.3 Bulbar palsy with deafness (Brown–Vialetto–van Laere syndrome) (mostly ar)
5. Spinobulbar muscular atrophy type, Kennedy's disease (XI)

ar, autosomal recessive; ad, autosomal dominant; XI, X-linked.

because two or more *SMN1* copies are present on normal chromosomes in 3–4% of carriers.

Patients with SMA type IV are only exceptionally associated with alterations of the *SMN1* gene. The genetic basis is largely unknown, and most patients have no family history. Kennedy syndrome is X-linked and easily diagnosed by the presence of a CAG repeat expansion in the *androgen receptor* gene (*8*).

127.3 DISTAL SMA

Distal SMA comprises a group of genetically and clinically heterogeneous disorders with a broad spectrum of clinical manifestations. Distal SMAs are also denoted as distal hereditary motor neuronopathies (HNMs) and are frequently listed among the Charcot–Marie–Tooth (CMT) neuropathies (*9*). Both autosomal dominant and recessive genes cause childhood- and adult-onset forms, and the course is usually a chronic and benign. Most patients show a clinical picture of distal weakness and atrophy similar to *CMT neuropathy*. As this is the main differential diagnosis, it is important to exclude peripheral nerve involvement by electroneurography before diagnosing distal SMA. The diagnostic yield with the known genes causing distal SMA does not exceed 15% by the end of 2010 (*10*).

127.4 THERAPY AND FUTURE PROSPECTS

With increasing knowledge of genes that cause anterior horn cell degeneration, our understanding for the pathogenesis and possible therapeutic interventions has much improved (*11*). Despite promising results from genetic studies, preliminary clinical trials and experiments with animal models, a curative treatment of SMA is not yet available. Therefore, symptomatic treatment is the mainstay in the management of patients (*12*).

REFERENCES

1. Lefebvre, S.; Bürglen, L.; Reboullet, S., et al. Identification and Characterization of Spinal Muscular Atrophy—Determining Gene. *Cell* **1995**, *80*, 155–165.
2. Ogino, S.; Wilson, R. B. Genetic Testing and Risk Assessment for Spinal Muscular Atrophy (SMA). *Hum. Genet.* **2002**, *111*, 477–500.
3. International, S. M. A. Consortium: Workshop Report. *Neuromuscul. Disord.* **1999**, *9*, 272–278.
4. Lunn, M. R.; Wang, C. H. Spinal Muscular Atrophy. *Lancet* **2008**, *371*, 2120–2133.
5. Zerres, K.; Rudnik-Schöneborn, S. Natural History in Proximal Spinal Muscular Atrophy (SMA): Clinical Analysis of 445 Patients and Suggestions for a Modification of Existing Classifications. *Arch. Neurol.* **1995**, *52*, 518–523.
6. Zerres, K.; Rudnik-Schöneborn, S. 93rd ENMC International Workshop: Non 5q-Spinal Muscular Atrophies (SMA)—Clinical Picture (6–8 April 2001, Naarden, The Netherlands). *Neuromuscul. Disord.* **2003**, *13*, 179–183.
7. Feldkötter, M.; Schwarzer, V.; Wirth, R., et al. Quantitative Analyses of SMN1 and SMN2 Based on Real-Time Light Cycler PCR: Fast and Highly Reliable Carrier Testing and Prediction of Severity of Spinal Muscular Atrophy. *Am. J. Hum. Genet.* **2002**, *70*, 358–368.
8. Rhodes, L. E.; Freeman, B. K.; Auh, S., et al. Clinical Features of Spinal and Bulbar Muscular Atrophy. *Brain* **2009**, *132*, 3242–3251.
9. Irobi, J.; Dierick, I.; Jordanova, A., et al. Unraveling the Genetics of Distal Hereditary Motor Neuropathies. *Neuromolecular Med.* **2006**, *8*, 131–146.
10. Dierick, I.; Baets, J.; Irobi, J., et al. Relative Contribution of Mutations in Genes for Autosomal Dominant Distal Hereditary Motor Neuropathies: A Genotype-Phenotype Correlation Study. *Brain* **2008**, *131*, 1217–1227.
11. Sendtner, M. Therapy Development in Spinal Muscular atrophy. *Nat. Neurosci.* **2010**, *13*, 795–799.
12. Wang, C. H.; Finkel, R. S.; Bertini, E. S., et al. Consensus Statement of Standard of Care in Spinal Muscular Atrophy. *J. Child Neurol.* **2007**, *22*, 1027–1049.

CHAPTER

128

Hereditary Muscle Channelopathies

Frank Lehmann-Horn, Reinhardt Rüdel and Karin Jurkat-Rott

Neurosciences Division of Neurophysiology, Ulm University, Ulm, Germany

During a muscle's major tasks, namely during contraction and relaxation, the well-regulated intracellular concentrations of the major ions Na^+, Cl^-, K^+, Ca^{2+} undergo transient changes, which—with the exception of the change of the Ca^{2+} concentration—are relatively small, yet decisive. The changes are brought about by the flux of ions through various parts of the muscle fiber membrane system (subsynaptic membrane, plasmalemma, triads of the sarcoplasmic reticulum), and in every case, the transmembraneous flux is mediated through proteins embedded in the membrane's lipid matrix. These proteins are called *ion channels* because they contain a pore with a filter that is more or less specific for each particular ion species. Channels are usually composed of several subunits that are encoded by separate genes. Many of these channels are tissue specific, for example are the Na^+ channels for cardiac and skeletal muscle encoded by different genes.

A typical feature of most of the relevant channel species is that their ion conductance can be modified by "gates" that may assume several states, for example "open" or "closed." A change from one state to the other is called "gating" and may be effected by either changes of the transmembrane voltage (in voltage-gated channels, such as the Na^+ channels responsible for excitation) or by chemical ligands (in ligand-gated channels, such as the Na^+ and K^+ conducting acetylcholine receptor in the subsynaptic membrane which is essential in neuromuscular transmission).

Channelopathies are hereditary diseases in which the function of a particular species of ion channels is disturbed as a consequence of mutations in one of its encoding genes. According to the above-mentioned membrane specification, this chapter treats first the disorders of the subsynaptic membrane, then those of the plasmalemma and then those of the triads.

If the disturbance affects the acetylcholine receptor in the subsynaptic membrane, the patients suffer from one of the various forms of *congenital myasthenic syndrome*, with muscle weakness as the main symptom. Other deficits in these patients depend on the specific mutation.

If the disturbance affects channels responsible for the resting and/or the action potential of the plasmalemma, this usually changes the excitability of the muscle fibers. Increased excitability or reduced excitability may result, and these changes are experienced by the patients as muscle stiffness (myotonia) or weakness (paralysis), respectively. The symptoms are usually transient and may be of short duration as in the *congenital myotonias*, or of longer duration as often in the *familial dyskalemic periodic paralyses*. In some of the diseases, for example in *paramyotonia congenita*, a transition from hyperexcitability to a lack of excitability may occur. The patients then experience stiffness that gives way to weakness. In *hyperkalemic* and *hypokalemic periodic paralysis*, an increasing fraction of the muscle fibers may stay depolarized for very long times. This then results in a permanent weakness. The depolarized fibers remain in a disturbed homeostasis of their electrolytes, which in turn provokes edemas and finally muscle degeneration. This section of the chapter also elaborates in detail the complicated implications of various channel defects, changes in the serum potassium level and episodes of local or generalized paralysis in the various forms of *dyskalemic periodic paralysis*.

If the ryanodine receptor, a ligand-gated Ca^{2+} release channel of the sarcoplasmic reticulum, which is under the voltage control of the L-type calcium channel of the tubular system, is affected, muscle stiffness and increased metabolism *or* weakness may be resulting. Mutations that increase the sensitivity of the channel to activating ligands channels like some inhalational anesthetics predispose to *malignant hyperthermia*, a life-threatening anesthesia-related event that is characterized by generalized muscle stiffness and increased metabolism. Other mutations increase the open probability of the Ca^{2+} release channel, leading to depleted Ca^{2+} stores and weakness as in central-core and multi-minicore myopathy.

The majority of the muscle channelopathies had been thoroughly investigated both clinically and electrophysiologically when—about 20 years ago—additional results from molecular genetics provided fantastic progress in the understanding of the pathology of these diseases. Today, virtually for all diseases of this group the specific sites of the mutations and their consequences for the particular channel functions are known. Puzzling variants of symptoms have been elucidated and new nosological entities have been delineated, for example the *sodium channel myotonias* from the classical *chloride channel myotonias*. The different modes of inheritance (dominant, recessive) in *myotonia congenita* are now explained by the fact that the chloride channel is a dimer where with certain mutations a single affected monomer can block the channel. The contrasting symptoms of stiffness and paralysis in the two sodium channel diseases *hyperkalemic* and *hypokalemic periodic paralysis* are explained by mutations in the same gene creating gain-of-function or loss-of-function, respectively.

The pathogenetic mechanism of *hypokalemic periodic paralysis* was particularly resistant to elucidation, because single-channel measurements with mutant Na^+ or Ca^{2+} channels in expression systems always yielded almost normal currents. The eventual explanation was that here, the mutations do not substantially influence the gating function of the pore, but rather lead to the creation of an additional Na^+ conducting pore that is permanently open. To make things even more complicated, homologous mutations in the gene encoding the homologous Ca^{2+} channel evoke the same symptoms by the same mechanism!

Two types of Ca^{2+} channels can be mutated in muscle channelopathies affecting excitation–contraction coupling: (1) the so-called dihydropyridine receptor of the tubular membrane and (2) the ryanodine receptor of the reticular membrane of skeletal muscle (RyR1). More important than the few disease-causing mutations are those mutations that play a decisive role in dangerous events that may occur in general anesthesia, the already mentioned *malignant hyperthermia*. Mutations in the genes of either channel have been found to lead to susceptibility to malignant hyperthermia whereby the RyR1 mutations are located in the protein part which bridges the two membrane (the so-called protein foot). In contrast dominant or recessive mutations in the transmembraneous segments spanning the reticular membrane cause a leaky channel, which depletes the Ca^{2+} stores and leads to weakness and a core myopathy. The pathogenetic mechanism of the central cores or the multi-mini cores, which lack mitochondrial activity, is still unclear.

One of the pleasant experiences with channelopathies is the fact that a rather great number of drugs influence the function of ion channels, some in the direction of increasing conduction, others in decreasing conduction, so that for instance excitation can be promoted or reduced, as desired. Also, for the various forms of *dyskalemic periodic paralysis*, methods are described as to how the serum potassium value can be shifted in one or the other direction. Therefore, the chapter rather extensively treats the many possibilities available to the medical practitioner with virtually all the various clinical pictures described.

FURTHER READING

1. Beeson, D.; Higuchi, O.; Palace, J.; Cossins, J.; Spearman, H.; Maxwell, S.; Newsom-Davis, J.; Burke, G.; Fawcett, P.; Motomura, M., et al. Dok-7 Mutations Underlie a Neuromuscular Junction Synaptopathy. *Science* **2006,** *313,* 1975–1978.
2. Brownlow, S.; Webster, R.; Croxen, R.; Brydson, M.; Neville, B.; Lin, J. P., et al. Acetylcholine Receptor Delta Subunit Mutations Underlie a Fast-Channel Myasthenic Syndrome and Arthrogryposis Multiplex Congenita. *J. Clin. Invest.* **2001,** *108,* 125–130.
3. Fontaine, B.; Khurana, T. S.; Hoffman, E. P.; Bruns, G. A.; Haines, J. L.; Trofatter, J. A.; Hanson, M. P.; Rich, J.; McFarlane, H.; Yasek, D. M., et al. Hyperkalemic Periodic Paralysis and the Adult Muscle Sodium Channel Alpha-Subunit Gene. *Science* **1990,** *250,* 1000–1002.
4. Jurkat-Rott, K.; Mitrovic, N.; Hang, C.; Kouzmenkine, A.; Iaizzo, P.; Herzog, J.; Lerche, H.; Nicole, S.; Vale-Santos, J.; Chauveau, D., et al. Voltage-Sensor Sodium Channel Mutations Cause Hypokalemic Periodic Paralysis Type 2 by Enhanced Inactivation and Reduced Current. *Proc. Natl. Acad. Sci. U.S.A.* **2000,** *97,* 9549–9554.
5. Jurkat-Rott, K.; Weber, M. A.; Fauler, M.; Guo, X. H.; Holzherr, B. D.; Paczulla, A.; Nordsborg, N.; Joechle, W.; Lehmann-Horn, F. K+-dependent Paradoxical Membrane Depolarization and Na+ Overload, Major and Reversible Contributors to Weakness by Ion Channel Leaks. *Proc. Natl. Acad. Sci. U.S.A.* **2009,** *106,* 4036–4041.
6. Koch, M. C., et al. The Skeletal Muscle Chloride Channel in Dominant and Recessive Human Myotonia. *Science* **1992,** *257,* 797–800.
7. McCarthy, T. V.; Healy, J. M. S.; Heffron, J. J. A.; Lehane, M.; Deufel, T.; Lehmann-Horn, F.; Farrall, M.; Johnson, K. J. Localisation of the Malignant Hyperthermia Susceptibility Locus to Human Chromosome 19q12-13.2. *Nature* **1990,** *343,* 562–564.
8. Ohno, K.; Engel, A. G., et al. Rapsyn Mutations in Humans Cause Endplate Acetylcholine-Receptor Deficiency and Myasthenic Syndrome. *Am. J. Hum. Genet.* **2002,** *70,* 875–885.
9. Plaster, N. M.; Tawil, R.; Tristani-Firouzi, M.; Canun, S.; Bendahhou, S.; Tsunoda, A., et al. Mutations in Kir2.1 Cause the Developmental and Episodic Electrical Phenotypes of Andersen's Syndrome. *Cell* **2001,** *105,* 511–519.
10. Ricker, K.; Moxley, R. T.; Heine, R.; Lehmann-Horn, F. Myotonia Fluctuans. A Third Type of Muscle Sodium Channel Disease. *Arch. Neurol.* **1994,** *51,* 1095–1102.
11. Senderek, J.; Müller, J. S.; Dusl, M.; Strom, T. M.; Guergueltcheva, V.; Diepolder, I., et al. Hexosamine Biosynthetic Pathway Mutations Cause Neuromuscular Transmission Defect. *Am. J. Hum. Genet.* **2011,** *88,* 162–172.
12. Zhou, H.; Brockington, M.; Jungbluth, H.; Monk, D.; Stanier, P.; Sewry, C. A.; Moore, G. E.; Muntoni, F. Epigenetic Allele Silencing Unveils Recessive RYR1 Mutations in Core Myopathies. *Am. J. Hum. Genet.* **2006,** *79,* 859–868.

CHAPTER

129

The Myotonic Dystrophies

Chris Turner

MRC Centre for Neuromuscular Disease, National Hospital for Neurology and Neurosurgery, London, UK

There are currently two forms of myotonic dystrophy: myotonic dystrophy type 1 (DM1), also known as "Steinert's disease," and myotonic dystrophy type 2 (DM2), also known as proximal myotonic myopathy. DM1 and DM2 are progressive multisystem genetic disorders with several clinical and genetic features in common. DM1 is the most common form of adult-onset muscular dystrophy and is caused by an expansion of unstable CTG trinucleotide repeats in the 3′ untranslated region of the dystrophia myotonica protein kinase (*DMPK*) gene on chromosome 19q13.3 (*1*). DM1 has a European prevalence of 3–15 per 100,000 and demonstrates genetic anticipation, which is strongest through the maternal lineage (*2*). There is a diverse range of clinical phenotypes associated with DM1. The severity of the phenotypes is partially correlated with the length of the expansion of the CTG repeat (*3*). Patients with congenital-onset disease have large CTG expansions greater than 1000. Patients with congenital DM1 almost exclusively have inherited the disease from their mother. Congenital DM1 often presents before birth as polyhydramnios and reduced fetal movements. After delivery, the main features are severe generalized weakness, hypotonia and respiratory compromise and mortality is high. Failure to thrive, clubfeet and feeding difficulties are common problems, but surviving infants experience gradual improvement in motor function, can swallow and independently ventilate. Cognitive and motor milestones are nevertheless delayed and all patients with congenital DM1 develop learning difficulties and require special needs schooling. A progressive myopathy and the other features seen in the adult-onset form of DM1 can develop, although this does not start until early adulthood and usually progresses slowly. Patients often develop severe problems from cardiorespiratory complications in their third and fourth decades and die in their mid-forties (*2*). A childhood-onset form is less common and tends to have clinical features between those of the adult onset and surviving congenital patients with muscle weakness and myotonia in the context of psychosocial problems and low intelligence.

The commonest form of DM1 is adult-onset disease. This is a multisystem disease with typical onset in early adulthood often with myotonia of the hands. The muscular dystrophy progresses over several decades and typically affects the face, swallowing, eyelids, neck flexion, finger flexion, and ankle dorsiflexion. The conduction system of the heart is particularly affected and leads to potentially fatal brady and tachyarrhythmias. The combination of dysphagia associated with aspiration and a poor cough often leads to aspiration pneumonia, which is often fatal. Patients also develop early cataracts, frontal balding, insulin resistance, excessive daytime sleepiness, apathy with poor motivation and IBS-like bowel symptoms (*1*). Over 70% of patients die prematurely from cardiorespiratory complications (*4*).

A mild "late-onset" form of DM1 is often asymptomatic until old age although many patients develop cataracts in their forties.

DM2 is caused by an expansion of an intronic CCTG repeat in the zinc finger 9 (*ZNF9*) gene on chromosome 3q13.3-q24. DM2 tends to have a milder, but similar, phenotype compared to DM1 with later onset of symptoms. It is rarer than DM1 and may be more common in Northern Germany. DM2 is called "proximal myotonic myopathy" because of the predilection for the proximal muscles. Patients may experience worse myalgia and insulin resistance than DM1 patients. Many patients are minimally affected until they are older. There is no congenital or childhood-onset forms of DM2 (*5*).

The screening and treatment of complications remains the mainstay of management of DM1 and DM2. Patients, at a minimum, should have a yearly ECG and a low threshold for treatment of arrhythmias. Excessive daytime sleepiness can be helped with the psychostimulant, modafinil. Obstructive sleep apnoea and central apnoea may be helped with noninvasive ventilation (NIV), although many patients do not tolerate NIV. Bowel symptoms may be treated with bile acid sequestrants and antibiotics. The presence of cataracts should be monitored and treated appropriately. Monitoring for diabetes mellitus is important but the incidence of overt diabetes may be over-estimated. Patients benefit from a multidisciplinary approach to their care and in particular from assessment for dysphagia, orthotics for foot drop and community occupational therapy and social services support (*1*).

The most probable molecular mechanism underlying the pathogenesis of DM1 and DM2 is a toxic RNA gain of function. The mutant pre-mRNA of DMPK accumulates in ribonuclear foci within the nucleus. These foci interfere with RNA splicing factors such as CUG-BP1 and MBNL-1, which cause many pre-mRNA species to undergo abnormal splicing leading to more fetal isoforms being made. This explains the multisystem nature of the disease. Other mechanisms may also be important and include transcriptional dysregulation, altered RNA breakdown and possibly reduced DMPK protein activity and altered expression of genes neighboring the *DMPK* gene. Some of the molecular and clinical effects of mutant CUGs, such as reduced MBNL-1 levels and myotonia, have been reversed using antisense oligonucleotides and small molecules in transgenic mouse models and in vitro cell models of DM1. These findings have heralded a new era of potential genetic treatments for the myotonic dystrophies (*6*).

REFERENCES

1. Turner, C.; Hilton-Jones, D. The Myotonic Dystrophies: Diagnosis and Management. *J. Neurol. Neurosurg. Psychiatr.* **2010,** *81,* 358–367.
2. Harper, P. S. *Myotonic Dystrophy*, 3rd ed.; Saunders: London, 2001.
3. Monckton, D. G.; Ashizawa, T. Molecular Aspects of Myotonic Dystrophy: Our Current Understanding. In *Myotonic Dystrophy: Present Management and Future Therapy*; Harper, P. S.; van Engelen, B.; Eymard, B.; Wilcox, D. E., Eds.; Oxford University Press: Oxford, 2004; pp 14–36.
4. Mathieu, J.; Allard, P.; Potvin, L., et al. 10-Year Study of Mortality in a Cohort of Patients with Myotonic Dystrophy. *Neurology* **1999,** *52,* 1658–1662.
5. Day, J. W.; Ricker, K.; Jacobsen, J. F., et al. Myotonic Dystrophy Type 2: Molecular, Diagnostic and Clinical Spectrum. *Neurology* **2003,** *60,* 657–664.
6. Osborne, R.; Thronton, C. RNA-Dominant Diseases. *Hum. Mol. Genet.* **2006,** *15* (2), R162–R169.

CHAPTER

130

Hereditary and Autoimmune Myasthenias

David Beeson

Nuffield Department of Clinical Neuroscience, University of Oxford, Oxford, UK

The neuromuscular junction is a synapse where symptoms resulting from impaired function are readily apparent. The phenotypic characteristic of both congenital myasthenic syndromes (CMSs) and autoimmune myasthenia gravis (MG) is fatigable muscle weakness. This characteristic weakness derives from the underlying threshold effect that is a key feature of neuromuscular transmission. The postsynaptic membrane must undergo sufficient depolarization to activate voltage-gated sodium channels, which then propagate the signal into muscle contraction. Normally, in humans, a nerve impulse depolarizes the muscle membrane from about −70 to about −40 mV, which only modestly exceeds the threshold of −50 mV. The safety margin is around 10 mV and any pathogenic mechanism that causes the depolarization not to reach −50 mV will result in myasthenic fatigable weakness.

Signal transmission at the neuromuscular junction and in particular the role of the muscle nicotinic acetylcholine receptor (AChR) have been intensely studied over many years and this has provided the basis for dissection of the many differing pathogenic mechanisms that cause the myasthenias (*1*). A series of tests are available for identifying the causes of impaired signal transmission, but a good case history and neurophysiology still provide important initial pointers for diagnosis. If an inherited (congenital) myasthenic syndrome is suspected, there are a series of identified genes that can be screened for mutations. Subtle differences in phenotype can provide valuable pointers for the likely gene defect. Similarly, a number of different diagnostic tests are available to detect the targets of pathogenic autoantibodies, including tests for antibodies to the muscle AChR, muscle-specific kinase (MuSK), and presynaptic voltage-gated calcium channels (VGCC). Although for both inherited and autoimmune forms of myasthenia, the AChR is the most frequently targeted, newly identified genes and neuromuscular junction proteins associated with novel myasthenic disorders continue to emerge.

130.1 CONGENITAL MYASTHENIC SYNDROMES

CMSs are rare inherited disorders of neuromuscular transmission with no underlying autoimmune basis. They are heterogeneous, reflecting the diversity of underlying genetic defects, but share the characteristic fatigable muscle weakness with autoimmune MG. CMSs account for only around 5% of all myasthenias at about 1:100,000, but this number is increasing with improved diagnosis. Mutations in at least 15 different genes have been found associated with CMS, although some are exceptionally rare and at least two are restricted to single case reports (*2*). All disorders (except for the slow channel syndrome) are recessive. They may involve presynaptic, synaptic, or postsynaptic proteins, although postsynaptic defects make up the majority of cases (*1,2*).

The postsynaptic CMS can be divided into three broad categories: those that lead to a deficiency of AChR on the postsynaptic membrane; those that cause a kinetic abnormality of AChR function; and those that alter the stability/maintenance of the synaptic structure. The AChR is a pentamer made up of $\alpha_2\beta\delta\varepsilon$ subunits in the adult form and $\alpha_2\beta\gamma\delta$ subunits in the fetal form. AChR deficiency is most commonly associated with mutations in *CHRNE*, the gene that encodes the adult-specific AChR ε-subunit. Many AChRs are ε-subunit null mutations and it is likely that in these cases residual expression of the fetal AChR, which occurs in human muscle but not mouse muscle, is able to partially compensate for loss of the adult receptors (*3*). Weakness is usually evident at birth or in the first few years of life and is characterized by feeding difficulties, ptosis, impaired eye movements, and delayed motor milestones. Ophthalmoplegia is a prominent and early feature that may help in differential diagnosis. The disorder tends to be stable over time and shows a good but only partial response to therapy with either pyridostigmine or 3,4-diaminopyridine, or a combination of the two. A second common cause of a deficiency of AChR at the endplate is caused by mutations in *RAPSN*. This protein is responsible for the aggregation of the receptors beneath the nerve terminal. In the majority of cases, a common mutation, N88K, is found either on both alleles or as a compound heterozygote with a different mutation on the second allele. By contrast with the *CHRNE* deficiency mutations, patients with mutations in *RAPSN* have little eye muscle involvement, but do suffer from severe bouts of respiratory insufficiency in early childhood often brought on by infections. As for *CHRNE* mutations, there is a good response to pyridostigmine and 3,4-diaminopyridine, and usually, once past the early years, patients improve over time (*1*).

Patients with kinetic abnormalities of the AChR form a second category of postsynaptic disorders. Here, prolonged or attenuated activations of the ion channels lead to slow or fast channel syndromes. The slow channel syndrome is the only CMS that shows dominant inheritance and it may due to mutations in any of the four subunits that make up the adult AChR (*CHRNA, CHRNB, CHRND, CHRNE*). Age of onset is variable as is disease severity. The prolonged channel activations are thought to cause excess calcium entry leading to excitotoxic damage within the immediate postsynaptic regions. The disorder may respond to treatment with drugs that block the receptor ion channel when it is in the open state, such as quinidine or fluoxetine (*4*). In the fast channel syndrome, which shows recessive inheritance, it is shortened channel activations that cause impaired signal transmission. This is often one of the more severe forms of CMS, with patients suffering severe bouts of respiratory insufficiency in early childhood, and with treatments of pyridostigmine and 3,4-diaminopyridine only partially effective (*5*).

A third category of postsynaptic disorders involves the AChR clustering pathway. In this pathway, Agrin is released from the nerve terminal into the synaptic cleft. In combination with the protein LRP4, it activates MuSK and with the help of the cytoplasmic adapter protein Dok-7 initiates a kinase signaling pathway that results both in the clustering of the AChR and the maintenance of the synaptic structures. Mutations in *AGRN* and *MUSK* are rare, but mutations in *DOK7* are relatively common. This disorder tends to cause a predominantly limb-girdle pattern of muscle weakness and responds remarkably to treatment with either ephedrine or salbutamol (*4,6*), both of which are β2 adrenergic receptor agonists.

Within the synapse ColQ forms, the tail of the asymmetric endplate form of acetylcholinesterase anchors it at the synapse. Mutations in *COLQ* cause a severe reduction in the amount of endplate acetylcholinesterase, and result in overstimulation of the postsynaptic membrane in a similar fashion to the slow channel syndrome. On the presynaptic side of the neuromuscular junction, mutations in cholineacetyltransferase, *CHAT*, result in reduced neurotransmitter release, and a disorder that again is characterized by severe apnoeic attacks in early childhood usually brought on by infections. Patients with mutations in *CHAT* respond to treatment with pyridostimine, though this will depend on how severely the mutation disrupts CHAT enzymic function.

The CMSs can be viewed as a spectrum of disorders varying from lethal forms that cause death in utero through to those with only mild weakness. Definitive diagnosis of CMS often requires screening of candidate gene loci, followed by functional studies of the identified mutations in cellular systems. Experienced clinicians can often make a provisional diagnosis from the clinical and electromyography (EMG) features. Clinical features such as mild arthrogryposis at birth, the pattern of muscle affected, and response to anticholinesterase medication may suggest which gene is defective.

130.2 ACQUIRED MYASTHENIAS

MG, in which autoantibodies are directed against the muscle AChR, has an overall prevalence of MG of approximately 15 per 100,000 in Europe and North America (*7*). MG is characterized by weakness of skeletal muscle, which is typically fatigable and thus usually at its worst in the evenings. It commonly first affects movements of the eyes and, at a later stage, may involve bulbar, neck, hand, limb, or respiratory muscles. In severe cases, difficulties in swallowing and breathing can be life threatening. However, its mortality has declined greatly since the advent of assisted ventilation, plasma exchange, and particularly, corticosteroids. The autoimmune disorder targeting AChR can typically be divided into early-onset and late-onset cases. The clinical course in early-onset cases is very variable; 20–30% of patients experience long-term remissions, especially after thymectomy. The myasthenia may fluctuate, and in another 20–30%, the disease may progress and require long-term immunosuppressive therapy. Patients with late-onset disease often show more stable behavior, but MG associated with thymoma tends to deteriorate, especially after thymectomy. Between 10% and 20% of MG worldwide is associated with the presence of a thymoma. Patients with pure extraocular muscle involvement for at least 3 years rarely progress to generalized MG.

The diagnosis of MG depends principally on the clinical picture, detection of anti-AChR antibodies by radioimmunoassay, and EMG, which shows typical decrement and jitter. A number of patients with clear clinical features of myasthenia are negative on the standard radioimmunoprecipitation assay using ^{125}I-α-bungarotoxin. A proportion of these patients are found to have antibodies to the neuromuscular junction protein MuSK. MuSK antibodies are more prevalent on the latitude of Mediterranean countries and are rarer further from the equator (*8*). A further subset can be shown to have antibodies that bind AChRs expressed in a cell lines in which AChR are aggregated on the cell surface by co-expression with the AChR-clustering protein, rapsyn. It is likely that additional key proteins at the neuromuscular junction will be autoantibody targets, and indeed recent reports suggest that antibodies to the MuSK-associated protein LRP4 are present in a few patients.

First-line therapy in MG is with anticholinesterases and in many patients this agent alone affords adequate control, although ocular weakness may respond poorly. Patients with anti-MuSK antibodies may be difficult to treat since they tend to show a poor response to anticholinesterases. In young patients with seropositive generalized MG, most neurologists favor early thymectomy, because about two-thirds of patients can subsequently be maintained on pyridostigmine alone. Many patients will require immunosuppressive therapy, for example, with alternate day corticosteroids, perhaps supplemented with azathioprine (to reduce the steroid dose). Although the thymomas are usually slow growing, their local spread can be dangerous and they should be sought and removed except in very frail cases. Plasma exchange and intravenous immunoglobulin are usually reserved for crises and to improve strength preoperatively (*10*).

Autoantibodies may also target the presynaptic side of the neuromuscular junction. The Lambert–Eaton myasthenic syndrome (LEMS) is a rare disorder with proximal muscle weakness associated with additional autonomic symptoms. It is due to antibodies directed against presynaptic VGCC, which cause reduced ACh release. Post-tetanic potentiation on EMG and the associated autonomic symptom are helpful for the differential diagnosis. The majority of LEMS cases are associated with small cell lung cancer. It is considered a classic paraneoplasmic condition since the autoantibodies that restrict transmission at the neuromuscular junction are thought to be secondary to the immune response against the VGCC expressed by the small cell tumors. 3,4-Diaminopyridine will help enhance transmission in these patients, and immunosuppressive therapy may also be used.

REFERENCES

1. Engel, A. G. Current Status of the Congenital Myasthenic Syndromes. *Neuromuscul. Disord.* **2012**, *22*, 99–111.
2. Chaouch, A.; Beeson, D.; Hantaï, D.; Lochmüller, H. 186th ENMC International Workshop: Congenital Myasthenic Syndromes 24–26 June 2011, Naarden, The Netherlands. *Neuromuscul. Disord.* **2012 Jan 7**, [Epub ahead of print].
3. Ohno, K.; Quiram, P. A.; Milone, M.; Wang, H. L.; Harper, M. C.; Pruitt, J. N., 2nd; Brengman, J. M.; Pao, L.; Fischbeck, K. H.; Crawford, T. O.; Sine, S. M.; Engel, A. G. Congenital Myasthenic Syndromes due to Heteroallelic Nonsense/Missense Mutations in the Acetylcholine Receptor Epsilon Subunit Gene: Identification and Functional Characterization of Six New Mutations. *Hum. Mol. Genet.* **1997**, *6*, 753–766.
4. Engel, A. G. The Therapy of Congenital Myasthenic Syndromes. *Neurotherapeutics* **2007**, *4*, 252–257.
5. Palace, J.; Lashley, D.; Bailey, S.; Jayawant, S.; Carr, A.; McConville, J.; Robb, S.; Beeson, D. Clinical Features in a Series of Fast Channel Congenital Myasthenia Syndrome. *Neuromuscul. Disord.* **2012**, *22*, 112–117.
6. Lashley, D.; Palace, J.; Jayawant, S.; Robb, S.; Beeson, D. Ephedrine Treatment in Congenital Myasthenic Syndrome due to Mutations in DOK7. *Neurology* **2010**, *74*, 1517–1523.
7. Phillips, L. H. The Epidemiology of Myasthenia Gravis. *Ann. N. Y. Acad. Sci.* **2003**, *998*, 407–412.
8. Vincent, A.; Leite, M. I. Neuromuscular Junction Autoimmune Disease: Muscle Specific Kinase Antibodies and Treatments for Myasthenia Gravis. *Curr. Opin. Neurol.* **2005**, *18*, 519–525.
9. Leite, M. I.; Jacob, S.; Viegas, S.; Cossins, J.; Clover, L.; Morgan, B. P.; Beeson, D.; Willcox, N.; Vincent, A. IgG1 Antibodies to Acetylcholine Receptors in "Seronegative" Myasthenia Gravis. *Brain* **2008**, *131*, 1940–1952.
10. Spillane, J.; Beeson, D. J.; Kullmann, D. M. Myasthenia and Related Disorders of the Neuromuscular Junction. *J. Neurol. Neurosurg. Psychiatr.* **2010**, *81*, 850–857.

CHAPTER

131

Motor Neuron Disease

Teepu Siddique, H X Deng and Senda Ajroud-Driss

Department of Neurology and Clinical Neurosciences and Department of Cell and Molecular Biology, Northwestern University Feinberg School of Medicine, Chicago, IL, USA

Amyotrophic lateral sclerosis (ALS), also known as Lou Gehrig disease, motor neuron disease, or Charcot disease, is an adult-onset, progressive neurodegenerative disorder involving the large motor neurons of the brain and the spinal cord. It causes a characteristic clinical picture with weakness and wasting of the limbs and bulbar muscles leading to death from respiratory failure within 5 years. It was in 1874 that the French neurologist Jean Martin Charcot defined the clinical and pathological feature of the disease and gave it its name of amyotrophic lateral sclerosis, and Sir William Osler recognized the hereditary component of the condition in 1880. More than a century later, the first gene linked to adult-onset autosomal dominant ALS, *SOD1*, was identified and that led to the engineering of the first transgenic animal model of the disease. This mouse model became the backbone of ALS research and is used worldwide to investigate pathogenic mechanisms and to test experimental therapies. In the past decade, the list of genes causing motor neuron disease has grown significantly, with first the identification of *ALSIN* causing a rare form of recessive juvenile-onset ALS with predominant upper motor neuron features or ALS2, followed by the discovery of *SENATAXIN* mutations causing ALS4, and *VAPB* causing ALS8. The identification of TDP43 as a major disease protein in the ubiquinated inclusions in ALS and frontotemporal dementia led to the detection of mutations in *TDP43* in a subset of familial ALS patients or ALS 10. The function of TDP43 as a DNA/RNA binding protein shifted ALS research to the role of DNA/RNA metabolism in motor neuron degeneration. Mutations in *FUS/TLS* encoding a protein of similar function to TDP43 were found to cause ALS6 and further implicated DNA/RNA fate in the pathogenesis of ALS. Optineurin gene and valosin-containing protein gene, both involved with Paget's disease of bone, were recently linked to autosomal dominant ALS, whereas SGP11-associated gene *SPATACSIN* was found to cause autosomal recessive juvenile ALS or ALS5. Most recently, ubiquilin 2 *(UBQLN2)* was found to cause X-ALS or ALS/dementia, and an expanded hexanucleotide repeat in C90RF72 was identified in familial and sporadic ALS and familial ALS with dementia.

A few other genes, such as Angiogenin, *FIG4*, and D amino acid oxidase make a small contribution to familial ALS, expanding the pathogenic mechanisms leading to motor neuron death. Despite all this genetic progress, a few loci remain challenging, like the ALS7 locus on 20pter, and mainly the ALS-FTD locus on 9p21, where extensive genetic analysis of all the genes in the candidate region by several independent investigators has failed to reveal any pathogenic mutation.

Although the genetics of familial ALS has helped us to better understand some of the molecular pathways involved in motor neuron degeneration, 90% of ALS is sporadic and of unknown etiology. The completion of the Human Genome Project and the International Hap Map Project, together with the development of high-throughput sequencing technology of the whole exome or whole genome, opened the door to genome-wide association studies (GWAS) to identify genetic risk factors for sporadic ALS or important clues for disease pathogenesis. The interpretation of GWAS results and the translation of the genetic findings have proved to be challenging, since results have not been replicated in different populations. Of all the susceptibility genes revealed by the different GWAS studies, the paraoxanase gene cluster has been extensively examined in the past years and has emerged as the most robust genetic risk factor for sporadic disease.

Despite the genetic progress, the majority of the genes that cause familial ALS continue to be unknown and the cause of sporadic ALS remains elusive. One can only expect that the application of the tools of modern genetics to the study of familial and sporadic disease will identify novel causative genes that will open new therapeutic horizons for this devastating disease.

Ophthalmologic Disorders

Genetics of Color Vision Defects, 493
Optic Atrophy, 496
Glaucoma, 498
Defects of the Cornea, 501
Congenital Cataracts and Genetic Anomalies of the Lens, 503
Hereditary Retinal and Choroidal Dystrophies, 505
Strabismus, 507
Retinoblastoma and the RB1 Cancer Syndrome, 508
Anophthalmia, Microphthalmia, and Uveal Coloboma, 511

CHAPTER

132

Genetics of Color Vision Defects

Samir S Deeb and Arno G Motulsky

Division of Medical Genetics, Departments of Medicine and Genome Sciences, School of Medicine, University of Washington, Seattle, WA, USA

132.1 INTRODUCTION

The human retina contains two classes of photoreceptors, rods and cones. Rods are responsible for vision in dim light, whereas cones mediate vision in bright light and enable the perception of color. Individuals with normal color vision have three types of cone photoreceptors, the short-wave-sensitive or blue, middle-wave-sensitive or green, and long-wave-sensitive or red cones. Such individuals have normal trichromatic color vision. Red–green color vision defects are common among males. Individuals with severe color vision defects usually either have nonfunctional red (protanopes) or green (deuteranopes) cone photoreceptors. Such individuals have dichromatic rather than trichromatic color vision. Those with milder color vision defects usually have either their red or green cone photoreceptor pigments replaced by an anomalous pigment with altered spectral sensitivity. Such individuals are classified as having protanomalous or deuteranomalous trichromatic color vision (Table 132–1).

A wide variation in the ability to discriminate between colors exists among individuals with defective color vision. Subtle variation in color perception also exists among individuals with normal color vision. Significant advances have been made during the last 20 years toward understanding the molecular and genetic bases of variation in both defective and normal color vision. In this chapter, genotype to phenotype relationships in normal color vision and in the various classes of defective color vision will be reviewed. Rare color vision defects include tritanopia (loss of function of the blue ospin), blue-cone monochromacy or incomplete achromatopsia (loss of function of both the red and green opsins), and complete achromatopsia (loss of function of all three cone types). Emphasis in this chapter will be placed on the common red–green color vision defects. Genetics plays a central role in red–green color vision defects. The common defects result from unequal homologous recombinations between the highly homologous and adjacent red and green opsin genes on the X-chromosome.

132.2 RED–GREEN COLOR VISION DEFECTS

Normal color vision in humans is trichromatic, due to the absorption of light by three classes of retinal cone photoreceptors, the short-wave-sensitive (S, or blue), middle-wave-sensitive (M, or green) and long-wave-sensitive (L, or red) cones. Neural comparisons of quantal catches by the three classes of cones give the sensation of color. Each cone contains a single type of photopigment, composed of a protein moiety called opsin, linked to the chromophore 11-*cis*-retinal, a derivative of vitamin A.

In humans, great apes, and Old World monkeys, the genes encoding the red (*OPN1LW*) and green (*OPN1MW*) pigments are arranged in a head-to-tail tandem array on the long arm of the X-chromosome (Xq28). The first gene in the array is an *OPN1LW* gene, followed by one or more red opsin genes; however, only the first two genes in an array are expressed in photoreceptors. The blue opsin gene (*OPN1SW*) is located on chromosome 7q32. The red opsin and green opsin genes are about 96% identical in sequence, whereas they are about 40% identical to the blue opsin gene. The red and green opsin genes arose via gene duplication about 40 MYA in the Old World after separation from the New World. The near identity of the red and green opsin genes resulted in relatively frequent unequal crossing over and gene conversion events that are responsible for the common occurrence (~8% in males and ~0.5% in females of European origin) of red–green color vision deficiencies.

Differences at three amino acids sites (Ala180Ser, Phe277Tyr, and Ala285Thr) account for the major difference in spectral sensitivity between the L- and M-opsins. We discovered a common Ser180Ala polymorphism (62% Ser, 38% Ala among males of European origin) of the L-opsin that is associated with a change in both spectral tuning and variation in perception of the normal range of color perception. This Ala allele is derived from the green opsin gene by gene conversion or unequal homologous recombination.

Unequal crossing over between the red and green opsin genes can lead to dichromatic color vision, a relatively severe form of red–green color vision deficiency, due to loss of either the green opsin (deuteranopia) or the red opsin (protanopia). Such individuals can perceive only blue to yellow hues. Unequal crossing over events can also lead to relatively milder color vision deficiencies (anomalous trichromacy) due to conversion of normal green opsin to red-like opsin (deuteranomaly) or vice versa (protanomaly). In anomalous trichromacy, the differences in spectral sensitivities between the red opsin and the red-like opsin in deuteranomaly, and the green opsin and green-like opsin in protanomaly contribute to the severity of the color vision defect (*1*).

Color vision deficiency in females (~5%) is based on homozygosity or compound heterozygosity for abnormal gene arrays. Approximately 15–16% of the female population of European origin is heterozygote for arrays associated with red–green color vision defects, but most have normal color vision.

Rare red–green color vision deficiencies are also caused by inactivating point mutations, such as the Cys203Arg in the red and/or green opsin genes. Tritanopia is caused by mutations in the blue opsin gene on chromosome 7, and is transmitted in an autosomal dominant manner with incomplete penetrance.

132.2.1 Potential Gene Therapy of Red–Green Color Vision Defects

Improvement of red–green color blindness by gene therapy was explored in the dichromatic adult squirrel monkey (*Saimiri sciureus*) that is missing the L-opsin gene. Sub-retinal injection of a recombinant adeno-associated virus expressing the L-opsin resulted in trichromatic color vision behavior (*2*). This provides the potential for gene therapy in humans to improve defective color vision in the foreseeable future.

132.2.1.1 Blue Cone Monochromacy (or Incompleter Achromatopsia). In this rare (1/100,000 males) X-linked recessive disorder most L- and M-cones are nonfunctional, causing color blindness, reduced visual acuity and pendular nystagmus. Blue cone monochromacy may be caused by either deletion of the regulatory region controlling expression of both the red and green opsin genes, or deletion of all but one of opsin genes plus the presence of inactivating mutations (Cys203Arg) in the one remaining pigment gene (*3*).

132.2.1.2 Complete Achromatopsia. Achromatopsia is a rare autosomal recessive (1/30,000) disorder characterized by reduced visual acuity and complete loss of color discrimination due to loss of function of all three types of cone photoreceptors. Mutations in two genes encoding the α- and β-subunits of the cyclic nucleotide GMP (CNG)-gated channel (*CNGA3*, *CNGB3*) that are critical for light-evoked signal transduction in cones account for the majority of cases. Mutations in two other genes, *GNAT2* (the α-subunit of the cone α-transducin *(4,5)*), and *PDE6C* (cone phosphodiesterase), have also been reported. Gene therapy in a canine model of a CNGB3 defect showed rescue of cone photoreceptor function (*6*).

132.2.1.3 Enhanced Blue Cone Syndrome. Enhanced blue-cone syndrome is a rare autosomal recessive disease characterized by hyper-differentiation of blue cones at the expense of rods, resulting in blindness in the late stages. This human syndrome is caused by mutations in the orphan nuclear receptor NR2E3 (*7,8*).

TABLE 132-1 Classification of the Common X-linked Recessive Color Vision Defects

Class	Retinal Cones			Frequency (European Men)	Color Discrimination
Normal	B	G	R		
Protanopia			No	~1%	
Protanomaly			G'	~1%	
Deuteranopia		No		~1%	
Deuteranomaly		R'		~5%	

R', red-like; G', green-like types of cones. Color detection patterns in the visible spectrum are shown for normal trichromats and dichromats with protanopia or deuteranopia. Anomalous trichromats (protanomalous and deuteranomalous) have a wide range of color discrimination capacity.

FURTHER READING

1. Deeb, S. S.; Motulsky, A. Disorders of Color Vision. *In Genetic Diseases of the Eye. Ophthalmic Disorders;* Traboulsi, E. L. Ed.; 2011 (in press).
2. The Genetics of Normal and Defective Color Vision (*8*).

Internet Review

3. Deeb, S. S.; Motulsky, A. In *Red-Green Color Vision Defects. GeneReviews [Internet] Seattle (WA)*; Pagon, R. A.; Bird, T. D.; Dolan, C. R.; Stephens, K., Eds.; University of Washington: Seattle, May, 2011, updated.

Websites

4. Deeb, S. S.; Motulsky, A. *Red-Green Color Vision Defects. GeneReviews Seattle (WA);* University of Washington: Seattle, May, 2011, updated.
5. Color Vision: http://webvision.med.utah.edu/book/part-vii-color-vision/color-vision.

REFERENCES

1. Deeb, S. S. Genetics of Variation in Human Color Vision and the Retinal Cone Mosaic. *Curr. Opin. Genet. Dev.* **2006,** *16* (3), 301–307.
2. Mancuso, K.; Hauswirth, W. W.; Li, Q.; Connor, T. B.; Kuchenbecker, J. A.; Mauck, M. C.; Neitz, J.; Neitz, M. Gene Therapy for Red-Green Colour Blindness in Adult Primates. *Nature* **2009,** *461* (7265), 784–787.
3. Gardner, J. C.; Michaelides, M.; Holder, G. E.; Kanuga, N.; Webb, T. R.; Mollon, J. D.; Moore, A. T.; Hardcastle, A. J. Blue Cone Monochromacy: Causative Mutations and Associated Phenotypes. *Mol. Vis.* **2009,** *15,* 876–884.
4. Trankner, D.; Jagle, H.; Kohl, S.; Apfelstedt-Sylla, E.; Sharpe, L. T.; Kaupp, U. B.; Zrenner, E.; Seifert, R.; Wissinger, B. Molecular Basis of an Inherited Form of Incomplete Achromatopsia. *J. Neurosci.* **2004,** *24* (1), 138–147.
5. Kohl, S.; Baumann, B.; Rosenberg, T.; Kellner, U.; Lorenz, B.; Vadala, M.; Jacobson, S. G.; Wissinger, B. Mutations in the Cone Photoreceptor G-Protein Alpha-Subunit Gene GNAT2 in Patients with Achromatopsia. *Am. J. Hum. Genet.* **2002,** *71* (2), 422–425.
6. Komaromy, A. M.; Alexander, J. J.; Rowlan, J. S.; Garcia, M. M.; Chiodo, V. A.; Kaya, A.; Tanaka, J. C.; Acland, G. M.; Hauswirth, W. W.; Aguirre, G. D. Gene Therapy Rescues Cone Function in Congenital Achromatopsia. *Hum. Mol. Genet.* **2010,** *19* (13), 2581–2593.
7. Kanda, A.; Swaroop, A. A Comprehensive Analysis of Sequence Variants and Putative Disease-Causing Mutations in Photoreceptor-Specific Nuclear Receptor NR2E3. *Mol. Vis.* **2009,** *15,* 2174–2184.
8. Neitz, J.; Neitz, M. The Genetics of Normal and Defective Color Vision. *Vision Res.* **2011,** *51* (7), 633–651.

CHAPTER

133

Optic Atrophy

Grace C Shih and Brian P Brooks

Ophthalmic Genetics and Visual Function Branch, National Eye Institute, Bethesda, MD, USA

Optic atrophy results from loss of the retinal ganglion cell axons that form the optic nerve, the anatomic and physiologic output of the eye. If the axons that subserve central vision are lost—the so-called papillomacular bundle—visual acuity is lost. The two most common inherited forms of primary optic atrophy are dominant optic atrophy (DOA, MIM#165500, *OPA1,* *605290) and Leber hereditary optic atrophy (LHON, MIM#535000) (*1*). While the clinical presentation of these two diseases is usually quite different in the following table, they are both due to genetic changes that affect mitochondrial metabolism and cellular bioenergetics.

	Leber Hereditary Optic Atrophy (LHON)	**Dominant Optic Atrophy,Kjer Type**
Inheritance	Mitochondrial DNA, maternal	Autosomal Dominant
Typical age of onset	Second–third decade	First–second decade
Presentation	Acute, painless visual loss	Chronic, painless, insidious visual loss
Gene(s)	*MTND4* (m.11778G>A) *MTND1* (m.3460G.A) *MTND6* (m.14484T>C) Collectively account for >90% cases	*OPA1*, 3q28-q29
End stage	Optic nerve pallor, legal blindness	Optic nerve pallor, legal blindness
Treatment	None	None

DOA is characterized by slow, insidious, painless vision loss, often beginning within the first two decades of life. Patients present with loss in visual acuity, abnormal color vision (dyschromatopsia), central/paracentral visual field loss, nerve fiber layer loss (particularly in the maculopapillary fibers), and optic nerve pallor (particularly in the temporal quadrant). As its name implies, inheritance is dominant with nearly 100% penetrance and variable expressivity. In 2000, two groups simultaneously reported on mutations in the *OPA1* gene in DOA (*2*). A nuclear gene, *OPA1*, codes for a dynamin-related GTPase that localizes to the inner mitochondrial membrane. Disruption of OPA1 function impairs ATP production, results in mitochondrial fragmentation, and may produce other, secondary, mitochondrial genome changes. While the pathogenesis is probably most often related to haploinsufficiency, missense-mutation containing alleles may exert a dominant-negative effect if expressed. Approximately 20% of individuals with *OPA1* mutations will develop a "DOA-plus" phenotype, which can include sensorineural hearing loss, peripheral neuropathy, ataxia, myopathy and progressive external ophthalmoplegia (*3*). At present, there is no proven treatment for DOA, although avoidance of potential mitochondrial toxins (tobacco, alcohol, certain drugs) is likely prudent.

Leber hereditary optic neuropathy, in contrast, usually presents as sudden, painless vision loss. Although bilateral vision loss can initially occur, patients more often present with unilateral disease; involvement of the second eye almost always follows, within weeks to months. The typical age of presentation is the second to third decade, although LHON can present at any age. The acute phase is characterized by swelling and hyperemia of the optic disc with telangiectatic and tortous blood vessels (Figure 133-1). This appearance resolves in weeks, leaving optic nerve pallor, particularly temporal. Visual acuity is usually permanently affected, usually resulting in legal blindness.

LHON is the prototypical mitochondrial DNA disease with three mutations (m.3460G>A, m.11778G>A, and m.14484T>C) in genes coding for mitochondrial complex I subunits accounting for >90% of cases. As such, all the children of a carrier female are predicted to carry the disease mutation, while males do not pass the mutation onto their progeny. These mutations are most often homoplasmic and disrupt complex I activity in the mitochondrial electron transport chain, resulting in reduced ATP production and fragmentation of mitochondria. A rough genotype–phenotype correlation has been described, with m.11778G>A having the worst visual prognosis, m.14484T>C having the best prognosis, and m.3460G>A having an intermediate prognosis. Incomplete penetrance is typical, with approximately 50% of males and 10% of females carrying the mitochondrial DNA mutations exhibiting visual signs/symptoms over their lifetime. The precise reasons for incomplete penetrance and male predominance are not known, although certain haplogroups of mitochondrial DNA and environmental

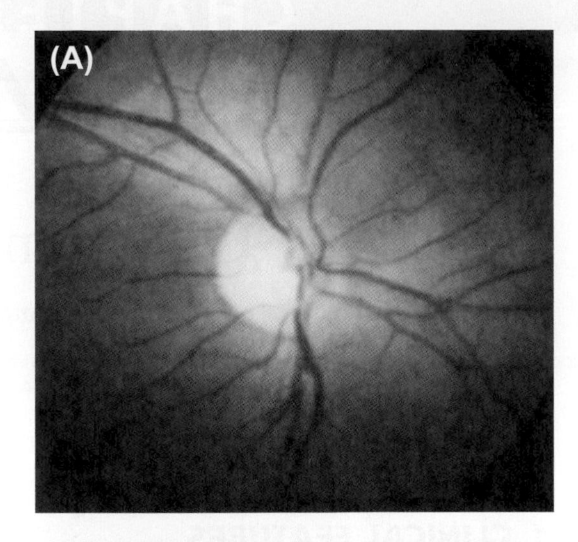

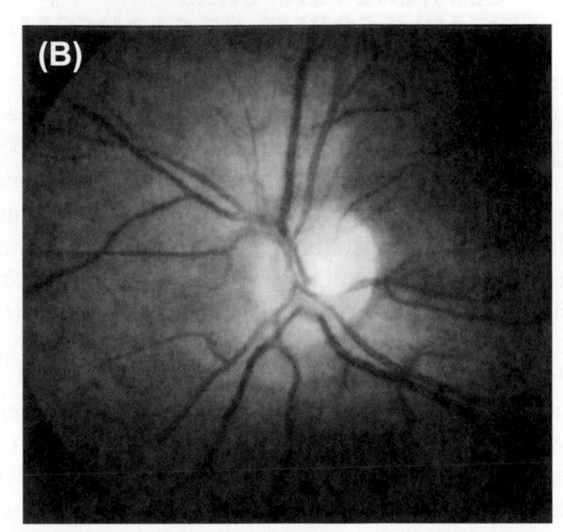

FIGURE 133-1 Dominant optic atrophy. (A) Right eye (OD). (B) Left eye (OS). This 37-year-old man has 20/30 vision in both eyes, a tritan color defect, and exhibits the characteristic temporal pallor and atrophy of the disc.

factors (e.g. smoking, excessive drinking) may play some role. Nuclear genetic factors (X-linked and autosomal) have been postulated to modify this phenotype. A subset of patients may also present with cardiac and neurologic symptoms. Of note, the m.11778G>A mutation appears to predispose affected females to develop a multiple sclerosis-like syndrome.

Like DOA, there is no proven treatment for LHON and care is supportive. Because many individuals carrying the mitochondrial mutation do not go on to develop disease, careful pretest and posttest genetic counseling should be pursued. Electrocardiogram screening of affected patients is a noninvasive way of detecting potentially significant cardiac conduction defects.

REFERENCES

1. Yu-Wai-Man, P.; Griffiths, P. G.; Chinnery, P. F. Mitochondrial Optic Neuropathies—Disease Mechanisms and Therapeutic Strategies. Prog. Retin. Eye Res. 2011, 30 (2), 81–114.
2. (a) Alexander, C.; Votruba, M.; Pesch, U. E.; Thiselton, D. L.; Mayer, S.; Moore, A.; Rodriguez, M.; Kellner, U.; Leo-Kottler, B.; Auburger, G., et al. OPA1, Encoding a Dynamin-Related GTPase, Is Mutated in Autosomal Dominant Optic Atrophy Linked to Chromosome 3q28. Nat. Genet. 2000, 26 (2), 211–215.
   (b) Delettre, C.; Lenaers, G.; Griffoin, J. M.; Gigarel, N.; Lorenzo, C.; Belenguer, P.; Pelloquin, L.; Grosgeorge, J.; Turc-Carel, C.; Perret, E., et al. Nuclear Gene OPA1, Encoding a Mitochondrial Dynamin-Related Protein, is Mutated in Dominant Optic Atrophy. Nat. Genet. 2000, 26 (2), 207–210.
3. Hudson, G.; Amati-Bonneau, P.; Blakely, E. L.; Stewart, J. D.; He, L.; Schaefer, A. M.; Griffiths, P. G.; Ahlqvist, K.; Suomalainen, A.; Reynier, P., et al. Mutation of OPA1 Causes Dominant Optic Atrophy with External Ophthalmoplegia, Ataxia, Deafness and Multiple Mitochondrial DNA Deletions: A Novel Disorder of mtDNA Maintenance. Brain 2008, 131 (Pt 2), 329–337.

CHAPTER

134

Glaucoma

Janey L Wiggs

Department of Ophthalmology, Harvard Medical School, Massachusetts Eye and Ear Infirmary, Boston, MA, USA

The glaucomas are a leading cause of blindness in the United States. One to two percent of Americans older than age 40 years are affected with glaucoma, and each year over 10,000 are blinded by the disease. The glaucomas all share a bilateral, progressive degeneration of the optic nerve. This glaucomatous optic neuropathy causes an irreversible loss of vision that can lead to complete blindness without proper therapy. In most cases, glaucoma is associated with elevated intraocular pressure (IOP); however, approximately one-third of cases have optic nerve degeneration despite intraocular pressures in the normal range. The anatomy and physiology of the optic nerve and the structures regulating aqueous humor dynamics are important elements in the biology of glaucoma and serve as a framework for investigating its pathogenesis. The optic nerve forms a link between the neurosensory retina and the brain, and is an important segment of the visual pathway.

Elevated IOP is associated with most forms of glaucoma and is currently the only known modifiable risk factor for these diseases. IOP is determined by the balance between the rate of aqueous humor production and the rate of outflow from the eye. Aqueous humor is produced by the ciliary body, a tissue located posterior to the peripheral iris. The majority of the aqueous humor drains from the eye through the trabecular meshwork, a sponge-like tissue located at the iridocorneal angle, the junction of the iris and the cornea. Current treatment strategies are all directed at lowering IOP, and in many cases, these interventions are effective at slowing or halting the progression of glaucoma. The early signs of the disease are often subtle and significant nerve damage and vision loss may occur before glaucoma is recognized by either patients or clinicians. Genetic studies of glaucoma could lead to the development of new diagnostic and therapeutic tools. This chapter is divided into four sections: (i) Clinical presentation of heritable forms of glaucoma, including nomenclature and phenotypic descriptions; (ii) identification of glaucoma genes using human linkage studies; (iii) glaucoma genes identified from studying naturally occurring animal models; and (iv) genes identified using genomic approaches.

134.1 CLINICAL FEATURES OF HERITABLE FORMS OF GLAUCOMA

The glaucomas are classified as two groups: primary and secondary glaucomas. Primary glaucoma is defined as isolated, idiopathic disease of the anterior chamber of the eye and the optic nerve, whereas secondary glaucoma is associated with known predisposing events including developmental abnormalities, systemic diseases, drug therapy, or trauma. Primary glaucomas are further characterized by the status of the iridocorneal angle (open or closed) and by the age of onset of disease (congenital (before age 3), juvenile (between ages 3 and 40), and adult). Adult-onset primary open angle glaucoma is the most common type of glaucoma. Both adult and juvenile open angle glaucoma have anatomically normal appearing ocular anterior segments. Developmental glaucomas, including primary congenital glaucoma and Axenfeld–Rieger syndrome, have abnormal anterior segment anatomy resulting from dysfunctional developmental processes. Common forms of secondary glaucoma include pigment dispersion syndrome and pseudoexfoliation syndrome. In pigment dispersion syndrome, pigment granules released from the iris are dispersed throughout the ocular anterior segment. Pseudoexfoliation is characterized by the deposition of a fibrillar material on the lens surface as well as in other ocular structures.

134.2 IDENTIFICATION OF GLAUCOMA GENES USING LINKAGE ANALYSIS

Early-onset forms of glaucoma are amendable to linkage studies. Gene discovery using a positional cloning/linkage approach requires large affected pedigrees with well-defined inheritance. The early-onset glaucomas are inherited as either autosomal dominant or autosomal recessive traits. Disease onset during childhood results in sufficiently large affected pedigrees that a linkage-based approach is likely to identify a genomic region harboring the causative gene. Some adult-onset (age >40) glaucoma pedigrees have familial disease distribution suggestive of autosomal dominant or autosomal recessive inheritance patterns.

Linkage studies have identified over 20 loci for Mendelian forms of primary and secondary glaucomas. Important genes that have been identified at these mapped loci include MYOC (juvenile onset primary open angle glaucoma), OPTN (normal tension glaucoma), CYP1B1 (primary congenital glaucoma), PITX2 (Axenfeld–Rieger syndrome), and FOXC1 (anterior segment dysgenesis).

134.3 GENES CAUSING GLAUCOMA IN ANIMAL MODELS

One approach for finding candidate genes for human glaucoma is to study animals with hereditary forms of glaucoma. The assumption of this approach is that genes important to animal forms of glaucoma will also contribute to human glaucoma. There are also valuable animal models relevant to glaucoma with inducible phenotypes. Important genes responsible for naturally occurring forms of glaucoma are described in mice, dogs and zebrafish.

The DBA/2 lineage of mice develops age-related forms of glaucoma that include iris disease, increased IOP, and retinal ganglion cell loss. The disease causing mutations have been identified and result from a digenic interaction of mutations in two genes encoding melanosomal proteins, Tyrp1 and Gpnmb. A colony of Beagle dogs with the anatomic characteristics of primary open angle glaucoma (POAG), including elevation of intraocular pressure and optic nerve disease, was used for a series of genetic linkage experiments that identified a homozygous missense change in ADAMTS10 as the responsible mutation. A forward-genetic screen for adult ocular abnormalities in zebrafish identified bugeye, with phenotypic features related to glaucoma phenotype including enlarged eyes with myopia, elevated IOP, and damage to retinal ganglion cells. Using linkage analysis, nonsense mutations in low density lipoprotein receptor-related protein 2 (lrp2) were identified in mutant fish.

134.4 IDENTIFICATION OF GLAUCOMA GENES USING GENOMIC APPROACHES

Most human patients with glaucoma are affected by the adult-onset forms of the disease, POAG, exfoliation glaucoma, and angle-closure glaucoma. The genetic and phenotypic complexity of these conditions complicates genetic linkage approaches used to identify causative genes. Recent advances in molecular genetics and genomics have supported the genetic study of common forms of adult-onset glaucoma. In the particular, the annotation of the human genome sequence and the development of the HapMap have facilitated genome-wide association studies (GWAS) identifying genes contributing to ocular quantitative traits related to glaucoma pathogenesis (cup/disc ratio, optic nerve size and central corneal thickness), as well as genes associated with exfoliation syndrome and POAG. Over 20 genes have been identified using GWAS as associated with common forms of glaucoma, including LOXL1 (pseudoexfoliation syndrome), CAV1/CAV2 (adult-onset POAG), CDKN2BAS (adult-onset POAG), and a number of genes contributing to ocular quantitative traits.

During the past decade, a number of new genes associated with glaucoma have been identified and future studies will likely identify many more. The identification of glaucoma genes will be the first step toward the development gene-based screening and diagnostic tests allowing for the identification of individuals at risk for disease before irreversible blindness occurs. The identification of genes contributing to glaucoma will also provide new insight into the molecular events underlying the disease pathophysiology suggesting novel therapeutic approaches. In particular, genes that predispose to optic nerve disease in glaucoma could identify new targets for neuroprotective therapies which may delay or even prevent blindness associated with glaucoma.

FURTHER READING

1. Abiola, O.; Angel, J. M.; Avner, P., et al. The Nature and Identification of Quantitative Trait Loci: A Communities View. *Nat. Rev. Genet.* **2003**, *4*, 911–916.
2. Fan, B. J.; Wiggs, J. L. Glaucoma: Genes, Phenotypes, and New Directions for Therapy. *J. Clin. Invest.* **2010 Sep 1,** *120* (9), 3064–3072.
3. Fingert, J. H. Primary Open-Angle Glaucoma Genes. *Eye (Lond)* **2011 May,** *25* (5), 587–595.
4. Gelatt, K. N.; Brooks, D. E.; Samuelson, D. A. Comparative Glaucomatology. I: The Spontaneous Glaucomas. *J. Glaucoma* **1998,** *7,* 187–201.
5. Gordon, M. O.; Beiser, J. A.; Brandt, J. D., et al. The Ocular Hypertension Treatment Study: Baseline Factors that Predict the Onset of Primary Open-Angle Glaucoma. *Arch. Ophthalmol.* **2002,** *120,* 714–720, discussion 829–830.
6. Leske, M. C.; Connell, A. M. S.; Schachat, A. P., et al. The Barbados Eye Study—Prevalence of Open Angle Glaucoma. *Arch. Ophthalmol.* **1994,** *112,* 821–829.
7. Libby, R. T.; Gould, D. B.; Anderson, M. G., et al. Complex Genetics of Glaucoma Susceptibility. *Annu. Rev. Genomics. Hum. Genet.* **2005,** *6,* 15–44.
8. Liu, Y.; Allingham, R. R. Molecular Genetics in Glaucoma. *Exp. Eye Res.* **2011 Oct,** *93* (4), 331–339.
9. McMahon, C.; Semina, E. V.; Link, B. A. Using Zebrafish to Study the Complex Genetics of Glaucoma. *Comp. Biochem. Physiol. C Toxicol. Pharmacol.* **2004,** *138,* 343–350.
10. Menaa, F.; Braghini, C. A.; Vasconcellos, J. P.; Menaa, B.; Costa, V. P.; Figueiredo, E. S.; Melo, M. B. Keeping an Eye on Myocilin: A Complex Molecule Associated with Primary Open-Angle Glaucoma Susceptibility. *Molecules* **2011 Jun 27,** *16* (7), 5402–5421.
11. Munroe, R. J.; Bergstrom, R. A.; Zheng, Q. Y., et al. Mouse Mutants from Chemically Mutagenized Embryonic Stem Cells. *Nat. Genet.* **2000,** *24,* 318–321.
12. Nadeau, J. H. Modifier Genes in Mice and Humans. *Nat. Rev. Genet.* **2001,** *2,* 165–174.
13. Quigley, H. A. Glaucoma. *Lancet* **2011 Apr 16,** *377* (9774), 1367–1377.

14. Sanfilippo, P. G.; Hewitt, A. W.; Hammond, C. J.; Mackey, D. A. The Heritability of Ocular Traits. *Surv. Ophthalmol.* **2010 Nov–Dec**, *55* (6), 561–583.
15. Sarfarazi, M.; Stoilov, I.; Schenkman, J. B. Genetics and Biochemistry of Primary Congenital Glaucoma. *Ophthalmol. Clin. North Am.* **2003**, *16*, 543–554, vi.
16. Tielsch, J. M.; Sommer, A.; Katz, J., et al. Racial Variations in the Prevalence of Primary Open-Angle Glaucoma. The Baltimore Eye Survey. *JAMA* **1991**, *266*, 369–374.
17. Kwon, Young H.; Fingert, John H.; Kuehn, Markus H.; Alward, Wallace L. M. Primary Open-Angle Glaucoma. *N. Engl. J. Med.* **2009**, *360*, 1113–1124.

CHAPTER

135

Defects of the Cornea

R Krishna Sanka, Elmer Tu, and Joel Sugar

University of Illinois at Chicago, Department of Ophthalmology and Visual Sciences, Chicago, IL, USA

Defects of the cornea can play an important role in clinical genetics because of the relative ease with which abnormalities can often be detected on a clinical ophthalmic examination. This allows for efficient screening, phenotypic categorization, and correlation with the molecular genetic findings.

Corneal dystrophies tend to be autosomal dominant in inheritance, bilateral, not usually associated with other ocular or systemic abnormalities, are often limited to a single layer in the cornea, commonly involve the central cornea, and progress with advancing age. Some changes are usually present in the first few decades of life. The International Committee for Classification of Corneal Dystrophies (IC3D) recently organized the dystrophies by the corneal layer that is chiefly affected and includes information regarding genetic analysis, histopathology, and phenotypic description (*1*). Each dystrophy is categorized by how well defined it is as a distinct entity. Classification of the dystrophies is complex, as many display genetic heterogeneity and/or phenotypic variability.

The corneal epithelium and basement membrane are typically affected in epithelial and subepithelial dystrophies, and decreased vision and painful recurrent erosions can occur. The Bowman layer dystrophies can present with recurrent erosions and epithelial changes, but also demonstrate scarring at the level of Bowman's layer, the "anterior membrane" of the corneal stroma. Stromal dystrophies are highlighted by the accumulation of opaque substances in the corneal stroma, either within the keratocytes or between the stromal collagen fibers. The endothelial dystrophies affect the corneal endothelium, which produces the Descemet membrane as its basement membrane and actively maintains the cornea in a state of normal hydration. These dystrophies cause endothelial dysfunction, which results in corneal stromal and epithelial over hydration, corneal edema, clouding, and decreased vision.

Two of the Bowman layer dystrophies include Reis–Bucklers corneal dystrophy (RBCD) and Thiel–Behnke corneal dystrophy, which are genetically distinct and distinguishable on electron microscopy. The discovery of mutations in TGFBI, the gene for keratoepithelin, a protein expressed in the cornea, skin, and connective tissue, has led to the closer grouping of these dystrophies with two stromal dystrophies, lattice corneal dystrophy (LCD), granular corneal dystrophy, and their variants. Varying defects in this gene, especially the mutational hotspot codons Arg124 and Arg555, have been associated with these dystrophies (*2*). Some defects may also be associated with epithelial basement membrane dystrophy, an epithelial dystrophy (*3,4*). Over 30 variants of LCD have been discovered, and while the mutations have phenotypic variety, with one resembling RBCD, almost all of these mutations lie in the Fas-1 domain 4 of TGFBI. The mutations in this domain include conserved sequences and may affect protein structure leading to phenotypic abnormalities (*5*).

Recent discoveries in molecular genetic defects have enhanced our understanding of several other dystrophies. For example, two dystrophies previously thought to be separate, Schnyder corneal dystrophy and central discoid corneal dystrophy, have been determined to be caused by mutations in the same gene, UBIAD1 (*6*). The discovery of mutations in common genes, such as COL8A2 and SLC4A11, may suggest that the endothelial dystrophies are on a spectrum of clinical disorders with genetic overlap.

Keratoconus is a highly prevalent disorder with unknown pathogenesis, and is characterized by a conical protrusion of the cornea associated with central stromal thinning and irregularity of the corneal surface. While several chromosome loci have been implicated, no candidate gene has been identified, and it is likely that no single locus is responsible for all observed disease. Keratoconus can be associated with systemic diseases, such as atopic eczema and trisomy 21, and ocular diseases, such as vernal conjunctivitis and pigmentary retinopathy. It usually begins in the teenage years with bilateral, often asymmetrical, progressive myopia and astigmatism. Frequent spectacle changes can address refractive change until irregular astigmatism prevents optimal visual correction. Rigid contact lenses can then allow good visual acuity by providing a regular refractive surface to the eye. As keratoconus progresses, contact lenses may not be well tolerated and visual acuity may be inadequate. Surgical intervention, usually in the form of penetrating keratoplasty (PK), then becomes necessary to restore good vision when other nonsurgical approaches are not an option. Alternatives to PK have been utilized, with optimism surrounding two options: deep anterior lamellar keratoplasty and corneal collagen cross linking. Corneal collagen cross linking involves the application of riboflavin and ultraviolet light A to promote cross linking of corneal collagen to stiffen and provide more structural strength within the cornea. Studies have suggested that treatment could safely halt progression, reduce corneal steepening, and improve vision (7). It is unclear how long the treatment lasts and whether retreatment will be needed.

Metabolic disorders are associated with accumulation of an abnormal substance in the cornea. If the substance is produced by the corneal tissues, it may be found throughout the cornea. If it is found in elevated amounts in the blood, it is more commonly accumulated in the corneal periphery. Mucopolysaccharidoses, sphingolipidoses, and mucolipidoses are associated with the accumulation of excess material in the lysosomes of the corneal cells. In lipid metabolic abnormalities, cholesterol esters, phospholipids, and/or triglycerides can be deposited into the cornea. Protein and amino acid metabolic abnormalities can yield a variety of corneal findings such as crystal deposition with tyrosinemia and cystinosis or copper deposition with Wilson disease.

Corneal scarring and vascularization can be seen with various cutaneous disorders such as epidermolysis bullosa. Several congenital anomalies, such as cornea plana (commonly autosomal dominant) and megalocornea (usually X-linked recessive), can involve the cornea.

Many inherited corneal defects can occur both as isolated findings and in association with systemic disease. An awareness of these is important to the general physician and geneticist as well as to the ophthalmologist.

REFERENCES

1. Weiss, J. S.; Moller, H. U.; Lisch, W.; Kinoshita, S.; Aldave, A. J.; Belin, M. W.; Kivela, T.; Busin, M.; Munier, F. L.; Seitz, B.; Sutphin, J.; Bredrup, C.; Mannis, M. J.; Rapuano, C. J.; Van Rij, G.; Kim, E. K.; Klintworth, G. K. The IC3D Classification of the Corneal Dystrophies. *Cornea* **2008,** *27* (Suppl 2), S1–S83.
2. Munier, F. L.; Frueh, B. E.; Othenin-Girard, P.; Uffer, S.; Cousin, P.; Wang, M. X.; Heon, E.; Black, G. C.; Blasi, M. A.; Balestrazzi, E.; Lorenz, B.; Escoto, R.; Barraquer, R.; Hoeltzenbein, M.; Gloor, B.; Fossarello, M.; Singh, A. D.; Arsenijevic, Y.; Zografos, L.; Schorderet, D. F. BIGH3 Mutation Spectrum in Corneal Dystrophies. *Invest. Ophthalmol. Vis. Sci.* **2002,** *43* (4), 949–954.
3. Boutboul, S.; Black, G. C.; Moore, J. E.; Sinton, J.; Menasche, M.; Munier, F. L.; Laroche, L.; Abitbol, M.; Schorderet, D. F. A Subset of Patients with Epithelial Basement Membrane Corneal Dystrophy have Mutations in TGFBI/BIGH3. *Hum. Mutat.* **2006,** *27* (6), 553–557.
4. Paliwal, P.; Sharma, A.; Tandon, R.; Sharma, N.; Titiyal, J. S.; Sen, S.; Kaur, P.; Dube, D.; Vajpayee, R. B., TGFBI Mutation Screening and Genotype–Phenotype Correlation in North Indian Patients with Corneal Dystrophies. *Mol. Vis. 16*, 1429–1438.
5. Kannabiran, C.; Klintworth, G. K. *TGFBI* Gene Mutations in Corneal Dystrophies. *Hum. Mutat.* **2006,** *27* (7), 615–625.
6. Weiss, J. S.; Wiaux, C.; Yellore, V.; Raber, I.; Eagle, R.; Mequio, M.; Aldave, A., Newly Reported p.Asp240Asn Mutation in UBIAD1 Suggests Central Discoid Corneal Dystrophy Is a Variant of Schnyder Corneal Dystrophy. *Cornea 29*(7), 777–780.
7. Wollensak, G.; Spoerl, E.; Seiler, T. Riboflavin/Ultraviolet-a-Induced Collagen Crosslinking for the Treatment of Keratoconus. *Am. J. Ophthalmol.* **2003,** *135* (5), 620–627.

CHAPTER 136

Congenital Cataracts and Genetic Anomalies of the Lens

Arlene V Drack and Yaron S Rabinowitz

Dept. of Ophthalmology and Vision Science, University of Iowa, Iowa City, IA, USA

Edward Cotlier

Former Professor of Ophthalmology and Visual Sciences, Yale University School of Medicine, New Haven, CT, USA

The lens is a transparent crystalline structure in the eye, which transmits and focuses light on the retina. The retina transmits these signals to the optic nerve, and from there they reach the visual cortex and are perceived as vision. Any perturbation of the lens has a severe impact on the quality of vision. Since the visual cortex is not fully developed at birth, and requires excellent visual input to develop as it should, congenital anomalies of the lens can cause lifelong loss of vision if not recognized and treated promptly. Since most early onset disorders of the lens are genetic, correct diagnosis is necessary to give appropriate counseling to families about recurrence risk.

The most common serious abnormality of the lens is cataract, which refers to opacity of the lens. This may be partial or complete. Other more rare anomalies include aphakia (absence of the lens), colobomas, abnormalities of the embryonic vascular system, anomalies of shape (lenticonus, spherophakia), anomalies of size (microspherophakia), and dislocation (ectopia lentis).

In the category of noncataractous anomalies, aphakia is perhaps the most severe manifestation. Both autosomal dominant, for example, as part of BOFS, and autosomal recessive inheritance, for example, due to mutations in FOXE3, have been reported. Lens coloboma, which may or may not be related to chorioretinal coloboma, may deform the lens. Ectopia lentis, or a displaced lens, can be seen as an isolated condition, or, more commonly as part of systemic syndromes such as Marfan syndrome, homocystinuria, and 283235 syndrome. Microspherophakia is most commonly associated with Weill–Marchesani syndrome, and may be inherited either as an autosomal recessive or as an autosomal dominant condition. Persistent papillary membranes and Mittendorf dots are remnants of the fetal vasculature that usually do not interfere with vision, but may in some cases.

Lenticonus refers to an outpouching of the lens, either anteriorly or posteriorly, which typically leads to myopia, and then opacity. Anterior lenticonus may be associated with Alport syndrome. Isolated, small flat anterior polar cataracts may be autosomal dominant and may not cause visual disturbance other than hyperopia. Posterior lenticonus may be inherited and has been associated with the *TDRD7* gene.

Congenital cataracts, when bilateral, are usually genetic. Many mutations in more than 100 genes have been reported to cause congenital cataracts, and even more genes are involved when one looks at all hereditary cataracts. Most hereditary cataracts have their onset in the presenile age range, i.e. before 45 years old. However, even age-related cataracts, which are extremely common with advancing years, have a genetic predisposition associated with their occurrence.

Most hereditary congenital cataracts, defined as onset before 6 months of age, are inherited in an autsomal dominant fashion. Autsomal recessive and X-linked forms are not uncommon, however. The morphology of the cataracts may sometimes be helpful in diagnosis, but as with many genetic disorders there can be wide phenotypic variability even within the same family with the same genotype. Conversely, patients with differing genotypes may look identical clinically.

Of the isolated hereditary congenital cataracts, most of the genes responsible code for either crystallins, membrane or cytoskeletal proteins, or transcription factors.

Many congenital cataracts are associated with other ocular anomalies and malformations. These cataracts are often due to mutations in homeobox genes such as *PAX6* or *PITX3*. These eyes are especially prone to glaucoma and other anomalies and have a more guarded prognosis for vision.

Some congenital cataracts are associated with syndromes such as Lowe syndrome, Nance–Horan syndrome, metabolic disorders, or others. For this reason, a complete work up by the pediatrician and geneticist, as well as the ophthalmologist, is of paramount importance.

Congenital cataracts are a true emergency for pediatric ophthalmologists since vision in the 20/40 or normal range can only be achieved if the opacity is surgically removed and optical correction (spectacles or contact lenses to replace the focusing power of the crystalline lens) is in place by 4 months of age. For this reason, early detection in the nursery is key. Likewise, for the family dealing with this very labor-intensive condition, genetic diagnosis and recurrence risk are of great interest. However, at this writing, routine fee for service testing for isolated congenital cataracts is not available. Testing is available for a number of syndromic ocular and systemic related cataracts.

FURTHER READING

1. Bergman, J. E.; Janssen, N.; Hoefsloot, L. H., et al. CHD7 Mutations and CHARGE Syndrome: The Clinical Implications of an Expanding Phenotype. *J. Med. Genet.* **2011,** *48* (5), 334–342.
2. Edwards, M. J.; Challinor, C. J.; Colley, P. W., et al. Clinical and Linkage Studies of a Large Family with Simple Ectopia Lentis Linked to FBN1. *Am. J. Med. Genet.* **1994,** *53,* 65–71.
3. Faivre, L.; Dolfus, H.; Lyonnet, S., et al. Clinical Homogeneity and Genetic Heterogeneity in Weill–Marchesani Syndrome. *Am. J. Med. Genet.* **2003,** *123A* (2), 204–207.
4. Hanson, I. M.; Fletcher, J. M.; Jordan, T., et al. Mutations at the PAX 6 Locus Are Found in Heterogenous Anterior Segment Malformations Including Peter's Anomaly. *Nat. Genet.* **1994,** *6,* 168–173.
5. Hejtmancik, J. F. Congenital Cataracts and Their Molecular Genetics. *Semin. Cell Dev. Biol.* **2008,** *19,* 134–149.
6. Gestri, G.; Osborne, R. J.; Wyatt, A. W., et al. Reduced TFAP2A Function Causes Variable Optic Fissure Closure and Retinal Defects and Sensitizes Eye Development to Mutations in Other Morphogenetic Regulators. *Hum. Genet.* **2009,** *126* (6), 791–803.
7. Lachke, S. A.; Alkuraya, F. S.; Kneeland, S. C., et al. Mutations in the RNA Granule Component TDRD7 Cause Cataract and Glaucoma. *Science* **2011,** *331* (6024), 1571–1576.
8. Shiels, A.; Bennett, T. M.; Hejtmancik Cat-Map: Putting Cataract on the Map. *Mol. Vis.* **2010,** *16,* 2007–2015.
9. Valleix, S.; Niel, F.; Nedelec, B., et al. Homozygous Nonsense Mutation in the *FOXE3* Gene as a Cause of Congenital Primary Aphakia in Humans. *Am. J. Hum. Genet.* **2006,** *79* (2), 358–364.

CHAPTER

137

Hereditary Retinal and Choroidal Dystrophies

Suma P Shankar

Department of Human Genetics and Ophthalmology, Emory University School of Medicine, Decatur, GA, USA

Inherited retinal and choroidal dystrophies are a clinically and genetically heterogeneous group of disorders and can occur isolated or in association with other systemic disorders. The clinical presentation of these dystrophies is variable. They can present at birth, during childhood, or adulthood; be stationary or progressive; cause profound or mild visual loss; predominantly involve rods or cones or both; be a central receptor disease; or primarily involve the choriocapillaries. The photoreceptors, retinal pigment epithelium (RPE) and choriocapillaries are in close relationship and are interdependent on one another in their functioning and maintenance of structure. Any disorder, genetic or environmental, that causes degeneration of one will likely eventually result in involvement of the others making accurate diagnosis difficult. The classical diagnostic approach for retinal/choroidal diseases has been a clinical examination by ophthalmologists assisted by electrophysiologic and psychophysical tests such as the electroretinogram (ERG), electrooculogram (EOG), dark adaptation, visual field test and ocular coherence tomography (OCT). In this chapter we describe a number of inherited retinal and choroidal dystrophies. The mode of inheritance for these disorders can be autosomal dominant, autosomal recessive, X-linked, or mitochondrial. Incomplete penetrance and variable expressivity, digenic inheritance pattern and role of modifying factors have been discussed. The clinical presentation, methods of examination and genes causing a number of inherited retinal disorders are described. Generally, retinal diseases causing primarily rod photoreceptor dysfunction/degeneration are referred to as pigmentary retinopathy/retinitis pigmentosa accounts for the most common retinal disorder. This can also occur in association with a number of systemic diseases such as mucopolysaccharidosis, Batten disease, and nephronopthisis. Usher syndrome presents with congenital hearing loss and later with retinitis pigmentosa. Leber congenital amaurosis presents with poor vision since birth, non-recordable ERG and nystagmus. Diseases predominantly affecting cone function or structure cause cone dystrophy, color vision problems including achromatopsia and L-cone nonochromatism. Best disease primarily is a disease of a chloride channel leading to deposition of lipofuscin-like material beneath the RPE.

Stargardt disease is caused by mutations in the gene *ABCA4*, and the same gene also causes retinitis pigmentosa, fundus flavimaculatus and cone dystrophy. Mutations in a single gene peripherin2/RDS cause pattern dystrophies, retinitis pigmentosa-like picture, cone dystrophy, as well as choroidal dystrophy-like appearance. This gene has also been implicated in digenic retinal disease along with gene *ROM1*. Flecked retinal disorders such as fundus albipunctatus and punctata albescens (RPA) are described. Another group of retinal diseases in which there is hypopigmentation of the iris and retina, resulting in foveal hypoplasia, misrouting of optic tracts and skin and hair hypopigmentation is oculocutaneous albinism. Retinoschisis is a condition with splitting of the neurosensory retina. Primary diseases of the choroid include choroideremia and gyrate atrophy of the choroid. These often have distinctive appearance on exam. Genetic as well as phenotypic heterogeneity is a feature of most retinal diseases. When clinical picture is not distinctive for an accurate diagnosis, other pychophysical tests such as ERG, EOG, flourescein angiography and OCT have been used but may not be sufficient for final diagnosis. With the recent advances in molecular genetics, the genetic etiology for a large number of retinal/choroidal diseases has been identified. Where single gene testing is not an option, multiple retinal gene testing panels are becoming an option. Clinical genetic testing provides the option for more accurate diagnosis, genetic counseling, prognosis, therapeutic and research options to individuals and families with inherited diseases. Clinical trials are being offered for a number of retinal diseases and many of them necessitate accurate molecular diagnosis.

Genetherapy for Leber congenital amaurosis caused by *RPE65* gene is in phase 3 clinical trial and opened options for other gene therapies for other diseases. We are in an exciting era looking forward to personalized medicine and gene-directed therapies.

CHAPTER

138

Strabismus

J Bronwyn Bateman and Sherwin J Isenberg

Strabismus, or "squint," is a misalignment of the visual axes of the eyes and occurs commonly. For the lay population, it may be termed "lazy-," "wall-," or "cross-eyed." Nomenclature and classification systems are based on the amount (measured in prism diopters); direction of the deviation—horizontal, vertical, or torsional; laterality of deviating eye; frequency; age of onset; and the amount of deviation in the different fields of gaze (comitance or concomitance). The deviation can be latent (phoria; see below) or intermittent, both of which are based on fusional or binocular capabilities of the individual or, alternatively, constant (tropia). The monofixation syndrome, a subset of strabismus defined as a small deviation (usually 10 prism diopters or less), may not be evident on external inspection and is associated with reduced but measurable fusion (binocularity; use of the two eyes together). Strabismus may be categorized on the basis of the age at onset of the deviation: "infantile" is generally defined as occurring in the first 6 months of life, and "acquired" develops at any age thereafter. Typically, as opposed to cranial nerve palsies or restrictive disorders such as dysthyroid ophthalmopathy, deviations that develop in childhood do not vary substantially in different fields of gaze (comitant or concomitant) such as right/left and up/down. The appearance of strabismus with aligned visual axes, as commonly occurs in infants and young children, with epicanthal folds is termed pseudostrabismus. Additionally, facial configuration may result in the appearance of strabismus even if the eyes are normally aligned.

Most forms of developmental delay are associated with a higher incidence of strabismus. Some forms of strabismus are associated with cranial nerve aplasia/malinnervation. Strabismus is common and occurs in 1–6% of all children. Although refractive error is a variation of normal and has a genetic component, it is not a single gene phenotype. Both myopia (nearsightedness) and hyperopia (farsightedness) are important risk factors for the development of strabismus. The pathophysiologic basis(es) of most cases is (are) poorly understood. Historically, strabismus has been divided into phorias and tropias. A tropia is a "manifest" deviation that is evident under binocular conditions (both eyes uncovered). Horizontal deviations are termed exotropia or exophoria (divergent), and esotropia or esophoria (convergent); vertical deviations are described as hyper- or hypotropias or phorias. As most individuals with strabismus have one eye that has better vision than the other, the deviating or nonfixating eye is designated, for example left exotropia. The horizontal and vertical deviations are quantified in prism diopters; a cyclotropia in which one or both eyes are torsionally deviated is measured in degrees. Horizontal and vertical deviations may be intermittent, namely the strabismus may be evident at one moment but not the other. A phoria is a deviation that develops under monocular viewing conditions; to identify a phoria or an intermittent (latent) strabismus (heterophoria), one eye is covered to eliminate the binocular stimulus to the central nervous system for ocular alignment. The deviation under the occluder is measured and is termed as phoria.

The causes of the common forms of strabismus are poorly understood and the natural history differs in children as compared to adults. The genetic forms are less common and the bases are numerous; they may be isolated ocular disorders or associated with systemic diseases. Mutations of a single gene can cause a variety of clinical manifestations and a clinical disorder can be caused by defects of multiple genes. The most common single gene disorders include ptosis and a cluster of well-documented cranial nerve disorders, Duane anomaly/syndrome and congenital ocular fibrosis. Strabismus is common in the genetic forms of craniofacial dysostoses such as Crouzon disease and the Apert and Pfeiffer syndromes as well as the mandibulofacial dysostoses. Chronic progressive external ophthalmoplegia is an acquired form of strabismus that may be associated with a mitochondrial myopathy, or may be inherited in an autosomal dominant or recessive pattern. The mitochondrial bases of strabismus are broad. Moebius syndrome may be inherited by a single gene or be caused by intrauterine insults such as thalidomide, and the mechanism of intrauterine insults such as thalidomide may be genetic.

CHAPTER

139

Retinoblastoma and the RB1 Cancer Syndrome

A Linn Murphree, Robin D Clark, Linda M Randolph, Uma M Sachdeva, Dan S Gombos and Joan M O'Brien

Retinoblastoma was the first tumor suppressor gene to be discovered and has defined the paradigm of the two-hit model of tumorigenesis. Because heritable retinoblastoma is associated with the RB1 cancer syndrome and puts offspring at risk for retinoblastoma development, it is important for patients presenting with retinoblastoma to receive genetic counseling.

139.1 TWO HIT HYPOTHESIS

In 1971, Alfred Knudson separated this childhood tumor into heritable and nonheritable forms (*1*). Knudson's two-hit hypothesis, by which tumors develop from a single cell that has undergone two separate mutational events, M1 and M2, was based on the observed difference in the ages at diagnosis of bilateral and unilateral retinoblastoma patients and has provided the framework for our understanding of recessive tumor suppressor genes (*1–4*). The retinoblastoma tumor suppressor gene (RB1) is located on chromosome 13q14 and encodes the retinoblastoma tumor suppressor protein, pRB (*5,6*). Regardless of the genetic subgroup to which it may belong, retinoblastoma develops when an insufficient amount of the active form of pRB is available within a retinoblast to prevent repeated entry into the cell cycle (*5–7*). pRb plays a pivotal role in regulating the exit of cells from the cell cycle and in preventing continuous cell cycle progression.

139.2 THE RB1 CANCER SYNDROME

Patients with a germline RB1 mutation possess the RB1 cancer syndrome, putting them at risk for multiple types of tumors, in addition to retinoblastoma. These patients are heterozygous for one inactive or deleted RB1 allele in every cell type of the body.

The two major consequences of the RB1 cancer syndrome are the lifelong increase in risk for developing certain cancers, including retinoblastoma, bone, and soft tissue sarcomas, and the risk of transmitting this cancer predisposition to subsequent generations.

When retinoblastoma affects both eyes or there is a positive family history of this disease, the patient demonstrates a heritable form. Unfortunately, unequivocal clinical features that are diagnostic of the RB1 cancer syndrome (primarily bilateral retinoblastoma) are present in fewer than 40% of all retinoblastoma cases. The other 60% of patients with retinoblastoma present unilaterally and the genetics of the disease cannot be clinically deduced. In order to detect the 15% of unilaterally affected patients who have the germline mutation and have the RB1 cancer syndrome, genetic investigation is critical.

139.3 CLINICAL FEATURES

Common presenting signs of retinoblastoma in the developed world include leukocoria (white pupil) and strabismus (misalignment of the eyes). Retinoblastoma shows no gender predisposition. The tumor occurs with equal incidence in right and left eyes, and is present in both eyes in about 20–40% of all cases. In bilateral patients, relatively small tumors often cause visual symptoms and lead to early detection and treatment. The survival of retinoblastoma exceeds 95% North America and Western Europe. In contrast, survival may be as low as 50% is less developed countries (*8*).

139.4 CLASSIFICATION OF RETINOBLASTOMA

Retinoblastoma is unique among the solid childhood tumors in that a tissue diagnosis is not a requirement before treatment initiation. In 2005, a new grouping system (the ABC or International Classification of Intraocular Retinoblastoma) based on the natural history of the disease, was introduced by Murphree (9). He observed that the most significant feature of the intraocular tumor that impacted ocular salvage was the presence or absence of tumor seeding (*9,10*) His system grouped eyes from A to E.

Group A eyes are limited to those containing small tumors (less than 3mm in greatest diameter) that are located away from the fovea and optic nerve. Group B is assigned to all other eyes with no seeding. Group C eyes have minimal localized seeding, whereas group D eyes have massive, diffuse seeding. Group E includes those eyes that present at diagnosis with such extensive intraocular disease that their function and/or structure have been destroyed (Table 139-1).

139.5 TREATMENT OF INTRAOCULAR RETINOBLASTOMA

139.5.1 Unilateral Retinoblastoma

Primary treatment generally consists of primary enucleation for group E eyes. Groups A through C eyes are often salvaged with systemic chemotherapy and/or local therapies.

139.5.2 Bilateral Retinoblastoma

For eyes in which salvage of some useful visual acuity is likely (bilateral groups A through D), systemic chemoreduction with local consolidation is commonly used.

The most commonly used protocol is six cycles of three agent chemotherapy given over 5—6 months. This is followed by sequential local therapy (laser, cryotherapy).

139.6 TESTING ISSUES

Although not always available, DNA from fresh frozen tumor can be useful in a genetic testing strategy. If a mutation is detectable in the tumor, a negative blood result is more reliable. When the patient has bilateral retinoblastoma, a mutation in RB1 is virtually certain. DNA testing can be carried out on blood alone. All patients with bilateral retinoblastoma are presumed to have a germ line mutation. When the proband has sporadic unilateral retinoblastoma, a fresh frozen sample of the tumor tissue as well as a blood sample may be sent for analysis. In unilateral cases, the tumor tissue can be used first to identify both RB1 mutations before the blood studies are initiated.

TABLE 139-1 Risks of Non-ocular Cancers in 1-Year Survivors of Heritable Retinoblastoma. Radiation (*n* = 849; Person-yr at Risk, 21,706), No Radiation (*n* = 114; Person-yr at Risk, 3602)

	Observed	Expected	SIR	95% CI	Observed	Expected	SIR	95% CI
All Sites	241	11	22	19–24	19	2.77	6.9	4.1–11
Heavily irradiated sites (>1 Gy)								
Bone	73	0.18	406	318–511	2	0.03	69	8.4–250
Soft tissue	33	0.23	140	96–196	1	0.04	23	0.6–131
Nasal cavities	32	0.02	1364	933–1925	0	0.01	0	0.0–688
Eye and orbit	17	0.05	312	181–499	0	0.01	0	0.0–392
Brain, CNS	10	0.62	16	7.7–29	0	0.11	0	0.0–33
Pineoblastoma	5	0.05	104	34–244	0	0.01	0	0.0–509
Buccal cavity	7	0.27	26	10–53	0	0.07	0	0.0–54
Thyroid	2	0.5	4	0.5–15	0	0.11	0	0.0–35
Moderately irradiated sites (0.4–1.0 Gy)								
Female breast	8	1.91	4.2	1.8–8.2	2	0.61	3.3	0.4–12
Skin melanoma	26	0.85	30	20–45	3	0.2	15	3.1–44
Lung	2	0.63	3.2	0.4–11	3	0.21	30	3.0–42
Leukemia	1	0.76	1.3	0.03–7.3	1	0.13	7.8	0.2–43
Lightly irradiated sites								
Uterus	5	0.25	20	6.4–46	2	0.1	20	2.5–74
Bladder	2	0.25	7.9	0.9–28	0	0.07	0	0.0–52
Excess absolute risk per	105.9				45.1 10,000 person-yr			

The SIR, or standardized incidence ratio, is the number calculated by dividing the number of observed cases (non-ocular tumors) by the number of non-ocular tumors expected in the general population. The larger the SIR, the greater the effect of the factor being measured.
CI, confidence interval; CNS, central nervous system; SIR, standardized incidence ratio.
Source: Adapted from *Kleinerman, R. A.; Tucker, M. A.; Tarone, R. E.; et al. J. Clin. Oncol.* **2005**, *23*, 2272–2279.

REFERENCES

1. Knudson, A. G., Jr. Mutation and Cancer: Statistical Study of Retinoblastoma. *Proc. Natl. Acad. Sci. U.S.A.* **1971**, *68*, 820–823.
2. Knudson, A. G. Hereditary Cancer: Two Hits Revisited. *J. Cancer Res. Clin. Oncol.* **1996**, *122*, 135–140.
3. Knudson, A. G. Hereditary Predisposition to Cancer. *Ann. N. Y. Acad. Sci.* **1997**, *833*, 58–67.
4. Knudson, A. G., Jr.; Hethcote, H. W.; Brown, B. W. Mutation and Childhood Cancer: A Probabilistic Model for the Incidence of Retinoblastoma. *Proc. Natl. Acad. Sci. U.S.A.* **1975**, *72*, 5116–5120.
5. Cobrinik, D.; Dowdy, S. F.; Hinds, P. W., et al. The Retinoblastoma Protein and the Regulation of Cell Cycling. *Trends Biochem. Sci.* **1992**, *17*, 312–315.
6. Weinberg, R. A. The Retinoblastoma Protein and Cell Cycle Control. *Cell* **1995**, *81*, 323–330.
7. Cobrinik, D.; Francis, R. O.; Abramson, D. H.; Lee, T. C. Rb Induces a Proliferative Arrest and Curtails Brn-2 Expression in Retinoblastoma Cells. *Mol. Cancer* **2006**, *12* (5), 72.
8. Chantada, G. Leal-Leal C International Perspective: Delayed Diagnosis and Advanced Disease. In *Clinical Ophthalmic Oncology*; Singh, A.; Damato, B.; Pe'er, J., Eds.; Elsevier: Philadelphia, 2006.
9. Murphree, A. L. The Case for a New Group Classification of Intraocular Retinoblastoma. *Ophthalmol. Clin. North Am.* **2005**, *18*, 41–53.
10. Murphree, A. L. The Robert M. Ellsworth Memorial Lecture: The Case for a New Group Classification of Intraocular Retinoblastoma. Presented at the Joint Meeting of the Xth International Retinoblastoma Symposium and the Xth International Conference of Ocular Oncology, Whistler, BC, Canada, 2005.

CHAPTER

140

Anophthalmia, Microphthalmia, and Uveal Coloboma

Brian P Brooks

Ophthalmic Genetics and Visua l Function Branch, National Eye Institute, Bethesda, MD, USA

Ocular development visibly begins in the fourth week of human gestation, when the rostral neural tube evaginates bilaterally to form the optic vesicles (*1*). The optic vesicle, which consists of a single layer of neuroepithelium connected to the neural tube via the optic stalk, invaginates to form the bilayered optic cup as it approaches the overlying surface ectoderm. The inner layer of the optic cup forms the future retinal pigment epithelium, while the outer layer will form the neural retina. This invagination is asymmetric, such that there is a ventral opening through which periocular mesenchyme migrates into the developing eye to, in part, form the primordial blood supply of the developing lens and eye—the tunica vasculosis lentis. Beginning in the fifth week of gestation, the edges of this opening—known as the optic fissure—approximate and fuse so that the eye becomes closed around its perimeter. This complex organogenesis is orchestrated via a series of critical transcription factors and growth factors—many of which, when mutated, lead to congenital ocular malformations.

Failure of the formation or maintenance of the optic vesicle results in a clinical absence of ocular tissue, commonly called anophthalmia (*2*). Failure of the growth and/or normal development of the eye results in a small eye or microphthalmia. An eye that is small but otherwise normally formed is said to have "simple microphthalmia," whereas an eye that is small and has other congenital malformations is said to have "complex microphthalmia." Although it is helpful to divide the clinical presentation of an individual into "anophthalmia" vs. "microphthalmia," in reality, these two designations are two points along a continuum of phenotypes. Uveal coloboma results when the edges of the optic fissure fail to close, in whole or in part (*3*). Uveal coloboma may or may not be accompanied by microphthalmia and microphthalmia may or may not be accompanied by coloboma. All these congenital ocular malformations may be isolated or associated with other systemic findings (i.e. syndromic; Table 140-1). The most common syndromic cause of uveal coloboma is the CHARGE syndrome (an acronym of Coloboma, Heart abnormalities, Atresia of the choanae, Retardation of growth and development, Genitourinary abnormalities, and characteristic Ear abnormalities). Approximately two-thirds of patients with CHARGE syndromes have mutations in the chromodomain gene, *CHD7* (*4*). In general, the degree of vision loss associated with microphthamia–anophthalmia–coloboma (MAC) depends upon the severity of the malformation—anophthalmia having the worst visual prognosis. Patients with coloboma may have normal vision if only the distal portions of the optic fissure failed to close; vision loss usually results from optic nerve and/or macular involvement. Visual rehabilitation includes treating any refractive error, possible overlying amblyopia, and referral to low vision services, as needed.

Individuals on the MAC spectrum often appear sporadically within pedigrees, although autosomal dominant, autosomal recessive and X-linked pedigrees have been reported. Substantial progress has been made in the identification of genes responsible for these phenotypes (Table 140-1), including mutations in several transcription factor genes (e.g. *SOX2*, *CHD7*, *OTX2*, *RX*, *VSX2/CHX10*) and several growth factor genes (*BMP4*, *GDF6*, *GDF3*). Most cases of isolated MAC, however, do not have mutations in or near the coding regions of these known genes, making molecular diagnosis and genetic counseling a challenge.

The approach of the clinical geneticist to a patient with MAC depends upon the clinical presentation. No consensus exists on what systemic and/or molecular testing ought to be done on a patient with anophthalmia, microphthalmia and/or coloboma. In general, more severe phenotypes, dysmorphic features and abnormalities on physical exam will prompt further work-up.

By the time of a term birth, the fetal vasculature originating from the optic nerve and surrounding the lens (the tunica vasculosis lentis) has largely regressed and is replaced by a retinal circulation originating from the central retinal artery and an anterior segment circulation originating from the ciliary arteries. If the tunica vasculosis fails to regress, persistence of the fetal vasculature (PFV) results and may be accompanied by microphthalmia, cataract, and vision loss (*5*). PFV is largely a unilateral condition whose genetics are unclear. PFV is sometimes referred to as "persistent hyperplastic primary vitreous" or PHPV.

TABLE 140-1 Reported Loci/Genes for Microphthalmia

Designation	MIM	Locus (Gene)	Inheritance	Associated Abnormalities
MCOP1	%251600	14q32	AR	
MCOP2 MCOPCB3	#610093	14q24.3 (*CHX10/VSX2*)	AR	
MCOP3	#611038	18q21.3 (*RX/RAX*)	AR	
MCOP4	#613094	8q22.1 (*GDF6*)	AD	Preaxial polydactyly, vertebral anomalies
MCOP5	#611040	11q23 (*MFRP*)	AR	Posterior microphthalmia/nanophthalmos, retinal degeneration, foveoschisis, optic disc drusen
MCOP6	%613517	2q37.1	AR	Posterior microphthalmia
MCOP7 MCOPCB1	#613704	12p13.1 (*GDF3*)	AD	Klippel–Feil
MCOPCB1	%300345	X	XL	
MCOPCB2	%605738	15q12-q15	AD	
MCOPCB4				No locus or gene. Microphthalmia with colobomatous cyst
MCOPCB5	#611638	7q36 (*SHH*)	AD	
MCOPCT1	%156850	16p13.3	AD	Mental retardation in some patients
MCOPCT2	#212550	14q23 (*SIX6*)	AD	
MCOPCT3	302300		X (?)	
MCOPCT4	#610426	22q11.2-q13.1 (*CRYBA4*)	AD	
MCOPS1	%309800	Xq27-q28	XL	Lenz microphthalmia (cleft lip/palate; dental abnormalities, syndactyly, abnormal ear helices, pectus excavatum)
MCOPS2	#300166	Xp11.4	XL	Oculofaciocardiodental syndrome, Lenz microphthalmia (microcephaly, mental retardation, GU abnormalities, abnormal ear helices)
MCOPS3	#206900	3q26.3-q27 (*SOX2*)	AD	Anophthalmia, developmental delay, CNS abnormalities, esophageal atresia, tracheoesophageal fistula, GU abnormalities
MCOPS4	%301590	Xq27-q28	XL	Ankyloblepharon, same region as MCOPS1
MCOPS5	#610125	14q21-q22 (*OTX2*)	AD	Cataract, retinal degeneration, hypoplasia/aplasia of optic nerves, agenesis of corpus callosum, developmental delay, seizures, hypotonia
MCOPS6	#607932	14q21-q22 (*BMP4*)	AD	Developmental brain abnormalities, hypopituitarism, polysyndactyly, retinal dystrophy
MCOPS7	#309801	Xp22 (*HCCS*)	XL	microphthalmia, dermal aplasia, sclerocornea (MIDAS) syndrome, microphthalmia with linear skin defects, CHD/cardiomyopathy
MCOPS8	%601349	6q21	AD (?)	Mental retardation, borderline microcephaly, prognathism, cleft lip/palate, ectrodactyly, premature aging of skin
MCOPS9	#601186	15q24.1 (*STRA6*)	AR	Matthew–Wood syndrome, diaphragmatic hernia with pulmonary hypoplasia, CHD, IUGR
MCOPS10	%611222		AR (?)	Microcephaly, progressive spasticity, seizures, mental retardation

Although many genes are associated with complex microphthalmia, at least two have been associated with simple microphthalmia. Note that although some loci are designated as "non-syndromic," systemic abnormalities have been reported in some of these patients.
MCOP, microphthalmia, non-syndromic; MCOPCB, microphthalmia, isolated with coloboma; MCOPCT, microphthalmia, isolated, with cataract; MCOPS, microphthalmia, syndromic; AD, autosomal dominant; AR, autosomal recessive; XL, X-linked; CHD, congenital heart disease; GU, genitourinary; IUGR, intrauterine growth abnormalities.

REFERENCES

1. Harada, T.; Harada, C.; Parada, L. F. Molecular Regulation of Visual System Development: More Than Meets the Eye. *Genes Dev.* **2007**, *21* (4), 367–378.
2. Verma, A. S.; Fitzpatrick, D. R. Anophthalmia and Microphthalmia. *Orphanet J. Rare Dis.* **2007**, *2*, 47.
3. Chang, L.; Blain, D.; Bertuzzi, S.; Brooks, B. P. Uveal Coloboma: Clinical and Basic Science Update. *Curr. Opin. Ophthalmol.* **2006**, *17* (5), 447–470.
4. Bergman, J. E.; Janssen, N.; Hoefsloot, L. H.; Jongmans, M. C.; Hofstra, R. M.; van Ravenswaaij-Arts, C. M. CHD7 Mutations and CHARGE Syndrome: The Clinical Implications of an Expanding Phenotype. *J. Med. Genet.* **2011**, *48* (5), 334–342.
5. Ceron, O.; Lou, P. L.; Kroll, A. J.; Walton, D. S. The Vitreo-Retinal Manifestations of Persistent Hyperplasic Primary Vitreous (PHPV) and Their Management. *Int. Ophthalmol. Clin.* **2008**, *48* (2), 53–62.

Deafness

Hereditary Hearing Impairment, 517

CHAPTER

141

Hereditary Hearing Impairment

Rena Ellen Falk

Medical Genetics Institute and Department of Pathology and Laboratory Medicine, Cedars Sinai Medical Center, Los Angeles, CA, USA

Arti Pandya

Departments of Human and Molecular Genetics, Pediatrics and Pathology, Virginia Commonwealth University, Richmond, VA, USA

Hearing impairment (HI) comprises the most common sensory deficit in man. A genetic etiology accounts for at least 35% of HI across the age spectrum and is responsible for about half of the cases of congenital, prelingual, or early-onset HI. The stunning heterogeneity seen in nonsyndromic hereditary HI is matched only by the diversity of gene functions which, when disrupted, lead to abnormal development, maintenance, or structure in the inner ear or auditory nerve pathway. These genes represent virtually the entire range of biologic functions found in human cells and organs, including structural cellular and extracellular elements, ion channels, enzymatic activity, energy production, cell–cell communication, protein modification, regulation of transcription and transmembrane signaling, as well as others. In addition to the greater than 400 syndromes associated with hearing loss, nearly 140 nonsyndromic deafness loci have been identified, and at least 67 nuclear genes have been characterized to date. These include 40 genes involved in autosomal recessive HI (designated DFNB), 24 involved in autosomal dominant HI (DFNA) and 3 X-linked *HI* genes (DFNX). Of particular importance, mutations in one gene, *GJB2*, which encodes connexin 26, account for about 20% of childhood deafness and closer to 50% of autosomal recessive prelingual HI worldwide. Identification of founder mutations and a mutation hot spot in this gene raises the possibility of population-based carrier screening. Of note, mutations in *GJB2* and a common deletion in a related gene, *GJB6*, encoding connexin 30, provide one of a number of examples of digenic inheritance of hereditary hearing loss.

In addition to *GJB2* and *GJB6*, the other known autosomal recessive, autosomal dominant, X-linked and mitochondrial genes involved in nonsyndromic SNHL are discussed, including mitochondrial mutations associated with aminoglycoside ototoxicity. A section on auditory neuropathy (AUN) addresses the genes known to cause AUN and the limitations of neonatal hearing screening in children whose HI is associated with mutations in the gene encoding otoferlin, which causes prelingual HI that often presents as an AUN. Also, several recently identified modifier genes are described. Of note, mutations in the same gene are known to cause both nonsyndromic and syndromic hearing loss, or can act either in an AR or AD manner. In many cases, particularly for the genes involved in Usher syndrome, clear genotype–phenotype correlation has been established, though the finding of more severe or early-onset HI in some patients than in family members who carry the same mutations suggests that other genetic and/or environmental factors play a role in gene expression leading to HI.

This chapter reviews major environmental causes of teratogenic and postnatal HI, with particular attention to prenatal cytomegalovirus infection, which is a major cause of both congenital and late-onset HI in children, some of whom may pass a neonatal hearing screen. Genetic and environmental factors that play a role in several adult-onset forms of HI are discussed, including the relationship between measles virus, genes involved in T-cell function and otosclerosis, and the effect of cigarette smoking and noise exposure on presbycusis. In addition, selected chromosomal, metabolic and syndromic causes of HI are reviewed with detailed discussion of current information about the Alport, branchio-oto-renal, Jervell and Lange-Nielsen, Pendred, Stickler, Usher and

Waardenburg syndromes. In each section, the clinical presentations, molecular aspects and relevant management issues of the syndrome are emphasized.

General evaluation and management of the hearing impaired individual is addressed with attention to unusual historical or physical clues to a particular diagnosis, as well as the indication for specific types of testing. Strategies to determine which specific gene tests are indicated are likely to be replaced by lower priced, chip-based, or other molecular technologies, allowing identification of mutations in a large number of genes for a cost equivalent to evaluating a single gene in the past. The rapid rate of discovery of deafness genes, coupled with increased understanding of the importance of early intervention in achieving better educational and social outcomes, has led to widespread neonatal hearing screening. The history of neonatal hearing screening and early intervention is discussed, along with examples of both successes and problems in global screening efforts. Molecular discoveries have provided new insights to the remarkable complexity and heterogeneity of genetic deafness, though many genes remain to be identified. Our emerging understanding of the complex pathways involved in genetic HI, and the pathogenic mechanisms associated with dysfunction of specific genes, has already provided a basis for more specific, targeted approaches to therapy. Cochlear implants have been effective in children with a variety of genetic diagnoses. In addition, some efforts to promote regeneration of the auditory hair cells, which undergo degeneration in a variety of conditions with HI, raise hope for other treatment to restore hearing.

FURTHER READING

1. American Academy of Pediatrics Joint Committee on Infant Hearing. Year 2007 Position Statement. *Pediatrics* **2007**, *120* (4), 898–921.
2. Azaiez, H.; Chamberlin, G. P.; Fischer, S. M.; Welp, C. L.; Prasad, S. D.; Taggart, R. T.; del Castillo, I.; Van Camp, G.; Smith, R. J. GJB2: The Spectrum of Deafness-Causing Allele Variants and Their Phenotype. *Hum. Mutat.* **2004**, *24* (4), 305–311.
3. Bizhanova, A.; Kopp, P. Genetics and Phenomics of Pendred Syndrome. *Mol. Cell. Endocrinol.* **2010**, *322* (1–2), 83–90.
4. Dahle, A.; Fowler, K. B.; Wright, J. D.; Boppana, S. B.; Britt, W. J.; Pass, R. F. Longitudinal Investigation of Hearing Disorders in Children with Congenital Cytomegalovirus. *J. Am. Acad. Audiol.* **2000**, *11* (5), 283–290.
5. Di Domenico, M.; Ricciardi, C.; Martone, T.; Mazzarella, N.; Cassandro, C.; Chiarella, G.; D'Angelo, L.; Cassandro, E. Towards Gene Therapy for Deafness. *J. Cell Physiol.* **2011**, *226* (10), 2494–2499.
6. Dollard, S. C.; Schleiss, M. R.; Grosse, S. D. Public Health and Laboratory Considerations Regarding Newborn Screening for Congenital Cytomegalovirus. *J. Inherit. Metab. Dis.* **2010**, *33* (Suppl 2), S249–S254, Epub June 8.
7. Dror, A. A.; Avraham, K. B. Hearing Impairment: A Panoply of Genes and Functions. *Neuron* **2010**, *68* (2), 293–308.
8. Gates, G. A.; Mills, J. H. Presbycusis. *Lancet* **2005**, *366* (9491), 1111–1120.
9. ACMG Statement Genetics Evaluation Guidelines for the Etiologic Diagnosis of Congenital Hearing Loss. Genetic Evaluation of Congenital Hearing Loss Expert Panel. *Genet. Med.* **2002**, *4* (3), 162–171.
10. Hennekam, R. C. M.; Allanson, J. E.; Krantz, I. D.; Gorlin, R. J. *Gorlin's Syndromes of the Head and Neck*, 5th ed.; Oxford University Press: Oxford; New York, 2010.
11. Hertz, J. M. Alport Syndrome. Molecular Genetic Aspects. *Dan. Med. Bull.* **2009**, *56* (3), 105–152.
12. Hudspeth, A. J. How the Ear's Works Work: Mechanoelectrical Transduction and Amplification by Hair Cells. *C. R. Biol.* **2005**, *328* (2), 155–162.
13. Keats, J. B.; Savas, S. Genetic Heterogeneity in Usher Syndrome. *Am. J. Med. Genet. A.* **2004**, *130A* (1), 13–16.
14. Kimberling, W. J.; Hildebrand, M. S.; Shearer, A. E.; Jensen, M. L.; Halder, J. A.; Trzupek, K.; Cohn, E. S.; Weleber, R. G.; Stone, E. M.; Smith, R. J. H. Frequency of Usher Syndrome in Two Pediatric Populations: Implications for Genetic Screening of Deaf and Hard of Hearing Children. *Genet. Med.* **2010**, *12* (8), 512–516.
15. Kochhar, A.; Fischer, S. M.; Kimberling, W. J.; Smith, R. J. Branchio-oto-Renal Syndrome. *Am. J. Med. Genet. A.* **2007**, *143A* (14), 1671–1678.
16. Kopecky, B.; Fritzsch, B. Regeneration of Hair Cells: Making Sense of All the Noise. *Pharmaceuticals (Basel)* **2011**, *4* (6), 848–879.
17. Martinez, A.; Linden, J.; Schimmenti, L. A.; Palmer, C. G. Attitudes of the Broader Hearing, Deaf, and Hard-of-Hearing Community Toward Genetic Testing for Deafness. *Genet. Med.* **2003**, *5* (2), 106–112.
18. Morton, C. C. Genetics, Genomics and Gene Discovery in the Auditory System. *Hum. Mol. Genet.* **2002**, *11* (10), 1229–1240.
19. Morton, C. C.; Nance, W. E. Newborn Hearing Screening—A Silent Revolution. *N. Engl. J. Med.* **2006**, *354* (20), 2151–2164.
20. Pandya, A., GeneReviews [Internet]; In *Nonsyndromic Hearing Loss and Deafness, Mitochondrial*; Pagon, R. A.; Bird, T. D.; Dolan, C. R., Eds.; University of Washington: Seattle (WA), 2011.
21. Schrauwen, I.; Van Camp, G. The Etiology of Otosclerosis: A Combination of Genes and Environment. *Laryngoscope* **2010**, *120* (6), 1195–1202.
22. Schwartz, P. J.; Spazzolini, C.; Crotti, L.; Bathen, J.; Amlie, J. P.; Timothy, K.; Shkolnikova, M.; Berul, C. I.; Bitner-Glindzicz, M.; Toivonen, L.; Horie, M.; Schulze-Bahr, E.; Denjoy, I. The Jervell and Lange–Nielsen Syndrome: Natural History, Molecular Basis, and Clinical Outcome. *Circulation* **2006**, *113* (6), 783–790.
23. Toriello, H. V.; Reardon, W.; Gorlin, R. J. *Hereditary Hearing Loss and Its Syndromes*, 2nd ed.; Oxford University Press: Oxford; New York, 2004.
24. White, K. R.; Forsman, I.; Eichwald, J.; Munoz, K. The Evolution of Early Hearing Detection and Intervention Programs in the United States. *Semin. Perinatol.* **2010**, *34*, 170–179.
25. Yan, D.; Liu, X. Z. Genetics and Pathological Mechanisms of Usher Syndrome. *J. Hum. Genet.* **2010**, *55* (6), 327–335.
26. Yoshinaga-Itano, C. Levels of Evidence: Universal Newborn Hearing Screening (UNHS) and Early Hearing Detection and Intervention Systems (EHDI). *J. Commun. Disord.* **2004**, *37* (5), 451–465.

USEFUL WEBSITES

27. MRC Institute of Hearing Research: http://www.ihr.mrc.ac.uk/index.php/research/.
28. Online Mendelian Inheritance in Man (OMIM): http://omim.org/.

29. National Institute of Deafness and Other Communication Disorders: www.nidcd.nih. www.geneclinics.org.
30. National Institute on Deafness and Other Communication Disorders, NIDCD Presbycusis health statistics. http://www.nidcd.nih.gov/health/hearing/presbycusis. htm 2010.
31. National Institute on Deafness and Other Communication Disorders, NIDCD Hearing health statistics. www.nidcd. nih. gov/health/statistics/quick. htm 2010.
32. Van Camp, G.; Smith, R. J. H. Hereditary Hearing Loss Homepage. http://hereditaryhearingloss.org/.

Craniofacial Disorders

Clefting, Dental, and Craniofacial Syndromes, 523
Craniosynostosis, 527

CHAPTER

142

Clefting, Dental, and Craniofacial Syndromes

Jeffrey C Murray and Mary L Marazita

The University of Iowa, Iowa City, IA, USA

Craniofacial (CF) anomalies comprise a significant component of human birth defects requiring surgical, nutritional, dental, speech, medical and behavioral interventions, and imposing a substantial economic and personal burden. The most common CF anomalies are orofacial clefts (OFCs) of the lip with or without the palate (CL/P) or of the cleft palate only (CPO). OFCs affect about 1/700 births. In general, Asian or Amerindian populations have the highest frequencies, 1/500 or higher, with Caucasian populations intermediate, and African-derived populations having the lowest at 1/2500. OFCs are caused by a mixture of genetic and environmental triggers that disrupt the process of normal embryogenesis. Mendelian and chromosomal causes are found in more than 500 specific anomaly syndromes, but most OFCs are complex traits caused by a combination of multiple genetic and environmental factors.

Epidemiologic studies of CL/P probands provide recurrence risks for first-, second-, and third-degree relatives of 3.5%, 0.8% and 0.6%, respectively. Individuals affected by the most severe oral cleft have a significantly higher recurrence risk among both offspring and siblings; for example, the recurrence risk for siblings of a proband with isolated bilateral CL/P is 4.6% versus 2.5% for a proband born with a unilateral defect. Thus anatomical severity does have an effect on recurrence in first-degree relatives and the type of cleft is predictive of the recurrence type.

The complex etiology of OFC affords opportunities to identify genes and gene–gene (G×G) or gene–environment (G×E) interactions to learn more about human embryology and its disturbances. A major limitation in studying clefting (and most complex traits) has been the lack of a population that is sufficiently large to have adequate power to detect G×G and G×E interactions and that is sufficiently well characterized across the lifespan to identify the long-term effects of OFC on health. Recent successes in genome-wide linkage and association studies for CL/P have led to five well-replicated loci/genes for CL/P. Prior to genome-wide association, only one gene, *IRF6*, had been shown to have common associated variants that contribute to the isolated forms of CL/P. Four studies identified one new locus at 8q24 with a highly associated SNP (rs987525; *p*-value 3.3×10^{-24}) in a gene-poor region, and SNPs associated with at least three new genes (*MAFB*, *ABCA4/ARHGAP29* and *VAX1*) in CL/P.

Among environmental exposures, smoking has been repeatedly associated with clefting and a meta-analysis strongly supported an overall OR of ~1.3 for smoking with clefting risk. Increased risks from exposures can suggest metabolic pathways in which disruptions may play a role in the development of CL/P and investigations into the role of alcohol or micronutrient deficiencies are ongoing.

Clinical care for children with OFCs involves multidisciplinary teams of surgeons, speech pathologists, dentists, pediatricians, psychologists and other to provide optimal outcomes. Evolving data on sub-phenotypes in speech patterns, orbicularis oris defects, or dental anomalies in relatives of OFC cases suggests a role for decreased penetrance of some phenotypes and opens new doors to investigations. Genetic and environmental findings can provide useful counseling tools in cases of Mendelian disorders or teratogenic exposures but is not yet of clinical validity for the isolated or nonsyndromic forms of CL/P.

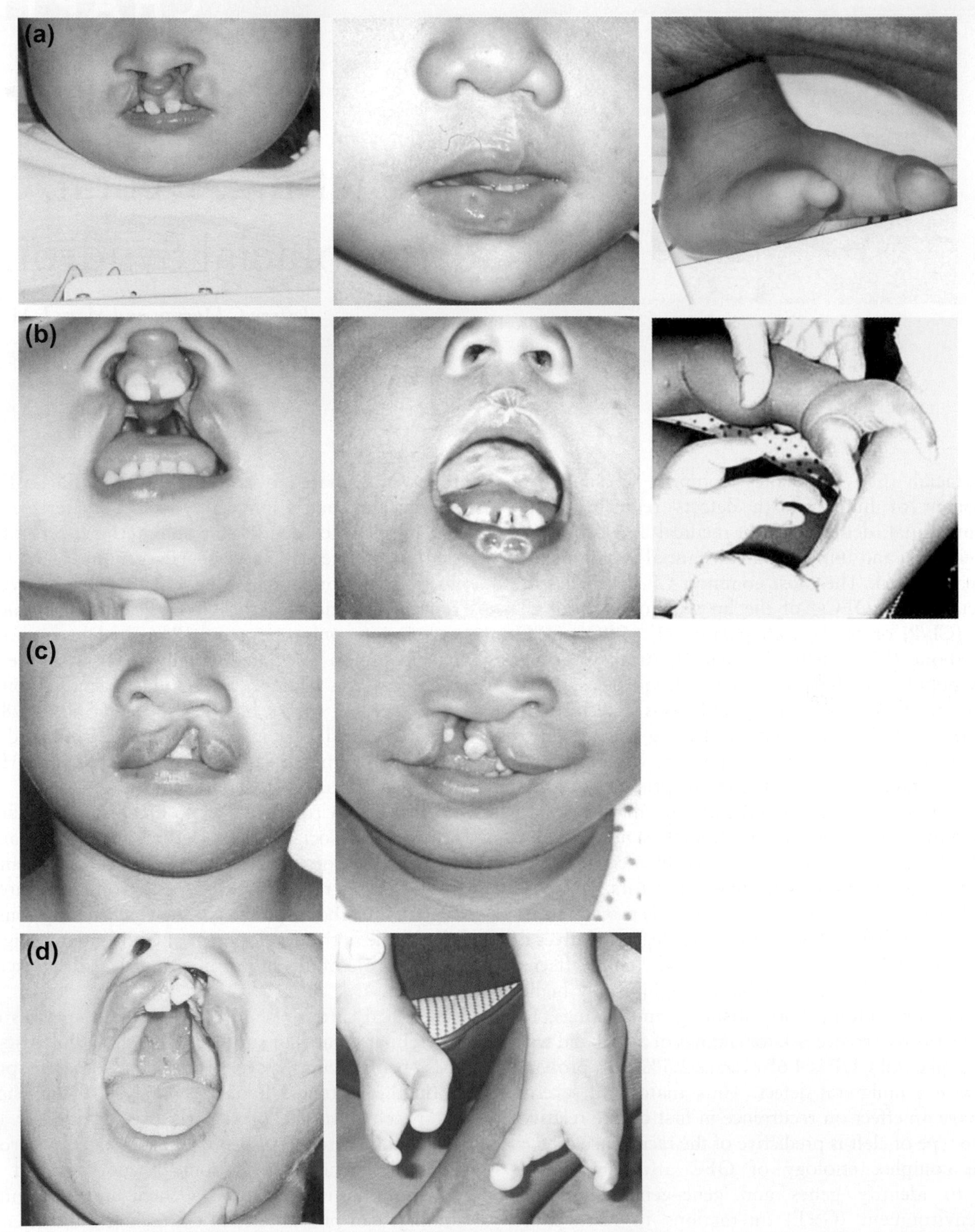

FIGURE 142-1 (A) A child with bilateral cleft lip and palate. (B) A child with unilateral cleft lip and palate. (C) Lip pits in the lower lip as seen in two cases of van der Woude syndrome. (D) Facial, limb, and hand defects seen in the EEC syndrome.

TABLE 142-1 Candidate Genes from Mendelian Disorders

Gene Symbol	Syndrome Name	CL	CP	Dental	Associated Findings
FGFR1	Kallman	+	+	+	Anosmia, infertility
IRF6	van der Woude/PPS	+	+	+	Lip pits
MSX1	Witkop	+	+	+	Hypodontia, nasal dysplasia
P63	EEC	+	+	−	Ectrodactyly
PTCH	Gorlin basal cell nerves	Rare	+	+	Bifid ribs, jaw cysts, macrocephaly
PVRL1	Margarita Island	+	+	−	None in heterozygotes
TBX22	Cleft/ankyloglossia	−	+	−	Ankyloglossia
TGFBR1	Loeys–Dietz	−	+	−	Aortic aneurysm
TGFBR2					Craniosynotosis

EEC, ectrodactyly-ectodermal dysplasia and clefting; PPS, popliteal pterygium syndrome.

TABLE 142-2 Phenotypic Features in Nonsyndromic Clefting

Region	Specific Features and Comments	Representative References
Craniofacial	Face shape differences: More than 20 studies have investigated heritable aspects of facial shape; in general, each found numerous differences between unaffected relatives of cleft cases and controls. The most consistent findings across studies are increased upper- and mid-facial widths	McIntyre and Mossey (2002), Ward et al. (2002)
	Elevated directional asymmetry (DA): Several cephalometric studies found significantly more DA in unaffected parents of NS cleft cases than controls. There is also some evidence that excess soft tissue nasal asymmetry may represent a cleft microform	Farkas and Cheung (1979), McIntyre and Mossey (2002)
	Orbicularis oris muscle (OOM) defects: Using high-resolution ultrasound, significantly more subclinical defects (e.g. breaks) in the OOM have been identified in relatives of NS cleft cases than in controls	Martin et al. (2000), Weinberg et al. (2003)
Dentition	Developmental anomalies: Numerous reports suggest that NS cleft cases are characterized by an increased frequency of hypodontia, supernumerary teeth, asymmetrical growth and eruption, enamel formation defects, delayed dental age, and tooth size reduction compared to controls. Evidence for elevated rates of such defects in unaffected family members has been inconsistent	Ranta (1986), Harris (2002)
	Elevated fluctuating asymmetry (FA): Four studies to date have reported elevated dental FA in NS cleft cases compared to controls, with a more pronounced effect in familial cases than in sporadic cases. Further, the FA appears to be pervasive, affecting multiple teeth in both the deciduous and the permanent dentition of the maxilla and mandible	SoFaer (1979), Werner and Harris (1989)
Brain	Structural differences: Recent neuroimaging studies suggest that the brains of adult men with NS clefting are characterized by significantly larger frontal and parietal lobes, smaller temporal and occipital lobes, decreased cerebellar volumes, and an increased frequency of midline brain anomalies than matched controls. Further, cases demonstrated significantly more posterior cerebral and cerebellar asymmetry than controls	Nopoulos et al. (2000, 2002)
	Excess non-righthandedness (NRH): Elevated rates of NRH in NS cleft cases compared to controls have been reported in several studies, but with no consistent findings related to cleft laterality. Two studies have also reported increased NRH in unaffected parents of cleft cases	Rintala (1985), Wenzlaff et al. (1997)
	Atypical hair whorls: Recent pilot data suggest that NS CL/P unaffected family members are characterized by an increased frequency of counterclockwise parietal hair whorls	Scott et al. (2005)
Vertebrae	Developmental anomalies: A number of studies have reported a higher incidence of various anomalies in the cervical vertebrae (e.g. fusion anomalies, posterior arch deficiencies) in NS cleft cases compared to controls	Šmahel and Škvarilova (1993)
Dermatoglyphics	Pattern type differences: Assessed in seven NS cleft populations to date. In general, cleft cases tend to possess a higher frequency of arches and ulnar loops and a lower frequency of wharfs compared to controls. Similar trends are reported for unaffected relatives of cleft cases	Balgir (1993), Neiswanger et al. (2002)
	Elevated fluctuating asymmetry: Increased levels of both metric and non-metric dermatoglyphic FA have been reported in NS cleft cases compared to controls, with a more pronounced effect on familial cases than on sporadic cases. There is also some evidence for elevated dermatoglyphic FA in unaffected case relatives	Adams and Niswander (1967), Woolf and Gianas (1977)
Growth	Low birth weight/reduced body size: A number of studies have shown NS cleft cases to be smaller than normal at birth, even after gestational age and maternal factors are accounted for. Additional evidence suggests that this growth deficit may persist, as children with CL/P have been shown to have reduced trunk and limb dimensions	Wyszynski et al. (2003)

See www.pubmed.gov for references.

TABLE 142-3 Empiric Recurrence Risk Estimates (in Percentages) for Isolated Orofacial Clefts

Affected Family Member	Cleft Lip with or without Cleft Palate	Cleft Palate
One sibling	1.2–5.1	2–5
Two siblings	9.0–14.0	10–20
Mother	1.9–6.8	–
Father	2.0–4.5	–
One parent (not specified)	1.7–5.0	3.0–7.0
First cousin	0.2–0.5	–
Aunt/uncle	0.4–0.8	–
Grandparent	0.2–0.8	–

Source: Data from literature reviews by Wyszynski et al. (1998) and Stadter.

FURTHER READING

1. Birnbaum, S.; Ludwig, K. U.; Reutter, H.; Herms, S.; Steffens, M.; Rubini, M.; Baluardo, C.; Ferrian, M.; Almeida de Assis, N.; Alblas, M. A., et al. Key Susceptibility Locus for Nonsyndromic Cleft Lip with or without Cleft Palate on Chromosome 8q24. *Nat. Genet.* **2009,** *41* (4), 473–477.
2. Dixon, M. J.; Marazita, M. L.; Beaty, T. H.; Murray, J. C. Cleft Lip and Palate: Understanding Genetic and Environmental Influences. *Nat. Rev. Genet.* **2011,** *12* (3), 167–178.
3. Genisca, A. E.; Frias, J. L.; Broussard, C. S.; Honein, M. A.; Lammer, E. J.; Moore, C. A.; Shaw, G. M.; Murray, J. C.; Yang, W.; Rasmussen, S. A. Orofacial Clefts in the National Birth Defects Prevention Study, 1997–2004. *Am. J. Med. Genet. A* **2009,** *149A* (6), 1149–1158.
4. Grosen, D.; Chevrier, C.; Skytthe, A.; Bille, C.; Molsted, K.; Sivertsen, A.; Murray, J. C.; Christensen, K. A Cohort Study of Recurrence Patterns Among More Than 54,000 Relatives of Oral Cleft Cases in Denmark: Support for the Multifactorial Threshold Model of Inheritance. *J. Med. Genet.* **2010,** *47* (3), 162–168.
5. Hochheiser, H.; Aronow, B. J.; Artinger, K.; Beaty, T. H.; Brinkley, J. F.; Chai, Y.; Clouthier, D.; Cunningham, M. L.; Dixon, M.; Donahue, L. R., et al. The Facebase Consortium: A Comprehensive Program to Facilitate Craniofacial Research. *Dev. Biol.* **2011,** *355* (2), 175–182.
6. Jugessur, A.; Farlie, P. G.; Kilpatrick, N. The Genetics of Isolated Orofacial Clefts: From Genotypes to Subphenotypes. *Oral Dis.* **2009 Oct,** *15* (7), 437–453.
7. Letra, A.; Bjork, B.; Cooper, M. E.; Szabo-Rogers, H.; Deleyiannis, F. W.; Field, L. L.; Czeizel, A. E.; Ma, L.; Garlet, G. P.; Poletta, F. A., et al. Association of AXIN2 with Non-Syndromic Oral Clefts in Multiple Populations. *J. Dent. Res.* **2012 May,** *91* (5), 473–478.
8. Mangold, E.; Ludwig, K. U.; Nöthen, M. M. Breakthroughs in the Genetics of Orofacial Clefting. *Trends Mol. Med.* **2011.**
9. Rahimov, F.; Marazita, M. L.; Visel, A.; Cooper, M. E.; Hitchler, M. J.; Rubini, M.; Domann, F. E.; Govil, M.; Christensen, K.; Bille, C., et al. Disruption of an Ap-2alpha Binding Site in an Irf6 Enhancer Is Associated with Cleft Lip. *Nat. Genet.* **2008,** *40* (11), 1341–1347.
10. Sakai, D.; Dixon, J.; Dixon, M. J.; Trainor, P. A. Mammalian Neurogenesis Requires Treacle-plk1 for Precise Control of Spindle Orientation, Mitotic Progression, and Maintenance of Neural Progenitor Cells. *PLoS Genet.* **2012 Mar,** *8* (3), e1002566.
11. Weinberg, S. M.; Brandon, C. A.; McHenry, T. H.; Neiswanger, K.; Deleyiannis, F. W.; de Salamanca, J. E.; Castilla, E. E.; Czeizel, A. E.; Vieira, A. R.; Marazita, M. L. Rethinking Isolated Cleft Palate: Evidence of Occult Lip Defects in a Subset of Cases. *Am. J. Med. Genet. A* **2008,** *146A* (13), 1670–1675.

CHAPTER

143

Craniosynostosis

Ethylin Wang Jabs

Department of Genetics and Genomic Sciences, Mount Sinai School of Medicine, New York, NY, USA

Amy Feldman Lewanda

Division of Genetics and Metabolism, Children's National Medical Center, Washington, DC, USA

Craniosynostosis is defined as the premature fusion of the skull bones. While most often isolated, craniosynostosis is a feature of more than 100 genetic syndromes. Great progress has been made in identifying the genes responsible for both isolated and syndromic craniosynostoses, helping to explain the underlying process of skull and sutural development.

Cranial bones overlap at birth to allow passage through the birth canal; sutures remain open in childhood to allow for growth of the developing brain. Premature closure of one or more sutures causes an abnormal skull shape, which can be predicted based on which suture or sutures are involved. Craniosynostosis occurs in approximately 1/2500 births. In addition to genetic factors/syndromes, craniosynostosis can result from various metabolic, teratogenic, and hematologic conditions, underlying brain malformations or iatrogenic causes. The suture most commonly involved is the sagittal, followed by coronal and metopic, which each occur about half as frequently as sagittal synostosis.

The most common syndromic craniosynostosis is Crouzon syndrome, involving premature fusion of the coronal sutures, typical facial features, and little involvement outside of the head/neck region. It is most often caused by missense mutations in the fibroblast growth factor receptor (*FGFR2*) gene. Jackson–Weiss syndrome is allelic to Crouzon (also caused by *FGFR2* mutations) but involves bony anomalies of both the skull and the feet. Pfeiffer syndrome is caused by mutations in *FGFR1* or *FGFR2* and may have dramatic features such as kleeblattschadel or cloverleaf skull deformity. In addition to craniosynostosis, broad thumbs and great toes are typical. Apert syndrome, the most easily recognizable of these conditions, causes highly consistent features of craniosynostosis and "mitten type" syndactyly of the hands and the feet. The etiology of the skull deformity is unique, as the sagittal suture is completely absent—leaving a wide gap which eventually fills in with bony islands that fuse together, preventing additional skull growth or expansion in this area. Compared with the other craniosynostosis syndrome, Apert syndrome is more often associated with internal organ malformations, CNS defects, and mental deficiency, and is associated with just two specific mutations in *FGFR2*. Other less common craniosynostoses have also been associated with the FGFR family of genes, including Muenke and Crouzonodermoskeletal syndromes (each associated with *FGFR3* mutations), Beare–Stevenson Cutis Gyrata (*FGFR2*), and Osteoglophonic dysplasia (*FGFR1*). All of these FGFR-related syndromes are autosomal dominant, involve the coronal suture, and present with brachycephaly.

A number of genes outside of the FGFR family have also been implicated in syndromic craniosynostosis, including *TWIST1* (Saethre–Chotzen), *MSX2* (Boston-type Craniosynostosis), *EFNB1* (Craniofrontonasal), *RECQL4* (Baller–Gerold), *POR* (POR syndrome), *RAB23* (Carpenter), and *I11RA* (Interleukin 11 receptor, alpha) (Craniosynostosis and Dental Anomalies syndrome). The last three conditions are autosomal recessive and the latter two differ from the others because the midline sutures are also involved. Few cases of nonsyndromic craniosynostosis have mutations in these known craniosynostosis genes. Research into copy number variation, genome-wide association studies, and high throughput sequencing are elucidating additional loci and genes for both syndromic and nonsyndromic craniosynostosis.

The *FGFR* genes code for a family of proteins consists of an extracellular domain with three immunolglobulin-like domains, a transmembrane (TM) domain, and an intracellular domain with tyrosine kinase activity. They act as receptors for more than 20 FGF ligands. The complexes formed by the FGFR, FGF, and necessary heparin-sulfate containing proteoglycans are critical for signaling distinct developmental pathways. FGFRs 1–3 are known to be expressed during calvarial ossification. Mutations in the *FGFR* genes have been identified in at least nine different craniosynostosis syndromes. FGFR2 mutations typically cause craniosynostosis due to a gain of function.

Some FGFR mutations are quite phenotype specific, such as the two well-documented FGFR2 Ser252Trp and Pro253Arg mutations associated with Apert syndrome. However, a number of craniosynostosis syndromes such as Crouzon, Pfeiffer, and Jackson–Weiss syndromes are now known to be allelic, and do share some identical mutations. The FGFR3 Pro250Arg in Muenke syndrome is associated with variable expression, and some cases of nonsyndromic coronal synostosis have the identical mutation. This suggests that additional factors are involved which can influence craniofacial and limb development. A number of autosomal dominant craniosynostoses are associated with advanced paternal age. Such conditions are known to be caused by mutations arising exclusively in the paternal cell line.

The *TWIST1* gene codes for a transcription factor with a basic DNA binding and helix-loop-helix domains, and is expressed during development of the head and limbs. Calvarial development is apparently dosage sensitive to *TWIST1*, as evidenced by early closure/synostosis when it is deleted or disrupted by translocation, and delayed closure of a large anterior fontanelle when the gene is duplicated. Interestingly, loss of TWIST1 expression leads to activation of FGFRs. Dosage sensitivity is also evidenced by the *MSX2* gene, a transcription factor protein containing a homeobox domain. A gain of function mutation in the large family in which this gene was reported led to highly variable calvarial features ranging from mild fronto-orbital recession to a cloverleaf skull. Msx2 deficient mice, however, have defects in skull ossification and persistent parietal foramina. The *Ephrin B1* gene is responsible for craniofrontonasal syndrome, and in fact may function together with *TWIST1* and *MSX2* in sutural boundary formation and the pathogenesis of coronal synostosis. A number of additional genes and their functional activities as enzymes, transcription factors, and receptors have been associated with particular craniosynostosis syndromes, including POR, RAB23, and I11RA. Causative genes have also been identified for several other syndromes in which craniosynostosis is only an occasional feature, such as Greig cephalopolysyndactyly syndrome and Alagille syndrome.

Clinical evaluation of a patient with craniosynostosis should include a careful dysmorphology exam, radiographs, pregnancy and family history, and appropriate genetic testing. The dysmorphology exam should evaluate the particular cranial shape, which is very helpful in predicting which sutures may be prematurely fused. Noting facial features such as proptosis, beaked nose, facial asymmetry, and palatal abnormalities are helpful in predicting a syndromic diagnosis. Evaluation of hands and feet may reveal additional findings, or they may not be detected without radiographic studies. Skin abnormalities in pigmentation or texture can also be significant. A CT scan with 3D reconstruction is the most accurate way of diagnosing specific sutural synostoses; X-rays of hands and feet can provide additional diagnostic information.

As craniosynostosis can also be due to noninherited causes, a thorough pregnancy history should include factors such as uterine crowding and maternal medication exposure. Family history should include questioning parents about any family members with abnormal head shape, hand or foot abnormalities, vision or hearing problems, or early death. Both parents should be evaluated as well for mild head shape abnormality due to the variable expression of many autosomal dominant disorders. Isolated craniosynostosis can still be inherited even if it is not syndromic, and parents should be made aware of any potential recurrence risks in future pregnancies.

There are many options for genetic testing in a patient with craniosynostosis. If additional malformations and developmental delay are present, one may start with a chromosome or microarray analysis to detect deletions/duplications that may explain the findings. Gene testing can be ordered for a particular suspected syndrome. For some, evaluation of a single gene may be sufficient. For others, a panel of genes may give the best chance of identifying a causative mutation. Once identified, a particular genetic diagnosis can be used to anticipate and potentially avoid complications such as increased intracranial pressure, hydrocephalus, vision or hearing loss, dental malocclusion, and psychosocial difficulties.

TABLE 143-1 Craniosynostosis Conditions and Mutations[a]

Condition	Gene		Protein			
Diagnosis OMIM Number	Symbol and Location OMIM Number	Nucleotide Change	Type	Region	Amino Acid Change	Mechanism of Action
FGFR craniosynostosis syndromes						
Crouzon syndrome #123500 Jackson–Weiss syndrome #123150 Pfeiffer syndrome #101600	FGFR2 exon IIIa (8 or U) exon IIIc (10 or D) [a]176943	Missense, splice site, in-frame insertion and deletion	Tyrosine kinase receptor	Extracellular IgII-IgIII domain	p.C278F, p.Q289P p.T320GfsX5, p.C342R, p.C342S, p.C342W, p.C342Y, p.A344A, p.A344G, p.S347C, p.S351C, p.S354C c.940-2A>G	Constitutive activation
Pfeiffer syndrome #101600	FGFR1 exon IIIa [a]136350	Missense	Tyrosine kinase receptor	Extracellular IgII-IgIII linker	p.P252R	Enhanced ligand affinity
Apert syndrome #101200	FGFR2 exon IIIa [a]176943	Missense, splice site, Alu repeat insertion, intragenic deletion	Tyrosine kinase receptor	Extracellular IgII-IgIII domain	p.S252W, p.P253R, p.S252F, c.1041-1042insAlu, c.940-2AA>G, 1372bp deletion resulting in chimeric IIIb and IIIc domain	Ectopic ligand-dependent activation
Muenke syndrome; Craniosynostosis, Adelaide Type #602849; %600593	FGFR3 exon IIIa #134934	Missense	Tyrosine kinase receptor	Extracellular IgII-IgIII linker	p.P250R	Enhanced ligand affinity
Crouzonodermoskeletal syndrome #612247	FGFR3 exon IIIa #134934	Missense	Tyrosine kinase receptor	Transmembrane domain	p.A391E	Stabilization of receptor dimerization by hydrogen bonding
Beare-Stevenson cutis gyrata syndrome #123790	FGFR2 exon 11 [a]176943	Missense	Tyrosine kinase receptor	Juxta- or trans-membrane domain	p.S372C, p.Y375C	Constitutive activation
Osteoglophonic dysplasia #166250	FGFR1 [a]136350	Missense	Tyrosine kinase receptor	Extracellular IgIII domain, juxta- or trans- membrane domain	p.N330I, p.Y372C, p.C379R	Constitutive activation
Thanatophoric dysplasia Type II #187601	FGFR3 #134934	Missense	Tyrosine kinase receptor	Intracellular tyrosine kinase domain	p.K650E	Constitutive activation
Saethre-Chotzen syndrome; Robinow-Sorauf syndrome #101400; #180750	TWIST exon 1 and non-coding region [a]601622	Nonsense, in-frame duplication, missense, microdeletion, large chromosome deletion 7p15.3-p22	Helix-loop-helix transcription factor	Entire protein	p.Q71X, p.Y103X, p.P139dup7, p.R154fsX237,	Haploinsufficiency

Continued

TABLE 143-1 Craniosynostosis Conditions and Mutations[a]—*Cont'd*

Condition	Gene		Protein			
Diagnosis OMIM Number	**Symbol and Location OMIM Number**	**Nucleotide Change**	**Type**	**Region**	**Amino Acid Change**	**Mechanism of Action**
Craniosynostosis, Boston type #604757	MSX2 exon 2 [a]123101	Missense	Homeobox domain transcription factor	Homeodomain	p.P148H	Enhanced DNA binding affinity
Craniofrontonasal syndrome #304110	EFNB1 [a]300035	Missense, splice site, frameshift insertion and deletion, nonsense	Transmembrane protein with ephrin and PDZ domains	Entire protein	p.R66X, p.P119H, p.G151S	Haploinsufficiency
Baller–Gerold syndrome #218600	RECQL4 [a]603780	Splice site, missense, frameshift deletion	DNA helicase		p.R102W, p.D779LfsX57, c.2335del22, g.2886delT, IVS17-2A>C,	
Antley–Bixler syndrome with genital abnormalities and disordered steroidogenesis #207410	POR [a]124015	Missense, frameshift insertion, in-frame deletion and insertion, splice site, nonsense	Multidomain with FMN, FAD, NADPH binding domain	Entire protein	p.R457H in Japanese	Recessive loss of function
Carpenter syndrome #201000	RAB23 [a]606144	Missense, frameshift insertion or deletion, nonsense	RAB guanosine triphosphatase		p.L145X	Recessive loss of function
Craniosynostosis and dental anomalies syndrome (CRSDA) #614188	IL11RA [a]600939	Missense, duplication, nonsense	Cell surface receptor of interleukin 11	Fibronectin-type domain III	p.Gln159X, p.Pro221Arg, p.Ser245Cys, p.Arg296Trp, c.916_924dup	Recessive loss of function with decreased stimulation of STAT3 with IL11
Greig cephalopolysyndactyly #175700	GLI3 [a]165240	Large chromosomal deletion 7p13, intragenic deletion in-frame insertion and deletion, missense, splice site; frameshift, nonsense in first or third of gene	Zinc finger transcription factor	Entire protein	p.R290X, p.R792X	Loss of function
Alagille syndrome #118450	JAGGED1 [a]601920	Frameshift, nonsense	Ligand for NOTCH receptors		p.E553X, c.delGAAAG	Loss of function
Sphrintzen–Goldberg syndrome #182212	FBN1 [a]134797	Missense, splice site	Extracellular microfibril protein	Repeat EGF-like domain	p.C1223Y, p.C1221Y	
Furlong syndrome #609192	TGFBR1 [a]190181	Missense	TGFB receptor superfamily protein	Kinase domain	p.S241L	
C syndrome, Opitz trigonocephaly syndrome #211750	CD96 [a]606037	Missense	Immunoglobulin superfamily protein	IgIII domain	p.T280M	Loss of cell adhesion and growth

9p- syndrome with trigonocephaly #158170	FREM1, contiguous gene syndrome [a]608944	Large chromosome deletion 9p22-p24, missense	Extracellular matrix protein		p.Y285C, p.R498Q, p.G1500V	
11q- syndrome with trigonocephaly; Jacobsen syndrome #147791	Contiguous gene Syndrome	Large chromosome deletion 11q23-q24				
Hypophosphatasia, infantile type #241500	ALPL [a]171760	Missense, frameshift deletion, splice site	Alkaline phosphatase	Entire protein	p.M45I, p.M45L, p.M45V	Reduced enzyme function
Craniosynostosis, anal anomalies, porokeratosis syndrome (CAP) %603116	Pathway including RUNX2 [a]600211					
Nonsyndromic coronal synostosis, associated with Crouzon; or normal phenotype	FGFR2 [a]176943	Missense	Tyrosine kinase receptor	Extracellular and intracellular domains	p.S252L, p.A362S, p.K526E; p.A315S, p.A337T	
Nonsyndromic coronal synostosis, associated with Muenke syndrome; or normal phenotype	FGFR3 [a]134934	Missense	Tyrosine kinase receptor	Extracellular IgII-IgIII linker	p.P250R	Enhanced ligand affnity
Nonsyndromic coronal synostosis	TWIST1 [a]601622	Missense	Helix-loop-helix transcription factor	TWIST1 Box domain	p.A186T	Potential loss of binding to RUNX2
Nonsyndromic coronal synostosis	EFNA4 [a]601380	Missense, frameshift	Ephrin ligand	Frameshift mutation in C-terminus	p.H60Y, p.P117T, c.471_472delCCinsA	Alternative isoform disrupts ephrin signaling; reduced receptor binding;
Trigonocephaly, Metopic synostosis with facial skin tags #190440	FGFR1 [a]136350	Missense	Tyrosine kinase receptor	Extracellular IgIII domain	p.I300W	
Nonsyndromic sagittal synostosis	FGFR1 [a]136350	Missense	Tyrosine kinase receptor	Extracellular domain	p.T261M[b]	
Nonsyndromic sagittal and/or unilambdoid synostosis	FGFR2 [a]176943	Missense	Tyrosine kinase receptor	Extracellular domain	p.C278W, p.A315T[b]	
Nonsyndromic sagittal synostosis	TWIST1 [a]601622	Missense	Helix-loop-helix transcription factor	TWIST1 Box domain	p.S188L, p.S201Y	Potential loss in binding to RUNX2

[a]For some disorders, there is no definitive information in the literature for each heading, to date. Either the most common mutations if there are multiple mutations associated with a condition or most of the known mutations if there are only a few mutations associated with a condition.
[b]There are also variations in the noncoding regions of these genes in nonsyndromic sagittal synostosis and their significance is unknown.

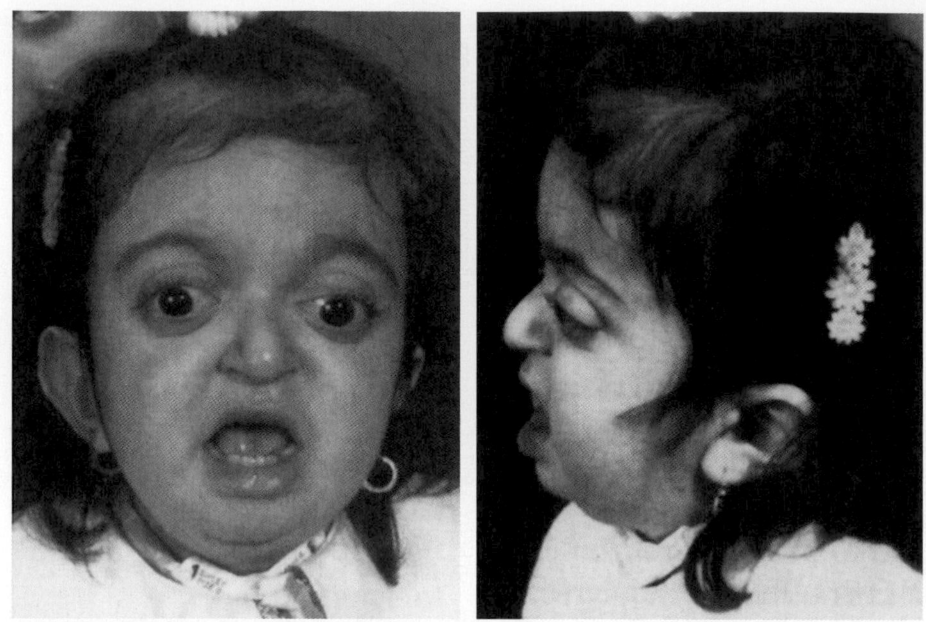

FIGURE 143-1 Crouzon syndrome. Brachycephaly, ocular proptosis, beaked nose, and midface hypoplasia. *(From Tewfik, T. L.; Teebi, A. S.; der Kaloustian, V. M. Selected Syndromes and Conditions. In: Congenital Anomalies of the Ear, Nose, and Throat; Tewfik, T. L., der Kaloustian, V. M., Eds.; 1st edn. Oxford University Press, Oxford, 1997, pp 461.)*

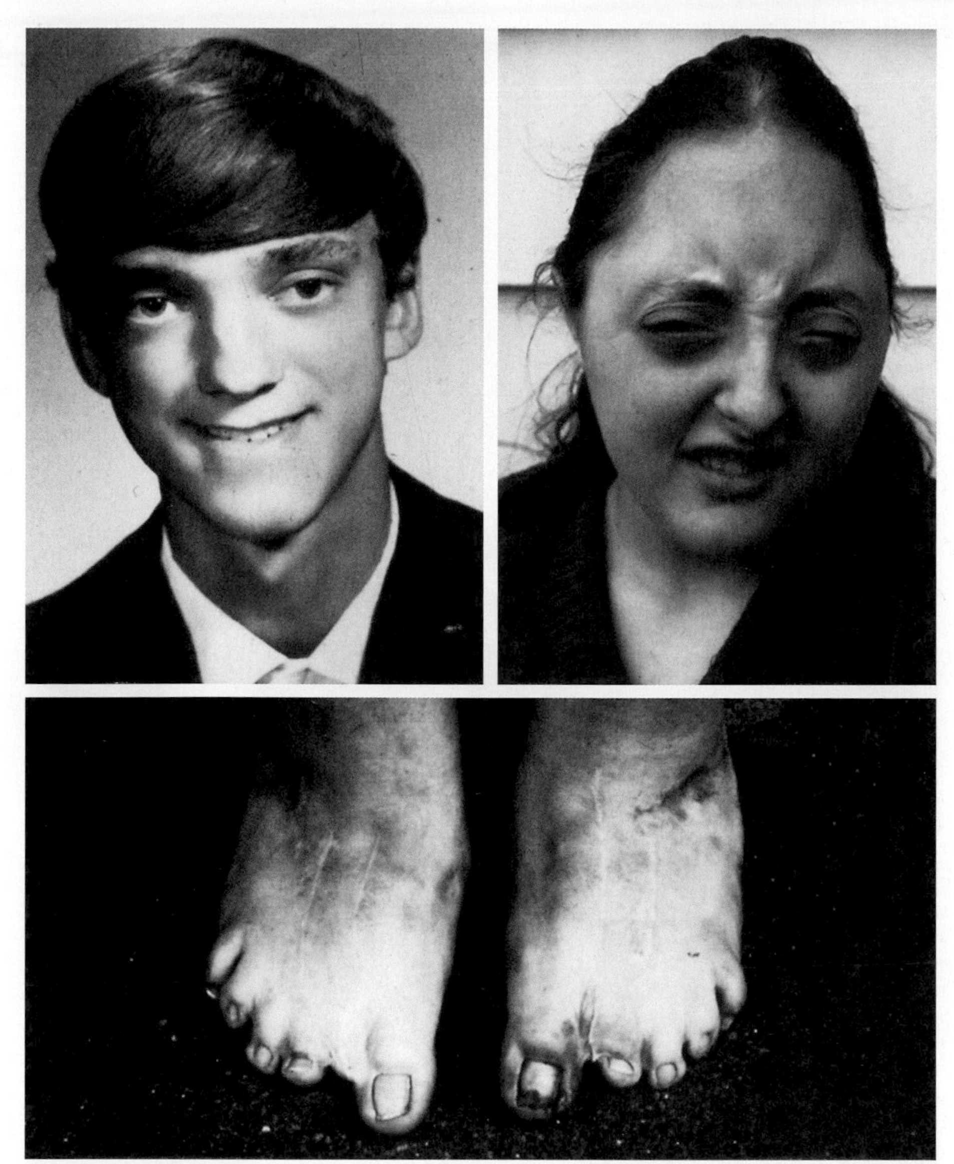

FIGURE 143-2 Jackson–Weiss syndrome. Variable craniosynostosis with brachycephaly (left) and acrocephaly (right) and large great toes and syndactyly. *(From Jabs, E. W.; Li, X; Scott, A. F., et al. Jackson-Weiss and Crouzon Syndromes Are Allelic with Mutations in Fibroblast Growth Factor Receptor 2. Nat. Genet.* ***1994****, 8 (3), 275–279; Jackson, C. E.; Weiss, L.; Reynolds, W. A., et al. Craniosynostosis, Midfacial Hypoplasia and Foot Abnormalities: An Autosomal Dominant Phenotype in a Large Amish Kindred. J. Pediatr.* ***1976****, 88 (6), 963–968.)*

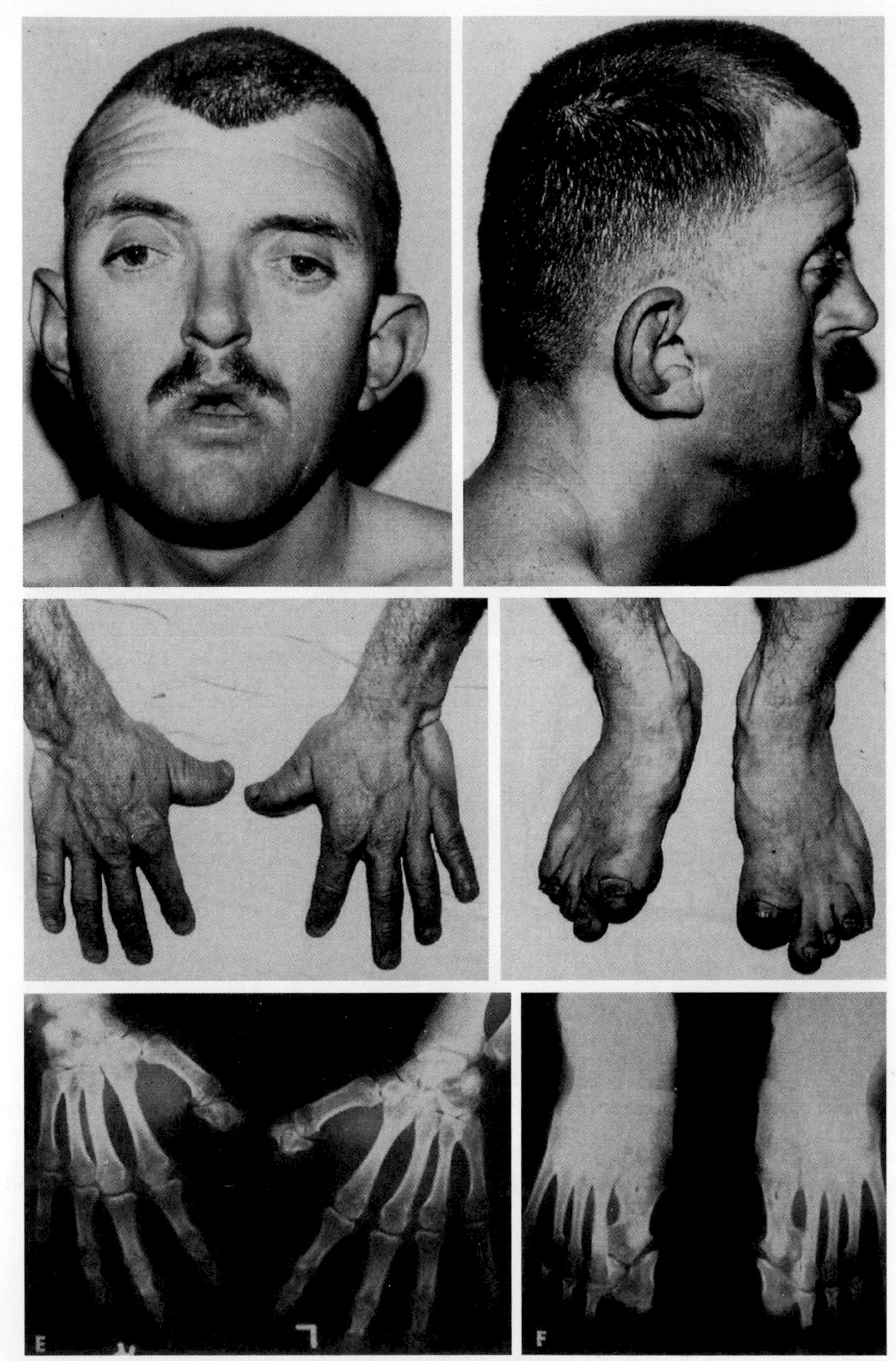

FIGURE 143-3 Pfeiffer syndrome type I. Brachycephaly, midface hypoplasia, mandibular prognathism, brachydactyly, broad thumbs that deviate radially, and short broad great toes with varus deformity and partial cutaneous syndactyly. *(From Cohen, M. M. Jr. An Etiologic and Nosologic Overview of Craniosynostosis Syndromes. In Malformation Syndromes, Birth Defects Original Article Series; Bergsma, D., Ed.; 1975; 11(2), pp. 137–189.)*

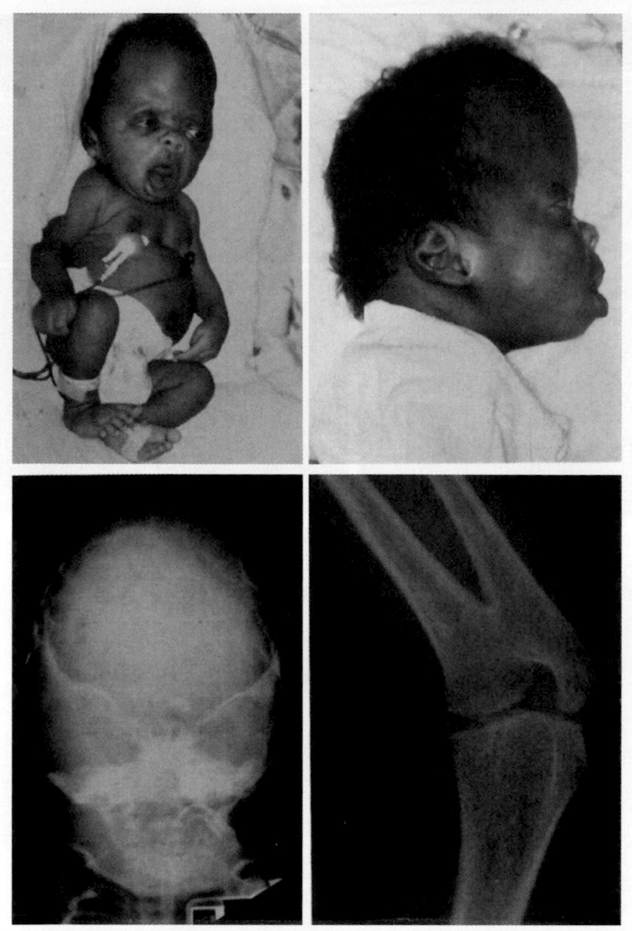

FIGURE 143-4 Pfeiffer syndrome type II. Brachy-turricephaly, ocular proptosis, hypertelorism, depressed nasal bridge, midface hypoplasia, and low-set posterior rotated ears. Radiographs show mild cloverleaf skull deformity, increased digital markings, and radioulnar synostosis. *(From Okajima, K.; Robinson, L. K.; Hart, M. A., et al. Ocular Anterior Chamber Dysgenesis in Craniosynostosis Syndromes with a Fibroblast Growth Factor Receptor 2 Mutation. Am. J. Med. Genet.* ***1999****, 85(2), 160–170.)*

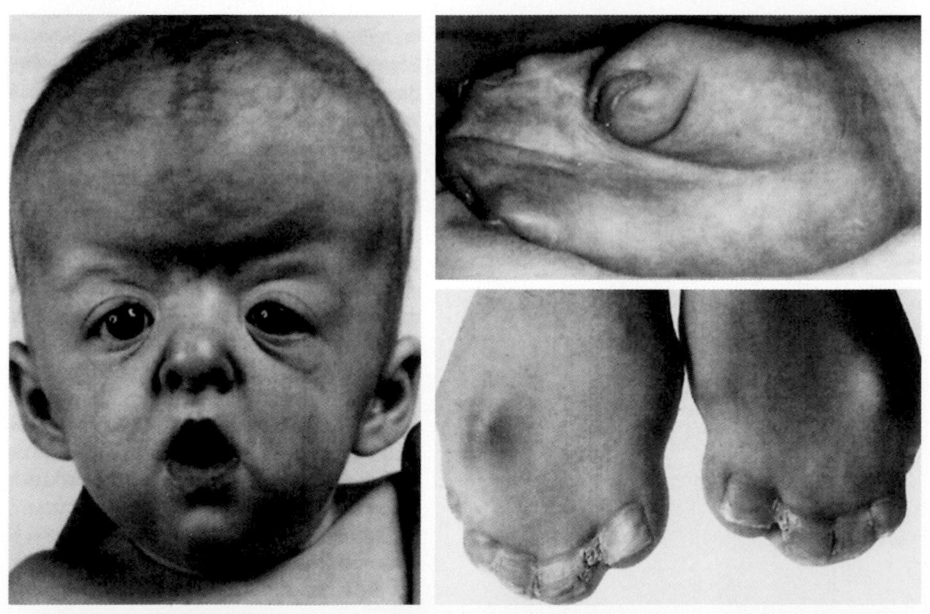

FIGURE 143-5 Apert syndrome. High prominent forehead, downslanting palpebral fissures, midface hypoplasia, and mitten type syndactyly of digits and severe syndactyly of toes. *(From Lessard, M-L.; Mulliken, J. B. Major Craniofacial Anomalies. In Congenital Anomalies of the Ear, Nose, and Throat; Tewfik, T. L., der Kaloustian, V. M., Eds. 1st ed. Oxford University Press, Oxford, 1997, pp. 307; Mann, T.P. Ed. Colour Atlas of Pediatric Facial Diagnosis, Kluwer Academic Publishers: London, 1989, pp. 88.)*

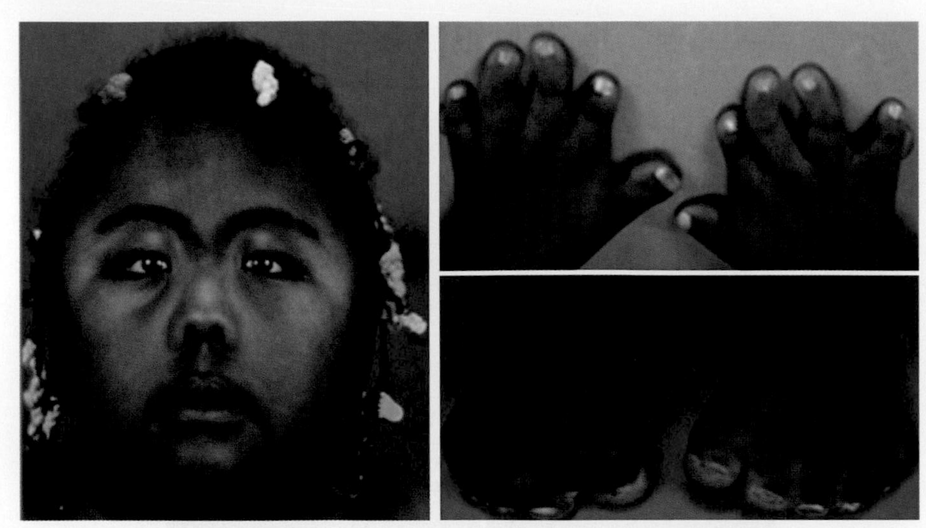

FIGURE 143-6 Carpenter syndrome. Metopic ridging and lateral bulge with multiple suture synostosis, broad thumbs and syndactyly, brachydactyly, clinodactyly, and polydactyly. *(From Jenkins, D., Seelow, D.; Jehee, F. S., et al. RAB23 Mutations in Carpenter Syndrome Imply an Unexpected Role for Hedgehog Signaling in Cranial Suture Development and Obesity. Am. J. Hum. Genet.* ***2007****, 80, 1162–1170.)*

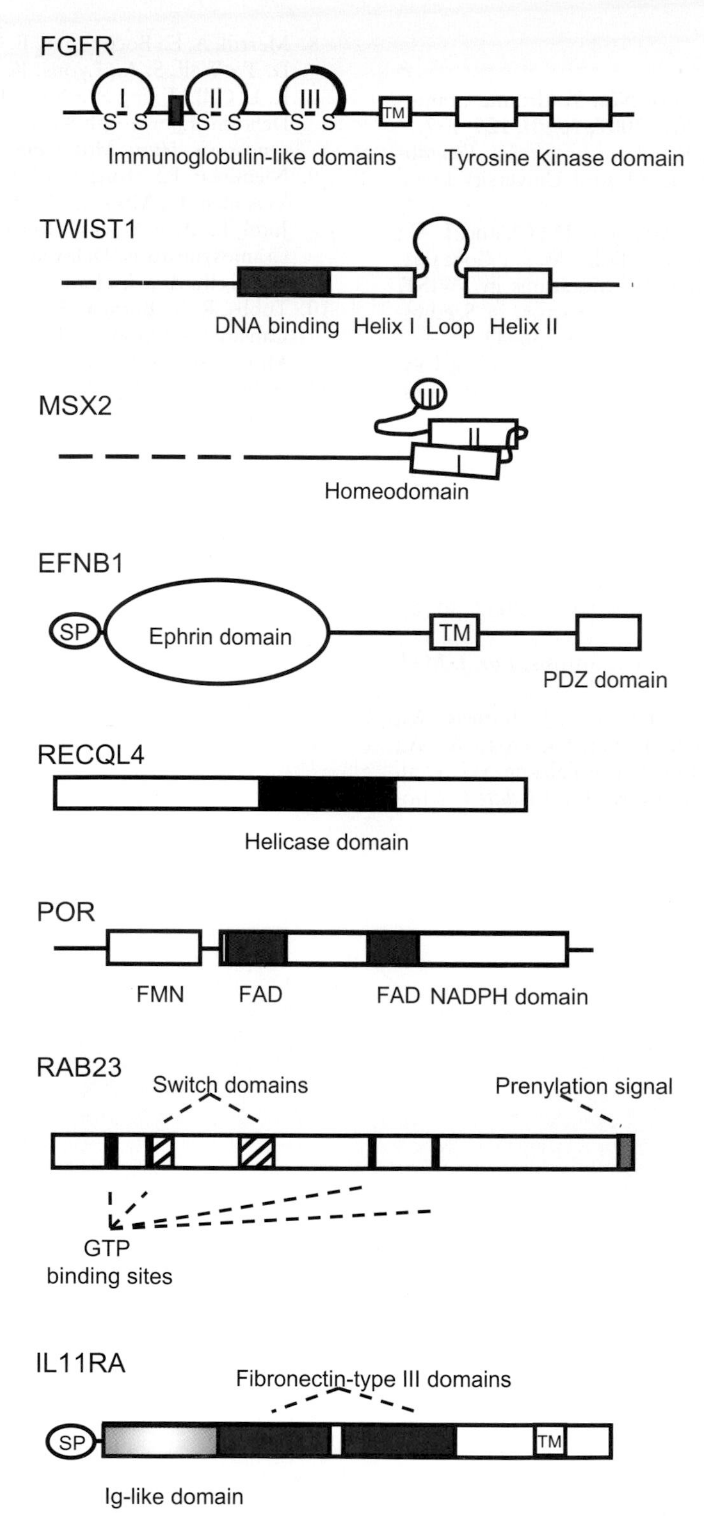

FIGURE 143-7 Schematic protein diagrams. FGFR has an extracellular domain composed of three immunoglobulin-like domains (I, II, III) with disulfide bonds (S–S) and heparin binding sequence of heparan sulfate-containing proteoglycans binding domain (gray box), TM domain, and intracellular split tyrosine kinase domain (two rectangular boxes). The second half of IgIII is coded by a variably spliced exon (thickened portion of IgIII). TWIST1 has a DNA binding domain and helix-loop-helix region. The MSX2 homeodomain is composed of a flexible N-terminal extension followed by helices I, II, and III. EFNB1 has an extracellular domain composed of a signal peptide (SP) and ephrin domain, TM domain, and intracellular tyrosine residues and PDZ domain. RECQL4 has a helicase domain. POR has an electron donating flavin mononucleotide domain (FMN), flavin adenine dinucleotide domains (FAD), and nicotinamide adenine dinucleotide phosphate (NADPH) binding domain. RAB23 has GTP binding sites, switch domains, and prenylation signal. IL11RA has a SP, immunoglobulin-like domain, fibronectin-like III domain, and TM domain.

FURTHER READING

1. Boyadjiev, S. A. Genetic Analysis of Non-Syndromic Craniosynostosis. *Orthod. Craniofac. Res.* **2007,** *10* (3), 129–137.
2. Cohen, I. M. M. J.; Maclean, R. *Craniosynostosis: Diagnosis, Evaluation, and Management*; Oxford University Press: Oxford, 2000.
3. Howard, T. D.; Paznekas, W. A.; Green, E. D.; Chiang, L. C.; Ma, N.; Ortiz de Luna, R. I.; Garcia Delgado, C.; Gonzalez-Ramos, M.; Kline, A. D.; Jabs, E. W. Mutations in TWIST, a Basic Helix-Loop-Helix Transcription Factor, in Saethre-Chotzen Syndrome. *Nat. Genet.* **1997,** *15* (1), 36–41.
4. Jabs, E. W. Toward Understanding the Pathogenesis of Craniosynostosis Through Clinical and Molecular Correlates. *Clin. Genet.* **1998,** *53* (2), 79–86.
5. Jenkins, D.; Seelow, D.; Jehee, F. S.; Perlyn, C. A.; Alonso, L. G.; Bueno, D. F.; Donnai, D.; Josifova, D.; Mathijssen, I. M.; Morton, J. E.; Orstavik, K. H.; Sweeney, E.; Wall, S. A.; Marsh, J. L.; Nurnberg, P.; Passos-Bueno, M. R.; Wilkie, A. O. RAB23 Mutations in Carpenter Syndrome Imply an Unexpected Role for Hedgehog Signaling in Cranial-Suture Development and Obesity. *Am. J. Hum. Genet.* **2007,** *80* (6), 1162–1170.
6. Johnson, D.; Wilkie, A. O. M. Craniosynostosis. *Eur. J. Med. Genet.* **2011,** *19* (4), 369–376.
7. McCarthy, J. G.; Warren, S. M.; Bernstein, J.; Burnett, W.; Cunningham, M. L.; Edmond, J. C.; Figueroa, A. A.; Kapp-Simon, K. A.; Labow, B. I.; Peterson-Falzone, S. J., et al. Parameters of Care for Craniosynostosis. *Cleft Palate Craniofac. J.* **2012,** *49* (Suppl), 1S–24S.
8. Merrill, A. E.; Bochukova, E. G.; Brugger, S. M.; Ishii, M.; Pilz, D. T.; Wall, S. A.; Lyons, K. M.; Wilkie, A. O. M.; Maxson, R. E. Cell Mixing at a Neural Crest-Mesoderm Boundary and Deficient Ephrin-Eph Signaling in the Pathogenesis of Craniosynostosis. *Hum. Mol. Genet.* **2006,** *15* (8), 1319–1328.
9. Nieminen, P.; Morgan, N. V.; Fenwick, A. L.; Parmanen, S.; Veistinen, L.; Mikkola, M. L.; van der Spek, P. J.; Giraud, A.; Judd, L.; Arte, S., et al. Inactivation of IL11 Signaling Causes Craniosynostosis, Delayed Tooth Eruption, and Supernumerary Teeth. *Am. J. Hum. Genet.* **2011,** *89* (1), 67–81.
10. Tubbs, R. S.; Bosmia, A. N.; Cohen-Gadol, A. A. The Human Calvaria: A Review of Embryology, Anatomy, Pathology, and Molecular Development. *Child Nerv. Syst.* **2012,** *28* (1), 23–31.
11. Wilkie, A. O. Bad Bones, Absent Smell, Selfish Testes: The Pleiotropic Consequences of Human FGF Receptor Mutations. *Cytokine Growth Factor Rev.* **2005,** *16* (2), 187–203.
12. Wilkie, A. O.; Byren, J. C.; Hurst, J. A.; Jayamohan, J.; Johnson, D.; Knight, S. J.; Lester, T.; Richards, P. G.; Twigg, S. R.; Wall, S. A. Prevalence and Complications of Single-Gene and Chromosomal Disorders in Craniosynostosis. *Pediatrics* **2010,** *126* (2), e391–e400.

Dermatologic Disorders

Abnormalities of Pigmentation, 541
Ichthyosiform Dermatoses, 544
Epidermolysis Bullosa, 546
Ectodermal Dysplasias, 551
Skin Cancer, 557
Psoriasis, 559
Cutaneous Hamartoneoplastic Disorders, 561
Inherited Disorders of the Hair, 563

CHAPTER

144

Abnormalities of Pigmentation

Richard A Spritz

Human Medical Genetics and Genomics Program, University of Colorado School of Medicine, Aurora, CO, USA

Vincent J Hearing

Laboratory of Cell Biology, National Cancer Institute, National Institutes of Health, Bethesda, MD, USA

Because of their visually evident phenotypes, disorders of pigmentation were the first genetic diseases ever documented, with clinical descriptions in some cases dating back hundreds or even thousands of years. Phenotypically similar abnormalities were recognized early in many different mammalian species, and the breeding of mice with interesting coat pigmentation patterns became a popular hobby in the Victorian era, forming the basis of one of the most important animal models of modern scientific research. With the rediscovery of Mendel's principles of genetics early in the twentieth century, the inheritance of normal human pigmentary variation was one of the first traits studied, though with limited success.

Many of the genes that regulate mammalian pigmentation have been associated with specific human diseases that include decreased pigmentation (hypopigmentation), increased pigmentation (hyperpigmentation), or abnormal pigmentary patterning. In general, disorders of pigmentation can be categorized on functional grounds as (1) disorders of melanocyte development, differentiation, and/or migration; (2) disorders of melanocyte function; and (3) disorders of melanocyte survival (*1–3*). This chapter summarizes those disorders and the corresponding genes known at this time.

The process of producing and distributing melanin pigment (also termed melanogenesis) in the skin and other tissues is surprisingly complex. Pigment cells, known as melanocytes, must develop from the neural crest, differentiate, migrate to their various destination tissues and distribute in appropriate patterns, proliferate, survive, produce melanin pigment in specialized organelles known as melanosomes, transfer pigmented melanosomes to keratinocytes, and respond to various environmental stimuli to achieve final levels of visible pigmentation. It is no wonder then that more than 378 different genes play roles in the pigmentation process (with more being identified continuously), those genes and their encoded proteins operating at various steps in the pathway to regulate eventual pigment patterns. Variation and mutation in many of these genes can result in individual differences in pigmentary phenotypes ranging from normal variation in pigment quantity, quality, or pattern to severe diseases with fatal consequences. In the normal context, melanins serve a number of important evolutionary functions (e.g. species recognition and sexual dimorphism), play essential roles in survival (e.g. camouflage, UV photoprotection, regulation of body temperature, free radical scavenging), and can provide visible output assays of key biological processes (e.g. signaling and transcription).

144.1 DISORDERS OF MELANOCYTE DEVELOPMENT, DIFFERENTIATION, AND/OR MIGRATION

Examples of such developmental diseases include Waardenburg syndrome, Hirschsprung's disease, and Piebaldism (*4,5*). Pigmentary anomalies in these diseases typically manifest as congenital, stable white spotted areas, frequently on the forehead and abdomen, which failed to receive sufficient melanoblasts during development or in which melanoblasts failed to survive or differentiate to functional melanocytes to populate the skin.

Melanocytes are dendritic cells that originate as melanoblasts in the neural crest during embryologic development. Melanoblasts migrate throughout the developing organism to several principal locations—the skin (at the epidermal/dermal border), the eyes (in the choroid and iris), and the hair bulbs—as well as to several other minor areas, such as the stria vascularis of the inner ear. In contrast, the retinal pigment epithelium derives from neural ectoderm origin. Genes that function in early melanoblast development, differentiation, and/or migration encode, among other things, transcription factors (e.g. *FOXD3*, *PAX3*, *SOX10*, *MITF*), receptors and their ligands (e.g. *EDNRB*/*EDN3*, *KIT*/*KITL*, *MC1R*/*POMC*, *FZD4*/*WNT3A*), and other factors important to the start/continue/stop signals for melanoblast differentiation and migration (e.g. *ADAMTS20*, *MCOLN3*, *ITGB1*). Furthermore, once melanoblasts have arrived at their final destinations in the tissues, they must differentiate to become functional melanocytes, and then must survive and proliferate to populate the tissue to achieve the correct eventual density of melanocytes required for normal function. Again, a relatively large number of genes are involved in these processes, some of which are the same as those involved in melanoblast development, such as *MITF*, *SOX10*, *KIT*, and *EDNRB*, which thus function at multiple levels.

144.2 DISORDERS OF MELANOCYTE FUNCTION

Examples of such functional diseases include hypopigmentary diseases such as oculocutaneous albinism (OCA), ocular albinism type 1 (OA1), Hermansky--Pudlak syndrome (HPS), Chediak–Higashi syndrome (CHS), and Griscelli syndrome (GS). Pigmentary anomalies in these diseases typically manifest as congenital global hypopigmentation of skin, hair, and irises, though in some patients there may be a slight increase in pigmentation over time, particularly during puberty in males. There are also many hyperpigmentary diseases, for example melasma, lentigo senilis, post-inflammatory hyperpigmentation and UV-induced melanosis, wherein pigmentation in tissues is increased above normal levels, frequently due to environmental stresses.

Melanin is a large pigmented biopolymer that is produced only within specific subcellular organelles termed melanosomes. The melanins produced, and the melanosomes wherein they are deposited, can be of several types, with differing visible colors and presumably with distinct functional properties; the major classes of melanins are the brown–black eumelanins and the yellow–red pheomelanins. Melanosomes are members of the lysosome-related organelle (LRO) family, which also includes lysosomes, platelet dense bodies, and synaptosomes, and they contain a limited number of enzymatic and structural proteins that are uniquely expressed by melanocytes (e.g. TYR, TYRP1, DCT, PMEL17, GPR143). Defects of genes encoding such melanocyte-specific functional proteins typically result in altered pigmentation phenotype and stereotypic abnormalities of optic tract development that result from deficient melanin pigment, with few or no effects on other, nonpigmented tissues; examples of such diseases include the four forms of oculocutaneous albinism: OCA1 (*TYR*), OCA2 (*OCA2*), OCA3 (*TYRP1*), OCA4 (*SLC45A2*) (*6–9*), and ocular albinism type 1 (*GPR143*) (*10*). In contrast, defects of other genes that encode proteins necessary for functions of melanosomes as well as other LROs (e.g. protein trafficking, signaling) often affect not only pigmentation but also other tissues and cells that depend on these other LROs, such as platelets and natural killer (NK) cells; examples of such diseases include the various forms of Hermansky–Pudlak syndrome (*11*): HPS1 (*HPS1*), HPS2 (*AP3B1*), HPS3 (*HPS3*), HPS4 (*HPS4*), HPS5 (*HPS5*), HPS6 (*HPS6*), HPS7 (*DTNBP1*), HPS8 (*BLOC1S3*), HPS9 (*PLDN*) and Chediak–Higashi syndrome (*LYST*) (*12,13*). Finally, defects of genes that encode proteins involved in the transport of melanosomes to the dendrites of melanocytes and/or their subsequent transfer to adjacent keratinocytes can also affect coloration, but also typically affect the functions of other LROs and thus can have quite pleiotropic phenotypic manifestations; examples of such diseases include the various forms of Griscelli syndrome (*14*): GS1 (*MYO5A*), GS2 (*RAB27A*), GS3 (*MLPH*).

144.3 DISORDERS OF MELANOCYTE SURVIVAL

Examples of such diseases include the various manifestations of generalized vitiligo and segmental vitiligo. Pigmentary anomalies in these diseases typically manifest as one or more acquired, progressive, waxing and waning, or stable white spotted areas, frequently on the face, trunk, and extremities, in which melanocytes have been destroyed and in which there is insufficient repopulation from adjacent areas to restore normal pigmentation.

Generalized vitiligo is a relatively common autoimmune disease in which melanocytes are targeted for destruction by the immune system (*15*). There is strong epidemiological association with other autoimmune diseases, in both vitiligo patients and their close relatives, suggesting that many susceptibility genes are shared in common among these disorders. Generalized vitiligo is a complex, polygenic, multifactorial disorder, with over 30 susceptibility loci identified thus far (*16,17*). Most of these genes encode immunoregulatory proteins (e.g. *HLA-A*, MHC class II, *NLRP1*, *PTPN22*, *RERE*, *FOXP1*, *CTLA4*, *FOXP3*, *IFIH1*, *LPP*, *CD44*, *CD80*, *CLNK*, *BACH2*, *CCR6*, *SLA*, *IL2RA*, *CASP7*, *SH3B3*, *GZMB*, *TICAM1*, *UBASH3A*, *C1QTNF6*, *XBP1*, *TSLP*), many of which are also involved in susceptibility to the other autoimmune diseases that are epidemiologically associated with generalized vitiligo. However, several (e.g. *TYR*, *OCA2*, *MC1R*) encode melanocyte-specific proteins that may play roles in triggering the melanocyte-specific autoimmune response and/or in mediating melanocyte targeting and destruction by immune effector cells. One unique family with autosomal dominant generalized vitiligo has a mutation that affects transcription of *FOXD3*, a master regulator of melanoblast development, likely reducing the number of skin melanoblasts and thus the ultimate population of skin melanocytes in affected individuals.

Segmental vitiligo typically manifests as a single acquired white spot, typically on the face, trunk, or an extremity, often occurring in childhood and typically with very rapid onset and progression but limited extent. Recent evidence suggests that segmental vitiligo may also have an autoimmune basis, which for some reason affects only a localized area.

REFERENCES

1. Tomita, Y.; Suzuki, T. Genetics of Pigmentary Disorders. *Am. J. Med. Genet.* **2004,** *131C,* 75–81.
2. Hornyak, T. J. The Developmental Biology of Melanocytes and Its Application to Understanding Human Congenital Disorders of Pigmentation. *Adv. Dermatol.* **2006,** *22,* 201–218.
3. Fistarol, S. K.; Itin, P. H. Disorders of Pigmentation. *J. Dtsch. Dermatol. Ges.* **2010 Mar,** *8* (3), 187–201.
4. Nordlund, J. J.; Boissy, R. E.; Hearing, V. J.; Oetting, W. S.; King, R. A.; Ortonne, J. P. *The Pigmentary System: Physiology and Pathophysiology*, 2nd ed.; Blackwell Science: Edinburgh, 2006, 1–1200 p.
5. Pingault, V.; Ente, D.; Dastot-Le, M. F.; Goossens, M.; Marlin, S.; Bondurand, N. Review and Update of Mutations Causing Waardenburg Syndrome. *Hum. Mutat.* **2010 Apr,** *31* (4), 391–406.
6. Gronskov, K.; Ek, J.; Brondum-Nielsen, K. Oculocutaneous Albinism. *Orphanet. J. Rare. Dis.* **2007,** *2,* 43, PMCID: PMC2211462.
7. Hutton, S. M.; Spritz, R. A. Comprehensive Analysis of Oculocutaneous Albinism Among Non-Hispanic Caucasians Shows that OCA1 is the Most Prevalent OCA Type. *J. Invest. Dermatol.* **2008 Oct,** *128* (10), 2442–2450.
8. Hutton, S. M.; Spritz, R. A. A Comprehensive Genetic Study of Autosomal Recessive Ocular Albinism in Caucasian Patients. *Invest. Ophthalmol. Vis. Sci.* **2008 Mar,** *49* (3), 868–872.
9. Suzuki, T.; Tomita, Y. Recent Advances in Genetic Analyses of Oculocutaneous Albinism Types 2 and 4. *J. Dermatol. Sci.* **2008 Jul,** *51* (1), 1–9.
10. Schiaffino, M. V.; Tacchetti, C. The Ocular Albinism Type 1 (OA1) Protein and the Evidence for an Intracellular Signal Transduction System Involved in Melanosome Biogenesis. *Pigment. Cell Res.* **2005 Aug,** *18* (4), 227–233.
11. Huizing, M.; Helip-Wooley, A.; Westbroek, W.; Gunay-Aygun, M.; Gahl, W. A. Disorders of Lysosome-Related Organelle Biogenesis: Clinical and Molecular Genetics. *Annu. Rev. Genomics Hum. Genet.* **2008,** *9,* 359–386, PMCID: PMC2755194.
12. Spritz, R. A. Chediak-Higashi Syndrome. In *Primary Immunodeficiency Diseases*, 2nd ed.; Ochs, H. D.; Smith, C. I. E.; Puck, J. M., Eds.; Oxford: New York, 2007; pp 570–577.
13. Kaplan, J.; De Domenico, I.; Ward, D. M. Chediak-Higashi Syndrome. *Curr. Opin. Hematol.* **2008 Jan,** *15* (1), 22–29.
14. Van Gele, M.; Dynoodt, P.; Lambert, J. Griscelli Syndrome: A Model System to Study Vesicular Trafficking. *Pigment Cell Melanoma Res.* **2009 Jun,** *22* (3), 268–282.
15. Picardo, M.; Taieb, A. *Vitiligo*; Springer: Heidelberg, 2010.
16. Spritz, R. A. Recent Progress in the Genetics of Generalized Vitiligo. *J. Genet. Genomics.* **2011 Jul 20,** *38* (7), 271–278.
17. Spritz, R. A. Six Decades of Vitiligo Genetics: Genome-Wide Studies Provide Insights into Autoimmune Pathogenesis. *J. Invest. Dermatol.* **2012 Feb,** *132* (2), 268–273.

CHAPTER 145

Ichthyosiform Dermatoses

Howard P Baden

Department of Dermatology, Cutaneous Biology Research Center, Harvard Medical School, Massachusetts General Hospital, Boston, MA, USA

John J DiGiovanna

Center for Cancer Research, National Cancer Institute, National Institutes of Health, Bethesda, MD, USA

The ichthyoses include a spectrum of common to rare disorders characterized by scaling of the skin. Clinical manifestations may be limited to the skin or involve other organ systems. Both acquired and inherited forms occur and onset may be congenital or later in life. Approaching a diagnosis for an ichthyosis patient involves understanding the clinical features of the various disorders and matching those to the patient's presentation including skin findings, family history (pedigree), age of onset, and associated features. The most common forms are ichthyosis vulgaris, X-linked, autosomal recessive congenital ichthyosis and epidermolytic hyperkeratosis, but there are many less common forms.

The epidermis is a multilayered tissue which is firmly attached to the dermis by a series of interlocking structures, and the keratinocytes are linked to each other to form a cohesive membrane. The cells die and cornify at the top to form the stratum corneum. The thickness of the epidermis and its various layers differ in various areas of the body but is generally about 100 mm. Keratinocytes contain the usual complement of subcellular organelles essential for energy production, synthesis of proteins, etc. In addition, one finds tonofilaments, desmosomes, keratohyalin granules, and cornified envelopes which are essential for full expression of epidermal differentiation.

Ichthyosis vulgaris is the commonest ichthyosis, with at least 1% of the population involved to some degree. Fine light scales are widespread, but spare the flexural areas and the turned-up margins give a rough feel to the skin. Atopy is frequently associated and in some patients it is a major problem. The pattern of inheritance is semi-dominant in that a mutation of one copy of the filaggrin gene (*FLG*) causes mild disease and if both alleles are mutated the condition is more severe. The barrier function of the epidermis is reduced which allows enhanced penetration into the skin.

X-linked ichthyosis may present at birth or early infancy. There is generalized involvement with the extensor surfaces most severely affected including the neck, but not the palms and soles. Asymptomatic corneal opacities are frequently associated and the incidence of cryptochidism is elevated. Patients have a deficiency of steroid sulfatase activity usually resulting from a deletion of a segment of DNA on the X chromosome containing the gene.

Ichthyosis of the newborn may present at birth or shortly after and may be the only organ affected or be part of a syndrome. Formulating a diagnosis involves assessment of clinical appearance, pedigree, and associated findings.

The collodion baby is born encased in a parchment-like membrane and may be the presentation of several different disorders which vary from mild to severe. Most have either autosomal recessive congenital ichthyosis or represent one of the congenital ichthyoses associated with abnormalities of other organs. However, some of these infants have a self-healing collodion phenotype which dramatically clears over a few weeks to leave minimal involvement. This type has been described with mutations in *TGM1* or *ALOX12B*. The harlequin fetus is a distinct genetic entity with a strikingly grotesque appearance. The skin has a thick armor-like covering of an off-white color which has deep fissures running in different directions and marked ectropion and eclabium. It is due to mutations in *ABCA12*.

Hyperkeratosis of the palms and soles (keratoderma) accompanies many ichthyosiform disorders but also occurs as a group of primary conditions where the keratoderma is the major clinical manifestation.

Trichothiodystrophy is an autosomal recessive disorder characterized by sulfur-deficient brittle hair and a broad spectrum of clinical manifestations including photosensitivity, short stature, intellectual impairment, recurrent infections, ichthyosis, and others. Mutations in *XPB*, *XPB*, *TTDA* or *TTDN1* have been found. Once the disease is suspected, a simple examination of hair shafts under polarizing light microscopy can confirm the diagnosis.

Clinical features of phytanic acid storage disease include ichthyosis, retinitis pigmentosa, peripheral polyneuropathy and cerebellar ataxia. The onset is usually before 10 years of age but can occur later. It is caused by a mutation in either the gene for phytanoyl-CoA hydoxylase gene (*PHYH)* or peroxin-7 (*PEX7*). The infantile form of the disease has biochemical abnormalities not restricted to phytanic acid and results from mutations in the peroxisome proteins PEX1, PEX2 or PEX26.

Chanarin-Dorfman syndrome is a systemic lipoidosis associated with scaling skin beginning in infancy and multisystem organ involvement. It is caused by mutations in the *CGI58* gene (abhydrolase domain-containing 5, *ABHD5*).

Infants with Sjögren–Larsson syndrome tend to be preterm and at birth or soon after develop hyperkeratotic ichthyosiform skin. Neurologic symptoms begin as delay in reaching motor milestones and develop into spastic diplegia or quadriplegia. Leukodystrophy may be seen on MRI. Characteristic glistening white dots in the retina may help suggest the diagnosis, which is a result of mutations in *ALDH3A2*, the gene encoding fatty aldehyde dehydrogenase.

Erythrokeratodermia variabilis is characterized by ichthyosiform scaling or fixed plaques with migratory red patches. This autosomal dominant disorder is the result of mutations in the genes encoding connexin-31 (*GJB3*) or connexin-30.3 (*GJB4*).

Chondrodysplasia punctata is a heterogeneous group of disorders with ichthyosis and punctate calcifications that show up as stippling of the endochondral bone in radiographs. Several subtypes have been described with mutations in genes of lipid metabolism or peroxisome function.

Patients with Netherton syndrome have an ichthyosiform dermatosis, atopy, and a distinctive hair shaft abnormality called trichorrhexis invaginata. Some have ichthyosis linearis circumflexa, which appears as serpiginous lesions with a distinctive, double-edged scale. Failure to thrive, enteropathy and infections can be fatal in infants. It is caused by mutation of the *SPINK5* gene.

CHILD syndrome is characterized by unilateral erythema and scale, with a distinct demarcation in the middle of the trunk and ipsilateral limb defects. Mutations have been described in the *NSDHL* gene, which affects cholesterol synthesis or in *EPB*.

KID syndrome, with keratitis, ichthyosis, and deafness, is caused by dominant mutations in *GJB2*, encoding connexin-26. Skin involvement may appear as distinctive fixed plaques and corneal neovascularization is progressive.

While the ichthyosiform dermatoses encompass a broad spectrum of disorders, specific dermatologic features, age of onset, pedigree, and associated findings can help formulate a specific diagnosis.

FURTHER READING

DiGiovanna, J. J. Robinson-Bostom. Ichthyosis: Etiology, Diagnosis, and Management. *Am. J. Clin. Dermatol.* **2003**, *4* (2), 81–95.

Elias, P. M.; Williams, M. L.; Crumrine, D.; Schmuth, M. Ichthyoses: Clinical, Biochemical, Pathogenic and Diagnostic Assessment. In *Series: Current Problems in Dermatology*; Itin, P., Ed., Vol. 39, Karger: Basel, 2010.

Fleckman, P.; DiGiovanna J. J. Ichthyoses. In *Fitzpatrick's Dermatology in General Medicine*, 8th ed.; Wolff K., et al., Eds.; McGraw-Hill: New York, (in press).

Oji, V.; Tadini, G.; Akiyama, M.; Blanchet Bardon, C.; Bodemer, C.; Bourrat, E.; Coudiere, P.; DiGiovanna, J. J.; Elias, P.; Fischer, J., et al. Revised Nomenclature and Classification of Inherited Ichthyoses: Results of the First Ichthyosis Consensus Conference in Sorèze 2009. *J. Am. Acad. Dermatol.* **2010,** *63* (4), 607–641.

Traupe, H. *The Ichthyoses: A Guide to Clinical Diagnosis, Genetics Counseling and Therapy*; Springer-Verlag: Berlin, 1989.

CHAPTER

146

Epidermolysis Bullosa

Cristina Has and Leena Bruckner-Tuderman

Department of Dermatology, University of Freiburg, Freiburg, Germany

Jouni Uitto

Department of Dermatology and Cutaneous Biology, Jefferson Medical College, Thomas Jefferson University, Philadelphia, USA

Epidermolysis bullosa (EB) is a clinically and genetically heterogeneous group of genodermatoses characterized by mucocutaneous blistering and chronic epithelial fragility (*1*). The most typical symptom, skin blistering, results from dermo-epidermal tissue separation caused by minor friction or trauma. EB is encountered in all populations throughout the world, but no accurate numbers exist on its global prevalence or incidence. The cutaneous basement membrane zone, which attaches the epidermis to the dermis in the skin, is structurally and functionally altered in EB. In rare cases, intraepidermal cell–cell adhesion is perturbed. At least 15 genes encoding proteins of the epidermis or the cutaneous basement membrane zone are involved in the pathogenesis of EB. Detailed knowledge on the molecular components and their supramolecular organization forms the basis for diagnostic evaluation of EB skin biopsy specimens by immunofluorescence staining or transmission electron microscopy.

The latest international consensus classification defines four major EB types based on the ultrastructural level of tissue separation (*1*): EB simplex (EBS), junctional EB (JEB), dystrophic EB (DEB), and the Kindler syndrome (KS). These can be further separated into subtypes. In addition to the level of tissue separation, clinical severity and the inheritance pattern were applied as criteria for the major subtypes. Further clinical and molecular genetic characteristics could distinguish more than 30 minor subtypes of EB (*1*), although many of these represent allelic diseases and it remains questionable whether they should be considered as distinct variants.

In EBS, the cleavage occurs within the epidermis. This category is subdivided into two major subtypes: (1) suprabasal with cleavage within the suprabasal epidermal layers and (2) basal with cleavage within the basal epidermal layer. Severe suprabasal EBS variants can be caused by mutations in the genes encoding desmoplakin or plakoglobin, whereas mutations in the gene for plakophilin 1 underlie the milder ectodermal dysplasia—skin fragility syndrome.

Basal EBS is the most common EB subtype, accounting for about one half of all cases (*2*). Blisters result from disintegration of the basal keratinocytes in the lowermost epidermis. Most cases of EBS are inherited in an autosomal dominant manner. Blistering is noted at birth or soon thereafter, and the clinical spectrum ranges from mild localized to severe generalized blistering. The blisters heal without scarring, although secondary trauma can cause milia, mild scars and nail dystrophy. The most common subtypes are EBS localized, EBS Dowling-Meara, EBS-other generalized. In addition, EBS with mottled pigmentation and migratory circinate EBS represent rare, genetically and clinically distinct subtypes. Autosomal recessive EBS due to *KRT14* null mutations accounts for about one-third of EBS cases in the Middle East (*2*). Plectin deficiency may cause three different, rare subtypes of basal EBS: EBS with muscular dystrophy, EBS with pyloric atresia, both inherited in an autosomal recessive manner, and EBS Ogna, which is an autosomal dominant disease (*3*).

In JEB, dermo-epidermal separation takes place along the lamina lucida within the basement membrane zone, and the hemidesmosome-anchoring filament complex appears altered or rudimentary. Transmembrane and extracellular proteins of the hemidesmosomes and the anchoring filaments are affected: collagen XVII, laminin-332, or integrin α6β4. The separation into major JEB subtypes relies on the severity of the disorder. JEB-Herlitz, caused by lack of laminin-332, has a lethal course due to extreme fragility of the skin and mucous membranes, whereas non-Herlitz JEB (also called JEB-other) exhibits milder, yet variable degree of skin fragility but an essentially normal life span (*4*). JEB with pyloric atresia is a relatively rare EB subtype, characterized by skin blistering and pyloric atresia, caused by mutations in the genes coding for β4 and α6 integrin, *ITGB4* and *ITGA6*.

In DEB, blistering takes place at the level of the anchoring fibrils in the uppermost dermis, and the fibrils, if present, are ultrastructurally altered. Mutations in the gene for collagen VII, *COL7A1*, underlie both dominant and recessive forms of DEB. The clinical spectrum of the dominant DEB is broad and can range from localized acral blistering to generalized involvement. Blisters always heal with scarring, and albopapuloid, scar-like lesions on the trunk, prurigo-like nodules on the lower legs, nail dystrophy and involvement of the gastrointestinal tract are also typical features (*5*). The severe generalized recessive DEB (RDEB) is associated with profound disability, and has a profound impact on the quality of life of the patients and their families. Generalized blistering is present at birth and progresses with advancing age leading to poorly healing wounds, extensive scarring and mitten deformities of hands and feet relatively early in life. Oral and gastrointestinal involvement with blistering, scarring and strictures leads to malnutrition, which impairs food intake, and in combination with protein loss through the wounds result in anemia and growth retardation. Aggressive, rapidly metastasizing squamous cell carcinoma is a devastating complication of severe generalized RDEB, with a cumulative risk of 70% by the age of 35 years (*6*). The group of RDEB-other comprises a wide range of clinical manifestations extending from localized erosions and blistering to generalized blistering and scarring of the skin and mucous membranes. Scarring, mucosal involvement, dental and nail dystrophy, or loss of nails, are common; however, mitten deformities of hands and feet do not develop (*5*).

KS is considered as a distinct EB subtype because, in contrast to the other major types above, there are typically multiple cleavage planes—intraepidermal, junctional, or sub-lamina densa (*1*). Mutations in the *FERMT1* gene, which encodes kindlin-1, a keratinocyte protein in the focal adhesion complexes, cause KS. It is characterized by congenital skin blistering and mild photosensitivity, which improve with age, and a progressive generalized poikiloderma with extensive skin atrophy (*7*). Palmoplantar keratoderma, nail dystrophy, webbing of the fingers and joint contractures may occur. Oral, ocular, esophageal, intestinal, anal, and urogenital mucous membranes are severely affected in the majority of the patients, leading to impaired quality of life. Squamous cell carcinoma is a relatively common complication of KS in adults (*7*).

Although EB should be included in the differential diagnosis of a newborn with blisters, the clinical features in a newborn are not characteristic for any particular subtype. The differential diagnosis of a newborn must include, apart from all EB types and subtypes, other genetic skin diseases manifesting with skin erosions or blisters. In the postnatal period, infections (herpes simplex, staphylococcal infections, candidiasis) should be excluded, as well as blistering due to maternal autoimmune diseases (transplacental antibodies in herpes gestationis, pemphigus). The differential diagnosis of acquired skin fragility in children and adults is quite broad.

Not until very recently, there has been relatively little progress in developing effective and specific treatments for EB. However, identification of specific mutations in the candidate genes and elucidation of the consequences of such mutations have provided a basis for development of novel therapeutic approaches, taking advantage of the progress in molecular and cell biology in general areas. These approaches consist of gene therapy, protein replacements, or cell-based therapies (*8*). Some of these approaches, such as allogeneic bone marrow transfer and fibroblast therapy, are in early clinical trials for patients with RDEB (*9,10*).

TABLE 146-1 Classification of EB, Causative Genes, Affected Proteins and Mode of Inheritance

Major EB types	Major EB Subtypes	Gene, MIM	Protein	Inheritance
EBS	Suprabasal	*DSP, 125647*	desmoplakin	AR
		PKP1, 601975	plakophilin 1	AR
		JUP, 173325	plakoglobin	AR
	Basal	*KRT5, 148040*	keratin 5	AD
		KRT14, 148066	keratin 14	AD, AR
		PLEC, 601282	plectin	AR, AD
		DST, 113810	BPAG1	AR
		ITGB4, 147557	α6β4 integrin	AR
		COL17A1, 113811	collagen XVII	AR
JEB	Herlitz	*LAMA3, 600805* *LAMB3, 150310* *LAMC2, 150292*	laminin-332	AR
	Non-Herlitz (also called JEB-other)	*LAMA3, LAMB3, LAMC2*	laminin-332	AR
		COL17A1, 113811	collagen XVII	AR
		ITGA6, 147556 *ITGB4, 147557*	α6β4 integrin	AR
	With renal and respiratory involvement	*ITGB3*	integrin α3 subunit	AR
DEB	Dominant	*COL7A1,120120*	collagen VII	AD
	Recessive			AR
KS	–	*FERMT1, 607900*	kindlin-1	AR

AR, autosomal recessive; AD, autosomal dominant.

TABLE 146-2 Main Complications and Extracutaneous Manifestations in EB

Symptom	Mainly seen in	Comments
Cutaneous complications		
• Infections	All	Sepsis in JEB and lethal congenital EB
• EB nevi	All	
• Webbing of fingers and toes, pseudosyndactylies	DEB, KS	In KS only partial webbing, but pseudoainhum possible; loss of dermatoglyphics
• Skin cancer at a later age	DEB, KS	In DEB after the age of 20, in KS after the age of 40 years
Ocular	JEB, DEB, KS	Can cause severe pain
• Blisters		
• Blepharoconjunctivitis		
• Corneal erosions		
• Exposure keratitis		
Upper respiratory airways		
• Blisters, erosions, scarring of the oropharynx	JEB, DEB	
• Hoarse cry or voice	JEB	
Gastrointestinal tract		
• Oral blistering and erosions	All	May impede food intake
• Oral scarring, microstoma	DEB, KS	May impede food intake, speech and dental care
• Dysphagia	DEB, KS	Impedes swallowing
• Esophageal stenosis	DEB, KS	Impedes swallowing
• Pyloric atresia	EBS and JEB w. PA	
• Colitis	KS	Rare
• Constipation	JEB, DEB, KS	
Genitourinary tract		
• Urethral strictures	JEB, DEB, KS	
Dental		
• Enamel defects	JEB	
• Increased prevalence of dental caries	JEB, DEB	
Musculoskeletal system		
• Mitten deformities of hands and feet	DEB sev. gen.	
• Joint contractures	DEB	
• Osteopenia and osteoporosis	DEB	
• Muscular dystrophy	EBS w. MD	Later-onset muscular dystrophy
Cardiomyopathy	DEB sev. gen.	
Anemia	JEB, DEB	

JEB, a junctional EB; DEB, dystrophic EB; KS, Kindler syndrome; JEB w. PA, junctional EB with pyloric atresia; EBS w. MD, EB simplex with muscular dystrophy; DEB sev. gen., dystrophic EB severe generalized.

TABLE 146-3 Differential Diagnosis of EB: Other Genetic Skin Diseases

Disease	Mutated Genes
Acral peeling skin syndrome[a]	*TGM5*
Keratinopathic ichthyosis[a]	
Epidermolytic ichthyosis	*KRT1, KRT10*
Superficial epidermolytic ichthyosis	*KRT2e*
Pachyonychia congenital	*KRT16, KRT6A, KRT6B, KRT17*
Incontinentia pigmenti[a]	*NEMO*
Hereditary porphyries	*UROD, UROS,PPOX, HMBS, FECH*
AEC syndrome[a]	*p63*
Hailey–Hailey disease	*ATP2C1*
Darier disease	*ATP2A2*

[a]In newborn.

REFERENCES

1. Fine, J. D.; Eady, R. A.; Bauer, E. A.; Bauer, J. W.; Bruckner-Tuderman, L.; Heagerty, A.; Hintner, H.; Hovnanian, A.; Jonkman, M. F.; Leigh, I., et al. The Classification of Inherited Epidermolysis Bullosa (EB): Report of the Third International Consensus Meeting on Diagnosis and Classification of EB. *J. Am. Acad. Dermatol.* **2008**, *58*, 931–950.
2. Sprecher, E. Epidermolysis Bullosa Simplex. *Dermatol. Clin.* **2010**, *28*, 23–32.
3. Pfendner, E.; Rouan, F.; Uitto, J. Progress in Epidermolysis Bullosa: The Phenotypic Spectrum of Plectin Mutations. *Exp. Dermatol.* **2005**, *14*, 241–249.
4. Kiritsi, D.; Kern, J. S.; Schumann, H.; Kohlhase, J.; Has, C.; Bruckner-Tuderman, L. Molecular Mechanisms of Phenotypic Variability in Junctional Epidermolysis Bullosa. *J. Med. Genet.* **2011**, *48*, 450–457.
5. Bruckner-Tuderman, L. Dystrophic Epidermolysis Bullosa: Pathogenesis and Clinical Features. *Dermatol. Clin.* **2010**, *28*, 107–114.
6. Fine, J. D.; Johnson, L. B.; Weiner, M.; Li, K. P.; Suchindran, C. Epidermolysis Bullosa and the Risk of Life-Threatening Cancers: The National EB Registry Experience, 1986–2006. *J. Am. Acad. Dermatol.* **2009**, *60*, 203–211.
7. Has, C.; Castiglia, D.; del Rio, M.; Diez, M.G.; Piccinni, E.; Kiritsi, D.; Kohlhase, J.; Itin, P.; Martin, L.; Fischer, J.; Zambruno, G.; Bruckner-Tuderman, L.; Kindler Syndrome: Extension of Mutational Spectrum and Natural History. *Hum. Mut.* in press.
8. Uitto, J.; McGrath, J. A.; Rodeck, U.; Bruckner-Tuderman, L.; Robinson, E. C. Progress in Epidermolysis Bullosa Rresearch: Toward Treatment and Cure. *J. Invest. Dermatol.* **2010**, *130*, 1778–1784.
9. Wagner, J. E.; Ishida-Yamamoto, A.; McGrath, J. A.; Hordinsky, M.; Keene, D. R.; Woodley, D. T.; Chen, M.; Riddle, M. J.; Osborn, M. J.; Lund, T., et al. Bone Marrow Transplantation for Recessive Dystrophic Epidermolysis Bullosa. *N. Engl. J. Med.* **2010**, *363*, 629–639.
10. Wong, T.; Gammon, L.; Liu, L.; Mellerio, J. E.; Dopping-Hepenstal, P. J.; Pacy, J.; Elia, G.; Jeffery, R.; Leigh, I. M.; Navsaria, H., et al. Potential of Fibroblast Cell Therapy for Recessive Dystrophic Epidermolysis Bullosa. *J. Invest. Dermatol.* **2008**, *128*, 2179–2189.

CHAPTER

147

Ectodermal Dysplasias

Dorothy Katherine Grange

Department of Pediatrics, Washington University School of Medicine, St. Louis, MO, USA

The ectodermal dysplasias are a complex and heterogeneous group of disorders characterized by anomalies of ectodermal structures, including abnormalities of the teeth, hair, nails, sweat glands, or other eccrine glands (*1*). These malformations result from developmental defects in tissues originally derived from the ectoderm of the developing embryo. Two or more systems must be affected in order for a condition to be defined as an ectodermal dysplasia syndrome. The ectodermal dysplasia syndromes may be associated with congenital anomalies in additional organ systems. The mammary gland, thyroid gland, adrenal medulla, anterior pituitary, central nervous system, external ear, cornea, conjunctiva, lacrimal gland and lacrimal duct are other structures that are derived from the embryonic ectoderm. Congenital malformations of teeth, hair, nails, or sweat glands may also occur as single isolated malformations. Developmental disorders involving only one type of structure (teeth, hair, nails, or sweat glands), even if associated with other malformations, have arbitrarily been classified as not constituting an ectodermal dysplasia syndrome.

Almost 200 conditions have been classified as ectodermal dysplasias. All of the ectodermal dysplasias appear to be genetic in etiology, with every mode of Mendelian inheritance encountered among the syndromes. The most common disorder is X-linked hypohidrotic ectodermal dysplasia (XLHED) due to mutations in the gene for ectodysplasin. There are no accurate population-based studies of the prevalence of this group of disorders. Various classification schemes have been proposed over the past 30 years. Friere-Maia and Pinheiro published a monograph in 1984 with later updates that classified the disorders according to clinical features and associated defects (*2–4*). They divided the ectodermal dysplasias into groups, based on which of the basic ectodermal structures of hair, teeth, nails, or eccrine sweat glands were involved. All possible combinations of two or more structures were described, resulting in 11 subgroups. However, the investigators realized that this taxonomic system is arbitrary, without taking into account the pathogenesis or genetics of the specific disorders. Identification of the genes responsible for some of these disorders in humans and other animals, and the understanding of the developmental biology of such structures as hair, skin and teeth have advanced markedly during the past 15 years. These scientific advances have lead to efforts to reclassify the ectodermal dysplasias based on molecular and developmental biology. The ectodermal dysplasias can be grouped according to their molecular basis, including the categories of defects in the NF-κB signaling pathway genes, transcription factors, gap junctions, structural molecules and adhesive molecules (*5,6*). The development of epithelial appendages depends on intricate networks mediating epithelial–mesenchymal interactions during development. Genetic studies of the ectodermal dysplasias have helped define these networks, and have led to the elucidation of developmental pathways. Hopefully, these scientific discoveries will lead to molecular-based therapies, with the possibility of treating or replacing defective and missing teeth, hair follicles, or eccrine sweat glands. In the future, new tissue engineering techniques using adult stem cells may be able to treat hypodontia postnatally. Gene delivery and therapy to existing hair follicles and skin is also being actively explored. Finally, even in utero therapy may be possible in the future to replace deficient proteins or genes needed for normal development. This has recently been done for the XLHED in the *tabby* mouse.

XLHED, also known as anhidrotic ectodermal dysplasia or Christ–Siemens–Touraine syndrome (MIM 305100), is the most common ectodermal dysplasia syndrome. The majority of individuals with hypohidrotic ectodermal dysplasia have the X-linked form. It is characterized by hypotrichosis (sparseness of scalp and body hair), hypohidrosis (reduced ability to sweat), and hypodontia (congenital absence of teeth) and characteristic facial features. Related forms of HED are caused by mutations in the ectodysplasin receptor (EDAR) and other genes in the NF-κB signaling pathway (EDARADD and IKBKG). Mutations in EDAR and EDARADD cause autosomal recessive and autosomal dominant forms of HED which are often indistinguishable from XLHED. Hypomorphic mutations in IKBKG in males, the same gene which is abnormal in females with incontinenetia pigmenti, cause an X-linked recessive form of hypohidrotic ectodermal dysplasia associated with immunodeficiency (*7*).

There are multiple well-recognized ectodermal dysplasia syndromes that are associated with mutations in p63 (*8*). These disorders are a heterogeneous group with some overlapping clinical features, but there are also distinctive features that allow them to be separated into different syndromes. The disorders include ectrodactyly–ectodermal dysplasia–clefting syndrome (MIM 604292), ankyloblepharon–ectodermal defects–cleft lip/palate or Hay–Wells syndrome (MIM 106260), limb–mammary syndrome (MIM 603543), acro-dermato-ungual-lacrimal-tooth syndrome (MIM 103285), Rapp–Hodgkin syndrome (MIM 129400) and split hand–split foot malformation syndrome (MIM 605289). The differences in the phenotypes are linked to the domain of the protein in which the mutation is located.

Hidrotic ectodermal dysplasia, also known as Clouston syndrome, is an autosomal dominant disorder that affects the hair, nails, and skin. It is caused by mutations in the gap junction protein beta-6 gene (*GJB6*) (*9*).

Witkop syndrome is an autosomal dominant ectodermal dysplasia syndrome, also called tooth and nail syndrome. Affected individuals have hypodontia and nail dysplasia, caused by heterozygous mutations in the *MSX1* gene (*10*).

Goltz syndrome, also known as focal dermal hypoplasia, is a rare X-linked dominant disorder associated with defects in tissues of both ectodermal and mesodermal origin. Goltz syndrome is due to mutations or deletions of the *PORCN* gene located at Xp11.23 (*11,12*).

Odonto-onycho-dermal dysplasia syndrome, a form of hypohidrotic ectodermal dysplasia, has been shown to be caused by mutations in the *WNT10A* (wingless-type MMTV integration site family, member 10A) gene (*13*). There is a broad spectrum of clinical features in these patients, typically including significant oligodontia as well as abnormal fingernails and toenails (*14*). Most cases are inherited in an autosomal recessive fashion, associated with homozygous or compound heterozygous *WNT10A* mutations, but up to 50% of heterozygotes may display some clinical features.

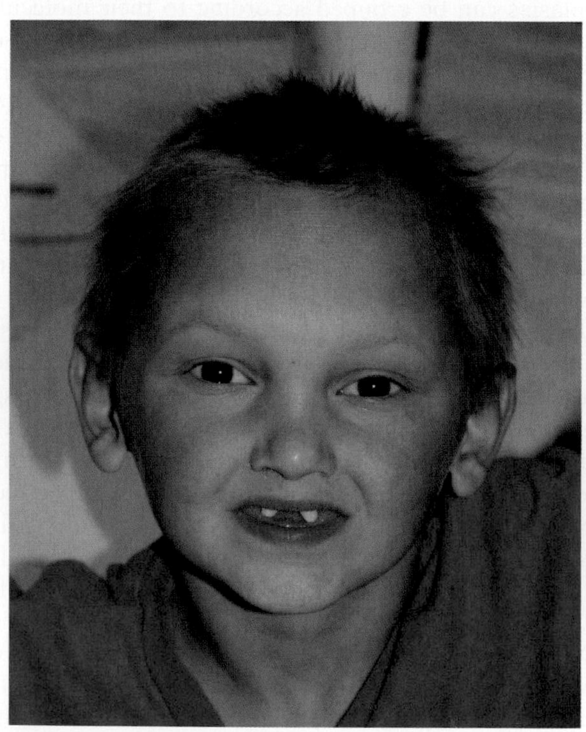

FIGURE 147-1 Boy with XLHED.

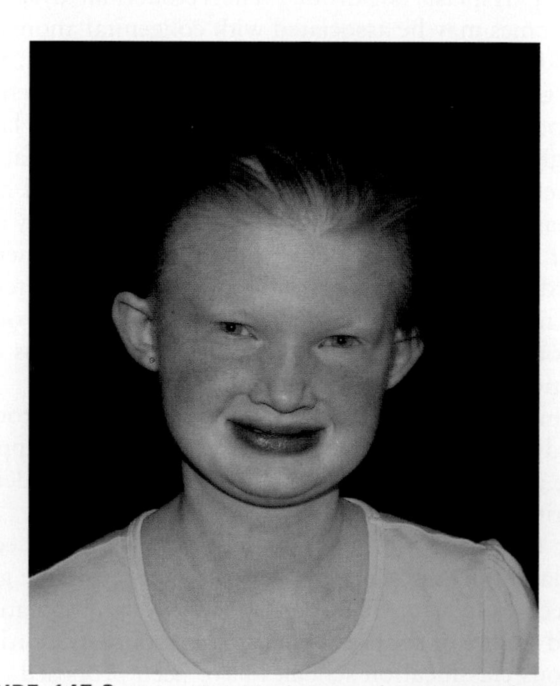

FIGURE 147-2 Girl with HED showing characteristic facies and sparse hair.

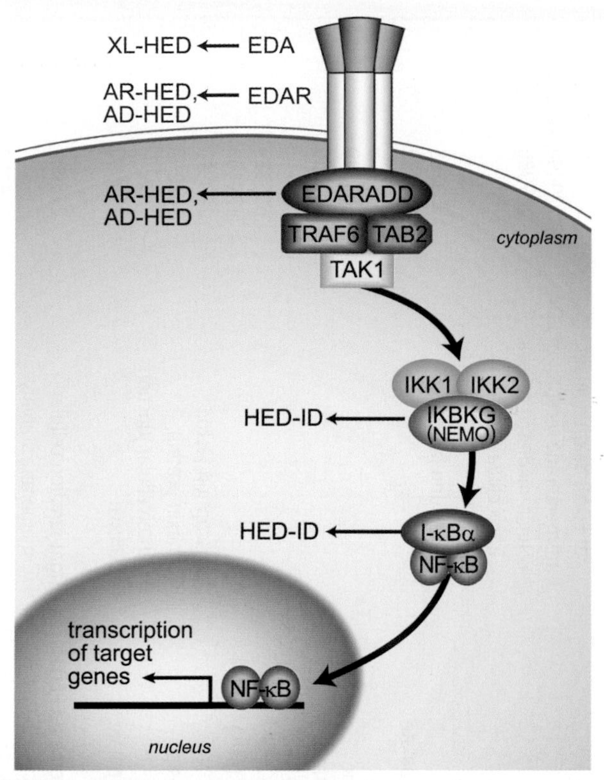

FIGURE 147-3 Ectodysplasin signal transduction pathway. *(Adapted from Mikkola, AJMG, 2009.)*

FIGURE 147-5 Goltz syndrome.

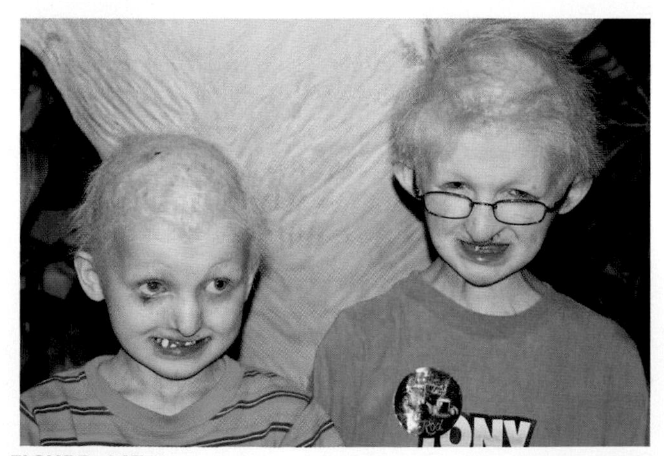

FIGURE 147-4 Boys with AEC syndrome showing the scarring alopecia.

TABLE 147-1 Genes and Chromosomal Regions for Selected Ectodermal Dysplasias

Chromosome	Gene	Protein or Gene Product	Inheritance	Ectodermal Dysplasia (ED)	Protein Function	OMIM Number
Xp11.23	*PORCN*	Five isoforms (PORCA-PORCE)	XLD	Focal dermal hypoplasia (Goltz syndrome)	Membrane targeting and secretion of Wnt proteins necessary for embryonic tissue development	305600
Xq12-q13.1	*EDA1*	Ectodysplasin-A	XLR	XLHED	Triggering ligand molecule	305100
Xq28	*IKBKG (NEMO)*	IKK- γ (NF-kappa-B essential modulator)	XLD	Incontinentia pigmenti 2	NF-κB cytoplasmic inhibitor	308300
			XLR	OL–HED–ID syndrome		300301
			XLR	Hypohidrotic ED with immune deficiency		300291
1q32	*PKP1*	Plakophilin 1	AR	ED/skin fragility syndrome	Desmosomal plaque accessory protein	604536
1q42.2-q43	*EDARADD*	Ectodysplasin-A receptor adapter	AD or AR	ADHED and ARHED	Intracellular molecule adaptor of EDAR death domain	129490, 224900
2q11-q13	*EDAR*	Ectodysplasin-A receptor	AD or AR	ADHED and ARHED	Transmembrane receptor of EDA	129490, 224900
2q35	*WNT10A*	Wingless-type MMTV integration site family, member 10A	AR	Odonto–onycho–dermal	β catenin-mediated specific intracellular signaling	257980
3q27	*TP63*	p63	AD	ADULT syndrome	Transcription factor	103285
			AD	Ectrodactyly, ED, cleft lip/palate syndrome 3 (EEC3)		604292
			AD	Limb- mammary syndrome		603543
			AD	Ankyloblepharon-ectodermal defects-clefting (AEC)		106260
			AD	SHFM4 syndrome		605289
			AD	Rapp-Hodgkin syndrome (RHS)		129400
4p16.1	*MSX1*	Msx1	AD	Witkop syndrome	Transcription factor	189500
6q21-q23.2	*GJA1*	Connexin 43	AD	Oculodentodigital dysplasia (ODDD)	Connexin protein, intercellular junction	164200
7q11.2-q21.3	*EEC1*	Unknown	AD	Ectrodactyly, ED, cleft lip/palate syndrome 1 (EEC1)	Unknown	129900
11q23-q24	*PVRL1*	Nectin 1	AR	Cleft lip/palate–ED syndrome (CLPED1)	Tight junction cellular membrane stability	225060
			AR	Rosselli-Gulienetti syndrome		225000

12q13	*KRT6A*	Keratins 6A and 6B	AD	Pachyonychia congenita 1 and 2	Structural component of hair and nails	167200
	KRT6B				Structural component of hair and nails	167210
12q13	*KRTHBS*	Keratin 85	AD	ED, 'pure' hair-nail type	Structural component of hair and nails	602032
13qll-q12	*GJB2*	Connexin 26	AD	Palmoplantar keratoderma, with deafness	Connexin protein, intercellular junction	148350
			AD	Keratitis–ichthyosis–deafness syndrome, AD (KID, AD)		148210
			AD	Ichthyosis, hystrix-like, with deafness (HID syndrome)		602540
13q12	*GJB6*	Connexin 30	AD	Clouston syndrome	Connexin protein, intercellular junction	129500
14q13	*IKBA*	IκBα	AD	Hypohidrotic ED with immune deficiency	NFκB cytoplasmic inhibitor	164008
16q22.1	*CDH3*	Cadherin·3	AR	ED, ectrodactyly, and macular dystrophy (EEM)	Adhesion molecule cell–cell binding function	225280
17q12-q21	*KRT14*	Keratin 14	AD	Naegeli–Franceschetti–Jadassohn syndrome	Structural component of hair and nails	161000
17q12-q21	*KRT16*	Keratins 16 and 17	AD	Pachyonychia congenita 1 and 2	Structural component of hair and nails	167200
	KRT17				Structural component of hair and nails	167210
17q21.3-q22	*DLX3*	Homeobox protein DLX-3	AD	Tricho–dento–osseous syndrome	Transcription factor	190320
19	*EEC2*	Unknown	AD	Ectrodactyly, ED, cleft lip/palate syndrome 1 (EEC2)	Unknown	

REFERENCES

1. Visinoni, A. F., et al. Ectodermal Dysplasias: Clinical and Molecular Review. *Am. J. Med. Genet. A* **2009,** *149A* (9), 1980–2002.
2. Freire-Maia, N.; Pinheiro, M. Ectodermal Dysplasias–Some Recollections and a Classification. *Birth Defects Orig. Artic. Ser.* **1988,** *24* (2), 3–14.
3. Pinheiro, M.; Freire-Maia, N. Ectodermal Dysplasias: A Clinical Classification and a Causal Review. *Am. J. Med. Genet.* **1994,** *53* (2), 153–162.
4. Freire-Maia, N.; Lisboa-Costa, T.; Pagnan, N. A. Ectodermal Dysplasias: How Many? *Am. J. Med. Genet.* **2001,** *104* (1), 84.
5. Wright, J. T., et al. Classifying Ectodermal Dysplasias: Incorporating the Molecular Basis and Pathways (Workshop II). *Am. J. Med. Genet. A* **2009,** *149A* (9), 2062–2067.
6. DiGiovanna, J. J.; Priolo, M.; Itin, P. Approach Towards a New Classification for Ectodermal Dysplasias: Integration of the Clinical and Molecular Knowledge. *Am. J. Med. Genet. A* **2009,** *149A* (9), 2068–2070.
7. Doffinger, R., et al. X-Linked Anhidrotic Ectodermal Dysplasia with Immunodeficiency Is Caused by Impaired NF-kappaB Signaling. *Nat. Genet.* **2001,** *27* (3), 277–285.
8. Rinne, T., et al. Pattern of p63 Mutations and Their Phenotypes–Update. *Am. J. Med. Genet. A* **2006,** *140* (13), 1396–1406.
9. Baris, H. N., et al. A novel GJB6 Missense Mutation in Hidrotic Ectodermal Dysplasia 2 (Clouston Syndrome) Broadens Its Genotypic Basis. *Br. J. Dermatol.* **2008,** *159* (6), 1373–1376.
10. Jumlongras, D., et al. A Nonsense Mutation in MSX1 Causes Witkop Syndrome. *Am. J. Hum. Genet.* **2001,** *69* (1), 67–74.
11. Wang, X., et al. Mutations in X-Linked PORCN, a Putative Regulator of Wnt Signaling, Cause Focal Dermal Hypoplasia. *Nat. Genet.* **2007,** *39* (7), 836–838.
12. Leoyklang, P., et al. Three Novel Mutations in the *PORCN* Gene Underlying Focal Dermal Hypoplasia. *Clin. Genet.* **2008,** *73* (4), 373–379.
13. Adaimy, L., et al. Mutation in WNT10A Is Associated with an Autosomal Recessive Ectodermal Dysplasia: The Odonto-Onycho-Dermal Dysplasia. *Am. J. Hum. Genet.* **2007,** *81* (4), 821–828.
14. Megarbane, H., et al. Further Delineation of the Odonto-Onycho-Dermal Dysplasia Syndrome. *Am. J. Med. Genet. A* **2004,** *129A* (2), 193–197.

CHAPTER 148

Skin Cancer

Julia A Newton Bishop and Rosalyn Jewell

Section of Epidemiology and Biostatistics, Leeds Institute of Molecular Medicine, University of Leeds, Leeds, UK

Skin cancer can arise from all cell types within the skin and is becoming increasingly common (*1*). Rare individuals and families exist in which there is an increased risk of skin cancer due to inherited highly penetrant genes. However, most increased susceptibility results from a combination of inherited less penetrant genes and environmental exposures.

Basal cell carcinomas (BCCs) arise from the epidermis and are the commonest form of cutaneous malignancy. The main etiologic factor is sun exposure (*2*) and tumors develop as slow-growing nodules mainly on the central part of the face. BCCs are usually sporadic although there is genetic variation between individuals in terms of susceptibility. The strongest determinant of risk is skin color as BCCs are seen predominantly in white skinned peoples. Genome-wide association and candidate gene studies have identified pigment genes as BCC susceptibility genes. These include the melanocortin 1 receptor gene (*MC1R*), agouti signaling protein locus (*ASIP*) and the oculocutaneous albinism A2 gene (*OCA2*). There is also evidence that nonpigment genes are also associated with risk, such as a gene coding for keratin 5 and *TERT* (*3*). There are two inherited syndromes in which there is a marked propensity to BCC: the nevoid basal cell carcinoma syndrome (Gorlin–Goltz syndrome) and the Bazex–Dupre–Christol syndrome.

The nevoid basal cell carcinoma syndrome is inherited as an autosomal dominant trait with nearly complete penetrance and variable expressivity. Affected individuals have a combination of congenital skeletal defects plus a tendency to develop tumors, including multiple basal cell carcinomas, ovarian fibromas, and medulloblastomas. The commonest first symptomatic presentation is of jaw cysts (keratocysts). The skin signs are of small pits in the skin of the palms and soles (*4*). Basal cell nevi start to arise around puberty. Only a small proportion of these lesions evolve into BCC, generally beginning in the second decade. Mutations in the *PTCH* gene at 9p22.3 have been shown to be the cause of the nevoid basal cell carcinoma syndrome (*5*). Somatic mutations in genes in the same pathway (*SMO* and *Gli 1*) have also been linked to the development of familial and sporadic basal cell carcinomas. Diagnosis is made on the basis of clinical criteria and mutation analysis can be offered. In a mutation positive patient, sun avoidance is recommended. Basal cell nevi are observed in patients with this syndrome but basal call carcinomas are removed.

Bazex–Dupre–Christol syndrome is a very rare disorder of the hair follicle associated with early development of BCC (in childhood to middle age). Affected children are born with hypotrichosis (thin and abnormal hair) and milia. Subsequently, deep and lax hair follicles become obvious particularly on the dorsae of the hands (follicular atrophoderma). Inheritance of Bazex–Dupre–Christol syndrome is by the X-linked dominant route and the gene has been mapped to Xq 24–27.1 (*6*).

Squamous cell carcinoma (SCC) is a malignancy derived from the epidermal keratinocyte. Most are postulated to result from excessive chronic sun exposure (*7*) and the clinical appearance of the classic SCC is that of a nonhealing nodule on any body site. Fair skinned people are more susceptible, this phenotype is in turn controlled by the pigmentation genes (for example, *OCA-2* and *MC1R*) and these are the most important low penetrance susceptibility genes (*8*). In terms of somatic genetic events, there is clear evidence of mutation of the *p53* gene being present in nearly half of the tumors (*9*). Other somatic changes such as abnormalities of chromosomes 3p, 9p, 13q, 17p and 17q and loss of the 9p tumor suppressors at the *CDKN2A* locus have been observed. There are in addition a number of rare hereditary conditions predisposing to SCC of the skin such as xeroderma pigmentosum, self-healing epitheliomas of Ferguson Smith, epidermodysplasia verruciformis, dystrophic epidermolysis bullosa (Hallopeau–Siemens type), and porokeratoses.

Melanoma is responsible for the great majority of skin cancer deaths. It is predominantly a cancer of fair skinned people and higher incidence where such fair skinned people live in hotter countries indicates that sun exposure is the main cause (*10*). Melanoma commonly arises from melanocytic nevi, or moles, but it may arise from normal appearing skin. The most common genetic determinants of susceptibility to melanoma are the skin pigmentation genes. Genes associated with skin pigmentation such as *MC1R* are established as common low penetrance melanoma susceptibility genes (*11*). Melanoma can also occur in genetically determined family cancer syndromes such as Li Fraumeni and familial retinoblastoma. Approximately 5% of melanoma patients have a clear history of melanoma in a first-degree relative; 40% of families with three or more cases of melanoma carry causal mutations in the *CDKN2A* gene at 9p21 which codes for a cell cycle control protein, p16 (*12*). In some geographical areas of the world, inherited *CDKN2A* mutations also increase the risk of pancreatic cancer. Individuals with multiple (three or more) primary melanomas and those with one melanoma and/or family history of two or more cases of melanoma or pancreatic cancer may carry a *CDKN2A* mutation and benefit from genetic counseling and potentially genetic testing.

REFERENCES

1. Diepgen, T. L.; Mahler, V. The Epidemiology of Skin Cancer. *Br. J. Dermatol.* **2002,** *146* (Suppl 61), 1–6.
2. Steding-Jessen, M.; Birch-Johansen, F.; Jensen, A.; Schuz, J.; Kjaer, S. K.; Dalton, S. O. Socioeconomic Status and Non-Melanoma Skin Cancer: A Nationwide Cohort Study of Incidence and Survival in Denmark. *Cancer Epidemiol.* **2010,** *34* (6), 689–695.
3. Gerstenblith, M. R.; Shi, J.; Landi, M. T. Genome-Wide Association Studies of Pigmentation and Skin Cancer: A Review and Meta-Analysis. *Pigment Cell Melanoma Res.* **2010,** *23* (5), 587–606.
4. Kimonis, V. E.; Goldstein, A. M.; Pastakia, B.; Yang, M. L.; Kase, R.; DiGiovanna, J. J.; Bale, A. E.; Bale, S. J. Clinical Manifestations in 105 Persons with Nevoid Basal Cell Carcinoma Syndrome. *Am. J. Med. Genet.* **1997,** *69* (3), 299–308.
5. Pan, S.; Dong, Q.; Sun, L. S.; Li, T. J. Mechanisms of Inactivation of *PTCH1* Gene in Nevoid Basal Cell Carcinoma Syndrome: Modification of the Two-Hit Hypothesis. *Clin. Cancer Res.* **2010,** *16* (2), 442–450.
6. Parren, L. J.; Abuzahra, F.; Wagenvoort, T.; Koene, F.; Van Steensel, M. A.; Steijlen, P. M.; Van Geel, M.; Frank, J. Linkage Refinement of Bazex-Dupre-Christol Syndrome to an 11.4-Mb Interval on Chromosome Xq25-27.1. *Br. J. Dermatol.* **2011,** *165* (1), 201–203.
7. Hemminki, K.; Zhang, H.; Czene, K. Time Trends and Familial Risks in Squamous Cell Carcinoma of the Skin. *Arch. Dermatol.* **2003,** *139* (7), 885–889.
8. Duffy, D. L.; Box, N. F.; Chen, W.; Palmer, J. S.; Montgomery, G. W.; James, M. R.; Hayward, N. K.; Martin, N. G.; Sturm, R. A. Interactive Effects of MC1R and OCA2 on Melanoma Risk Phenotypes. *Hum. Mol. Genet.* **2004,** *13* (4), 447–461.
9. Bolshakov, S.; Walker, C. M.; Strom, S. S.; Selvan, M. S.; Clayman, G. L.; El-Naggar, A.; Lippman, S. M.; Kripke, M. L.; Ananthaswamy, H. N. p53 Mutations in Human Aggressive and Nonaggressive Basal and Squamous Cell Carcinomas. *Clin. Cancer Res.* **2003,** *9* (1), 228–234.
10. Chang, Y. M.; Barrett, J. H.; Bishop, D. T.; Armstrong, B. K.; Bataille, V.; Bergman, W.; Berwick, M.; Bracci, P. M.; Elwood, J. M.; Ernstoff, M. S., et al. Sun Exposure and Melanoma Risk at Different Latitudes: A Pooled Analysis of 5700 Cases and 7216 Controls. *Int. J. Epidemiol.* **2009,** *38* (3), 814–830.
11. Bishop, D. T.; Demenais, F.; Iles, M. M.; Harland, M.; Taylor, J. C.; Corda, E.; Randerson-Moor, J.; Aitken, J. F.; Avril, M. F.; Azizi, E., et al. Genome-Wide Association Study Identifies Three Loci Associated with Melanoma Risk. *Nat. Genet.* **2009,** *41* (8), 920–925.
12. Goldstein, A.; Chan, M.; Harland, M.; Gillanders, E.; Hayward, N.; Avril, M. -F.; Azizi, E.; Bianchi-Scarra, G.; Bishop, D.; Bressac de Paillerets, B., et al. The Lund Melanoma Study Group; (GenoMEL), t. M. G. C. High-Risk Melanoma Susceptibility Genes and Pancreatic Cancer, Neural System Tumours, and Uveal Melanomas Across GenoMEL. *Cancer Res.* **2006,** *66,* 9818–9828.

CHAPTER

149

Psoriasis

Johann E Gudjonsson and James T Elder

University of Michigan Health System, Ann Arbor, MI, USA

Psoriasis is a chronic inflammatory skin disease that affects around 2% of the Caucasian population. It is clinically characterized by erythematous scaly plaques which are sharply demarcated from normal skin and predominantly located on the scalp and extensor surfaces of the extremities, although in some cases it can involve the entire skin surface. The majority of cases arise in individuals younger than 30 years of age with more than 10,000 cases per year arising in children less than 10 years old. The cutaneous manifestations of psoriasis are visibly obvious, with a strong negative impact on quality of life. Up to 40% of patients develop an associated arthritis, which can be severe and deforming. The disease has a strong genetic component and tends to have a fluctuating course that can sometimes be linked to environmental factors.

Psoriasis has been appreciated as a genetic disease for nearly 100 years. However, only about one-third of patients have an affected first-degree relative. The strongest evidence for the genetic basis of psoriasis was previously provided from twin-studies with concordance in monozygotic twins ranging from 35% to 73% in various studies, which is amongst the highest of all inflammatory diseases.

Multiple environmental factors have been implicated in the pathogenesis of psoriasis and the impact of these factors is demonstrated by the less-than-perfect disease concordance in monozygotic twins. The specific roles of the various environmental factors are likely to differ with some being important in triggering the disease while others exacerbate or modify the disease. Streptococcal throat infections are the best characterized environmental factor in psoriasis and have been shown to both trigger the disease and exacerbate preexisting chronic plaque psoriasis. Interestingly, primary skin infections by streptococci generally do not trigger or exacerbate psoriasis.

Psoriasis, particularly severe disease, has been shown to be associated with an increased risk of myocardial infarction and death consistent with what has been observed in other systemic inflammatory diseases, suggesting that a state of systemic inflammation, perhaps promoted by shared genetic susceptibility variants, may predispose to atherosclerosis in these disorders.

In recent years, considerable progress has been made in unraveling the genetic complexity of psoriasis through the use of genome-wide association studies. Currently, over twenty genetic susceptibility loci have been identified and while in most cases the true causal variants responsible for these association signals remain to be identified, it is intriguing that many of these loci map close to genes that fall into a few distinctive biological pathways. These include antigen processing and presentation, the Th1 and Th17 inflammatory axes, TNF-α and NF-κB signaling, dendritic cell and macrophage activation, and epidermal differentiation and innate defense.

Multiple lines of evidence suggest that psoriasis is a T-cell mediated disease driven at least in part by a positive feedback loop from activated T-cells to antigen-presenting cells with contributions from innate immune mechanisms involving the epidermis, macrophages vascular endothelium and possibly mast cells. Th1 IFN-γ secreting cells have for long thought to play a major role, but recently the focus has shifted toward a novel subset of CD4+T-lymphocytes producing IL-17 (Th17 cells) and/or IL-22 (Th22 cells). These cells are increased in psoriatic lesions and their maintenance and expansion is dependent on IL-23 which is mainly derived from dendritic cells, macrophages. The importance of these cells in the pathogenesis of psoriasis is supported by the rapid decrease in Th17 responses seen with effective anti-TNF treatment. In psoriatic skin there is a distinct compartmentalization of T-cells, with CD4+T-cell predominating in the upper dermis while CD8+T-cells mostly localize to the epidermis. Importantly, entry of T-cells into the epidermis appears to be a crucial event in the development of psoriatic lesions. CD8+T-cells have similar functional division as CD4+T-cells, with the majority producing TNF-α along with IFN-γ (Tc1), IL-17 (Tc17) and/or IL-22 (Tc22), all of which are found in increased numbers in the psoriatic epidermis. These subsets have similar cytokine requirements as the T-helper cells with the Tc17 being similarly being dependent on IL-23 for maintenance and expansion. Although CD8+T-cells are thought to be primarily cytotoxic in nature, the injury might be sublethal as psoriatic keratinocytes are relatively resistant to apoptotic damage. Thus, instead of causing cytotoxic death, these cells may promote the hyperplastic and proinflammatory psoriatic epidermal response through the release of cytokines. Some CD8+T-cells in psoriasis are clonally expanded, suggesting that they may be responding to a limited set of antigens in the context of HLA Class I molecules, notably HLA-Cw6. The identification of HLA-Cw6, *ERAP1* and *ZAP70* as psoriasis susceptibility loci implicates antigen processing and presentation by CD8+T-cells as a central element in psoriasis pathogenesis.

Other cell types contribute to the inflammatory network in psoriasis including keratinocytes, which produce many of the effects of innate defense known to be highly overexpressed in psoriasis, including beta-defensins and various Th1 and Th17 chemokines. Macrophages and dendritic cells are important cells of the innate immune system and are capable of antigen presentation and activating T-cell during stimulation, particularly after IFN-γ priming as is observed with these cells in psoriasis. These cells co-express inducible nitric oxide synthase (iNOS) and contribute to the inflammatory network in psoriasis through production and release of TNF-α and IL-23. Beyond the cellular players described above, mast cells, neutrophils, and endothelial cells have all been suggested to play a role in the inflammatory network. Based on recent genetic findings, it is now evident that genetic defects in several of these cellular compartments play a role in the pathogenesis of psoriasis. Thus, genes involved in the IL-23/IL-17 signaling such as *IL12B*, *IL23A*, *IL23R*, *TYK2* and *TRAF3IP2* are likely to have a major role in amplifying the effect of the Th17/Tc17 inflammatory responses. The NF-κB regulatory variants *TNIP1*, *TNFAIP3*, *FBXL19*, *NFKBIA*, and *REL* may play active roles downstream of cytokines such as TNF-α and IL-17. Interferon signaling, as endorsed by *IFIH1*, and Th2 balance, through the *IL4/IL13* locus, appear to be important for both triggering and maintaining the disease, while macrophages and dendritic cells may amplify the inflammatory network through the release of nitric oxide (*NOS2*). As a skin disease, it is not surprising to find risk genes such as *LCE3B* and *LCE3C* that are involved in epidermal barrier formation.

Although it is foreseeable that this blueprint is going to become even more comprehensive with the elucidation of additional causative variants, we are now finally able to begin to understand how such a distinctive pattern of cutaneous inflammation develops in this fascinating yet enigmatic disease.

CHAPTER

150

Cutaneous Hamartoneoplastic Disorders

Katherine L Nathanson

Perelman School of Medicine at the University of Pennsylvania, Philadelphia, PA, USA

Cutaneous hamartoneoplastic disorders encompasses a wide variety of syndromes, among which are neurofibromatosis types 1 (NF1) and 2 (NF2), schwannomatosis, tuberous sclerosis complex, nevoid basal cell carcinoma syndrome or Gorlin syndrome, PTEN hamartomatous syndrome (encompassing Cowden disease and Bannayan–Riley–Ruvalcaba syndrome), hereditary leiomyomatosis and renal cancer (HLRCC), and Birt–Hogg–Dubé syndrome (BHD). These syndromes are addressed in detail in other chapters, except HLRCC and BHD, which are covered in detail below.

150.1 HEREDITARY LEIOMYOMATOSIS AND RENAL CANCER

Hereditary leiomyomatosis and renal cancer (HLRCC; OMIM 150800) is a syndrome characterized by the development of cutaneous and uterine leiomyomas and renal cancer, papillary type 2. Mutations in fumarate hydratase (*FH*) were identified as the cause of this autosomal dominant disorder through linkage studies. Mutations in *FH* are generally missense, and lead to a decrease of 50% or more in FH activity. The decrease in function leads to a "pseudohypoxic" state, such as that associated with other cancer susceptibility syndromes, including von Hippel–Lindau disease and susceptibility to pheochromocytoma and paraganglioma due to mutations in the *SDHx* genes. Recent studies also have implicated loss of FH as leading to activation of Nrf2-dependent activation of antioxidant pathways.

The major phenotypic manifestations of HLRCC are renal cancer, papillary type 2, and cutaneous and uterine leiomyomas. The frequency of these three findings can vary between individuals. The renal cancers in patients with HLRCC have a characteristic pathological appearance with large nuclei with inclusion-like orangiophilic or eosinophilic nucleoi surrounded by a clear halo. They tend to be very aggressive renal cancers, with a high rate of metastatic disease, which can be very early onset. The frequency of renal cancer varies depending on the patient series from 10% to 65% of families, with bias of ascertainment in that families ascertained on the basis of having renal cancer have higher rates than those ascertained due to cutaneous leiomyomas. Uterine fibromas also are early onset, and severe, frequently leading to early hysterectomy. They are described in 75–98% women with HLRCC. The cutaneous leiomyomas, described as pink–purple nodules, are seen in almost 100% of patients, tend to present in the mid-twenties. Patients can have from less than five to over 100; they also range in terms of pain from being not noticeable to very painful.

No standard screening guidelines exist for HLRCC. However, given the early onset and aggressive renal cancer, annual screening with CT or MRI is generally recommended starting at age 15. Unlike other hereditary renal cancers, removal of a renal cancer upon identification is recommended. Cutaneous and uterine leiomyomas are managed symptomatically.

150.2 BIRT HOGG DUBÉ SYNDROME

BHD is an autosomal dominant syndrome characterized by the development of fibrofolliculomas, renal cancer, predominantly hybrid oncocytic tumors, and lung cysts and pneumothoraxes. BHD is caused by mutations in *FLCN* (folliculin), which are mainly truncating. The function of folliculin has been difficult to elucidate, due to the fact it has no homology with other known proteins, but recent data have implicated it associated with TGFβ signaling. Similarly to HLRCC, while with a family, all the major phenotypic manifestations of BHD are seen, individuals can only have one or two components of the phenotype. *FLCN* mutations have been identified in patients with multiple lung cysts, and no family history, and families with multiple case of isolated pneumothorax or renal cancer without other findings of BHD.

Fibrofolliculomas, dysplastic growths of the hair follicle, most commonly appear on the cheeks as white domed lesions, but can extend around the neck and upper back. They do not have implicated beyond cosmetic.

Renal cancers are generally benign of an unusual hybrid oncocytic/chromophobe type, but other types of renal cancers have been observed, as well as metastatic disease. Again like HLRCC, the rate of renal cancer varies depending on the type of ascertainment and can range from 16% to 35%. Lung cysts are almost universally present. Pneumothoraces cause the highest morbidity for these patients; they are estimated to have a 50-fold increased relative risk.

There are no standard screening recommendations for BHD. Based on screening recommendations for other hereditary renal cancers, annual to biannual screening of the kidneys by MRI starting in the early twenties, given the earliest age of cancer diagnosis in patients with BHD, seems reasonable. Resection is not recommended until the tumor reaches 3 cm. Baseline screening of the lungs by chest CT should be done, with follow-up tailored to initial findings.

CHAPTER

151

Inherited Disorders of the Hair

Mazen Kurban

Department of Dermatology, Columbia University, NY, USA

Angela M Christiano

Department of Dermatology and Department of Genetics and Development, Columbia University, NY, USA

Recent advances in molecular genetics have enabled the identification of many genes and pathways that are involved in HF morphogenesis and cycling. Furthermore, mutations in some of these genes were shown to be associated with hereditary hair diseases in humans. Identification of causative genes for hair diseases has provided a better understanding of the crucial roles of these genes in HF morphogenesis, development, and hair growth in humans. Mutations in some of these genes have been shown to underlie hereditary hair disorders, such as atrichia with papular lesions, T-cell immunodeficiency, congenital alopecia, and nail dystrophy, localized autosomal-recessive hypotrichosis, autosomal-dominant hypotrichosis simplex of the scalp, autosomal-dominant hypotrichosis simplex, autosomal-recessive woolly hair, autosomal-dominant woolly hair, and hypotrichosis with recurrent skin vesicles which are caused by mutations in hairless, *FOXN1*, *DSG4*, *CDSN*, *APCDD1*, *P2RY5/LPAR6 and LIPH*, *Keratin 74* and *DSC3* genes, respectively. In addition, it has been shown that mutations in *CDH3* gene underlie both hypotrichosis with juvenile macular dystrophy and ectodermal dysplasia, ectrodactyly, and macular dystrophy. On the other hand, the genetic basis of hypertrichosis is largely unknown. To date, several candidate regions for congenital hypertrichosis have been identified. For example, X-linked generalized CH has previously been mapped to chromosome Xq24-q27.1. However, causative genes for CH of any type have not yet been identified. Nevertheless, recent studies have gradually disclosed the molecular mechanisms responsible for congenital hypertrichosis. Over the past two decades, there have been several reports on hypertrichosis in association with chromosomal rearrangements and copy number variations (CNVs). Although the gene(s) involved in the pathophysiology of this disorder have not been identified, molecular characterization of patients with hypertrichosis in association with chromosomal abnormalities is likely to prove useful in understanding the pathogenesis of this disorder, and potentially lead to the identification of causal genes. Recently, it has been shown in three Chinese families and one isolated Chinese patient that CNVs including both deletions and duplications in the region of ch.17q24.2-24.3 is associated with hypertrichosis with and without gingival hyperplasia. Several reciprocal translocations have been associated with hypertrichosis syndromes; in particular, we have reported several patients with Ambras syndrome in which affected individuals also have associated craniofacial abnormalities and involvement of chromosome 8q22-24. Although no mutations were associated with the breakpoints, we found evidence of a position effect on the *TRPS1* gene as a causative event in Ambras syndrome.

As for complex hair diseases such as alopecia areata and androgenetic alopecia, genome-wide association studies have identified several loci that are implicated in the pathogenesis of such conditions, setting the ground for potential targeting of specific genes or gene products for achieving treatment.

Connective Tissue Disorders

Marfan Syndrome and Related Disorders, 567
Ehlers – Danlos Syndrome, 575
Heritable Diseases Affecting the Elastic Fibers: Cutis Laxa, Pseudoxanthoma Elasticum, and Related Disorders, 579

CHAPTER

152

Marfan Syndrome and Related Disorders

Reed E Pyeritz

Perelman School of Medicine at the University of Pennsylvania, Philadelphia, PA, USA

Several 100 distinct heritable disorders of connective tissue (HDCT) have been described phenotypically. Each is presumed to be due to a mutation in a single gene because of the inheritance pattern in families or evidence that a component of the extracellular matrix (ECM) is abnormal. Mutations could occur in a structural gene for a macromolecule, a gene specifying one of the many posttranslational modifications that components of the ECM undergo, a gene encoding a growth factor or a growth factor receptor, and so forth. The upper bounds of the number of HDCT and the number of constituents of the ECM are unknown but are clearly not identical; numerous clinically distinct disorders are due to different mutations in the same gene. This chapter reviews the composition, structure, function, and pathobiology of one particular class of components of the ECM, the extracellular microfibrils, and the common disorder, the Marfan syndrome (MFS).

152.1 MARFAN SYNDROME (OMIM 154 700)

152.1.1 Prevalence

Based on crude calculations of the size of the catchment area and the number of Marfan patients in the files of Johns Hopkins Hospital, the prevalence was estimated as four to six per 100,000. Since the manifestations of MFS may extend from the limits of normal to the floridly "classic", in which the diagnosis is unquestionable, the actual prevalence of MFS clearly exceeds this estimate.

MFS occurs in all races and major ethnic groups who reside in the United States. Relative prevalences in ethnic groups elsewhere are unknown, but cases of MFS have been reported from around the world.

152.1.2 Marfan Phenotype and Natural History

In 1972, the records of 257 Marfan patients at Johns Hopkins Hospital were examined for life expectancy and causes of death. The study was performed when medical and surgical therapy had virtually no beneficial impact on patient survival. Survival had fallen to 50% for men at age 40 years and for women at age 48 years, a reduction in expected life span of about 30–40% for both sexes. The mean age of death of the 72 deceased patients was 32 years. The immediate cause of death in more than 90% of cases was due to cardiovascular complications. Dissection or rupture of the aorta and chronic aortic regurgitation with congestive heart failure accounted for the vast majority of deaths. MFS can usually be defined on the basis of clinical features and the mode of inheritance, despite discovery of the fundamental defect. Increasingly, clinicians resort to searching for mutations in the gene encoding fibrillin-1 (*FBN1)*.

The skeletal features are due to the combination of overgrowth of tubular bones (arachnodactyly, dolichostenomelia, anterior chest deformity) and ligamentous laxity (joint hypermobility, pes planus, scoliosis).

In the eye, subluxation (partial displacement) and dislocation (complete displacement from the pupil) of the lens (ectopia lentis) occur in a proportion of cases variously estimated at 50–80%, is usually bilateral, and is often not evident at birth. The lens is most commonly displaced superotemporally and the zonules usually remain intact. The range of refractive errors detected in MFS is extremely broad and is not limited to myopia. However, myopia is frequent and may appear early and be severe.

The two most common cardiovascular features of MFS are mitral valve prolapse (MVP) and dilatation of the ascending aorta. The former may result in mitral regurgitation, while the latter may result in aortic regurgitation and predispose to aortic dissection and rupture. The prolapse is often pansystolic, with exaggerated leaflet excursion and thickness, suggesting redundancy of the valvular tissue. Dilation of the mitral annulus is common, and calcification occurs in a minority. MVP may not be clinically or echocardiographically present during infancy but may be noted several years later. The degree of prolapse may worsen with age, and mitral regurgitation may appear and progress hemodynamically in some patients who initially had only prolapse. Even in children, the mitral regurgitation can become severe enough to warrant valve repair or replacement. The diameter of the aortic root (sinuses of Valsalva (SoV)) is usually, but not always, greater than the upper limit of the normal range, even in young children. Aortic dilatation that begins in the SoV rarely involves the annulus and may remain confined there or progress into the proximal ascending aorta after effacement of the sinotubular junction. Except when dissection occurs, dilation rarely progresses as far as the innominate artery. About 90% of acute dissections in MFS begin in the aortic root, just distal to the coronary ostia. Many dissections of the ascending aorta extend, either immediately or in a stuttering pattern over hours or days, to involve the arch and descending aorta, often extending into the iliac vessels. The dissection event is usually sudden, with all the classic signs and symptoms reflected in series of patients with MFS. However, some dissections occur with few or atypical symptoms and if the patient survives the acute event, he or she will be left with a chronic dissection of some portion of the aorta. The elastic properties of the Marfan aorta are distinctly abnormal. The wall is considerably stiffer than expected from an early age; this is associated with heightened systemic pulse wave velocities. MRI and tissue Doppler imaging enable determination of aortic distensibility. The more stiffer the aorta, and the larger the diameter, the more likely that the aortic root will dilate progressively.

Ectasia of the caudal dural sac is a common finding, evident on radiographs of the lumbosacral spine as bony erosions of the neural foramina and anteroposterior scalloping of vertebrae, and on axial CT or MRI scans as a widened neural canal. An extreme manifestation is an intrapelvic meningocele, which may be present as a pelvic mass and confused with an ovarian cyst or tumor. Dural ectasia is positively associated with reports of back pain in MFS.

The predominant abnormality of the skin is the stria atrophica, which is most commonly found over the anterior shoulders, lumbar region, and lateral hips.

Spontaneous pneumothorax occurs in about 5% of Marfan patients.

152.1.3 Genetics

In the vast majority of reported families, segregation of MFS is consistent with autosomal dominant inheritance. Reports of multiple affected sibs with ostensibly normal parents are rare. Based on the pedigree analysis, germinal mosaicism is extremely uncommon, but has been confirmed by molecular studies. While various biases influence who attends a clinic, sporadic cases account for 15–30% of all patients. The average age of the fathers of sporadic cases exceeds by some 7 years that of the fathers in the general population. The extreme interfamilial variability in the Marfan phenotype largely reflects the extensive genetic heterogeneity in mutations of *FBN1*. However, intrafamilial variability can be marked; this undoubtedly explains the claims of "nonpenetrance" that appear in the older literature.

152.1.4 Etiology

Mutations in the gene that encodes the large glycoprotein, fibrillin-1 (*FBN1*) cause classic MFS. Cases resembling MFS, but not meeting the revised Ghent criteria, can be due to mutations in genes for TGF-β receptors. Genotype–phenotype correlations have been few, with the most well established being the missense mutations of the calcium-binding epidermal growth factor (EGF)-like domains in the middle region of the coding sequence that are more likely to cause a severe phenotype evident at birth. An international database of mutations in *FBN1* is available (http://www.umd.be:FBN1; www.biobase-international.com).

152.1.5 Molecular Pathology

The most common mutation is missense, and most affect calcium-binding EGF-like motifs. A frequent class of mutation is substitution for one of the cysteine residues of the EGF-like motifs. There have been few large deletions. Little can be deduced about genotype–phenotype correlations. Mutations near the C-terminus were initially predicted to cause a milder phenotype, even isolated ectopia lentis. However, severe MFS can result from mutations in this region. Severe MFS evident in infancy tends to involve missense mutations in calcium-binding EGF-like motifs encoded by exons 24 through 27 and 31 through 32. However, mutations in the same exons cause much less severe forms of MFS. Mutations that likely result in early degradation of the mutant protein (e.g. premature chain termination mutations) tend to cause a less severe disorder, including MASS (mitral valve, aorta, skin, and skeleton) phenotype.

152.1.6 Pathogenesis

For over a decade from the discovery of mutations in *FBN1* as the cause of MFS, pathogenesis was assumed to be due to a "weakness" in the ECM. However, microfibrils play an important role in regulating the activity of TGF-β, and abnormal or deficient FBN1 appears to lead to increased TGF-β activity during pre- and postnatal development. Most of the features of MFS, with the possible exceptions of ectopia lentis and osteopenia, are due to failure in regulating cytokine signaling. The strongest evidence for the role of altered TGF-β signaling came from the studies of development of the lung and mitral valve in mice bearing MFS mutations. There is a failure of distal alveolar septation, leading to an emphysematous-like pathology. The abnormal lung shows abnormally high levels of active TGF-β; treatment of perinatal MFS mice with and anti-TGF-β antibody prevents the abnormal lung phenotype. Mice deficient in FBN1 show postnatal development of MVP, associated with abnormally increased TGF-β signaling. As in the lung, inhibition of TGF-β activity "rescues" the MVP phenotype. Enhanced TGF-β signaling through both the canonical (Smad-facilitated) and noncanonical pathways occurs.

Serum from MFS mice and humans with MFS contain increased levels of circulating TGF-β.

Additional evidence for the role of excessive TGF-β signaling in Marfan syndrome derives from the therapeutic efficacy of angiotensin receptor blockade in preventing, and even reversing, the cardiovascular phenotype in the mouse model of MFS. Losartan, which is in widespread use as an antihypertensive due to its blockage of the angiotensin II type 1 receptor, also interferes with the action of TGF-β. Prenatal treatment of mice pregnant with MFS normalizes aortic histopathology. Additionally, MFS mice that were are not treated until 2 months of age (adolescence in human terms) have little aortic or lung pathology when sacrificed at 8 months of age. Human trials of angiotensin receptor blockade are in progress.

152.1.7 Diagnosis

Each of the clinical features of MFS occurs with variable frequency in the general population. Occasionally, several will occur together by chance alone. In determining which of these individuals are affected by MFS, some other systemic connective tissue disorder, or no clear syndrome, more diagnostic reliance is placed on the presence of manifestations that are at once uncommon in the population but common in MFS (subluxed lenses, aortic dilation or dissection, dural ectasia) than on soft features (myopia, mitral prolapse, tall stature, joint laxity, and arachnodactyly). The clinical diagnostic criteria last proposed in 2010 by an international committee and referred to as the "Revised Ghent Criteria" are summarized in Table 152-1. In comparison with the original Ghent criteria, more emphasis is placed on aortic dilatation, ectopia lentis and a mutation in *FBN1*. Less emphasis is placed on dural ectasia, mainly because a separate radiologic imaging study is necessary to detect it. The major weakness of the new criteria is the reliance on the "Z-score" to define aortic dilatation with a score greater than +2.0 being required; this represents a measurement greater than two standard deviations above the mean adjusted for age, sex, and body surface area. Without any change in the aortic diameter, the interpretation of whether it is dilated or not in the same patient can vary as the patient grows, or gains or loses weight. Thus, there is a possibility that the diagnosis of MFS can vary for purely trivial reasons. Additionally, very large individuals can have an aortic Z-score <2 despite the actual diameter being larger than 40 mm, which is considered pathologically dilated in the minds of most cardiologists. The first step in establishing the diagnosis should always be a careful medical history, family history, and physical examination. Determination or rejection of the diagnosis of MFS on purely clinical grounds often must await the results of the detailed ophthalmologic examination and echocardiogram. Understanding the molecular basis of MFS has greatly aided diagnosis in specific instances, but *FBN1* mutations cause conditions other than MFS, very often the conditions that clinicians are interested in excluding.

152.1.8 Management

No specific therapy exists for the underlying defect in MFS. Therapeutic efforts are directed at first establishing an accurate diagnosis, determining which problems are present at diagnosis, anticipating the problems that will probably arise in the future, and pursuing certain prophylactic measures for specific problems. These patients should have one physician who is knowledgeable about the syndrome, who may or may not serve as the primary caregiver, but who refers to specialists as the need arises.

The use of adrenergic blockade to delay or prevent severe aortic complications was suggested in 1980s, but convincing evidence of its utility appeared much later. Any patient with classic MFS should be considered for prophylactic therapy. Atenolol is the drug we currently use; it is relatively cardioselective, is less lipophilic (so central nervous system side effects are minimal), and has a longer half-life than propranolol. In adults, we begin treatment at 50 mg in the morning and advance the dose (giving an evening dose as the next step), to keep the resting heart rate at less than 60 bpm and the heart rate during moderate exercise at less than 100 bpm. Because of marked variation in individual responsiveness to all β-adrenergic blocking agents, the dose must be titrated for patients of any age. We begin treatment in children as soon as possible. Circulating levels of TGF-β are increased in people with MFS. Chronic treatment with a β-blocker (or an angiotensin receptor blocker) reduce the serum level of TGF-β. Based on the mouse model of MFS, angiotensin receptor blockade is being compared to β-adrenergic blockade in the on-going human trials.

Prophylactic surgery of the dilated aortic root has had dramatic impact on improving mortality. Various techniques to preserve the native aortic valve while replacing the dilated SoV and proximal ascending aorta have emerged. Results of valve sparing employing the "reimplantation" approach have been quite encouraging. The notion of preserving the aortic valve is appealing for anyone who face lifelong anticoagulation, but particularly very important for young women who wish to attempt pregnancy and also for anyone with a relative contraindication to warfarin. After any repair of the aorta, patients should be maintained on chronic β-adrenergic blockade. Some patients require mitral valve surgery alone, or in addition to aortic surgery. In most cases, repair rather than replacement of the mitral valve is possible, with acceptable long-term outcome.

All women with MFS are at further increased risk of aortic dissection during pregnancy, especially if the aortic root dimension is greater than 40 mm.

152.1.9 Counseling

One major issue in genetic counseling is straightforward, each affected person having a 50% probability of passing the allele to any offspring. When MFS is diagnosed in a young child and the family history is negative, the parents must be carefully examined to ensure that neither has signs of MFS. If both parents are phenotypically normal, considerable reassurance can be given that the risk of future affected offspring is slight. Women affected with MFS must deal with two concerns. The first is the 50% risk that any offspring will inherit the syndrome. The second concern is the risk of cardiovascular problems during pregnancy. Parents should be counseled about the range of intrafamilial variability in MFS. Affected offspring may be more or less severely involved than their parents.

The complex issues in transitioning responsibility of care from parent to affected child should be addressed longitudinally over a period of years.

Support groups for MFS have been established in many countries. The National Marfan Foundation in the United States maintains web-based resources (http://www.marfan.org), and sponsors national meetings for patients and families and periodic international gatherings of investigators. The International Federation of Marfan Syndrome Organizations (http://marfan world.org) also exists.

152.1.10 Life Expectancy

Compared to the natural history study in 1990s, an examination of contemporary clinical history showed a marked improvement of about 25% in life expectancy for men and women. This gain is primarily due to both better survival after cardiovascular surgery and early, prophylactic surgical intervention. Viewed another way, over the past 30 years, life expectancy has increased about 30 years.

152.1.11 Related Conditions

Conditions that share phenotypic features, mutations in *FBN1*, or other involvement of microfibrils are listed in Table 152-2. Conditions that affect the aorta are listed in Table 152-3 along with any relevant animal models.

ACKNOWLEDGMENTS

This revision was accomplished during the time of support by the National Heart, Lung and Blood Institute (GenTAC) and the National Human Genome Research Center (Center of Excellence in ELSI Research P50-HG-004487), both of the US National Institutes of Health, and while in residence at the Brocher Foundation, Hermance, Switzerland. I also thank all of my colleagues who shared their published and unpublished work and commented on sections of the chapter.

TABLE 152-1 Features of Marfan Syndrome in 50 Consecutive Clinic Patients

	No. of Patients Demonstrating Clinical Features
Ocular	35/50
Ectopia lentis	30/50
Myopia	17/50
Cardiovascular	49/50
Mid-systolic click only	15/50
Mid-systolic click and late systolic murmur	9/50
Aortic regurgitant murmur	5/50
Mitral regurgitant murmur only	3/50
Abnormal echocardiogram	48/50
Aortic enlargement	42/50
MVP	29/50
Prosthetic aortic valve	5/50
Musculoskeletal	50/50
Arachnodactyly	44/50
US/LS 2 SD below mean for age	36/47
Pectus deformity	34/50
High, narrow palate	30/50
Height >95th percentile for age	29/50
Hyperextensible joints	28/50
Vertebral column deformity	22/50
Pes planus	22/50
Family history	40/47
Additional documented cases of syndrome	40/47
Sporadic cases (presumed new mutations)	7/47
Unclear or unknown pedigree	3/50

SD, standard deviations: US/LS, upper segment-to-lower segment ratio.

TABLE 152-2 Disorders Related to Marfan Syndrome Through Phenotype, Etiology or Pathogenesis

MASS phenotype (OMIM157700)
Ectopia lentis (OMIM129600; 225100)
Weill–Marchesani syndrome (OMIM277600; 608328)
Stiff skin syndrome (OMIM184900)
Familial tall stature
Familial kyphoscoliosis
Shprintzen–Goldberg syndrome (OMIM182212)
Familial aortic aneurysm and dissection (OMIM132900; 607086; 607087; 611788; 613780)
Congenital contractural arachnodactyly (OMIM121050)
Loeys–Dietz syndrome (OMIM609192)
Arterial tortuosity syndrome (OMIMM208050)
PHACE(S) association (OMIM606519)
Bicuspid aortic valve (OMIM109730)
Familial intracranial aneurysm (OMIM105800)
Familial pneumothorax (OMIM135150)
Lujan–Fryns syndrome (OMIM309520)
Abdominal aortic aneurysm (OMIM100070)

TABLE 152-3 Human Hereditary Aneurysm Conditions and Murine Models of Aneurysm

Gene (Protein)	Human Aneurysmal Syndrome	Animal Model Phenotype	Pathway Implicated
ECM Protein			
FBN1 (fibrillin-1)	Marfan Syndrome—Fully penetrant ascending aortic aneurysm	KO: Perinatal lethality, pulmonary hypoplasia, arteriopathy Hypomorphic: Arteriopathy, aneurysm, dissection	TGFβ
EFEMP2 (fibulin-4)	Cutis Laxa with aneurysm—ascending aortic aneurysm and tortuosity	KO: Ascending aortic aneurysm, defective elastogenesis, perinatal lethality Sm22-Cre (smooth muscle) KO: Ascending aortic aneurysm	TGFβ
ELN (Elastin)	Cutis laxa with aneurysm—low penetrance ascending aortic aneurysm and dissection	Haploinsufficient: Obstructive arterial disease with increased VSM proliferation, increased lamellae number KO: Accentuated phenotype	Unknown
COL1A1 (Collagen 1 alpha-1)	Osteogenesis imperfecta—extremely rare aortic aneurysm Ehlers–Danlos Syndrome, type 7A—dissection of medium size arteries	KO: Adult onset aortic aneurysm and dissection	Collagen metabolism
COL1A2 (Collagen 1 alpha-2)	Osteogenesis imperfecta—extremely rare aortic aneurysm Ehlers–Danlos syndrome—cardiac valvulodystrophy type 7B—borderline aortic root enlargement with aortic regurgitation	Homozygous LOF: Decreased body weight, bony abnormalities, no arterial phenotype reported	Collagen metabolism
COL3A1 (Collagen 3 alpha-1)	Ehlers–Danlos syndrome, type 4—frequent arterial dissection with infrequent aneurysm	KO: Frequent neonatal mortality, aortic rupture, intestinal rupture	Collagen metabolism
COL4A1 (Collagen 4 alpha-1)	Hereditary angiopathy, nephropathy, aneurysms, and muscle cramps—infrequent aneurysms	KO: Embryonic lethal (E10.5–11.5), basement membrane failure	Collagen metabolism
COL4A5 (Collagen 4 alpha-5)	X-linked Alport syndrome—ascending aortic and abdominal aneurysms and dissections	Nonsense mutation: No overt aortic disease noted	Collagen metabolism
LOX1 (lysyl oxidase 1)	No human phenotype described	KO: Low penetrance aortic aneurysm, perinatal lethality	Collagen metabolism TGFβ
PLOD1 (lysyl hydroxylase 1)	Ehlers–Danlos syndrome, type 6—rare aneurysm	KO: Spontaneous aneurysm and dissection, gait abnormalities	Collagen metabolism
PLOD3 (lysyl hydroxlase 3)	Bone fragility with contractures, arterial rupture, and deafness—frequent medium sized arterial aneurysms	KO: Embryonic lethality (E9.5) and basement membrane fracture	Collagen metabolism
Transmembrane Protein			
TGFBR1 (tgfbr1)	Loeys–Dietz syndrome—highly penetrant root and diffuse large and medium arterial aneurysm	KO: Midgestational death with yolk sac defects M318R heterozygous knock in: Aortic root and diffuse aneurysm (D. Loch, unpublished observations)	TGFβ
TGFBR2 (tgfbr2)	Loeys–Dietz syndrome—highly penetrant root and diffuse large and medium arterial aneurysm FTAAD—highly penetrant root and medium arterial aneurysm	KO: Defects in hematopoiesis and vasculogenesis, lethal (E10.5) *TGFBR2* flox: Impaired elastogenesis, decreased lysyl oxidase in aorta G457W heterozygous knock in: Aortic root and diffuse aneurysm (D. Loch, unpublished observations)	TGFβ

TABLE 152-3 Human Hereditary Aneurysm Conditions and Murine Models of Aneurysm—cont'd

Gene (Protein)	Human Aneurysmal Syndrome	Animal Model Phenotype	Pathway Implicated
SLC2A10 (Glucose transporter 10)	Arterial tortuosity syndrome—diffuse arterial tortuosity, stenoses, aneurysms	Homozygote missense: Arterial thickening with increased elastin deposition, elastin fractures at advanced age	TGFβ
NOTCH1 (notch1)	Bicuspid valve with ascending aortic aneurysm	KO: Embryonic lethal (E9.5) required for somite segmentation, defects in angiogenesis.	NOTCH/ Jagged1
JAG1 (jagged1)	Alagille syndrome—intracranial aneurysms, coarctation of the aorta, aortic aneurysm	KO: Embryonic lethal (E9.5) with diffuse hemorrhages	NOTCH/ Jagged1
PKD1 (polycystin-1)	Polycystic kidney disease with intracranial aneurysms	KO: Embryonic lethal (E14.5) with polycystic kidneys Hypomorphic expression: Adult onset aortic aneurysm and dissection	Unknown
PKD2 (polycystin-2)	Polycystic kidney disease with intracranial aneurysms	KO: Defects in cardiac septation and left–right axis determination, kidney and pancreatic cysts	Unknown
GJA1 (connexin 43)	Hypoplastic left heart syndrome	Nonsense (W45X): Coronary artery aneurysms	Unknown
ENG, ACVRL1	Hereditary hemorrhagic telangiectasia—dilated ascending aorta in some adults; arteriovenous malformations common	*Eng* and *Acvrl1* KO: aorta not evaluated	? BMP
Cytoplasmic Proteins			
ACTA2 (α-smooth muscle actin)	Familial aortic aneurysm with levido reticularis and iris flocculi	KO: Viable offspring with normal life span and impaired vascular contractility	IGF-1, Ang2
MYH11 (smooth muscle myosin)	Familial aortic aneurysm with patent ductus arteriosis	KO: Neonatal lethality, urinary retention, dilated cardiomyopathy	IGF-1, Ang2
FLNA (Filamin A)	Periventricular nodular heterotopia with Ehlers–Danlos features—ascending aortic aneurysm and valvular dystrophy	KO: Neonatal lethality, Persistent truncus arteriosis, endothelial cell–cell contact defects	Unknown
NF1 (Neurofibromin 1)	Neurofibromatosis—medium sized arterial aneurysm and stenosis	KO: Enlarged head, pale liver, cardiac malformations	Ras/MEK/ ERK
*PTPN11 (*SH2 domain-containing protein tyrosine phosphatase-2)	Noonan and LEOPARD Syndrome—coronary artery aneurysms and rare ascending aortic aneurysm	KO: Embryos die preimplantation Missense (D61G): Cardiac defects, defective valvulogenesis, skeletal anomalies, myeloproliferative disorder	Ras/MEK/ERK
NPHP3 (Nephrocystin-3)	Nephronophthisis	KO: Low penetrance intracranial aneurysms	Unknown
NOS3 (Nitric Oxide Synthetase 3)	Refractory hypertension	KO: Abnormal aortic development with BAV, in combination with ApoE-/-, mice show abdominal arterial aneurysm and dissections	NO
TSC2 (tuberin)	Tuberous Sclerosis—diffuse thoracoabdominal aneurysms	Heterozygous KO: Increased proliferation of VSMCs upon injury	mTOR/ AKT
GAA (alpha-1,4-glucosidase)	Acid maltase deficiency, adult onset—intracranial aneurysms	KO: Lysosomal accumulation in heart, aorta, skeletal muscle	Unknown
S100A12 (S100A12)	No human phenotype, increased *S100A12* expression in human MYH11 aneurysmal tissues	Sm22α promoter-S100A12 transgenic mouse: VSM disarray, elastin fragmentation, thoracic aneurysm	TGFβ, IL-6

Continued

TABLE 152-3 Human Hereditary Aneurysm Conditions and Murine Models of Aneurysm—cont'd

Gene (Protein)	Human Aneurysmal Syndrome	Animal Model Phenotype	Pathway Implicated
Nuclear Proteins			
MED12 (mediator complex subunit 12)	Lujan–Fryns Syndrome—extremely rare aneurysm	Hypomorphic mutants: Embryonic lethal (E10), defects in neural tube closure, somatogenesis, heart formation	Wnt/β-catenin, Wnt/PCP
KLF15 (Kruppel-like factor 15)	No human phenotype, *Kruppel-like factor 15* downregulated in human abdominal aortic aneurysm	KO: Aortic aneurysm and cardiomyopathy	TSP-1, p53
KLF2 (Kruppel-like factor 2)	No human phenotype	KO: Embryonic aortic aneurysm and dissection	Unknown
Inborn Errors of Metabolism			
GLA (α-galactosiadase A	Fabry disease—dilatation of ascending aorta common	Gla-deficient mice	Unknown
GAA (alpha-1,4-glucosidase)	Pompe disease, adult onset—intracranial aneurysms	KO: Accumulation of glycogen in lysosomes in heart, aorta, skeletal muscle	Unknown
Chromosomal Anomaly			
45 X,0	Turner Syndrome—bicuspid aortic valve, coarctation of the aorta, ascending aneurysm	XO mice demonstrate no phenotypic heart disease	Unknown
Chemical Models			
	No human phenotype	Ang2 infusion model	Ang2, TGFβ, MCP-1, IL-6
	No human phenotype	Elastase infusion model	Unknown
	No human phenotype	Periarterial calcium application	JNK1

Adapted from Lindsay et al. with permission and appreciation.

FURTHER READING

David, T. E.; Armstrong, S.; Maganti, M.; Colman, J.; Bradley, T. J. Long-term Results of Aortic Valve-Sparing Operations in Patients with Marfan Syndrome. *J. Thorac. Cardiovasc. Surg.* **2009**, *138*, 859–864.

Dietz, H. C. Marfan Syndrome. In: *GeneReviews*; Pagon, R. A, Bird, T. D, Dolan, C. R., Stephens, K., Eds.; University of Washington: Seattle. http://www.ncbi.nlm.nih.gov/books/NBK1335/ updated 2009 Jun 30; accessed 2011 Mar 19.

Erkula, G.; Jones, K. B.; Sponseller, P. D., et al. Growth and Maturation in Marfan Syndrome. *Am. J. Med. Genet.* **2002**, *109*, 100–115.

Keane, M. G.; Pyeritz, R. E. Medical Management of Marfan Syndrome. *Circulation* **2008**, *117*, 2802–2813.

Lacro, R. V.; Dietz, H. C.; Wruck, L. M., et al. Rationale and Design of a Randomized Clinical Trial of Beta Blocker Therapy (Atenolol) vs Angiotensin II Receptor Blocker Therapy (Losartan) in Individuals with Marfan Syndrome. *Am. Heart J.* **2007**, *154*, 624–631.

Lindsay, M. E.; Dietz, H. C. Lessons on the Pathogenesis of Aneurysm from Heritable Conditions. *Nature* **2011**, *473*, 308–316.

Loeys, B. L.; Dietz, H. C. Loeys–Dietz Syndrome. In: *Gene Reviews*; Pagon, R. A., Bird, T. D., Dolan, C. R., Stephens, K. Eds.; University of Washington: Seattle. http://www.ncbi.nlm.nih.gov/pubmed/20301312/ updated 2008 Apr 29; accessed 2011 Mar 19.

Milewicz, D. M.; Kwartler, C. S.; Papke, C. L., et al. Genetic Variants Promoting Smooth Muscle Cell Proliferation Can Result in Diffuse and Diverse Vascular Diseases: Evidence for a Hyperplastic Vasculomyopathy. *Genet. Med.* **2010**, *12*, 196–203.

Pyeritz, R. E. Marfan Syndrome: 30 years of Research Equals 30 years of Additional Life Expectancy. *Heart* **2009**, *95*, 173–175.

Ramirez, F.; Dietz, H. C. Extracellular Microfirils in Vertebrate Development and Disease Processes. *J. Biol. Chem.* **2009**, *284*, 14677–14681.

CHAPTER

153

Ehlers–Danlos Syndrome

Peter H Byers

Department of Pathology, University of Washington, Seattle, WA, USA

The Ehlers–Danlos syndromes (EDS) are a group of genetically heterogeneous disorders rather than a "single" entity, in which skin, joint and internal organ fragility are the hallmarks (Table 153-1). Clinical descriptions that emphasized soft doughy skin with easy bruising and unusually broad scars after minor lacerations date from the turn of the last century by Ehlers, a Danish dermatologist and Danlos, a French dermatologist. As other individuals were identified who had features that included easy bruising, acrogeria, and a propensity for vascular tears, they were included under the rubric of Ehlers–Danlos. In a landmark series of papers and a book, Beighton reported clinical descriptions of 100 individuals with different skin, joint and systemic findings and provided a numerical classification of five types of EDS. These included the gravis and mitis forms (EDS type I and II) characterized by soft, doughy, hyperextensible skin, increased bruising, and marked joint hypermobility; hypermobile EDS (EDS type III) with significant joint hypermobility in multiple joints; the ecchymotic for of EDS (EDS type IV) with thin skin, easily visible veins, marked bruising and a propensity to arterial rupture and early death; and fifth form similar to EDS type I but apparently due to mutations on the X-chromosome and including muscle hematomas. At the time, the molecular bases of none of these disorders were known.

In the early 1970s, biochemical studies of collagens began to point to genetic defects in subsets of individuals previously thought to have the more typical forms of EDS. This development created both the clinical tools and then the biochemical or molecular genetic means to distinguish unique forms of EDS. Pinnell and colleagues identified sib with unaffected parents, whose type I collagen had decreased hydroxylation of lysyl residues in the triple helical domain. Analysis of hydroxylated cross-links derived from type I collagen in urine provided a cheap and effective test for this disorder, which became known as EDS type VI, before the isolation of the gene and characterization of bi-allelic mutations in *PLOD1*. Biochemical analysis of type I collagen from cattle with a recessively inherited disorder, dermatosparaxis, characterized by marked skin fragility and joint hypermobility determined that the amino-terminal propeptides of type I procollagen were ineffectively removed and led to defective collagen fibrillogenesis in skin. This discovery, coming at a time of renewed interest in the biosynthesis of collagens, provided some of the first insights into the complex pathway that was needed to produce stable extracellular matrix collagens. A seventh type of EDS was identified through the study of a single individual with bilateral congenital hip dislocation and marked joint hypermobility which was originally thought to be a human model of dermatosparaxis. Instead, she was shown to be heterozygous for a donor splice site mutation in intron 6 of *COL1A2* that encodes the proα2(I) chains of type I procollagen. As a result, exon 6 was skipped, and the proteolytic conversion site and an intermolecular cross-link site were deleted from the mature protein and the amino-terminal propeptide was retained in a subset of type I collagen molecules. It took another 20 years for the autosomal recessive human form of dermatosparaxis to be identified and shown to result from mutations in *ADAMTS2*, which encodes the protease that removes the amino-terminal propeptides of both proα1(I) and proα2(I) chains.

The growing knowledge of the heterogeneity of collagens and rapid recognition of additional genes that encoded fibrillar and nonfibrillar collagens, coupled with the maturation of tissue culture techniques, provided the tools to characterize the collagens produced by dermal fibroblasts and to identify underlying defects. For example, cells from individuals with the ecchymotic form (also known as the Sack–Barabas form or EDS type IV) make either defective or markedly reduced type III procollagen. Initially proposed to be a recessive disorder, both clinical studies and the subsequent characterization of

TABLE 153-1 Types of Ehlers–Danlos Syndrome

Numerical Type	Descriptive Type	Gene	Mode of Inheritance	Clinical Features
I (Gravis) and II (Mitis)	Classical	*COL5A1, COL5A2*	AD	Marked joint hypermobility, skin hyperextensibility, bruising, abnormal scar formation
III	Hypermobile	Most unknown	AD	Marked joint hypermobility, minor skin findings
		TNXB	AD	Rare, mostly women
IV	Vascular	*COL3A1*	AD	Thin translucent skin, marked bruising, small joint hypermobility. High risk for arterial and bowel rupture and uterine rupture during pregnancy
VI	Kyphoscoliotic	*PLOD1*	AR	Joint hypermobility, kyphoscoliosis that is recalcitrant to surgical intervention, risk for arterial rupture
VIIA and B	Arthrochalasis	*COL1A1, COL1A2*	AD	Very marked joint hypermobility, bilateral congenital hip dislocation that is difficult to repair
VIIC	Dermatosparaxis	*ADAMTS2*	AR	Soft, very fragile skin, blue sclerae, joint hypermobility, late onset of skin redundancy
VIII	Periodontal	Chromosome 12p3	AD	Periodontal loss, soft skin, joint hypermobility, characteristic plaque on anterior shin
	Classical	*COL1A1*	AD	Marked joint hypermobility, skin hyperextensibility, bruising, abnormal scar formation, aortic aneurysm formation
	Tenascin X type	*TNXB*	AR	Joint hypermobility, sleeve like character to skin which is hyperextensible, bruising, scarring is normal
	Cardiac valvular	*COL1A2* null mutations	AR	Joint hypermobility, skin hyperextensibility, cardiac valvular insufficiency
	Periventricular heterotopias with EDS	*FLMNA*	XLR	Significant joint hypermobility with periventricular heterotopias
	Spondylocheiro dysplastic	*SLC39A13*	AR	Hyperelastic, thin, and bruisable skin, hypermobility of the small joints with a tendency to contractures, protuberant eyes with bluish sclerae, hands with finely wrinkled palms. Spondyloepiphyseal dysplasia with mild short stature

heterozygous mutations in *COL3A1* confirmed its autosomal dominant character.

Further clinical studies distinguished individuals with joint hypermobility, a unique plaque-like darkened pretibial region, and early onset periodontal loss that appeared to be noninflammatory. Called the periodontal form of EDS (EDS type VIII), the causative locus in some families was mapped to 12p, but the causative gene awaits identification.

The profusion of single families that shared some of the features of different forms of EDS led to an attempt in the late 1990s to bring order into the clinical classification system that resulted in "renaming" with "descriptive" rather than numerical names. This classification, developed in 1998, is now showing its age as additional conditions that share at least some of the hallmark features of different forms of EDS and their mutant genes are identified. The use of a descriptive nomenclature has de-emphasized some features of the different forms of EDS (e.g. the use of "vascular EDS" in place of EDS type IV may draw attention away from other presentations in this group, that include bowel rupture, rupture of the gravid uterus, and spontaneous pneumothorax) and potentially delay clinical recognition.

Most of the genes in which mutations that lead to forms of EDS encode proteins that become part of the extracellular matrix, modify those proteins, or facilitate interaction between those proteins in the matrix and other intracellular proteins. One clinical feature common to the different types of EDS is joint hypermobility, which may be limited to a few joints or be widespread. It is striking that isolated joint hypermobility as a clinical condition (EDS type III) has not yielded to the usual forms of genetic attack (candidate gene testing or use of families to define linked regions and sequence analysis to identify mutations) that have defined the genes for many other types of EDS.

153.1 TYPES OF EDS

EDS type I/II, now referred to as the "Classical" type, is a dominantly inherited form characterized by very marked skin extensibility and fragility with abnormal scar formation, and joint hypermobility. Most people have mutations in type V collagen genes, *COL5A1* and

COL5A2, but there may be genetic heterogeneity. The frequency of this disorder is not known, but it probably affects about 1/50,000 (there are no good epidemiologic data to support this or any estimate). Most individuals do not have major problems with other organs although a few individuals have mild aortic enlargement. A subset of such individuals has mutations in the *COL1A1* gene that result in substitutions in the triple helical domain of the proα1(I) chains that introduce cysteine residues, ordinarily absent from this domain. It is not clear when or in whom screening echocardiography should be employed.

There are two recessively inherited forms of EDS that overlap with the classical form or EDS types I and II. The first results from homozygosity or compound heterozygosity for mutations in *COL1A2* that result in loss of stable mRNA from both alleles. These individuals, in addition to joint hypermobility and soft, hyperextensible skin have multivalvular cardiac involvement. These phenotypes contrast with *COL1A2* mutations that allow stable expression of an mRNA that encodes a proα2(I) chain that fails to incorporate into type I procollagen trimers. Those mutations result in a rare recessive form of osteogenesis imperfecta. A recessively inherited form of an EDS type I-like condition, characterized by marked joint hypermobility, bruising, and skin extensibility without scarring, results from mutations in the *tenascin X* gene (*TNXB*). It is apparently rare but the structure of the gene has made it difficult to develop reliable diagnostic testing, a significant lack in as much as heterozygous women can have significant joint hypermobility.

EDS type III, the hypermobile type, is almost certainly a heterogeneous group within itself. In some, pain is a dominant feature while in others dysautonomia including orthostatic hypotension is a major feature, and in some the hypermobility may be limited to a few joints. It is probably dominantly inherited but the clinical features may vary widely and women appear far more likely to be symptomatic than men. One clue to causative mutations is that some women who are heterozygous for mutations in *TNXB* are hypermobile.

EDS type IV, the vascular type, is dominantly inherited and results from mutations in the *COL3A1* gene, which encodes type III collagen. The major complications of this disorder are arterial and bowel rupture and rupture of the uterus during late pregnancy. Early death may result from these complications. Diagnosis by analysis of these genes provides information for guiding both medical decision making and counseling. One class of mutations—those that result in inactivation of one copy of *COL3A1*—is associated with prolonged survival and a far lower risk of major complications.

EDS type VI, also known as the kyphoscoliotic or ocular scoliotic type, results from bi-allelic mutations in *PLOD1*, which encodes lysyl hydroxylase 1, which hydroxylates lysyl residues in the triple helical domain collagen, a step critical to the formation of stable intermolecular cross-links. Affected individuals have very marked joint laxity, especially when young, and scoliosis that is resistant to surgical correction. Some individuals develop large- and medium-sized artery aneurysms and dissection, which can lead to death. A subset seems to respond to high doses of vitamin C, a cofactor for the missing or defective enzyme, by a measurable increase in urinary cross-links.

EDS type VIIA and B, known also as the arthrochalasis type, characterized by very marked joint laxity and bilateral congenital hip dislocation is recalcitrant to surgical intervention. Very marked joint hypermobility persists into adult years and some adults require wheelchairs for mobility. Skin findings are not as dramatic as in the classical form of EDS. This form of EDS results from heterozygous mutations in *COL1A1* and *COL1A2*, and in both affect the splicing of exon 6, a sequence that encodes a short segment of the protein that contains the substrate site for the enzyme that removes the amino-terminal propeptide of type I procollagen and a lysyl residue involved in intermolecular cross-link formation. Loss of the sequence results in retention of a portion of the propeptide and interferes with collagen fibrillogenesis and fibril stabilization.

A recessively inherited form, dermatosparaxis (EDS type VIIC), results from lack of the enzyme, encoded by *ADAMTS2*, that cleaves the amino-terminal propeptides of all three chains of a procollagen trimer during extracellular processing. The affected children have very marked skin fragility, blue sclerae, marked joint hypermobility, and very soft and fragile skin that tears and bruises easily. With time, they develop skin laxity apparent in their face and hands. There is moderate variation in the phenotype that depends on the extent to which there is complete enzymatic loss. The long-term natural history is not known.

EDS type VIII, the periodontal type, is probably uncommon and genetically heterogeneous. The clinical features include joint hypermobility, hyerextensible, soft skin that scars abnormally, anterior shin pigmented regions, and the characteristic early periodontal loss. One gene is located on the short arm of chromosome 12. Other loci have not been identified.

Several other genes have recently been identified, mutations in which give rise to conditions in which joint hypermobility and skin alterations can be part of the presentation. Periventricular heterotopia with joint laxity result from mutations in the X-chromosomal gene, *FLNA*, which encodes an actin binding protein filamin A, that functions to organize the cytoskeleton. Mutations are generally lethal in males so there is a marked predominance of women with this condition. Aortic root dilatation can be seen and seizures are part of the phenotype. The extent to which mutations in this gene could contribute to the excess of women with joint hypermobility has not been examined.

Mutations in *SLC39A1*, which encodes a zinc transporter gene, result in a rare recessively inherited form of EDS, called the spondylocheiro-dysplastic type, characterized by hyperelastic, thin, and bruisable skin, hypermobility of the small joints with a tendency to contractures, protuberant eyes with bluish sclerae, hands with finely wrinkled palms, and a spondyloepiphyseal dysplasia with mild short stature.

153.2 CONCLUDING REMARKS

The last decade and a half has witnessed dramatic advances in the understanding of the molecular bases and translation to therapeutic trials in some disorders of the extracellular matrix, including Marfan syndrome. EDS has remained a poor cousin in this journey, perhaps because of the difficulty in devising therapies for disorders like EDS type IV and because of the enormous difficulty in discriminating genetic entities within the EDS type III or hypermobile group.

ACKNOWLEDGMENTS

Supported in part by funds from the Freudmann Fund for Translational Research in Ehlers–Danlos syndrome at the University of Washington.

FURTHER READING

1. Beighton, P.; De Paepe, A.; Steinmann, B.; Tsipouras, P.; Wenstrup, R. J. Ehlers-Danlos Syndromes: Revised Nosology, Villefranche, 1997. [Article]; *Am. J. Med. Genet.* **1998 Apr,** *77* (1), 31–37.
2. Malfait, F.; Coucke, P.; Symoens, S.; Loeys, B.; Nuytinck, L.; De Paepe, A. The Molecular Basis of Classic Ehlers-Danlos Syndrome: A Comprehensive Study of Biochemical and Molecular Findings in 48 Unrelated Patients. *Hum. Mutat.* **2005 Jan,** *25* (1), 28–37.
3. Schwarze, U.; Atkinson, M.; Hoffman, G. G.; Greenspan, D. S.; Byers, P. H. Null Alleles of the *COL5A1* Gene of Type V Collagen are a Cause of the Classical Forms of Ehlers-Danlos Syndrome (Types I and II). *Am. J. Hum. Genet.* **2000 Jun,** *66* (6), 1757–1765.
4. Malfait, F.; Symoens, S.; De Backer, J.; Hermanns-Lê, T.; Sakalihasan, N.; Lapière, C. M., et al. Three Arginine to Cysteine Substitutions in the Pro-Alpha (I)-Collagen Chain Cause Ehlers-Danlos Syndrome with a Propensity to Arterial Rupture in Early Adulthood. *Hum. Mutat.* **2007 Apr,** *28* (4), 387–395.
5. Schwarze, U.; Hata, R.; McKusick, V. A.; Shinkai, H.; Hoyme, H. E.; Pyeritz, R. E., et al. Rare Autosomal Recessive Cardiac Valvular Form of Ehlers-Danlos Syndrome Results from Mutations in the *COL1A2* Gene that Activate the Nonsense-Mediated RNA Decay Pathway. *Am. J. Hum. Genet.* **2004 May,** *74* (5), 917–930.
6. Burch, G. H.; Gong, Y.; Liu, W.; Dettman, R. W.; Curry, C. J.; Smith, L., et al. Tenascin-X Deficiency Is Associated with Ehlers-Danlos Syndrome. *Nat. Genet.* **1997 Sep,** *17* (1), 104–108.
7. Zweers, M. C.; Bristow, J.; Steijlen, P. M.; Dean, W. B.; Hamel, B. C.; Otero, M., et al. Haploinsufficiency of TNXB is Associated with Hypermobility Type of Ehlers-Danlos Syndrome. *Am. J. Hum. Genet.* **2003 Jul,** *73* (1), 214–217.
8. Pepin, M.; Schwarze, U.; Superti-Furga, A.; Byers, P. H. Clinical and Genetic Features of Ehlers-Danlos Syndrome Type IV, the Vascular Type. *N. Engl. J. Med.* **2000 Mar,** *342* (10), 673–680.
9. Schwarze, U.; Schievink, W. I.; Petty, E.; Jaff, M. R.; Babovic-Vuksanovic, D.; Cherry, K. J., et al. Haploinsufficiency for One COL3A1 Allele of Type III Procollagen Results in a Phenotype Similar to the Vascular Form of Ehlers-Danlos Syndrome, Ehlers-Danlos Syndrome Type IV. *Am. J. Hum. Genet.* **2001 Nov,** *69* (5), 989–1001.
10. Krane, S. M.; Pinnell, S. R.; Erbe, R. W. Lysyl-Protocollagen Hydroxylase Deficiency in Fibroblasts from Siblings with Hydroxylysine-Deficient Collagen. *Proc. Natl. Acad. Sci. U.S.A.* **1972 Oct,** *69* (10), 2899–2903.
11. Giunta, C.; Superti-Furga, A.; Spranger, S.; Cole, W. G.; Steinmann, B. Ehlers-Danlos Syndrome Type VII: Clinical Features and Molecular Defects. *J. Bone Joint Surg. Am.* **1999 Feb,** *81* (2), 225–238.
12. Smith, L. T.; Wertelecki, W.; Milstone, L. M.; Petty, E. M.; Seashore, M. R.; Braverman, I. M., et al. Human Dermatosparaxis: A Form of Ehlers-Danlos Syndrome That Results from Failure to Remove the Amino-Terminal Propeptide of Type I Procollagen. *Am. J. Hum. Genet.* **1992 Aug,** *51* (2), 235–244.
13. Malfait, F.; De Coster, P.; Hausser, I.; van Essen, A. J.; Franck, P.; Colige, A., et al. The Natural History, Including Orofacial Features of Three Patients with Ehlers-Danlos Syndrome, Dermatosparaxis Type (EDS type VIIC). *Am. J. Med. Genet. A.* **2004 Nov,** *131* (1), 18–28.
14. Sheen, V. L.; Jansen, A.; Chen, M. H.; Parrini, E.; Morgan, T.; Ravenscroft, R., et al. Filamin A Mutations Cause Periventricular Heterotopia with Ehlers-Danlos Syndrome. *Neurology* **2005 Jan,** *64* (2), 254–262.
15. Fukada, T.; Civic, N.; Furuichi, T.; Shimoda, S.; Mishima, K.; Higashiyama, H., et al. The Zinc Transporter SLC39A13/ZIP13 Is Required for Connective Tissue Development; Its Involvement in BMP/TGF-Beta Signaling Pathways. *PLoS One* **2008,** *3* (11), e3642.

CHAPTER

154

Heritable Diseases Affecting the Elastic Fibers: Cutis Laxa, Pseudoxanthoma Elasticum, and Related Disorders

Jouni Uitto

Department of Dermatology and Cutaneous Biology, Jefferson Medical College, Philadelphia, PA, USA

The elastic fibers provide elasticity and resilience to a number of organs, including the skin, the lungs and the arterial blood vessels. These fibers are assembled through a multistep, enzymatically mediated process that requires coordinated expression of at least a dozen of genes. Also, maintenance of the physiological steady-state levels of elastic fibers in tissues requires a careful balance between the rate of synthesis and the degradative processes mediated by elastases. Errors in these processes can lead to phenotypic manifestations in a number of both heritable and acquired disorders.

The major components of the elastic fibers are (i) elastin, a well-characterized connective tissue protein and (ii) elastin-associated microfibrils which are composed of a dozen of less well characterized proteins. In a mature elastic fiber, elastin is the predominant component providing the elastic properties to the tissues. Elastin-associated microfibrils, even though a lesser component in mature elastic fibers, play a critical role in assembly of the elastic fibers during development and elastic fiber regeneration by providing a scaffold into which the elastin molecules align. Thus, formation of properly functional elastic fibers requires critical interactions and a physiologic balance between elastin and the microfibrillar components. Subsequent to the assembly of the elastin molecules into a fiber structure, the fibers become stabilized by synthesis of covalent, lysine derived cross-links, desmosines, in a process that requires lysyl oxidase, a copper-dependent enzyme. Thus, proper assembly of the elastic fibers requires a number of critical steps, and abnormalities in the primary structures of the elastic fiber components or perturbations in the steps required for fibrillogenesis can result in pathology involving the elastic fibers.

Heritable disorders affecting the elastic fibers can be divided into three broad categories, based primarily on clinical findings and histopathological observations on elastic fibers. These groups include (i) diseases manifesting primarily with lax skin and associated features due to loss of functional elastic fibers, such as the cutis laxa syndromes; (ii) those where elastin accumulation in tissues is a predominant histopathologic feature, resulting in generalized or localized lesions with associated pathologic findings, as exemplified by cutaneous elastomas and pseudoxanthoma elasticum; and (iii) those in which abnormalities in elastin fibrillogenesis can explain the clinical phenotypes due to lack of stabilization of functional elastic fibers, as in copper deficiency syndromes.

Cutis laxa, a heterogeneous group of disorders, demonstrates loose and sagging skin with loss of recoil with associated clinical features in the lungs and vascular connective tissues. Histopathology of the skin of these patients reveals loss and fragmentation of the elastic fibers. This group of disorders is also genetically heterogeneous, and both autosomal dominant and autosomal recessive inheritance have been documented. In these cases, mutations have been identified in the genes encoding elastin (ELN), fibulins 4 and 5 (FBLN4 and FBLN5), latent TGF-β binding protein4 (LTBP4), and in the gene encoding the A2 subunit of vacuolar H+ATPase (ATP6V082). Mutations in the *PYCR1* gene encoding pyrroline-5-carboxylate reductase 1 are also associated with cutis laxa-like phenotypes. Increased elastase-like activities have been documented in an autosomal recessive variant of cutis laxa, and a distinct variant, acquired cutis laxa, can develop as a result of inflammatory reactions in the skin, sometimes associated with allergic drug reactions with little or no evidence of internal organ involvement.

Cutis laxa-like phenotypes, either localized or generalized, have been encountered in a number of related disorders, including anetoderma, mid-dermal elastolysis, the wrinkly skin syndrome, and the Michelin tire baby syndrome. Also, de Barsy syndrome, an autosomal recessive developmental disorder displays cutis laxa-like features accompanied by progeria-like appearance, dwarfism, neurologic abnormalities, and corneal opacities. Elastin mutations can also explain the cardiovascular involvement in the Williams syndrome and in supravalvular aortic stenosis.

A prototype of the elastin accumulation diseases is pseudoxanthoma elasticum, which clinically involves the skin, the eyes and the cardiovascular system. Characteristic lesions in the skin are yellowish papules which tend to coalesce into larger plaques of inelastic and leathery skin. Histopathology of skin demonstrates accumulation of pleomorphic elastotic material which becomes progressively mineralized by calcium phosphate deposition. The eye manifestations consist primarily of angioid streaks which lead to loss of visual acuity and rarely blindness. The cardiovascular involvement results from mineralization of the arterial blood vessels, clinically manifesting with hypertension, intermittent claudication, occasional arterial bleeding, and rare early myocardial infarcts. The classic forms of PXE are inherited in an autosomal recessive pattern, and are caused by mutations in the *ABCC6* gene which encodes a putative transmembrane efflux transporter expressed primarily in the liver and the kidneys. Based on these and related observations, PXE is now considered to be a metabolic disorder. PXE-like skin lesions have also been encountered in unrelated heritable conditions, including β-thalassemia and sickle cell anemia without mutations in the *ABCC6* gene, as well as in patients with vitamin K dependent coagulation factor deficiency due to mutations in the *GGCX* gene.

Disorders of elastin fibrillogenesis include the Menkes syndrome and occipital horn syndrome, allelic conditions due to mutations in *ATP7A* gene. This gene encodes a copper transport protein and its functional deficiency results in reduced serum copper levels. The spectrum of connective tissue disorders and neurological and hair manifestations in these diseases can be explained by variably reduced activity of a number of copper-dependent enzymes. Elastic fiber abnormalities are also evident in a number of other syndromes, including the Buschke–Ollendorff syndrome, the Costello syndrome, and elastoderma, but the pathomechanisms and the precise consequences of the gene defects in these disorders are currently unknown.

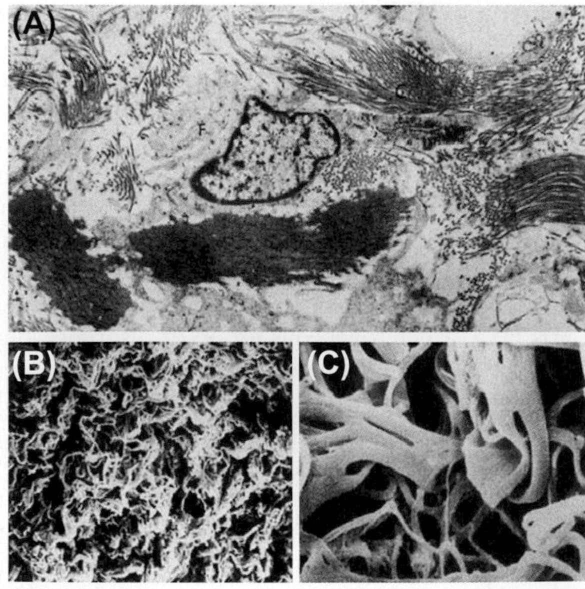

FIGURE 154-1 Ultrastructure of elastic fibers in human skin: A. Transmission electron microscopy demonstrates a fibroblast (F) surrounded by elastic sheets (E) and collagen fibers (C). B and C. Scanning electron microscopy reveals the presence of interconnecting fibers of variable diameters. *(Modified from Uitto, J; Rosenbloom, J. Elastic Fibers.* In Dermatology in General Medicine; *Freedberg, I. M.; Eisen, A. Z.; Wolff, K., et al. Eds.; 5th ed. McGraw Hill, New York, 1999, pp. 260.)*

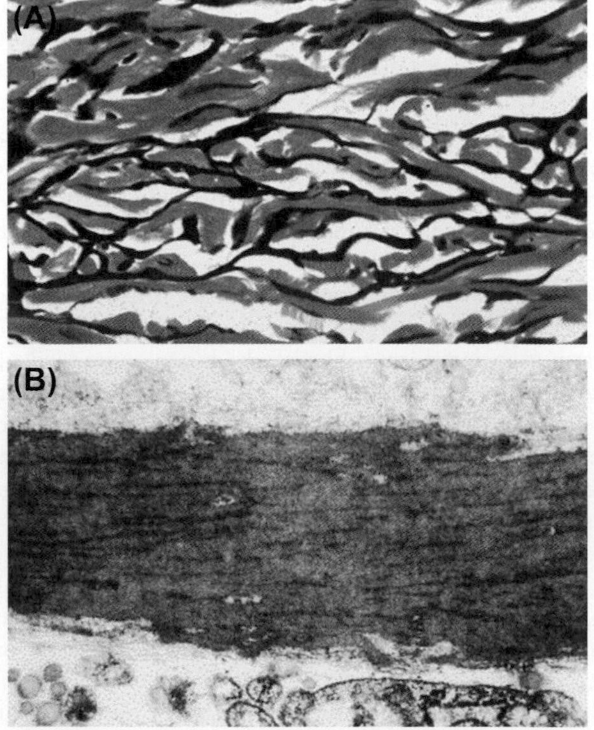

FIGURE 154-2 Morphology of the elastic fibers in human dermis. A. Elastic fibers, visualized by histopathology using an "elastin-specific" stain (Verhoeff–van Gieson), form an interconnecting network (black). B. Transmission electron microscopy of an individual elastic fiber reveals the presence of an electron-lucent central core consisting of elastin, and electron-dense (darker) elastin-associated microfibrils, which are superimposed on the core.

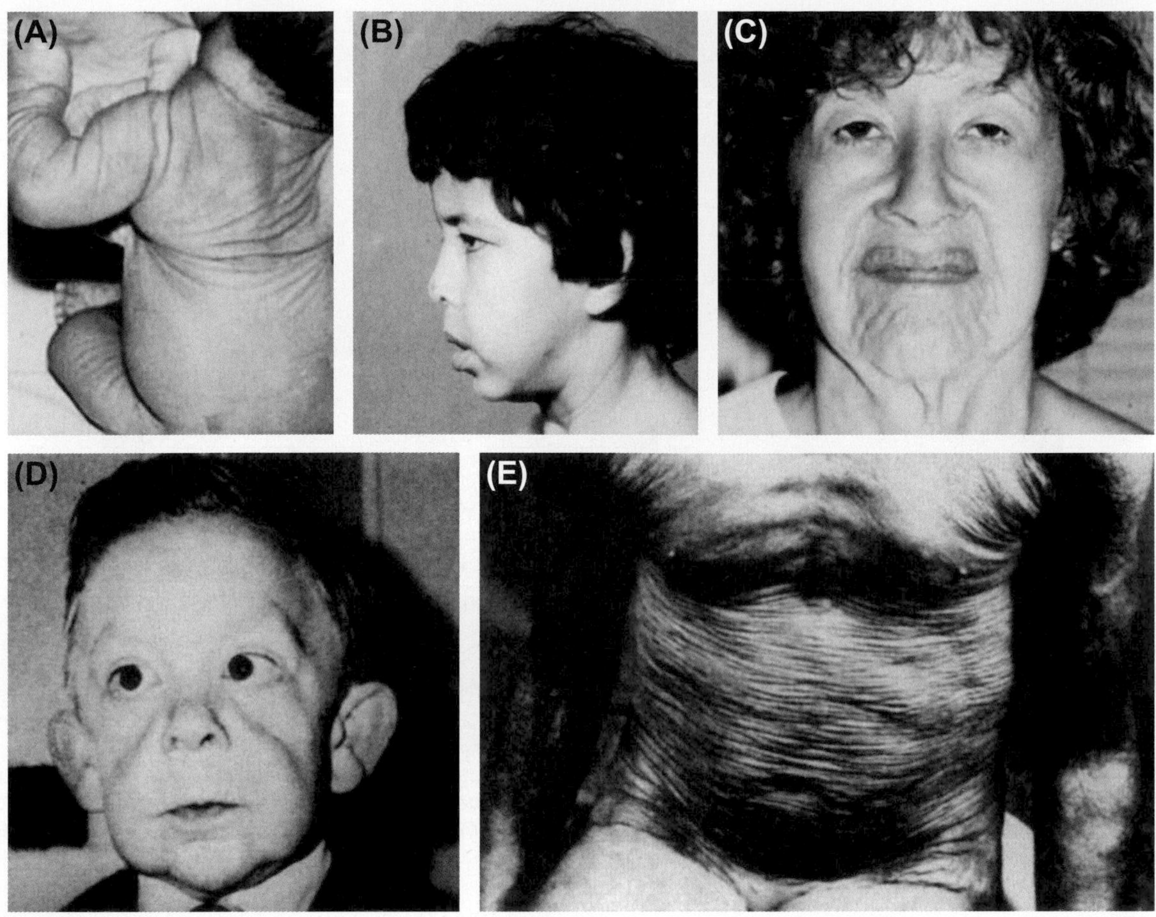

FIGURE 154-3 Clinical features of cutis laxa. Note loose and folded skin in an infant (A), sagging jowl in a 4-year-old boy (B), premature aging appearance in a 30-year-old female (C), characteristic facial features in a patient with autosomal recessive cutis laxa (D), and generalized mid-truncal wrinkling as a consequence of an inflammatory urticarial reaction (E). *(D is from Pope, F. M. Pseudoxanthoma Elasticum, Cutis Laxa, and Other Disorders of Elastic Tissue. In* Principles and Practice of Medical Genetics*; Rimoin, D. L., Connor, J. M., Pyeritz, R. E., Eds.; 3rd ed. Churchill Livingstone, New York, 1997; p. 1083. E is from Verhagen, A. R., Woerdeman, M. J. Post-inflammatory elastolysis and cutis laxa.* Br. J. Dermatol. ***1975****, 92, 183–190; with permission.)*

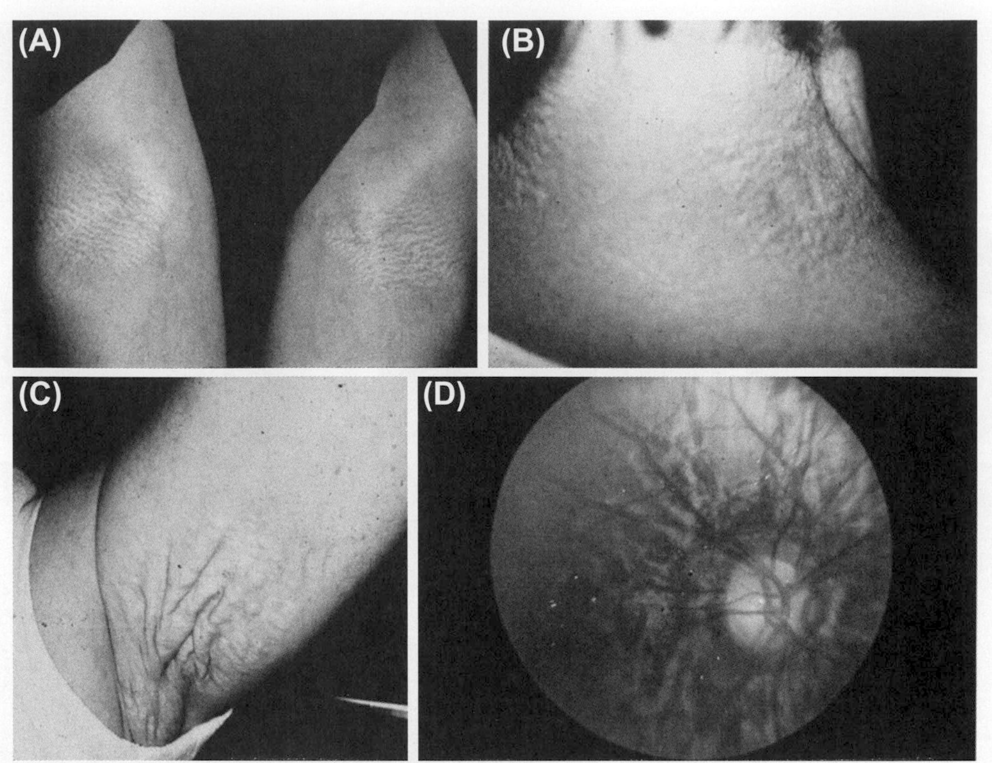

FIGURE 154–4 Characteristic clinical findings in pseudoxanthoma elasticum. Note small yellowish papular lesions on the antecubital fossa (A) and on the site of the neck (B). These primary lesions tend to coalesce into large plaques rendering skin redundant and inelastic (C). Ophthalmologic examination reveals the presence of angioid streaks (D). *(Modified from Uitto, J.; Pulkkinen, L.; Ringpfeil, F. Molecular Genetics of Pseudoxanthoma Elasticum—a Metabolic Disorder at the Environment/Genome Interface?* Trends Mol. Med. ***2000****, 7, 13–17.)*

FIGURE 154-5 Typical facial features of a patient with Williams syndrome. *(From Pope, F. M. Pseudoxanthoma Elasticum, Cutis Laxa, and Other Disorders of Elastic Tissue. In* Principles and Practice of Medical Genetics, *3rd edn; Rimoin, D. L., Connor, J. M., Pyeritz, R. E. Eds.; Churchill Livingstone: New York, 1997; p. 1083; reproduced with permission.)*

FURTHER READING

1. Kielty, C. M. Elastic Fibres in Health and Disease. *Expert Rev. Mol. Med.* **2006,** *8* (19), 1–23.
2. Lewis, K. G.; Bercovitch, L.; Dill, S. W.; Robinson-Bostom, L. Acquired Disorders of Elastic Tissue: Part I. Increased Elastic Tissue and Solar Elastotic Syndromes. *J. Am. Acad. Dermatol.* **2004,** *51* (1), 1–21.
3. Lewis, K. G.; Bercovitch, L.; Dill, S. W.; Robinson-Bostom, L. Acquired Disorders of Elastic Tissue: Part II. Decreased Elastic Tissue. *J. Am. Acad. Dermatol.* **2004,** *51* (2), 165–185.
4. Uitto, J.; Li, Q.; Jiang, Q. Pseudoxanthoma Elasticum: Molecular Genetics and Putative Pathomechanisms. *J. Invest. Dermatol.* **2010,** *130* (3), 661–670.
5. Pfendner, E. G.; Vanakker, O. M.; Terry, S. F.; Vourthis, S.; McAndrew, P. E.; McClain, M. R.; Fratta, S.; Marais, A. S.; Hariri, S.; Coucke, P. J., et al. Mutation Detection in the *ABCC6* Gene and Genotype-Phenotype Analysis in a Large International Case Series Affected by Pseudoxanthoma Elasticum. *J. Med. Genet.* **2007,** *44* (10), 621–628.
6. St Hilaire, C.; Ziegler, S. G.; Markello, T. C.; Brusco, A.; Groden, C.; Gill, F.; Carlson-Donohoe, H.; Lederman, R. J.; Chen, M. Y.; Yang, D., et al. NT5E Mutations and Arterial Calcifications. *N. Engl. J. Med.* **2011,** *364* (5), 432–442.
7. LaRusso, J.; Li, Q.; Jiang, Q.; Uitto, J. Elevated Dietary Magnesium Prevents Connective Tissue Mineralization in a Mouse Model of Pseudoxanthoma Elasticum (Abcc6(–/–)). *J. Invest. Dermatol.* **2009,** *129* (6), 1388–1394.
8. Hellemans, J.; Preobrazhenska, O.; Willaert, A.; Debeer, P.; Verdonk, P. C.; Costa, T.; Janssens, K.; Menten, B.; Van Roy, N.; Vermeulen, S. J., et al. Loss-of-Function Mutations in LEMD3 Result in Osteopoikilosis, Buschke-Ollendorff Syndrome and Melorheostosis. *Nat. Genet.* **2004,** *36* (11), 1213–1218.
9. Ramirez, F.; Sakai, L. Y.; Rifkin, D. B.; Dietz, H. C. Extracellular Microfibrils in Development and Disease. *Cell Mol. Life Sci.* **2007,** *64* (18), 2437–2446.
10. Yanagisawa, H.; Davis, E. C. Unraveling the Mechanism of Elastic Fiber Assembly: The Roles of Short Fibulins. *Int. J. Biochem. Cell Biol.* **2010,** *42* (7), 1084–1093.
11. Urban, Z.; Michels, V. V.; Thibodeau, S. N.; Donis-Keller, H.; Csiszar, K.; Boyd, C. D. Supravalvular Aortic Stenosis: A Splice Site Mutation within the Elastin Gene Results in Reduced Expression of Two Aberrantly Spliced Transcripts. *Hum. Genet.* **1999,** *104* (2), 135–142.
12. Lin, A. E.; Alexander, M. E.; Colan, S. D.; Kerr, B.; Rauen, K. A.; Noonan, J.; Baffa, J.; Hopkins, E.; Sol-Church, K.; Limongelli, G., et al. Clinical, Pathological, and Molecular Analyses of Cardiovascular Abnormalities in Costello Syndrome: A Ras/MAPK Pathway Syndrome. *Am. J. Med. Genet. A* **2011.**

Skeletal Disorders

Osteogenesis Imperfecta (and Other Disorders of Bone Matrix), 587
Disorders of Bone Density, Volume, and Mineralization, 590
Chondrodysplasias, 593
Abnormalities of Bone Structure, 595
Arthrogryposes (Multiple Congenital Contractures), 597
Common Skeletal Deformities, 599
Hereditary Noninflammatory Arthropathies, 601

CHAPTER

155

Osteogenesis Imperfecta (and Other Disorders of Bone Matrix)

Craig Munns

Department of Endocrinology and Diabetes, Sydney Children's Hospital Network (Westmead), Westmead, NSW, Australia

David Sillence

Connective Tissue Dysplasia Service, Sydney Children's Hospital Network (Westmead), Westmead, NSW, Australia

Osteogenesis Imperfecta (OI) is a collective term for a heterogeneous group of connective tissue dysplasia syndromes characterized by liability to fractures throughout life. The nomenclature and classification have evolved over at least 200 years, but the 2010 recommended nosology is based on that published in 1979 by Sillence and colleagues which grouped OI syndromes by primary clinical characteristics and pattern of inheritance into four groups, OI types I–IV. The syndrome of OI with calcification in interosseous membranes ± hyperplastic callus previously designated OI type V is included in the 2010 nosology as the fifth of these clinical syndromes.

In the last decade, the genetic complexity in OI is evidenced in the molecular investigation of OI cases, which has identified 10 genes in addition to *COL1A1* and *COL1A2* that contribute to the phenotypic spectrum of different types of OI. These genes code for proteins that are either enzymes, chaperone proteins or proteins involved in mineralization. There are three proteins required for prolyl 3-hydroxylation and an enzyme that codes for bone collagen lysyl hydroxylation. Five of these genes also code for proteins that are important chaperone proteins at the level of either the rough or smooth endoplasmic reticulum or involved further in trafficking. Osterix, a protein involved in mineralization coded for by the gene *SP7*, is one of a growing number of proteins involved in the process of mineralization. In this regard, another major group of disorders characterized by bone fragility in the newborn, hypophosphatasia also result from a defect in mineralization due to deficiency of the enzyme tissue non-specific alkaline phosphatase (TNSALP). Mutations in bone morphogenetic protein 1, which also acts as the type I collagen carboxy peptidase, result in a brittle bone disorder characterized by high bone mass.

While the severity of the bone fragility and osteoporosis leading to skeletal deformity are important factors in the clinical assessment of patients with OI, "severity" in itself is not a genetic type of OI. A grading of severity was proposed and adopted by the International Nosology Group in 2009. It provides an assessment based on clinical and historical data, fracture frequency, bone densitometry, level of mobility impairment and level of social and educational integration.

In European populations, a type of OI that is virtually always associated with distinctly blue–gray coloration of the scleras when viewed in daylight is the most common disorder. This disorder usually is not associated with progressive deformity of long bones or spine, nor of marked reduction in stature. It is associated with susceptibility to hearing loss, which commences in adolescence and progresses in young adult life. Its birth frequency is in the order of 1 in 25,000 lives birth and its clinical features are well described such that it has become known as the "classic" type of OI. It results from heterozygous nonsense or frameshift mutations in *COLIA1* or *COLIA2*. The distinctly blue–gray color of the scleras results from accumulation of an electron dense granular material between the collagen lamellae which result in an abnormal diffraction of reflected light. Almost equally common in population frequency, autosomal dominantly inherited common variable OI with normal scleras is associated with a low frequency of hearing impairment. These patients may have bluish scleras when they are young. This type of OI is variable within families and variable among families. In many affected, height is normal but there may be considerable variability within the same family resulting in progressive skeletal deformity. Usually the upper limbs are normal. This type of OI, which was previously designated OI type IV, is due to missense mutations in type I collagen genes and the phenotypic findings are continuous with those patients with progressively deforming OI with normal scleras and patients with the perinatally lethal OI syndromes.

The group of OI patients with progressive deformity shows intense genetic heterogeneity. This disorder may result from heterozygous mutations in *COL1A1* or *COLA2* or from autosomal recessive inheritance of at least nine other genes. There is strong evidence that further genes whose gene product codes for either enzymes involved in posttranslational modifications of collagens or proteins involved in chaperone and assembly functions of the type I collagen polypeptides will soon be discovered. Furthermore, with the discovery of mutations in SP7/osterix, it is likely that this will result in the discovery of a further class of proteins involved in the biogenesis of matrix vesicles and proteins involved in the process of bone mineralization. Prolyl 3-hydroxylation is accomplished by a multi-protein enzymatic unit consisting of Leprecan, Cartilage Associated Protein and peptidyl-prolyl isomerase B. Mutations in the gene coding for bone lysyl hydroxylation (*PLOD2*) also result in a distinct form of progressively deforming OI with joint contractures which is one of the two syndromes known as Bruck syndrome. The chaperone protein FKBP65 coded by the gene *FKPB10* results in both the Bruck syndrome phenotype as well as non-syndromic progressively deforming OI and a pattern of bone fragility and skeletal deformity with club feet at birth and progressive protrusio acetabulae.

OI with progressive ossification in interosseous membranes plus or minus hyperplastic callus formation is a familial autosomal dominant disorder first described over 100 years ago. While 100% of patients clinically develop the calcification between the radius and ulnar leading to dislocation of radial heads, calcification between the tibia and fibula does not always occur. The true frequency of hyperplastic callus is unknown. Massive callus formation leading to swelling and pain at the side of the fracture may be a medical emergency. The syndrome is found in approximately 5% of individuals with OI seen in a hospital setting, but up to 10% of patients with moderate to severe fracture frequency and disability. Diagnosis can be confirmed by histomorphometry of specially prepared trans-iliac bone biopsies. At present, the genetic basis for OI type V is unknown. Anti-inflammatory medication may prevent progression and reduce the severity of the painful swelling of the limb that precedes the hyperplastic callus development. The disorder may be manifested as mild through severe disease within the same family and many individuals go through life without development of hyperplastic callus.

Although rare, several syndromic forms of OI including the Bruck syndromes with congenital contractures and syndromes with eye disease including osteoporosis with pseudoglioma are known. The pathogenesis of the Bruck syndromes may have been elucidated with the discovery that Bruck type I may result from mutations in *FKBP10* and Bruck type II from mutations in *PLOD2*. The Cole Carpenter type of OI is accompanied by progressive craniosynostosis and otherwise is phenotypically a severe progressively deforming type of OI. The Hypophosphatasia syndromes while clinically heterogeneous appear at the present time to have a single pathogenesis reflecting mutations in the TNSALP. Syndromes that can be distinguished range from a perinatally lethal form through to a non-skeletal form with premature exfoliation of teeth known as odontohypophosphatasia. The observation that there appeared to be cases with radiographic findings of hypophosphatasia yet a normal fibroblast TNSALP, normal serum and urinary phosphoethanolamine indicate that hypophosphatasia is also pathogenetically heterogeneous. Treatment with enzyme replacement holds great potential for affected individuals. Some disorders of the bone matrix including chronic recurrent multifocal osteomyelitis and SAPHO syndrome are also associated with generalized or regional osteoporosis, but have unknown pathogenesis. However, in some communities, there is a form of disorder associated with congenital dyserythropoietic anemia and palmar-plantar pustulosis, which is known as Majeed syndrome.

The last 20 years have witnessed a revolution in the multidisciplinary care of children with OI and related disorders. This program of care, which includes physical rehabilitation, timely orthopedic surgery and the early introduction of medical therapy with bisphosphonates, has revolutionized care. Although other more potent bisphosphonate have been used in selected patients, for reasons of safety and cost, the two drugs known as intravenous pamidronate and intravenous neridronate will continue to be the drugs of choice for the treatment of these disorders on a worldwide basis.

FURTHER READING

1. Munns, C. F.; Sillence, D. O. *Disorders Predisposing to Bone Fragility and Decreased Bone Density*, 5th ed.; Churchill Livingstone: Philadelphia, 2007, p. 3671–3691.
2. Warman, M. L.; Cormier-Daire, V.; Hall, C., et al. Nosology and Classification of Genetic Skeletal Disorders: 2010 Revision. *Am. J. Med. Genet. A.* **2011**, *155A*, 943–968, Epub 2011/03/26.

3. Rauch, F.; Lalic, L.; Roughley, P., et al. Genotype-Phenotype Correlations in Nonlethal Osteogenesis Imperfecta Caused by Mutations in the Helical Domain of Collagen Type I. *Eur. J. Hum. Genet.* **2010,** *18,* 642–647.
4. Forlino, A.; Cabral, W. A.; Barnes, A. M., et al. New Perspectives on Osteogenesis Imperfecta. *Nat. Rev. Endocrinol.* **2011,** *7,* 540–557.
5. Pyott, S. M.; Schwarze, U.; Christiansen, H. E., et al. Mutations in Ppib (Cyclophilin B) Delay Type I Procollagen Chain Association and Result in Perinatal Lethal to Moderate Osteogenesis Imperfecta Phenotypes. *Hum. Mol. Genet.* **2011,** *20,* 1595–1609.
6. Christiansen, H. E.; Schwarze, U.; Pyott, S. M., et al. Homozygosity for a Missense Mutation in SERPINH1, Which Encodes the Collagen Chaperone Protein Hsp47, Results in Severe Recessive Osteogenesis Imperfecta. *Am. J. Hum. Genet.* **2010,** *86,* 389–398.
7. Becker, J.; Semler, O.; Gilissen, C., et al. Exome Sequencing Identifies Truncating Mutations in Human SERPINF1 in Autosomal-Recessive Osteogenesis Imperfecta. *Am. J. Hum. Genet.* **2011,** *88,* 362–371.
8. Lapunzina, P.; Aglan, M.; Temtamy, S., et al. Identification of a Frameshift Mutation in Osterix in a Patient with Recessive Osteogenesis Imperfecta. *Am. J. Hum. Genet.* **2010,** *87,* 110–114.
9. Russell, R. G. G. Bisphosphonates: The First 40 Years. *Bone* **2011,** *49,* 2–19.
10. Rauch, F.; Munns, C. F.; Land, C., et al. Risedronate in the Treatment of Mild Pediatric Osteogenesis Imperfecta: A Randomized Placebo-Controlled Study. *J. Bone Miner. Res.* **2009,** *24,* 1282–1289.
11. Horwitz, E. M.; Prockop, D. J.; Gordon, P. L., et al. Clinical Responses to Bone Marrow Transplantation in Children with Severe Osteogenesis Imperfecta. *Blood* **2001,** *97,* 1227–1231.
12. Whyte, M. P.; Greenberg, C. R.; Salman, N. J., et al. Enzyme-Replacement Therapy in Life-Threatening Hypophosphatasia. *N. Engl. J. Med.* **2012,** *366,* 904–913.

CHAPTER

156

Disorders of Bone Density, Volume, and Mineralization

Maria Descartes

Department of Genetics, University of Alabama at Birmingham, Birmingham, AL, USA

David O Sillence

Connective Tissue Dysplasia Management Service and Centre for Children's Bone Health, Sydney Children's Hospital Westmead Campus, Westmead, NSW, Australia

More than 30 disorders are known with generalized or localized increase in the density or size of the skeleton or individualized skeletal elements.

Bone is a dynamic tissue in which osteoblasts synthesize bone matrix while osteoclasts resorb bone. Therefore, bone density is dependent on the relative function of these two types of cells. Osteopetrosis and related disorders are a heterogeneous group of heritable conditions in which there is a defect in bone resorption by osteoclasts characterized by increased bone density on radiographs. Osteopetrosis was categorized clinically based on presentation and severity. Mutations in at least 10 causative genes have been identified. The understanding of the genetic basis of the disease remains incomplete but recent advances have provided a broad grouping of osteoclasts defects into defects in early osteoclast differentiation, defects in receptor activation of nuclear factor-kB and related proteins, defects in osteoclast function, and defects in osteoclast acidification.

The osteopetrotic conditions can be inherited as autosomal recessive, dominant, or X-linked traits. They vary in their clinical presentation and severity. The autosomal recessive forms are the most severe. The overall incidence of the autosomal recessive osteopetrosis (ARO) is 1 in 250,000 births, and autosomal dominant osteopetrosis (ADO) has an incidence of 1 in 18,000 births in Denmark.

In classic ARO, symptoms may be recognized at birth but are most frequently discovered during the first few months of life. They may present with failure to thrive, short stature, compressive neuropathies, hypocalcemia with seizures, and anemia with thrombocytopenia. Life expectancy is diminished in the severe infantile forms of osteopetrosis. Mutations in *TCIRG1* are responsible for 50–60% of children with severe osteopetrosis. Mutations in *CLCN7* are responsible for 15% of the cases. Fractures, osteomyelitis, and compressive neuropathies presenting in childhood, adolescence, or young adult life are more typical of ADO. Families with autosomal recessive inheritance have been reported with this milder presentation. Mutations in *CLCN7* account for approximately 70–75% of the ADO cases. Treatment for these conditions is mostly symptomatic. Hematopoietic stem cell transplantation can cure some patients with ARO.

The presence of primary neurodegeneration, mental retardation, fractures in utero, dysmorphic features, short limb dwarfism, skin and immune system involvement, or renal tubular acidosis is suggestive of other forms of osteopetrosis.

Osteopetrosis with renal tubular acidosis is an autosomal recessive type of infantile osteopetrosis that presents with fractures and/or short stature, visual impairment, and mental retardation in the first few years of life. The defect is in carbonic anhydrase II, an enzyme important for acidification by osteoclasts in the process of mineral resorption and in the renal tubules. These patients generally show mixed (proximal and distal) renal tubular acidosis. Raine dysplasia is an autosomal recessive disorder that combines diffuse osteosclerosis with distinctive facial dysmorphism and intracranial calcification. Mutations in *FAM20C* that code for the human homolog of DMP4, a dentin matrix protein, cause Raine syndrome. Pyknodysostosis is a rare generalized hyperostotic bone disease with disproportionate short stature, characteristic facies, and wide anterior fontanelle that persists into adult life. There is progressive osteolysis of the distal phalanges in older patients. Mutations in the cathepsin *K* gene are responsible for pyknodysostosis. Dysosteosclerosis is characterized by generalized increase in bone density and postnatal onset of short stature. Radiographic evidence of platyspondyly, irregularity of vertebral ossification, and high incidence of developmental defects of the teeth help differentiate this condition from osteopetrosis and pyknodysostosis. Reports of consanguinity indicate autosomal recessive inheritance, yet more affected males than females suggest X-linked recessive inheritance. Its etiology remains unknown.

Osteopoikilosis and melorheostosis are commonly asymptomatic. In osteopoikilosis, osteodense foci are seen in the epiphyses and the carpal and tarsal centers. In melorheostosis, irregular linear osteodense lesions are seen along the axis of the tubular bones. The coexistence of cutaneous lesions, such as disseminated lenticular dermatofibrosis, and skeletal changes (osteopoikilosis and/or melorheostosis) characterize Buschke–Ollendorf syndrome. Mutations in *LEMD3* cause Buschke–Ollendorff syndrome and osteopoikilosis, with or without melorheostosis.

Linear regular bands of increased density throughout the skeleton are characteristics of osteopathia striata with autosomal dominant inheritance. Typical changes of osteopathia striata are seen in diverse syndromes. Osteopathia striata with cranial sclerosis is an X-linked dominant disorder. It is characterized by psychomotor retardation, hearing deficiency, and congenital heart defects. Macrocephaly with frontal bossing and cranial hyperostosis are present in virtually all patients. There is an excess of affected females and the affected females have more severe presentation with more complications. Germline mutations in *WTX*, a gene that encodes a repressor of canonical WNT signaling, cause osteopathia striata congenita with cranial sclerosis.

The craniotubular dysplasias are a heterogeneous group characterized by abnormal modeling error as well as increased bone density of the craniofacial and tubular bones. All these disorders result from excess bone deposition versus resorption with specifically different patterns of skeletal involvement. Essentially, they are all disorders in which there is minimal involvement of the spine compared with osteopetrosis, pyknodysostosis, and dysosteosclerosis, in which increased osteodensity is seen throughout the spine and the rest of the skeletal with minimal changes in the cranial vault. The cranial involvement leads to cranial nerve compression, particularly of the optic and auditory nerves, due to internal bony overgrowth and facial deformities involving/affecting the cranial foramina and fissures. Craniotomy and cranial nerve decompression can be beneficial to some patients. Nonsurgical treatment is supportive.

Diaphyseal dysplasia (Camurati–Engelmann disease) is a rare craniotubular remodeling disorder with significant neuromuscular involvement. Presenting symptoms start in childhood and include failure to thrive, fatigability, abnormal gait, and pain in the legs of increasing severity. The hallmark of the disorder is cortical thickening of the diaphysis of long bones but the skull and pelvis could also be affected. It is inherited as autosomal dominant with variable penetrance and wide clinical variability. Camurati–Engelmann disease is caused by mutations in *TGFB1*.

Craniodiaphyseal dysplasia is a rare craniotubular remodeling disorder characterized by massive hyperostosis and sclerosis of the skull and facial bones and hyperostosis and defective modeling of the shafts of the tubular bones. This disorder is inherited as an autosomal dominant trait in most cases. Familial recurrence in few cases suggests autosomal recessive inheritance but the nature of the basic defect is not known. In the autosomal dominant form of craniodiaphyseal dysplasia heterozygous mutations in *SOST* have been documented in two affected children.

Endosteal hyperostosis, Van Buchem disease and sclerosteosis are a group of disorders characterized by marked accretion of osseous tissue at the endosteal (inner) surface of bone, leading to narrowing of the medullary canal or obliteration of the medullary space. Endosteal hyperostosis, Worth type is a rare dominantly inherited variety, is frequently associated with the presence of torum palatinus. Gain of function mutations in the *LRP5* gene are responsible for the dominantly inherited Worth type of endosteal hyperostosis. Van Buchem disease is recessively inherited variety of endosteal hyperostosis characterized by progressive mandibular enlargement and increased density of the cortices of tubular bones. Sclerosteosis is autosomal recessive disorder clinically and radiologically almost indistinguishable from Van Buchem disease. Sclerosteosis has a high frequency of hyperostosis of the nasal and facial bones and syndactyly of the second and third fingers. Sclerosteosis and Van Buchem disease are caused by mutations in *SOST*.

Pachydermoperiostosis is an unusual condition characterized by progressive thickening of the skin and clubbing of the fingers. Radiographic findings are similar to those observed in hypertrophic osteoarthropathy associated with pulmonary disease. Pachydermoperiostosis may be genetically heterogeneous. Families with vertical transmission consistent with autosomal dominant transmission have been observed. The molecular basis of the autosomal dominant form is still unknown. Autosomal recessive inheritance has been documented. The autosomal recessive form is molecularly heterogeneous. Mutations in *HPGD* and *SLCO2A1*, genes involved in prostaglandin E2 metabolism and transport respectively, have been documented.

Craniometaphyseal dysplasia (CMD) is a genetically heterogeneous group of disorders with both autosomal and recessive inheritance with overlap in the clinical and radiographic findings. They both have progressive broadening of the osseous prominence of the nasal root. The essential radiographic features are hyperostosis of the skull, nasal, and maxillary bones extending bilaterally across the zygoma. The long bones show flaring and decreased density of the metaphyses due to failure of remodeling of the metaphyses during growth. Mutations in *ANK* are responsible for the dominantly inherited CMD cases.

Craniotubular remodeling disorders resulting from mutations in filamin A include frontometaphyseal dysplasia (FMD), otopalatodigital syndrome type I (OPDI), otopalatodigital syndrome type II (OPDII), and Melnick–Needles syndrome (MNS). All four disorders are characterized by sclerosis of the skull base and to a varying extent, facial bones. X-linked inheritance has been demonstrated in all four disorders.

Tubular stenosis (Kenny–Caffey syndrome) is characterized by narrowing of the medullary cavity and cortical thickening. Features include short stature, tetanic seizures due to hypocalcemia, hypoparathyroidism, and small hands and feet. Radiographically, there is widening of the cortex of the long bones and short tubular bones of the hands and feet, without overall widening of the diaphyses, leading to reduction of the medullary cavity. Both autosomal dominant and recessive forms have been reported. The autosomal recessive Kenny–Caffey syndrome results from mutations in the *TCBE* gene.

Hyperphosphatasia with osteoectasia (juvenile Paget disease) is a progressive skeletal deformation with associated marked elevation of alkaline phosphatase. Manifestations occur early in life and include large head and enlarged bones, and painful deformity of the extremities. The clinical symptoms are similar to Paget disease in adult but are more generalized and symmetrical in distribution. The inheritance is autosomal recessive. Mutations in *TNFRSF11B* result in hyperphosphatasia with osteoectasia.

Disorders of parathyroid hormone resistance or pseudohypoparathyroidism (PHP) describe a group of disorders characterized by biochemical hypoparathyrodism, increased serum concentrations of parathyroid hormone (PTH), and end-organ resistance to the biological actions of PTH. The commonly associated clinical findings with PHP disorders are disproportionate short stature with acrodysplasia, round face, and obesity. Collectively these features are termed Albright hereditary osteodystrophy (AHO). The clinical manifestations were variable and range from decreased bone density to osteosclerosis. It is usual to subdivide AHO into subtypes which have significant overlap in phenotypic features, but distinctive endocrine and biochemical findings: PHP type I (Type Ia, Type Ib, and Type Ic); Pseudo-PHP; PHP type II. The genetic defect associated with different forms of PHP involves the α-subunit of the stimulatory G protein (Gsα008). A defect in the renal response to PTH is the hallmark of all forms of PHP. The genetic basis of PHP is complex. The different subtypes result from tissue-specific imprinting and imprinting mutation in the GNAS locus.

X-linked hypophosphatemic rickets (XLH) is the most common heritable form of rickets with a prevalence of 1:20,000. It is a renal phosphate wasting disorder characterized by rickets resistant to vitamin D therapy, low serum phosphate, and inappropriately normal 1,25-dihydroxy vitamin D [1,25(OH)2D]. XLH results from impaired phosphate reabsorption, inadequate 1-25 dihydroxy vitamin D response to hypophosphatemia, and an intrinsic osteoblast defect. Children with XLH exhibit short stature, rickets, genu varum or valgum, and dental abscesses. Adults typically have osteomalacia. Enthesopathy may result in joint limitations and neurologic compressive syndrome. Affected females may have less severe manifestations and tend to respond better to therapy. XLH is inherited as an X-linked dominant trait. Mutations in *PHEX* are associated with XLH.

FURTHER READING

1. Del Fattore, A. P. B. Clinical, Genetic, and Cellular Analysis of 49 Osteopetrotic Patients: Implications for Diagnosis and Treatment. *J. Med. Genet.* **2006**, *43,* 315–325.
2. Janssens, K. V. F. Camurati-Engelmann Disease: Review of the Clinical, Radiological, and Molecular Data of 24 Families and Implications for Diagnosis and Treatment. *J. Med. Genet.* **2006**, *43,* 1–11.
3. Jenkins, Z. A. vK Germline Mutations in WTX Cause a Sclerosing Skeletal Dysplasia But Do Not Predispose to Tumorigenesis. *Nat. Genet.* **2009**, *41,* 95–100.
4. Kim, S. J. B. T. Identification of Signal Peptide Domain SOST Mutations in Autosomal Dominant Craniodiaphyseal Dysplasia. *Hum. Genet.* **2011**, *129,* 497–502.
5. Mantovani, G. Pseudohypoparathyroidism: Diagnosis and Treatment. *J. Clin. Endocrinol. Metab.* **2011**, *96,* 3020–3030.
6. Simpson, M. A. Mutations in FAM20C Are Associated with Lethal Osteosclerotic Bone Dysplasia (Raine Syndrome), Highlighting a Crucial Molecule in Bone Development. *Am. J. Hum. Genet.* **2007**, *81,* 906–912.
7. Stark, Z.; Savarirayan, R. Osteopetrosis. *Orphanet J. Rare Dis.* **2009**, *4,* 5.
8. Tolar, J.; Teitelbaum, S. L.; Orchard, P. J. Osteopetrosis. *N. Engl. J. Med.* **2004**, *351,* 2839–2849.
9. Uppal, S.; Diggle, C. P.; Carr, I. M., et al. Mutations in 15-Hydroxyprostaglandin Dehydrogenase Cause Primary Hypertrophic Osteoarthropathy. *Nat. Genet.* **2008**, *40,* 789–793.
10. Whyte, M. P.; Wenkert, D.; McAllister, W. H., et al. Dysosteosclerosis Presents as an "Osteoclast-Poor" Form of Osteopetrosis: Comprehensive Investigation of a 3-Year-Old Girl and Literature Review. *J. Bone Miner. Res.* **2010**, *25,* 2527–2539.
11. Zhang, Y.; Castori, M.; Ferranti, G., et al. Novel and Recurrent Germline LEMD3 Mutations Causing Buschke-Ollendorff Syndrome and Osteopoikilosis But Not Isolated Melorheostosis. *Clin. Genet.* **2011**, *79,* 556–561.
12. Zhang, Z.; Xia, W.; He, J., et al. Exome Sequencing Identifies SLCO2A1 Mutations as a Cause of Primary Hypertrophic Osteoarthropathy. *Am. J. Hum. Genet.* **2012**, *90,* 125–132.

CHAPTER

157

Chondrodysplasias

David L Rimoin[†]

Medical Genetics Institute, Cedars-Sinai Medical Center, Los Angeles, CA, USA

Ralph Lachman and Sheila Unger

Service of Medical Genetics, Lausanne, Switzerland

The skeletal dysplasias are a heterogeneous group of disorders associated with abnormalities in the size and shape of the limbs, trunk, and/or skull that frequently result in disproportionate short stature. Until the 1960s, most disproportionate dwarfs were considered to have either achondroplasia (short limbs) or Morquio disease (short trunk). It is now apparent that there are well over 450 distinct skeletal dysplasias that have been classified primarily on the basis of their clinical or radiographic characteristics.

Current nomenclature for the chondrodysplasias is somewhat confusing and is based on: (1) the part of the skeleton that is affected radiographically (e.g. the metaphyseal dysplasias) or clinically and/or radiographically (e.g. the acromesomelic dysplasias; (2) a Greek term that describes the appearance of the bone or the course of the disease (e.g. diastrophic [twisted] dysplasia, thanatophoric [death-seeking] dysplasia); (3) an eponym (e.g. Kniest dysplasia, Ellis-van Creveld syndrome); or (4) a term that attempts to describe the pathogenesis of the condition (e.g. achondroplasia, osteogenesis imperfecta) or the gene implicated (e.g. SEMD, Aggrecan type). Other clinical classifications have been based on the age of onset of the disorder and those disorders that manifest themselves at birth (achondroplasia) versus those that first manifest in later life (e.g. pseudoachondroplasia). Although this information is occasionally still useful in describing various skeletal dysplasias and in arriving at a diagnosis, the advent of nearly universal prenatal ultrasound examinations has altered our perception of the timeline and revealed this criterion to be often subjective in nature. Other disorders have been classified on the basis of their apparent mode of inheritance, for example the dominant and X-linked varieties of spondyloepiphyseal dysplasia. The most widely used method of differentiating the skeletal dysplasias has been the detection of skeletal radiographic abnormalities. Radiographic classifications are based on different parts of the long bones that are abnormal (epiphyses, metaphyses, or diaphyses). Thus, there are epiphyseal and metaphyseal dysplasias, which can be further divided depending on whether or not the spine is also involved (spondyloepiphyseal dysplasias, spondylometaphyseal dysplasias).

Clinical evaluation: As in the differential diagnosis of most other disorders, an accurate history, family history, and physical examination may lead one to the correct diagnosis. Certain skeletal dysplasias have prenatal onset and the more severe forms are detected by fetal ultrasound and are evident at birth. However, others may not manifest until late infancy or early childhood. Thus, a child who was normal until 2 years of age and then develops disproportionate short-limbed dwarfism is more likely to have pseudoachondroplasia or multiple epiphyseal dysplasia (MED) than achondroplasia or spondyloepiphyseal dysplasia congenita. A detailed physical examination may reveal the correct diagnosis or point to the likely diagnostic category. Affected individuals usually present with the complaint of disproportionate short stature and the abnormality in stature must first be documented by the use of the appropriate growth curves with adjustment for ethnic background and parental heights. In general, patients with disproportionate short stature have skeletal dysplasias, and those with relatively normal body proportions have endocrine, nutritional, prenatal, or other nonskeletal defects. There are exceptions to these rules, as congenital hypothyroidism can lead to disproportionate short stature, and a variety of skeletal dysplasias, such as osteogenesis imperfecta and hypophosphatasia, may result in normal body proportions. A disproportionate body habitus may not be readily apparent on casual physical examination. Thus, anthropometric measurements, such as upper to lower segment ratio, sitting height, and arm span must be obtained. First, one must establish whether the disproportionate shortening affects primarily the trunk or the limbs and, if the latter, whether it is proximal (rhizomelic), middle segment (mesomelic), or distal (acromelic), or a combination of these. The term rhizomelia is frequently inadvertantly abused and should not be applied without measuring.

A variety of dysmorphisms and malformations can be seen in the skeletal dysplasias. A disproportionately large

[†]Deceased.

head with frontal bossing and flattening of the bridge of the nose suggests achondroplasia. Multiple joint dislocations suggest Larsen syndrome, Ehlers–Danlos syndrome type VII, or otopalatodigital syndrome; less severe degrees of joint laxity, particularly of the hands, may be seen in other types of skeletal dysplasia (e.g. cartilage-hair hypoplasia and pseudoachondroplasia). Bone fractures may occur in all of the osteogenesis imperfecta syndromes and several types of hypophosphatasia, osteopetrosis, dysosteosclerosis, and achondrogenesis type IA (Houston–Harris).

Congenital cardiac defects are seen in several of the skeletal dysplasias. In chondroectodermal dysplasia, the most common lesion is an A-V cushion defect with a common atrium. In short rib-polydactyly syndrome I (Saldino–Noonan), a variety of very complex lesions involving the great vessels, transposition or double outlet right or left ventricle, and ventricular septal defect have been reported, and in short rib-polydactyly syndrome II (Majewski), the most common cardiac defect is transposition of the great vessels.

Thus, careful physical examination with delineation of all the skeletal and nonskeletal abnormalities can be quite helpful in arriving at a diagnosis.

The next step in the evaluation of the disproportionately short patient is to obtain a full set of skeletal radiographs. A full series of skeletal views, including anteroposterior (AP) and lateral views of the skull, AP and lateral views of the spine, and AP views of the pelvis and extremities, with separate views of the hands and feet, is optimal. A lateral radiograph of the knees can be helpful in diagnosing the recessive form of MED. Due to their intrinsic natural variability, radiographs of the feet are not generally helpful. An abridged skeletal survey should include at a minimum AP view of the pelvis, AP view of the knees, left hand, and lateral spine. Skeletal radiographs alone are often sufficient to make an accurate diagnosis, since the classification of skeletal dysplasias has been based primarily on radiographic criteria. Attention should be paid to the specific parts of the skeleton involved (spine, limbs, pelvis, skull) and within each bone to the location of the lesion (epiphysis, metaphysis, diaphysis). The skeletal radiographic features of many of these diseases change with age, and it is usually beneficial to review radiographs taken at different ages when possible. In some disorders, the radiographic abnormalities following epiphyseal plate fusion are nonspecific, so that the accurate diagnosis of an adult disproportionate dwarf may be impossible unless prepubertal films are available.

Radiologic diagnosis is also based on recognition of unique patterns of abnormal skeletal ossification. Some radiographic features characterize certain disorders. For example, in achondroplasia the acetabulae are flat with tiny sacrosciatic notches, rather square iliac wings with rounded corners, and an oval translucent area in the proximal femora and humeri in infants. The finding of a decreasing interpedicular distance from L1 to L5 is seen nearly uniformly in people with achondroplasia and hypochondroplasia.

Marked decrease in or absence of ossification of the vertebral bodies suggests a diagnosis of achondrogenesis. Deficient ossification of several vertebral bodies with good preservation of the pedicles is fairly specific for hypophosphatasia. Coronal clefts of the vertebrae can be seen in Kniest dysplasia, CHST3-related dysplasia, and rhizomelic chondrodysplasia punctata.

These examples are representative of a few of the many typical radiographic features seen in the skeletal dysplasias. In many instances, an accurate diagnosis can be made by simply examining the skeletal radiographs, but in other disorders only the general type of dysplasia, such as spondyloepiphyseal dysplasia, can be readily classified, and further information may be required to diagnose its exact form. Furthermore, only part of the heterogeneity of the skeletal dysplasias has been delineated to date and there are many disorders that will require morphologic, biochemical, or molecular studies for their exact delineation.

FURTHER READING

1. Jones, K. L. *Smith's Recognizable Patterns of Human Malformation*, 6th ed.; Elsevier: Philadelphia, 2005.
2. Kozlowski, K.; Beighton, P. *Gamut Index of Skeletal Dysplasias*; Springer–Verlag: Berlin, 2001.
3. Krakow, D.; Lachman, R. S.; Rimoin, D. L. Guidelines for the Prenatal Diagnosis of Fetal Skeletal Dysplasias. *Genet. Med.* **2009**, *11* (2), 127–133.
4. Maroteaux, P. *Les Maladies Ossèoses de L'Enfant*, 3rd ed.; Flamanon: Paris, 1995.
5. Ornoy, A.; Borochowitz, Z.; Lachman, R.; Rimoin, D. *Atlas of Fetal Skeletal Radiology*; Year Book Medical Publishers: Chicago, 1988.
6. Rimoin, D. L.; Lachman, R. S. Genetics Disorders of the Osseous Skeleton. In *McKusick's Heritable Disorders of Connective Tissue*; Beighton, P., Ed.; Mosby: St Louis, 1993; pp 557–689.
7. Spranger, J. Pattern Recognition in Bone Dysplasia. In *Endocrine Genetics and Genetics of Growth*; Papadatos, C. J.; Bartsocas, C. S., Eds.; Wiley: NY, 1985; pp 315–342.
8. Superti-Furga, A.; Bonafé, L.; Rimoin, D. L. Molecular-Pathogenetic Classification of Genetic Disorders of the Skeleton. *Am. J. Med. Genet.* **2001**, *106* (4), 282–293.
9. Taybi, H.; Lachman, R. *Radiology of Syndromes, Metabolic Disorders, and Skeletal Dysplasias*, 5th ed.; Mosby Elsevier: Philadelphia, 2006.
10. Warman, M. L.; Cormier-Daire, V.; Hall, C., et al. Nosology and Classification of Genetic Skeletal Disorders: 2010 Revision. *Am. J. Med. Genet. A* **2011**, *155A* (5), 943–968.

CHAPTER 158

Abnormalities of Bone Structure

William A Horton

Director of Research, Shriners Hospital for Children, Professor of Molecular and Medical Genetics, Oregon Health and Science University, Portland, OR, USA

The disorders discussed in this chapter are due to localized disturbances of the endochondral growth plate development or localized disturbances in the maturation of bone and related fibrous tissues. The clinical manifestations are typically asymmetric in distribution and less predictable than constitutional bone dysplasias in terms of ages when specific manifestations arise. In some cases, the disease manifestations reflect nonclassical Mendelian inheritance mechanisms such as somatic mosaicism and somatic loss of heterogeneity. A genetic basis has been uncovered for several of these conditions.

Dysplasia epiphysealis hemimelica is a developmental disorder of childhood characterized by asymmetrical growth of epiphyseal cartilage. It presents most often in 2–14-year-old boys with joint deformity of the knees and ankles with palpable lesions that appear on X-ray as lobulated multicentric masses adjacent to epiphyses or bones. Histologically, they resemble secondary ossification centers. The lesions and secondary deformities may increase during childhood; they exhibit little change after puberty. Treatment is individualized. Almost all cases are sporadic.

Hereditary multiple exostosis is characterized by the formation of many cartilage-capped exostoses that give rise to deformities of the growing skeleton. It usually presents in the first decade when palpable masses are detected and exostoses are documented by skeletal X-rays. The lesions typically increase in number and size during childhood most often near the ends of bones where they interfere with normal bone growth and produce local deformities. No new lesions form after completion of puberty, and the activity of existing lesions ceases, although asymptomatic ones may be detected by radiograph at any age. The most serious complication is malignant degeneration of the exostoses, which occurs in 3–10% of patients, although higher risks have been reported. Tumors most often arise in the pelvic region. Hereditary multiple exostosis is inherited as an autosomal dominant trait and mutations map to one of two EXT loci, mostly EXT1. *EXT* genes encode a glycosyltransferase required for synthesis and polymerization of heparan sulfate, which is essential for signaling many growth factors, such as Indian hedgehog in the growth plate. Recent evidence suggests that exostoses develop within clones of growth plate chondrocytes following loss of heterozygosity for EXT 1 or 2, which causes clonal loss of glycosyltransferase activity, deficient heparan sulfate production and aberrant growth factor signaling within the exostoses.

The Langer–Giedion syndrome, also designated as tricho-rhino-phalangeal (TRP) syndrome type II, combines multiple exostoses with features of the TRP syndromes and nonspecific manifestations including mild mental retardation and hearing loss. It is a contiguous gene syndrome resulting from small deletions of chromosome 8 (q24.11–q24.13). The deleted region contains the *EXT1* gene and the gene mutated in TRP syndrome types I and III.

Enchondromatosis (Ollier disease) is a rare condition resulting from the presence of multiple cartilaginous tumors in the metaphyses of growing bones, especially tubular bones. Characteristic deformities result from direct expansion of the tumors and from reduced linear growth of the affected bones. The tumors arise during childhood and the severity of deformities is quite variable. The tumors usually become quiescent after puberty. The tumors may undergo malignant degeneration. Enchondromatosis has occurred in a sporadic fashion in almost all cases reported to date. Mutations in the PTH/PTHrP type I receptor (PTHR1) have been identified in a few patients or tumors raising the possibility that the enchondromas result from somatic PTHR1 mutations.

Multiple enchondromas occur with cutaneous hemangiomas and other, primarily skin lesions in Maffucci syndrome. The hemangiomas are often detected at or shortly after birth. Adults with this syndrome are predisposed to neoplasia. The highest risk is for sarcomatous degeneration of the enchondromas, that is, 5–30% of patients, but malignant degeneration of hemangiomas and lymphangiomas also occurs, and patients may develop multiple primary tumors. Maffucci syndrome occurs sporadically.

The combination of multiple exostoses and enchondromas is found in the autosomal dominant condition, metachondromatosis. The lesions have unique radiographic characteristics and exhibit a tendency to regress or disappear in adulthood. Loss of function mutations of PTPN11 has been detected. PTPN11 encodes (PTP)

SHP2, a regulator of RAS/MAPK signaling downstream of many receptor tyrosine kinases.

Fibrous dysplasia of bone is characterized by the replacement of bone by dysplastic fibrous tissue. It can be divided into the much more common monostotic form and a polyostic form, which includes a number of extraskeletal manifestations, mostly cutaneous and endocrine, under the rubric of the McCune–Albright syndrome. Monostotic fibrous dysplasia occurs sporadically most often affecting the craniofacial bones, femur, tibia and humerus. The lesions typically present in adolescents with localized swelling, which may be painful, and characteristic radiographic changes. They usually become inactive after puberty; malignant degeneration occurs but is rare.

The bone lesions are the same in the McCune–Albright syndrome in which they arise at multiple locations, usually present before age 10 years and may lead to severe deformity. The skin lesions of this syndrome typically consist of irregular flat patches of brown pigmentation that are frequently evident at birth. Sexual precocity occurs in about one-third of patients, mostly females. Hyperthyroidism is also common. Both forms of fibrous dysplasia result from somatic mutations of the heterotrimeric signal transducer guanine nucleotide-binding protein, alpha-stimulating activity polypeptide 1, which couples cell surface receptors to adenyl cyclase–dependent downstream pathways. If the mutation occurs early in development, mutation-bearing cells are likely to reside in many locations giving rise to the McCune–Albright syndrome. Monostotic lesions reflect isolated somatic mutations in bone cells that occur later in life.

Cherubism is an autosomal dominant trait with considerable clinical variability. Affected children usually present with painless symmetrical swelling of the jaws between the ages of 18 months and 7 years. The swelling progresses rapidly over the next 2–3 years, after which it slows until puberty; the cherubic features may normalize in adulthood. Gain-of-function mutations have been detected in the adapter protein SH3BP2, which promotes inflammatory bone loss mediated substantially through tumor necrosis factor alpha.

CHAPTER 159

Arthrogryposes (Multiple Congenital Contractures)

Judith G Hall

UBC and Children's and Women's Health Centre of BC, Department of Medical Genetics, BC's Children's Hospital, Vancouver, BC, Canada

Arthrogryposis multiplex congenita (AMC) is a term that has been used for almost a century to describe conditions with nonprogressive multiple congenital joint contractures. The conditions that have been called arthrogryposis range from well-known syndromes to nonspecific combinations of joint contractures. The term has become descriptive rather than diagnostic and is now used in connection with a very heterogenous group of patients and disorders, all of which have in common multiple congenital joint contractures. The term arthrogryposis implies a generalized involvement, with multiple joints having congenital contractures, and is usually reserved for conditions that involve more than one part of the body. We have begun to realize that over 350 specific disorders are or maybe associated with multiple contractures in the newborn, and that, although most improve over time, some are progressive.

Arthrogryposis has been said to be a rare condition, but in fact it has an incidence somewhere between 1 in 3000 and 1 in 5000 live births. The medical literature on arthrogryposis is very confusing. Over the past 50 years, more than 3000 articles have been published describing various types of arthrogryposis, but the term has been used very loosely, initially as a diagnostic term but more recently as a clinical sign or as a general category of disorders. In addition to the imprecise use of the term, the medical literature on arthrogryposis is confusing, because many authors fail to fully describe the clinical features of their cases, or to separate subgroups. In the past, authors have often lumped together several patients with congenital contractures, who in actuality represent different specific entities, and have then made generalizations about recurrence, management, prognosis, and treatment. More recently, responsible genes have been identified for many specific types of arthrogryposis and the range of variability related to a specific gene is being defined.

This chapter attempts to sort out some of the major clinical entities with congenital contractures, to describe a clinical approach to distinguishing heterogeneity, to discuss the investigation of the individual with congenital contractures, and to make some comments about genetics, recurrence risk, prenatal diagnosis, and therapy.

It has become increasingly apparent, both from animal studies and from human work, that anything that leads to decreased movement in utero may also lead to congenital contractures or fixation of joints at birth. Even prolonged hypotonia in utero may lead to congenital contractures. Swinyard has called this a "collagenic response," with thickening of the joint capsule and fibrous development in muscle tissue. Undoubtedly, lack of movement either produces a set of cytokines or fails to produce a set of cytokines that ultimately leads to this connective tissue response. The growing embryo or fetus that has superimposed limitation of movement may develop even more marked contractures because the process of growth may compound the contractures leading to additional deformation.

The potential causes of limitation of movement in utero include:

1. Myopathic processes, including myopathies and abnormal muscle structure or function, such as absence or loss of muscle tissue and congenital myopathies
2. Neuropathic processes, including abnormalities in nerve structure or function, either central or peripheral; failure of nerves to form, migrate, or myelinate; and congenital neuropathies
3. Neuromuscular end-plate abnormalities, including abnormal structure receptors and transmitters
4. Abnormal connective tissue, including bone, joint and tendon abnormalities; skin abnormalities (such as restrictive dermopathy); and abnormal tendon attachments
5. Limitation of space or restriction of movement within the uterus, as in the case of twins, structural anomalies

of the uterus, amniotic bands, fibroids, or decreased amniotic fluid, as with leakage
6. Problems related to maternal illness, including maternal infections, interference with fetal neurotransmitters as by antibodies, etc.
7. Maternal exposures related to medications, drugs, and environmental factors, such as muscle relaxants, high temperature, fever, etc.
8. Compromise of the blood supply to the placenta and/or the embryo/fetus

An approach to sorting out various types of congenital contractures includes taking a careful history of the pregnancy and delivery; a full family history; a detailed physical examination with documentation of what parts of the body are involved in the process; photographs at different ages; measurements, including range of movement of various joints; a natural history of complications and response to therapy; laboratory data, such as muscle biopsies; imaging studies; autopsy results, including central nervous system (CNS) histopathology; chromosome studies; and molecular studies as appropriate, including comparative genomic hybridization microarray. This kind of evaluation has proven to be important and necessary in distinguishing different types of arthrogryposis.

The clinical approach found to be most useful has been to first distinguish three categories of congenital contractures on a clinical basis: (1) primarily limb involvement; (2) musculoskeletal involvement plus other system malformations or anomalies; and (3) musculoskeletal involvement plus lethality, CNS dysfunction, and/or significant mental retardation/mental disability.

Every effort to achieve a specific diagnosis should be made; however, for those families in which a specific diagnosis cannot be made, the empiric recurrence risk to unaffected parents of an affected child or to the affected individual with arthrogryposis is in the 3–5% range.

Multiple congenital contractures are relatively frequent and often part of recognizable syndromes. Marked heterogeneity exists. Careful investigation should lead to a specific diagnosis in more than half the cases, allowing more specific prognostication, counseling, and therapy.

RELEVANT WEBSITES

Arthrogryposis: A Text Atlas. http://www.globalhelp.org/publications/books/book_arthrogryposis.html.

Avenues. A National Support Group for Arthrogryposis Multiplex Congenital. http://www.avenuesforamc.com/.

CHAPTER

160

Common Skeletal Deformities

William A Horton

Director of Research, Shriners Hospital for Children, Professor of Molecular and Medical Genetics, Oregon Health and Science University, Portland, OR, USA

Familial aggregation occurs frequently among common skeletal deformities even after simply inherited syndromes have been excluded. Environmental factors have been implicated in some cases, but genetic factors are also thought to contribute as well. Advances in genomic technologies have begun to identify chromosomal regions and even specific genes that may contribute to these disorders, although in most cases the mechanisms through which they act remain elusive. The genetics of several common conditions are examined in this chapter.

The incidence of idiopathic scoliosis—curvature of the spine—varies from 2% to 8% in the population. It usually becomes evident during periods of rapid growth, and is divided into a juvenile form that occurs most often in infant boys, and an adolescent form that occurs typically in girls during their pubertal growth spurt. Twin studies have shown a substantially higher rate of concordance for monozygotic compared to dizygotic twins. Similarly, family surveys have revealed a higher incidence of scoliosis in first-degree relatives, especially girls. Most evidence to date supports a multifactorial model for genetic predisposition to scoliosis. It also suggests female gender is a strong determinant in converting the genetic predisposition to clinical disease. Recent genome-wide linkage and association studies have pointed to several different chromosomal locations and gene loci that may contribute to the predisposition; however, none have been firmly established in this role.

Spondylolisthesis refers to slippage of a vertebral body over another, most often in the lumbar spine. It is thought to be preceded by a defect in the posterior inferior process of the relevant vertebral arch, spondylolysis, which occurs in 4–8 % of the population. Spondylolysis with and without slippage sometimes aggregates within families, and radiographic surveys of family members have detected spondylolysis in as many as 27% of near relatives. Type IX collagen has been implicated in spondylolisthesis based on detection of genetic polymorphisms in type IX collagen genes in patients undergoing surgery for spinal stenosis with spondylolisthesis.

Congenital dislocation of the hip (CDH), sometimes referred to as developmental dysplasia of the hip, is characterized by the displacement of the femoral head outside the acetabulum before or slightly after birth. CDH is a common birth defect and many predisposing factors have been identified including time of birth in the year, intrauterine posture and especially female gender. Familial aggregation of CDH has long been recognized. Twin studies have consistently shown a higher concordance rate for monozygotic than for dizygotic twins and family surveys have revealed a higher incidence of CDH in close relatives. The genetic predisposition is thought to have two components. The first involves the configuration of the acetabulum or pelvic socket into which the femoral head fits. Multiple genes determine the acetabular shape including depth. The shallow end of the depth range is called acetabular dysplasia; it predisposes to CDH. The other factor is generalized joint laxity, which is inherited mainly as a polygenic trait. Genome-wide investigations have identified two candidate genes for CDH predisposition, TBX4 and GDF5, both of which are regulators of early skeletal development.

Idiopathic isolated clubfoot is a common birth defect. Three different forms are recognized and all three exhibit familial aggregation. The most common form, talipes equinovarus, is characterized by adduction of the forefoot, inversion of the heel, and plantar flexion of the forefoot and ankle. It occurs more often in boys than in girls and in first-degree relatives compared to second- and third-degree relatives. Genetic analyses have suggested that both Mendelian and non-Mendelian factors influence the predisposition. There is recent evidence for contributions from *TBX4* and *HOX* genes, although the nature of these contributions is unknown. In the second form of clubfoot, talipes calcaneovalgus, there is dorsal flexion of the forefoot and the plantar surface of the foot faces laterally. It is mild, often correcting spontaneously, and it occurs more commonly in girls often in first-born infants suggesting that uterine constraint might be an etiologic factor. The incidence of affected sibs is 4.5%, suggesting multifactorial inheritance. Inversion and adduction of the forefoot are found alone in metatarsus varus. It resembles talipes calcaneovalgus, is often mild, and may go unnoticed. It has been postulated that it could represent the very mild end of the diastrophic dysplasia clinical spectrum, but this notion is controversial.

The juvenile osteochondroses are a group of disorders in which localized noninflammatory arthropathies result from regional disturbances of skeletal growth. There is ischemic necrosis of either primary or secondary endochondral ossification centers. Most of the abnormalities occur sporadically, but familial forms have been described. Dominant inheritance of Legg–Perthes disease, osteonecrosis of the capital femoral epiphysis, has been observed; however, most of these cases have turned out to be mild multiple epiphyseal dysplasia (Ribbing type). When the latter cases are excluded, this condition appears to not be inherited. Blount disease, a growth disturbance of the medial aspect of the proximal tibial growth plate, has also been reported to show autosomal dominant inheritance. But most evidence suggests that common environmental factors are largely responsible for the familial aggregation. Osteochondritis dissecans involving multiple sites, especially the knees, hips, elbows, and ankles, appears to behave as an autosomal dominant trait in some families. There also appears to be a poorly defined genetic predisposition to Scheuermann disease, which affects the spine.

FURTHER READING

1. Miller, N. H. Genetics of Familial Idiopathic Scoliosis. *Clin. Orthop. Relat. Res.* **2007,** *462,* 6–10.
2. Ogilvie, J. Adolescent Idiopathic Scoliosis and Genetic Testing. *Curr. Opin. Pediatr.* **2010,** *22* (1), 67–70.
3. Matsui, Y.; Mirza, S. K.; Wu, J. J.; Carter, B.; Bellabarba, C.; Shaffrey, C. I., et al. The Association of Lumbar Spondylolisthesis with Collagen IX Tryptophan Alleles. *J. Bone Joint Surg. Br.* **2004,** *86* (7), 1021–1026.
4. Lee, M. C.; Eberson, C. P. Growth and Development of the Child's Hip. *Orthop. Clin. North Am.* **2006,** *37* (2), 119–132, v.
5. Osarumwense, D.; Popple, D.; Kershaw, I. F.; Kershaw, C. J.; Furlong, A. J. What Follow-Up Is Required for Children with a Family History of Developmental Dysplasia of the Hip? *J. Pediatr. Orthop. B* **2007,** *16* (6), 399–402.
6. Stevenson, D. A.; Mineau, G.; Kerber, R. A.; Viskochil, D. H.; Schaefer, C.; Roach, J. W. Familial Predisposition to Developmental Dysplasia of the Hip. *J. Pediatr. Orthop.* **2009,** *29* (5), 463–466.
7. Dietz, F. R.; Cole, W. G.; Tosi, L. L.; Carroll, N. C.; Werner, R. D.; Comstock, D., et al. A Search for the Gene(s) Predisposing to Idiopathic Clubfoot. *Clin. Genet.* **2005,** *67* (4), 361–362.
8. Engesaeter, L. B. Increasing Incidence of Clubfoot: Changes in the Genes or the Environment? *Acta Orthop.* **2006,** *77* (6), 837–838.
9. Kruse, L. M.; Dobbs, M. B.; Gurnett, C. A. Polygenic Threshold Model with Sex Dimorphism in Clubfoot Inheritance: The Carter Effect. *J. Bone Joint Surg. Am.* **2008,** *90* (12), 2688–2694.
10. Damborg, F.; Engell, V.; Andersen, M.; Kyvik, K. O.; Thomsen, K. Prevalence, Concordance, and Heritability of Scheuermann Kyphosis based on a Study of Twins. *J. Bone Joint Surg. Am.* **2006,** *88* (10), 2133–2136.

CHAPTER

161

Hereditary Noninflammatory Arthropathies

Mariko L Ishimori

Division of Rheumatology, Cedars-Sianai Medical Center, Los Angeles, CA, USA

161.1 INTRODUCTION

Noninflammatory arthropathy is a major manifestation of a number of genetic disorders. There is a great variety of phenotypes, which present a diagnostic challenge in the clinical setting. In some of these conditions, this arthropathy is generalized, while in others it predominates in the hip joints, with or without significant involvement of the spine. Degenerative osteoarthritis (OA), especially of the weight-bearing joints, is a common complication in a large number of genetic skeletal dysplasias and disorders (*1*). This chapter reviews the following hereditary noninflammatory arthropathies, focusing on the most recent genetic linkages: spondyloepiphyseal dysplasia (SED), familial osteoarthropathy with a focus on hand osteoarthritis, primary osteoarthropathy of the hip, and Mseleni joint disease (MJD).

161.2 SPONDYLOEPIPHYSEAL DYSPLASIAS

SEDs are characterized by predominant involvement of the vertebral bodies and epiphyses of the proximal joints, as well as considerable phenotypic and genetic heterogeneity; in some forms dwarfism with a characteristics shortened trunk is severe, while in others stature approaches normality. Myopia and hearing loss are variable syndromic components. The classical severe form of SED congenital (SEDC) is inherited as an autosomal dominant trait, and is recognizable at birth, but other milder autosomal dominant forms may only become evidence in late childhood.

Linkage to the type II collagen gene (*COL2A1*) has been demonstrated in some families with classical SEDC but not in others (*2–8*). In the same way, some of the milder, late-onset forms of autosomal dominant SED have been shown to be linked to type II collagen. A mutation in aggrecan has been described in an autosomal recessive form of spondyloepimetaphyseal dysplasia (SEMD), aggrecan type (*9*). Characterization of mutations in *COL2A1* has been undertaken, and the Cardiff University Human Gene Mutation Database (*10*) lists at least 33 human mutations that have been described. There is great phenotypic variation in SED, making genotype–phenotype correlations more challenging (*11*), and it is possible that there are hot spots in this gene with some forms of mild SED in which OA occurs (*12*).

In the context of SED, it is noteworthy that in pseudoachondroplasia and some forms of multiple epiphyseal dysplasia (AD-MED), the disease-associated genes are allelic on chromosome 19. The phenotype in these conditions is the result of defective cartilage oligomeric matrix protein. AD-MED may also result from mutations in matrillin-3 (MAT3) and type IX collagen (*13*).

161.3 FAMILIAL OSTEOARTHROPATHY

The major problem in the elucidation of the genetic determinants of OA is the considerable heterogeneity of the disorder and the great difficulty in precise phenotypic delineation of autonomous entities within this general category. Primary OA, or degenerative arthropathy, is a very common disorder of middle and old age. Risk factors include obesity and trauma, but this is not solely an age-related disorder and there is some evidence for a genetic component (*14*). The findings of the large-scale Framingham offspring investigation (*15*) and the Baltimore longitudinal study (*16*) have also indicated that genetic factors are involved. Twin studies have been suggestive of a significant genetic component in OA of the hand and knee (*17*). It is possible, though unproven, that some "normal polymorphisms" of type II collagen might convey an increased propensity to the development of OA. It is also possible that subchondral bone may be primarily involved in the pathogenesis (*18*).

The best example of familial OA is probably Heberden's arthropathy. Stecher in 1941 noted a hereditary disposition for hand OA expression, with a twofold excess of disease in mothers and a threefold excess in sisters in patients with Heberden's nodes compared to unrelated controls (*19*). The form of generalized OA that is accompanied by Heberden's nodes may represent a sex-influenced or sex-limited autosomal dominant trait (*16*). The results of familial aggregation and twin studies provide strong evidence for an inherited predisposition to hand OA. The exact mode of inheritance remains unclear and multiple studies have resulted in reports of association or linkage with a variety of genetic loci (*20*). Hunter et al. have concluded that a joint-specific approach to hand OA genetics may provide greater linkage as evaluating hand OA as a general entity may decrease the strength of association (*21*). Mutations in *MATN3* have been reported in association with first carpometacarpal joint OA (*22,23*).

161.4 PRIMARY OSTEOARTHROPATHY OF THE HIP

In the context of primary OA, the question arises as to whether or not OA of the hip joint, in the absence of significant involvement of other joints, is an independent genetic entity. Family clustering has been documented (*14,24,25*), and twin studies have yielded positive results (*26*). Family clustering was identified in a large investigation in Iceland (*27*) after which a susceptibility locus on chromosome 16p was identified in a large family (*28*). At the clinical, radiologic, histologic, and phenotypic levels, this familial OA of the hip was indistinguishable from the idiopathic, nonfamilial OA of the hip (*29*).

The determinant genes for the α_1 chain of type II collagen and the vitamin D receptor are adjacent loci on 12q. Granchi and colleagues (*30*) investigated polymorphic sites in these genes in 143 persons in whom hip joint replacement had been undertaken for primary or secondary OA. The findings were interpreted as providing evidence for a genetic component for the risk of OA in persons with severe hip dysplasia.

It is possible that collagen polymorphisms might be involved in the pathogenesis of OA of the hip. The Genetics, Osteoarthritis and Progression study identified the deiodinase, iodothyronine, type II (*D2*) gene (*DIO2*) as a susceptibility gene for hip OA and found that a mutation in this gene is more likely to increase the vulnerability of cartilage to nonoptimal hip morphology instead of causing these shapes (*31*).

161.5 MSELENI JOINT DISEASE

MJD is a remarkable disorder that presents with widespread generalized degenerative OA in late childhood and causes severe crippling handicap in adulthood (*32*). The condition occurs in high frequency in an isolated area in northern KwaZulu-Natal, South Africa, and was first described in 1970 (*33*). A condition very similar to MJD, Kasin–Beck disease occurs in the Urov Valley of Siberia, parts of China and North Korea (*34*). Selenium deficiency, aflatoxins in foodstuffs and hypoxia (*7*) have been incriminated in the latter, but despite more than two decades of intensive investigations, no environmental determinants have been identified for MJD.

To date, extensive research has yet to uncover the etiology of MJD. There is no evidence for Mendelian inheritance patterns and epigenetic changes in response to the environment have been postulated (*35*). By the end of 2004, there was anecdotal evidence that the incidence of MJD in the Mseleni region was diminishing rapidly. This observation, if substantiated, would be suggestive of the fluctuation of the influence of an unrecognized environmental agent.

TABLE 161-1 The Hereditary Noninflammatory Arthropathies
Defects of the cartilage matrix
SED group of disorders, including Kniest and Stickler syndromes, and the familial hip joint dysplasias
(Proven or possible defects of type II collagen)
Alkaptonuria
(Abnormal binding of polymers of homogentisic acid to cartilage collagen)
Gout and pseudogout
(Deposition of calcium pyrophosphate and hydroxyapatite crystals in the cartilage matrix)
Infiltration and/or aseptic necrosis (femoral head)
Storage disorders
Wilson disease
Hemochromatosis
Gaucher, Fabry, and Farber diseases
Hemoglobinopathies
Mechanical collapse due to interaction of primary defects in cartilage and external forces
(notably the femoral head)
Skeletal dysplasias
Hypermobility syndromes
Primary hip joint dysplasias
Perthes disease
Slipped femoral capital epiphyses
Neuropathic arthropathy
Amyloidosis
Charcot–Marie–Tooth syndrome
Déjérine–Sottas syndrome
Familial dysautonomia

SED, spondyloepiphyseal dysplasia.

REFERENCES

1. Cicuttini, F. M.; Spector, T. D. Genetics of Osteoarthritis. *Ann. Rheum. Dis.* **1996,** *55,* 665–667.
2. Anderson, I. J.; Goldberg, R. B.; Marion, R. W., et al. Spondyloepiphyseal Dysplasia Congenita: Genetic Linkage to Type II Collagen (COL2A1). *Am. J. Hum. Genet.* **1990,** *46,* 896–901.
3. Murray, L.; Bautista, J.; James, P. L.; Rimoin, D. L. Type II Collagen Defects in the Chondrodysplasias: 1. Spondyloepiphyseal Dysplasias. *Am. J. Hum. Genet.* **1989,** *45,* 5–15.
4. Wordsworth, P.; Ogilvie, D.; Priestly, L., et al. Structural and Segregation Analysis of the Type II Collagen gene (*COL2A1*) in Some Heritable Chondrodysplasias. *J. Med. Genet.* **1988,** *25,* 521–527.
5. Zhang, Z.; He, J.-W.; Fu, W.-Z., et al. Identification of Three Novel Mutations in the COL2A1 Gene in four Unrelated Chinese Families with Spondyloepiphyseal Dysplasia Congenita. *Biochem. Biophys. Res. Commun.* **2011,** *413,* 504–508.
6. Mark, P. R.; Torres-Martinez, W.; Lachman, R. S.; Weaver, D. D. Association of a p.Pro786Leu Variant in COL2A1 with Mild Spondyloepiphyseal Dysplasia Congenita in a Three-Generation Family. *Am. J. Med. Genet. A* **2011,** *155,* 174–179.
7. Zhang, F.; Guo, X.; Wang, W., et al. Genome-Wide Expression Analysis Suggests an Important Role of Hypoxia in the Pathogenesis of Endemic Osteochondropathy Kashin-Beck Disease. *PLoS One* **2011,** *6,* e22983.
8. Hoornaert, K. P.; Dewinter, C.; Vereecke, I., et al. The Phenotypic Spectrum in Patients with Arginine to Cystein Mutations in the COL2A1 Gene. *J. Med. Genet.* **2006,** *43,* 406–413.
9. Tompson, S.; Merriman, B.; Funari, V., et al. A Recessive Skeletal Dysplasia, SEMD Aggrecan Type, Results Forom a Missense Mutation Affecting the C-Type Lectin Domain of Aggrecan. *Am. J. Hum. Genet.* **2009,** *84,* 72–79.
10. Institute of Medical Genetics in Cardiff. The Human Gene Mutation Database. http://www.hgmd.org.
11. Kannu, P.; Bateman, J.; Savarirayan, R. Clinical Phenotypes Associated with Type II Collagen Mutations. *J. Paediatr. Child Health* **2011**10.1111/j.1440–1754.2010.01979.x, [Epub ahead of print].
12. Bleasel, J. F.; Holderbaum, D.; Mallock, V., et al. Hereditary Osteoarthritis with Mild Spondyloepiphyseal Dysplasia: Are There “Hot Spots” on COL2AI? *J. Rheumatol.* **1996,** *23,* 1594–1598.
13. Jackson, G. C.; Mittaz-Crettol, L.; Taylor, J. A. Pseudoachondroplasia and Multiple Epiphyseal Dysplasia: A 7-year Comprehensive Analysis of the Known Disease Genes Identify Novel and Recurrent Mutations and Provides an Accurate Assessment of their Relative Contribution. *Hum. Mutat.* **2011**10.1002/humu.21611, [Epub ahead of print].
14. Chitnavis, J.; Sinsheimer, J. S.; Clipsham, K., et al. Genetic Influences in End-Stage Osteoarthritis: Sibling Risks of Hip and Knee Replacement for Idiopathic Osteoarthritis. *J. Bone Joint Surg. Br.* **1997,** *79,* 660–664.
15. Felson, D. T.; Couropmitree, N. N.; Chaisson, C. E., et al. Evidence for a Mendelian Gene in a Segregation Analysis of Generalized Radiographic Osteoarthritis: The Framingham Study. *Arthritis Rheum.* **1998,** *41,* 1064–1071.
16. Hirsch, R.; Lethbridge-Cejku, M.; Hanson, R., et al. Familial Aggregation of Osteoarthritis: Data from the Baltimore Longitudinal Study on Ageing. *Arthritis Rheum.* **1998,** *41,* 1227–1232.
17. Spector, T. D.; Cicuttini, F.; Baker, J., et al. Genetic Influences on Osteoarthritis in Women: A Twin Study. *BMJ* **1996,** *312,* 940–943.
18. Dieple, P. Editorial. Osteoarthritis: Time to Shift the Paradigm. *BMJ* **1999,** *318,* 1299–1300.
19. Stecher, R. M. Heberden’s Nodes: Heredity in Hypertrophic Arthritis of the Finger Joints. *Am. J. Med. Sci.* **1941,** *201,* 801–809.
20. Livshits, G.; Kalichman, L.; Cohen, Z.; Kobyliansky, E. Mode of Inheritance of Hand Osteoarthritis in Ethnically Homogenous Pedigrees. *Hum. Biol.* **2002,** *74,* 849–860.
21. Hunter, D. J.; Demissie, S.; Cupples, L. A., et al. A Genome Scan for Joint-Specific Hand Osteoarthritis Susceptibility: The Framingham Study. *Arthritis Rheum.* **2004,** *50,* 2489–2496.
22. Stefansson, E.; Jonsson, H.; Ingvarsson, T., et al. Genomewide Scan for Hand Osteoarthritis: A Novel Mutation in Matrillin-3. *Am. J. Hum. Genet.* **2003,** *72,* 1448–1459.
23. Min, J. L.; Meulenbelt, I.; Riyazi, N., et al. Association of Matrillin-3 Polymorphisms with Spinal Disc Degeneration and Osteoarthritis of the First Carpometacarpal Joint of the Hand. *Ann. Rheum. Dis.* **2006,** *65,* 1060–1066.
24. Lanyon, P.; Doherty, S.; Muir, K.; Doherty, S. Strong Genetic Predisposition to Hip Asteoarthritis. *Arthritis Rheum.* **1998,** *41,* S351.
25. Lindberg, H. Prevalence of Primary Coxarthritis in Siblings of Patients with Primary Coxarthritis. *Clin. Orthop.* **1986,** *203,* 273–275.
26. MacGregor, A. J.; Spector, T. D. Twins and the Genetic Architecture of Osteoarthritis. *Rheumatology (Oxford)* **1999,** *38,* 583–588.
27. Ingvarsson, T.; Stefansson, S. E.; Hallgrimsdottir, I. B., et al. The Inheritance of Hip Osteoarthritis in Iceland. *Arthritis Rheum.* **2000,** *43,* 2785–2792.
28. Ingvarsson, T.; Stefansson, S. E.; Gulcher, J. R., et al. A Large Icelandic Family with Early Osteoarthritis of the Hip Associated with a Susceptibility Locus on Chromsome 16p. *Arthritis Rheum.* **2001,** *44,* 2548–2555.
29. Ingvarsson, T. Prevalence and Inheritance of Hip Osteoarthritis in Iceland. *Acta Orthop. Scand. Suppl.* **2000,** *298,* 1–46.
30. Granchi, D.; Stea, S.; Sudanese, A., et al. Association of Two Gene Polymorphisms with Osteoarthritis Secondary to Hip Dysplasia. *Clin. Orthop.* **2002,** *403,* 108–117.
31. Viljoen, D.; Fredlund, V.; Ramesar, R.; Beighton, P. Brachydactylous Dwarfs of Mseleni. *Am. J. Med. Genet.* **1993,** *46,* 636–640.
32. Du Toit, G. T. Hip Disease of Mseleni. *Clin. Orthop.* **1979,** *141,* 223–236.
33. Wittman, W.; Fellingham, S. A. Unusual Hip Disease in Remote Part of Zululand. *Lancet* **1970,** *295,* 852–853.
34. Sokoloff, L. Endemic forms of Osteoarthritis. *Clin. Rheum. Dis.* **1985,** *11,* 187–202.
35. Gibbon, V. E.; Harington, J. S.; Penny, C. B.; Fredlund, V. Mseleni Joint Disease: A Potential Model of Epigenetic Chondrodysplasia. *Joint Bone Spine* **2010,** *77,* 399–404.

Pathways

Pathways—Cohesinopathies, 607
Genes and Mechanisms in Human Ciliopathies, 612

CHAPTER

162

Pathways—Cohesinopathies

Matthew A Deardorff

The Perelman School of Medicine at the University of Pennsylvania, Philadelphia, PA, USA

Ian D Krantz

Division of Human Molecular Genetics, The Children's Hospital of Philadelphia, Philadelphia, PA, USA

The group of diagnoses linked together by their common disruption of the chromosomal cohesion complex, cohesin, or its regulators have been collectively termed "cohesinopathies." Multiple proteins in the cohesin pathway are also involved in additional fundamental biological events such as double strand DNA break repair transcriptional regulation. It is cohesin's role in transcriptional regulation that likely plays a key role during development and results in several congenital disorders (collectively termed the "cohesinopathies") when disrupted.

162.1 CORNELIA DE LANGE SYNDROME

Cornelia de Lange syndrome (CdLS; OMIM #122470, #300590, and #610759), the first disorder found to be caused by alterations in a cohesin regulatory protein, is currently the best recognized and most well studied. CdLS, also known as Brachmann–de Lange syndrome, is a rare genetically heterogeneous disorder affecting multiple organs and systems during development. CdLS has been estimated to occur in about 1:10,000 to 1:50,000 individuals. However, the actual incidence is likely higher as the clinical presentations are quite variable. With recent molecular diagnostic capabilities, we now realize a higher prevalence of very mildly affected individuals that have been underappreciated and historically were less likely to be diagnosed as CdLS. Almost all cases are sporadic resulting from dominantly acting de novo mutations, although recurrence in siblings due to parental (germ line) mosaicism has been reported, as has direct transmission from a mildly affected individual to their children. Somatic growth, central nervous system, craniofacial, musculoskeletal, and gastrointestinal systems are the most commonly affected. Additional systemic involvement includes auditory, genitourinary, cardiac, integumentary, hematopoietic and ophthalmologic. Mutations in three genes, *NIPBL* on chromosome 5p13, *SMC1A* on chromosome Xp11, and *SMC3* on chromosome 10q25, can be identified (collectively) in 70% of individuals with clinically diagnosed CdLS, with *NIPBL* being the major contributor. All of the identified causative genes to date are structural or regulatory components of cohesin.

Approximately 60% of CdLS probands have a heterozygous mutation in *NIPBL*. Genotype–phenotype correlations among a large cohort of CdLS probands indicate that presumably haploinsufficient *NIPBL* mutations (truncating mutations, splice site mutations, or frame-shifting indels) often result in a more severe cognitive and structural phenotype than do missense mutations. Approximately 5% of probands with a clinical diagnosis of CdLS were found to have missense or small in-frame deletion mutations in *SMC1A*, and one individual was found to have an in-frame 3-bp deletion in the *SMC3* gene. Individuals with *SMC1A* and *SMC3* mutations tend to have mild to moderate mental retardation without significant impairments in growth or structural abnormalities of the limb or other organ systems. Individuals with *SMC1A* mutations tend to have a more prominent nasal bridge than is typically seen in patients with *NIPBL* mutations, and the majority have normal growth parameters at birth and even later in life. For most individuals with *SMC1A* mutations, walking and speech are often acquired, and overall this group exhibits a much milder level of cognitive involvement. The genotype–phenotype correlations seen in CdLS are summarized in Figure 162-3.

Overall, approximately 65% of CdLS probands with a confident clinical diagnosis have mutations in one of the cohesin-associated genes (*NIPBL*, *SMC1A*, or *SMC3*). Among individuals with a more severe or "classic" CdLS phenotype, the mutations detection rate is even higher (~80%) and much more likely to involve the *NIPBL* gene. The molecular etiology of the remaining 35% of probands is unknown at this time.

162.2 COHESIN BIOLOGY

While mitotic chromosomes were first described in the late 1800s, it was not until 1985 that the first yeast mutant involved in maintaining the stability of chromosomes

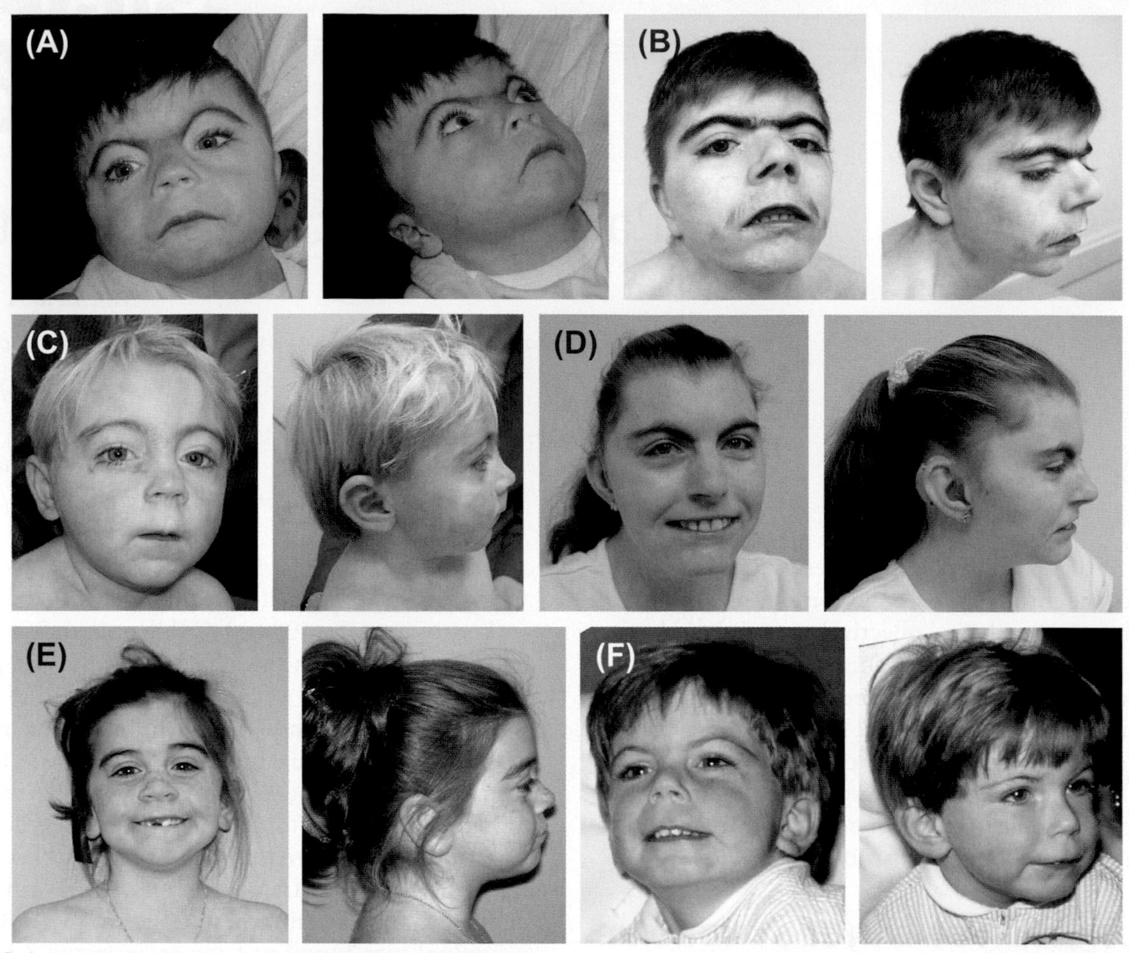

FIGURE 162-1 Facial Characteristics in Cornelia de Lange Syndrome. (A) Typically affected 1-year-old male, and (B) 19-year-old male, both with severe truncating mutations in *NIPBL*. Note characteristic facial features (arched eyebrows, synophrys, ptosis, anteverted nares, long philtrum, thin upper lip with down-turned corners and micrognathia) and severe asymmetrical defects of the forearms. (C) More mildly affected 3-year-old boy and (D) 19-year-old girl with missense mutations in *NIPBL*. Note characteristic but more subtle facial features. (E) 7-year-old girl and (F) 5-year-old boy with *SMC1A* mutations and mild facial characteristics.

through mitosis was described. Termed *SMC1* for stability of mini-chromosomes, the gene was subsequently cloned in 1993. With the identification of additional structurally related proteins, "SMC" proteins were renamed "structural maintenance of chromosome" proteins. The term "cohesins" was coined to clarify the involvement of yeast Smc1, Smc3 and Scc1(Rad21) in a "common cohesion apparatus for all eukaryotic cells in mitosis and meiosis." Since that time, over 25 proteins have been implicated in cohesin's canonical role of controlling appropriate sister chromatid segregation during the mitotic cell cycle.

Cohesin is a dynamic complex regulated at various cell cycle stages by multiple mechanisms. The canonical role of cohesin, as first described in yeast, is to control sister chromatid segregation during both mitosis and meiosis. Four evolutionarily conserved subunits form the core structural component of the cohesin complex, two SMC proteins SMC1A and SMC3, a kleisin protein RAD21 (also known as MCD1 or SCC1), and STAG1/STAG2 (also known as SA1/SA2). Paralogs SMC1B, REC8 and STAG3, respectively, are cohesin subunits with specialized roles in meiosis. Homologs of the cohesin complex and its regulatory genes have been identified in all eukaryotic model systems including *Saccharomyces cerevisiae*, *Schizosaccharomyces pombe*, *Drosophila melanogaster*, *Xenopus laevis*, mouse and in humans.

NIPBL heterodimerizes with hSCC4 (MAU-2) and is required for loading of cohesin onto chromatin in mitosis, and to all chromosome regions currently under study such as heterochromatin, CARs, centromeres, and DSBs. Despite its importance, the mechanism of NIPBL loading of cohesin onto DNA is poorly understood. The NIPBL/MAU-2 complex seems to be involved in all cohesin activities including SMC ATPase activation, hinge dimerization, chromatin binding, and chromatin remodeling. Cells elaborately regulate chromatin binding of cohesin, both temporally and spatially, although most of these mechanisms are unknown at present. This complex regulation enables cohesin to perform diverse biological functions and it is anticipated that mutations in associated proteins are expected to contribute to multiple human diseases, both known or unknown.

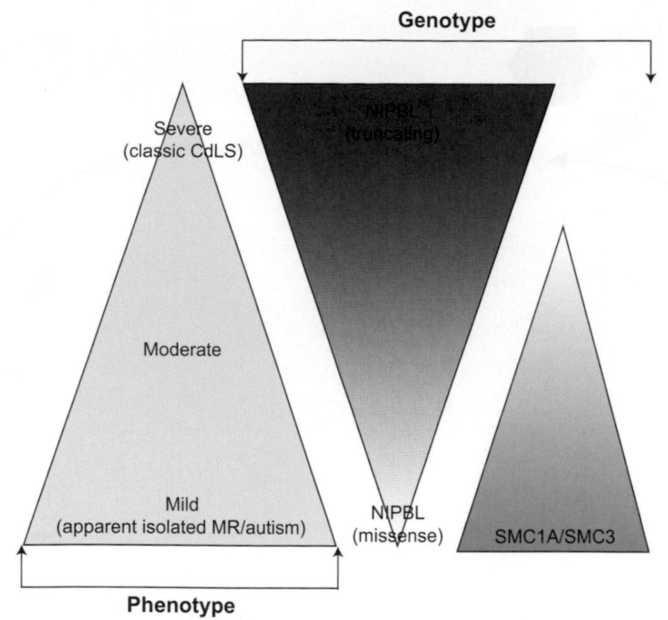

FIGURE 162-2 Genotype–Phenotype Correlations in CdLS. Left triangle represents distribution of severity seen in CdLS with the classic/severe phenotype being seen in the minority of probands (~30%) and the mild phenotype representing the majority of ascertained probands (some of which have been ascertained as apparent isolated mental retardation and/or autism, who upon referral to genetics were recognized as having subtle features of CdLS). The middle inverted triangle represents the distribution of *NIPBL* mutations identified with the majority of probands, with the classic/severe phenotype having truncating mutations and the majority of probands with a mild phenotype having missense or other frame-preserving mutations. The triangle on the right represents the distribution of mutations seen in *SMC1A* and *SMC3*, with the majority being seen in mildly to moderately affected probands and not seen in the severe/classic form of CdLS.

162.3 OTHER DISORDERS OF COHESIN AND SISTER CHROMATID COHESION

162.3.1 Roberts/SC Phocomelia Syndrome

Roberts/SC phocomelia syndrome (RBS/SCP OMIM #268300) is an autosomal recessive genetic disorder caused by homozygous or compound heterozygous mutations in the *ESCO2* gene. The clinical features of this syndrome are distinct from CdLS but display systemic overlap. The consistent features seen in RBS include pre- and postnatal growth retardation, microcephaly, bilateral cleft lip and palate, symmetric mesomelic limb shortening ("phocomelia"), genitourinary abnormalities, and congenital heart defects.

162.3.2 α-Thalassemia/Mental Retardation Syndrome, X-Linked

α-Thalassemia/mental retardation syndrome, X-linked (ATRX) (OMIM #301040), is a multisystem disorder characterized by postnatal growth and mental deficiency, microcephaly, dysmorphic craniofacial features (hypertelorism, midface hypoplasia, anteverted nares and full lips with protruding tongue), lack of speech, seizures and abnormal genitalia in males. Affected individuals usually have a mild form of hemoglobin H (Hb H) disease. ATRX is caused by mutations in the *ATRX* gene on the X chromosome.

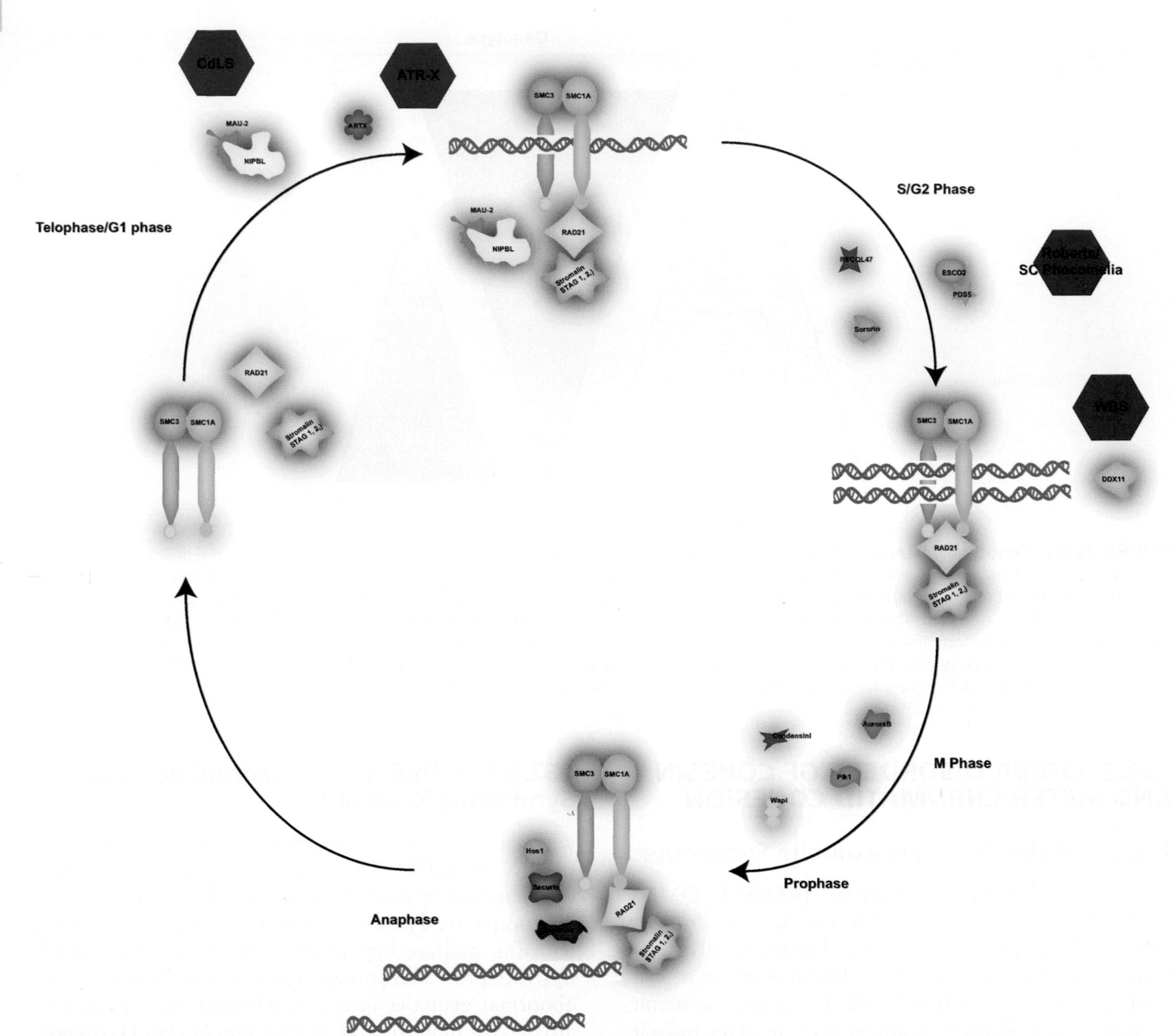

FIGURE 162-3 Cohesinopathies. Human disorders associated with mutations in cohesin and its accessory proteins. SMC1A, SMC3, RAD21, and either SA1 or SA2 are the four major subunits of cohesin in somatic vertebrate cells. RAD21 crosslinks the head domains of SMC1A and SMC3, in an ATP-dependent manner, whereas RAD21 also binds to the fourth cohesin subunit, SA. NIPBL and MAU-2 form a complex and facilitate cohesin loading and unloading. Sister chromatid cohesion is established during S phase, mediated by sororin and ESCO1/ESCO2, after which PDS5 maintains cohesion through G2 phase. The removal of cohesin from the chromosome arms starts at prophase and is regulated by PLK1, aurora B kinase, condensin I, and WAPL. During the metaphase-to-anaphase transition, the separase inhibitor securin is degraded by APC thereby activating separase, which in turn cleaves centromeric cohesin as well as residual cohesin on the chromosome arms. Recently, ESCO1 and its yeast homolog Eco1 were identified as acetyltransferases required for the acetylation of human SMC3, a necessary step to establish the cohesiveness of chromatin-loaded cohesin during S phase. This acetylation counteracts the effects of WAPL and PDS5, two additional regulatory proteins. SMC3 is subsequently deacetylated during anaphase by yeast Hos1. Cohesin cleavage by separase at the onset of anaphase triggers SMC3 deacetylation, and SMC3 molecules that remain acetylated after mitosis due to Hos1 inactivation are unable to generate cohesion during the subsequent S phase. Human disorders are listed corresponding to their disease-causing genes in the cohesion pathway and at which point in the cell cycle these respective genes are active.

TABLE 162-1 Cohesinopathies

Diagnosis	Gene	Chromosome Location	Phenotype Features	Inheritance
Cornelia de Lange Syndrome	*NIPBL*	5p13.1	Severe, moderate and mild CdLS	Autosomal dominant
	SMC1A	Xp11.2	Physically milder CdLS, significant cognitive impairment	X-linked dominant
	SMC3	10q25	Physically milder CdLS, significant cognitive impairment (only one proband identified)	Autosomal dominant
Roberts /SC phocomelia syndrome	*ESCO2*	8p21.1	No established genotype-phenotype correlation	Autosomal recessive
Warsaw breakage syndrome	*DDX11*	12p11	Only one proband identified to date	Autosomal recessive
α-Thalassemia/mental retardation syndrome, X-linked	ATR-X	Xq13		X-linked recessive

FURTHER READING

1. Barisic, I.; Tokic, V.; Loane, M., et al. Descriptive Epidemiology of Cornelia de Lange Syndrome in Europe. *Am. J. Med. Genet. A.* **2008,** *146A,* 51–59.
2. Deardorff, M. A.; Kaur, M.; Yaeger, D., et al. Mutations in Cohesin Complex Members SMC3 and SMC1A Cause a Mild Variant of Cornelia de Lange Syndrome with Predominant Mental Retardation. *Am. J. Hum. Genet.* **2007,** *80,* 485–494.
3. Jackson, L.; Kline, A. D.; Barr, M. A., et al. de Lange Syndrome: A Clinical Review of 310 Individuals. *Am. J. Med. Genet.* **1993,** *47,* 940–946.
4. Kagey, M. H.; Newman, J. J.; Bilodeau, S., et al. Mediator and Cohesin Connect Gene Expression and Chromatin Architecture. *Nature* **2010,** *467,* 430–435.
5. Mannini, L.; Liu, J.; Krantz, I. D., et al. Spectrum and Consequences of SMC1A Mutations: The Unexpected Involvement of a Core Component of Cohesin in Human Disease. *Hum. Mutat.* **2010,** *31,* 5–10.
6. Peters, J. M.; Tedeschi, A.; Schmitz, J. The Cohesin Complex and Its Roles in Chromosome Biology. *Genes. Dev.* **2008,** *22,* 3089–3114.
7. van der Lelij, P.; Chrzanowska, K. H.; Godthelp, B. C., et al. Warsaw Breakage Syndrome, a Cohesinopathy Associated with Mutations in the XPD Helicase Family Member DDX11/ChlR1. *Am. J. Hum. Genet.* **2010,** *86,* 262–266.

CHAPTER

163

Genes and Mechanisms in Human Ciliopathies

Dagan Jenkins and Philip L Beales
Molecular Medicine Unit, UCL Institute of Child Health, London, UK

The term "ciliopathy" was first coined in 1984 by Cornillie and colleagues to describe the varied ultrastructural abnormalities of motile (9+2) cilia observed in the respiratory tracts of children with recurrent respiratory tract infections. This term was subsequently adopted in its current form, in reference to the varied phenotypes caused by mutations that disrupt cilia or basal body formation, when it was noted that a subset of clinical features are shared by each of the five clinically distinct syndromes that fall into this category, implicating ciliary dysfunction in their pathogenesis. We present an analysis of ciliopathies that is limited to those diseases for which the causative gene(s) has been identified. We also limit our description of the phenotypes associated with each gene to the actual phenotypes listed for molecularly proven cases in original publications, and we have made every effort to document individual clinical features precisely, and to give some indication about the frequencies with which they occur. This analysis shows that ciliopathies represent a spectrum of disorders, with variability in clinical presentation both within and between families. Individual clinical features that characterize ciliopathies include retinitis pigmentosa, renal and hepatic cystic disease, situs inversus/laterality defects, polydactyly and specific brain malformations, including the molar tooth sign.

At the level of entire syndromes, ciliopathies can be broadly categorized into groups of related disorders, including Bardet–Biedl syndrome (for which at least 17 causative genes have been identified), the Meckel–Joubert spectrum disorders, the short-rib polydactylies and related skeletal ciliopathies, as well as primary ciliary dyskinesia. There is considerable genetic heterogeneity and allelism within ciliopathies, likely reflecting the formation of large protein complexes involved in intracellular trafficking and intraflagellar transport (IFT). There are also clear examples of genotype–phenotype correlations. For example, whereas homozygous/compound heterozygous nonsense or frameshift mutations cause the more severe embryonic-lethal Meckel–Grüber syndrome, patients with Joubert syndrome caused by mutations in either *CC2D2A*, *INPP5E*, *MKS3* or *RPGRIP1L* always carry at least one missense mutation which are likely to represent hypomorphic alleles. A similar scenario is also found in patients with mutations in *DYNC2H1*, *IFT80*, *IFT122*, or *WDR35*, causing either Short-rib polydactyly or Sensenbrenner syndrome, who also always carry at least one missense/point mutation, suggesting that homozygous/compound heterozygous nonsense or frameshift mutations would be lethal or cause a more severe phenotype. Further evidence for complex inheritance in ciliopathies has been generated, particularly from examples of possible triallelic inheritance in Bardet–Biedl syndrome, whereby three mutant alleles in two genes are required to cause disease. Families segregating three mutations in two genes have also been described for other pairs of ciliopathy genes in nephronophthisis, Leber congenital amaurosis, Joubert, Meckel and Senior–Løken syndromes, suggesting that this may be a more general mechanism.

Cilia are microtubular protrusions present on the apical surface of most cells. Some cilia also have a central pair of microtubules, and are therefore known as "9+2" cilia, while others do not have a central pair and are referred to as "9+0" cilia. Classically, "9+2" cilia are motile, exhibiting very regular patterns of beating to generate fluid flow in structures such as the respiratory epithelium. In contrast, "9+0" cilia are generally nonmotile, and are referred to as primary cilia to account for the observation that almost all cells within the body will at some point form a single "9+0" cilium. The ciliary axoneme is anchored within the cell by a cytoplasmically located structure known as the basal body. Cilia formation is intimately linked to the cell cycle. Following docking of the basal body and initial nucleation of the ciliary axoneme, so-called "anterograde" movement of various cargo to the tip of the cilium is driven by kinesin-2, a molecular motor that travels toward the "minus" end of microtubules; a second molecular motor known as dynein travels toward the "plus" end of microtubules toward the cytoplasm to drive retrograde IFT.

Some of the mechanisms leading to ciliopathies are beginning to be understood. One mechanism of pathogenesis relates to the essential role of cilia in regulating signal transduction pathways, of which the hedgehog pathway is perhaps the best example. Hedgehog pathway components have been shown to be enriched in cilia, and dynamic trafficking of these proteins through the cilium is essential for proper regulation of the pathway. Defective hedgehog signaling is associated with characteristic defects in humans, including polydactyly and skeletal defects, which relate to the roles of hedgehog signaling in limb bud patterning and endochondral ossification, respectively. Evidence from mice with mutations in ciliary proteins suggests that defective hedgehog signaling causes these defects in ciliopathies.

Cilia are also important for sensing fluid flow. This is exemplified by work on the polycystins, which are proteins that form ion channels on the ciliary membrane and which are mutated in autosomal dominant polycystic kidney disease. It is likely that bending of the cilium in response to fluid flow leads to the influx of calcium into the cell, which is mediated by polycystins and may ultimately feed into a variety of cellular processes such as mTOR signaling and transcriptional regulation via molecules such as Foxj1, Mef2c and HDAC5. Two features of ciliopathies that likely relate to roles of cilia in fluid flow sensation are renal cyst formation and laterality defects. In the latter, it is likely that fluid flow within the embryonic node leads to bending of cilia on the left of the node, and that the induction of calcium influx initiates an asymmetric signaling cascade that determines normal situs, as well as sensing fluid flow, directed beating of motile cilia generate fluid flow, which is essential for clearance of mucus from respiratory epithelia and for movement of sperm along the reproductive tract. Mutations affecting ciliary beating, therefore, lead to phenotypes such as bronchiectasis and male infertility. Motile cilia also generate fluid flow within the embryonic node, and so defects in this process are also likely to account for laterality defects in ciliopathies.

Finally, certain ciliopathies are also likely to reflect defects in the formation of specialized cilia. Photoreceptor cells in the retina are one example of specialized sensory cilia, whereby the connecting cilium is essential for the transport of various essential proteins from the inner to the outer segments. It is generally considered that defects in this process lead to retinal degeneration and retinitis pigmentosa, which are cardinal ciliopathies. Specialized ciliary axonemes, known as kinocilia, are also a striking feature of sensory hair cells within the inner ear of mammals. Each kinocilium (one per cell) is associated with many stereocilia, actin filament bundles encased by membrane. Collectively, kinocilia and stereocilia form a chevron-shaped bundle of cytoskeletal components that sense auditory vibrations, and defects in their formation can cause sensorineural hearing loss and balance disturbance. It is noteworthy that several Bardet–Biedl syndrome proteins are required to maintain the correct planar cell polarity of hair cells in mice.

Collectively, the identification of many genes mutated in ciliopathies is illuminating the genetic architecture of human inherited disease, and the many mechanisms of ciliary biology that have been described is providing new insight into the pathogenesis of many important human conditions.

Index

Note: Page numbers with "b" denote boxes; "f" figures; "t" tables.

A
Abetalipoproteinemia (ABL), 382–383
Abnormal body size, 131–137
Aceruloplasminemia, 393
N-Acetylglucosamine phosphotransferase deficiency, 399–400
Achromatopsia
  complete, 494
  incomplete, 494
Acid lipase deficiency, 404–405
Acquired myasthenias, 487–488
Acro-dermato-ungual-lacrimal-tooth syndrome, 552
ACT Sheets, 90
Acute lymphoid leukemia (ALL), 300
Acute myeloid leukemia (AML), 300–301
Acute promyelocytic leukemia (APL), 300
Acyl-CoA dehydrogenase deficiency
  medium chain, 385
  short chain, 385–386
Addictive disorders, 431–432
Adipose tissue, development and function of, 365
Adolescents, nephrotic syndrome in, 256
Adrenal gland, disorders of, 350–351
Adrenocorticotropic hormone (ACTH), 325
Adults
  common diseases of, genetic evaluation for, 69–73
  nephrotic syndrome, 256
Adverse drug reactions (ADRs), 64
Affective disorders, 429–430
Agenesis of the ureters, 241
Aging, 60–62
  defined, 60
  process, 60
  reasons for, 60
Agreeableness, 415
AIRE syndrome, 354
Alagille syndrome (AGS), 267–268, 271
Allelic disorders, 259–260
Alloimmune thrombocytopenia, 299
Alloimmunization
  etiology of, 298
  management of, 299
  prevention of, 299
  treatment for, 299
Alpha 1-antitrypsin (AAT) deficiency, 234
Alzheimer disease (AD), 18, 36, 427–428
  early-onset, 427
  gene therapy for, 93
  late-onset, 427–428
Ambras syndrome, 563
Amino acid metabolism, 372–373
Amniocentesis, 85, 87f
Amniotic band disruption sequence, infant with, 119f
Amyloidosis, 309–311
  apolipoprotein AI, 309
  apolipoprotein AII, 309–310
  cystatin C, 310
  DNA diagnosis for, 310
  gelsolin, 310
  hereditary systemic, 309
  lysozyme, 309
  transthyretin, 309
  treatment for, 310
Amyotrophic lateral sclerosis (ALS), 489
Anemia
  Diamond–Blackfan, 290
  dyserythropoietic, 291
  Fanconi, 148, 290–291
  and hereditary hemorrhagic telangiectasia, 187
  megaloblastic, 291
  sideroblastic, 291
Aneuploidy
  maternal serum screening for, 83
  prenatal screening for, 83
  ultrasound screening for, 83–84
Angelman syndrome, 17–18, 161, 440–441
Angiogenesis, 94
Anhidrotic ectodermal dysplasia. *See* X-linked hypohidrotic ectodermal dysplasia
Animal models, 365
  genes causing glaucoma in, 499
Ankylosing spondylitis, 302–303, 306–307
Anophthalmia, 511–514
Antenatal screening, 96–97
Anterior pituitary, 325–328
  developmental anomalies and syndromes, 326
  development of, 325–326
  function of, 325–326
Anthropometry, 441
Anti-citrullinated protein antibodies (ACPA), 306–307
Antithrombin, 210
Apert syndrome, 507, 527–528, 535f
Aphakia, 503–504
Aplacia of Müllerian ducts, 354
Apolipoprotein AI (apo-AI) amyloidosis, 309
Apolipoprotein AII (apo-AII) amyloidosis, 309–310
Array-based genomic tests, 76–77
Array comparative genome hybridization (aCGH), 31, 169
  whole genome, 418
Arrhythmogenic right ventricular dysplasia, 176
Arterial hypertension, 203f
Arterial thrombosis, genome-wide association studies for, 210
Arterial tortuosity syndrome, 374
Arthritis
  enteropathic, 306–307
  juvenile idiopathic, 307
  psoriatic, 306–307
  reactive, 306–307
  rheumatoid, 306
Arthrogryposes, 597
Arthrogryposis multiplex congenita (AMC), 597
Artificial reproductive technology (ART), 18
Aspartylglucosaminuria (AGU), 401
Assisted reproductive technologies (ARTs), 53
Asthma, 229–233
  pathogenesis of, 233f
  susceptibility, 233f
Ataxia
  Friedreich's, 445
  hereditary, 445–446
  spinocerebellar, 445
Ataxia telangiectasia (AT), 147–148, 469
Ataxia telangiectasia-like disorder (ATLD), 148
Atherosclerotic cardiovascular disease, 213–214
Atlantoaxial subluxation, 156
Atransferrinemia, 393
Attention deficit/hyperactivity disorder (ADHD), 417, 423–424
  behavioral genetics, 423
  molecular genetics, 423–424
Auditory neuropathy (AUN), 517
Autism spectrum disorder (ASD), 124–125, 417, 425–426
  metabolic screening test for, 130t
  X-chromosome genes causing, 129t
Autoimmune myasthenia gravis (MG), 486–488
Autoimmune polyendocrinopathy-candidiasis-ectodermal dystrophy (APECED). *See* Type 1 autoimmune polyglandular syndrome
Autoimmune polyglandular syndromes (APS), 332
Autoimmune thyroid disease (AITD), genetic basis of, 332
Autonomic disorders, 454–457
Autosomal dominant inheritance, 25
  defined, 25
  recurrence risks, 25
Autosomal dominant neurohypophyseal diabetes indipidus (ADNDI), 328
Autosomal dominant osteopetrosis (ADO), 590
Autosomal dominant polycystic kidney disease (ADPKD), 252
Autosomal recessive disorders, 81t
  genomic carrier screening for, 97
Autosomal recessive inheritance, 25–26
  defined, 25
  recurrence risks, 25–26
Autosomal recessive osteopetrosis (ARO), 590
Autosomal recessive polycystic kidney disease (ARPKD), 252
Autosomal recessive Turcot syndrome, 147
Autosomal trisomies, 155–156

Autosomes
deletions, 161–166
structural abnormalities of, 161–166
Axenfeld–Rieger syndrome, 498

B
Bannayan–Riley–Ruvalcaba syndrome, 273
Bardet–Biedl syndrome, 109, 612–613
Barth syndrome, 321–322
Basal ganglia disorders, 443–444
Base excision repair, disorders of, 146–147
Basel cell carcinoma (BCC), 557
nevoid, 557
Batten disease, 406–407, 505
Bazex–Dupre–Christol syndrome, 557
B-cell lymphoma, 300–301
Beare–Stevenson Cutis Gyrata, 527
Becker muscular dystrophy, 473
Beckwith–Wiedemann syndrome (BWS), 17–18, 132, 260
Behavioral genetics, 423
Bernard–Soulier syndrome, 293
Bilateral retinoblastoma, 509
Bile pigment metabolism, 270–271
Bilirubin, 270–271
Bipolar disorder (BPD), 429
Birt–Hogg–Dube (BHD) syndrome, 260–261, 561–562
Bladder tumors, 261
Blastogenesis, 47–48
Blepharophimosis-ptosis-epicanthus (BPE) type II syndrome, 353–354
Blood pressure regulation, 201–205
dysregulated blood pressure, personalized management of, 202
loci associated with abnormal, 203t
Bloom syndrome, 148–149
Blount disease, 599–600
Blue cone monochromacy, 494
Blue rubber bleb naevus (BRBN) syndrome, 217
Body fat stores, regulation of, 365
Body mass disorders, 365–366
Body mass index (BMI), 365
Bone, 590–592
structure, abnormalities of, 595–596
Bone morphogenetic protein (BMP), 181–182
signaling, 183f
Bowman layer dystrophies, 501
Brachmann–de Lange syndrome. *See* Cornelia de Lange syndrome
Brain
development, timing of, 51t
primary/secondary vesicles, 52f
BRCA1, 361
BRCA2, 361
Breast cancer, 361–362
familial risk for, 71f
Breast–ovarian cancer, 148
Bronchopulmonary dysplasia (BPD), 144
Burkitt lymphoma, 300–301
Buschke–Ollendorff syndrome, 580, 591

C
Calciferols
excess state, 388
normal physiology of, 387
Camurati–Engelmann disease, 591
Canadian Collaborative Project on Genetic Susceptibility to MS (CCPGSMS), 464
Cancer
breast, 71f, 361–362
breast–ovarian, 148
colon, 146–147, 272–280
colorectal. *See* Colorectal cancer
female reproductive tract, 361–362
gene therapy for, 93–94
genetic counseling for, 75
hereditary diffuse gastric, 273
hereditary leiomyomatosis and renal, 561
hereditary non-polyposis colon, 147
kidney, 259–264
medullary thyroid, 332
molecular biology of, 58–59
pancreatic, 273
prostate, 261–262
skin, 557–558
urinary tract, 259–264
Cancer Genome Project, 58
Candidate gene association study, 201, 229, 366
loci identification through, 231t
Candidate genes and type 1 diabetes mellitus, 338–339
Capillary malformation–arteriovenous malformation (CM–AVM), 219–222
cerebral cavernous malformation, 220
Carbohydrate metabolism disorders, 374–375
Carbonic anhydrase II deficiency, 258
Carcinoma
basel cell, 557
of the prostate, 261–262
renal cell, 260–261
familial non-VHL, 261
hereditary papillary, 261
of the renal pelvis, 261
squamous cell, 557
thyroid, 332
Cardiac condition disturbances, Mendelian conditions associated with, 198t
Cardiac dysrhythmia, 196
conduction disturbance and, familial forms of, 197
Mendelian conditions associated with, 199t
Cardiac electrophysiology, in humans, 196–200
personalized disorders management of, 197–198
Carnegie Stage, 47, 49t–51t
Carney complex, 262, 328
Carpenter syndrome, 536f
Carrier screening, 80–81, 100
for autosomal recessive disorders, 97
Cartilage hair dysplasia, 321–322
CATCH-22 syndrome, 98
Celiac disease, 302–303
Cell division, 30–31
Cell proliferation, 30–31
Central core disease (CCD), 478
Central nervous system (CNS)
hereditary hemorrhagic telangiectasia in, 186
involvement in muscular dystrophies, 474
primary tumors of, 469–470
Central precocious puberty, 327
Centromere-banding, 31
characteristics of, 32t
Centronuclear myopathy (CNM), 478–479
Cerebral arteriovenous malformations (CAVMs), 185–186
Cerebral autosomal dominant arteriopathy with subcortical infarcts and leukoencephalopathy (CADASIL), 467–468
Cerebral autosomal recessive arteriopathy, 467–468
Cerebral cavernous malformation (CCM), 220
Cerebral cortical development disorders, 438–439
Cerebrotendinous xanthomatosis (CTX), 405
Chanarin-Dorfman syndrome, 545
Channelopathies, 482
hereditary muscle, 482–483
Charcot disease. *See* Amyotrophic lateral sclerosis
Charcot–Marie–Tooth (CMT) neuropathy, 476–477, 481
CHARGE association syndrome, 17–18
Chediak–Higashi syndrome (CHS), 293, 321–322, 542
Cherry-red spot myoclonus syndrome. *See* Juvenile normosomatic sialidosis
Cherubism, 596
Childhood absence epilepsy, 440–441
Children, nephrotic syndrome in, 255–256
CHILD syndrome, 545
Chloride channel myotonias, 482–483
Cholecalciferol activation, metabolic pathways for, 389f
Cholesteryl ester storage disease (CESD), 404–405
Chondrodysplasia punctata, 545
Chondrodysplasias, 593–594
clinical evaluation for, 593
Chorionic villus sampling (CVS), 85–86
transcervical, 88f
Choroidal dystrophy, 505–506
Choroideremia, 505–506
Christ–Siemens–Touraine syndrome. *See* X-linked hypohidrotic ectodermal dysplasia
Chromatin, 30
epigenetic organization of, 19f
Chromosome abnormalities, 12, 27
and fetal loss, 113
and hypergonadotropic hypogonadism, 105–106
and male infertility, 108
thyroid disease associated with, 332
Chromosome microarray (CMA), 161–162
Chronic granulomatous disease (CGD), 322
Chronic myeloid leukemia (CML), 300–301
Chronic obstructive pulmonary disease (COPD), 234
Chronic progressive external ophthalmoplegia, 507
CHST3-related dysplasia, 594
Ciliopathy, 612–613
Classic cystinuria, 257
Clinical risk assessment, 74–75

Clinical teratology, 120–123
  categories of, 120
  characterization of, 122t
    for counseling, 122t
  mechanisms of, 122t
Clone-by-clone sequencing, 15
Clot formation, 211f
Clouston syndrome. *See* Hidrotic ectodermal dysplasia
Coagulation cascade, 209–210, 212f, 296f
  components of, 294t
  factor V, 209
  factor VII, 209
  factor VIII, 209
  factor XIII, 210
  fibrinogen, 209–210
  inherited disorders of, 292
  proteins, defects in, 293
  prothrombin, 209
  von Willebrand factor, 209
Cochlear implants, 518
Cockayne syndrome, 146
Cohen syndrome, 321–322
Cohesin biology, 607–608
Cohesinopathies, 607–611, 610f, 611t
Collodion baby, 544
Colon cancer, 146–147, 272–280
Colorectal cancer
  genomic landscape of, 277f
  microsatellite instability in, 278f
  sequential, multistep, 276f
Color vision defects, 493–495
  red–green, 493–494
  X-linked recessive, 494t
Combined pituitary hormone deficiency (CPHD), 105, 326
Comparative genome hybridization (CGH), 27, 31, 76–77
  array, 31, 169
  whole genome array, 418
Complement regulatory proteins, 316t
Complement system, 317f
Complete androgen insensitivity syndrome (CAIS), 106
Complex diseases, genomic screening for, 98
Conduct disorder, 423
Confidentiality, 99
Congenital adrenal hyperplasia (CAH), 350
Congenital anomalies of the kidney and urinary tract (CAKUT), 241–251
  chromosomal disorders associated with, 243t
  frequency of, 242t
  migration error, 242
  obstruction, 242–251
  organogenesis, errors of, 241–242
  position error, 242
  with teratogen exposures, 244t
Congenital bilateral absence of the vas deferens, 109
Congenital bone marrow failure syndromes, 290–291
Congenital cardiac defects, 594
Congenital cataract, 503–504
Congenital dislocation of the hip (CDH), 599
Congenital disorders of protein glycosylation (CDG), 376–378
Congenital fiber type disrproportion (CFTD), 479
Congenital galactosialidosis, 399
Congenital generalized lipodystrophy (CGL), 367, 369f
Congenital heart defect (CHD), 169–174
  management of, 171t–173t
  nonsyndromic, 170t–171t
Congenital hypogonadotropic hypogonadism (cHH), 109
  genes classification and phenotypes associated with, 110t–111t
Congenital intrinsic factor deficiency, 291
Congenital muscular dystrophies (CMDs), 473–474
Congenital myasthenic syndromes (CMSs), 482, 486–487
Congenital myopathies, 478–479
Congenital myotonias, 482
Congenital ocular fibrosis, 507
Congenital segmental urethral agenesis, 241
Congenital sensory neuropathy with anhidrosis, 454–455
Congenital unilateral absence of vas deferens (CUAVD), 109
Connective tissue disorders, 567–574
Conscientiousness, 415
Consent, 98–99
Contiguous gene deletion syndromes, 332
Copy number variation (CNVs), 38–39, 132, 424, 440–441, 527, 563
Corneal defects, 501–502
Cornea plana, 502
Cornelia de Lange syndrome (CdLS), 607
  facial characteristics in, 608f
  genotype–phenotype correlations in, 609f
Coronary heart disease (CHD), 213–214
Costello syndrome, 580
Counseling
  genetic. *See* Genetic counseling
  for hereditary hemorrhagic telangiectasia, 187
Cowden syndrome, 273, 469
Craniodiaphyseal dysplasia, 591
Craniofacial (CF) anomalies, 523
Craniometaphyseal dysplasia (CMD), 591–592
Craniosynostosis, 527–538
  conditions and mutations, 529t–531t
Craniotubular remodeling disorders, 592
Cri du Chat syndrome, 161
Crigler–Najjar syndrome type 1 (CN1), 270–271
Crigler–Najjar syndrome type 2 (CN2), 270–271
Cross-cultural genetic counseling, 97–98
Crouzonodermoskeletal syndrome, 527
Crouzon syndrome, 507, 527–528, 532f
Cutaneous hamartoneoplastic disorders, 561–562
  Birt–Hogg–Dubé syndrome, 561–562
  hereditary leiomyomatosis and renal cancer, 561
Cutis laxa, 579–580
  clinical features of, 581f
Cystatin C (ACys) amyloidosis, 310
Cystic fibrosis (CF), 80, 225–228, 354
  incidence of, 227f
  phenotypic features of, 226t
  signs and symptoms of, 226t
Cystic fibrosis transmembrane conductance regulator (CFTR), 225
  mutations, cellular consequences of, 228f
  phenotypes associated with, 226t
  structural and functional domains of, 227f
Cystic renal disease, 252–254
Cytogenetic analysis, 76–77

D
Danon's disease, 375
Dark adaptation, 505
Defective double strand break repair (DSBR) disorder, 147–148
Degenerative arthropathy, 601–602
Dentatorubro-pallidoluysian atrophy (DRPLA), 440–441
Denys–Drash syndrome, 255, 259–260
DeSanctis–Cacchione syndrome, c
Developmental dysplasia of the hip. *See* Congenital dislocation of the hip
Developmental fields, 47–48
Diabetes mellitus, 338–349
  type 1, 338–339
    candidate genes and, 338–339
    genes, identification of, 339
    genetic approaches in, 338
    genetic contribution to, evidence of, 338
    susceptibility genes for, 341t–344t
  type 2, 339–340
    candidate genes and, 339
    genes, identification of, 339–340
    genetic approaches in, 339
    genetic contribution to, evidence of, 339
    susceptibility genes for, 345t–348t
Diagnostic genetic tests, 95
Diagnostic molecular genetics, 78–79
Diamond–Blackfan anemia (DBA), 290
Diaphyseal dysplasia, 591
Diffuse mesangial sclerosis (DMS), 255–256
Dihydroorotate dehydrogenase deficiency, 380
Dihydropyrimidinase deficiency, 380
Dihydropyrimidine dehydrogenase deficiency, 380
Dilated cardiomyopathy (DCM), 175–176
Disaccharidase deficiencies, 374
Disorders of sex development (DSD), 262, 352–360
  ovotesticular, 353
  46,XY, 353
    due to androgen action defects, 353
Distal RTA (type I RTA), 258
Dizygotic twins, 53–54. *See also* Twin(s/ning)
DNA
  diagnosis, for amyloidosis, 310
  forensic testing, 100
  free fetal, diagnosis of, 97
  long noncoding, 17
  methylation, 17–18, 64
  micro, 17–18
  mitochondrial, 35–37
  MMR system, 278f
  nuclear, 35–36
  repair and metabolism, disorders of, 146–150

DNA (*Continued*)
sequence information, patenting, 101
sequencing, 58
small interfering, 17
Dominant optic atrophy (DOA), kjer type, 496, 496t
Down syndrome, 76, 83–84, 117, 155–156
antenatal screening for, 96–97
management of, 171t–173t
Duane anomaly, 507
Dubin–Johnson syndrome (DJS), 271
Duchenne muscular dystrophy (DMD), 86, 473–474
diagnosis of, 95
Duty to recontact, 98
Dyschromatosis symmetrica hereditaria, 48–49
Dyserythropoietic anemias, 291
Dyskalemic periodic paralysis, 482–483
Dyslexia, 420–422
Dysmorphic child, clinical approach to, 117–119, 118f
Dysosteosclerosis, 590–591
Dysplasia epiphysealis hemimelica, 595
Dysregulated blood pressure, personalized management of, 202
Dystonia, 443
Dystrophic epidermolysis bullosa (DEB), 546–547, 548t
Dystrophinopathies, 473

E
Ectodermal dysplasias, 551–556
genes and chromosomal regions for, 554t–555t
X-linked hypohidrotic, 551–552
Ectopia lentis, 503
Ectrodactyly–ectodermal dysplasia–clefting syndrome, 552
Ehlers–Danlos syndrome (EDS), 193–194, 468, 575–578, 593–594
types of, 576–578, 576t
Elastic fibers, affected by heritable diseases, 579–584
Elastoderma, 580
Electrooculogram (EOG), 505
Electroretinogram (ERG), 505
Embryogenesis, 47–48
Emery Dreifuss muscular dystrophies (EDMDs), 473–474
Emilin, 569
Enchondromatosis, 595
Encyclopedia of DNA Elements (ENCODE) Project, 63
Endoscopic retrograde cannulation of the pancreas (ERCP), 186
Endosteal hyperostosis, 591
End stage renal disease (ESRD), 255–256
Energy expenditure, 365
Enhanced blue-cone syndrome, 494
Enteropathic arthritis, 306–307
Enzyme disorders, red blood cell, 290
Enzyme replacement therapy (ERT)
for Gaucher disease, 91–92, 405
for lysomal storage disease, 91–92
for mucopolysaccharidoses, 398
Ependymomas, 470
Epidermolysis bullosa (EB), 502, 546–550
classification of, 548t
complications of, 549t
differential diagnosis of, 549t
extracutaneous manifestations in, 549t
Epidermolysis bullosa simplex (EBS), 546, 548t
Epigenetics, 17–21, 64
trait, 17
Epilepsy, 440–441
childhood absence, 440–441
juvenile absence, 440–441
juvenile myoclonic, 440–441
Epistaxis, 186
Ergocalciferol activation, metabolic pathways for, 389f
Erythrokeratodermia, 545
Erythropoetic protoporphyria (EPP), 391–392
Ethical issues, in clinical genetics, 95–99
antenatal screening, 96–97
autosomal recessive disorders, genomic carrier screening for, 97
complex diseases, genomic screening for, 98
cross-cultural genetic counseling, 97–98
diagnostic genetic tests, 95
duty to recontact, 98
free fetal DNA and noninvasive prenatal diagnosis, 97
genetic services, goals and outcomes of, 95
labeling of syndromes, 98
naming of syndromes, 98
newborn screening, 96
population screening, 96
predictive genetic tests, 96
research, 98–99
confidentiality, 99
consent, 98–99
eugenics, 99
feedback, 98–99
geneticization, 99
Ethylmalonic aciduria, 384
Euchromatin, 30
Eugenics, 99
Eugonadal causes, of infertility, 106
Eukaryotic initiation factors (eIFs), 48–49
Evaluation of COPD Longitudinally to Identify Predictive Surrogate End points (ECLIPSE), 234
Extracellular matrix (ECM), 567–568
Extracellular microfibrils, structure and composition of, 567–568
constituents of, 571t
Extraversion, 415

F
Fabry disease (FD), 237–238, 406, 468
Facioscapulohumeral muscular dystrophy (FSHD), 473–474
Factor V (FV), 209
domain structure and processing of, 297f
Factor VII (FVII), 209
Factor VIII (FVIII), 209
gene inversion model, 296f
Factor V Leiden, 294, 467
Factor XIII (FXIII), 210
Familial adenomatous polyposis (FAP), 272–273
Familial dysautonomia, 454
Familial dyskalemic periodic paralyses, 482
Familial hypercalcemic syndromes, 335t
Familial hypobetalipoproteinemia (FHBL), 382–383
Familial interstitial pneumonia (FIP), 237
Familial isolated hypoparathyroidism (FIH), 334
Familial isolated pituitary adenoma (FIPA), 327
Familial nephronophthisis, 252
Familial non-VHL renal cell carcinoma, 261
Familial osteoarthropathy, 601–602
Familial partial lipodystrophy (FPL), 367, 371f
Familial retinoblastoma, 558
Families, genes in, 25–27
autosomal dominant inheritance, 25
autosomal recessive inheritance, 25–26
sex-linked inheritance, 26
X-linked dominant inheritance, 26
X-linked recessive inheritance, 26
Y-linked (holandric) inheritance, 26–27
Fanconi anemia (FA), 148, 257–258, 290–291
Fanconi–Bickel syndrome, 374–375
Farber's disease, 404
Fasciculus gracilis, 447
Fatty acid oxidation, disorders of, 384–386
Feedback, 98–99
Feeding, control of, 365
Female infertility in humans, genetics of, 105–107
diagnosis of, 106f, 107t
eugonadal causes of, 106
hypergonadotropic hypogonadism, 105–106
chromosome abnormalities and, 105–106
ovarian failure, single gene disorders associated with, 106
hypogonadotropic hypogonadism, 105
hypothalamic causes of, 105
pituitary causes of, 105
Female reproductive tract cancer, 361–362
Fetal blood sampling, 86
Fetal loss, 113–116
early
evaluation and management of, 114t
factors associated with, 114t
Fetal tissue sampling, 86
Fetomaternal hemorrhage, detection of, 298
Fibrillins, 568
Fibrinogen, 209–210
cascade, 210
Fibrinogen Aα-chain (AFibA), 309
Fibrinolytic system defects, 293
Fibrofolliculomas, 561–562
Fibrous dysplasia of bone, 596
Fibulin-2, 569
Field defects, 47–48
First- and Second-Trimester Evaluation of Risk (FASTER) trial, 84
Fitzgerald factor deficiencies, 293
Five factor model, 415
Fletcher factor deficiencies, 293
Fluorescence in situ hybridization (FISH), 27, 31, 76–77
telomere, 27

Focal dermal hypoplasia. *See* Goltz syndrome
Focal segmental glomerulosclerosis (FSGS), 255
Follicle-stimulating hormone (FSH), 325
Formylglycine generating enzyme (FGE), 401–402
Founder effect, 41
Fragile X full mutation, 79
Fragile X Mental Retardation Protein (FMRP), 417–418
Fragile X syndrome (FXS), 417–418
Frasier syndrome, 255, 259–260
Friedreich ataxia, 393, 445
Fructose-1,6-diphosphatase deficiency, 375
Fructose metabolism disorders, 374
Fucosidosis, 400–401
Fundus albipunctatus, 505–506

G
Galactose-1-phosphate uridyltransferase deficiency, 353–354, 374
Galactose metabolism disorders, 374
Galactosialidosis
  congenital, 399
  juvenile, 399
  late infantile, 399
Gastrointestinal carcinoid tumors, 273
Gastrointestinal hemorrhage and hereditary hemorrhagic telangiectasia, 187
Gastrointestinal stromal tumors (GISTs), 273
Gastrointestinal tract (GI), 267–269
  cancer, 272–280
  hamartomatosis syndromes, 273
Gaucher disease (GD), 405
Gaucher disease, enzyme replacement therapy for, 91–92
Gelsolin amyloidosis (AGel), 310
Gene–environment correlation, 416
Gene–environment interaction, 366, 416, 523
Gene expression, regulation of, in human development, 48–49
Gene–gene interaction, 523
Gene mutation analysis, 132
Generalized vitiligo, 542–543
Gene structure, 16
Gene therapy, 93–94
Genetic counseling, 74–75
  for adulthood diseases, 70
  applications of, 75
  cross-cultural, 97–98
  defined, 74
  future of, 75
  goals of, 74
  for lymphatic system disorders, 194
  for multiple sclerosis, 464, 466t
  tasks of, 74–75
Genetic discrimination, 100
Genetic disease
  chromosomal disorders, 12
  multifactorial disorders, 13
  nature and frequency of, 12–14
  single-gene disorders, 12–13
  somatic cell genetic disorders, 13–14
Genetic hyperthyroidism, 331–332
Genetic information, legal issues in use of, 100
Genetic Information and Non-Discrimination Act of 2008, 100
Geneticization, 99
Genetic services, goals and outcomes of, 95
Genetic testing
  diagnostic, 95
  predictive, 96
Genome, 16
  recombination, 16
  replication, 16
  transcription, 16
  translation, 16
Genome-wide association studies (GWAS), 38–39, 45, 63–64, 98, 138, 201–202, 209–211, 213–214, 229–230, 306–307, 339, 379, 527
  of addictive disorders, 431
  of Alzheimer disease, 427–428
  for arterial thrombosis, 210
  of attention-deficit/hyperactivity disorder, 423–424
  of autism spectrum disorders, 426
  glaucoma genes, identification of, 499
  loci identification through, 231t–232t
  for motor neuron disease, 489
  of personality, 416
  for stroke, 468
  type 1 diabetes genes, identification of, 338
  type 1 diabetes genes, identification of, 340
  for venous thrombosis, 210
Genome-wide linkage studies
  of attention-deficit/hyperactivity disorder, 423
  type 1 diabetes genes, identification of, 339
Genomic approaches
  glaucoma genes, identification of, 499
Genomic imprinting, 17–18, 31
Genomics, 9, 15
Giemsa-banding, 31, 34f, 76–77
  characteristics of, 32t
Glaucoma, 498–500
  gene identification
    using genomic approaches, 499
    using linkage analysis, 498–499
  genes causing, in animal models, 499
  heritable forms of, clinical features of, 498
Glioblastomas, 470
Gliomas, 469
Global developmental delay, medical genetics evaluation for, 125t
Globin
  chains, primary and secondary structures of, 286f
  genes complexes, 287f
  genes, 287f
  qualitative/quantitative changes in, during human development, 287f
  synthesis, 288f
Glomuvenous malformation (GVM), 215–216
Glucose-6-phosphate dehydrogenase (G6PD) deficiency, 290, 322
Glucosyltransferase 1 deficiency (ALG6-CDG), 376
Glutaric academia type 1 deficiency, 384
Glutathione reductase deficiency, 322
Glutathione synthetase deficiency, 322
Glycine, 257
Glycogen storage cardiomyopathies, histopathology of, 178f
Glycogen storage diseases (GSDs), 374–375
Glycogen synthase deficiency, 375
Glycosaminoglycans (GAGs), 397
Glycosylation, congenital disorders of, 48–49
N-Glycosylation disorders, 376–377, 377t–378t
O-Glycosylation disorders, 376–377, 378t
GM1-gangliosidoses, 403
GM2-gangliosidoses
  chronic, 403–404
  infantile acute, 403–404
  late onset, 403–404
  subacute, 403–404
GM3-synthase, loss of function mutation of, 404
Goltz syndrome, 26, 552
Gonadoblastoma, 262
Gonadotropin pathway disorders, 327
Gorlin–Goltz syndrome, 557
Gorlin syndrome, 469
Gower's maneuver, 473
Graft versus host disease (GVHD), 142
Granulocyte chemotaxis disorders, 322
Gray platelet syndrome, 293
Griscelli syndrome (GS), 321–322, 542
Growth hormone (GH), 325–326
  pathway disorders, 326–327, 329t
  resistance, 327
Gyrate atrophy of the choroid, 505–506

H
H3 N-terminal tails, modifications of, 20f
H4 N-terminal tails, modifications of, 20f
Hageman factor deficiencies, 293
Hair, inherited disorders of, 563–564
Hamartoma tumor syndrome, 273
Hand–foot–genital (HFG) syndrome, 354
HapMap project, 38, 499
Hardy–Weinberg law, 41
Hartnup disease, 257–258
Hay–Wells syndrome, 552
Hb variants, molecular basis of, 285t
Hearing impairment (HI), 517–520
Heberden's arthropathy, 602
Hematopoietic stem cell transplantation (HSCT)
  for Krabbe disease, 405–406
  for mucopolysaccharidoses, 397–398
Hemoglobinopathies, 283–289
Hemoglobins
  human, 284t
  mutants, clinical manifestations of, 285t
  variants, 285t
Hemophilia A, clinical classification of, 295t
Hemophilia B, clinical classification of, 295t
Hemophilias, 292
Hemorrhagic stroke, 467
Hemorrhagic telangectasia
  management of, 171t–173t

Hemostasis, 295f
disorders of, 292–297
Hepatobiliary duct system, 267–269
Hereditary ataxias, 445–446
Hereditary diffuse gastric cancer (HDGC), 273
Hereditary elliptocytosis (HE), 290
Hereditary hemorrhagic telangiectasia (HHT), 184–191, 468
anemia and, 187
counseling for, 187
diagnosis of, 185–186
gastrointestinal hemorrhage and, 187
genetics of, 185
life expectancy, 187
management of, 186–187
central nervous system, 186
epistaxis, 186
liver, 186–187
lung, 186
mucocutaneous telangiectases, 186
natural history of, 184–185
pathogenesis of, 185
phenotype, 184–185
prevalence of, 184
Hereditary leiomyomatosis and renal cancer (HLRCC), 561
Hereditary lymphedema I, 193
Hereditary lymphedema II, 193
Hereditary multiple exostosis, 595
Hereditary muscle channelopathies, 482–483
Hereditary non-polyposis colon cancer, 147
Hereditary non-polyposis colorectal cancer, 362
Hereditary papillary renal carcinoma, 261
Hereditary persistence of fetal hemoglobin (HPFH) disorders, 284
Hereditary pulmonary emphysema, 234–236
Hereditary sensory and autonomic neuropathies (HSANs), 454
mutations, 457t
type III, 454
type IV, 454–455
Hereditary spastic paraplegias (HSPs), 447–453
genetic heterogeneity of, 447
genetic types of, 448t–452t
molecular basis of, 447–452
neuropathology of, 447
treatment for, 452
Hereditary spherocytosis, 290
Hereditary systemic amyloidosis, 309
Hereditary vitamin D–dependent rickets type 1 (VDDR-1), 388
Hereditary vitamin D–dependent rickets type 2 (VDDR-2), 388
Hermansky–Pudlak syndrome (HPS), 237–238, 293, 542
Heterochromatin, 30
Heterozygote testing, 80–81
Heterozygous familial hypercholesterolemia (HeFH), 382
Hidrotic ectodermal dysplasia, 552
High throughput sequencing, 527
Hirschsprung disease, 267, 268f, 541
syndromic forms of, 269t
Histogenesis, 47
Holandric inheritance. *See* Y-linked inheritance
Holt-Oram syndrome, management of, 171t–173t
Homocystinuria, 503
Homozygous familial hypercholesterolemia (HoFH), 382
Horseshoe kidneys, 242
Houston–Harris syndrome, 593–594
hPpM disorders, 64
Human copper metabolism, inherited disorders of, 393
Human developmental genetics, 47–52
developmental fields, 47–48
field defects, 47–48
gene expression, regulation of, 48–49
normal human development, timing of, 47
organogenesis, 49
repertoire of, 48
Human epilepsy, genetic aspects of, 440–442
Human Gene Mutation Database, 58
Human Genome Project, 28, 43, 63, 77, 79, 86, 489
Human imprinted genes, ideograms of, 21f
Human leukocyte antigen (HLA), 141–142, 306
class Ia associated diseases, 139t
class II associated diseases, 139t
Humans, progeroid syndromes of, 61
Hunter syndrome, 397–398
Huntington disease (HD), 443
genetic testing for, 96
Hurler–Scheie syndrome, 397
Hurler syndrome, 397–400
Hutchinson–Gilford syndrome, 61
Hydronephrosis, 242
4-Hydroxybutyric aciduria, 384
11β-Hydroxylase deficiency, 351
21-Hydroxylase deficiency, 350–351
27-Hydroxylase deficiency, 405
Hyperbilirubinemia, 270
Hypergonadotropic hypogonadism, 105–106
chromosome abnormalities and, 105–106
ovarian failure, single gene disorders associated with, 106
Hyper-IgE syndrome, 322
Hyperkalemic periodic paralysis, 482–483
Hyperkeratosis of the palms, 544
Hyperkeratosis of the soles, 544
Hyperkeratotic cutaneous capillarovenous malformation, 216–217
Hyperphosphatasia with osteoectasia, 592
Hypertension
arterial, 203f
Mendelian forms of, 201–202, 204t
mitochondrial conditions associated with, 204t. *See also* Blood pressure regulation
Hyperthyroidism, genetic, 331–332
Hypertriglyceridemia (HTG), 383
Hypertrophic cardiomyopathy (HCM), 175
Hypogonadism
hypergonadotropic, 105–106
hypogonadotropic, 105, 327
Hypogonadotropic hypogonadism, 105, 327
hypothalamic causes of, 105
pituitary causes of, 105
Hypokalemic periodic paralysis, 482–483
Hypoketotic hypoglycemia, 385
Hypoparathyroidism
biochemical characteristics of, 336t
forms of, 336t
idiopathic, 334–335
Hypotension, Mendelian forms of, 202, 204t
Hypothalamic-pituitary congenital hypothyroidism, 330
Hypothalamus–pituitary–gonadal axis, 107f
Hypothyroidism, 96, 155–156
hypothalamic-pituitary congenital, 330
thyroidal congenital, 330–331
Hypoxanthine guanine phosphoribosyltransferase (HGPRT) deficiency, 379–380

I

Ichthyosiform dermatoses, 544–545
Ichthyosis vulgaris, 544
Idiopathic hepatic copper toxicosis, 393
Idiopathic hypogonadotropic hypogonadism (IHH), 105
Idiopathic hypoparathyroidism, 334–335
Idiopathic interstitial pneumonias (IIPs), 237
Idiopathic isolated clubfoot, 599
Idiopathic pulmonary fibrosis (IPF), 237
Idiopathic scoliosis, 599
Imerslund–Grasbeck syndrome, 291
Imino acid proline, 373
Imino acids, 257
Immunodeficiency-centromeric instability-facial anomalies (ICF) syndrome, 48–49
Immunologic disorders, 302–303
Incomplete Müllerian fusion (IMF), 354
malformation syndromes in, 358t
Incontinentia pigmenti, 26
Infertility, 18
eugonadal causes of, 106
Inheritance, 331
chromosomal basis of, 30–34
Inherited bleeding disorders, clinical findings in, 294t
Inherited cardiomyopathies, 175–180
arrhythmogenic right ventricular dysplasia, 176
dilated cardiomyopathy, 175–176
hypertrophic cardiomyopathy, 175
ventricular noncompaction, 176
Inherited complement deficiencies, 315–318
clinical aspects of, 315
diagnosis of, 316
genetics of, 315–316
Inherited disease, human gene mutation in, 22–24
Inherited porphyrias, 391–392
Inherited venous malformation, 216
Intellectual disability, 124–125
dosage-sensitive genes causing, 126t
medical genetics evaluation for, 125t
metabolic screening test for, 130t
monogenic causes of, 127t–128t
recurrent interstitial CMA deletions and duplications in, 125t–126t
X-chromosome genes causing, 129t

Intellectual disability (ID), 17–18, 417
X-linked, 417
Internal genital duct anomalies, 354
Interstitial lung diseases (ILDs), 237–238
Intestinal polyposis syndromes, 275t
Intraocular retinoblastoma, 509
Intraventricular hemorrhage (IVH), 144–145
Ion channels, 482
Iron metabolism, 394–396
disorders
clinical conditions associated with, 393
hereditary, 393
Ischemic stroke, 467
Isolated distal motor neuropathy, 393
Isolated growth hormone deficiency (IGHD), 326–327
Isovaleric academia, 384

J
Jackson–Weiss syndrome, 527–528, 533f
Job syndrome, 322
Joubert syndrome, 612
Jugular lymphatic obstruction sequence, 192
Junctional epidermolysis bullosa (JEB), 546, 548t
Juvenile absence epilepsy, 440–441
Juvenile galactosialidosis, 399
Juvenile idiopathic arthritis (JIA), 307
Juvenile myoclonic epilepsy, 440–441
Juvenile normosomatic sialidosis (JNS), 399
Juvenile osteochondroses, 599–600
Juvenile Paget disease, 592
Juvenile polyposis syndrome, 273

K
Kabuki syndrome, 17–18
Kallmann syndrome (KS), 105, 109
Kearns–Sayre syndrome, 291, 335
Kenny–Caffey syndrome, 334–335, 592
Keratoconus, 501–502
Keratoplasty, 501–502
Kidney
cancer, 259–264
congenital anomalies of, 241–251
KID syndrome, 545
Kindler syndrome (KS), 547, 548t
Kleefstra syndrome, 17–18
Kleihauer–Betke acid elution test, 298
Klinefelter syndrome (KS), 108, 157
abnormalities associated with, 159b
assessment and follow-up program for, 160b
Klippel–Trenaunay syndrome (KTS), 193, 217
Klippel–Trenaunay–Weber syndrome, 458
Kniest dysplasia, 594
Kostmann syndrome, 321
Krabbe disease (KD), 405–406
adult form, 405–406
infantile form, 405–406
juvenile form, 405–406
late infantile form, 405–406
Kufs disease, 408

L
Labeling of syndromes, 98
Lambert–Eaton myasthenic syndrome (LEMS), 488
Laminin β2, 256
Langer–Giedion syndrome, 161, 595
Larsen syndrome, 593–594
Late infantile galactosialidosis, 399
Leber congenital amaurosis, 505–506, 612
Leber hereditary optic atrophy (LHON), 496–497, 496t
Leber hereditary optic neuropathy (LHON), 26, 35–36
Left ventricular hypertrophy (LVH), unexplained, 175, 178f
Legal issues, in genetic medicine, 100–102
Legg–Perthes disease, 599–600
Lens coloboma, 503
Lenticonus, 503
LEOPARD syndrome, 98
Leptin, 365
Leukocyte adhesion deficiencies (LAD), 322
Leukocyte function disorders, 321–324
granulocyte chemotaxis disorders, 322
leukocyte adhesion deficiencies, 322
microbicidal activity disorders, 322–323
monocyte/macrophage function disorders, 323
neutrophil number disorders, 321–322
Leydig cells, 352
Liddle syndrome, 202
Li–Fraumeni syndrome, 558, 469
Limb girdle muscular dystrophies (LGMDs), 473–474
Limb–mammary syndrome, 552
Linkage analysis, 28–29, 235t, 304, 366
glaucoma genes, identification of, 498–499
Lipid metabolism, 382–383
Lipodystrophies, 367–371
classification of, 368t
congenital generalized, 367
familial partial, 367
Lipoprotein, 382–383
Lipoprotein lipase (LPL) deficiency, 383
Lissencephaly, 438
Liver
hereditary hemorrhagic telangiectasia in, 186–187
transplantation, for amyloidosis, 310
Long QT syndrome (LQTS), 197, 199t
Lou Gehrig disease. *See* Amyotrophic lateral sclerosis
Lung, hereditary hemorrhagic telangiectasia in, 186
Luteinizing hormone (LH), 325
Lymphangioleiomyomatosis (LAM), 237–238
Lymphatic system, 192–195
development of, 192
disorders of, 192–193
genetic counseling for, 194
mendelian disorders affecting, 193, 195t
Lymphedema–distichiasis syndrome, 193
Lymphoma
B-cell, 300–301
Burkitt, 300–301
T-cell, 300–301
Lynch syndrome, 147, 272, 362
clinical criteria for, 274t
clinical features of, 273t–274t
Lysomal storage disease
enzyme replacement for, 91–92
pharmacologic chaperone therapies for, 91–92
Lysosomal acid α-D-mannosidase (LAMAN) deficiency, 400
Lysosomal enzyme deficiencies, 323
Lysosomal *N*-acetyl-α-galactosaminidase (NAGA) deficiency, 401
Lysozyme amyloidosis, 309
Lysyl oxidase, 569

M
Maffucci syndrome, 595–596
Major depression disorder (MDD), 429
Major histocompatibility gene complex (MHC), 141
Male infertility, genetics of, 108–112
chromosome anomalies, 108
endocrine forms of, gene defects involved in, 108–109
post-testicular/primary testicular forms of, monogenic defects in, 109
syndromic monogenic defects, 109
Male psudohermaphroditism. *See* 46,XY disorders of sex development
Malignant hyperthermia, 482–483
Mandibulofacial dysostoses, 507
Maple syrup urine disease (MSUD), 373
Marfan syndrome, 193–194, 237–238, 503, 567–574
management of, 171t–173t
Maroteaux–Lamy syndrome, 397–398
Maternal serum alpha-fetoprotein (MSAFP), 82–83
Maternal serum screening, for aneuploidy, 83
McArdle disease, 375
McCune–Albright syndrome, 596
McKusick–Kaufman syndrome (MKS), 354
Meacham syndrome, 259–260
Meckel–Grüber syndrome, 612
Meckel–Joubert spectrum disorders, 612
Medical genetics
history of, 3–8
timeline for, 3, 4t–7t
Medicine
in genetic context, 9–11
mitochondrial, 35–37
Medium chain acyl-CoA dehydrogenase deficiency (MCADD), 89–90
Medullablastomas, 470
Medullary cystic kidney disease (MCKD), 252
Medullary thyroid cancer (MTC), 332
Megaloblastic anemias, 291
Megalocornea, 502
Meiosis, 30–31
stages of, 33f
Melanocyte disorders
development, 541–542
differentiation, 541–542
function, 542
migration, 541–542
survival, 542–543
Melanocyte-stimulating hormone (MSH), 325
Melanoma, 558

Melnick–Needles syndrome (MNS), 592
Membrane disorders, red blood cell, 290
Membrane transporters, 331
Mendelian disorders, 262
  associated with ovarian failure, 356t
  candidate genes from, 525t
  and obesity, 366
  pathogenetics of, 44–45
Meningiomas, 470
Menkes syndrome, 393, 580
Mental retardation. *See* Intellectual disability
Metabolic disorders, 502
Metabolomics, 64–65
Metachromatic leukodystrophy (MLD), 406
3-Methylcrotonyl-CoA carboxylase deficiency, 386
Methylmalonic academia, 384
Microarray analysis, 27
  of autism spectrum disorders, 425
Microbicidal activity disorders, 322–323
Microcephaly, 438
Microfibril-associated glycoprotein (MAGP), 568
Microfibril-associated protein-1 (MFAP1), 568
Microfibril-associated protein-3 (MFAP3), 568–569
Microfibrils
  extracellular, 567–568, 571t
  functions of, 570
Microphthalmia, 511–514, 512t
Microspherophakia, 503
Miller–Dieker syndrome, 161
Miller syndrome, 38–39
Mismatch repair deficiency syndrome (MMR-D), 147
Mitochondrial encephalomyopathy, lactic acidosis and stroke-like episodes (MELAS), 35
Mitochondrial fatty acid oxidation, 385
Mitochondrial inheritance, 26
Mitochondrial medicine, 35–37
Mitochondrial myopathy, encephalopathy, lactic acidosis and stroke-like episodes (MELAS) disorder, 468
Mitosis, 30–31
  stages of, 32f
Moebius syndrome, 507
Molecular genetics, 423–424
Monocyte/macrophage function disorders, 323
Monogenic pharmacogenetic disorders, 65t
Monozygotic twins, 53–54
  anomalies exclusive to, 55t
  discordance in, 56t
  placentation, types of, 55t
  during postfertilization, 54f
  structural defects in, 55t. *See also* Twin(s/ning)
Morphogenesis, 47
Morquio A syndrome, 397
Morquio B syndrome, 397
Motor neuron disease, 489–490
Motor neuropathies, 476–477
Mseleni joint disease (MJD), 602
Mucocutaneous telangiectases, 186
Mucolipidosis I. *See* Sialidoses
Mucolipidosis IV, 402
Mucopolysaccharidoses (MPS), 397–398, 505
Mucosulfatidosis. *See* Multiple sulfatase deficiency
Muenke syndrome, 527–528
Müllerian aplasia (MA), 106
  malformation syndromes in, 357t
Multicystic kidneys, disorders associated with, 249t–250t
Multifactorial disorders, 13
Multifactorial inheritance, 27
  and complex diseases, 38–40
Multi-minicore disease (MmD), 478–479
Multiple carboxylase deficiency, 384
Multiple congenital contractures, 597
Multiple enchondromas, 595
Multiple endocrine neoplasia type 1 (MEN1), 327–328
Multiple endocrine neoplasia type 4 (MEN4), 328
Multiple epiphyseal dysplasia (MED), 593
Multiple malformation syndrome, 117–118
Multiple sclerosis (MS), 302–303, 464–466
  genetic counseling for, 464, 466t
  recurrence risk of, 466t
Multiple sulfatase deficiency (MSD), 401–402, 406
Muscular dystrophies (MDs), 473–475
  Becker, 473
  clinical course of, 474
  CNS involvement in, 474
  congenital, 473
  diagnosis of, 474
  Duchenne, 473
  Emery Dreifuss, 473
  facioscapulohumeral, 473
  limb girdle, 473
  natural history of, 474
Myasthenias
  autoimmune, 486–488
  hereditary, 486–488
Myelodysplastic syndrome (MDS), 300
Myeloperoxidase (MPO) deficiency, 322
Myeloproliferative neoplasm (MPN), 300–301
Myoclonic epilepsy with ragged red fibers (MERRF), 440–441
Myofibrillar myopathies (MFMs), 473–474
Myotonia congenita, 482–483
Myotonic dystrophies, 484–485
Myotonic dystrophy, 48–49, 109
  type 1 (DM1), 484–485
  type 2 (DM2), 484–485
Myotubular myopathy (MTM), 478

N
NACHT leucine-rich-repeat protein 5 (NALP5), 335
Naming of syndromes, 98
Narcolepsy, 302–303
Natowicz syndrome, 397
Natural anticoagulants, 210
  deficiency of, 294
Necrotizing enterocolitis (NEC), 145
Nemaline myopathy (NM), 478
NEMO deficiency, 323
Neonatal screening, 89–90
Neoplasia, 327–328
Nephrin, 255
Nephronophthisis, 612
Nephronopthisi, 505
Nephrotic disorders, 255–256
  in adolescents, 256
  in adults, 256
  in children, 255–256
  diagnosis of, 256
  management of, 256
  in newborns, 255–256
Netherton syndrome, 545
Neural tube defects (NTDs), 82–84, 435–437
  defined, 435
  diagnosis of, 82–83, 436–437
  outcome of, 436–437
  prenatal screening for, 82–83
  risk factors associated with, 435–436
  treatment for, 436–437
Neurodevelopmental disabilities, 124–130
Neuroferritinopathy, 393
Neurofibromatose 1 (NF1), 469
  age of onset, 459t–460t
  clinical features of, 459t–460t
  diagnostic criteria for, 459t–460t
Neurofibromatosis 2 (NF2), 469
  clinical features of, 461t
  diagnostic criteria for, 461t
Neurogenetics
  genetic counseling in, 75
Neuronal ceroid lipofuscinosis syndromes, 406–408
Neuroticism, 415
Neutral amino acids, 257–258
Neutrophil granulocytes, 321
Neutrophil number disorders, 321–322
Nevoid basal cell carcinoma syndrome, 557
Newborn(s)
  nephrotic syndrome, 255–256
  screening, 96
    legal issues associated with, 100
Niemann–Pick disease, 237–238, 404
  type A, 404
  type B, 404
  type C, 91–92, 404
  type D, 404
NIH
  Epigenomics of Health and Disease Roadmap Program, 18
  Roadmap Epigenomics Mapping Consortium, 18
Nijmegen breakage syndrome (NBS), 148
Nodulosis–arthropathy–osteolysis syndrome, 399
NOG syndrome, 354
Non-*HLA* microbe-gene susceptibility associations, 140t
Non-Hodgkin lymphoma (NHL), 300
Noninflammatory arthropathy, 601–604, 602t
Noninvasive prenatal diagnosis, 97
  rhesus, 299
Non-mammalian models, 366
Nonsyndromic clefting, phenotypic features in, 525t

Nonsyndromic congenital heart defect, 170t–171t
Noonan syndrome, 109, 193
  management of, 171t–173t
NOR-banding, 31
Normal human development, timing of, 47
Nuchal translucency (NT), 84
Nucleotide diversity, in genomic regions, 23f

O
Obesity
  in animals, polygenic models of, 366
  genetic architecture of, 365
  problems, addressing, 366
Obsessive compulsive disorder (OCD), 417
Occipital horn syndrome (OHS), 393, 580
Ocular albinism type 1 (OA1), 542
Ocular coherence tomography (OCT), 505
Oculocutaneous albinism (OCA), 542
Odonto-onycho-dermal dysplasia, 552
$1\alpha,25(OH)_2D$, transcriptional and nongenomic effects of, 387–388
Oligodendrogliomas, 470
Oligosaccharidoses, 399–402
  disorders allied to, 401–402
Ollier's disease, 335, 595
Omenn syndrome, 147
Online Mendelian Inheritance in Man (OMIM), 15
Openness to experience, 415
Optic atrophy, 496–497
  dominant, 496, 496t
  Leber hereditary, 496–497, 496t
Organic acidemias, 384–386
Organogenesis, 47–49
  errors of, 241–242
Orofacial clefts (OFCs)
  of cleft palate only, 523
  of lip with or without palate, 523
Orphan Drug Act, 101
Orthogenetics, 43
Osler–Weber–Rendu syndrome, 184–191
  anemia and, 187
  counseling for, 187
  diagnosis of, 185–186
  gastrointestinal hemorrhage and, 187
  genetics of, 185
  life expectancy, 187
  management of, 186–187
    central nervous system, 186
    epistaxis, 186
    liver, 186–187
    lung, 186
    mucocutaneous telangiectases, 186
  natural history of, 184–185
  pathogenesis of, 185
  phenotype, 184–185
  prevalence of, 184
Osteochondritis dissecans, 599–600
Osteogenesis imperfecta (OI), 587–589
Osteoglophonic dysplasia, 527
Osteopetrosis, 323, 590–591
Osteopoikilosis, 591
Otitis media, 155–156
Otopalatodigital syndrome type I (OPDI), 592
Otopalatodigital syndrome type II (OPDII), 592
Ovarian failure
  Mendalian disorders associated with, 356t
  single gene disorders associated with, 106
Overgrowth disorders, 132, 260
Ovotesticular disorders of sex development, 353
Oxidative phosphorylation (OXPHOS), 35–36

P
p14/ROBLD3 deficiency, 321–322
Pachydermoperiostosis, 591
Pallister–Killian syndrome, 162
Pancreatic cancer, 273
Pancreatic islet cell tumors, 273
Paramyotonia congenita, 482
Parathyroid disorders, 334–337
Parathyroid hormone (PTH), 334
Parkes Weber syndrome, 193
Parkinson disease (PD), 36, 443
  gene therapy for, 93
Patent ductus arteriosus (PDA), 144
Pathogenetics, 43–46
  of Mendelian disorders, 44–45
Pathway analysis, 39
Pearson syndrome, 291, 321–322, 335
Pediatric genetics, genetic counseling in, 75
Pemphigus vulgaris, 302–303
Perlman syndrome, 260
Peroxisomal disorders, 409–412, 410t
Peroxisome biogenesis disorder (PBD), 409
Perrault syndrome, 354, 409
Personality, 415–416
Personalized disorders management, of cardiac electrophysiology, 197–198
Pervasive developmental disorder, not otherwise specified (PDD-NOS), 417
Peutz–Jegher syndrome, 262, 273
Pfeiffer syndrome, 507, 527–528
  type I, 534f
  type II, 535f
Phakomatoses, 458–463, 459t
Pharmacogenetics, 63–66
  defined, 63
Pharmacogenomics, 63–66
  defined, 63
Pharmacologic chaperone therapies, for lysomal storage disease, 91–92
Phenylalanine hydroxylase (PAH), 372
Phenylketonuria, 96
Phosphoenolpyruvate carboxykinase deficiency, 375
Phospholipase CE1 (PLCE1), 255–256
Phosphomannomutase 2 deficiency (PMM2-CDG), 376
Phosphoribosyl pyrophosphate (PRPP) synthetase disorders, 380t
Phytanic acid storage disease, 545
Piebaldism, 541
Pigmentation, abnormalities of, 541–543
Pituitary gland, genetic disorders of, 325–329
Pituitary hypersecretion disorders, 327–328
Placenta, vascular problems in, 114t
Placental growth factor (PlGF), 207
Plasminogen activator inhibitor-1 (PAI-1), 210
Platelet
  disorders, 293
  function, 210
    inherited disorders of, 293
  glycoprotein receptors, 210
Podocin, 255
Polycystic kidneys, disorders associated with, 249t–250t
Polycystic ovarian syndrome (PCOS), 351
Polymerase chain reaction (PCR), 78–79
Polymicrogyria, 438–439
Polymorphonuclear leukocytes (PMNs), 321
PolyPhen, 38–39
Pompe disease, 375
Population genetics, 41–42
Population screening, for genetic disorders, 96
Porphyria cutanea tarda (PCT), 391
Posterior pituitary disorders, 328
Posterior urethral valves (PUVs), 241–242
Prader–Willi syndrome, 17–18, 109, 161
Predictive genetic tests, 96
Preeclampsia, 206–208
  recurrence risk of, 208t
Premature newborn, disorders affecting, 144–145
  bronchopulmonary dysplasia, 144
  intraventricular hemorrhage, 144–145
  necrotizing enterocolitis, 145
  patent ductus arteriosus, 144
  respiratory distress syndrome, 144
  retinopathy of prematurity, 145
Premature ovarian failure (POF), 353–354
Prenatal diagnosis techniques, 85–88
  amniocentesis, 85
  chorionic villus sampling, 85–86
  fetal blood sampling, 86
  fetal tissue sampling, 86
Prenatal-onset growth deficiency disorders, 133t–135t
Prenatal-onset overgrowth syndromes, 136t–137t
Prenatal Rh genotyping, 298–299
Prenatal screening, 100
  for aneuploidy, 83
  for neural tube defects, 82–83
Primary ciliary dyskinesia, 109, 612
Primary hyperparathyroidism (PHPT), 334
Primary osteoarthropathy of the hip, 602
Primary pulmonary hypertension (PPH). *See* Pulmonary arterial hypertension
Primary tumors, of central nervous system, 469–470
Primitive neuroectodermal tumors (PNETs), 470
Progeroid syndromes, of humans, 61
Progressive familial intrahepatic cholestasis syndromes, 271
Progressive osseous heteroplasia (POH), 335
Prolactin (PRL), 325
Pro-longevity genes, allelic variants homologous to, 61
Propionic academia, 384
Prostate cancer, 261–262
Protein C, 210
Protein S, 210
Proteomics, 15, 65, 79
Prothrombin, 209

Proximal myotonic myopathy. *See* Myotonic dystrophy type 2
Proximal RTA (type II RTA), 258
Pseudoachondroplasia, 593
Pseudohypoparathyroidism (PHP), 592
  biochemical characteristics of, 336t
  PHP-1a, 335
  PHP-1b, 335
Pseudopseudohypoparathyroidism (PPHP), 335
Pseudovasculogenesis, 206
Pseudoxanthoma elasticum, 580
  clinical features of, 582f
Psoriasis, 559–560
Psoriatic arthritis, 306–307
Psuedohypoaldosteronism type 2, 202
Ptosis, 507
Puberty, central precocious, 327
Pulmonary alveolar microlithiasis, 237–238
Pulmonary arterial hypertension (PAH)
  heritable, 181–183
  idiopathic, 181–183
Pulmonary arteriovenous malformations (PAVMs), 185–186
  right upper lobe, 190f
Pulmonary disorders
  interstitial, 237–238
  restrictive, 237–238
Pulmonary Langerhans cell histiocytosis, 237–238
Punctata albescens, 505–506
Purine metabolism, 379–381
  pathways of, 381f
Pyknodysostosis, 402, 590–591
Pyloric stenosis, in families, risk of, 269t
Pyrimidine metabolism, 379–381
Pyrimidine 5′-nucleotidase deficiency, 380
Δ^1-Pyrroline-5-carboxylate synthetase deficiency, 373
Pyruvate carboxylase deficiency, 375
Pyruvate dehydrogenase complex deficiency, 375
Pyruvate kinase (PK) deficiency, 290
22q11.2 deletion syndrome
  management of, 171t–173t

R
Raine dysplasia, 590–591
Random genetic drift, 41
RANKL system, 387–388
Rapp–Hodgkin syndrome, 552
RB1 cancer syndrome, 508
  testing of, 509
Reactive arthritis, 306–307
Recombination, 16
  unequal homologous, 24f
RecQ helicase disorders, 148–149, 149t
Red blood cell (RBC) disorders, 290–291
  congenital bone marrow failure syndromes, 290–291
  dyserythropoietic anemias, 291
  enzyme disorders, 290
  megaloblastic anemias, 291
  membrane disorders, 290
  sideroblastic anemias, 291
Red–green color vision defects, 493–494
  gene therapy for, 494
Refsum disease, 409
Reis–Bucklers corneal dystrophy (RBCD), 501
Renal agenesis, 241
Renal cell carcinoma (RCC), 260–261
  familial non-VHL, 261
  hereditary papillary, 261
Renal dysplasia, 241
  disorders associated with, 245t–248t
Renal glycosuria, 258
Renal-hepatic-pancreatic dysplasia, 249t–250t
Renal hypodysplasia (RHD), 241
Renal transplantation, 256
Renal tubular acidosis, 258
Renal tubular disorders, 257–258
Replication, 16
Reproductive genetics, genetic counseling in, 75
Respiratory distress syndrome (RDS), 144
Response to infection, 138–140
Reticular dysgenesis, 321–322
Retinal dystrophy, 505–506
Retinal pigment epithelium (RPE), 505
Retinoblastoma, 508–510, 558
  bilateral, 509
  classification of, 508–509
  clinical features of, 508
  testing of, 509
  unilateral, 509
Retinopathy of prematurity (ROP), 145
Retinoschisis, 505–506
Rett syndrome, 17–18, 45
Reverse-banding, 31
  characteristics of, 32t
Reverse genetics, 28
Rhesus (Rh), 298–299
  antigens, molecular basis of, 298
  blood group system, 298
  noninvasive prenatal diagnosis, 299
  prenatal genotyping, 298–299
Rheumatoid arthritis (RA), 302–303, 306
Rhizomelic chondrodysplasia punctata, 594
Right to abortion, 100
Riley–Day syndrome, 454
RIVUR study, 242–251
RNA
  micro, 48–49, 58–59
Roberts/SC phocomelia syndrome (RBS/SCP), 609
Robin malformation sequence, infant with, 119f
Roe v. Wade, 100
Rothmund–Thomson syndrome, 149
Rotor syndrome, 271
Rubella embryopathy, 122f
Rubinstein–Taybi syndrome, 48–49
Russell Silver syndrome, 17–18

S
Sanfilippo syndrome, 397–398
Sanjad–Sakati syndrome, 334–335
Schizencephaly, 438–439
Schizophrenia, 429–430
Schwachmann–Diamond syndrome, 321–322
Schwannomas, 470
Sclerosteosis, 591
Segmental vitiligo, 542–543
Senior–Løken syndrome, 612
Sensenbrenner syndrome, 612
Sensory disorders, 454–457, 456t
Sensory neuropathies, 476–477
Septo-optic dysplasia (SOD), 326
Sequencing
  clone-by-clone, 15
  shot-gun, 15
  whole genome, 15
Seronegative spondyloarthropathies, 306–307
Severe combined immunodeficiency syndrome (SCID), 147
Severe congenital neutropenia (SCN), 321
Sex chromosome abnormalities, 157–160
Sex-linked inheritance, 26
Short QT syndrome, 197
Short-rib polydactyly, 612
Short stature, 131, 155–156
  classification of, 132f
Shot-gun sequencing, 15
Sialic acid storage disorders (SASDs), 401
Sialidoses, 399
  juvenile normosomatic, 399
Sialuria, 401
Sickle-cell disease, 468
Sideroblastic anemias, 291
SIFT, 38–39
Single-gene disorders, 12–13
Single nucleotide polymorphisms (SNPs), 28, 38–39, 58, 98, 304, 361–362, 467
Single primary defect in development, 117, 119t
Sitosterolemia, 383
Sjögren–Larsson syndrome, 545
Skin cancer, 557–558
Sly syndrome, 397–398
Smith–Magenis syndrome, 161
Social issues, in clinical genetics, 95–99
Sodium channel myotonias, 482–483
Somatic cell genetic disorders, 13–14
Somitogenesis, clock-and-wavefront model of, 52f
Sotos syndrome, 17–18, 132
Specific language impairment (SLI), 420–422
Specific reading disability (SRD), 420–422
  gene localization studies with, 421t
Speech sound disorder (SSD), 420–422
Spielmeier–Vogt–Batten–Mayou disease, 407
Spinal muscular atrophies (SMA), 480–481
  classification of, 481t
  diagnosis of, 480
  distal, 481
  future of, 481
  genetics of, 480–481
  proximal, 480–481
  treatment for, 481
Spinocerebellar ataxias (SCAs), 445
Split hand–split foot malformation syndrome, 552
Spondyloepiphyseal dysplasia (SED), 601
Spondylolisthesis, 599
Sporadic venous malformation, 216
Squamous cell carcinoma (SCC), 557
Squint. *See* Strabismus

Stargardt disease, 505–506
Steinert's disease. *See* Myotonic dystrophy type 1
Storage disorders, 445
Strabismus, 507
Stroke, 467–468
  family history of, 467
  genetic polymorphisms, 467
  genome-wide association studies for, 468
  hemorrhagic, 467
  ischemic, 467
  monogenic disorders and, 467–468
  as primary manifestation, 467–468
  as secondary manifestation, 468
Structural nondirectiveness, 96
Sturge–Weber syndrome, 458
Substrate reduction therapy, 91–92
  for Niemann–Pick disease, 404
Succinic semialdehyde dehydrogenase deficiency, 384
Sucrase-isomaltase deficiency, 374
Sugar transport disorders, 258
Sulfatase modifying factor 1 (SUMF1), 401–402
Susceptibility, 138–140
Syndromic monogenic defects, 109
Systemic lupus erythematosus (SLE), 304–305

T
Tall stature, 131
Tandem mass spectrometry (MS/MS), 89
Tarui disease, 375
Tay–Sachs disease (TSD), 80
T-cell leukemia, 300–301
T-cell lymphopma, 300–301
Telomere fluorescence in situ hybridization, 27
Teratogen exposures, urinary tract anomalies associated with, 244t
Teratogenicity, 120
Teratogens, 171t
Testicular neoplasms, 262
TGF-β receptor syndromes, management of, 171t–173t
Thalassemias, 283–289
  α-thalassemias, 283–284
  β-thalassemias, 284
α-Thalassemia/mental retardation syndrome, X-linked (ATRX), 609
Thalidomide embryopathy, 121
Thiel–Behnke corneal dystrophy, 501
Thrombocytopenia and absent radii (TAR) syndrome, 293
Thrombomodulin, 210
Thrombosis, inherited disorders predisposing to, 293–294
Thrombotic thrombocytopenic purpura (TTP), 294
Thymidine kinase 2 deficiency, 380
Thyroidal congenital hypothyroidism, 330–331
  inheritance, 331
Thyroid carcinoma, genetic basis of, 332
Thyroid disorders, 330–333
  associated with chromosomal disorders, 332
  autoimmune thyroid disease, genetic basis of, 332
  contigous gene deletion syndromes, 332
  genetic hyperthyroidism, 331–332
  hypothalamic-pituitary congenital hypothyroidism, 330
  thyroid carcinoma, genetic basis of, 332
  thyroid hormone transport proteins, disorders of, 331
  thyroidal congenital hypothyroidism, 330–331
Thyroid follicular cell tumors, 332
Thyroid hormone action, 331
Thyroid hormone transport proteins, disorders of, 331
Thyroid stimulating hormone (TSH), 325
Tissue factor pathway inhibitor (TFPI), 210
Tissue-type plasminogen activator (tPA), 210
Tooth and nail syndrome. *See* Witkop syndrome
Torsades de pointes, 197
Tourette syndrome (TS), 443–444
Transaldolase deficiency (TALDO), 374
Transcobalamin II deficiency, 291
Transcription, 16
Transcriptomics, 65
Transient myeloproliferative disease (TMD), 156
Translation, 16
Transplantation genetics, 141–143
Transthyretin (TTR) amyloidosis, 309
Triacylglycerol biosynthesis, pathways for, 370f
Tricho-rhino-phalangeal (TRP) syndrome type II. *See* Langer–Giedion syndrome
Trichothiodystrophy, 544–545
Trontometaphyseal dysplasia (FMD), 592
Tuberous sclerosis, 469
Tuberous sclerosis complex (TSC), 458
  clinical features of, 461t
Tubular stenosis, 592
Turcot syndrome, 469
Turner syndrome (TS), 157, 193
  clinical out-patient program for, 159b
  management of, 171t–173t
Twin(s/ning), 53–57
  dizygotic, 53–54
  monozygotic, 53–54
  sex ratio in, 55t
283235 syndrome, 503
Two hit hypothesis, 508
Type 1 autoimmune polyglandular syndrome (APS-1), 335
Type 1 carbohydrate-deficient-glycoprotein (CDG) deficiency, 353–354
Type 1 diabetes mellitus, 302–303, 338–339
  candidate genes and, 338–339
  genes, identification of, 339
  genetic approaches in, 338
  genetic contribution to, evidence of, 338
  susceptibility genes for, 341t–344t
Type 2 diabetes mellitus, 339–340
  candidate genes and, 339
  genes, identification of, 339–340
  genetic approaches in, 339
  genetic contribution to, evidence of, 339
  susceptibility genes for, 345t–348t
Type 2 Hermansky Pudlak syndrome, 321–322
Typological thinking, 9

U
Ultrasound screening, for aneuploidy, 83–84
UMP synthase deficiency, 380
Unilateral retinoblastoma, 509
Uniparental disomy (UPD), 161
Urea cycle disorders, 373
Urinary tract
  cancer, 259–264
  congenital anomalies of, 241–251
Usher syndrome, 517
Uveal coloboma, 511–514

V
Van Buchem disease, 591
Varicose veins, 193–194
Vascular endothelial growth factor (VEGF), 207
VATER association, 47–48
Venous system, mendelian disorders affecting, 193
Venous system disorders, 215–218
  blue rubber bleb naevus syndrome, 217
  glomuvenous malformation, 215–216
  hyperkeratotic cutaneous capillarovenous malformation, 216–217
  inherited, 216
  Klippel–Trenaunay syndrome, 217
  sporadic, 216
Venous thrombosis, genome-wide association studies for, 210
Ventricular noncompaction, 176
Vesicoureteral reflux (VUR), 242–251
Villin disease, 271
Visual field test, 505
Vitamin D metabolism, 387–390
  $1\alpha,25(OH)_2D$, transcriptional and nongenomic effects of, 387–388
  calciferol excess state, 388
  calciferols, normal physiology of, 387
  history of, 387
von Hippel–Lindau (VHL) disease, 260–261, 458, 469
  age at presentation, 463t
  clinical features of, 463t
von Willebrand Disease (VWD), 292–293
von Willebrand factor (vWF), 209

W
Waardenburg syndrome, 541
Weill–Marchesani syndrome, 503
Werner syndrome, 61, 149, 469
WHIM syndrome, 321–322
Whole exome sequencing (WES), 201–202
Whole genome array-comparative genomic hybridization, 418
Whole genome sequencing (WGS), 15, 201–202
Williams-Beuren syndrome, management of, 171t–173t
Williams syndrome, 161, 582f
Wilms' tumor, 259–260
  WT1 gene and, 259
Wilms tumor factor 1 (WT1), 255
Wilson disease, 393, 445
Wiskott-Aldrich syndrome (WAS), 293, 321

Witkop syndrome, 552
Wolf–Hirschhorn syndrome, 161
Wolf–Parkinson–White (WPW) syndrome, 196
Wolfram syndrome (WFS), 328
Wolman disease (WD), 404–405
WT1 gene and Wilms' tumor, 259

X
Xeroderma pigmentosum (XP), 146
X-linked adrenoleukodystrophy (X-ALD), 409
X-linked dominant inheritance, 26
  defined, 26
  recurrence risks, 26
X-linked dominant lethal alleles, 26
X-linked hypohidrotic ectodermal dysplasia (XLHED), 551–552
X-linked hypophosphatemic rickets (XLH), 592
X-linked ichthyosis, 262, 544
X-linked intellectual disability (XLID), 417
X-linked recessive color vision defects, classification of, 494t
X-linked recessive inheritance, 26
  defined, 26
  recurrence risks, 26
X-linked sideroblastic anemia, 291
  with ataxia, 393
46,XX ovarian dysgenesis, 353–354
46,XX sex reversal (XX males), 352
46,XY disorders of sex development, 353
  due to androgen action defects, 353
  malformation syndromes in, 355t
XY sex reversal (46,XY females), 352–353

Y
Y-linked (holandric) inheritance, 26–27
  chromosomal disorders, 27
  mitochondrial inheritance, 26
  multifactorial inheritance, 27

Z
Zellweger syndrome (ZS), 409